FOUNDATION PAPERS IN LANDSCAPE ECOLOGY

EDITORS

John A. Wiens
Michael R. Moss
Monica G. Turner
David J. Mladenoff

COLUMBIA UNIVERSITY PRESS
New York

Columbia University Press

Publishers Since 1893

New York   Chichester, West Sussex

Copyright © 2007 Columbia University Press

Library of Congress Cataloging-in-Publication Data

Foundation papers in landscape ecology / editors, John A. Wiens ... [et al.].

  p. cm.

Includes bibliographical references (p.   ).

ISBN 0-231-12680-8 (cloth) — ISBN 0-231-12681-6 (pbk.)

1. Landscape ecology.   I. Wiens, John A.   II. Title.

QH541.15.L35F68   2006

577.27— dc22

2006049183

FOUNDATION PAPERS *in* LANDSCAPE ECOLOGY

Contents

FOUNDATION

PAPERS *in*

LANDSCAPE

ECOLOGY

Introduction

New disciplines or fields of study do not spring to life fully formed, replete with arsenals of ideas, questions to be researched, theories to be tested, methods and tools to be used, or applications for their findings ready and waiting. Instead, they usually begin with glimmerings of new ideas or different perspectives, often developed as part of some seemingly unrelated discipline. These ideas or perspectives may lie dormant for some time, perhaps decades, while work in the mainstream field continues apace or evolves in different directions. At some point the ideas emerge again, prompted perhaps by thinking in other disciplines, new methods or technological advances, or simply someone reading the old papers in a fresh context and seeing things in a new light. The ideas and approaches begin to coalesce into something that has its own identity.

The fusion of new or forgotten ideas and approaches is most likely to occur when different disciplines collide, drawing energy from the fringes that share a fascination with a common set of phenomena or problems. At least this is how landscape ecology came into being. Physical and human geography, soil science, ecology, land-use planning, landscape architecture—all of these disciplines in some way focus on the landscape as a discrete object in its own right and on spatial phenomena, relationships, and configurations within these landscapes. Landscape ecology has drawn from all of these disciplines to develop a spatially explicit perspective on the relationships between ecological patterns and processes that can be applied across a range of scales. The focus is on spatial heterogeneity, the notion that things vary among locations and that where things are, and where they are relative to other things, can have important consequences for a wide range of phenomena. Because most landscapes are modified by human actions, landscape ecology also integrates humans with natural ecosystems, and it brings a spatial perspective to fields such as natural resource management, conservation, and urban planning.

Landscape ecology is now widely recognized as an important and identifiable area of research and application. Concepts from landscape ecology now permeate ecological research across most levels of ecological organization and many scales and increasingly influence land-use policy, conservation, landscape architecture, and geography. In other words, landscape ecology is now giving back to some of the disciplines from which it was born. Although the term *paradigm shift* may be too strong to characterize the influence of landscape ecology, the discipline nonetheless embodies new ways of thinking about both natural and human-dominated systems and has helped to shape efforts in other fields.

To fully understand the current thinking and approaches of landscape ecology and why they have become influential, one must understand something of its history. And there is no better way to do that than by reading some of the works that inspired or exemplify that history. Our goal in this volume is to foster an increased appreciation of the earlier creative work that led to the emergence of landscape ecology and of the varied and irregular pace of development of the discipline. Landscape ecology grew in fits and spurts, not in some smooth, orderly, and logical progression of new thinking. We aim to provide an easy entrée into key papers, papers that in one way or another helped shape what landscape ecology has become.

The determination of what represents a foundation or classic paper is always somewhat arbitrary. We have restricted consideration to papers published prior to 1990, partly because the literature grew explosively during the 1990s and partly because this allows sufficient time to determine whether a paper really has had an impact or represents an early version of what has become a major theme in landscape ecology. Although landscape ecology is a relatively young discipline, its antecedents go back to the early part of the twentieth century. Several of these early papers that describe and define a landscape perspective have not previously been available to readers who do not know German or Russian. Others encompass the range of disciplines that influenced the first uncertain steps toward landscape ecology and thus are difficult to track down from within our own disciplines. Some classics could not be included because of their length (classics in any discipline often seem to be long).

The papers in the book are organized into seven thematic areas. **Part I** includes contributions that illustrate the diverse disciplinary roots of landscape ecology and show that so-called modern thinking about landscapes may not always be so modern after all. The papers in **part II** exemplify a growing awareness of the importance of spatial pattern, particularly in ecology. Landscape ecology began to take on an identifiable form in the late 1970s and 1980s, and the papers in **part III** show both this coalescence and the divergent views about what landscape ecology was or should be. One of the main conceptual themes of landscape ecology as it developed was the focus on scale and its consequences; the papers in **part IV** highlight the different ways that landscape ecologists incorporated scale into their work. The two papers in **part V** illustrate ways to quantify and analyze spatial patterns, an essential step in the progression of landscape ecology from a qualitative to a quantitative discipline. **Part VI** contains a collection of papers that serve as examples of how landscape thinking and methods began to be applied to a host of issues and questions. By the late 1980s, the discipline had reached the point at which a synthesis of the field could be undertaken; the paper included in **part VII** both reviews earlier literature and points out how this emerging discipline can contribute to research and thinking in mainstream ecology.

We have written short introductions to each of these sections. These introductions highlight the contributions and context of each of the papers and provide (at least indirectly) some of the reasons for its inclusion in this volume. These introductions are not intended to review or summarize the various themes of landscape ecology; several recent texts do that well, and we include references to some key background volumes at the end of this introduction.

This book developed from the merger of two pairs of collaborators who had independently identified the need for developing a collection of readings of foundation papers in landscape ecology. John Wiens and Michael Moss worked together in the late 1990s as president and secretary-general, respectively, of the International Association for Landscape Ecology (IALE). David Mladenoff and Monica Turner jointly taught a graduate course in landscape ecology during the 1990s at the University of Wisconsin–Madison, where both are on the faculty. When each group learned of the shared goals and potential duplication, it only made sense to combine efforts and edit a single volume. This book is the result of that combined effort.

Landscape ecology has grown into a vital area of research and practice, seeking better understanding of the reciprocal interactions between spatial patterns and ecological processes, testing the generality of its concepts across systems and scales, and extending the incorporation of human actions on the landscape. This, in turn, should enhance the capacity of landscape ecology to effect better management and conservation of natural resources, especially in the places where people live and work. It is our hope that the window into the past provided by the papers collected here will help focus the landscape ecology of the future.

A note on the translations: Several papers in this collection were originally published in Russian or German, at a time when the style of scientific writing was different from what it is now. In translating these papers, the translators have presented the science and ideas accurately, while retaining the flavor of the original writing. The translators of these papers deserve our special thanks and appreciation: Alexander V. Khoroshev (Department of Physical Geography, Moscow State University) and Sergei Andronikov (Department of Geography, George Mason University) for the Berg and Solnetsev papers; Conny Davidsen (Institüt für Internationale Forst und Holzwirtschaft, Dresden) for the Troll article; and Olaf Bastian (Saxon Academy of Sciences, Dresden) for the Neef paper.

References

Bastian, O., and U. Steinhardt, eds. 2001. *Development and perspectives of landscape ecology.* Dordrecht, Netherlands: Kluwer.

Bissonette, J. A., ed. 1997. *Wildlife and landscape ecology.* New York: Springer-Verlag.

Bissonette, J. A., and I. Storch, eds. 2003. *Landscape ecology and resource management: Linking theory with practice.* Washington, DC: Island Press.

Burel, F., and J. Baudry. 2003. *Landscape ecology: Concepts, methods, and applications.* Enfield, NH: Science Publishers.

Farina, A. 1998. *Principles and methods in landscape ecology.* London: Chapman & Hall.

Forman, R. T. T. 1995. *Land mosaics: The ecology of landscapes and regions.* Cambridge: Cambridge University Press.

Gutzwiller, K., ed. 2002. *Applying landscape ecology in biological conservation.* New York: Springer-Verlag.

Haines-Young, R. H., D. R. Green, and S. Cousins. 1993. *Landscape ecology and geographic information systems.* London: Taylor & Francis.

Hansson, L., L. Fahrig, and G. Merriam, eds. 1995. *Mosaic landscapes and ecological processes.* London: Chapman & Hall.

Klopatek, J. M., and R. H. Gardner, eds. 1999. *Landscape ecological analysis.* New York: Springer-Verlag.

Kolasa, J., and S. T. A. Pickett, eds. 1991. *Ecological heterogeneity.* New York: Springer-Verlag.

Krönert, R., U. Steinhardt, and M. Volk, eds. *Landscape balance and landscape assessment.* New York: Springer-Verlag.

Liu, J., and W. W. Taylor. 2002. *Integrating landscape ecology into natural resource management.* Cambridge: Cambridge University Press.

Mander, Ü., and R. H. G. Jongman, eds. 2000. *Consequences of land use changes.* Southampton, UK: WIT Press.

Mladenoff, D. J., and W. L. Baker, eds. 1999. *Spatial modeling of forest landscape change: Approaches and applications.* Cambridge: Cambridge University Press.

Nassauer, J. I., ed. 1997. *Placing nature: Culture and landscape ecology.* Washington, DC: Island Press.

Naveh, Z., and A. Lieberman. 1994. *Landscape ecology: Theory and application.* 2nd ed. New York: Springer-Verlag.

Reiners, W. A., and K. L. Driese. 2004. *Transport processes in nature.* Cambridge: Cambridge University Press.

Turner, M. G., and R. H. Gardner, eds. 1991. *Quantitative methods in landscape ecology.* New York: Springer-Verlag.

Turner, M. G., R. H. Gardner, and R. V. O'Neill. 2001. *Landscape ecology in theory and practice.* New York: Springer-Verlag.

Wiens, J. A., and M. R. Moss, eds. 2005. *Issues and perspectives in landscape ecology.* Cambridge: Cambridge University Press.

Zonneveld, I. S. 1995. *Land ecology.* Amsterdam: SPB Academic.

1 The Early Antecedents of Landscape Ecology

Introduction and Review

As with many of the natural sciences, the study of landscapes had its origins in the nineteenth-century scientific explorations of Africa and South America by Europeans and in the mapping and resource surveys of the American West. Although the primary aim was to describe and categorize different environments and vegetation types, some attention also was given to the environmental determinants and consequences of human occupancy of these places. Some of these early natural scientists also began to seek underlying principles to explain the broad-scale spatial variations in environments and the distributions of plants and animals.

As recognizable scientific disciplines began to emerge from these beginnings, landscapes came to be viewed from several largely independent perspectives. Moreover, because these disciplinary approaches to landscapes adopted different objectives in different parts of the world, distance and language hindered communication. The roots of modern landscape ecology, therefore, varied. Nonetheless, four major cornerstone themes of this diverse foundation can be identified:

1. A Russian/Soviet biophysical approach based on soil science, physical geography, and geology. Papers by **Berg** and **Solnetsev** illustrate this theme.

2. In the development of rapid land-resource inventories in which ideas derived from physical geography were applied at broad scales, including whole continents. **Christian**'s paper illustrates this development in the context of Australia.

3. The regional concepts of the geographer, particularly the approaches of cultural/historical geographers, initially in Germany and France but also in North America in the first half of the twentieth century. **Sauer** and **Troll** exemplify this theme.

4. A biologically based approach that emerged within ecology. **Watt** provides the basis for this theme's development.

Any review of the literature relating to the scientific foundations of landscape analysis would be drawn immediately to the paper by **L. S. Berg.** First presented in 1913, it is still the keystone paper of the geoecological tradition within landscape ecology.

Berg placed the study of landscapes as *the* focus around which the emerging field of geography should coalesce. His review of geography up to that time relied heavily on the earlier methodological writings of the key German founders of that discipline, Ferdinand von Richthofen (1833–1905) and Alfred Hettner (1859–1941), and incorporated the thinking of important early

French geographers such as Emmanuel de Martonne (1873–1955). Berg also cited the work of one of the key founders of American geography, W. M. Davis (1850–1934), although he used a curiously obscure and unrepresentative illustration of Davis's argument.

Berg emphasized the interdisciplinary nature of geography. Rather than seeing geography merely as a compilation of multidisciplinary information, however, he saw the need to use methods and knowledge from other specializations to categorize the landscape from a distinctly geographic point of view. This viewpoint, he suggested, considered the landscape through four categories: distribution (the chorological category), form (the morphological category), material (the hylological category), and change over time (the chronological category). These categories are identical to those proposed for landscape ecology by Zonneveld (1990).

Berg viewed interrelated groups of humans, animals, plants, and landforms as the key features of the inorganic and organic worlds on the earth's surface; they form landscapes, and, therefore, "geography is the science of landscapes...a chorological science" (Berg 1915:14). He differentiated between natural landscapes (those formed without human activity) and cultural landscapes (those sharply affected by humans) and described natural landscapes in terms that would suffice for many landscape ecologists today. Cultural landscapes were not discussed to the same degree, and subsequent writings by both Sauer and Troll noted the difficulties of incorporating the cultural dimension of landscapes into an integrated landscape system.

N. Solnetsev's paper, published in 1948 in the USSR, developed in a somewhat uncritical manner the proposals outlined by Berg, placing them into a more practical applied context that aligned with the stated objectives of landscape work during the Soviet era. Although Solnetsev's focus was in the field of geography, physical geography within the Soviet Union had a wider, more inclusive grasp of such fields as geohydrology, soil science, and geochemistry than did its counterparts in the West. In Solnetsev's view, the landscape remained the focus for geography, although the scale of enquiry was refined to that of the *geographic individuum* as the basic landscape unit. The geographic individuum corresponds to the ideas of the geocomplex developed later in the Soviet Union and Central Europe (Preobrazhenskiy 1983; Snytko 1983) and elsewhere as land systems (see Christian, part I).

The value of this approach was in the identification of functional, process-based landscape units. Solnetsev recognized these units as having "internal integrity, completeness and unity" for which there is "no possibility of further subdivision" (Solnetsev 1948:19). These are the natural units of the earth's surface that are transformed by economic activities and human intervention. In this view, land-unit processes were determined by the interrelationships between heat (energy) and moisture. The recognition by Soviet scientists of these fundamental driving forces was later adopted as the basis for much of the landscape work undertaken during the decade of the International Biological Program (1964–1974; see part II). This approach had a short-lived acceptance within physical geography in the United States (Carter et al. 1972). Elsewhere, it became more broadly accepted as an essential base of applied landscape ecology that examined the efficiencies of land-system processes such as primary productivity and decay and decomposition and that used these natural processes as benchmarks for evaluating land-use practices, particularly in agro-ecosystems (see, e.g., Moss and Davis 1981; Meentemeyer 1984; Ryszkowski and Kedziora 1987).

The geoecological focus outlined by Berg and Solnetsev required the identification of individual, integrated land(scape) units at increasingly fine scales. By the early 1940s, aerial photography had been developed to the point that it could be used for rapid reconnaissance of unmapped areas. The paper by **C. S. Christian** summarized the state of the art of this broad scale of landscape analysis. Christian supported his approach using previously published regional land-system reports, illustrating how this information fit into a scheme of a hierarchy of land units and discussing the relative scale of these units and the controlling variables at each level in his integrated land-system hierarchy. With minor variations, this approach was developed further in many parts of the world during the same period. Canadian systems of land classification have parallels (see Rubec et al. 1988). Subsequent developments in this aspect of landscape ecology fall into two broad areas: inventories of landscape component information and schemes that attempt to incorporate land-system process information in their structure. Moss (1975, 1985) reviewed this problem for several widely accepted schemes, and Klijn (1994) provided a detailed analysis of the problems and limitations of ecosystematic land-classification procedures, even in the age of geographic information technologies.

Of the four cornerstone themes for landscape ecology, this approach of land-unit classification showed little further development. Beyond reference to geosystems in the work of Central European landscape ecologists in the 1970s and 1980s (part III), few advances have been made. Nevertheless, the importance of this work as a foundation for conceptualizing a systematic approach to land-unit identification cannot be denied.

The third cornerstone theme for landscape ecology is found in cultural or historical geography. Although regional analysis was a strength of geography within much of Western Europe (see Troll, part I), cultural geography developed its strongest base in the United States, particularly from 1920 to 1960. Much of the driving force for

this development came from the work of **Carl Sauer** and his many disciples from the University of California–Berkeley School of Geography. Sauer's paper is still regarded as a foundation paper for geography.

Born in the United States, Sauer received a significant part of his early education in Germany and was one of few U.S. geographers familiar with the ideas developed by the German geographers. Sauer viewed the landscape as being the "unit concept of geography" (Sauer 1925:25). He translated and equated the German *Landschaftskunde* and *Länderkunde* to be the equivalents of geography. The study of landscape—or geography—was a chorological science, one dealing with so-called place facts, with the landscape being an association of these facts having an identifiable structure and function. Sauer's perspective focused on the links between the geological base through geognostic factors, specifically factors of rock character and condition, with specific climatic factors that were expressed in the vegetation of a landscape. These interrelated factors, over time, formed natural landscapes. Sauer viewed landscapes as being also cultural entities, with forms arising from the local culture, influenced by time, within the medium of the natural landscape. These relationships would produce the various characteristics of the cultural landscape: for example, population density, housing structures, and production. By recognizing interrelationships rather than causal relationships between natural and cultural elements, Sauer eschewed the established deterministic viewpoint within U.S. geography, which emphasized physiography or landform as the critical element defining the nature of the landscape.

For the next forty years, Sauer's work and that of his students profoundly influenced the thinking about landscape by geographers, particularly in North America. By the 1960s, however, the focus of geography had shifted, particularly in North America, and the influence of Sauer's Berkeley School and of the cultural/historical dimension of geography waned. Indeed, some viewed the concept of landscape as espoused by Sauer as a "chaotic conception" (Sayer 1984). We return to this shift in the focus of geography in the introduction to part II.

Although the landscape focus faded in prominence in North American geography, European geographers had been much more diligent in refining the landscape focus as their raison d'être. Chief among these was **Carl Troll,** whose 1950 paper is generally regarded as *the* foundation paper for landscape ecology. This is the publication in which the term *landschaftsökologie* (landscape ecology) was first extensively used, although Troll had coined the term in a presentation in 1938 and used it in print in 1939 (Troll 1939), albeit merely as a passing reference. Troll's writing had a strong conceptual foundation, with a clear link to the theoretical bases of landscape geography outlined by Berg and Solnetsev. The value of Troll's paper to landscape ecology was in his refinement of these earlier ideas from geography in combination with Tansley's (1935) ecosystem concept. Troll subsequently applied these new ideas to both continental areas and small areas in parts of the Rhine Plateau of Germany.

It has been suggested that in 1938 Troll was unaware of Tansley's ecosystem concept (Haber 2004), although he was clearly aware of the great potential of the emerging techniques of aerial photography (Schreiber 1990). Whatever the circumstances, Troll came to his position of bridging geography and ecology from a varied background, having graduated in natural sciences (including geography and botany) from the University of Munich in 1919, obtaining a doctorate in botany in 1921, and subsequently conducting phytogeographic work in the Alps and the Andes. This experience gave him a keen awareness of the interdependence of soil, slope, and vegetation and, subsequently, an appreciation of the value of aerial photography in his work

on past glaciations, soils, and native agriculture in Latin America.

To Troll, the geographic region and the units that make up the earth's surface were landscape units: natural landscapes of interrelated phenomena with both inner cohesion and outer spatial relationships (Dickinson 1969). Units at different scales make up hierarchies of regional units—again fitting into the geographers' earlier concepts of natural regions. The strength of Troll's approach, however, was in his focus on landscape units at a scale relevant to regional planning and development. The smallest landscape units, the key elements in his landscape ecology, were those where further breakdown in the areal extent of individual landscape elements do not occur in combination with other areal elements, but as individual distributions. These units are the landscape individuals or ecotopes, the spatial coincidence and interdependence of natural or physical conditions. Although Troll's ecotopes excluded human activities, he did consider the problems of defining and mapping cultural and economic factors of human occupancy of the land. He was particularly concerned with the disharmony in regional (or landscape) structure where systems of land occupancy did not coincide with natural units, asking what forces lie behind these differentials between human occupancy and site. In the 1950s, this was viewed as a need to understand environmental interrelationships.

Today's landscape ecologists will find in Troll's paper virtually all of the topics that are currently discussed in landscape ecology (see Wiens and Moss 2005). Although written primarily as a methodology and focus for geography, the familiar key words and phrases of landscape ecology are all there: scale, hierarchy, spatial distribution, integration, natural versus cultural landscapes, and so on.

The previous three cornerstone themes for landscape ecology were derived from the field of geography. This contrasts with the situation in North America and Britain, where the field of ecology had been growing in importance from the late 1800s. Although this growth was to become significant to the development of landscape ecology within North America (part II), its relevance was presaged in Britain, first by Tansley's elaboration of the ecosystem concept (Tansley 1935). Tansley stressed natural, functional interrelationships within what were essentially land units, thereby recognizing the importance of a spatial perspective. **Alex Watt's** presidential address to the British Ecological Society in 1947 played a major role in placing spatial patterns and dynamics on a firmer footing in ecology, establishing the foundation for what later would become *spatial ecology*. Watt attempted to articulate a middle ground in plant community ecology between the work of Frederic Clements, who viewed plant communities as environmentally proscribed, compositionally fixed, and deterministic in their directional change, and Henry Gleason, who declared that plant species responded individualistically to the environment, resulting in variable and indeterminate assemblages of species (see McIntosh 1975). Watt described studies of several plant communities, extending the dynamic nature of the temporal succession of Clements into a framework that interacted in space as well. Watt defined plants within a bounded community as being distributed in patches that form mosaics. Although Watt saw vegetation change as quite proscribed, he differed from Clements in describing it as a cyclic series of changes that had spatial as well as temporal dynamics (this has come to be known as the *shifting-mosaic steady-state* view of vegetational dynamics; Bormann and Likens 1979).

In the 1950s and 1960s, research in the United States by Robert Whittaker and John Curtis would reinforce Gleason's concept, showing that environmental and disturbance gradients of varying steepness can create community patchiness in space with varying species composition and contrast. In combination with Watt's emphasis on the spatial and temporal dynamics of vegetation, this community work showed the potential for integration with the more explicitly spatial concepts being formulated within geography, soil science, and the earth sciences at this time.

By the 1950s, then, the cornerstones of modern landscape ecology were in place. The greater part of these foundations came from geography, a geography that recognized the interconnected natural and cultural elements of landscapes. Although the importance of this integrative and spatial perspective subsequently diminished in geography, particularly within North America, it combined with the ecological perspective illustrated by Watt to become the dominant approach to landscape ecology as it is now recognized, not only in North America but worldwide. Calls for the convergence of the geographic and the ecological approaches (e.g., Bastian 2001; Moss 2000) need only go back to these foundations to find early attempts to develop a method and focus for understanding the nature of landscapes.

References

Bastian, O. 2001. Landscape ecology—Towards a unified discipline? *Landscape Ecology* 16:757–766.

Berg, L. S. 1915. The objectives and tasks of geography. *Proceedings of the Russian Geographical Society* 15 (9):463–475.

Bormann, F. H., and G. E. Likens. 1979. *Pattern and process in a forested ecosystem*. New York: Springer-Verlag.

Carter, D. B., T. H. Schmudde, and D. M. Sharpe. 1972. *The interface as a working environment: A purpose for physical geography*. Tech. Paper 7. Washington, DC: Association of American Geographers.

Dickinson, R. E. 1969. *The makers of modern geography*. London: Routledge and Kegan Paul.

Haber, W. 2004. Landscape ecology as a bridge from ecosystems to human ecology. *Ecological Research* 19:99–106.

Klijn, F. 1994. Spatially nested ecosystems: Guidelines for classification from a hierarchical perspective. In

Ecosystem Classification for Environmental Management, ed. F. Klijn, 85–116. Dordrecht, Netherlands: Kluwer.

Meentemeyer, V. 1984. The geography of organic decomposition rates. *Annals Association of American Geographers* 74:551–560.

McIntosh, R. P. 1975. H. A. Gleason, "individualistic ecologist," 1882–1975: His contributions to ecological theory. *Bulletin of the Torrey Botanical Club* 102:253–273.

Moss, M. R. 1975. Biophysical land classification schemes: Their relevance and applicability to agricultural development in the humid tropics. *Journal of Environmental Management* 3:287–307.

——. 1985. Land processes and land classification. *Journal of Environmental Management* 20:295–319.

——. 2000. Interdisciplinarity, landscape ecology, and the "Transformation of Agricultural Landscapes." *Landscape Ecology* 15:303–311.

Moss, M. R., and S. Davis. 1981. The actual and potential primary productivity of southern Ontario's agro-ecosystem. *Applied Geography* 2:17–37.

Preobrazhenskiy, V. S. 1983. Geosystem as an object of study. *GeoJournal* 7:131–134.

Rubec, C. D. A., E. B. Wiken, J. Thie, and G. R. Ironside. 1988. Ecological land classification and the role of the CCELC and the formation of CSLEM. In *Landscape Ecology and Management*, ed. M. R. Moss, 51–54. Montreal: Polyscience.

Ryszkowski, L., and A. Kedziora. 1987. Impact of agricultural landscape structure on energy flow and water cycling. *Landscape Ecology* 1:85–94.

Sauer, C. O. 1925. The morphology of landscape. *University of California Publications in Geography* 2:19–53.

Sayer, R. A. 1984. *Method in social science: A realist approach.* London: Hutchinson.

Schreiber, K.-F. 1990. The history of landscape ecology in Europe. In *Changing Landscapes: An Ecological Perspective*, eds. I. S. Zonneveld and R. T. T. Forman, 21–35. New York: Springer-Verlag.

Snytko, V. A. 1983. Substance dynamics in geosystems. *GeoJournal* 7:135–138.

Solnetsev, N. A. 1948. The natural geographic landscape and some of its general rules. *Proceedings of the All-Union Geographical Congress.* Vol. 1. Moscow: State Publishing House for Geographic Literature.

Tansley, A. G. 1935. The use and abuse of vegetational concepts and terms. *Ecology* 16:284–307.

Troll, C. 1939. *Luftbildplan und ökologische Bodenforschung*, 241–298. Berlin: Zeitschrift der Gesellschaft für Erdkunde.

——. 1950. The geographic landscape and its investigation. *Studium Generale* 3 (4/5):163–181.

Wiens, J. A., and M. R. Moss, eds. 2005. *Issues and perspectives in landscape ecology.* Cambridge: Cambridge University Press.

Zonneveld, I. S. 1990. Scope and concepts of landscape ecology. In *Changing Landscapes: An Ecological Perspective*, eds. I. S. Zonneveld and R. T. T. Forman, 3–20. New York: Springer-Verlag.

The Objectives and Tasks of Geography

L. S. Berg

Proceedings of the Russian Geographical Society, *15, no. 9: 463–475*

A Historical Review of the Problem

One could hardly name any other science that has generated so much discussion concerning its content, objectives, and methods as has geography. Some people even argue that geography is not a science at all but an assemblage of facts collected from various other fields of knowledge, both from the natural sciences and the humanities. Even those who recognize geography as a separate discipline often disagree about its goals and objectives.

As an example of one extreme opinion on this topic let us cite the famous and respected geographer Davis. From his point of view, geography deals with interrelationships existing between the organic and inorganic worlds. As an illustration of actual geographic relationships, Davis cites the following example: Aquatic plants are sustained in a vertical position because of differences between the density of water and that of air; therefore, they do not need solid or ligneous stems, which are essential for subaerial plants. Other examples of geographic phenomena cited by Davis are the flight of birds, the relationship between specific weights of water and fish, and the vital requirements of living organisms for oxygen (Davis 1906).

From my point of view, there is nothing geographic in the examples cited above. This is a list of phenomena from the field of morphology and physiology of plants and animals and has no relationship to geography. The definition of geography given by Davis, on the one hand, has no roots in the historical development of our science; on the other hand, it does not take a logical position that geography has, or must take, with respect to other sciences.

Many scientists are of the opinion that geography deals with the study of the earth as a whole in all its aspects. They take as a basis the Greek definition of the term *geography*, "the description of the earth." This point of view was shared, by the way, by Professor Petri, who said: "The whole Earth belongs to it (to Geography—L. S. Berg). The task of Geography is to understand the essence and life of the Earth. Geography receives its ideas, documents, facts, and data from a number of other sciences including the natural, historical, economic, philosophical disciplines. Its objective is to combine these into one focus and to apply this to understanding the total characteristics of the Earth" (1892:34). In Petri's opinion it is impossible to separate geography from the other sciences from which it takes its information. Geographers use the methods of other specialties, but from a different point of view (Petri 1892:40); they analyze data concerning its geographic *distribution* (the chorological category), *shape* (the morphological category), *material composition* (the hylological category), and *change in time* (the chronological category). Genetic, dynamic, and anthropogeographic categories are also relevant (Petri 1892:41).

The opinion that geography studies the globe as a whole (some add "as the habitat of human beings") is widespread. This viewpoint is shared by Geiki (1900:1), Wagner (1913:37), Davis (1911), and others. Wagner qualifies this definition by noting that geography pays particular attention to "the reasons for the spatial distribution of bodies and phenomena on the Earth's surface."

However, we do not see any unifying principle in such a definition. To state that geography studies all the earth in all its aspects confirms the opinion that geography is not a science but a compilation of disconnected knowledge about the earth and its inhabitants.

The lack of credibility of this viewpoint is nicely demonstrated by Hettner in his splendid article "Das Wesen und die Methode der Geographie" (On essence of geography and its methods; 1905:546). Even if it were possible to unite, in a single discipline, all knowledge related to the physical nature of the earth, to do so as far as the organic world is concerned would be quite impossible. Actually, in textbooks of geography we often see more or less complete information on the field of so-called physical geography, but with respect to plants, animals, and human beings, geographers leave this theme for the specialized sciences—botany, zoology, anthropology, ethnography, demography, and so on—and limit themselves only to presenting facts about geographic distribution.

Contradiction is obvious. On the one hand, geography should study the earth as a whole; on the other hand, geographers select only the objects and phenomena that relate to the influence of the inorganic world on the organic. The systematic description of the organic world is left for other sciences. Such a contradiction indicates clearly that the task of encompassing all the earth sciences is beyond the scope of geography as it is of any other scientific discipline.

Richthofen and other geographers limit the scope of geography to the surface of the earth: Geography is not a simple description of the earth (*Erdkunde*) but a science of the earth's surface (*Erdoberflächenkunde*). Geography must take as its principal viewpoint the causal interrelationships between phenomena and objects, on the one hand, and the surface of the earth, on the other hand (Richthofen 1883:25). The tasks of scientific geography, according to Richthofen, are (1) the investigation of the lithosphere, hydrosphere, and atmosphere regarding its shape, material composition, transformation, and genesis; (2) the study of the plant and animal worlds on the basis of the same topics; and (3) the study of human beings and the effects of social life and culture from the same basic principles.

Although the limits proposed by Richthofen make the domain of geography narrower, geography would still include, as before, those disciplines that claim a separate existence. For example, the study of the form and material composition of the plant world is a domain of botany; there is no reason to include it in geography. The same holds true for zoology, anthropology, ethnography, and so on.

In another paragraph in his article, Richthofen states that the first-order task for geography is to investigate the relationships between man and both organic and inorganic nature: landforms, distribution of water, atmosphere (1893:24).

The starting point for the geographer is not the interrelationships between objects and phenomena but the geographic distribution of these objects and phenomena.

This viewpoint of Richthofen, who limited the domain of geography to the earth's surface, was opposed by Gerland (1888). In his opinion, geography deals with the study of the earth as a whole, namely, the study of interaction between the earth's surface and its interior (Gerland 1888:20). Geography should investigate the results of the interaction of forces linking the earth's materials, formation, transformation, and development. Because anthropogeography, anthropology, ethnology, and historical geography do not provide data for investigation of physical properties of the earth, Gerland excludes these sciences from scope of geography. The content of the latter is represented by (1) mathematical geography and cartography (the study of the size and movements of the earth); (2) geophysics (the study of the general properties and forces of the earth's matter, including studies of the earth as a whole, the land surface, oceans, and atmosphere); (3) regional studies, defined by Gerland as the study of the properties and development of certain parts of the globe (this section could be more correctly named *physical land studies* because Gerland excludes from it all information from biogeography and anthropology; Gerland [1888:46] names this section local or special geophysics); and (4) the geography of organisms.

Thus, only the ideas of Hettner (which are shared by the author) create a solid basis for geography as a distinct science.

Hettner (1905:552), relying on Ritter's definition of geography as a "science of space and its material content," argues that geography is a chorological science of the earth's surface (the term *chorological* has as its root the Greek word *choros,* meaning "place"), its subject being the attributes and properties of countries.

Hettner determined the true domain of geography to be regional studies. A. A. Yarilov came to the same conclusion. This scientist argued that the so-called general geography (i.e., separate sections from related sciences: geology, meteorology, oceanography, etc.) should be excluded from the domain of geography; the latter should become "a discipline, placing human-beings in a specific habitat as determined by nature."

As a conclusion to this short historical review, it is necessary to say a few words concerning the viewpoints of P. P. Semyonov. He states, "Geography—the science of the earth—is a word with a lot of meaning attached to it." The late geographer distinguishes "geography in its general and in its narrow sense. In its general sense its subject is the complete study of the globe.... Hence it is not actually a science but a set of the natural sciences." In its narrower meaning, geography is "the description of both constant properties created by nature and unchanged for centuries, and changeable ones created by human activities" (Semyonov 1856:6–11) The term *regional studies* is not used by P. P. Semyonov, but obviously his "narrow meaning of Geography" is nothing other than regional studies.

A Definition of Geography

Generally speaking, geography is a science about the horizontal and vertical distribution of various objects and phenomena on the earth's surface within the upper lithosphere, in the atmosphere, and in the hydrosphere in the past, the present, and the future.

Geography, as Hettner has revealed, is a chorological science. It studies the spatial distribution of objects and phenomena. At the same time, however, geography studies the distribution not only of individual objects but also of the combination and interaction of these objects and phenomenon. Geography is not a chorology of separate objects and phenomena but a chorology of their assemblages, that is, a chorology of interrelated groups of humans, animals, plants, landforms, and so forth, on the earth.

What are these groups and communities of the organic and inorganic world on the earth's surface? These are landscapes. Thus, geography is the science of landscapes.

It follows from the definition given previously that the scope of geography does not take in the composition or form of objects or phenomena; rather, it is interested in their distribution in space. Geography is a chorological science.

Similarly, history is a chronological science: It studies the distribution of phenomena and objects in time but not the origin of these objects and phenomena.

Actual information on topics of interest to the geographer is obtained from related disciplines: geomorphology, hydrology, meteorology, soil science, geology, botany, zoology, anthropology, ethnography, statistics, political economy, and so on.

Let us summarize our definition in detail. We asserted that geographic distribution of various phenomena is the object of geography. By this, we say that both the phenomena of a physical nature and the material and spiritual activities of organisms (including human beings) are the subjects of geographic research, the distribution of these phenomena being a principal objective. For example, geography can include, with equal importance, the distribution of mountains, rivers, thunderstorms, coral reefs, coniferous and marsupial species, races and religions, industry and the consumption of sugar, customs (e.g., cannibalism), fairy tales, judicial norms, criminal activity, and so on.

In a similar way, history deals with all the facts of the past, not only political and military facts but also the results of human activities in science, fine art, technology, and socioeconomic relations.

Because it is impossible and useless to embrace everything, the geographer should focus attention on objects of great geographic significance. The problem of what is or is not significant is quite subjective and can be resolved within the context of the spirit of the times and the state of science. In any case, as Hettner underlined, for the geographer, the objects and phenomena that are important are those that act, not separately, but which interact with others.

Many are of the opinion that the scope of geography should be limited by the earth's surface. This is not correct. Geography should consider all the earth's crust to the depth affected by those exterior forces that transform the surface. In other words, the geographer should study the lithosphere to the depth of the weathering crust and the groundwater table and in the sea to the ocean floor to a depth at which the influences of the ocean waters are no longer manifest.

A simple listing of facts concerning the distribution of any phenomena or objects—in other words, chorography—does not bring much to geography; this is the material for geographic description. As an example of such geographic description, note *Kniga Bol'shomu Chertezhu* (The book of great drawing, 1570). The purpose of geographic research is to reveal links and regularities that exist between separate objects of interest to the geographer, the main question being, "How do particular sets of objects and phenomena affect each other, and what are the results of this spatially?" In other words, the final goal is the investigation and description of both natural and cultural landscapes. By natural landscapes we mean those formed without human activity, differing from cultural landscapes, which are strongly affected by humans and human activities. Cities, towns, and villages, by our definition, are components of the cultural landscape.

The natural landscape is the region in which the character of landforms, climate, and plant and soil cover are integrated into one harmonic whole, typically repeated within a certain zone of the earth. A study of the causes leads to the fact that relief, climate, and plant and soil cover form one unique landscape system or organism resulting from the interactions of different factors that affect each other and that constitute the natural landscape. This is the task of scientific geography. All that has been said previously is also true for cultural landscapes.

If we translate the word *geography* as "land studies," the meaning that should be assigned to it is not "the study of the earth" but "the study of lands" *(Länderkunde,* not *Erdkunde)* or even better, "the study of landscapes."

The task of great significance for geographers is the division of the earth's entire surface, or its parts, into regions based on their natural attributes.

We now have quite a few works of scientific regional studies. Usually, however, authors describe landforms, climate, flora, fauna, and population of the areas investigated but do not integrate the results obtained into a united, harmonic picture; the interaction of various factors is not analyzed, and the general image of the natural landscape remains unclear.

From the previous statements, it follows that the geographer, in borrowing information from other sciences, then uses this information for his own purposes—the study of landscapes—by applying the chorological method. Thus, geography has both its own particular research objectives and methods.

It follows, therefore, that the frequent reproaches to geographers concerning the lack of originality of their science are quite unfair. With the same argument, one could say that geology is a combination of facts borrowed from geomorphology, geophysics, petrography, zoology, and botany or that soil science is derived from geological, geomorphological, chemical, botanical, and other facts. We do not even discuss history, sociology, or philosophy; such a discussion requires a much more encyclopedic knowledge from the researcher than does geography. Nowadays, however, it is much easier to gain an understanding of a group of sciences than it was before, when knowledge of Aristotle or Humboldt was in demand. Now many books on related sciences—geomorphology, meteorology, geology, biogeography—are available to ensure a general acquaintance with the subject.

In contrast to the viewpoint developed here, Hettner (1905:559) underlines that geography is a science not of the spatial distribution of various objects (this being a branch of corresponding specialized sciences) but of the composition of areas or countries. As an example, Hettner mentions that the study of the distribution of particular species, genera, families, and so on of plants and animals is one of the branches of systematic zoology or botany (geographic zoology or geographic botany), whereas the study of the faunal or floral species composition of different countries is a branch of geography, namely, zoological geography (zoogeography) or botanical geography (phytogeography).

The difference between the viewpoints of the author and Hettner, however, is more apparent than real if we take into consideration the necessity of studying not only the distribution of separate phenomena and facts but their assemblages. Hettner speaks about geography as a science studying countries; we consider the study of landscapes or natural regions as the first-order task.

Our definition is similar to that of Martonne: Geography is a science of physical, biological, and social phenomena considered from the viewpoint of their distribution on the surface of the earth, their causes and mutual interactions (1909:23).

A Subdivision of Geography and Its Relationships to Other Sciences

Geography deals with the description of countries or, to be more exact, landscapes.

The distribution of objects and phenomena on the earth's surface falls within the realm of general geography (e.g., Suess 1883). If the description concerns any part of the earth, we name it special geography or regional studies. There is no reason to speak much about the description of regions being delimited on the basis of administrative or political boundaries as being nonscientific: Regional studies deal with natural landscapes.

The following sciences are related to geography: selenography (describing the moon in the same way that geography describes the earth), heliography (dealing with the sun), and astronomy (an assemblage of sciences interpreting the distribution of objects and phenomena in the universe in general).

Geography deals with problems of distribution, both in the present and in the past. Dealing with present-day chorology, we are within the domain of general geography. Dealing with chorology of the historical past, we enter the domain of historical geography; and the chorology of prehistoric periods comes within the terms of reference of geology. Geology is the geography of the prehistoric past; it studies the distribution of various facts and phenomena in prehistoric times.

The field of dynamic or physical geology is not related to geology at all; it is a part of geomorphology, outlined below. Because geography deals with the study of the distribution and grouping of objects and phenomena, it has no reason to study the objects and phenomena themselves. Instead it permits other sciences to do this. Consequently, geography involves neither the study of the agents transforming the earth's surface at the present time and in the past nor the investigation and classification of the resulting landforms. This is the task of a specialized science, namely geomorphology. An investigation of the water envelope of the earth falls in the realm of hydrology; the air envelope, meteorology; rocks, petrography; flora, botany and phytopaleontology; and human beings, anthropology, archeology, ethnography, demography, political economy, and so on.

The investigation of the causes of mountain building, the classification of mountain chains, and the investigation of their change under various conditions are not the tasks of the geographer. However, the study of the distribution of mountain chains on the earth and the investigation of the causes of regularities in this distribution fall within the realm of the geographer. The well-known work by Suess, *Das Antlitz der Erde,* provides the best example of geographic work in this field.

Climatology, or the study of the distribution of climatic phenomena on the earth's surface, is one of the most important parts of geography. The same can be said concerning soil geography, phytogeography, zoogeography, and anthropogeography. The study of soil-forming processes, the study of the soil as a physical body, and the classification of soils are within the competence of the soil scientist, but the distribution of soil types comes within the province of geography.

To show the difference between this viewpoint and that of Richthofen, let us give one more example. The celebrated geographer says: "If, at this moment, a new volcano comes into existence, an investigation of the causes of its origin in connection with that of other volcanoes should be included within the scope of the geologist, but a study of the transformation of its form, once it has appeared, would be a task for the geographer" (1893:21). In contrast, to this we say that the causes of this volcanic activity and changes to its morphology should be studied by a special science—geomorphology—but an investigation of the location of the new volcano in relation to older ones, as well as the study of the new natural landscape, belongs to the field of geography.

Because geography is a chorological science, the first task for a geographer is the precise definition of the site and location by latitude, longitude, and altitude above sea level. Therefore, the geographer should know the methods of astronomical and topographic definition of locations on the earth's surface by topographic survey, mapping, interpretation, and measurement of maps. Mathematical geography—interpreting these topics—should be a practical introduction to geography.

Moreover, it is necessary for the geographer to study landforms on the earth's surface because that is the very place where all the phenomena he is interested in occur. Thus, knowledge of geomorphology is a necessary premise for every geographer working in the science. Geomorphology is not geography but the principal basis for geography.

The sciences of the earth's landforms (geomorphology), waters (hydrology), and atmospheric envelope (meteorology) are sometimes united under the umbrella term *physical geography,* but they are distinct sciences and, to avoid confusion, should not be called geography. If a common name is necessary, *physiography* (physiogeography) would be better. It unites geomorphology, hydrology, and meteorology and is widely accepted in the United States.

References

Davis, W. M. 1906. An inductive study of the content of geography. *Bulletin of the American Geographical Society* 38.

——. 1911. *Grundzuge der Physiogeographie.* Leipzig, Germany.

——. 1912. *Die erklärende Beschreibung der Landformen.* Leipzig, Germany.

Geiki, A. 1900. O prepodavanii geografii [On teaching geography]. *Zemlevedeniye* [Land studies]. Suppl. 2–3.

Gerland, G. 1888. Zadachi i razdeleniye geografii [Tasks and subdivision of geography]. *Izvestiya Russkogo Geograficheskogo Obshchestva* [Proceedings of the Russian Geographic Society] Suppl. 24.

Hettner, A. 1905. Das Wesen und die Methode der Geographie. Geogr. Zeitschr., Jg. 11.

Marthe, F. 1877. Begriff, Ziel und Methode der Geographie. Zeitschr. Gesellsch. Erdkunde zu Berlin, Bd. 12.

Martonne, E. 1909. *Traité de géographie physique.* Paris.

Petri, E. 1892. *Metody i printsipy geografii* [Methods and principles of geography]. St. Petersburg.

Richthofen, F. 1893. *Aufgaben und Methoden der heutiggen Geographie.* Leipzig, Germany.

Semyonov, P. P. 1856. Predisloviye perevodchika k knige: K. Ritter [Introduction by the translator to the book: K. Ritter]. *Zemlevedeniye Azii* [Land Studies of Asia]. St. Petersburg.

Suess, E. 1883. *Das Antlitz der Erde.* Vols. 1–3. Vienna.

Wagner, H. 1913. Lehrbuch der Geographie. 9-te Aufl., Bd. 1, Hanover, Germany.

Yarilov, A. A. 1905. Pedologiya kak samostoyatel'naya yestestvenno-nauchnaya distsiplina o zemle. Opyt istoriko-metodologicheskogo issledovaniya. Chast' 2. Mesto pedologii sredi nauk o zemle. [Pedology as an individual natural-scientific discipline about the earth. The experience of historical-methodological study. Part 2. The place of pedology among the earth sciences]. Yuryev.

——. 1913. Geografiya i "geleografiya" [Geography and "heliogeography"]. *Zemlevedeniye* (Land studies). Vols. 1–2.

The Natural Geographic Landscape and Some of Its General Rules

N. A. Solnetsev

Proceedings of the All-Union Geographical Congress.
Vol. 1. Moscow: State Publishing House for Geographic Literature

Definition of Geographic Landscape

In my opinion, which is in agreement with that of academician L. S. Berg, the geographic landscape is the main taxonomic territorial unit. In this sense, the term *geographic individuum,* which has been frequently used, is quite acceptable. By its use, authors who share such an idea of landscape wish to underline, that on the one hand, there is internal integrity, completeness, and unity, and, on the other hand, there is no possibility of further division into its original constituents. Naturally, such territorial integrity is determined by its genetic coherence as well as by its morphological and dynamic homogeneity. All three of these conditions are closely interrelated, so a change in one (and especially in all three) means transition to another geographic landscape.

Definitions of the idea of landscape have been presented by numerous authors with varying degrees of success. In this paper I will not be concerned with the analysis and criticism of such definitions. I shall only note that the definition of the geographic landscape must provide the researcher with the possibility of differentiating, in the field, one landscape from another. This is why I propose the following definition, which is similar to that of Berg: The geographic landscape is defined as such genetically homogenous territory in which a regular recurrence of the same interrelated combination of factors takes place; these factors are geological composition, forms of relief, surface and subterranean waters, microclimates, soil units, phytocoenoses, and zoocoenoses.

It is necessary to add that, in many cases, such natural combinations can be transformed, to varying degrees, by the economic activities and the intervention of human beings. In these cases, however, not all components are subject to transformation; those affected are usually plant cover and soils. The radical transformation of all the landscape components is a rather rare phenomenon, even in densely populated cultural regions. The landscape of major cities with populations of many millions may serve as an example of such significantly changed landscapes.

From the previous definition, it can be seen that without claiming to be original it does claim some specificity. In earlier definitions of landscape, attention was generally paid to the interrelated combinations of landscape components in general: relief, climate, soil cover, vegetation, and so on. The definition proposed points to combinations of the same regularly recurring forms of relief, to microclimates but not to climate in general, to combinations of soils and soil units but not to soil cover in general, to combinations of phytocoenoses but not to vegetation in general, and so on. As the geographer observes combinations of the same forms of relief, the same biocoenoses, the same bodies of water, the same soils and soil units, he can be sure he is still in the same landscape. As soon as new forms of relief appear, however, which have not been seen before, or as soon as he notices other kinds of soil and phytocoenoses, it means the researcher has entered a new geographic landscape.

As an example, I would like to mention some simple landscapes of the Soviet Union. The simplest hilly moraine landscapes with lakes, in the northwestern part of the Kalinin region, present combinations of morainic hills and intervening depressions. Some depressions are occupied by lakes, others by swamps, and the third type are drained by rivers and streams. Depending on position relative to the elements of relief, regularly recurring combinations of microclimates, soil units, biocoenoses, and so on can be observed.

In the southern part of the Kursk region, the simplest landscape is formed by a combination of flat or slightly inclined watershed areas with rounded depressions and incised ravines, river valleys, and other erosional forms. Corresponding to these forms of relief, the other landscape elements will recur regularly.

In the northwestern parts of the Aral Kyzyl Kum sands, the following regularly recurring combinations can be observed (data supplied by A. G. Gayel, personal communication): (1) ancient river channels with weeds on solonchak soils and areas of green grass steppes with *Poa* species and *Artemisia lercheana*, (2) terrace-like rolling lowlands with wormwood steppes, and (3) intervalley ridges with sand hillocks and sparse plant cover.

One may see from these examples that the geographic landscape consists of several simple units. At first this would appear to be inconsistent with the thesis that the geographic landscape is the principal geographic unit. In reality, however, there is no contradiction because each of the constituent units is not unique. Either a few or many such units always exist in the landscape, and all of them constitute interrelated and recurring combinations. This phenomenon will be quickly revealed after division of the whole (i.e., the geographic landscape) into its constituent units.

Urotshistshe as Part of the Landscape

The phenomenon of the landscape consisting of several parts has been noted for a long time by many geographers. L. S. Berg named such parts *landscapes of the second order*. S. Passarge named them simply *parts of landscape (Landschaftsteilen)*; A. Ponomarev, *elementary landscapes*; and E. Markus, *natural complexes*. Recently, L. S. Berg has proposed the term *facies* for these parts of the landscape.

However, the term *facies*, which has been carried over from geology, is unsatisfactory when applied to large parts of the landscape. In geology, a facies is usually characterized by the homogeneity of sedimentary conditions that are manifest in the lithologic homogeneity of the strata and the fossils they contain. Within large areas of the geographic landscape, however, diverse natural conditions can be observed. It is natural, for example, that the upper parts of morainic hill slopes and the depressions between them differ in depositional characteristics, humifying characteristics, and soil units as well as in the plant communities and zoocoenoses linked with them. For this reason, it would be advantageous to apply another term in preference to the Russian one.

Berg once proposed the term *urotshistshe* instead of the term *landscape*. It seems to me that the term could be used in our science, not as applied to landscape (we had already rejected this idea) but rather as applied to those areas that make up the landscape. In this interpretation, the term *urotshistshe* is close and may even correspond in meaning to its use in everyday language. *Depressions with lake, swamp basin, sandy massif used for grazing*, or *small mountain valley* are terms used everyday by people; these are urotshistshe. The term has been frequently used in other geographic publications. In the *Proceedings of the All-Union Geographical Society* (1946, no. 4), the term appears in the articles by S. V. Zonn about the Zagedan Valley and by K. I. Rubtsov about the Betpak-Dala desert.

The use of the term *facies*, as proposed by Berg, should be reserved in teaching about the geographic landscape when it is applied to parts of geographic urotshistshes. One may then speak of facies of ravine bottoms, ravine slopes, beach ridges, alluvial cones, and so on.

The principal conclusion is the following: The geographic landscape itself represents regularly recurring and interrelated combinations of a few or numerous geographic urotshistshes. Geographic urotshistshes are built from geographic facies that are systematically located with gradual transitions between them.

Thus, the landscape evidently has a clear structure. This phenomenon facilitates research. The geographer need only investigate the details of one urotshistshe from each group typical of a certain landscape. After revealing the interrelations between these, the geographer can then extrapolate conclusions about existing regularities to the whole landscape.

The Age of Landscape

To understand landscape better, as a methodological procedure it is possible to imagine that there is both material and a cutting edge. In each geographic landscape, the material is that inherited by the landscape from previous stages of development and with a sharply different structure reflecting different physical-geographic processes. The present-day geographic processes are the so-called cutting tools acting on this material. In reality, however, the geographic landscape and the processes do not exist independently of one another. On the contrary, they are linked and affect each other.

Also, just as the cutting edge of a lathe acts on material and shapes it step by step and little by little, geographic processes also gradually rebuild the material—the geographic heritage seeking to adjust to the new structure.

The latter is unattainable, however. Complete harmony between landscape morphology and the pattern of the impact of physical-geographic processes is impossible. The reasons are as follows: (1) the most significant physical-geographic process—climate—is to a greater or lesser extent changeable in time; (2) endodynamic changes (self-development) constantly take place within the landscape; and (3) because landscapes have an influence on the processes themselves, this, too, will lead to their transformation.

It is impossible to understand landscape if we do not understand fully how the material is influenced by the processes and if we do not study the pattern in detail and the way in which processes and material influence one another.

Let us explain this idea by an example. In the northwestern part of the Kalinin region, the geographer observes a young landscape with morainic hills and lakes formed by different geographic processes, namely a glacial one. Present-day geographic processes—with the most significant component, humid climate—aim to transform this landscape into a typical erosional one. Because of the young age of the landscape, however, the observer sees river valleys not yet perfectly formed; a network of ravines and small, hardly noticeable, flat-bottom valleys; watershed areas not yet transformed by the drainage network; and numerous large and small lakes with irregularly shaped beds and strongly dissected shorelines filling depressions in the morainic relief.

However, as soon as we cross the boundary of the last glaciation, a change of landscape character is noticeable. This phenomenon was clearly revealed by the Kalinin expedition of the Institute of Geography of Moscow State University. Why is there a landscape change? Because the present-day geographic processes are dealing with another geographic heritage: a more ancient, better transformed moraine landscape. That is why here we can observe a well-developed drainage network and river valleys with a complete complex of valley forms and fewer lakes. Small lakes have already disappeared, filled with deposits and peat. The larger ones have acquired a new morphological structure: a simple shoreline, a bed with simple relief, and surrounded by swamp areas.

By these two examples, I have shown only the difference in relief resulting from different landscape age. However, it is natural that these differences also cause transformation and change in all other components of the landscape: surface waters, microclimates, soil units, phytocoenoses, and zoocoenoses.

As a general rule we can state that the greater the discrepancy between the pattern of present-day geographic processes and the geographic heritage, the more intensive will be its transformation and the more dynamic will be the landscape.

Thus, the present-day landscape is a complicated phenomenon consisting of an ancient landscape heritage transformed, more or less, by present-day processes. That is why it is possible to discern in each landscape both ancient, dying features (relicts in the broadest sense of the word) and newly forming and developing features that are the products of present-day geographic processes. In young landscapes, relict features of different kinds are numerous and distinguishable; in old ones, they are the opposite—they have been removed, are rare, or are manifest as different relict fragments. This is why the abundance, or the sparsity, of different relict features is a reliable criterion for determining the relative age of a landscape. To be clearly understood, I underline once more that I do not mean simply relict features of one landscape element, such as floral relicts, but the combined presence of various relicts: geomorphological, soil, botanical, zoological, and so on.

In speaking on the topic of the age of landscape, the following is a natural question: "What moment should be taken as the time of its origin?" I believe this should be the moment of the appearance in a particular territory of all the components named in the definition of landscape that have never been present there before. This moment usually, but not always, is accompanied by a complete change of geographic processes. The reasons for this are always allied either to a display of endogenous forces or to reasons of extraterrestrial origin. For example, the rise of a section of the lithosphere relative to sea level is the birth of a new landscape. The release of territory from glacial cover is also the moment for the origin of several landscapes. The formation of a new lava sheet after a volcanic eruption is a third example of the birth of a new landscape. The latter is an example of a sharp change in landscape components but without the subsequent complete transformation of landscape processes.

During the subsequent life of the landscape, numerous changes in geographic processes may occur that cause changes to the landscape, but each such change will mark not the so-called birth date of the landscape but only a new stage in its life and a significant landmark in its development.

Types of Geographic Landscape and Stability

With the exception of tectonics, I would say that, among all the components of physical-geographic processes, the most important and determining one would be the relationship between heat and moisture. The great role of this component in the pattern of physical-geographic processes is well demonstrated in the work of Academician A. A. Grogoryev. Therefore, I do not have to explain how the relationship of heat and moisture changes in different geographic zones. Here I would like to discuss another question.

The relationship between heat and moisture, to a great extent, determines the type of geographic process, both on plains and in mountainous landscapes, though they belong to different structural classes. Unfortunately, there is no opportunity to discuss this here. The most obvious indicator of the type of geographic process is the nature of vegetation: tundra, forest, steppe, desert, and so on. This is why Berg chose geobotanical features as the basis for the division of land into geographic zones. This indicator had been used previously by climatologists when defining climatic zones and even provinces. It does not mean, however, that geographic zones are simply geobotanical zones because each one has a corresponding type of climate, types of geomorphological processes, types of soil-forming processes, biochemical processes, and so on. In general, however, each of these specific processes is always determined by a certain relationship between heat and moisture.

Proceeding from all this, we can justifiably state that to each geographic zone or altitudinal belt in mountains there is a corresponding landscape. Thus, the definition of the type of landscape is rather broad but at the same time very convenient. It seems to me that with such an understanding of the term, geographers will speak the same language as climatologists, geobotanists, pedologists, and so on because such a geographic idea does not contradict, nor is it discordant with the terminology used by allied specialists.

Each local deviation from the characteristic heat and moisture relationship for a particular zone leads to a local change in the type of geographic landscape. The most evident example of this concept is the occurrence of oases and tugais in deserts.

An understanding of this geographic rule has great practical significance. Human beings, artificially creating oases in deserts for their own needs, will change the relationship between heat and moisture in a limited area in a way that is alien to that zone.

Postglacial changes in climate can be reduced to changes in the relationship between heat and moisture. The Atlantic period is characterized by the climatic optimum, which caused the broad-leaved forests to extend far to the north, beyond the boundary of their present-day area. The Boreal period, which preceded the Atlantic, strongly differed from the latter because of the different relationship between heat and moisture. Other climatic epochs may be analyzed in the same way.

Naturally, each time such climatic changes occurred, a transformation of the pattern of physical-geographic processes took place. This was imprinted on the features of the landscape and may be left as relics.

According to one group of researchers, the last transformation of the landscapes of the Russian plain took place 2,500 to 3,000 years ago when the Sub-Boreal climate was substituted by the Sub-Atlantic. According to other researchers, this date was much earlier. This is not the essence of the issue, but the fact that from this time onward, within the geographic zones of the Russian plain, the climate has not changed substantially, and we can conveniently consider it to be stable. This enables us to state that for a long period of time, a great number of landscapes have developed under the same regime of geographic processes. Nowadays, changes in the types of natural landscape can only be observed in the border regions of present-day zones. A slow southward movement of all zones is now taking place.

Therefore, adhering to my definition of the term *landscape type,* it is clear that in the postglacial period, the landscapes of the Russian plain have often been changed, but their type did not change every time or in every place. Replacement of one type by another occurred only at some distance from the borders of geographic zones. In the Boreal climatic period, the prevalent steppe type of landscape penetrated much farther northward than today. In the Atlantic period, a forest type prevailed at the location of the present-day tundra landscape. On the basis of what has been written previously, the following law states: Under the relative stability of both the climate and the tectonic state of the earth's crust, the type of geographic process operating on the largest part of the zone will be constant, and this will inevitably lead to the stability of its geographic landscapes.

Thus, it is necessary to understand clearly that although each geographic landscape continuously changes and never repeats a stage of development, once that has been passed, the type of geographic landscape is much more stable. We may consider that throughout historical times the geographic landscape type remained almost unchanged and that the evolution of the majority of landscapes remained within the one type.

The understanding of this law is also important for human beings. It means that anthropogenic changes to the natural landscape cannot radically change its type; if we allow the landscape to develop over time, it will restore all the inherent features of that type.

The power of physical-geographic processes is so great that human beings must continuously expend energy to support the landscape in the desired state. Thus, in forest zones and, in particular, in wet tropical rainforest zones, abandoned arable lands and roads are quickly covered with forest.

G. I. Tanfilyev wrote: "Steppe ploughed for sowing wheat, later abandoned as long fallow, will be restored to its initial vegetation in a few years" ("Essay on the Geography and History of the Main Cultural Plants," p. 5).

It is widely known that in the abandoned oases of Central Asia, the soil cover very quickly reverts to its former type inherited from the surrounding desert areas. The stability of these landscape types can become, with rational use, powerful techniques in human hands. The "Plan of Agricultural Development for 1943," approved by the Central Committee of the Communist Party and Sovnarkom, recommended that arable lands be left fallow as one of the effective tools in fighting weeds on heavily infested fields; meaning, it is proposed to utilize the stability of the landscape type and to allow the natural vegetation inherent to this type to force out undesirable alien weed species.

The Concept of Natural Potential

Now I will deal with the problem of the so-called natural potential of landscape.

Each geographic landscape possesses within it certain inherent natural capabilities. These are determined by the geographic heritage of former times, on the one hand, and by the possibilities given by the present-day pattern of geographic processes on the other.

Human beings do not study the landscape impartially. On the basis of our knowledge we study landscapes to gain an understanding of the possibility of actively intervening in the natural trajectory of their development, to reconstruct landscapes in a desirable way, and to make their potential capabilities serve human beings. That is why the determination of the natural potential of each landscape is the most significant responsibility of each geographer. With this approach to landscape, geography will be able to promote successful solutions to practical problems within the economy.

One must differentiate two parts to this question, namely, that of *natural potential* and that of *cultural-technical potential*. Natural potential is that inherited capability created by nature, and geographers are obliged to assess and determine this correctly. Cultural-technical potential increases quickly and continuously; because of this, the possibility of humans utilizing landscape also grows. What was impossible or useless yesterday may be found to be quite possible, valuable, and necessary today. Thus, both natural potential and selection—that is, cultural-technical potential—can be introduced to agriculture. Thanks to cultural-technical potential, the limits to possible agriculture and to the possibilities of introducing cultural plants are expanded.

The concept of both the constancy of the nature of physical-geographic processes and the stability of the geographic landscape type, when applied to extensive areas of geographic zones, simplifies, to a great extent, the determination of the landscape's natural potential. This gives only general solutions to the problem, however. Within each geographic zone, there are many modifications of the general zonal type, that is, particular geographic landscapes. Each particular geographic landscape has its characteristic pattern of physical-geographic processes. This is why the solution to the problem of the natural potential for each landscape is possible only in the context of it being studied in detail. That is, the genesis, present-day morphology, and the characteristic pattern of geographic processes so revealed are the basis of both the present-day dynamics and the future trajectory for development.

Each geographic landscape conceals a number of potential capabilities: geobotanical, soil, geomorphological, and so on. That is, the natural potential of a landscape consists of a number of distinct potentials intimately linked and that influence one another. As a landscape eternally lives and changes, however, so, too, its natural potential is constantly changing because of this interconnected complex of specific potentials. That is why it is particularly important to take into consideration the dynamics of landscape and to determine its trajectory and development accurately.

Let us examine the following example. The soil conditions, as well as the combination of heat and moisture in the southern part of the middle Russian uplands, are favorable to grain and intertilled crops. Humans recognized this long ago and began to utilize the natural potential capabilities of this region. Plowing of the steppes, however, causes increased soil erosion and the rapid growth of a ravine network. To prevent these negative consequences, which potentially existed in the landscape before, it is necessary to use agrotechnical measures as well as the natural geobotanical potential. The geobotanical potential is manifest in attempts by trees, shrubs, and grasses to occupy the slopes and upper reaches of ravines, thus preventing further growth of the ravine network. Humans may substitute fruit trees for this wild vegetation on these worthless lands.

We can see from this example that, from a human point of view, the specific potentials are not equal. This is why I differentiate them into three groups: (1) positive, (2) negative, and (3) neutral.

As examples of the utilization of positive potential we can cite the enrichment of the Moscow regional fauna by muskrat from North America, the introduction of the fish *Gambusia* to malarial water bodies of Transcaucasia and Central Asia, the introduction of Australian eucalyptus into the swamps of the Riony lowland and *Acacia dealbata* on the worthless lands on the southern Black Sea shore, and so on.

Examples of negative potential, which should always be taken into consideration, are salinization of soils in arid regions under artificial irrigation, land subsidence of loess loams subject to artificial irrigation in some regions of the foothills of the Caucasus and in Central Asia, overlogging of watershed areas because of timber cutting in the northern part of the Russian plain, and the catastrophic expansion of rabbits and *Opuntia* introduced into Australia.

An example of neutral potential is the rapid spread of *Elodea* introduced into European bodies of water from America.

It is clear from these examples that a great deal of attention should be paid to the determination of natural potential. Humans have already revealed a lot about specific potentials. Recently, particular organizations have been engaged in this process (e.g., the Soviet Bureau of Plant Introductions), but no one has yet analyzed the natural potential of landscape as a whole. Here there are a great many possibilities for research by geographers.

I would like to conclude by referring to the fine words of Friedrich Engels, which bear a strong relationship to what has been written here: "Indeed, every day we study, in order to understand in a more perfect way, the laws of nature, and to realize the close and distant consequences of our active intervention in natural processes.... We shall be more and more able to predict, and thanks to this, to regulate the distant consequences, at least of our most common productive activities" (*The Dialectics of Nature*, 6th ed., 57).

The Concept of Land Units and Land Systems

Proceedings of the Ninth Pacific Science Congress, 20:74–81 (1958)

C. S. Christian

PROCEEDINGS OF THE NINTH PACIFIC SCIENCE CONGRESS

INSET MAPS

A number of inset maps may be placed along the margin of both the Land-Use Survey map and the Ecological map. The "Carte de la Végétation de la France" at the scale of 1:200,000 has the following six inset maps at 1:1,250,000:

(A) Climax vegetation

(B) Soils
(C) Land use and orchards
(D) Agriculture
(E) Precipitation and temperature
(F) Agricultural hazards

For maps at a scale of 1:1,000,000 the inset maps can be at the scale of 1:5,000,000. Only maps (A), (C), (D), and (E) will be useful. The pedological map has little value at this scale.

DISCUSSION

THE CONCEPT OF LAND UNITS AND LAND SYSTEMS

C.S. CHRISTIAN

Chief, Division of Land Research and Regional Survey, C.S.I.R.O., Canberra, A.C.T., Australia.

The characteristics of the land surface, and its suitability for supporting mankind, vary in an infinite number of ways, and the need for some kind of areal subdivision and categorization to enable problems of land use planning and economic assessment to be approached systematically is evident enough not to require emphasis.

Many kinds of subdivisions are used, either for a specific purpose or in relation to a specific characteristic. Thus, there are produced distribution maps for such characteristics as physiography, climate, vegetation, forestry, pastures, soils, water resources, present land use, land capabilities, and a host of other features, each of which serves to subdivide territory according to the classes adopted for that characteristic. In some

instances the specific feature mapped is the dominant characteristic influencing land use potential, but its importance may not be the same in different parts of the territory. In many instances it may be of relatively little significance by itself, and only of value if considered in conjunction with other features. Thus correlation between the mapped classes of one characteristic and land use potential in a region may be high, low, or variable, depending partly upon its range of variability and that of other characteristics and also to no small degree upon the method of classification which has been used for the particular feature. For example, there are the widely different physical reactions which can occur in similar soils under different climatic conditions leading to the

VOLUME 20   HUMID TROPICS

conclusions that in this instance it is really the soil-climate combination which is the important determining factor in land use rather than soil or climate alone. As numerous factors may in turn be significant, the number of combinations of characteristics which might be considered could be multiplied indefinitely. Any approach to the subdivision of territory into similar and dissimilar areas on the basis of like or unlike combinations of characteristics such as can be shown by the use of a system of overlay maps or by other means could lead to a degree of complexity which would be very difficult to put to practical use. It would also be subject to any inadequacy of the methods of classification used for the individual characteristics.

An ever present problem of classifying a variable characteristic arises from the fact that it is the observable and recordable aspect which must be used for classification. It may not follow that this is the most significant feature of the characteristic for the particular purpose of land use considerations. Alternatively there may not have been opportunity for the prior testing of correlations so that the classes used are established on an arbitrary or subjective basis and may not be the most satisfactory for the purpose of differentiating between different land use potentials. The many attempts which have been made to express climate, and especially one facet of it, rainfall incidence, are good examples of how inadequate classifications can be for particular applications. They also draw attention to a further point: a standard method of classification may not be suitable for all purposes, and classifications may have to be modified to suit the requirements of different areas.

A consequence of such difficulties is that while the many methods of subdivision of territory used may have significance on a wide continental basis, or at the other extreme in very small areas where intensive correlative work has already been done, they are usually of limited value when applied to the order of variation of an intermediate nature, such as within geographic regions—the sort of circumstance with which the land use planner is frequently concerned. This is especially true if there has been little land use experience in the area to serve as a guide.

This was the situation in Australia when the extensive series of regional surveys were first commenced in 1946 by the Northern Australia Regional Survey, now the Division of Land Research and Regional Survey of the Commonwealth Scientific and Industrial Research Organi-

zation. The objectives of these surveys were to describe, classify and map, and assess the land use, developmental possibilities and technical problems of large areas of country about which there was relatively little recorded scientific information and in which there was not any long term traditional form of land use other than hunting and gathering of native foods. The order of magnitude of these surveys is indicated by the fact that in the period 1946-1956 nearly half a million square miles of country was mapped by one survey unit on the Australian mainland, and between 1953-1956 an additional area of 7,500 sq. miles was mapped by a second unit in the Territories of Papua and New Guinea.

It was clearly not possible in the time available to map such large areas in detail. Therefore, a method was sought whereby the mass of detail could be mapped as complexes of country, the detail being indicated but not mapped as such. The concept of land units and land systems was the basis on which an appropriate technique for comprehensive surveys and land classification on this scale was developed.

The concept has been used in the field for a period of over ten years, in a wide range of conditions varying from arid Central Australia to the wet tropics of New Guinea, and has been found to have general applicability and to provide a satisfactory procedure for the systematic survey and application of large areas of country.

The actual methods of procedure have been described by Christian and Stewart (7) and Christian (4) and their particular applications to New Guinea by Taylor and Stewart (9).

THE CONCEPT OF LAND, LAND UNITS AND LAND SYSTEMS

LAND

The word land is used to refer to the land surface and all its characteristics of importance to man's existence and success. It is the integration of all such factors rather than mere likeness or unlikeness in some of the more obvious observable characteristics which determines the similarity or dissimilarity of areal subdivisions in respect to land use potential. The concept visualises that each part of the land surface is the end product of an evolution governed by parent geological material, geomorphological processes, past and present climates and time. During this period the land surface has been shaped to existing land forms, each developing in the process its own hydrological features, soil mantle, vegetation communities, animal populations, and range of microenvironments.

THE LAND UNIT

Where such parts of the land surface can be identified as having a similar genesis and can be described similarly in terms of the major inherent features of consequence to land use—namely, topography, soils, vegetation and climate—they are regarded as being members of the same land unit. It is almost axiomatic that there will also be a general similarity in climate, for otherwise differences in soils, vegetation, or land form would almost certainly occur. The common genesis of the various occurrences of one land unit also implies that they may be similar in many respects additional to those most easily observed, and this is likely to be expressed in the technical problems associated with land use development.

The degree of simplicity or complexity of a land unit is determined by the nature of the land form accepted as the unit of study. If this is a whole mountain structure it will have a variety of slopes and aspects, each with its own combination of soils and vegetation, and each representing a "site" as defined by Bourne (1). The land unit can range from a land form of this order of complexity down to the simplest topographic unit when it will be identical with Bourne's site.

It may be argued that considerations of genesis are not applicable as genesis can be deduced only from observations, and therefore it is the observation rather than the deduction from it which should be the basis of identification. However, the genesis of a land unit is interpreted not only from observations of a variety of characteristics of the land unit itself but also from the interrelationships of all land units in the landscape as a whole. Thus, external as well as internal evidence is used for interpretation, and in this instance, therefore, identification on the basis of genesis does have practical and scientific significance.

THE LAND SYSTEM

The land system is an assembly of land units which are geographically and genetically related. The term was initially defined by Christian and Stewart (6) in the advance report of the Katherine-Darwin region as "a region throughout which a recurring pattern of topography, soils and vegetation can be recognized." The definition and description was expanded in the printed revised report (7) as follows: "We have defined this unit which is a composite of related units, as an area, or group of areas, throughout which there is a

76

recurring pattern of topography, soils and vegetation. A change in the pattern determines the boundary of a land system," and "A simple land system is a group of closely related topographic units, usually small in number, that have arisen as the products of a common geomorphological phenomenon. The topographic units thus constitute a geographically associated series and are directly and consequentially related to one another."

The authors also recognized *Complex* and *Compound* Land Systems, the former being a group of related simple land systems, the latter a grouping made merely for mapping convenience. The kind of land systems most commonly mapped by the Division of Land Research and Regional Survey are either complex or compound. The simple land system is in much the same category as the "region" of Bourne(1).

In practice the boundaries of the land system are mapped, but the kind of detail which occurs within each is indicated by means of a diagrammatic cross section and tables. This shows the various land forms which comprise the land system and their associated characteristics and relative sizes. The actual areas of the land systems as a whole are calculated from the map so that there is a basis for judging the relative importance of the various land systems or of any particular land unit within them.

SOME GENERAL IMPLICATIONS OF THE CONCEPT

It should be emphasized that the land system is not just a convenient association of land units. Apart from the use of the compound land system, it is implied in the concept that, while the assembly of land units constitutes a recognisable and recurring pattern, all the component land units have been produced as the result of a particular channel of land surface evolution, that channel differing from that of adjacent country either in parent material, geomorphological processes, or in the length of the period over which these processes have been operating. The land system is therefore a natural aggregation of natural units. In the same way as land units may be dissimilar in unrecorded features irrespective of their appearance, because of different origins, so will the aggregation of land units into land systems imply that land systems as a whole may have unrecorded, but characteristic differences of significance to land use. As an example of this, Christian, Jennings and Twidale (5) quoting the findings of Hubble, describe two areas of heavy grey pedocal soils which cannot be differentiated by field characteristics of the soils themselves whereas the areas occupied by them can be differentiated on morphogenetic grounds. The phosphate status of one area is very low, but is satisfactory in the other. The boundary of these areas of soils of different phosphate status can be mapped geomorphologically after a relatively small amount of correlative work, but to map the boundary from field crop trials or by laboratory examination of soil samples would require a tremendous amount of effort. In such ways genesis can be expected to provide a basis of differentiation even though the significance of the differentiation cannot be interpreted at the present stage of knowledge at the time of the survey.

In practical application the significance of the land unit is that it enables a complex region to be analysed into all its component areas of different character or potentiality. The importance of the land system is that it aggregates these units into areas large enough to be mapped at the scale of working, but small enough in number to facilitate regional comprehension and planning.

The concept can be applied at any scale of working and it has the advantage that both the land unit and the land system are flexible units which can be adjusted in respect to degree of complexity, at the same time maintaining their logical relationship to one another. Thus, working on a very extensive scale, land units may represent such gross land forms as mountains, valleys, alluvial plains or plateaux, grouped according to their geomorphological relationships into land systems. On a moderately intensive scale these units may become the land systems, with the various slopes and aspects of the mountains or valleys, the various kinds of alluvial deposits of the flood plains, or the units of microtopography of the plateau, as the land units. On a very intensive scale further subdivision of parts of these units would provide the land units; and the survey would approach in nature a combination of a detailed ecological and soil survey, the land unit maintaining its land character as being a recurring topographic unit together with its characteristic soil or soils and its characteristic vegetation.

The concept is most easily applied to virgin country which has not been greatly modified by man, or to country so intensively developed that particular forms of land use have become characteristic of the land units. This is true of parts of Europe and some of the densely populated tropical countries. A more complex situation arises when land is developed only partly, and the association of land unit and land use is not complete.

The original nature of the land unit, particularly its vegetation, may be obscured by partial land use. In such circumstances the available evidence from remnants of the original vegetation associated with cultivated areas, from the land form itself, its soils, and its relationship to other land forms, will usually place an area in its correct category and enable the correlation between land units and the ways that experience has demonstrated they can be used to be recognized. Where shifting agriculture has been practised and various stages of regrowth occur, it is helpful if ecological study can relate these to and express them as stages of a particular mature vegetation.

SOME IMPLICATIONS OF THE CONCEPT TO THE STUDY OF SOILS AND VEGETATION

The concept provides a very useful approach to understanding both soils and vegetation. In both spheres there is the need for a classification based on characteristics of the subject being studied, but such a classification is often unsuitable for mapping purposes and may not be the most significant from an ecological point of view. In the past, this difficulty in the mapping of soils has been overcome largely by the use of the soil association which is a geographically associated group of soils. Major geomorphological subdivisions have often been used as convenient geographical subdivisions in soil surveys, but it is only relatively recently that pedologists have recognized the full significance of geomorphic origin to the derivation of landscapes and soil patterns (e.g., 2, 3). In the regional surveys conducted by the Division of Land Research and Regional Survey, the concept of land units and land systems and their genesis has substantially assisted towards a better understanding of the soils of the regions.

The mapping of vegetation has often been confused by varying emphasis on structure, floristics, and ecology. As the application of the concept of land units and land systems provides a natural subdivision into areas with distinctive environments, it automatically subdivides vegetation into ecological units which may then be aggregated into classes according to the objective in mind. Not only has the concept assisted in the ecological understanding of vegetation but it has also proved of value in the making of subdivisions of vegetation into floristic units. Where the changes in the land surface are sharp and of a major nature, the vegetation communities usually change just as markedly, and distinctive communities may be recognized and differentiated on a clear-cut basis. Where, however, the variations in the land surface are minor or gradual, the re-

PROCEEDINGS OF THE NINTH PACIFIC SCIENCE CONGRESS

arrangement or replacement of species in communities is often slight, and floristically constant communities are difficult to discern on a reconnaissance survey. The land unit can often provide a guide as to what subdivisions should be made. In such circumstances a truer picture is presented if the vegetation is regarded as a complex, with an indication of the variations associated with specified variations in the land surface, rather than if an attempt is made to separate constant floristic units which, if they occur at all, may occupy only a small proportion of the area relative to the gradations between them.

Insofar as one may expect the nature of the vegetation of any area to have been influenced by the history of the land surface, a knowledge of the geomorphic origin of land provides a basis for studies in the field of plant geography as exemplified by the work of Crocker and Wood *(8)*. The concept of land units and land systems offers scope for added precision in this type of study both of short and long term changes, but as yet this possibility has not been developed.

SOME WIDER APPLICATIONS

The use of the concept by the Division of Land Research and Regional Survey has been restricted to land use aspects, and mostly to virtually undeveloped regions. The maps produced by the surveys have been used in a variety of ways by other authorities, as for example the preliminary planning of access roads into new areas. It is believed that the basic classifications of environments provided by the concept could have applications in a variety of fields of which just a few may be quoted as examples. At one extreme one can visualize its assistance in anthropological studies of primitive peoples, for example where village habits, practices, and mode of living are influenced by the nature of the habitat. It would have its application to ecological studies of all kinds. In geographic and economic studies of advanced communities, the areal subdivision provided by land unit and land system mapping can serve as a very satisfactory framework for making a systematic sampling and establishing correlations. It has the advantage of being a subdivision into natural units of landscape and therefore provides a common basis for a wide range of studies.

ASSESSMENT OF LAND USE POTENTIALITIES

It is inevitable that whatever the stages of development of the area be at the time of land

classification, whether it be virgin land or intensively developed land, present knowledge of the land use problems and what constitutes the best form of land use, will be modified as more information is gained or as markets change. For this reason, it is desirable that the land classification of a region be based on fundamental qualities of the land itself and be independent of our present knowledge of land use. Land classification and assessment should, therefore, be regarded as separate phases of the investigation although in part the two may be done concurrently.

The present concept can be applied to land use studies in relatively unknown and undeveloped areas where a comprehensive assessment of possibilities has to be made or areas which are well developed but in which, for economic or other reasons, a reassessment is necessary to achieve a better adjustment of actual land use to land capability.

Successful planning for the optimum land use of an area requires a knowledge of:

1. its inherent potentialities and limitations.
2. the degree to which these may feasibly be modified.
3. the technical problems which must first be solved.
4. the difficulties which may be involved in achieving optimum development summarized in some form expressing labour, cost, and time factors.
5. the risks of deterioration in natural resources such as the permanent loss of forests, soil erosion, seepage, salting, and loss of fertility if a given land use policy is practised.

The investigator may have available to him experience of land use already gained in the area, or he may have to resort to the less reliable procedure of making comparisons with other areas near or far away. If neither of them are adequate, the information may have to be sought by experimentation, and it may be only through experimentation that the real limitations or technical problems will be defined.

The mapping of land systems will indicate those parts of the region which warrant the most attention and those which might well be left for later examination. Beyond this it is the productivity of the individual land unit and the feasibility of its use in relation to other land units associated with it that must first be determined. The application of knowledge gained about one land unit can confidently be extended to all occurrences of the same land unit in the same land system, but it should not be assumed that it applies to apparently similar land units of other land systems without confirmatory trials. Thus, the classification of land into land units and land systems not only permits the systematic planning of where further information should be sought, but it also indicates the extent and limits of application of new land use information as it is gained.

The theoretical potential of a given land unit may not represent a realistic picture of what is attainable. Because of transport difficulties, an agriculturally productive unit distributed as small areas throughout an otherwise unproductive land system may be very suitable for self-sufficient agriculture but entirely unsuitable for production for an export market irrespective of its productive capacity. On the other hand, intermittent land units with different potentialities will have advantages in respect to diversity of production, expressed as an ability to adjust production to changing market demands, or as the possibility for the internal consumption of the product of one industry by another for conversion into an exportable form. The use of grain from intensive agricultural areas by an animal industry based on an associated non-agricultural country, with the animal products being exported is an example. The distribution of land units and the sizes of their individual occurrences are therefore important factors in assessing regional potentialities and planning the kind of regional development. The mapping of land systems is a logical and satisfactory way of expressing and summarising this type of information for different parts of regions.

The potential of a land unit can be substantially changed through drainage or irrigation if it has agricultural possibilities, for example, through the provision of water for stock if it is an arid pasture area. The possibility of making such changes may be a function of the association of the land units, rather than the characteristics of the land unit itself. The land system will indicate the likelihood of irrigation schemes being engineering possibilities; the land unit will indicate the agriculture possibilities.

The survey of the Katherine-Darwin region provides an example of the sequence of events which may be followed in an undeveloped area. The region, which covers about 27,000 square miles, was subdivided by regional survey into 19 land systems ranging in area from 42 to 6,890 square miles. The number of land units identified in the land systems varied from 2 to 11.

One of the major recommendations concerning land use related to the Tipperary Land System in which there appeared to be prospects for dryland farming. This land system is about 7,000 square miles in area and has an annual rainfall ranging from about 30 to 45 inches. The soils are mostly of sandy to medium texture and have low to medium mineral fertility. The survey had to take note of this limitation and also of the limited distribution of rainfall, most of which occurs intermittently during a four-month period in the summer. Apart from limited fruit and vegetable production on favoured areas in the region, there was no agriculture in this climatic zone in Australia; and the nearest agricultural development of any magnitude was over 1,000 miles away. Direct comparison within the country was not possible, and therefore a Research Station was established at Katherine within the land system to determine the economic possibilities of a range of crops and pastures and any technical problems involved in their production. Investigations at this station demonstrated an acute deficiency in phosphorus on that land unit but reasonable production of adapted crops when this element was supplied. Yields were further enhanced in most instances by the addition of a nitrogenous fertilizer. A crop establishment problem was encountered due to a combination of certain physical properties of the soil and marked fluctuations in atmospheric conditions during the normal planting periods. This required a special study of land preparation to determine the best methods of seed bed preparation for reliable establishment. Other land preparation studies indicated the advantage of deep ploughing during the dry season over shallow or wet season ploughing. Crop and variety studies showed that sorghum and peanuts were the most reliable crops and indicated the best varieties for the district. Phenological studies established the best time of planting and certain relationships between climate and crop development. Observations on other land types with differing physical properties have indicated somewhat different crop responses and different potentialities. All these studies, and others, are contributing to establishing what is, and how to produce, a crop of reasonable standard on each land unit. Before this information can be applied to planned agricultural development, further information is required on costs of bringing land into production, farm capitalization costs, costs of crop production, labour requirements, crop returns, and any special farm problems arising from the cultivation for larger areas. This information which is being sought by a Demonstration Farm established by the Northern Territory Administration is essential before economic farm size or financial requirements for farm development and maintenance can be determined. When all this information is available, it should be possible to plan for a stable form of land use and to allot land to farms of a predetermined economic size, knowing the production potentialities of the various land units and the relative areas of each of which should be associated to ensure stability and economic success.

REFERENCES

(1) Bourne, R., 1931, Regional Survey and Its Relation to Stocktaking of the Agricultural and Forest Resources of the British Empire. *Oxford Forestry Memoirs,* **13.**

(2) Butler, B.E., 1950, A Theory of Prior Streams as a Causal Factor of Soil Occurrence in the Riverine Plain of Southeastern Australia, *Aust. J. Agric. Res.,* **1:** 231-52.

(3) ——————, 1956, Parna.—an Aeolian Clay, *Aust. J. Sci.,* **18:** 145-151.

(4) Christian, C.S., 1952, Regional Land Surveys, *J. Aust. Inst. Agric. Sci.,* **18:** 140-146.

(5) Christian, C.S., Jennings, J.N., and Twidale, C.R., 1957, Chapter on Geomorphology in *Guide Book to Research Data for Arid Zone Development.* Unesco, Paris.

(6) Christian, C.S. and Stewart, G.A., 1947, *North Australia Regional Survey, 1946, Katherine-Darwin Region.* General Report on land classification and development of land industries. Mimeo., Melbourne.

(7) Christian, C.S. and Stewart, G.A., 1952, Survey of Katherine-Darwin Region, 1946. *C.S.I.R.O. Aust. Land. Res. Ser.,* **1.**

(8) Crocker, R.L. and Wood, J.G., 1947, Some Historical Influences on the Development of the South Australian Vegetation Communities and their Bearing on Concepts and Classification in Ecology, *Trans. Roy. Soc. S. Aust.,* **71:** 91-136.

(9) Taylor, B.W. and Stewart, G.A., 1956, *Vegetation Mapping in the Territories of Papua and New Guinea Conducted by C.S.I.R.O.* (Paper presented at UNESCO Symposium on Methods of Study of Tropical Vegetation, Kandy, 1956.)

DISCUSSION

L.D. STAMP: Mr. Chairman, can minor land units as defined by Dr. Christian be regarded as ecosystems?

F.R. FOSBERG: Ecosystems can vary in size from a drop of water or smaller to the whole earth, or even the universe itself, but the size is usually determined by what is convenient for study and understanding. Certainly these land units can be treated as ecosystems.

A.W. KÜCHLER: Dr. Christian's ideas and methods are admirable. But his criticism that vegetation maps are too general and do not portray the site comprehensively seems unfounded. There are excellent vegetation maps by Buchwald, Jahn, Hohmeyer, and especially by Ellenberg that bring the very information which Dr. Christian is looking for in Australia. These maps are an integration of the work of various specialists and serve to establish the highest potential land use. The specialists, incidentally, are needed as much in known areas as in unknown areas.

C.S. CHRISTIAN: I agree that the same result can be obtained by detailed studies in small areas. I am more concerned with providing a systematic integrated picture over very large areas quickly so that there will be a permanent framework into which later information can be fitted.

F.R. FOSBERG: It must be emphasized that the excellent systems of integration presented by Dr. Christian and Professor Gaussen are not mutually exclusive. It would be very interesting to see the two systems applied to the same piece of land. Thus, perhaps, a degree of understanding would be brought out that would exceed that gained from the two systems used separately.

It should be pointed out that the examples used by both Professor Gaussen and Dr. Christian are arid zone areas.

In humid tropical areas the difficulties are increased — the geology is largely covered up by the rain forest. Even topography is obscured. The vegetation is much less well understood; the physical difficulties of travel are enormously greater. It is highly unlikely that the same team of specialists that mapped a half million square miles in northern Australia could have done nearly that much in a humid tropical area.

R.E. HUKE: Dr. Christian, on what scale are the slopes measured; are they regional or local slopes? How are they measured? I ask because in my work in Mindanao I found that areas corresponding to your peneplain local slopes, many of which would fail to show on 20-meter contour maps, were the limiting factor for future agriculture. On the other hand, local changes of slope were relatively unimportant.

C.S. CHRISTIAN: Slope is very important locally. Regional slopes are estimated from barometric readings checked by base records according to time of day. Local slopes are measured with an Abney level but are not mapped. A range of slopes characteristic of the land unit is incorporated in the description of the unit.

This raises the question of extent of detail which should be mapped. I am of the opinion that there should be a planned sequence of mapping, commencing with the broad reconnaissance, followed by more detailed work on selected areas, the remainder being disregarded at that stage. As information about selected areas is gained, so can surveys be in more detail and designed with more and more specific objectives. Thus following a reconnaissance and subsequent investigation, detailed surveys may be for the purpose of mapping soils, hydrological features, native pastures, slope, or whatever factor or group of factors the investigations on land use may have indicated are the most significant in any particular area.

C. TROLL: The discussion following Mr. Christian's paper on land units, their ecological character and cartographic representation concerns in a broad respect the problem of scale and of generalization which is necessary in changing from a larger to a smaller scale. The landscape pattern of the world comprehends a long series of natural land units at different scales, a "hierarchy of units," which begins with the climatic, vegetation, soil, and landscape belts on the one hand, and ends with unit areas which can be reproduced only on a large scale.

But the "Unit Area" as the smallest geographical individual land unit is subdivided into a certain number of smaller ecological units which exist in a certain area as a certain set of types, in a certain association or complex and in a certain pattern. These ecological units or "sites" (R. Bourne) or "stows" (Unstead) I denominated "ecotopes" broadening the purely biologic term "biotope." These ecotopes normally can be reproduced only on maps of a scale of about 1:25,000. Detailed vegetation maps in Central Europe are generally based on this scale.

However, each biotope or ecotope can be again subdivided into still smaller units which H. Gams calls "synusia." On a rock, for instance, which is often used by birds as a perch, we can distinguish a certain zonation of nitrophilous lichen associations, or on a tree trunk a zonation of algae, lichens, and mosses according to the exposure to rain and to the distance from the water evaporating soil surface. For a cartographic reproduction of such smallest units not even a scale of 1:1,000, such as used by Friedel recently for his vegetation map of the glacier foreland of the Pasterze (Austrian Alps), is adequate.

MAN'S INFLUENCE ON TROPICAL VEGETATION

E. AUBERT DE LA RÜE

16 rue Ribera, Paris 16e, France

INTRODUCTION

The physiognomy of the natural vegetation of the humid tropics, the only zones to be studied here and which cover some 28 million square kilometers or roughly 36% of the earth's land surface, until the present has remained somewhat better preserved than in temperate countries where man-made landscapes usually spread before the eye. While the face of the tropics has remained more natural, the plant cover is far from being preserved intact everywhere. Profound but irregularly distributed changes have taken place since very ancient times and have been increasing rapidly in the last hundred years.

Let us admit unhesitatingly that man's influence has often had catastrophic effects on the natural vegetation of the tropics, and that their consequences are, in the end, disastrous for man himself in many cases.

South Asia and the Malay Archipelago, over-populated and for the greater part deforested, are generally well cultivated and tend to become immense gardens, as have many tropical islands. Africa, on the contrary, devastated by dangerous agricultural practices and repeated brush fires tends toward the desert. Relatively spared so far, tropical America has long retained somewhat intact its primitive plant cover but it has been rapidly deteriorating in the last fifty years. Here is only, of course, a simplified sketch of the state of things, reality being much more complex.

The natural plant physiognomy of the humid tropics is above all, that of the great rain forest, associated with semi-deciduous forest in zones less humid and with a better defined dry season, with littoral mangrove and inland marshes, and with large areas of savanna of climatic or edaphic origin, usually containing trees, as in North Australia, various regions of Amazonia and of the Venezuelan Llanos, all areas which seem to be natural in origin and independent of any human intervention.

THE CAUSES OF DEFORESTATION

The primitive rain forest which in theory has never been affected by man's activities even back to the most ancient times is a survivor of the past and soon this designation will have no more meaning. While most of the other vegetation types have been deeply modified, this forest still remains relatively intact in a few privileged lands, but its exuberant and imposing aspect hides its great vulnerability. It has been able to persist until today only in sparsely inhabited territories. The hungry crowds cannot tolerate its existence. In tropical Asia and its great archipelagos, it has become almost unknown, except for a few great forest tracts remaining unchanged in Sumatra, Borneo, and New Guinea and located away from

UNIVERSITY OF CALIFORNIA PUBLICATIONS

IN

GEOGRAPHY

Vol. 2, No. 2, pp. 19–54   October 12, 1925

THE MORPHOLOGY OF LANDSCAPE

BY

CARL O. SAUER

_____

INTRODUCTION

Diverse opinions regarding the nature of geography are still common. The label, geography, as is the case with history, is no trustworthy indication as to the matter contained. As long as geographers disagree as to their subject it will be necessary, through repeated definition, to seek common ground upon which a general position may be established. In this country a fairly coherent series of viewpoints has been advanced, especially through presidential addresses before the Association of American Geographers, which may be accepted as mirror and mould of geographic opinion in America. They are sufficiently clear and well known that they need not be restated.[1] In European geography a somewhat different orientation appears to be developing. In various quarters significant activity is being displayed, probably in some measure influenced by anti-intellectualist currents. At any rate a shaking up of some vigor is under way. It may therefore be appropriate to reëxamine the field of geography, keeping current views abroad especially in mind, in order to attempt a working hypothesis that may serve to illuminate in some degree both the nature of the objective and the problem of systematic method.

_____

[1] In particular the following addresses are notable expressions of leading opinion: Davis, W. M., "An Inductive Study of the Content of Geography," Bull. Am. Geog. Soc., vol. 38, pp. 67–84 (1906); Fenneman, N. M., "The Circumference of Geography," Ann. Assoc. Am. Geog., vol. 9, pp. 3–12 (1919); Barrows, H. H., "Geography as Human Ecology," ibid., vol. 13, pp. 1–14 (1923).

THE FIELD OF GEOGRAPHY

The phenomenologic view of science.—All science may be regarded as phenomenology,[2] the term science being used in the sense of organized process of acquiring knowledge, rather than in the common restricted meaning of a unified body of physical law.   Every field of knowledge is characterized by its declared preoccupation with a certain group of phenomena, which it undertakes to identify and order according to their relations.   These facts are assembled with increasing knowledge of their connection; the attention to their connection denotes scientific approach.   "A fact is first determined when it is recognized as to limits and qualities, and it is understood when it is viewed in its relations.   Out of this follows the necessity of predetermined modes of inquiry and of the creation of a system that makes clear the relation of the phenomena. . . .   Every individual science is naïve as a special discipline, in so far as it accepts the section of reality which is its field *tel quel* and does not question its position in the general scene of nature; within these limits, however, it proceeds critically, since it undertakes to determine the connection of the phenomena and their order."[3]   According to such definition of the grounds of knowledge, the first concern is with the phenomena that constitute the "section of reality" which is occupied by geography, the next with the method of determining their connection.

Geography as a 'naïvely given section of reality.'—Disagreement as to the content of geography is so great that three distinct fields of inquiry are usually designated as geography: (1)   The study of the earth as the medium of physical processes, or the geophysical part of cosmologic science; (2) the study of life-forms as subject to their physical environment, or a part of biophysics, dealing with tropisms; and (3) the study of the areal or habitat differentiation of the earth, or chorology.   In these three fields there is partial accordance of phenomena, but little of relation.   One may choose between the three; they may hardly be consolidated into one discipline.

The great fields of knowledge exist because they are universally recognized as being concerned with a great category of phenomena. The experience of mankind, not the inquiry of the specialist, has made

---

2 v. Keyserling, H., Prolegomena zur Naturphilosophie, p. 11 (1910).

3 *Ibid.*, pp. 8 and 11.

the primary subdivisions of knowledge. Botany is the study of plants, and geology that of rocks, because these categories of fact are evident to all intelligence that has concerned itself with the observation of nature. In the same sense, area or landscape is the field of geography, because it is a naïvely given, important section of reality, not a sophisticated thesis. Geography assumes the responsibility for the study of areas because there exists a common curiosity about that subject. The fact that every school child knows that geography provides information about different countries is enough to establish the validity of such a definition.

No other subject has preëmpted the study of area. Others, such as historians and geologists, may concern themselves with areal phenomena, but in that case they are avowedly using geographic facts for their own ends. If one were to establish a different discipline under the name of geography, the interest in the study of areas would not be destroyed thereby. The subject existed long before the name was coined. The literature of geography in the sense of chorology begins with parts of the earliest sagas and myths, vivid as they are with the sense of place and of man's contest with nature. The most precise expression of geographic knowledge is found in the map, an immemorial symbol. The Greeks wrote geographic accounts under such designations as periplus, periodos, and periegesis long before the name geography was used. Yet even the present name is more than two thousand years old. Geographic treatises appear in number among the earliest printed books. Explorations have been the dramatic reconnaissances of geography. The great geographic societies, justly have accorded a place of honor to explorers. "Hic et ubique" is the device under which geography has stood always. The universality and persistence of the chorologic interest and the priority of claim which geography has to this field are the evidences on which the case for the popular definition may rest.

We may therefore be content with the simple connotation of the Greek word which the subject uses as its name, and which means most properly areal knowledge. The Germans have translated it as *Landschaftskunde* or *Länderkunde,* the knowledge of landscape or of lands. The other term, *Erdkunde,* the science of the earth in general, is falling rapidly into disuse.

The thought of a general earth science is impossible of realization; geography can be an independent science only as chorology, that is as knowledge of the varying expression of the different parts of the earth's surface. It is, in the first place, the study of lands; general geography is not general earth science, rather it presupposes the general properties and processes of the earth, or accepts them from other sciences; for its own part it is oriented about their varying areal expression.[4]

With this preference of synthetic areal knowledge to general earth science the entire tradition of geography is in agreement.

The interdependence of areal phenomena.—Probably not even the adherents of other, recent schools of geography would deny place for such a view of the subject, but they deem this naïvely given body of facts inadequate to establish a science, or at the most would consider it an auxiliary discipline which compiles fragmentary evidence, to find its place ultimately in a general geophysical or biophysical system. The argument then is shifted from the phenomenal content to the nature of the connection of the phenomena. We assert the place for a science that finds its entire field in the landscape on the basis of the significant reality of chorologic relation. The phenomena that make up an area are not simply assorted but are associated, or interdependent. To discover this areal "connection of the phenomena and their order" is a scientific task, according to our position the only one to which geography should devote its energies. The position falls only if the non-reality of area be shown. The competence to arrive at orderly conclusions is not affected in this case by the question of coherence or incoherence of the data, for their characteristic association, as we find them in the area, is an expression of coherence. The element of time is admittedly present in the association of geographic facts, which are thereby in large part non-recurrent. This, however, places them beyond the reach of scientific inquiry only in a very narrow sense, for time as a factor has a well-recognized place in many scientific fields, where time is not simply a term for some identifiable causal relation.

Historical development of chorologic relation into scientific system. —The older geography was troubled but little by critique. It was casually, even trivially, descriptive rather than critical. Yet though it is idle to seek in most of this literature a "system which makes clear the relation of the phenomena," we cannot dispose of all of it as accidental or haphazard in content. In some measure the notion of areal interdependence of phenomena as giving rise to areal reality

[4] Hettner, A., Methodische Zeit und Streitfragen, Geog. Ztschr., vol. 29, p. 37 (1923). Hettner is cited here in the latest statement of the position he has defended ably for many years. To American geographers Fenneman's address of 1919, cited above, is ever memorable for its spirited declaration of the same thesis.

was present, as any reader of Herodotus or Polybius knows. The *historia* of the Greeks, with its blurred feeling for time relations, had a somewhat superior appreciation of areal relations and represented a far from contemptible start in geography.[5] However much it may have been embroidered by geophysical, geodetic, and geologic notes, classical geography in general, not cosmology subsequently interpreted by some as geography, gave primary emphasis to areal description, with frequent observations on the interrelation of areal facts. The culminating school, of which Strabo was chief, was by no means entirely naïve, and rejected vigorously other definition of geography than as chorology, with express exclusion of cosmologic philosophy.

During the period of great discoveries a bona fide but uncritical geography attained its greatest development in the numerous travel relations and especially in the cosmographies of that time. An ever increasing body of facts about countries was at that time being brought before the Western World, which took keen interest in the rapidly widening horizon. With such a deluge of newly acquired facts about parts of the world, attempts at systematic ordering were numerous, but often grotesque rather than successful. It is not surprising that dynamic systems of geography should have emerged only as the furore of exploration became spent. Yet it is perhaps even more difficult for us to judge the thought of this period than that of classical antiquity. Yule has helped us to better appreciation of the geographic acumen of some of the men of this period. Of the cosmographers, at least Varenius has been accorded a higher rank than that of a compiler. One very great step in synthesis certainly took place at this time, that of the development of cartography into a real chorologic discipline. Only through a large amount of classification and generalization of geographic data was it possible to consolidate the scattered and voluminous data of exploration into the geographically adequate maps that characterize the latter part of the period. To this day many of the maps of the seventeenth and eighteenth centuries are in some respects monumental. However much may have been added since in precision of measurement, in many ways we have retained the chorologic content as formulated in the maps of this period beginning the 'Age of Surveys.'[6] "Every map which reproduces

---

[5] v. Humboldt, A., Kosmos (1845), vol. 1, pp. 64, 65: "In classical antiquity the earliest historians made little attempt to separate the description of lands from the narration of events the scene of which was in the areas described. For a long time physical geography and history appear attractively intermingled."

[6] Peschel's *Zeitalter der Messungen.*

the form of the earth's surface is a kind of morphologic representation.''[7] Not only for physical morphology, but for the cultural expression of landscape, these maps represented a highly successful series of solutions that are still used. Without such a preliminary synthesis of the facts of geography the work of the next period would have been impossible.

In the nineteenth century the contest between the cosmologic and the chorologic views became acute and the situation of geography was much in doubt. Rationalism and positivism dominated the work of geographers. The milieu became a leading doctrine and thus continued through the century. Divine law was transposed into natural law, and for geography Montesquieu and Buckle were prophets of major importance. Since natural law was omnipotent the slow marshaling of the phenomena of area become too tedious a task for eager adherents of the faith of causation. The areal complex was simplified by selecting certain qualities, such as climate, relief, or drainage, and examining them as cause or effect. Viewed as end products, each of these classes of facts could be referred back fairly well to the laws of physics. Viewed as agents, the physical properties of the earth, such as climate in particular with Montesquieu, became adequate principles for explaining nature and distribution of organic life. The complex reality of areal association was sacrificed in either case to a rigorous dogma of materialistic cosmology, most notably in American physiography and anthropogeography. About twenty years ago the most distinguished American geographer took the position ''that neither the inorganic nor the organic elements which enter into geographical relations are by themselves of a completely geographic quality; they gain that quality only when two or more of them are coupled in a relation of cause and effect, at least one element in the chain of causation being organic and one inorganic. . . . Any statement is of geographical quality if it contains a reasonable relation between some inorganic element of the earth, acting as a control, and some element of organic existence . . . . serving as a response.'' Indeed in this causal relation was, he said, ''the most definite, if not the only unifying principle that I can find in geography.'' Cause was a confident and alluring word and causal geography had its day. The *Zeitgeist* was distinctly unfavorable to those geographers who thought that the subject was in no wise committed to a rigidly deterministic formula.

[7] Penck, A., Morphologie der Erdoberfläche (1894), vol. 1, p. 2.

Latterly, Vidal de la Blache, in France, Hettner, Passarge, and Krebs, in Germany, and others have been reasserting more and more the classical tradition of geography as chorologic relation. It may be said that, after a period in which special, essentially physical disciplines were most in vogue, we are in process of returning to our permanent task and that this readjustment is responsible for the current activity of inquiry as to content of our field.

Summary of the objective of geography.—The task of geography is conceived as the establishment of a critical system which embraces the phenomenology of landscape, in order to grasp in all of its meaning and color the varied terrestrial scene. Indirectly Vidal de la Blache has stated this position by cautioning against considering "the earth as 'the scene on which the activity of man unfolds itself,' without reflecting that this scene is itself living."[8] It includes the works of man as an integral expression of the scene. This position is derived from Herodotus rather than from Thales. Modern geography is the modern expression of the most ancient geography.

The objects which exist together in the landscape exist in interrelation. We assert that they constitute a reality as a whole which is not expressed by a consideration of the constituent parts separately, that area has form, structure, and function, and hence position in a system, and that it is subject to development, change, and completion. Without this view of areal reality and relation, there exist only special disciplines, not geography as generally understood. The situation is analogous to history, which may be divided among economics, government, sociology, and so on; but when this is done the result is not history.

THE CONTENT OF LANDSCAPE

Definition of landscape.—The term 'landscape' is proposed to denote the unit concept of geography, to characterize the peculiarly geographic association of facts. Equivalent terms in a sense are 'area' and 'region.' Area is of course a general term, not distinctively geographic. Region has come to imply, to some geographers at least, an order of magnitude. Landscape is the English equivalent of the term German geographers are using largely and strictly has the same meaning, a land shape, in which the process of shaping is by no means thought of as simply physical. It may be defined, therefore,

---

[8] Principes de géographie humaine, p. 6 (1922).

as an area made up of a distinct association of forms, both physical and cultural.[9]

The facts of geography are place facts; their association gives rise to the concept of landscape. Similarly, the facts of history are time facts; their association gives rise to the concept of period. By definition the landscape has identity that is based on recognizable constitution, limits, and generic relation to other landscapes, which constitute a general system. Its structure and function are determined by integrant, dependent forms. The landscape is considered, therefore, in a sense as having an organic quality. We may follow Bluntschli in saying that one has not fully understood the nature of an area until one "has learned to see it as an organic unit, to comprehend land and life in terms of each other."[10] It has seemed desirable to introduce this point prior to its elaboration because it is very different from the unit concept of physical process of the physiographer or of environmental influence of the anthropogeographer of the school of Ratzel. The mechanics of glacial erosion, the climatic correlation of energy, and the form content of an areal habitat are three different things.

Landscape has generic meaning.—In the sense here used, landscape is not simply an actual scene viewed by an observer. The geographic landscape is a generalization derived from the observation of individual scenes. Croce's remark that "the geographer who is describing a landscape has the same task as a landscape painter"[11] has therefore only limited validity. The geographer may describe the individual landscape as a type or possibly as a variant from type, but always he has in mind the generic, and proceeds by comparison.

An ordered presentation of the landscapes of the earth is a formidable undertaking. Beginning with infinite diversity, salient and related features are selected in order to establish the character of the landscape and to place it in a system. Yet generic quality is nonexistent in the sense of the biologic world. Every landscape has individuality as well as relation to other landscapes, and the same is true of the forms that make it up. No valley is quite like any other valley; no city the exact replica of some other city. In so far as these qualities remain completely unrelated they are beyond the reach of

---

[9] Sölch, J., Auffassung der natürlichen Grenzen (1924), has proposed the term 'Chore' to designate the same idea.

[10] "Die Amazonasniederung als harmonischer Organismus," Geog. Ztschr., vol. 27, p. 49 (1921).

[11] Quoted by Barth, P., Philosophie der Geschichte (ed. 2), p. 10.

systematic treatment, beyond that organized knowledge that we call science. ''No science can rest at the level of mere perception..... The so-called descriptive natural sciences, zoology and botany, do not remain content to regard the singular, they raise themselves to concepts of species, genus, family, order, class, type.''[12] ''There is no idiographic science, that is, one that describes the individual merely as such. Geography formerly was idiographic; long since it has attempted to become nomothetic, and no geographer would hold it at its previous level.''[13] Whatever opinion one may hold about natural law, or nomothetic, genetic, or causal relation, a definition of landscape as singular, unorganized, or unrelated has no scientific value.

Element of personal judgment in the selection of content.—It is true that in the selection of the generic characteristics of landscape the geographer is guided only by his own judgment that they are characteristic, that is, repeating; that they are arranged into a pattern, or have structural quality, and that the landscape accurately belongs to a specific group in the general series of landscapes. Croce objects to a science of history on the ground that history is without logical criteria: ''The criterion is the choice itself, conditioned, like every economic art, by knowledge of the actual situation. This selection is certainly conducted with intelligence, but not with the application of a philosophic criterion, and is justified only in and by itself. For this reason we speak of the fine tact, or scent, or instinct of the learned man.''[14] A similar objection is sometimes urged against the scientific competence of geography, because it is unable to establish complete, rigid logical control and perforce relies upon the option of the student. The geographer is in fact continually exercising freedom of choice as to the materials which he includes in his observations, but he is also continually drawing inferences as to their relation. His method, imperfect as it may be, is based on induction; he deals with sequences, though he may not regard these as a simple causal relation.

If we consider a given type of landscape, for example a North European heath, we may put down notes such as the following:

The sky is dull, ordinarily partly overcast, the horizon is indistinct and rarely more than a half-dozen miles distant, though seen from a height. The upland is gently and irregularly rolling and descends to broad, flat basins. There are no

---

[12] Barth, *op. cit.*, p. 11.

[13] *Ibid.*, p. 39.

[14] On History, pp. 109, 110. The statement applies to the history that has the goal simply of ''making the past live again.'' There is, however, also a phenomenologic history, which may discover related forms and their expression.

long slopes and no symmetrical patterns of surface form. Watercourses are short, with clear brownish water, and perennial. The brooks end in irregular swamps, with indistinct borders. Coarse grasses and rushes form marginal strips along the water bodies. The upland is covered with heather, furze, and bracken. Clumps of juniper abound, especially on the steeper, drier slopes. Cart traces lie along the longer ridges, exposing loose sand in the wheel tracks, and here and there a rusty, cemented base shows beneath the sand. Small flocks of sheep are scattered widely over the land. The almost complete absence of the works of man is notable. There are no fields or other enclosed tracts. The only buildings are sheep sheds, situated usually at a distance of several miles from one another, at convenient intersections of cart traces.

The account is not that of an individual scene, but a summation of general characteristics. References to other types of landscape are introduced by implication. Relations of form elements within the landscape are also noted. The items selected are based upon "knowledge of the actual situation" and there is an attempt at a synthesis of the form elements. Their significance is a matter of personal judgment. Objective standards may be substituted for them only in part, as by quantitative representation in the form of a map. Even thus the personal element is brought only under limited control, since it still operates in choosing the qualities to be represented. All that can be expected is the reduction of the personal element by agreeing on a "predetermined mode of inquiry," which shall be logical.

Extensiveness of areal features.—The content of landscape is something less than the whole of its visible constituents. The identity of the landscape is determined first of all by conspicuousness of form, as implied in the following statement: "A correct representation of the surface form, of soil, and of surficially conspicuous masses of rock, of plant cover and water bodies, of the coasts and the sea, of areally conspicuous animal life and of the expression of human culture is the goal of geographic inquiry."[15] The items specified are chosen because the experience of the author has shown their significance as to mass and relation. The chorologic position necessarily recognizes the importance of areal extensiveness of phenomena, this quality being inherent in the position. Herein lies an important contrast between geography and physiography. The character of the heath landscape described above is determined primarily by the dominance of sand, swamp, and heather. The most important geographic fact about Norway, aside from its location, probably is that four-fifths of its surface is barren highland, supporting neither forests nor flocks, a condition significant directly because of its extensiveness.

---

[15] Passarge, Grundlagen der Landschaftskunde, vol. 1, p. 1 (1920).

Habitat value as a basis for the determination of content.—Personal judgment of the content of landscape is determined further by interest. Geography is distinctly anthropocentric, in the sense of value or use of the earth to man. We are interested in that part of the areal scene which concerns us as human beings because we are part of it, live with it, are limited by it, and modify it. Thus we select those qualities of landscape in particular that are or may be of use to us. We relinquish those features of area that may be significant to the geologist in earth history but are of no concern in the relation of man to his area. The physical qualities of landscape then are those that have habitat value, present or potential.

The natural and the cultural landscape.—''Human geography does not oppose itself to a geography from which the human element is excluded; such a one has not existed except in the minds of a few exclusive specialists.''[16] It is a forcible abstraction, by every good geographic tradition a *tour de force*, to consider a landscape as though it were devoid of life. Because we are interested primarily in ''cultures which grow with original vigor out of the lap of a maternal natural landscape, to which each is bound in the whole course of its existence,''[17] geography is based on the reality of the union of physical and cultural elements of the landscape. The content of landscape is found therefore in the physical qualities of area that are significant to man and in the forms of his use of the area, in facts of physical background and facts of human culture. A valuable discussion of this principle is given by Krebs under the title ''Natur- und Kulturlandschaft.''[18]

For the first half of the content of landscape we may use the designation 'site,' which has become well established in plant ecology. A forest site is not simply the place where a forest stands; in its full connotation, the name is a qualitative expression of place in terms of forest growth, usually for the particular forest association that is in occupation of the site. In this sense the physical area is the sum of all natural resources that man has at his disposal in that area. It

---

[16] Vidal de la Blache, P., *op. cit.*, p. 3.

[17] Spengler, O., Untergang des Abendlandes, vol. 1, p. 28 (1922–23): ''Kulturen die mit urweltlicher Kraft aus dem Schoose einer mütterlichen Landschaft, an die jede von ihnen im ganzen Verlauf ihres Daseins streng gebunden ist, erblühen.''

[18] Ztschr. Gesell. f. Erdkunde, Berlin (1923), p. 83. He states the content of geography as being ''in the area (Raum) itself with its surfaces, lines, and points, its form, circumference, and content. The relations to geometry, the pure areal science, become even more intimate, when not only the area as such, but its position with reference to other areas is considered.''

is beyond his power to add to them; he may 'develop' them, ignore them in part, or subtract from them by exploitation.

The second half of landscape viewed as a bilateral unit is its cultural expression. There is a strictly geographic way of thinking of culture; namely, as the impress of the works of man upon the area. We may think of people as associated within and with an area, as we may think of them as groups associated in descent or tradition. In the first case we are thinking of culture as a geographic expression, composed of forms which are part of geographic phenomenology. In this view there is no place for a dualism of landscape.

THE APPLICATION OF THE MORPHOLOGIC METHOD

Form of induction.—The systematic organization of the content of landscape proceeds with the repression of a priori theories concerning it. The massing and ordering of phenomena as forms that are integrated into structures and the comparative study of the data as thus organized constitute the morphologic method of synthesis, a special empirical method. Morphology rests upon the following postulates: (1) that there is a unit of organic or quasi-organic quality, that is, a structure to which certain components are necessary, these component elements being called 'forms' in this paper; (2) that similarity of form in different structures is recognized because of functional equivalence, the forms then being 'homologous'; and (3) that the structural units may be placed in series, especially into developmental sequence, ranging from incipient to final or completed stage. Morphologic study does not necessarily affirm an organism in the biologic sense, as, for example, in the sociology of Herbert Spencer, but only organized unit concepts that are related. Without being committed in any sense to a general biogenetic law, the organic analogy has proved most useful throughout the fields of social inquiry. It is a working device, the truth of which may perhaps be subject to question, but which leads nevertheless to increasingly valid conclusions.[19]

The term 'morphology' originated with Goethe and expresses his contribution to modern science. It may be well to recall that he turned to biologic and geologic studies because he was interested in the nature and limits of cognition. Believing that there were things "accessible and inaccessible" to human knowledge, he concluded:

---

[19] The assumption 'as if,' advanced by Vaihinger as "Philosophie des Als Ob."

"One need not seek for something beyond the phenomena; they themselves are the lore (Lehre)." Thus originated his form studies, and especially those of homology of form. His method of scientific inquiry rested on a definite philosophic position.

If therefore the morphologic method appears unpretentious to the student who is eager to come to large conclusions, it may be pointed out that it rests upon a deliberate restraint in the affirmation of knowledge. It is a purely evidential system, without prepossession regarding the meaning of its evidence, and presupposes a minimum of assumption; namely, only the reality of structural organization. Being objective and value-free, or nearly so, it is competent to arrive at increasingly significant results.

Application to social studies.—Morphologic method is not only the introduction to the biologic sciences but it is steadily growing in importance in the social fields. In biology it is the study of organic forms and their structure, or the architecture of organisms. In the social field the continued synthesis of phenomena by morphologic method has been employed with greatest success perhaps in anthropology. This science can claim an honor roll of workers who have had the patience and skill to approach the study of social institutions phenomenologically, by the classification of forms, ranging from the concrete materials of clothing, housing, and tools to the language and customs of a group, thereby identifying step by step the complex structure of cultures. Spengler's recent brilliant and highly controversial thesis of history is far and away the most pretentious application of the method to the human field. Disregarding its elements of intuitionalism, it is in effect comparative morphology as applied to history, the second volume bearing that title. He characterizes the forms that, to his mind, compose the great historic structures, subjects them to comparison for different periods as homologies, and traces developmental stages. By however much the author may have exceeded his and our knowledge in his daring synthesis, he has shown the possibilities of a morphology of history, or of the study of history on a scientific basis other than the causal formula of historical rationalism.[20]

[20] Untergang des Abendlandes. The mathematico-philosophical thesis of the cultural cycle, the complete antithesis of Buckle, in particular is of such importance that it should be known to every geographer, whatever his position may be with reference to Spengler's mysticism. There are at least three other similar views of the structure of history, apparently independently discovered; Frobenius' Paideuma (1921), Henry Adams' "Rule of Phase in History" (*in* The Degradation of the Democratic Dogma, 1919), and Flinders Petrie's Revolutions of Civilization (1911).

The introduction of morphology into geography and the results.— Method and term were first formally introduced into geography by Karl Ritter, whose restoration of geography succeeded finally, not in the idealistic cosmology he had espoused, but because after all he laid the foundations for comparative regional study. Thereafter, perhaps because there was so much to do, the morphologic studies were rapidly narrowed so as to regard only the surficial form of the land. Griesebach's classic definition that "the morphologic system illuminates, by regarding the relationship of forms, the obscurity of their descent"[21] was applied with fateful results to the field of geography. The restriction of forms to relief, and interest in the origin of these forms, shortly established, under the leadership of Peschel, v. Richthofen, and de la Noë, the genetic inquiry that was called geomorphology.[22] At first relying on the naïve descriptive classification of surface forms, as for example in Penck's Morphologie der Erdoberfläche, which is chorologic morphology, increasingly the trend was to classify on the basis of process, and to trace these forms back to more and more remote forms. The genetic historians of land form undertook increasingly the invasion of the field of geology. The final step was that some of these specialists lost sight almost completely of actual land forms and devoted themselves to the construction of theoretical forms deduced from individual physical processes. The defeat of geographic ends was therefore almost complete and such geomorphology became a separate branch of general earth science.

This autonomous genetic morphology inevitably led to an adverse reaction among the chorologically minded geographers, not because the work was not carefully done, nor because it failed to develop a valuable field of knowledge, but because it became unrecognizable as geography.[23] Unfortunately a very general name was applied to a very specialized discipline. Under a misapprehension of the term there has been a tendency to disregard in consequence the possibilities of the morphologic method. Vidal de la Blache perhaps earlier than any one else realized the situation and reëstablished morphology in its rightful position. The regional monographs that proceeded from his school expressed far more adequately than had been done before the full form content and structural relation of the landscape, finding in the cultural landscape the culminating expression of the organic area.

---

21 Vegetation der Erde, vol. 1, p. 10 (1884).
22 Penck, *op. cit.*, vol. 1, pp. 5, 6.
23 Hettner, *op. cit.*, pp. 41–46.

In these studies, for example, the position of man and his works explicitly is that of the last and most important factor and forms in the landscape.

The perversion of geographic ends in the definition of morphology as the causal study of relief forms appears from the following considerations: (1) Relief is only one category of the physical landscape and ordinarily not the most important one; it almost never supplies the complete basis of a cultural form. (2) There is no necessary relation between the mode of origin of a relief form and its functional significance, the matter with which geography is most directly concerned. (3) An inevitable difficulty with a purely genetic morphology of relief forms is that most of the actual relief features of the earth are of very mixed origin. Behind the present forms lie processal associations, previous or ancestral forms, and almost inscrutable expressions of time. For the present at least, therefore, genetic morphology isolates those form elements that yield to causal analysis. In the selection of those relief facts that are legible as to genesis, it neglects some, even many, of the features of relief and abandons therefore the structural synthesis of even this segment of the landscape in so far as chorology is concerned.

In the late enthusiasm for studies of relief forms the climatologists were crowded into a relatively obscure position. Yet they, most largely, escaped the geographically sterile pursuit of the pure genetic method. Climatology has been phenomenologic rather than genetic. In spite of very scant knowledge of the origin of climatic conditions, the facts of climate have been summarized in terms of their geographic significance most admirably. In particular Köppen's series of trials at climatic synthesis, carefully developed as to biotically critical values, admirably restrained as to genetic explanation, are among the most important if not the most important contribution in this generation to a general geographic morphology. Yet such is the force of associations that few doubtless would name such climatic synthesis as a fundamental part of geographic morphology. It is more than a matter of mere nomenclature to object to the misapplication of the term morphology; it is a rut into which we have slipped and which has limited our range. Perhaps some of the cross-purposes in present-day geography may be traced to the failure to recognize that all the facts of the subject are to be organized by a general system, through which alone their relation may be determined.

PREPARATORY SYSTEMATIC DESCRIPTION

The first step in morphologic study.—Historically "geography commenced by describing and registering, that is as a systematic study. It proceeded thereupon to . . . . genetic relation, morphology."[24] The geographic study is still thus begun. The description of observed facts is by some predetermined order that represents a preliminary grouping of the material. Such systematic description is for the purpose of morphologic relation and is really the beginning of morphologic synthesis. It is therefore distinguishable from morphology not at all in principle but in that it lies at a lower critical level. The relation is not dissimilar to that between taxonomy and biologic morphology.

Descriptive terminology.—The problem of geographic description differs from that of taxonomy principally in the availability of terms. The facts of area have been under popular observation to such an extent that a new terminology is for the most part not necessary. Salisbury held that the forms of landscape had generally received serviceable popular names and that codification might proceed from popular parlance without the coining of new terms. Proceeding largely in this manner, we are building up a list of form terms, that are being enriched from many areas and many languages. Very many more are still awaiting introduction into geographic literature. These terms apply as largely to soil, drainage, and climatic forms as they do to land surface. Also popular usage has named many vegetational associations and has prepared for us a still largely unprospected wealth of cultural form terms. Popular terminology is a fairly reliable warrant of the significance of the form, as implied in its adoption. Such names may apply to single form constituents, as glade, tarn, loess. Or they may be form associations of varying magnitude, as heath, steppe, piedmont. Or they may be proper names to designate unit landscapes, as, for example, the regional names that are in use for most parts of France. Such popular nomenclature is rich in genetic meaning, but with sure chorologic judgment it proceeds not from cause but from a generic summation; namely, from form similarities and contrasts.

If systematic description is a desideratum for geography, we are still in great need of enlarging our descriptive vocabulary. The meagerness of our descriptive terms is surprising by comparison with

---

24 Krebs, *op. cit.*, p. 81.

other sciences. Contributing causes may be the idiographic tradition of unrelated description, and the past predilection for process studies which minimized the real multiplicity of forms.

The predetermined descriptive system.—The reduction of description to a system has been largely opposed by geographers and not entirely without cause. Once this happens the geographer is responsible within those limits for any areal study he undertakes; otherwise he is free to roam, to choose, and to leave. We are not concerned here with geography as an art. As a science it must accept all feasible means for the regimentation of its data. However excellent the individualistic, impressionistic selection of phenomena may be, it is an artistic, not a scientific desideratum. The studies in geomorphology, in particular those of the school of Davis, represent perhaps the most determined attempt to oppose uncontrolled freedom of choice in observation by a strict limitation of observations and of method. Different observations may be compared as to their findings only if there is a reasonable agreement as to the classes of facts with which they deal. The attempt at a broad synthesis of regional studies by employing our existing literature immediately runs into difficulties, because the materials do not fit together. Findings on the most important theme of human destructiveness of natural landscape are very difficult to make because there are no adequate points of reference. Some observers note soil erosion systematically, others casually, and still others may pay no attention to it. If geography is to be systematic and not idiosyncratic, there must be increasing agreement as to items of observation. In particular this should mean a general descriptive scheme to be followed in the collection of field notes.[25]

A general descriptive scheme, intended to catalogue areal facts broadly, without proceeding at this stage from hypothetical origins and connections, has been recently proposed by Passarge under the name "*Beschreibende Landschaftskunde.*"[26] It is the first comprehensive treatment of this subject since v. Richthofen's Führer für Forschungsreisende, written just prior to the most flourishing period of geomorphology (1886). The work of Passarge is somewhat rough hewn and it is perhaps excessively schematic, yet it is the most adequate consideration by far that the whole matter of geographic description has had. Its express purpose is "first of all to determine the facts and to attempt a correct presentation of the significant,

---

[25] Sauer, C. O., The Survey Method in Geography, Ann. Assoc. Am. Geog., vol. 14, pp. 19 ff. (1924).

[26] Grundlagen der Landschaftskunde, vol. 1.

visible facts of area without any attempt at explanation and speculation."[27] The plan provides

"for the systematic observation of the phenomena that compose the landscape. The method resembles most closely the chrie, a device for the collection of material in theme writing. It helps to see as much as possible and to miss as little as possible and has the further advantage that all observations are ordered. If earlier geographers had been familiar with a method of systematic observation of landscape, it would have been impossible for the characteristic red color of tropical residual soils to have escaped attention until v. Richthofen discovered that fact."[28]

Passarge proceeds with an elaborate schedule of notes covering all form categories of the landscape, beginning with atmospheric effects and ending with forms of habitation. From these he proceeds to a descriptive classification of form associations into larger areal terms. For the further elaboration of the plan the reader is referred to the volume in question, as worthy of careful consideration.

The author has applied his system elsewhere to the 'pure' as against the 'explanatory' description of areas, as for example in his characterization of the Valley of the Okavango, in the northern salt steppe of the Kalahari.[29] That he succeeds in giving to the reader an adequate picture of the composition of area will probably be admitted.

One may note that Passarge's supposedly purely descriptive procedure is actually based on large experience in areal studies, through which a judgment as to the significant constituents of landscape has been formed. These are really determined through morphologic knowledge, though the classification is not genetic, but properly based on the naïvely generic forms. The capacious dragnet which Passarge has fashioned, though disclaiming all attempt at explanation, is in reality a device fashioned by experienced hands for catching all that may be wanted in an areal morphology and for deferring explanation until the whole material is sorted.

FORMS OF LANDSCAPE AND THEIR STRUCTURE

The division between natural and cultural landscapes.—We cannot form an idea of landscape except in terms of its time relations as well as of its space relations. It is in continuous process of development or of dissolution and replacement. It is in this sense a true appreciation of historical values that has caused the geomorphologists to tie the present physical landscape back into its geologic origins, and to derive it therefrom step by step. In the chorologic sense,

---

[27] *Ibid.*, p. vi.   [28] *Ibid.*, p. 5.   [29] Hamburg Mitt. Geog. Gesell., 1919, No. 1.

however, the modification of the area by man and its appropriation to his uses are of dominant importance. The area prior to the introduction of man's activity is represented by one body of morphologic facts. The forms that man has introduced are another set. We may call the former, with reference to man, the original, natural landscape. In its entirety it no longer exists in many parts of the world, but its reconstruction and understanding are the first part of formal morphology. Is it perhaps too broad a generalization to say that geography dissociates itself from geology at the point of the introduction of man into the areal scene? Under this view the prior events belong strictly in the field of geology and their historical treatment in geography is only a descriptive device employed where necessary to make clear the relationship of physical forms that are significant in the habitat.

The works of man express themselves in the cultural landscape. There may be a succession of these landscapes with a succession of cultures. They are derived in each case from the natural landscape, man expressing his place in nature as a distinct agent of modification. Of especial significance is that climax of culture which we call civilization. The cultural landscape then is subject to change either by the development of a culture or by a replacement of cultures. The datum line from which change is measured is the natural condition of the landscape. The division of forms into natural and cultural is the necessary basis for determining the areal importance and character of man's activity. In the universal, but not necessarily cosmologic sense, geography then becomes that part of the latest or human chapter in earth history which is concerned with the differentiation of the areal scene by man.

The natural landscape: geognostic basis.—In the subsequent sections on the natural landscape a distinction is implied between the historical inquiry into origin of features and their strictly morphologic organization into a group of forms, fundamental to the cultural expression of the area. We are concerned alone with the latter in principle, with the former only as descriptive convenience.

The forms of the natural landscape involve first of all the materials of the earth's crust which have in some important measure determined the surface forms. The geographer borrows from the geologist knowledge of the substantial differences of the outer lithosphere as to composition, structure, and mass. Geology, being the study of the history of these materials, has devised its classification on the basis of succession of formations, grouped as to period. In formations per

se the geographer has no interest. He is concerned, however, with
that more primitive phase of geology, called geognosy, which regards
kind and position of material but not historical succession. The name
of a geologic formation may be meaningless geographically, if it
lumps lithologic differences, structural differences, and differences in
mass under one term. Geognostic condition provides a basis of conversion of geologic data into geographic values. The geographer is
interested in knowing whether the base of a landscape is limestone
or sandstone, whether the rocks are massive or intercalated, whether
they are broken by joints or are affected by other structural conditions expressed in the surface. These matters may be significant to
the understanding of topography, soil, drainage, and mineral distribution.

The application of geognostic data in geographic studies is usual
in a sense, areal studies being hardly feasible without some regard for
the underlying materials. Yet to find the most adequate analysis of the
expression of the underlying materials in the surface, it is probably
necessary to go back to the work of the older American and British
geologists, such as Powell, Dutton, Gilbert, Shaler, and Archibald
Geikie. In the aggregate, of course, the geologic literature that
touches upon such matters is enormous, but it is made up of rather
incidental and informal items, because landscape is not in the central
field of interest of the geologist. The formal analysis of critical
geognostic qualities and their synthesis into areal generalizations has
not had a great deal of attention. Adequately comparable data are
still insufficient from the viewpoint of geography. In briefest form
Sapper has lately attempted a general consideration of the relation
of geologic forms to the landscapes of varying climates, to the illumination of the entire subject of regional geography.[30]

Rigorous methodologist that he is, Passarge has not failed to
scrutinize the geographic bearing of rock character and condition,
and has applied in intensive areal study the following observations
(somewhat adapted) :[31]

Physical resistance
    Soft, easily eroded formations
    Rocks of intermediate resistance
        much broken (*zerklüftet*)
        moderately broken
        little broken
    Rocks of high resistance
        as above

---
[30] Sapper, K., Geologischer Bau und Landschaftsbild (1917).
[31] Physiologische Morphologie, Hamburg Mitt. Geog. Gesell., vol. 26, pp. 133 ff.
(1912).

Chemical resistance and solubility
     Easily soluble
         highly permeable
         moderately permeable
         relatively impermeable
     Moderately subject to solution and chemical alteration
         as above
     Resistant

In a later study he added provision for rocks notably subject to creep (*fluktionsfähig*).[32] An interpretation of geologic conditions in terms of equivalence of resistance has never been undertaken for this country. It is probably possible only within the limits of a generally similar climatic condition. We have numerous classifications of so-called physiographic regions, poorly defined as to their criteria, but no truly geognostic classification of area, which, together with relief representation, and climatic areas, alone is competent to provide the base map of all geographic morphology.

The natural landscape: climatic basis.—The second and greater link that connects the forms of the natural landscape into a system is climate. We may say confidently that the resemblance or contrast between natural landscapes in the large is primarily a matter of climate. We may go farther and assert that under a given climate a distinctive landscape will develop in time, the climate ultimately cancelling the geognostic factor in many cases.

Physiography, especially in texts, has, largely, either ignored this fact or has subordinated it to such an extent that it is to be read only between the lines. The failure to regard the climatic sum of physiographic processes as differing greatly from region to region may be due to insufficient experience in different climatic areas and to a predilection for deductive approach. Most physiographic studies have been made in intermediate latitudes of abundant precipitation, and there has been a tendency to think of the agencies in terms of a standardized climatic milieu. The appreciation even of one set of phenomena, as for example drainage forms, is likely to be too much conventionalized by applying the schematism of standardized physiographic process and its results to New England and the Gulf states, to the Atlantic and the Pacific coasts, not to mention the deserts, the tropics, and the polar margins.

But, if we start from the areal diversity of climates, we consider at once differences in penetration of heat and cold diurnally and seasonally, the varying areal expression of precipitation as to amount, form, intensity, and seasonal distribution, the wind as a factor vary-

---

[32] Morphologie des Messtischblattes Stadtremda, *ibid.*, vol. 28, pp. 1 ff. (1914).

ing with area, and above all the numerous possibilities of combination of temperature, precipitation, dry weather, and wind. In short, we place major emphasis on the totality of weather conditions in the moulding of soil, drainage, and surface features. It is geographically much more important to establish the synthesis of natural landscape forms in terms of the individual climatic area than to follow through the mechanics of a single process, rarely expressing itself individually in a land form of any great extent.

The harmony of climate and landscape, insufficiently developed by the schools of physiography, has become the keystone of geographic morphology in the physical sense. In this country the emergence of this concept is to be sought largely in the studies in the arid and semi-arid West, though they did not result at once in the realization of the implied existence of a distinct set of land forms for every climate. In the morphologic form category of soils, the climatic factor was fully discovered first at the hand of Russian students, and was used by them as the primary basis of soil classification [33] in a more thorough-going manner than that which had been applied to topographic forms.[34] Under the direction of Marbut the climatic system has become basal to the work of the U. S. Bureau of Soils. Thus the ground was prepared for the general synthesis of physical landscape in terms of climatic regions.[35] Most recently, Passarge, using Köppen's climatic classification, has undertaken a comprehensive methodology on this basis.[36]

The relation of climate to landscape is expressed in part through vegetation, which arrests or transforms the climatic forces. We, therefore, need to recognize not only the presence or absence of a cover of vegetation, but the type of cover that is interposed between the exogenous forces of climate and the materials of the earth and that acts on the materials beneath.

Diagrammatic representation of the morphology of the natural landscape.—We may now attempt a diagram of the nature of physical morphology to express the relation of landscape, constituent forms, time, and connecting causal factors:

---

[33] Glinka, K., Typen der Bodenbildung (1914); revised and extended by Ramann, E., Bodenbildung und Bodeneinteilung (1918).

[34] For desert forms there was in existence the synthesis of Johannes Walther, Das Gesetz der Wüstenbildung, first published in 1900.

[35] Excellently done by Sapper, cited above, but also strongly emphasized by Davis, W. M., and Braun, G.; Physiogeographie (ed. 2, 1916), vol. 2, especially in the final chapters.

[36] Grundlagen der Landschaftskunde, vols. 2 and 3.

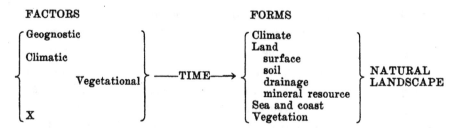

The thing to be known is the natural landscape. It becomes known through the totality of its forms. These forms are thought of not for and by themselves, as a soil specialist would regard soils, for example, but in their relation to one another and in their place in the landscape, each landscape being a definite combination of form values. Behind the forms lie time and cause. The primary genetic bonds are climatic and geognostic, the former being in general dominant and operating directly as well as through vegetation. The 'X' factor is the pragmatic 'and,' the always unequated remnant. These factors are justified as a device for the connection of the forms, not as the end of inquiry. They lead toward the concept of the natural landscape which in turn leads to the cultural landscape. The character of the landscape is determined also by its position on the time line. Whether this line is of determinate or infinite length does not concern us as geographers. In some measure, certainly, the idea of a climax landscape is useful, a landscape that, given a constancy of impinging factors, has exhausted the possibilities of autogenous development. Through the medium of time the application of factor to form as cause and effect relation is limited; time itself is a great factor. We are interested in function, not in a determination of cosmic unity. For all chorologic purposes the emphasis of the diagram lies at its right hand; time and factor have only an explanatory descriptive rôle.

This position with reference to the natural landscape involves a reaffirmation of the place of physical geography, certainly not as physiography nor geomorphology as ordinarily defined, but as physical morphology which draws freely from geology and physiography certain results to be built into a view of physical landscape as a habitat complex. This physical geography is the proper introduction to the full chorologic inquiry which is our goal.

Forms of the natural landscape: climate.—In the physical structure of landscape, climate is first in importance. In the diagram it appears at the head of the list of forms and also as the major factor behind the whole category of forms. As a form, climate is an areal expression, the sum of the atmospheric features of the area. This is the sense in which it is treated in climatology. In American literature climates were first introduced prominently as areal forms, fundamental to geography in general, through Tower's chapters on climate in Salisbury, Barrows, and Tower's Elements of Geography. The value of this view has been demonstrated by the steadily increasing rôle which climatology has played in the fundamental courses of instruction. In no respect are we as near to general agreement as in this.

Climatology is areal reality; meteorology is general process. The contrast is that between physical geography and physiography.

Land forms in the natural landscape.—The land includes four edaphic elements or properties analogous to the climatic elements; namely, surface or land form in the narrow sense, soil, drainage, and mineral forms. In the case of surface forms we are dealing with a body of fact that is of interest to geomorphology, to physiography, and to geographic morphology. The first is concerned with history, the second with process, the third with description and relation to other forms. For our purposes surface forms are to be regarded as climates are in climatology. Strictly we are concerned with the character of relief only, that is, with expressions of slope and exposure in relation to the other constituent forms of the landscape. The topographic map, interpreted in terms of use significance of different slopes, is in principle the complete chorologic representation of surface form. The relation of surface form to climate is so close that the grouping of surfaces by climates is warranted generally. Geognostic relation of surface also lends itself well to the areal grouping of land forms. The further penetration into genesis of forms leads increasingly away from geographic ends. Restraint in this respect is necessary and is attained through a proper realization of the goal of areal reality.

The areal differentiation of soils fundamentally is based on differences of productivity, or their habitat significance. Soils as areal form constituents are primarily grouped by climates; the secondary classification is geognostic and therefore also chorologically satisfactory. The placing of soils into the structure of landscape therefore presents little difficulty, the soil survey being in fact a highly specialized form of physical geography. Unlike some physiographers and geomorphologists, the field student of soils is not pursuing a nongeographic end, but is limiting himself to a small part of the geographic field.

Drainage forms are of course direct expressions of climate, and the most feasible classification of streams, swamps, and bodies of standing water is in climatic terms. For instance, moors are a type of high latitude swamp, permanent features under conditions of low evaporation. Their growth is especially favored by the presence of certain plants, such as sphagnum moss. Their position is not restricted to lowlands, but they extend over fairly irregular surfaces by the expansion of a marginal zone of spongy vegetation. These swamps illustrate the interrelation of physical areal forms. Under them a distinctive soil is developed and even the subsoil is altered. This swampy covering also protects the land surface it has occupied from the attacks of running water and wind and moulds it into broadly rounded forms. Where climate conditions are not favorable to the development of such swamps, both in still higher and in lower latitudes, the forms of drainage, soil, and surface change markedly.

Mineral resources belong among the physical forms under the view of the physical landscape as a human habitat. Here the geognostic factor dominates genetically. The diagrammatic relation still holds in a measure, because of the concentration of minerals due to underground waters both at and beneath the surface. It would be pedantic to urge this point strongly, nor are we desirous to urge genetic relation as a necessary principle.

Forms of the sea in the natural landscape.—The relation of sea to land is organizable on the same basis of climate and geognosy. The seacoasts are in the main an expression of tectonic history and of climatic setting. Areally, climates afford the broader basis of classification, because elevation and subsidence of coasts have varied and are changing, as to direction and amount, so greatly, over short distances, as to make a tectonic classification of seashores chorologically unsatisfactory. The seas themselves are obviously as intimately related to climate as is the land. Their currents, surface conditions, density, and temperatures are as certainly to be classified in climatic terms as land forms.

Vegetation forms in the natural landscape.—A. v. Humboldt was first to recognize through systematic observations, the importance of vegetation in the character of the landscape. "However much the character of the different parts of the world depends on the totality of external appearances; though outline of mountains, physiognomy of plants and animals, cloud forms and transparency of the atmosphere compose the general impression: yet it is not to be denied that the most important element in this impression is the cover of vegetation.''[37] The bonds between climate and vegetation are so direct and strong that a large measure of climatic grouping of vegetation forms is possible. Some plant geographers have found the classification of vegetational associations most desirable in terms of thermal or moisture belts.[38]

Summary of the form relations in the natural landscape.—The large emphasis on climate in the previous statements does not mean that geography is to be transformed into climatology. The physical area is fundamental to any geographic study because it furnishes the materials out of which man builds his culture. The identity of the physical area rests fundamentally on a distinctive association of physical forms. In the physical world, generic character of area and its genesis are coupled so closely that the one becomes an aid to the recognition of the other. In particular, climate, itself an areal form, largely obscure as to origin, so largely controls the expression of the other physical forms that in many areas it may be considered the determinant of form association. An express disclaimer may be entered, however, against the notion of the necessity of a genetic bond in order to organize the phenomenology of the natural landscape. The existence of such bonds has been determined empirically. By regarding the relationship of forms we have discovered an important light on ''the obscurity of their descent,'' but as geographers we are not enjoined to trace out the nature of this descent. This remains the problem of geomorphology, which indeed now appears more complicated than ever, the validity of climatic control and of great secular changes of climate being accepted.

Thus far the way is pretty well marked. We know the 'inorganic' composition of landscape fairly well, and, except for a somewhat excessive aloofness existing between plant and general geography, the place of vegetation in the landscape is properly cared for.[39]

---

[37] Ansichten der Natur, vol. 2, p. 20 (1849).

[38] The content of the preceding pages, dealing with the morphologic structure of physical area, is adapted from Sauer and Leighly, Syllabus for an Introduction to Geography (1924).

[39] Hettner, *op. cit.*, p. 39, comments as follows on biogeography: ''The great majority of studies in plant and animal geography have been made by botanists and zoologists, even though these works do not always completely satisfy our geographic needs. The botanist and zoologist are concerned with plants and animals, we with lands . . . . When they carry on plant and animal geography in this narrower sense, as for example, Griesebach, in his brilliant volume on the vegetation of the earth, they are doing geographic work, in the same manner as meteorologists who concern themselves with climatology; for the purpose is geographic, the results fit more closely into the structure of geographic instruction than into that of botany or zoology, and the whole process of thought and inquiry, oriented as it is about climate and soil, is geographic. We geographers are

The extension of morphology to the cultural landscape.—The natural landscape is being subjected to transformation at the hands of man, the last and for us the most important morphologic factor. By his cultures he makes use of the natural forms, in many cases alters them, in some destroys them.

The study of the cultural landscape is, as yet, largely an untilled field. Recent results in the field of plant ecology will probably supply many useful leads for the human geographer, for cultural morphology might be called human ecology. In contrast to the position of Barrows in this matter, the present thesis would eliminate physiologic ecology or autecology and seek for parallels in synecology. It is better not to force into geography too much biological nomenclature. The name ecology is not needed: it is both morphology and physiology of the biotic association. Since we waive the claim for the measurement of environmental influences, we may use, in preference to ecology, the term morphology to apply to cultural study, since it describes perfectly the method that is involved.

Among geographers in America who have concerned themselves with systematic inquiry into cultural forms, Mark Jefferson, O. E. Baker, and M. Aurousseau have done outstanding pioneering. Brunhes' ''essential facts of geography'' represent perhaps the most widely appreciated classification of cultural forms.[40] Sten DeGeer's population atlas of Sweden[41] was the first major contribution of a student who has concentrated his attention strictly on cultural morphology. Vaughan Cornish introduced the concepts of 'march,' 'storehouse,' and 'crossroads' in a most valuable contribution to urban

far from being jealous on that account; on the contrary, we acknowledge such aid gratefully; but rightly we have commenced also to do plant and animal geography, because certain problems concern us more than they do those who are not geographers and because we possess certain valuable preparations for such studies.'' The work of plant and animal geographers illustrates the partial artificiality of academic compartments. They require so specialized a training that ordinarily they are professionally classed as botanists and zoologists. Their method, however, is geographic to such an extent and their findings are so significant to geography that their work is more appreciated and perhaps even better evaluated by geographers than by biologists generally. Occasional field biologists, such as Bates, Hudson, and Beebe, have done work which encompasses so large a part of the landscape that they are really geographers of the highest accomplishment. It is, however, true that vegetation or fauna may be regarded somewhat differently as a part of the human habitat (economic plant and animal geography?) than as a part of botany or zoology. In this lies the justification of Hettner's recommendation for participation by the geographer in plant and animal studies. Now and then a geographer, as for instance Gradmann and Waibel, has mastered the field of biogeography to the enrichment of his whole position.

[40] Brunhes, J., Human Geography (1910, Am. ed., 1920).

[41] Befolkningens Fördeliing i Sverige (Stockholm, 1917).

problems.[42]   Most recently, Geisler has undertaken a synthesis of the urban forms of Germany, with the deserved subtitle: "A contribution to the morphology of the cultural landscape."[43] These pioneers have found productive ground; our periodical literature suggests that a rush of homesteaders may soon be under way.

Diagrammatic representation of the morphology of the cultural landscape.—The cultural landscape is the geographic area in the final meaning (*Chore*). Its forms are all the works of man that characterize the landscape. Under this definition we are not concerned in geography with the energy, customs, or beliefs of man but with man's record upon the landscape. Forms of population are the phenomena

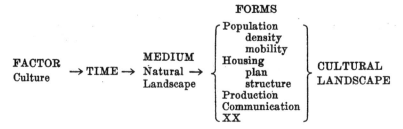

```
                              FORMS
                           ┌─ Population
                           │     density
                           │     mobility
        FACTOR     MEDIUM  │  Housing           CULTURAL
        Culture → TIME →  Natural →│     plan           LANDSCAPE
                          Landscape│     structure
                           │  Production
                           │  Communication
                           └─ XX
```

of mass or density in general and of recurrent displacement, as seasonal migration. Housing includes the types of structures man builds and their grouping, either dispersed as in many rural districts, or agglomerated into villages or cities in varying plans (*Städtebild*). Forms of production are the types of land utilization for primary products, farms, forests, mines, and those negative areas which he has ignored.

The cultural landscape is fashioned out of a natural landscape by a culture group. Culture is the agent, the natural area is the medium, the cultural landscape the result. Under the influence of a given culture, itself changing through time, the landscape undergoes development, passing through phases, and probably reaching ultimately the end of its cycle of development. With the introduction of a different, that is, alien culture, a rejuvenation of the cultural landscape sets in, or a new landscape is superimposed on remnants of an older one. The natural landscape is of course of fundamental importance, for it supplies the materials out of which the cultural landscape is formed. The shaping force, however, lies in the culture itself. Within the wide limits of the physical equipment of area lie many possible choices for man, as Vidal never grew weary of pointing out.

---

[42] The Great Capitals (London, 1923).

[43] Die deutsche Stadt (Stuttgart, 1924).

This is the meaning of adaptation, through which, aided by those suggestions which man has derived from nature, perhaps by an imitative process, largely subconscious, we get the feeling of harmony between the human habitation and the landscape into which it so fittingly blends. But these, too, are derived from the mind of man, not imposed by nature, and hence are cultural expressions.

MORPHOLOGY AS APPLIED TO THE BRANCHES OF GEOGRAPHY

The consolidation of the two diagrams gives an approximation of the total scientific content of geography on the phenomenologic basis by which we have proceeded.[44] They may readily be expressed so as to define the branches of geography. (1) The study of the form categories per se in their general relation, the system of the forms of landscape, is morphology in the purely methodologic sense and is the equivalent of what is called, especially in France and Germany, general geography, the propaedeutic through which the student learns to work with his materials. (2) Regional geography is comparative morphology, the process of placing individual landscapes into relation to other landscapes. In the full chorologic sense, this is the ordering of cultural, not of natural, landscapes. Such a critical synthesis of regions for the entire world is the latest contribution of Passarge, who has thereby nearly rounded out a critique of the entire field of geography.[45] (3) Historical geography may be considered as the series of changes which the cultural landscapes have undergone and therefore involves the reconstruction of past cultural landscapes. Of special concern is the catalytic relation of civilized man to area and the effects of the replacement of cultures. From this difficult and little touched field alone may be gained a full realization of the development of the present cultural landscape out of earlier cultures and the natural landscape. (4) Commercial geography deals with the forms of production and the facilities for distribution of the products of areas.

---

[44] The conclusions presented in this paper are substantially in agreement with Sten DeGeer's article On the Definition, Method, and Classification of Geography, Geog. Annaler, 1923, pp. 1–37, with the contrast that a 'concrete' *landscape* takes the place of DeGeer's 'abstract' *areal relation.*

[45] Vergleichende Landschaftskunde (Berlin, 1923); Landschaftsgürtel der Erde (Breslau, 1923).

BEYOND SCIENCE

The morphologic discipline enables the organization of the fields of geography as positive science. A good deal of the meaning of area lies beyond scientific regimentation. The best geography has never disregarded the aesthetic qualities of landscape, to which we know no approach other than the subjective. Humboldt's 'physiognomy,' Banse's 'soul,' Volz's 'rhythm,' Gradmann's 'harmony' of landscape, all lie beyond science. They seem to have discovered a symphonic quality in the contemplation of the areal scene, proceeding from a full novitiate in scientific studies and yet apart therefrom. To some, whatever is mystical is an abomination. Yet it is significant that there are others, and among them some of the best, who believe, that having observed widely and charted diligently, there yet remains a quality of understanding at a higher plane which may not be reduced to formal process.[46]

DIVERGENT VIEWS OF GEOGRAPHY

The geographic thesis of this article is so largely at variance with certain other views of the subject that it may be desirable to set forth in summary form what has been expressed and implied as to contrast in the several positions.

Geomorphology as a branch of geography.—German geographers in particular tend to regard geomorphology as an essential division of geography, and use largely the term *'Oberflächengestaltung,'* or the record of development of surficial form. The forms considered are ordinarily topographic only. The content of geomorphology has been most broadly defined by Penck,[47] who included the following forms: plains, hill surfaces, valleys, basins, mountains, cavernous forms, seacoasts, seafloors, islands. These descriptive topographic terms are studied by geomorphology as to their derivation, not as to use significance.

Geomorphology being the history of topography, it derives present surfaces from previous forms and records the processes involved. A

---

[46] A good statement of current searchings in this field is by Gradmann, R., Das harmonische Landschaftsbild, Ztschr. Gesell. Erdk., Berlin, 1924, pp. 129–147. Banse has been publishing since 1922 a non- or antiscientific journal, Die Neue Geographie, in which numerous good items are enclosed in a repellently polemic shell.

[47] Morphologie der Erdoberfläche (1894), vol. 2.

study of the geomorphology of the Sierra Nevada is a history of the sculpturing of the mountain massif, concerned with the uplift of the earth block, and the stages of modification in which erosional processes, secondary deformations, and structural conditions are in complex relations. Relief features in this sense are the result of the opposition of orogenic and degradational processes through geologic periods of time. Certain features, such as peneplains and terrace remnants, thus have high diagnostic value in reading the record of modification of surface. These elements of the landscape, however, may be of little or no significance in the chorologic sense. To geomorphology the peneplain has been extremely important; the trend of geography has not been notably affected by its discovery. Out of the topographic complex the geomorphologist may select one body of facts illustrative of earth history, the geographer will use a largely different set of facts which have habitat significance.

The geomorphologist, therefore, is likely to be a specialized historical geologist, working on certain, usually late, chapters of earth history. Conventional historical geology is mostly concerned with the making of rock formations. The geomorphologist directs attention to erosional and deformational surfaces in the record of the rocks. To such an extent has this been the American orientation that we have in our country little geomorphologic work of recent date that is consciously geographic in purpose, that is, descriptive of actual land surfaces.

The geomorphologist can and does establish a connection between the fields of geography and geology and his labors further our own work. He advances our studies of landscape materially where he has preceded the geographer, and we properly regard him potentially as much a collaborator in geography as in geology. One of the present needs in American geography is a greater familiarity with and application of geomorphologic studies.

Physiography and physical geography.—When Huxley reapplied the term physiography he disclaimed expressly the desire to reform physical geography. He was not lecturing, he said, "on any particular branch of natural knowledge, but on natural phenomena in general."[48] The subtitle of his treatise read: "An Introduction to the Study of Nature." He chose the Basin of the Thames as the area for his demonstration, not through chorologic interest, but in

---

[48] Physiography (1877), p. vi.

order to show that any area contained abundant material for the demonstration of the general laws of physical science. Huxley said:

> I endeavored to show that the application of the plainest and simplest processes of reasoning to any of these phenomena, suffices to show, lying behind it, a cause, which will again suggest another; until, step by step, the conviction dawns upon the learner that, to attain to even an elementary conception of what goes on in his parish he must know something about the universe; that the pebble he kicks aside, would not be what it is and where it is, unless a particular chapter of the earth's history, finished untold ages ago, had been exactly what it was.[49]

The two central ideas in his mind were the unity of physical law as shown by the features of the earth and the evolutionary march of the geologic record. It was the bright hour of dawn in scientific monism, with Huxley officiating at the observation of the lands. Physiography served in such a canonical rôle in elementary scientific education until a later age of machinery sent it into the discard in favor of 'general science.'

Physiography is still the general science of the earth, and concerns itself with the physical processes that operate at the surface of the earth and in the earth's crust. We still find the captions that Huxley introduced into his text: the work of rain and rivers, ice and its work, the sea and its work, earthquakes and volcanoes. These things have chorologic expression but they are studied as general processes. As an investigator the physiographer must be above all things a physicist, and increasing demands are made on his physical and mathematical knowledge. The way of the development of physiography as research is through geophysical institutes. Academically it fits in best as a part of dynamic geology. The geographer probably needs to know little more of it than he should know of historical geology.

One may question, therefore, the propriety of such terms as regional physiography and physiographic regions. They contradict the essential meaning of the subject and ordinarily mean rather a loose form of geomorphology, which of necessity has areal expression. Physiography was conceived as a purely dynamic relation and is categorically incapable of consistent areal expression unless it becomes also a name applied to physical geography or to geomorphology.

Geographic morphology vs. 'geographic influences.'—The study of the physical environment as an active agency has recently been subjected to trenchant criticism by L. Febvre, with an equally incisive

---

[49] *Ibid.*, pp. vii, viii.

foreword by Henri Berr.[50] Both thoroughly relish the chance to riddle this geographic ambition. Geography as they see it is "to give an example of the true task of synthesis. . . . The effort of synthesis is a directed activity; it is not a premature realization."[51] Questions of environment "may have for the geographer their interest; but they are not his end. He must guard well against acclaiming as 'scientific' verities theories of adaptation 'simpliste' in character which more competent people are in process of completing or correcting."[52] "What is, then, the commendable attitude in human geography? It can consist only in searching for the relations which exist between earth and life, the rapport which exists between the external milieu and the activity of the occupants."[53] Vidal de la Blache's thesis that in the relation of man to the earth there exists less of necessary adaptation than of 'possibilisme' is worked out with skill and conviction. Excepting for their spirited devotion to the master of French geography, the authors are not really familiar with geographic thought. They do not fairly represent the tenets of geography because they know chiefly the publicists of environmentalism, against whom they consider Vidal as the outstanding bulwark. Vidal will have an honored place in the history of geography, but we are no longer much impressed by his concern to establish decently good relations with rationalistic thought. Rationalism has seen better days than these; we no longer need to come to terms with it by diplomatic compromise. In spite of the deficient orientation in geographic thought, the volume directs a quality of dialectic at one geographic school which entitles it to high rank in geographic criticism.

In this country the theme that geography is the study of natural environment has been dominant in the present generation. It has come to be advertised abroad that such is the American definition of geography.[54] The earliest term was 'environmental control.' This was succeeded by 'response,' 'influence,' 'adjustment,' or some other word that does not change the meaning, but substitutes a more cautious term for the ringing declaration of control. All these positions are mechanistic. In some way they hope to measure the force that physical environment exerts over man. The landscape as such has no

[50] La terre et l'évolution humaine (Paris, 1922).

[51] *Ibid.*, p. ix.

[52] *Ibid.*, p. 11.

[53] *Ibid.*, p. 12.

[54] Van Valkenburg, Amsterdam Tijdschr., K. Ned. Aardr. Gesell., vol. 41, pp. 138, 139 (1924).

interest for them, but only those cultural features for which a causal connection with the physical environment can be established. The aim, therefore, is to make of geography a part of biophysics, concerned with human tropisms.

Geographic morphology does not deny determinism, nor does it require adhesion to that particular faith in order to qualify in the profession. Geography under the banner of environmentalism represents a dogma, the assertion of a faith that brings rest to a spirit vexed by the riddle of the universe. It was a new evangel for the age of reason, that set up its particular form of adequate order and even of ultimate purpose. The exposition of the faith could proceed only by finding testimonials to its efficacy. To the true believer there were visible evidences of the existence of what he thought should be, which were not to be seen by those who were weak in the faith. Unless one has the proper temperament, the continued elaboration of this single thesis with the weak instruments at his hand becomes dreadfully monotonous. In such a study one knows beforehand that one will encounter only variants of the one theme of 'influence.'

The narrowly rationalistic thesis conceives of environment as process and of some of the qualities and activities of man as products. The agency is physical nature; man responds or adapts himself. Simple as the thesis sounds, it incurs continually grave difficulties in the matching of specific response to specific stimulus or inhibition. The direct influence of environmental stimuli is purely somatic. What happens to man through the influence of his physical surroundings is beyond the competence of the geographer; at most he may keep informed as to physiologic research in that field. What man does in an area because of tabu or totemism or because of his own will involves use of environment rather than the active agency of the environment. It would, therefore, appear that environmentalism has been shooting neither at cause nor at effect, but rather that it is bagging its own decoys.[55]

---

[55] Kroeber, A. L., Anthropology (1923), pp. 180–193, 502–503, scrutinizes the ex parte nature of environmentalist tenets in their relation to culture.

CONCLUSION

In the colorful reality of life there is a continuous resistance of fact to confinement within any 'simpliste' theory. We are concerned with "directed activity, not premature realization" and this is the morphologic approach. Our naïvely selected section of reality, the landscape, is undergoing manifold change. This contact of man with his changeful home, as expressed through the cultural landscape, is our field of work. We are concerned with the importance of the site to man, and also with his transformation of the site. Altogether we deal with the interrelation of group, or cultures, and site, as expressed in the various landscapes of the world. Here are an inexhaustible body of fact and a variety of relation which provide a course of inquiry that does not need to restrict itself to the straits of rationalism.[56]

---

[56] Wissler, Clark, Ecology, vol. 5, p. 311 (1924): "While the early history of the concept is probably lost to us forever, there are not wanting indications that the ecological idea was conceived in the same atmosphere as the theory of design, or of purposeful adaptation. However that may be, the effort on the part of later professors of ecology has been to eschew all such philosophies except the fundamental assumption that plants and the rest of nature are intimately interdependent one upon the other." Thus "the anthropologist is not only trying to show what all the forms and forces of nature have done to man, but even with more emphasis what man has done to nature." (312) This definition of anthropology includes a very large part of the social field, and is also a good definition for geography. At present anthropology is the study of culture per se. If our studies of man and his work have large success in synthesis, a gradual coalescence of social anthropology and of geography may represent the first of a series of fusions into a larger science of man.

The Geographic Landscape and Its Investigation

C. Troll

Studium Generale, 3, no. 4/5: 163–181

For about the past three decades the spirit of our times has been reflected in geographic scholarship by synthesizing approaches. If we define geography as the science of features of the earth's surface—the lithosphere, hydrosphere, and atmosphere and their spatial differentiation and functional interrelations—then geographic synthesis means a shift from analysis directed at a single phenomenon to a contextual approach for these phenomena in their interrelated functions in the landscape. An understanding of the whole requires, of course, an understanding of the individual elements. A modern geographer, however, is not expected to deal with glacial movements, vegetation classification, or the trade output of countries but instead is expected to read and interpret the main character of a landscape and to explain the functional interrelationships of its elements. As Hassinger[1] stated in 1919, with landscapes as the natural regions, geography has finally found its own disciplinary focus, one that no other scientific discipline could claim as its own. On the other hand, research into individual factors such as geofactors, scenic elements, and the elements of a landscape requires continuous and close contact with the large number of related disciplines in the natural and the economic and social sciences. It is increasingly obvious that a landscape can be correctly considered as an "organic entity" or as a "harmonic individuum of space" and to perceive them "within the temporal and spatial rhythm of the interplay of their numerous and various factors."[2] The question has also been raised, "Does a landscape represent more than the sum and the unity of geographic objects, namely, a wholeness or a gestalt?" The gestalt and holistic philosophy that has developed since the end of the nineteenth century *(Chr. von Ehrenfels),*[3] but that has its conceptual roots farther in the past, especially in the works of Goethe, was developed in geography by K. Ritter. Initially, it was more precisely defined in psychology (H. Benusse, W. Köhler, K. Koffka, F. Krüger) and in biology (H. Driesch, W. Troll). Accordingly, German geographers[4] generally considered landscape regions as wholes or Gestalten, but not organisms or biological entities as described by W. Driesch, and limited the use of psychological wholeness to the conceptual construct of a landscape from a perspective of subjective experience and aesthetics, which E. Banse[5] strongly argued should become more of an issue for geography. But just as there are forms of living communities (biocoenoses) above the forms of diverse creatures and societal units such as family and people all carry their own conceptual gestalt,[6] spatial geographic features such as landscapes can also represent complex associations of very distinct objects and consequently can be considered geographic entities in themselves.

Landscape and Landscape Studies

In 1913, S. Passarge[7] coined the term *landscape geography* in German geographic literature and started promoting the term *landscape science* in a number of popular publications after 1919; that is, after it had appeared twice before, in 1884 and 1885.[8] He presented the latter term as a new subdiscipline of geography, which would now "win the position that it should have had much earlier." As Waibel[9] pointed out in 1936, because he failed to define precisely the term *landscape* prior to using it, particularly its relationship to the term *land,* a long discussion arose about the understanding and justification for this field. The German word *Landschaft* is more than a thousand years old and in popular usage has had a very instructive development.[10]

In Old High German, *Landschaft* means part of a country or region, approximately identical with the Latin *regio.* In the translation of the *Tatianischen Evangelienharmonie* (circa A.D. 830), the passage "Et pastores erant in regione eadem vigilantes" (Luk. 2:8) is interpreted as "Warun tho hirta in thero landskeffi wahhante," and in another passage "omnis regio circa Jordanem" (Matt. 3:4) is interpreted as "al thiu lantscaf umbi Jordanem." In Middle High German, the term is also used for the inhabitants of a region, as for example in Tristan (6501): "do kom al diu lantschaft und volkes ein so michel kraft." Only after the Renaissance was the term also used for the artistic portrayal of a region, enlarging the term to the nature and scenery of a landscape. In 1518, in Basel, instructions for the manufacture of an altar cloth asked for "die Landschaft in der Tafel verguldet oder versilbert und glasiert." Albrecht Dürer uses the term *landscape painter* in his *Tagebuch der niederlaendischen Reise* in 1521. The first description and poetic experience of a part of nature may be attributed to Hans Sachs[11] in his fable "Die ehrentreich fraw Miltigkeit" (1537). In this fable, the ascent to a mountain lookout with distant vistas is described as follows:

> Nachdem wir auf den thurn
> Bayde gelassen wurn
> Auff dem wir bayde sahen
> Die Landschaft ferr und nahen.

Twenty years later, the term again refers to the old meaning of *regio* in the title of a geographic travel description from Brazil in Hans Stadens von Homburg's *Wahrhaftige Historia und Beschreibung einer Landschaft der wilden, nacketen, grimmigen Menschenfresserleuthen, in der Neuen Welt Amerika gelegen* (Frankfurt 1556). Used in a cartographic context, for example, I found the term in a map by the Swiss naturalist J. J. Scheuchzer as "Die Landschaft Toggenburg," in which the northern part, the "Alte Landschaft" was identified, signifying "Fürstenland" (domain of the ruler). This was not under the control of the abbots of St. Gallen, and, therefore, reveals a reference to *Landschaft* in terms of political territories, as Luther and his contemporaries had in mind.

Today, the term *landscape* plays a role in diverse areas of the arts and sciences. Only geography has expanded its popular use to create a scientific term and establish a whole new subdiscipline based upon it. Starting from here, the nature protection movement and garden architecture have created the terms *landscape protection, landscape management,* and *landscape design* (see following section). The geographer approaches a landscape from its distinct physiognomy (landscape appearance) and certain functions. Individual parts, or landscape elements—whether visible and material or not—are arranged in a certain functional context. As long as no element—and, therefore, the whole landscape—is changed, this functional context is usually in equilibrium (Bürger's *dynamic equilibrium*). We can therefore distinguish between a physiognomic or formal and a functional (physiological, ecological) approach to a landscape observation. Oppel's landscape studies had this first approach in mind when defining a landscape as a "a space, which viewed from an arbitrary point, appears to be complete." Nowadays, people have the possibility of seeing landscapes from aircraft or by air photography in a geometrically revolutionary form, specifically with an aboveground perspective without the usual distortion of a terrestrial perspective. It is not surprising that aerial photographic research has triggered quite a push for landscape research.[12] In technical terminology, this became obvious in the use of the term *aerolandscape,* created by effective and sophisticated Russian air photography. Havemann and Faas define aerolandscape as a "complex of elements on the earth surface which are visible from the air above and which are characteristic of the physio-geographic regions in question."[13]

Because all scenic elements, animate and inanimate geofactors, and functions of the human economy and culture are connected with one another, interdependently, a functional approach is naturally advisable. Depending on the relative influence of humans, one can distinguish between natural landscapes and cultural landscapes. It is important to take into account, though, that a cultural landscape consists not only of its natural elements and the infrastructure of the economy, settlement, and transport but also the influence and output of its inhabitants: their traditions, language, nationality, societal structure, artistic development and feeling for art, and religion. A limited approach to purely economic features deals with what is called the *economic landscape,* with economic units ranging from entirely uninhabited natural landscapes to highly cultivated and densely populated agricultural, industrial, and urban landscapes *(Vollkulturlandschaften)* as, for example, in S. Passarge's and O. Jessen's *Raublandschaft* (pillaged landscapes).

All landscapes are also subject to temporal change, whereby they can retain characteristics of the past for long periods of time (relict forms). Although natural landscapes generally only change in geological dimensions of time, economic landscapes experience a very rapid change from generation to generation, even before the eyes of an observer.

Landscape, in scientific terms, is, therefore, a concept of regional and comparative geography. Even today, however, no consensus has been reached on the relationship of landscape science to regional studies, just as there is little consensus on the terms *landscape* and *land,* even though, by 1938, serious attempts had been made by scholars in various countries.[14] However, N. Krebs's statement that "landscapes are repeatedly recurring types, lands are individuals"[15] cannot be supported here. Continents, countries, and provinces today are usually subdivided into their natural regions that represent unique and individual areas (Passarge's *Reallandschaften*).[16] N. Krebs has subdivided southern Germany into landscapes and treated the whole mapped area accordingly.[17] The term *land* usually describes larger political or ethnic regions such as Scotland, Ireland, England, Greece, Finland, Lapland, Germany, or smaller, present-day or historic territories such as Braunschweig Land, Oldenburg Land, Ermland, Markgräfler Land, Münsterland, Hanauer Land, Lebus Land, and so on. Natural landscapes, in contrast, are Harz, Spessart, Schwarzwald, Ries, Münchener Ebene, Wiener Becken, Maifeld, Goldene Aue, Gutland, Ösling, and so on.

The commonly used term *Land*, however, is by no means demarcated or precisely defined. A whole list of "Lands" exist in Germany, the names of which initially referred to old territories or areas of sovereignty but that are now used for geographic landscapes with modified boundaries. Bergisches Land, as used today, includes not only the former Bergische Lande, Herzogtum Berg with Siegmündung, and Düsseldorf but also the elevated range east of the Rhine Valley. Jülicher Land does not stand for the entire area of the duchy but refers to the agricultural plains on both sides of the Rur River to where it meets the Erft River. We understand Sauerland, the old Westphalian Süderland, as a natural part of the Rheinische Schiefergebirge. Our usage of the name Siegerland, the former Princedom of Siegen, stands for the mining and industrial region at the headwaters of the Sieg River. In particular, the historically long-settled districts tend to carry the designation *Land* today (Bitburger Land = Bitgau, Oldenburgisches Ammerland = Ambergau, Samland as the old Pruzzengau, Bauland, etc.), and we perceive them as naturally defined cultural landscapes. This historic terminology apparently implies a notion of earth (land and water) in the German language, especially with a landscape-oriented perspective of cultivated land as opposed to meadow, marsh, or lea. This explains the names of marshlands, Wursten Land, Hadeln Land, Kehdingen Land, Altes Land, Vierlande, and particularly the small insular arable lands in the Havelländische Luch (Ländchen Friesack, Ländchen Berlin, Ländchen Glin). In southern Germany, some vegetable gardens and meadows in close proximity to villages are still sometimes called *Land*. These are undoubtedly cases of geographically defined landscapes for which we, nevertheless, prefer to keep alive the historical "Land" names. The Amt für Landeskunde project, an etymological-historical-geographic examination of popular landscape names in Germany, will certainly reveal a lot of interesting comparative material.

This being said, we should now modify K. Bürgers's[18]definition of a geographic landscape to the following statement, which would also be correct for the word *Länder*. A geographic landscape (landscape individual, natural region) is a part of the earth's surface that, in its outer appearance, by the interplay of its elements, and in its inner and outer spatial interdependencies, forms a spatial entity with a particular character and that becomes a landscape of a different character at its natural geographic boundaries. Lands are politically or administratively defined, sometimes historical territories or areas populated by particular settled populations.

According to the Amsterdam Discussion of 1938 and other statements, an international nomenclature could equate the following terms.

Table 5.1 International Nomenclature

German	French	English
Land	pays, région	country
Geographische/natürliche Landschaft	paysage géographique, paysage régional	natural region, landscape region
Naturlandschaft	paysage naturel, paysage spontané	natural landscape
Kulturlandschaft	paysage humanisé	cultural landscape, humanized landscape

In its usage in the German language, the term *natürliche Landschaft* (natural region) reflects an area with natural boundaries as opposed to artificial, arbitrary, political, or administrative boundaries that, nevertheless, allows cultural, as well as natural features, to be taken into account.[19] In order to eliminate possible confusion with the term *Naturlandschaft* (natural landscape), Waibel[20] recommends the term be discontinued. This would be possible by using the alternative, *geographische Landschaft* (geographic landscape). Uncertainties of terminology are intensified by differing opinions on the question of whether the subdivision of landscape regions is possible by natural features only or by human activity as well. In the future, this distinction will have to be made more carefully. In heavily human-influenced cultural landscapes like Central Europe, a spatial classification of natural features will always require a reconstruction of the natural landscape first. A subdivision that takes only current factors such as elevation, climate, soil composition, relief, hydrology, and vegetation into account is undoubtedly called a *naturräumliche Gliederung* (natural spatial classification).[21] Such a process is being conducted, for example, under the standardized rules of the *Amtes für Landeskunde* for the whole of Germany. W. Müller-Wille calls his similar subdivision of Westphalia[22] a *naturlandschaftliche Gliederung* (natural landscape classification) and the resulting units simply *Naturlandschaft*. This follows Waibel's suggestion to avoid the ambiguous term *natürliche Landschaft* but instead uses one term for which two precisely defined terms used to be used. When modified by a well-founded *kulturräumliche Gliederung* (cultural spatial classification), for which a methodology has yet to be developed, the *naturräumliche Gliederung* (natural spatial classification) will one day provide a complete geographic structure of Germany. In many regions of the earth, where human interference has changed only the vegetation but has not generated an intensive cultural landscape, it would probably be impossible to reconstruct the natural landscape without human interference. In such regions, for example the grasslands of the tropics, R. Gradmann's term *Naturlandschaft* would be more understood. His definition of *Naturlandschaft* is also physiogeographic but includes human influence on the character of a landscape.[23]

In the past decades, among schools of thought abroad, especially in American geography, there has been concern about the term *landscape*. In German, two extensive papers[24] are available, and in English there is the comprehensive work of R. Hartshorne.[25]

Problems and Tasks of Geographic Landscape Research

In terms of landscape, the tasks of modern geography are multifaceted.

Landscape Morphology

The initial task is to define the geographic boundaries and the concrete characteristics of the landscapes of the earth. These alternating views of both landscape content and landscape boundaries sharpen the understanding and perception of the substantive features of a landscape and its so-called landscape structure. Structuring the landscape results in an organization of groupings of different sizes. Every landscape is an individual at first but is intrinsically integrated into a spatial structure of interdependencies by its location among other landscapes. With regard to the cultural elements of a landscape, O. Schlüter[26] has already outlined a "morphology of the cultural landscape" and has also addressed generally "landscape composition studies."[27] E. Winkler[28] considers these issues of landscape morphology, or "structural studies" of the landscape, as the most important elements hierarchically within his concept of general geography, whereas he associates the spatial structure of landscapes with the subdiscipline of landscape chorology.

Landscape Ecology or Landscape Physiology

The second main task is a functional analysis of primary landscape content and a resolution of the numerous multilevel and reciprocal interdependencies that affect the parts of a landscape. In German geographic literature, for example, in the context of agricultural landscapes, reciprocal interdependencies have been addressed as physiological issues. E. Winkler[29] also calls this aspect of research *landscape physiology*. In the disciplinary source of these terms in biology, however, the study of relationships between organisms and their environment is called *ecology*. Precedence must be given, however, to the term *agricultural landscape* because it was defined thirty years ago by R. Krzymowski[30] and because the term *ecology* is widely used in international scientific research: *human ecology* in British and American geography, the Italian *ecologia agraria*, and the Russian *topoecology*. Therefore, we prefer to speak only of *landscape ecology*, whereby it can remain open whether this includes only the functional interrelations of the natural landscape or also includes the functional connection of human interference in the cultural landscape.

Landscape Typology or Landscape Systematics

A comparison of the infinite number of individual landscapes according to their character or their dominant features results in an organization of landscapes types, which Passarge calls *model landscapes* as opposed to *actual landscapes*. What criteria we prefer for the organization of types is dependent on our perspective. We can create criteria according to single factors of landscape that lead to special types, or we can create synthetic landscape types. For example, geomorphologic landscape types take into account only physical criteria, such as moraines, dunes, inselbergs, fjord landscapes, and rax-landscapes, whereas the climate and vegetation of the region, as well as landscape changes caused by cultural interference, are not taken into account at all. Other examples for similar landscape-type concepts are hydrographic landscape types such as lakes, deltas, or saline landscapes; or climatic classifications of vegetation such as tundra, taiga, savanna, or rainforest landscapes; and local vegetation types such as heath, swamp, river meadow, and parklike landscapes. If we, in turn, speak of rice-growing agriculture, pasture landscape, single-farm landscape, excavation landscape, or, more generally, agricultural, industrial, mining, urban, or cultural landscape, we ignore the natural foundations of a landscape. For the creation of a systematic concept, the grouping of cultural landscape types (domesticated landscapes), given by the biologist W. Schönichen,[31] appears to me to be useless: for example, agricultural landscape, resource landscape, protective landscape, transport landscape, experiential landscape. His introduced term *flirt landscape* ultimately comes very close to triteness. More interesting are the synthetic terms for landscape that combine a whole complex of landscape phenomena into one word.

Karst landscape immediately draws attention to the disposition of the geological subsurface, soil types, hydrology and hydrography, and partly to the vegetation and land use, whereby we can further distinguish between the karst landscapes of different climate zones (e.g., tropical conical karst or cockpit landscapes) and between chalk and gypsum karst. When talking about *high montane landscape* in scientific terms, it means a geomorphological-vegetation-oriented type of natural landscape, formed in all zones of the earth at certain elevations above the tree line and the glacial snow line.[32] *Bördelandschaft* is a type of a cultural landscape that for Central Europeans implies not only a loess surface, fertile farmland, and sparse forest but also a particular type of agriculture, even a particular agricultural system of wheat, barley, alfalfa, sugar beet, cattle herding, and sheep farming. Similar associations apply to marsh landscapes, sandy upland landscapes, and floodplain landscapes in which commonly used terms provide us with scientifically precise terms. Unfortunately, there are no commonly used terms provided for the Bunter sandstone landscape with its forest ridges, valley settlements, and water meadows; for the low mountain granite ranges with their spruce forests, swamps, and block fields; for the rolling Tertiary lands; or for the young gravel plains in the vicinity of the Alps.

Such synthetic landscape names exist in all parts of the world in the local language. In northern France, two types of French cultural landscape are distinguished as *Champagne* and *Bocage*. Romanians know that *Podgoria* stands for the pleasant, rolling land with vines at the foot of the mountains between the steppe plains and moist mountain forests. From this idea, the term *Podgoria* has been changed into a synonym for vineyard. Similar mechanisms apply for *Balta, Câmpia,* and *Lunca.* The Silurian limestone soils of the Baltic islands and Estonia are called *alvar,* describing a landscape type with characteristic geological, pedological, and vegetational features. *Mbuga* in East Africa stands for flooded savannas with glutinous black soils, little tree growth, limited options for land use, and seasonal contrasts that are remarkably obvious to drivers of vehicles in the region. Similarly, *Pampa, Sertâo, Chapada, Bolson, Puna, Paramo, Valle,* and *Cabezera de Valle* are complex landscape names from Latin America. *Paramo,* for example, includes not only the forest-free, ever-moist mountain landscape of the equatorial Andes that is used as pasture, but it also embraces the curious vegetation and its typical fog and drizzle *(paramillos)* with its impact on the state of the human mind and human body (Paramo sickness or *emparamarse*). The same term, *Paramo,* originally had a different meaning in its Spanish motherland but under different climatic conditions. Many Spanish landscape terms exist in Latin America to which very different meanings have been slowly attributed, across the different climatic zones from Mexico to Tierra del Fuego, and were adopted during colonization; for example, *Vega, Estero, Loma, Ciénaga, Pampa, Quebrada, Mesa,* and *Banco.* One of the tasks for future landscape research is to compare these local landscape terms and make them usable for a landscape chronology of the earth.

Landscape Chronology

Another task for landscape research is the analysis of the genesis of today's landscapes, both natural landscapes and cultural landscapes. This task requires the reconstruction of earlier landscape stages to understand the transformation processes that have led to the appearance and composition of today's landscapes. To some extent, it is generally possible to reconstruct today's cultural landscapes to their original natural landscape before any human interference. This can be done successfully with the help of plant sociology; some schools of plant geography are even reconstructing maps at large scales (1:25,000). R. Knapp[33] has shown the strongest awareness of this aspect of landscape research, emphasizing growth zones, growth areas, and growth landscapes. Over the past couple of decades, European scholars have increasingly utilized pollen analysis with remarkable success to trace the development of these natural landscapes through the postglacial period. In an important publication, F. Firbas[34] collected results for Central Europe. This provides precise knowledge of the original landscape; that is, "of the landscape as it existed shortly before the start of interference by farming and human settlement."[35] R. Gradmann had earlier examined this from the combined perspective of botanical, prehistoric, and demographic facts and published as an admirable contribution. Now, new knowledge from periglacial research has also enabled a relatively precise analysis of European landscapes during the last glacial period.[36]

Unfortunately, we do not yet have such assured methods that would allow, with confidence, the reconstruction of natural landscapes in nonforested regions of the earth, especially grasslands. This is because they do not provide the most important archives of landscape history: swamps. Instead, in these regions we have to study human interference in the destruction of vegetation under today's circumstances, dealing primarily with the impact of grass burning, overgrazing, migratory hoe farming, and related soil degradation. Literature on the so-called savanna problem in Africa, the pampas problem in Argentina, and the prairie problem in North America deals with these questions. There is yet no agreement as to how much, or whether at all, these grasslands were naturally forested; in doing this, I am leaving out such unscientific exaggerations that claim the Sahara to be a "desert created by humans in the distant past" or that "goats grazed the Sahara bare."

A second, different chapter of landscape genesis deals with changes to cultural and economic landscapes throughout human history to the present. This is part of historical geography and landscape research and is of special interest in Europe, with its continuously evolving culture. For German cultural landscapes, different schools of geographic thought have provided much, but still insufficient, material. This material covers primarily the past 150 years because the rationalization of agricultural and cultural technology since the late eighteenth century has brought with it cartographic records of land use. From these studies, which cannot be listed here in detail, we can conclude the following two points: (1) The transformation of agricultural landscapes took place following established rules triggered by economic development and a spatial shift of the resource and demand markets during the era of the railway, coal, and steamboat traffic (thus turning away from an economy of self-sufficiency, with the disappearance of primitive cultural methods except in certain remote areas, and with industrial relocation and the resulting establishment of new demand centers). Examples of such regularities include: a land use change from formerly extensive heath areas to farming and grazing areas or to coniferous forests (pine in lowlands, spruce in mountain areas); newly introduced agricultural land use in moist coastal and mountainous zones; a decrease of vine growing and an increase in modern horticulture; and the spread and repeated decline of sprinkler irrigation. (2) The structural changes listed, and the development of modern technology, by no means invalidate or override the geographic interconnections assumed by F. von Richthofen. Instead, it is the very change away from self-sufficiency to the market economy, following railroad traffic, that enhances the subtler natural, soil, and climatic distinctions and that makes them visible in the agrarian landscape. Today, agrarian landscapes are geographically structured by subtler distinctions than they were 150 years ago.

Landscape Management and Landscape Design

This retrospective view of landscapes and its recognition lead almost naturally to the last task of geographic landscape research: the influence of scientific research on future landscape development. This objective automatically emerges from the threat of the increasing degradation of landscape values by economic resource exploitation. Landscape protection is one of the current issues talked about that aims to prevent further serious landscape destruction. Landscape management aims to sustain the natural equilibrium and keep the cultural landscape in good health. Landscape design provides the active creation of the cultural landscape, not only to meet its functional requirements but also to create a harmonic structure in the appearance of the landscape, if not in its artistic design. Landscape protection emerged from environmental protection movements when the single-object approach to individual natural monuments and to nature preservation areas was widened to entire landscapes. Strong stimulation was provided in the written work of the art instructor and architect P. Schultze-Naumburg,[37] whereas its first incorporation into governmentally organized landscape preservation was actually achieved by H. Schwenkel[38] in Württemberg. This movement has been enthusiastically received in Germany in forest management[39] and especially in garden architecture and may develop in landscape architecture.[40] Also from this perspective, it claims its own academic program, leading from garden architecture right into the profession of landscape architect. The German Society for the Study of Garden Design and Landscape Management publishes a journal called *Garten und Landschaft*. Since 1936, numerous publications have appeared, almost all of them concerned with very general subject matter. This deficiency in the literature means that it cannot yet be grounded in the established, basic knowledge of landscape ecology. Furthermore, it treats landscape terms uncritically, it does not reflect the wide variety of geographic landscapes, it uses comparisons from very different climatic regions (deserts, steppes), and it utilizes the catchwords and sometimes embellishes the vague suggestions of the do-gooder. The questions of landscape preservation are intriguing, multifaceted, and have practical relevance, but they should be based on systematic research into the current ecological and past changes in landscapes and be supported by differentiated field experiments.

Landscape Structure and Landscape Classification

The landscape units on earth are of very different scales and extent. A global perspective across all continents helps convey the so-called landscape belts or landscape zones that reflect spatially the climate belts, climatic soil zones, and vegetation zones. Since 1921, S. Passarge has dedicated several books to this topic. A. J. Herbertson called them "major natural regions" in 1905, followed by other English geographers such as L. D. Stamp and J. F. Unstead.[41] On the other hand, the amazing spatial differentiation of earth can be advanced by grouping very small landscape individuals called *Kleinlandschaften* (small landscapes) or *Teillandschaften* (landscape components). In between exists a hierarchical order of landscapes of different dimensions. In the early decades of the nineteenth century, G. Hommeyer[42] had already developed a spatial hierarchy at four levels: *Ort—Gegend—Landschaft—Land* (place-area-landscape-country). Only a small distinction exists in P. E. James's[43] hierarchical concept: locality—district—subregion—region. J. F. Unstead[44] suggested a series of seven levels: stow—stow group—tract—tract group—subregion—minor region—major region, and K. H. Paffen[45] a series of eight: *Landschaftszelle—Kleinlandschaft—Einzellandschaft—Großlandschaft—Landschaftsgruppe—Landschaftsregion—Landschaftszone—Landschaftsgürtel* (landscape cell—small landscape—singular landscape—large landscape—landscape group—landscape region—landscape zone—landscape belt). The name and number are initially of little relevance.

If we begin to structure a climatic landscape belt into its different parts, solely according to its natural disposition, we would initially group the continental sections according to the principles of location. After the climatic aspect, the next grouping will also take into account structure, relief, soil, and hydrology.

A structuring of the boreal coniferous forest belt in Eurasia, for example, would be most expediently done from west to east, reflecting the increasing continentality of its climate and its distinctive structure (fig. 5). This results in at least six groupings: (1) Fennoscandian, reaching as far as the eastern boundary of Karelia from the White Sea to Lake Onega; (2) the North Russian coniferous forest areas to the Urals; (3) the West Siberian lowland plains; (4) the Central Siberian tableland between the Yenisey River and Lena River; (5) Transbaikalia; and (6) the northeast Siberian coniferous forest area. Fennoscandia is characterized by the glacially reshaped primordial mountain range of the Baltic Shield with its colorful mixture of small-scale landscape elements. The North Russian area is characterized by its continental climate (larch forests) and its much more monotonous landscape structure caused by its tableland structure. The West Siberian area is characterized by its low elevation and an increase of peat bogs and swamps, Central Siberia by a mountainous character and by the impacts of permafrost on the landscape, and northeast Siberia by its extreme continental climate and stronger topographic differentiation caused by young mountain ranges. The Scandinavian high mountains and the Urals, both crossing the entire coniferous forest belt in a meridional manner, also can be distinguished into parts. This reveals their montane character of horizontal landscape zones with the addition of a vertical landscape gradation that reflects the three-dimensional structure of the earth's landscapes.[46] Of the landscape boundaries mentioned previously, the one between Fennoscandia and North Russia has been examined by two groups.[47] Both come to the conclusion that the transition from the glacially reshaped basement range of East Keralia to the Russian tableland shows a clear division in terms of the spatial and material composition of the landscape elements because their flora and vegetation characteristics are determined by soil and relief.

Generally the following is true: The more we proceed to smaller structures, the more important becomes the relative importance of soil as opposed to climatic features because the former creates strict boundaries, the latter only gradual transitions. This brings up the questions: "How far should, and can, a geographic distinction of space go?" "What is the smallest dimension of a geographic landscape?" This can be tested in different landscape regions of the earth with the help of topographic and geological maps or air photos. I have tried to investigate this question in a case study of the Rhine Uplands.[48] The results showed that the elimination of smaller and smaller landscape individuals ended at a scale at which a further distinction led to landscape elements that were no longer identifiable as distinct entities but evident in large numbers. These elements, through characteristic combinations and particular spatial patterns, create the smallest landscape individuals. These are the smallest spatial structural types.

In the example shown (figs. 1 and 2), the *Kleinlandschaft* selected on the faulted, loess-covered plateau of the Bergische Land, on both sides of the lower Agger River, is constructed of the following landscape elements: loess-covered highland plains (mostly arable land), wet meadows or depressions of the high plains (mostly used for hamlets, orchards, and pastures close to farms), loess-covered and less steep slopes (mostly farmland), relatively steep rocky or stony slopes (forest), alluvial cones of tributary streams where they meet a main stream (meadows), moist valley bottoms of stream valleys (meadows), and wider bottoms of main valleys with gravel and terraces (meadows and farmland). These landscape benchmarks are, therefore, the type sites with a very distinct ecological quality with regard to rock composition, soil structure, exposure, weathering, and groundwater. A plant-sociological study of the area would reveal that every one of these site types naturally supports a particular uniform plant association. Similarly, agricultural economic values, which, in this case, are rather clear and distinctive, are reflected in the distribution of cultivated species. I prefer not to retain the term *landscape element* that I originally used, because it is already associated with other distinct landscape factors.

Since 1945, I have used the term *ecotope* for these smallest spatial entities, or elements of a geographic landscape, modifying the similar bio-

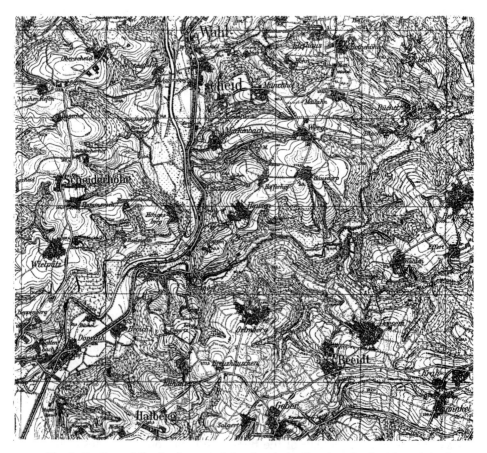

Fig. 1. Section of the landscape of the Bergische Land on both sides of the lower Agger Valley (from the Wahlscheid survey 5109)

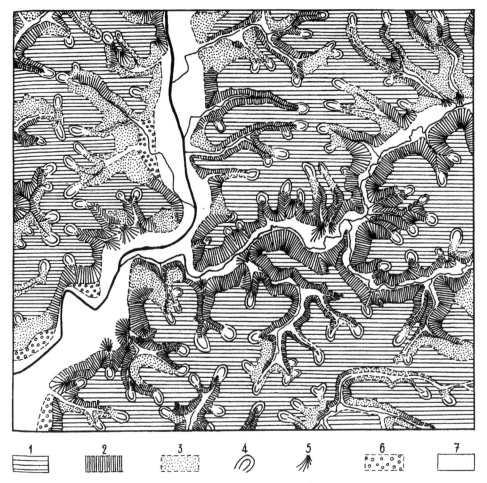

Fig. 2. Section of the landscape map in fig. 1 divided into ecotopes and landscape cells. 1. Loess-covered high plateau. 2. Stony or rocky hanging valleys. 3. Less steep hanging valleys covered by loessic loam. 4. Stream headwater areas of high plateaus and depressions. 5. Alluvial cones of the side valleys entering the main valley. 6. Gravel terraces of the Agger Valley. 7. Stream and Agger Valley floodplains.

logical term *biotope*. The British plant ecologist A. G. Tansley[49] had already used the term *ecotope* in 1939, whereas the Russians speak of *topoecological units*. K. H. Paffen suggested the German term *Landschaftszelle* (landscape cell) for the latter, which seems to me to be well chosen because different *Landschaftszellen* form a *Teillandschaft* (landscape component) in the same way the different cells of an organism form organs. This is not in any way intended to equate a landscape with an organism. It is defined as follows: The smallest-scale landscape individuals, or *Kleinlandschaften*, if spatially differentiated further, would form different, but ecologically homogeneous, local entities, existing in large number, which are called *ecotopes* or *landscape cells (Landschaftszellen)*. They have a defined composition and occur in characteristic associations (*Landschaftskomplex* [landscape complex]) and create a *Kleinlandschaft* because of a particular distribution pattern (*Landschaftsmuster* [landscape pattern], *Landschaftsmosaik* [landscape mosaic]).

At the same time, however, we are aware that a purely biological structuring of habitats has smaller-scale units. Within a plant population, for example, a tree trunk shows different, smallest-scale habitat units for moss and lichen cover, depending on its exposure, height above ground (the impacts of ground fog and snow cover), and type of bark. A block of stone, serving as a bird perch, shows different zones of nitrogen concentration, leading to different zones of lichen cover. But here we are dealing only with spatial structures created within a complete biocoenosis by its individual organisms (trees, birds). Examples of much more complex and interwoven habitats include epiphytic plant communities on rainforest tree branches and in the pits of bromeliad calyxes, and again the smallest-scale associations are found in still bodies of water. In the systematics of biocoenoses, these are called *synusae*.[50]

Kleinlandschaften and ecotopes, as the smallest members of the geographic landscape, are by far the most relevant for scholarly geography. Because they reflect the distribution of the ever-changing successional forces of landscapes and territories, they also have high practical relevance. Forestry uses them for the evaluation of the forest growth and for the identification and characterization of homogeneous growth areas. Soil science uses them for the agricultural assessment of locally variable soils. Vegetation mapping, hydrology, parasitology, and landscape contamination, as well as all the subdivisions of regional landscape planning, require information for analysis at a precise reference point. It is not surprising, therefore, that so many different scientific disciplines, even without cross-fertilization and enrichment and taking different approaches, are directed to the same goal. For the collective benefit of each, it is advisable to share research results and to harmonize terminology.

In the terminology of biology, ecotopes are the equivalent of Dahl's[51] *Lebensstätten* (living places) or *biotopes,* A. Kerner's (1891), M. Öttli's (1905) and H. Gams's (1918) *Standorte* (habitats), as in the recurring characteristic soil or land properties of a place. The British forester R. Bourne undertook a systematic spatial analysis of forest distribution in India's hinterland, Rhodesia, and England using aerial photography.[52] He structured a region into landscape individuals called *regions* or *scenic regions,* which in turn consist of *sites* that are "areas which provide for all practical purposes, and which have across their entire range, similar conditions in terms of climate, physiography, geology, soil and all edaphic factors generally." They usually recur again and again in a certain association and spatial structure. Applied soil science experienced similar results when attempting to gain knowledge about the distribution of topographic soil types on large-scale soil maps in lesser-developed countries.[53] Milne calls the representative character of the topographic structure of a physiographic soil complex a *soil type association.* Because every ecotope has a certain soil type, this edaphic complex also reflects the landscape mosaic. The soil types of the broken hill country between the bottom of the valley and the top of the mountain, which follow the strict rules of spatial distribution, have been called *catena-complexes* by Milne. P. Vagler[54] who expanded this term, in a generalized way, to the *catena-method* of presenting the topographic soil types mentioned previously. Mapping in plant sociology has also started moving in the same direction. The meaning of R. Tüxen's term *Klimaxschwarm* (climax cluster)[55] is the same as the association of ecotopes in a landscape. P. Knapp[56] already uses the term *landscape*. In his hierarchy of plant sociological areas (*Regionen—Wuchszonen—Wuchsgebiete—Wuchslandschaften* [region—growth zone—growth district—growth landscape]), the spatially and narrowly defined *Wuchslandschaften*, which carry a certain climax vegetation, are about equivalent to ecotopes. A. von Kruedener[57] calls ecotopes *Mikrolandschaftstypen* (microlandscape types); Passarge,[58] *Zwergräume* (minute areas) or *Mikroräume* (microareas); and J. G. Granö,[59] *Kleinräume* (small areas). In the terminology of nature and landscape conservation, the term *Landschaftsteil* (landscape component/element) has been introduced for small sections of a landscape that represent either natural landscapes or that have a high landscape value and should therefore be protected. *Landschaftsbestandteile* are the single structures of a landscape; for example, hedges, single trees, bank vegetation, gorges, and rock barriers (as determined according to a decree of November 6, 1935). E. Markus, whose thorough analysis of the term *geographic landscape*[60] we appreciate, uses the term *natural complex* for "the things and processes that are localized in a certain part of the earth's surface and that are causally interconnected." Accordingly, the smallest homogeneous landscape component of an ecotope is called the *smallest natural complex*. The term *unit area* seems to be used in rather different ways in British and American geography. P. W. Bryan[61] uses it for smallest landscape individuals, whereas Unstead refers to the old English term *stow*. At the same time, American geographers use the term for areas that appear homogeneous in terms of land use and physiogeographic features, with a total area depending on the landscape alone. Following this interpretation, the systematic structuring of land for the purpose of regional planning has been called the *Unit-Area Method*.[62] Because this term includes natural factors as well as other rather random factors of land use on newly available land, and because the author did not undertake a theoretical distinction, the Unit-Area Method does not permit a precise methodological evaluation. J. Sölch, in 1924, introduced term *chore*,[63] which has again and again been referred to as the term for the smallest landscape individuals, whereas I have been convinced, by reading his original work and by oral confirmation, that the author suggested it be used rather generally for geographic landscapes of different kinds and scales. He further distinguishes between *Physiochoren* (the generic natural appearance of the earth's surface and its natural spatial structure defined by natural spatial limits), *Kulturchoren*, and, if these are approximately equal, *Geochoren*.

The internationally recognized term *ecotope* has experienced a further special distinction, generated by the work of J. Schmithüsen,[64] regarding its abstract refinement concerning the contents of a landscape. To avoid any misunderstanding, he suggests different names for this basic topographic unit, depending on whether it refers either to its solely physiogeographic

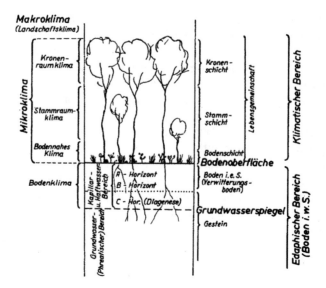

Fig. 3. Model of the Landscape Ecological Structure of an Ecotope

Makroklima (Landschaftsklima)	macroclimate (landscape climate)
Mikroklima	microclimate
Kronenklima	microclimate
Stammraumklima	trunk climate
Bodennahesklima	ground surface climate
Bodenklima	soil climate
Kapillar-u. Haftwasser Bereich	capillary and water retention zone
A-Horizont	A horizon
B-Horizont	B horizon
C-Hor. (Diagenese)	C hor. (diagenetic)
Grundwasser-(Phreatischer) Bereich	groundwater (phreatic) zone
Klimatischer Bereich	climatic zone
Lebensgemeinschaft	living community
Kronenschicht	canopy layer
Stammschicht	trunk layer
Bodenschicht	soil layer
Edaphischer Bereich (Boden i.w.S.)	edaphic zone (soil in the wider sense)
Bodenoberfläche	soil surface
Boden i.e.S. (Verwitterungsbereich)	soil in the narrower sense (weathering zone)
Grundwasserspiegel	water table
Gestein	rock

inorganic disposition or to its total natural disposition, including biological factors, or to its entire physio-, bio-, and cultural-geographic context. For the spatial area without biogeographic features, he suggests the Lower German term *Fliese*, although his understanding of the term *ecotope* includes bio- as well as cultural-geographic factors (natural landscapelike and cultural land-scapelike ecotopes). It should be noted though that the distinction between *Fliese* and ecotope cannot be used universally, at least not where the edaphic locational factors are created by animals, as in the case of moorland landscapes, termite savannas, and so on. Conversely, there should be a stricter distinction between the natural disposition and the cultural environment created by humans. The term *ecotope* should only be used for the smallest units of a landscape with regard to its natural disposition and function as a resource for human land use. In my understanding, the hierarchy physiotope—biotope—ecotope seems to make the introduction of the difficult term *Fliese* superfluous. The distinction between *Fliese* and *ecotope,* as suggested by Schmithüsen, is based on an idea similar to that of the dichotomy between *site* and *cover form* in P. James's[65] landscape analysis.

Landscape Ecology

Within an ecotope, the interplay between the different landscape elements is the closest and most immediate. We usually emphasize the harmonic interplay of climate—soil—vegetation in a rather generalized manner, but the equal importance and interconnectedness of animal life, including parasites and soil fauna, must be considered. Almost of greater importance than the soil is the nature of its moisture flow and the soil climate, which is also partly influenced by the active climatic budget, the microclimate, which is generated by the actual vegetation cover on the soil. Similarly, the soil's different layers do not develop merely from rock under the influence of soil moisture and climate alone, but also from flora and fauna. Hence, the spatial distribution of soil types, as expressed by their soil profile, reflects the spatial distribution of the vegetation cover almost perfectly. The interplay of vegetation and biocenotic life-forms with the soil brings geographic landscape research, soil science, and vegetation research into closest contact.[66] A. G. Tansley describes this intrinsically self-coherent entity of living community and environmental factors by the term *ecosystem.*

Figure 3 may provide an illustration of the relationships between climate, soil, and vegetation. We call all phenomena found beneath the soil surface, except for soil fauna and plant roots, the *soil* or *edaphic zone,* in the broader sense. It includes mineral and weathered soil, groundwater, areas of soil water, and soil climate. As the soil's counterpart, the climatic zone is structured into the macro- or landscape climate and the microclimate or *Kleinstklima,* which extends as far as the upper limit of the vegetation. In the case of normal forest vegetation, in agreement with R. Geiger, we can subdivide the microclimate further into ground surface, trunk, and crown zone climates, which are the most common determining factors (bark vegetation, epiphytes). Vegetation is the central point of the whole ecosystem because its existence determines the microclimate, its remains influence soil weathering, and its water regime regulates the characteristics of the soil water and the soil climate, either directly or through the microclimate. On the other hand, the vegetation as a whole is dependent on the macroclimate and the soil in a broader sense. Comprehensive and clear understanding of microclimatic conditions today is primarily because of R. Geiger.[67] Knowledge of the vegetation's dependence on edaphic conditions is provided by ecological plant geography, which focuses on ecological support and the water regime of vegetation. Ahead lies a vast amount of research into the budgetary regime of living communities as a whole. The main problem is that an individual researcher is hardly able to oversee these living communities, together with the natural cycles of all animals combined within one biocoenosis. Something of an agreement about research into biocoenology has been achieved only for living, freshwater communities in modern limnology (A. Thienemann, F. Ruttner, and others). For land biocoenoses, studies of tropical rain forests[68] may give direction to future research. The aims of landscape ecology will not be achieved before we are able to view the whole biochemical cycle between the producers, consumers, and reducers/decomposers of the major living communities of the earth.

If the vegetation cover in an ecosystem is modified, or even destroyed, by natural or human-induced interference, it of course changes the whole regime and the interplay of landscape elements, micro- and soil climate, characteristics of soil water, and the determinants of soil weathering. If nature is left alone, it tries to reestablish the disturbed equilibrium on its own by the slow regeneration of vegetation and soil through a long series of successions. Alluvial environments of rivers, dunes, deltas, landslides, springs, and avalanche tracks provide numerous such examples.

If we change our landscape ecology perspective from a single ecotope to the landscape complex of a whole landscape, a change in topographic position is obvious, not only for features of soil composition but also for climatic differentiation between the micro- and macroclimate scale. We summarize such climatic conditions as *local climate,* topographic or site climate, created by exposure to radiation, rain and wind, shade, snow accumulation, and so on in various settings such as basins, slopes, ravines, ridges, and peaks. The model landscape profile in fig. 4 illustrates this and should be evident without further explanation because it builds on figs. 1 and 2.

Looking into the inner structure of a landscape allows us to understand that we have to recognize the holistic or gestalt characteristics of a geographic landscape, not only of the landscape individuum but also, most notably, of an ecotope. A spatial synthesis of individual landscape elements into an ecotope creates not only an entity but a harmonic structure of the individual parts, which support one another and cannot exist on their own. Only with such a holistic perspective is it possible to grasp an ecotope entirely. This, however, also results in two methodological consequences for geography. The first deals with the method of landscape classification, the second with its graphic presentation. For a long time, there have been attempts to classify the landscape zones of the earth with a mechanical-analytical approach, either by delineating the spatial distribution of the individual landscape elements to turn their overlapping limits into well-defined boundaries or by synthesizing these overlapping areas into marginal belts (the cartographic method). This approach includes the long-established activities of Passarge's school, the fundamental work of Granö about the geographic areas of Estonia and Finland, and the boundary zone method of O. Maull. This enabled Maull to isolate and define the core area of Hungary from its marginal areas by using the example of the Pannonic Basin. He was also able to synthesize his results into recommendations on core and marginal areas of tropical zones.[69] Hassinger, Lautensach, and Bürger, however, have already shown the deficiencies of this method. Experience and the perception

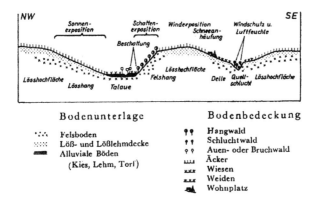

Fig. 4. Landscape Profile with Annotation for Site, Ground Cover, and Local Climatic Differentiation (referred to in figs. 1 and 2)

Bodenunterlarge	*subsoil*
Felsboden	*bedrock*
Löss- und Lösslehmdecke	*loess and loess loam cover*
Alluviale Boden (Kies, Lehm, Torf)	*alluvial soils (gravel, loam, peat)*
Bodenbedeckung	*ground cover*
Hangwald	*slope forest*
Schluchtwald	*ravine forest*
Auen- oder Bruchwald	*water meadows or marsh forest*
Äcker	*farmland*
Wiesen	*meadows*
Weiden	*pastures*
Wohnplatz	*dwelling*
Sonnenexposition	*adret slope*
Beschattung	*shaded*
Schattenexposition	*ubac slope*
Windexposition	*wind aspect*
Schneeanhäufung	*snow accumulation*
Windschutz und Luftfeuchte	*wind protection and humidity*
Lösshochflache	*upland loess areas*
Lösshang	*loess slopes*
Talaue	*flood plain*
Felshang	*rock slope*
Delle	*hollow*
Quellschlucht	*gorge*

of characteristic landscapes as a whole suggest a more intuitive approach to dealing with the idea and character of one or more neighboring landscapes, in looking at the essential landscape factors and their interplay, and in proceeding to a more specific definition of boundaries derived from the whole. From the same perspective, those methods that try to depict landscape character in simplified numerical or letter format appear unsatisfactory. These have been applied, initially by J. G. Granö and then in American geography.[70] Without doubt, they have their merits for the further development of knowledge and methods. They are not, however, sufficient to illustrate the inner interconnectedness of the elements and the functional structure of landscape.

Toward the Classification of Cultural and Economic Landscapes

The previous sections on landscape structure have consciously referred only to the organization of natural structures that try to classify the physical landscape. The cultural aspects of a landscape also follow certain rules for the grouping of patterns that must be understood to structure the cultural landscape. A complete structuring of cultural landscapes, however, requires recognition of both patterns in their mutual interplay. The synthesizing of natural and cultural landscape features into the complex structures of a cultural landscape creates a whole array of problems for geographers. This will probably require decades of research to acquire a satisfactory understanding. The distribution pattern of cultural features is, of course, subject to completely different causalities that are layered above the mechanical and biological ones of the natural environment. They result from the mental constructs of socially structured mankind (families, tribes, communities, cooperatives, entrepreneur associations, peoples, states) and have functional impacts on the supply of the material and nonmaterial needs of human groups. The characteristics of the land do not necessarily determine the forms of human labor; they provide only certain possibilities and suggestions. On the other hand, they also determine quite strict boundaries and limits. By using the opportunities provided by nature, humans are in an intimate interplay with the physical structure of the landscape. As Schmithüsen[71] stresses, cultural landscape research focuses on "the reciprocity and an adjustment in the harmonious interplay between both structural patterns and in the natural structure and culturally based activities, respectively" and on "the plan, as determined by natural settings, which provide—from a human perspective—a certain indication of its potential utilization." The cultural design of the earth, with its roots in human communities, is structured in a similar pattern to the natural structure; that is, in levels at different scales. The large landscape zones with a natural structure, and which stretch beyond continents, correspond to the global divisions of culture; the smallest units of the natural structure correspond approximately to the communities of a village population who shape their settlements and populated area both culturally and economically. A spatial understanding of the natural and cultural structural plan, at both large and small scales, is the key to understanding the whole landscape structure of the earth.

In his anthropogeographic structuring of the earth, based on cultural landscapes, F. Jaeger[72] has attempted also to portray cartographically the cultural-structural plan of the earth. For this he distinguishes between the following criteria: (1) affiliation to one of the fourteen large cultural divisions (culture area or culture group), which at the same time indicates affiliation to a certain type of state governance (European, Anglo-American, Latin American, Russian, East Asian, Indian, Indonesian, Oriental, Negro-African, etc. cultural realms); (2) a general form of soil and land use (arable,

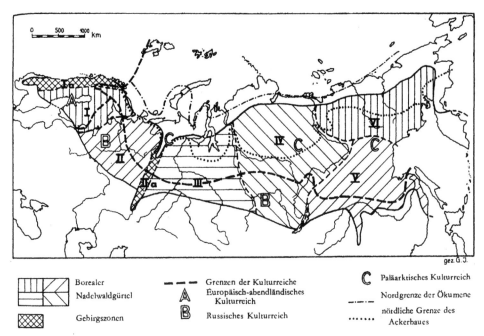

	Borealer Nadelwaldgürtel	– – –	Grenzen der Kulturreiche	**C**	Paläarktisches Kulturreich
		A	Europäisch-abendländisches Kulturreich	–··–	Nordgrenze der Ökumene
	Gebirgszonen	**B**	Russisches Kulturreich	·······	nördliche Grenze des Ackerbaues

I—VI Naturräumliche Landschaftsabschnitte des Nadelwaldgürtels

I Fennoskandien
i a Skandinavisches Gebirge (Skanden)
II Nordrussisches Nadelwaldgebiet
II a Ural

III Westsibirische Tiefebene
IV Mittelsibirisches Schollenland (Angaraland)
V Transbaikalien
VI Nordostsibirisches Nadelwaldgebiet

Fig. 5. The Natural and Cultural Spatial Structure of the Boreal Coniferous Forest of Eurasia

Borealer Nadelwaldgürtel — *boreal coniferous forest*
Gebirgszonen — *mountain zones*
Grenzen der Kulturreiche — *boundaries of cultural realms*
  A Europäisch-abendländisches Kulturreich — *A European-Occidental cultural realm*
  B Russisches Kulturreich — *B Russian cultural realm*
  C Paläarktisches Kulturreich — *C Palearctic cultural realm*
Nordgrenze der Ökumene — *northern boundary of the ecumene*
Nördliche Grenze des Ackerbaues — *northern boundary of cultivation*
I–VI Naturräumliche Landschaftsabschnitte des Nadelwaldgürtels — *Natural spatial division of the coniferous forest zone*
  I Fennoskandien — *Fennoscandian*
    Ia Skandinavisches Gebirge (Skanden) — *Scandinavian mountain range (Scandia)*
  II Nordrussisches Nadelwaldgebiet — *North Russian coniferous forest area*
    IIa Ural — *Urals*
  III Westsibirische Tiefebene — *West Siberian lowlands*
  IV Mittelsibirisches Schollenland (Angaraland) — *Mid-Siberian Massif (Angaraland)*
  V Transbaikalien — *Transbaikalia*
  VI Nordostsibirisches Nadelwaldgebiet — *Northeast Siberian coniferous forest area*

horticulture, rice growing, plantation agriculture, oasis agriculture, migratory hoe farming, nomadic livestock farming, stationary livestock farming, mining, and industry), which are closely dependent on natural landscape zones and can be compared to land use zones; (3) and the degree to which cultural structures have reshaped the earth and for which population density is an indicator of particular relevance. An important supplement to understanding global cultural divisions from their historical and dynamic perspective has been provided by H. Schmitthenner,[73] who explored interactive diffusion processes between cultural divisions, starting from the core regions of high civilizations.

If we apply this perspective to the boreal coniferous forest belt of Eurasia, whose natural environment we have already structured previously, we can see that the uniform landscape belt stretches over three different cultural divisions: the European-Occidental in Scandinavia and Finland, the Russian from East Karelia beyond the Dvina, and the remaining Palearctic, which is the largest part and is populated by different Arctic cultural groups. The boundaries of the cultural divisions by no means correspond to the structuring of the natural environment, nor does the boundary between the Occidental and the Russian division. East Karelia, though originally by nature as well as culture associated with Fennoscandia, has now been integrated into the North Russian cultural landscape through Russian colonization.

In accordance with the second point concerning economic land use, the coniferous forest belt is structured in its longitudinal range from a northern farming-free zone of caribou-hunting nomads and wildlife hunters to a southern zone of sporadic farming. At the same time, this zonal structuring relates to the third point, in that the degree of environmental reshaping through the cultural landscape has only created isolated spots of cultural landscape within the natural landscape so far.

This straightforward and simplified example may clarify how large-scale natural and cultural structures permeate one another and how their interactions and interconnections illuminate the separation of larger-scale cultural landscapes. Occasionally, natural landscape boundaries on the earth have such immense impact that they also become boundaries of cultural divisions. The boundary between the Oriental and Negro-African cultural division in the Sudan is probably the finest example of its kind.

Let us focus now on the small-scale structure of cultural landscapes. We previously chose an example of a cultural farming landscape from the Bergische Land because it still reveals a very close dependency of land use on its natural setting. Almost every ecotope is subject to land use that is particularly suitable for its natural assets and for landscape balance. The form of land use is by no means determined absolutely, and it can change according to developments in the farming economy. For example, pastures close to the farmsteads that extend to swales in high areas have only appeared in recent decades. However, the distributional pattern of land use, the entire rural agricultural landscape, settlement location, and transportation structures still have such an intimate harmony with the natural landscape structure that we are strongly inclined to speak of a term *harmonious cultural landscape.*[74] In such a case, there is no restriction on the use of the term *natural landscape*

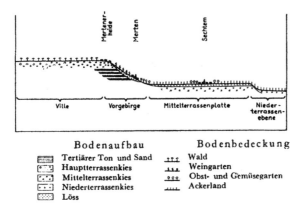

Fig. 6. Landscape Profile Through *Ville* and *Vorgebirge* Northwest of Bonn as Recorded in 1812

Bodenaufbau	soil structure
Tertiärer Ton und Sand	tertiary clay and sand
Haptterrassenkies	main gravel terrace
Mittelterrassenkies	middle gravel terrace
Niederterrassenkies	lower gravel terrace
Löss	loess
Bodenbedeckung	ground cover
Wald	forest
Weingarten	vineyard
Obst- und Gemüsegarten	orchard and vegetable gardens
Ackerland	farmland
Mittelterrassenplatte	middle terrace bench
Niederterrassenplatte	lower terrace bench

as a comprehensive term to include the cultural elements. F. Huttenlocher[75] suggests the term *geographic landscape* be limited to these cases and the term *länderkundliche Einheiten* (landscape units) for all other cases.

Generally, however, economically active humans entering the system induce new spatial interrelationships that are rooted in human social structures and in functional strategies for its economy and management. In those cases, in which this covers up the natural variations, a *disharmony of landscape structure* occurs.

This may be illustrated by the *Vorgebirge* in the lower Rhine landscape between Bonn and Cologne, for which H. Müller-Miny[76] provides an accurate land-use map (figs. 6–8). Starting from the high plains of the *Ville*, which are composed of terraces of crushed rock and covered by forests, the land falls away to a scarp that is covered by a thick layer of loess, then to the so-called promontory and across another loess-covered lower terrace to the low terrace of the Rhine, which is made up of moist channels (fig. 6). In 1810, the cultural landscape still showed a close relationship to its natural structures (fig. 7). The high crushed rock plain of the *Ville* was covered by forest, except for a few areas that were covered instead by a thin layer of loess. In these areas, a second line of villages developed at the upper bound-

ary of the loess-covered slope toward the table-rock outcrop, in addition
to the densely developed line of *Vorgebirge* villages at the foot of the loess
slope. The *Vorgebirge* villages were surrounded by vineyards and, close to
the houses, by small orchards and vegetable-growing areas. Between these,
farming took up the remaining land on the broad plains of the middle and
lower terraces. The uniform grain fields were interrupted only by the allot-
ment gardens of the widely spaced villages (Sechtem), by meadows in the
moist parts of Bornheim and Roisdorf, and by remnants of small woods on
sandy soils. This picture has undergone a radical change in the past one
hundred years. Viticulture has disappeared; instead, horticulture developed extensively, initially forming a small belt that eventually took over the
whole loess slope of the *Vorgebirge*. In more recent times (fig. 8), however,

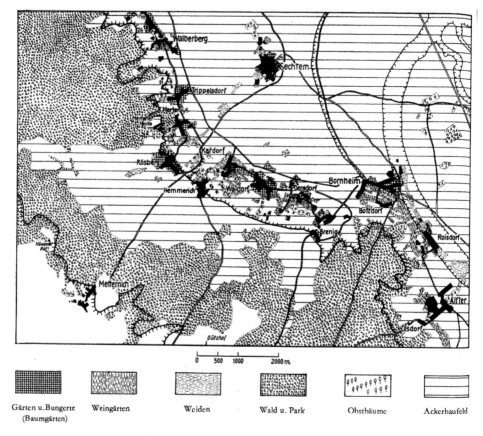

Fig. 7. Landscape Section of *Ville* and *Vorgebirge* Northwest of Bonn with the
Distribution of Cultivation Types in 1810 (after KUPAL)

Gärten u. Bungerte	*gardens and arboreta (orchards)*
Weingarten	*vineyards*
Weiden	*pastures*
Wald u. Park	*forest and parks*
Obstbäume	*fruit trees*
Ackerbaufeld	*farmland*

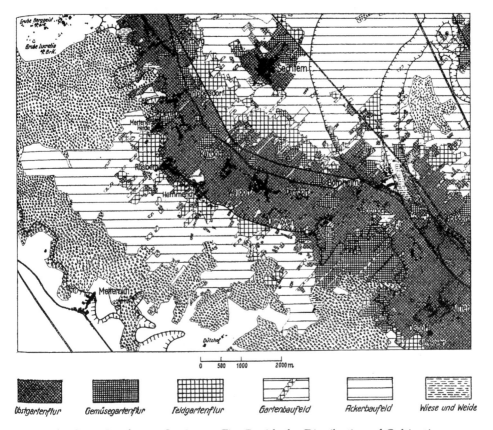

Fig. 8. The Same Landscape Section as Fig. 7 with the Distribution of Cultivation Types in 1944

Obstgartenflur	*horticultural land*
Gemüsegartenflur	*vegetable-growing land*
Feldgartenflur	*field cropland*
Gartenbaufeld	*market gardens*
Ackerbaufeld	*farmland*
Wiese und Weide	*meadow and pasture*

horticulture has experienced, and still experiences, such an upturn that it has expanded well beyond the plains and already reaches the expanding horticultural belt around Sechtem, thus increasingly narrowing the farmland. At the same time, horticulture has taken over the marginal soils of the crushed rock plain of the *Ville* plateau, where it even displaces the heath forest through the expansion of strawberry, raspberry, and cherry growing. Furthermore, starting from Walberberg northward, a cohesive area of mining landscape, due to the surface mining of the immense layers of brown coal located under the main terrace layer, has replaced the forest of the *Ville*.

For us, the most important question is "What forces have driven these landscape changes?" They are undoubtedly embedded in the general development of the economy, transport, and population in the last century. The growing population in urban and industrial areas has generated an ever-increasing demand for garden products, which generally must be supplied from a close proximity. Where social and business management conditions and traditions dictate, more and more farmland tends to be converted into more horticultural land. This encroachment onto the crushed rock soils of the *Ville* shows that location relative to the horticultural villages is a more important determining factor than are the natural properties of the soil. The whole spatial pattern of the cultural landscape, however, still reflects a subtle hint of the natural structure: the differentiation into the pebble highlands, the loess slope, and the terraced benches. The spatial extent of horticultural land, however, strictly follows the economic factors of relative location. In this context, it should not be forgotten that, in close proximity to the villages, horticulture originally developed from the long-established cultivation of vines. The corresponding factors on the mining landscape of the *Ville*, which initially focused on accessible surface mining for coal, were initially determined by natural factors, especially the rather deep-layered natural subsurface features of the geological landscape. This surface mining replaces the forests and the agricultural land of the villages, sweeping away the latter and leaving behind an emerging landscape of new settlements. The value of these settlements is entirely dependent on how careful the criteria of landscape design are initially taken into account in the reshaping of the terrain. The reshaping of the brown-coal mining landscape, planned for the near future, is partly subject to recent technical conditions (subsurface mining, underground engineering) and partly to agropolitical issues because their respective areas overlap with the richest agricultural landscapes.

The map shown in fig. 8, in L. Waibel's[77] terminology, displays three clearly distinctive economic formations. They form a clear spatial economic structure, as Hassinger[78] puts it. We see two different kinds of forces here, however: natural forces with a rather static impact and extremely dynamic cultural forces. J. O. Broek's[79] suggestions of viewing cultural landscapes as growing or shrinking spatial entities can be, therefore, easily recognized.

This particular, cartographically supported example may provide a starting point for a concluding exploration of economic and cultural landscapes in more general terms. Similar to natural structures, the spatial structures of economic landscapes of the earth also show hierarchical levels of quite different dimensions. Subsequent to a well-prepared methodology, H. Carol's[80] mapping of Switzerland has gone through various theoretical considerations concerning the structure of economic landscapes. For a landscape entity that has been partly determined by economic organization, he uses the term *functional unit of the economic landscape,* abbreviated to *functional,* as opposed to purely physiognomic entities, the *formal.* Every economic landscape is the result of interaction between the natural structure and the economic planning format. To clearly isolate the economic forces, we use a deductive approach, which assumes completely equivalent natural factors. This is the same assumption made by J. H. Thünen in his theory of the isolated state and later by W. Christaller in his theory of central functions. The smallest economic unit of an agricultural landscape is a single farm with its enclosed farmed space. The distribution of usable agricultural areas in it reflects both the natural potential and the relative distance from the farmhouse. The more distant parts can only be extensively managed, or at least not as intensively managed as those areas in close proximity to the farmhouse. Groups of similar farms usually coalesce to form settlements, hamlets, and villages. At this point, we have to distinguish between communal-living areas and the much larger agricultural area of a parish. The latter is much more differentiated than is the agricultural area of an individual farm because, as a result of greater distances, the principle of differential concentric rings of intensity will be much more noticeable, possibly separated into home gardens, pastures close to villages, inner and outer fields, and common lands. Another point is the distribution of property, which is dependent on the social structure of the village (strip- and block-shaped structures, parcels of land with and without a direct connection to the farmhouse, reallocation of land, etc.). In this case, there are already three overlapping organizational structures. W. Müller-Wille[81] has demonstrated a fine example of this in a Rhenish agricultural landscape. Further exploration reveals that the selection of location for housing and farming does not, by any means, show any tendency for land use management based on its natural disposition. Apart from locations where topography is taken into account (water supply, local climate), human economic activity tries to combine as many versatile areas of use (farmland, meadow, pasture, forest, bodies of water) around the location of a settlement, a preference that leads to a choice of location at the boundaries of the small-scale landscapes or ecotopes. Small or large groups of similarly structured economic units (districts) form the smallest economic landscape units. The economic functions, even of purely agricultural landscapes, are generally not closed. Neighboring economic landscapes enter into agricultural interrelationships; abundant grain fields and lush lowland meadows mutually trade their products, and Rhenish valley vineyards receive manure for their vineyards from the neighboring highlands. Cattle raising is diversified through breeding, feeding, and dairy operations, with frequent changes of location between different landscapes, and the agricultural system of the Alps moves cattle through different elevations to follow seasonal changes. Recently, F. Huttenlocher has portrayed the interplay of rural settlement and the natural structure of cultural landscapes in a methodologically refined case study of Württemberg.

Let us now leave the field of agriculture and focus on complex economic landscapes. A wide range of research on cultural landscapes deals with spatial sociogeographic interrelationships between urban and rural areas. Whereas J. H. von Thünen[82] had in mind only a structuring of the agricultural landscape with regard to one urban consumer center, W. Christaller[83] was the first to explore the location of larger settlements, which provide central economic, administrative, and cultural functions in relationships with their rural surroundings. In dealing with a regular distribution of central-place functions, the spatial structure of such economic landscapes reveals forces with both radiating and circular operating forces. Entire industrial landscapes enter into sociological interrelationships with neighboring or distant agricultural landscapes by offering additional labor opportunities, with daily or seasonal commuting for small-scale farms with insufficient farm output. An excellent example has been analyzed by G. von Siemens[84] in the Rhineland. The Oberbergische Land between the Kölnische Tieflandsbucht and the watershed divide along the Westfalian boundary, which also provided the examples in figs. 1 and 2, shows a clear natural structure from west to east with increasing elevation and a related drop in temperature, increasing precipitation, and a corresponding decrease in soil quality. Over this agroecological structure, however, is spread a system of sociogeographic forces; the close proximity of the industrial area of Siegburg in the west and the established iron and textile industry in the Agger River headwaters in the east enabled farmers to turn incidental labor opportunities into the combination of the two extremes of a wage economy and one of small-scale farming. A purely farming-oriented economy is maintained in between. In turn, this has an impact on the cultivation of crops grown; that is, between a more economically self-sufficient family production of potato and dairy products by part-time farmers, on the one hand, and a regular farming economy with a stronger focus on grain, on the other.

Even greater impacts of structural forces on the economic landscape are generated by the relationships between seaports or entire wharf areas and their hinterlands, where inland-bound traffic connections by road and river are of crucial importance. Because harbor cities develop overseas relationships with supply-and-demand markets and with marine fishing grounds at the same time, they manage to overcome the restrictions of the global economy. The economic landscapes of the earth constitute an economic spatial structure at the largest scale, where the forces of natural and cultural space overlap as much as do ecotopes and the agriculturally used areas of a farm at the small scale of a landscape mosaic. To both, H. Gutersohn's[85] statement is applicable: "Most of the different hypotheses of a natural, economic, technical, sociological and psychological form continuously provide new momentum which changes the landscape accordingly. Each one of these impulses is the starting point of a wave of development, stronger or weaker, proceeding rapidly or slowly, periodic or aperiodic, which results in all possible forms of interaction, mutual overlap, and which finally frames the picture of the cultural landscape as a whole." The functional analysis of its appearance and its intellectual rearticulation into a landscape entity is a central concern today for academic geography.

Notes

1. H. Hassinger, Über einige Aufgaben geographischer Forschung und Lehre. Kartor. und schulgeogr. Ztschr., 8, Wien, 1919.
2. W. Volz, "Das Wesen der Geographie in Forschung und Darstellung," *Schles. Jahrb. f. Geistes-u. Naturwiss.* 2 (1923).
3. W. Burkamp, *Die Struktur der Ganzheiten* (Berlin, 1929); J. C. Smuts, *Holism and Evolution* (London, 1927).
4. W. Volz, "Geographische Ganzheitlichkeit," *Ber. Sächs. Akad. Wiss. Leipz. Math.-Phys. Kl.* 84 (1932); E. Plewe, "Randbemerkungen zur geographischen Methodik," *Geogr. Ztschr.* (1935); K Bürger, "Der Landschaftsbegriff," *Dresden. Geogr. Stud.* 7 (1935); R. Wörner, Das geog-

raphische Ganzheits-problem vom Standpunkt der Psychologie aus. Geogr. Ztschr. 1938; H. Lautensach, "Über die Erfassung und Abgrenzung von Landschaftsräumen," *Comptes Rendus du Congrès Intern. Géog.*, vol. 2, sec. 5 (Leiden, 1938).

5. E. Banse, *Die Geographie und ihre Probleme* (Berlin, 1932).
6. A. Vierkandt, "Gesellschafts- und Geschichtsphilosophie," *Lehrb. D. Philos. v. M. Dessoir.* Berlin 1925.
7. S. Passarge, "Physiogeographie und vergleichende Landschaftsgeographie," *Mitt. Geogr. Ges. Hbg.* 27 (1913).
8. A. Oppel, *Landschaftskunde, Versuch einer Physiognomie der gesamten Erdoberfläche etc* (Breslau, Poland, 1884); I. Wimmer, *Historische Landschaftskunde,* (Innsbruck, Austria, 1885).
9. L. Waibel, "Besprechung von K. Bürger 'Der Landschaftsbegriff,'" *Dt. Lit. Ztg.* (1936).
10. Trübners Deutsches Wörterbuch, hrsg. V. A. Götze, 4 Bde. Berlin 1943.—Für freundl. Hilfe bei der philosophischen Seite danke ich Herrn Prof. W. Betz (Bonn herzlich). Manche Anregung gab auch die Dissertation meiner Schülerin M. Gurlitt-Jansen, "Die Geschichte der Alpenforschung und die Entwicklung des Landschaftserlebnisses im 18. Jahrhundert. Bonn 1948. (Manuskr.)
11. J. Messerschmidt-Schultz, "Zur Darstellung der Landscaft in der deutschen Dichtung des ausgehenden Mittelalters," PhD diss., Breslau 1938.
12. C. Troll, Luftbildplan und ökologische Bodenforschung. Ztschr. Ges. Erdk. Berlin 1939— Ders., Fortschritte der wissenschaften Luftbildforschung. Ebenda 1943. Methoden der Luftbildforschung. Sitz. Ber. Europ. Geograph., Würzburg 1942. Leipzig, 1943.
13. A. V. Havemann, and V. A. Faas, "On the Development of the Study of Aerolandscape," *Comt. Rend. (Doklady) Acad. Sciences USSR* 26 (1940).
14. *Comptes Rendus du Congrès International de Géographie,* vol. 1, sec. 5, Paysage géographique (Amsterdam, 1938; Leiden, 1938).
15. N. Krebs, Vom Wesen, and Wert der Länder, Abhdl. Preuß. Akad. Wiss. 1941, 4, 1941; ähnlich schon 1927 in Krebs, Die Entwicklung der Geographie in den letzten 15 Jahren. Frankf. Geogr. Hefte, I, 1, 1927.
16. S. Passarge, Die natürlichen Landschaften Afrikas. Pet. Mitt. 1908; C. Troll, Die natürlichen Landschaften des rechtsrheinischen Bayerns. Geogr. Anz., 27, 1926.
17. N. Krebs, Süddeutschland, Leipzig-Berlin 1923, S. 57.
18. K. Bürger, Der Landschaftsbegriff loc. cit. 1935, S. 29.
19. Also ganz im Sinne vom natürlichen und künstlichen System der Pflanzen.
20. L. Waibel, "Was verstehen wir unter Landschaftskunde?" *Geogr. Anz.* (1933).
21. E. Otremba, Die Grundsätze der naturräumlichen Gliederung Deutschlands. Erdk., II, Bonn 1948.
22. W. Müller-Wille, Die Naturlandscaften Westfalens. Westf. Forsch., Münster 1942.
23. R. Gradmann, Das harmonische Landschaftsbild. Ztsch. Ges. f. Erdk. Berlin 1924.
24. G. Pfeifer, Entwicklungstendenzen in Theorie und Methode der regionalen Geographie in den Vereinigten Staaten nach dem Kriege. Ebende 1938; H. Boesch, Beiträge zur Frage der geographischen Raumgliederung in der amerikanischen Literatur. Vierteljahr. Schr. Naturf. Ges. Zürich, 91, 1946.
25. R. Hartshorne, *The Nature of Geography,* 2nd ed. (Annals of the Association of American Geographers, 1946).
26. O. Schlüter, "Die Stellung der Geographie des Menschen in der erdkundlichen Wissenschaft," *Geogr. Abende* 5 (1919).
27. O. Schlüter, Die analytische Geographie der Kulturlandschaft. Sond. Bd. 100 Jahrfeier, Ztschr. Ges. F. Erdk. Berlin, 1928.
28. E. Winkler, Zur Frage der allgemeinen Geographie. Athenäumsschr. 1938, H. 2, Zürich 1938.
29. E. Winkler, Zur Frage der allgemeinen Geographie loc. cit. 1938.
30. R. Krzymowski, Die Agrargeographie Landwirtsch. Jahrb. 50 1917. Ders., Philosophie der Landwirtschaftlehre. Stuttgart 1919.
31. W. Schönichen, Biologie der Landschaft. Neudamm u. Berlin 1939.
32. C. Troll, Studien zur vergleichenden Geographie der Hochgebirge der Erde. Bonn 1941.
33. R. Knapp, Einführung in die Pflanzensoziologie, H, 1, Stuttgart 1948.

34. F. Firbas, Spät- und nacheiszeitliche Waldgeschichte Mitteleuropas nördlich der Alpen. Jena 1949.
35. Begriffsbestimmung nach W. Müller-Wille loc. cit. 1942.
36. J. Budel, Die räumlich und zeitliche Gliederung des Eiszeitklimas. Die Naturwiss., 36, 1949. H. Poser, Auftautiefe und Frostzerrung im Boden Mitteleuropas während der Wurmeiszeit. Ebenda, 34, 1947—Ders. Äolische Ablagerungen und Klima des Spätglazials im Mittel- und Westeuropa. Ebenda, 35, 1948.
37. P. Schultze-Naumburg, Entstellung unseres Landes. 1905—Ders., Die Gestaltung der Landschaft durch den Menschen. Kulturarbeit. Bd.7. München 1915.
38. H. Schwenkel, Grundzüge der Landschaftspflege. Neudamm u. Berlin 1938.
39. A. Vietinghoff-Reisch, Forstliche Landschaftsgestaltung. Neudamm 1939.
40. A. Seifert, Im Zeitalter des Lebendigen. Dresden u. Planegg 1941. H. F. Wiepking-Jurgensmann, Die Landschaftsfibel. Berlin 1942.
41. "Classification of Regions of the World. Report of a Committee of the Geographical Association," Geography 22 (1937).
42. H. G. Hommeyer, Beitrag zur Militär-Geographie der europäischen Staaten. I, Breslau 1805.
43. P. E. James, The terminology of regional description. Annals of the Association of American Geographers (1933).
44. J. F. Unstead, "A System of Regional Geography: Herbertson Memorial Lecture 1933," Geography (1935).
45. K. H. Paffen, Ökologische Landschaftsgliederung. Erdk., II, 1948.
46. C. Troll, Der asymmetrische Vegetations- und Landschaftsaufbau der Nord- und Südhalbkugel. Göttingen. Geogr. Abhdl., I, Gött. 1948.
47. A. Kalela, Die Ostgrenze Fennoskandiens in pflanzengeographischer Beziehung. Veröff. Geobot. Inst. Rübel, 20, Bern 1943. C. Regel, Die Bergrenzung von Fennoskandien nach Südosten und Osten. Peterm. Mitt. 1944.
48. C. Troll, Methoden der Luftbildforschung. Sitz. Ber. Europ. Geogr. Würzburg 1942; Leipzig 1945.
49. A. G. Tansley, The British Isles and Their Vegetation (Cambridge: Cambridge University Press, 1939).
50. H. Gams, Prinzipienfragen der Vegetationsforschung, Viertelj. Schr. Naturf. Ges. Zürich, 63, 1918.
51. F. Dahl, Grundsätze und Grundbegriffe der biozönologischen Forschung. Zool. Anz., 33, 1908.
52. R. Bourne, "Regional Survey and Its Relation to Stocktaking of the Agricultural and Forest Resources of the British Empire, Oxford Forest." Memoirs, 13, 1931.
53. G. Milne, A Provisional Soil Map of East Africa. Amani Memoirs, Amani 1936.
54. P. Vageler, Die Böden Westafrikas vom Standpunkt der Catena-Methode. Mitt. Gruppe Dt. Kolonialwirt. Unternehm., 2, Berlin 1940.
55. R. Tüxen, and H. Diemond, Klimagruppe und Klimaschwarm. 88.–89. Jahresber. Naturf. Ges. Hannover, 1937.
56. R. Knapp, loc. cit. 1948.
57. A. v. Kruedener, Waldtypen als kleinste Landschaftseinheiten bzw. Mikrolandschaftstypen. Pet. Mitt. 1926.
58. S. Passarge, Wesen, Aufgaben, und Grenzen der Landschaftskunde. Pet. Mitt., Erg. H. 209, Gotha 1930.
59. J. G. Granö, Reine Geographie. Helsinki 1929.
60. E. Markus, Geographische Kausalität (Tartu, Estonia, 1936).
61. P. W. Bryan, Mans Adaptation of Nature (London, 1933).
62. G. D. Hudson, "The Unit-Area Method of Land Classification," Annals of the Association of American Geographers 26 (1936); E. Kirchen, "Die Einheitsflächenmethode," diss. University of Zürich, 1949.
63. J. Sölch, Die Auffassung der "natürlichen Grenzen" in der wissenschaftlichen Geographie. Innsbruck 1924.
64. J. Schmithüsen, "Fliesengefüge der Landschaft" und "Ökotop." Ber. Z. Dt. Landeskde., 5, 1948.

65. P. E. James, "Blackstone Valley," *Annals of the Association of American Geography* (1929).

66. J. Schmithüsen, Vegetationsforschung und ökologische Standortslehre in ihrer Bedeutung für die Geographie der Kulturlandschaft. Ztschr. Ges. F. Erdk. Berlin 1942. C. Troll, Luftbildplan und ökologische Bodenforschung. Ebenda 1939.

67. R. Geiger, Das Klima der bodennahen Luftschicht—ein Lehrbuch der Mikroklimatologie. Braunschweig 2. Aufl., 1942.

68. H. Eidmann, Der tropische Regenwald als Lebensraum. Kolonialforstl. Mitt., 5, Neudamm u. Berlin 1942. Ders., Zur Ökologie der Tierwald des afrikanischen Regenswaldes. Beitr. Z. Kolonialforsch., 2, Berlin 1942. F. Mertens, Die Tierwelt des tropischen Regenswaldes. Frankfurt 1948.

69. O. Maull, "Allgemeine vergleichende Länderkunde," in Länderkunde Forsch., Festsch. N. Krebs, Berlin 1936.—Die Bestimmung der Tropen am Beispiel Amerikas. Festchr. Z. 100-Jahrfeier d. Ver. f. Geogr. u. Statistik zu Frankfurt a. M., 1936.

70. V. C. Finch, "Montford Area," *Bulletin of the Geographical Society of Chicago* 9 (1933).

71. Schmithüsen, "Vegetationsforschung," loc. cit. 1942.

72. F. Jager, Neuer Versuch einer anthropogeographischen Gliederung der Erdoberfläche. Pet. Mitt. 1943.

73. H. Schmitthenner, Lebensräume im Kampf der Kulturen. Leipzig 1938.

74. Im Sinne von R. Gradmann loc. cit. 1924, und H. Gutersohn, Harmonie in der Landschaft. Wesen und Ziel der Landesplanung. E. T. H. Zürich, Arbeit. a. d. Geogr. Inst., 4, 1946.

75. F. Huttenlocher, Versuche kulturlandschaftlicher Gliederung am Beispiel von Württemburg. Forsch. z. Dt. Landeskde., 47, Stuttgart. 1949.

76. H. Müller-Miny, Die linksrheinischen Gartenbaufluren der südlichen Kölner Bucht. Bericht z. Raumforschung u. Raumordnung, 5, Leipzig 1940.

77. L. Waibel, Probleme der Landwirtschaftsgeographie. Breslau 1933.

78. H. Hassinger, Die Geographie des Menschen (Anthropogeographie). Handb. Geogr. Wiss. Hrsg. V. F. Klute, Allgemeine Geographie, 2, Teil, Potsdam 1933.

79. J. O. Broek, "The Concept Landscape in Human Geography," *Comptes Rendus du Congrès International de Géographie,* vol. 2 (Amsterdam, 1938; Leiden, 1938).

80. H. Carol, "Die Wirtschaftslandschaft und ihre kartographische Darstellung," *Geogr. Helvetica* 1 (1946).

81. W. Müller-Wille, "Die Ackerfluren im Landesteil Birkenfeld," *Beitr. z. Landeskde. D. Rheinl.* 2, no. 5 (1936).

82. L. Waibel, Das Thünensche Gesetz und seine Bedeutung fur die Landwirtschaftsgeographie. In Waibel loc. cit. 1933.

83. W. Christaller, Die zentralen Orte in Süddeutschland. Jena 1933. Ders. Rapports fonctionnels entre les agglomérations urbaines et les campagnes. *Comptes Rendus du Congrès International de Géographie,* vol. 2 (Amsterdam, 1938; Leiden, 1938). R. E. Dickinson, *City Region and Regionalism* (London, 1947). E. Hoover, *The Location of Economic Activity* (New York, 1948).

84. G. v. Siemens, "Zur agrargeographischen Landschaftsgliederung," *Erdk.* 3 (1949).

85. H. Gutersohn, Geographie und Landesplanung. E. T. H. Zürich, Kultur- und Staatswiss. Schr., 31, Zürich 1942.

VOLUME 35, NOS. 1 AND 2   DECEMBER 1947

PATTERN AND PROCESS IN THE PLANT COMMUNITY*

By ALEX. S. WATT, *Botany School, University of Cambridge*

(*With eleven Figures in the Text*)

CONTENTS

THE PLANT COMMUNITY AS A WORKING MECHANISM

The plant community may be described from two points of view, for diagnosis and classification, and as a working mechanism. My primary concern is with the second of these. But inasmuch as the two aspects are not mutually exclusive, a contribution to our understanding of how a community is put together, and how it works, may contain something of value in description for diagnosis.

It is now half a century since the study of ecology was injected with the dynamic concept, yet in the vast output of literature stimulated by it there is no record of an attempt to apply dynamic principles to the elucidation of the plant community itself and to formulate laws according to which it maintains and regenerates itself. Pavillard's assessment of the dynamic behaviour of species comes very near it, but is essentially concerned with the 'influence (direct or indirect) of the species on the natural evolution of plant communities' (Braun-Blanquet & Pavillard, 1930). As things are, the current descriptions of plant communities provide information of some, but not critical, value to an understanding of them; how the individuals and the species are put together, what determines their relative proportions and their spatial and temporal relations to each other, are for the most part unknown. It is true that certain recent statistical work is stretching out towards that end, but the application of statistical technique, the formulation of laws and their expression in mathematical terms, will be facilitated if an acceptable qualitative statement of the nature of the relations between the components of the community is first presented. Such a statement is now made based on the study of seven communities in greater or less detail, for data of the kind required are seldom recorded.

The ultimate parts of the community are the individual plants, but a description of it in terms of the characters of these units and their spatial relations to each other is impracticable at the individual level. It is, however, feasible in terms of the aggregates of

* Presidential address to the British Ecological Society on 11 January 1947.

2 *Pattern and process in the plant community*

individuals and of species which form different kinds of patches; these patches form a mosaic and together constitute the community. Recognition of the patch is fundamental to an understanding of structure as analysed here.

In the subsequent analysis evidence is adduced to show that the patches (or phases, as I am calling them) are dynamically related to each other. Out of this arises that orderly change which accounts for the persistence of the pattern in the plant community. But there are also departures from this inherent tendency to orderliness caused by fortuitous obstacles to the normal time sequence. At any given time, therefore, structure is the resultant of causes which make for order and those that tend to upset it. Both sets of causes must be appreciated.

In describing the seven communities I propose in the first examples to emphasize those features which make for orderliness, in the later to content myself with little more than passing reference to these and to dwell specifically upon departures from it; in all examples to bring out special points for the illustration of which particular communities are well suited or for which data happen to be available.

For the present the field of inquiry is limited to the plant community, divorced from its context in the sere; all reference to relics from its antecedents and to invaders from the next state is omitted. I am assuming essential uniformity in the fundamental factors of the habitat and essential stability of the community over a reasonable period of time.

THE EVIDENCE FROM SEVEN COMMUNITIES
The regeneration complex

Regarded by some as an aggregate of communities, by others as one community, the regeneration complex fittingly serves as an introduction because its study emphasizes the underlying uniformity of the nature of vegetational processes. It consists of a mosaic of patches forming an intergrading series the members of which are readily enough assignable to a few types or phases. The samples of these phases are repeated again and again over the area; each is surrounded by samples of other, but not always the same phases.

As the work of Osvald (1923) and of Godwin & Conway (1939) shows, these phases are dynamically related to each other. For Tregaron Bog the sequence is briefly as follows. The open water of the pool is invaded by *Sphagnum cuspidatum* which in turn is invaded and then replaced first by *S. pulchrum*, then by *S. papillosum*, by whose peculiar growth a hummock is formed. This hummock is first crowned by *Calluna vulgaris*, *Erica tetralix*, *Eriophorum vaginatum* and *Scirpus caespitosus*, later by *Calluna vulgaris* with *Cladonia silvatica* forming a subsidiary layer. The proof of the time sequence lies in the vertical sequence of plant remains in the peat itself.

Each of the phases of the regeneration complex was at one time regarded as a community, and the whole as an aggregate of communities dynamically related to each other. Tansley (1939), for example, still calls them seral. The resemblance to a sere is close; each phase from the open water to the hummock with *Calluna* depends on its antecedents and is the forerunner of the next, the sequence depending on plant reaction.

The resemblance, however, is partial only, for a cycle of change is completed by the replacement of the hummock by the pool, a topographical change brought about by differential rates of rise in the different patches. That is, the immediate cause does not reside exclusively in the patch itself but in the spatial relation between patches and their relative changes in level. Thus the *Calluna* of the hummock dies or is killed and is followed in the place vacated by it by other species which have had no direct or indirect hand in its death. Thus in the full cycle we may distinguish an upgrade series and a downgrade.

Each patch in this space-time mosaic is dependent on its neighbours and develops under conditions partly imposed by them. The samples of a phase will in general develop under similar conditions but not necessarily the same, for the juxtaposition of phases will vary. This may be expected to affect the rate of development of the patches of a phase and their duration. But on the duration of the full cycle of change and its component phases there is inadequate information, although the *impression* is gained that in the upgrade series the net rate of production per annum is at first slow, then fast in the *Sphagnum papillosum* phase, then slow again in the final phases.

This brief summary presents the regeneration complex as a community of diverse phases forming a space-time pattern. Although there is change in time at a given place, the whole community remains essentially the same; the thing that persists unchanged is the process and its manifestation in the sequence of phases.

Dwarf Callunetum

It may well be argued that the regeneration complex is a special case. It is one of my objects to show that these dynamic phenomena are paralleled in a wide range of communities, all of which can hardly be set aside as special cases.

Occasionally in the regeneration complex there occur partial sequences in space which correspond with the time sequence. Complete correspondence between the space and time sequence is found, however, in certain communities under highly specialized conditions. Such conditions are found in the Arctic (Walton, 1922), where the prevailing winds determine the alinement of the plants in the community and their unidirectional vegetative spread. Similar phenomena are shown by the dwarf Callunetum of highly exposed places on the slopes of the Cairngorms.

The Callunetum consists of strips of *Calluna* separated by strips of bare wind-swept soil; all the *Calluna* plants lie side by side and in line, their apices spreading into the shelter created by the plant itself and their old parts dying away behind. In other places the strip of vegetation is double, consisting mainly of *Calluna* and *Arctostaphylos uva-ursi*; under the prevailing conditions both species have the same habit, but they differ so far in their specific make-up that *Arctostaphylos* is generally found to leeward of *Calluna* (Fig. 1). The spatial relation between these two species and their unidirectional growth suggest that as *Arctostaphylos* grows forward over the eroded soil, *Calluna* grows over the older parts of *Arctostaphylos*, suppressing the leaf-bearing shoots though not killing the old stems. This is, indeed, the case, and proof of the dynamic relation is found in the dead remains of *Arctostaphylos* below the *Calluna*.

The community just described is a three-phase (or more-phase depending on the refinement of the analysis) system with the phases arranged in linear series. In the dwarf Callunetum of more sheltered places unidirectional spread is replaced by centrifugal; each *Calluna* plant is free to spread until checked by other plants in its neighbourhood. The static pattern of the mature community is a background of *Calluna* with scattered patches of *Arctostaphylos*, *Cladonia silvatica* and bare soil in it. But the time sequence remains essentially the same, with four phases in the full cycle, which may be shortened to three or even to two phases.

4 *Pattern and process in the plant community*

Briefly summarized from details obtained by Dr G. Metcalfe (to whom I am also indebted for Fig. 1) during the Cambridge Botanical Expedition to the Cairngorms and kindly placed by him at my disposal, the history of the relations between the phases is as follows, the kind of evidence being the same (except for the positional relation) as that obtained from the linear series. The young vigorous shoots of *Calluna* are usually dense enough to exclude lichens; the older, fewer and less vigorous shoots are unable to

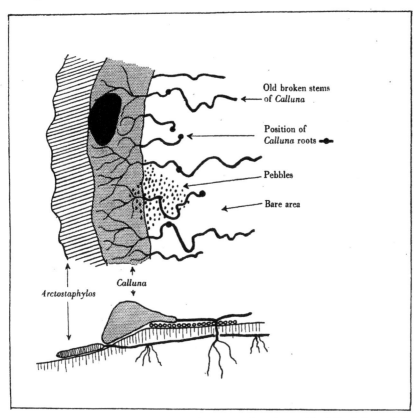

Fig. 1. Diagrammatic representation in plan and elevation to show the spatial relation between *Calluna* and *Arctostaphylos* in the 'double strips' separated by wind-swept bare soil in the Dwarf Callunetum of highly exposed places at approx. 2500 ft. on the northern slopes of the Cairngorms. The double strip moves forward in time; *Arctostaphylos* invades bare soil, and *Calluna* moves on and suppresses the adjacent *Arctostaphylos*.

do so and *Cladonia silvatica* is often abundant on the old parts of the plant. On its death *Cladonia* becomes dominant, anchored on the dead stems. (The counterpart of this phase in the linear series is the strip of dead *Calluna* stems behind the live; owing to the violence of the wind, *Cladonia*, although present, is unable to dominate.) In time the *Cladonia* mat disintegrates (in much the same way as it does in Breckland (Watt, 1937)) and bare soil is exposed, often with some remains of *Calluna* stems on it. If *Arctostaphylos* happens to be near such a gap (and a much ramified system of non-leaf bearing stems occurs under the *Calluna* mat) invasion is followed by complete occupation. In time by vegetative spread from the margin *Calluna* replaces the *Arctostaphylos*. The relations between the phases are indicated diagrammatically in Fig. 2, which also shows the short-circuiting in the absence of *Arctostaphylos*.

Clearly *Calluna* is the dominant plant; the locally dominant *Arctostaphylos* and *Cladonia* are allowed to occupy the ground which *Calluna* must temporarily vacate, *Arctostaphylos* as a phase in the upgrade series and *Cladonia* in the downgrade. Although no data are available it may be pointed out that from the annual rings of the woody stems of *Calluna* and *Arcto-staphylos* some estimate of the duration of the phases is made possible.

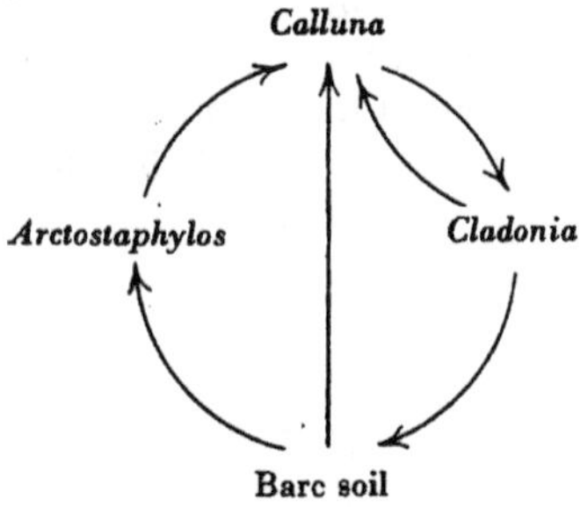

Fig. 2. Diagram illustrating the dynamic relations between the chief species in Dwarf Callunetum in less exposed places than those mentioned in Fig. 1. The arrows indicate the direction of change.

Eroded Rhacomitrietum

At somewhat higher altitudes on the Cairngorms an eroded community dominated by *Rhacomitrium lanuginosum* shows phenomena similar to those of the *Calluna* strips. For here, too, the direction of the prevailing wind, by determining the direction of spread of the plants, imposes on the patches of vegetation a space sequence which is also a time sequence.

The following very short account is based on details kindly supplied by Dr N. A. Burges, to whom I am also indebted for Fig. 3.

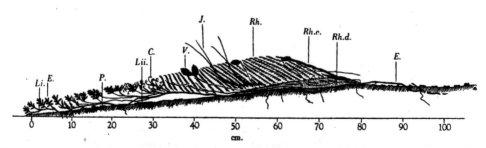

Fig. 3. Profile downwind or across a patch of vegetation in eroded Rhacomitrietum at approx. 3500 ft. on the northern slopes of Cairngorm. The spatial sequence is also a time sequence. *Li.*, *Lii.* and *P.* = bryophytes, *E.* = *Empetrum hermaphroditum*, *C.* = *Cladonia rangiferina*, *V.* = *Vaccinium myrtillus*, *J.* = *Juncus trifidus*, *Rh.* = *Rhacomitrium lanuginosum*, *Rh.e.* = *R. lanuginosum*, eroded face, *Rh.d.* = *R. lanuginosum*, dead.

The community consists of a network of patches of bare soil and of vegetation; each of the latter shows from the lee side to the exposed and eroded western side (Fig. 3) a series of phases characterized respectively by bryophytes, *Empetrum hermaphroditum*, a mixture of *Vaccinium myrtillus*, *V. uliginosum* and *Rhacomitrium lanuginosum*, and finally *Rhacomitrium* itself. Proof of the temporal sequence is found in the vertical layering of peaty remains found under the last phase. The *Rhacomitrium* phase is eroded by the wind, the exposed accumulated humus dispersed and mineral soil once again laid bare. Since the stems of *Empetrum* stretch from end to end of the patch the number of rings enables an estimate to be made of the duration of the whole and of its phases. The rate of advance is approximately 1 m. in 50 years.

6.   *Pattern and process in the plant community*

Bracken

The relation of *Arctostaphylos* to *Calluna* in the dwarf Callunetum of the less exposed places is deduced from internal evidence and supported by evidence provided by the study of their spatial and temporal relations in the specialized double strips. The Pteridietum (Fig. 4) which has been studied (Watt, 1947) provides a close parallel, for the area in which

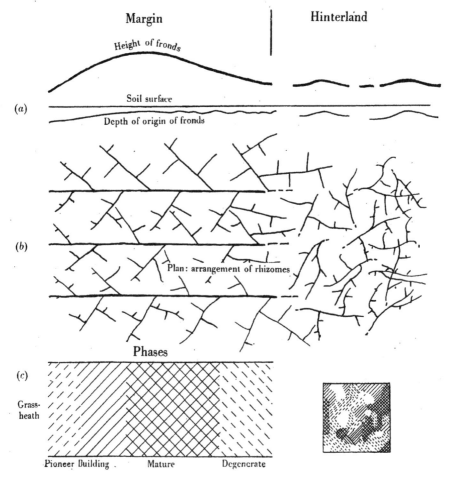

Fig. 4. Diagram to illustrate the spatial and temporal relations between the phases in the marginal belt and in the hinterland of a bracken community on Lakenheath Warren (Breckland). In (a) the change in height and in continuity of cover of the fronds are indicated: also the change in depth of origin of the fronds. In (b) the relative size and direction of growth of the main axes of the bracken plants are shown. In the marginal belt the axes are parallel, in the hinterland they form a network. In (c) the phases in the margin are in linear series: in the hinterland they are irregularly arranged.

it occurs consists of a hinterland (comparable with the continuous Callunetum) in which the fronds are patchily distributed and the axes of the bracken plants form a loose network; marginal to the hinterland is a belt of bracken invading grass-heath in which the individual plants lie side by side, with their main axes parallel to each other and their apices in line (cf. the strip of *Calluna*).

Such a marginal belt with unidirectional spread can be divided into a series of zones or phases by lines running parallel with the invading front. These phases can be distinguished by data from the frond and rhizome. The inherent circumstances make it clear that, from back to front, the space sequence of phases represents a sequence in time. Among the overdispersed fronds of the hinterland there are patches without fronds (the grass-heath phase) and patches with fronds which vary among themselves in features of frond and rhizome in much the same way as the sequence in the marginal belt; these are distinguished as the pioneer, building, mature and degenerate phases. These phases, together with the grass-heath phase, form a cycle of change in time at one place. Comparison of the two sets of data (Watt, 1945, 1947)—from the zones of the marginal belt and the patches of the hinterland—shows close agreement between the relations between the values within each set, an agreement all the more striking when allowance is made for the different circumstances. For in the marginal belt the phases are in linear series and bracken invades grass-heath free from bracken; in the hinterland the grass-heath phase is vacated by bracken but seldom completely before reinvasion takes place. There is thus some overlap between the beginning and end phases of the cyclic series.

The evidence for cyclic change derived from the specialized marginal belt is supported by the internal evidence from the hinterland itself. For in the grass-heath phase without fronds there are abundant remains of decaying rhizome, and under the litter in the later phases with bracken are the recognizable remains of *Festuca* and *Agrostis* of the grass-heath phase. Phases with, and a phase without, fronds alternate in time.

The processes involved in the marginal belt and the hinterland are the same. In the marginal belt the natural tendency of the bracken to spread centrifugally is checked by lateral competition; spread is thus directed by ecological opportunity into the grass-heath which is free from bracken. Once established the sequence of phases depends on the development of the bracken itself and development as affected by its own reaction. In the hinterland, where the axes form a loose network, rejuvenation is similarly directed by ecological opportunity and essentially restricted to areas vacated by the death of the bracken. The phases which follow are a biological consequence of established invasion and are limited to the area of that invasion.

Running parallel with the change in the vegetation of the cycle, there is change in the factors of the habitat and in total habitat potential. The variable factors of microclimate and soil, e.g. shelter, light intensity, temperature, the amount of litter, the state of the humus and its distribution in the soil, are differentiae of the phases superposed on the original foundation of a uniform habitat. They are closely linked with the phases because they are the effects of the plants themselves; the effects of one phase become part cause of the next. Thus the spatial variation in the habitat is primarily caused by the variable vegetational cover. Further, taking a broad view, we may note that in the upgrade series there is an accumulation of plant material and an increase in habitat potential; in the downgrade both are dissipated.

The significance of the differential action of the phases on microclimate may be illustrated by the effect of frost during the winter 1939–40. In the grass-heath or pioneer phase rhizome apices were killed at a maximum depth of 20 cm.; in the mature phase, protected by a blanket of litter, the maximum depth of lethal damage was 6 cm. Action of this kind retards development and prolongs the duration of a phase, upsetting the smooth course of the time sequence.

8   *Pattern and process in the plant community*

Grassland A (Breckland)

The vegetation of Grassland A (Watt, 1940) is obviously patchy. In its hollows and hummocks the habitat presents a striking resemblance to the regeneration complex, and here, too, vegetational variation is linked with variation in microtopography and soil habitat.

In a reinvestigation of the community upon dynamic lines four phases are recognized: the hollow, building, mature, degenerate. There are intermediates, but little difficulty is experienced in assigning all parts of the typical community to one or other of these four phases.

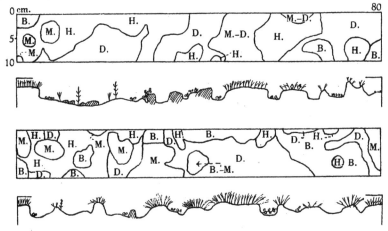

Fig. 5. The relative size and spatial relations of the phases in a plot of 160 × 10 cm. in Grassland A. The relation between the phases and the microtopography is seen in the profile taken along the upper edge of the plot.

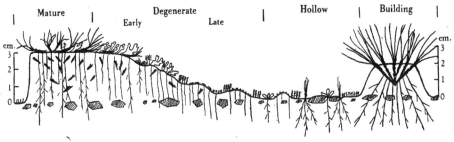

Fig. 6. Diagrammatic representation of the phases showing change in flora and habitat and indicating the 'fossil' shoot bases and detached roots of *Festuca ovina* in the soil.

Their spatial distribution is shown in Fig. 5; the patches are irregular in size and shape and their juxtaposition varies. These four phases form a time sequence, the evidence for which is emphasized in the following reconstruction (Fig. 6). The whole 'life' of the community centres round the reactions and life history of *Festuca ovina*, its growth and reproduction; it has every right to be called the dominant plant even although it occupies less than one-half of the total area (approximately 45 %). The seedling becomes established among the stones which floor the hollow phase; as the plant grows and spreads the level of the mineral soil inside the tussock rises. This soil is free from stones and its accumulation is due to the activities of ants and earthworms, very probably also to wind-borne particles and particles water-borne in the splash of heavy rain. The young vigorous fescue with relatively long leaves and many inflorescences on relatively long stalks constitutes the building phase; in it the shoots of the plants are still attached to the parent stock. By further accumulation of soil the hummock increases in height, attaining a maximum of about 4 cm., and then carrying much less vigorous fescue, with shorter leaves and fewer inflorescences with shorter stalks. The original many-branched tussock is replaced by numerous small plants each lying horizontally and consisting of one or two shoots with vertically descending roots; they have arisen from the larger plant by the separation of branches through the death and decay of the parent stock, the evidence for this being the presence throughout the soil profile of the hummock of the bases of lateral branches which are resistant to decay and thus persist.

In the spaces between the individual plants in the mature phase, the fruticose lichens *Cladonia alcicornis* and *C. rangiformis* become established. They spread and ultimately form a mat below which the remains of fescue are found. This is the early degenerate phase. 'Fossil' shoots and unattached roots are found in the soil. In the late degenerate phase, characterized by the crustaceous lichens *Psora decipiens* and *Biatorina coeruleo-nigricans*, 'fossil' shoots and unattached roots are again found. In places these roots project above the soil surface and bear witness to the erosion (already suggested by the pitted surface) which gradually wears down the hummock. Some vestiges of the hummock survive into the hollow phase, but eventually these disappear and the erosion pavement of flint and chalk stones is once again exposed.

Every part of the surface of the typical community can be assigned to one or other of these four phases. A single set of them is fully representative of the whole and summarizes in itself the processes at work and their manifestations. The understanding of the community as a working mechanism is based on the elucidation of the relations of the phases to each other. They are its minimum representative (minimum area), its very core.

Now the contributions which the phases make to the community as a whole may vary among themselves and also from year to year. To eliminate one source of variability, and at the same time to take account of it as an ecological phenomenon of great importance, I propose to base the description of the community upon areas of equal size in the several phases (and to call the whole the unit pattern) and to make a separate assessment of the relative areas occupied by the phases.

Some of the data for the unit pattern of Grassland A are presented graphically in Fig. 7; the diversity as well as the unifying continuity of the phenomena are clearly shown. The cover percentage of fescue and 'bare soil and stones' are inversely related. The holophytic bryophytes are virtually excluded from the building and mature phases and increase to a maximum in the hollow, where competition from fescue is least. For the lichens, on the other hand, the maximum appropriately lies in the phase where the fungal partners can utilize the organic remains accumulated during previous phases.

The selective effect of the phasic microenvironments on the distribution of seedlings and their subsequent fate is shown in Table 1. The total number of seedlings varies directly with the 'bare soil and stones'. The distribution of the adults suggests failure of the seedlings to reach maturity in the mature and degenerate phases and shows restriction to the building and hollow phase and virtually to the hollow phase because the great bulk of

10   *Pattern and process in the plant community*

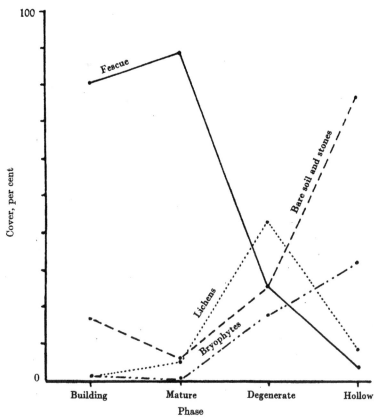

Fig. 7. Graphical presentation of some data for Grassland A. Note that most of the data for the building phase are from late stages.

Table 1. *Grassland A*

Species	Building	Mature	Degenerate	Hollow

	Building	Mature	Degenerate	Hollow
Species	Total no. in 40 plots, each 5 × 5 cm.			
Seedlings				
Arenaria serpyllifolia	0	0	0	1
Avena pratensis	0	0	0	2
Calamintha acinos	3	3	8	6
Cirsium lanceolatum	0	0	1	0
Crepis virens	24	7	10	21
Erigeron acre	25	17	31	68
Festuca ovina	2	0	10	66
Hieracium pilosella	9	5	6	7
Koeleria gracilis	0	0	1	0
Senecio jacobaea	0	0	0	1
Total	63	32	67	172
Adults				
Arenaria serpyllifolia	0	0	0	1
Calamintha acinos	0	0	0	3
Cerastium semidecandrum	1	0	0	8
Crepis virens	1	0	0	0
Erigeron acre	3	0	1	3
Galium anglicum	11	0	0	7
Hieracium pilosella	2	0	0	4
Saxifraga tridactylites	1	0	0	0
Total	19	0	1	26

the adults recorded from the building phase are in the immediate surround of the fescue tussock and not in it. This restriction to the hollow phase in all probability holds for the surviving fescue seedlings as well. Thus at any given time the initiation of the cycle of change is restricted in space, and at any given place to the hollow phase in the time sequence.

Estimated by two different methods the relative areas of the phases hollow, building, mature, degenerate are, respectively, 25·7, 15·1, 16·3 and 42·9. As far as I can judge from the annual charting of two plots since 1936, but without actually noting the spatial limits of the phases, Grassland A is remarkably stable both in its unit pattern and the areal extent of its phases.

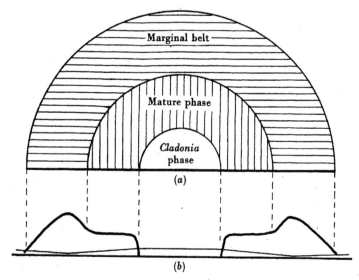

Fig. 8. Diagrammatic representation in plan (a) and profile (b) of an *Agrostis* ring set in a background of *Cladonia silvatica*. In (b) the darker line indicates the change in height (and/or number) of *Agrostis* shoots, the thinner line the change in thickness of the *Cladonia* mat.

Grass-heath on acid sands (Breckland)

The grass-heath on acid sands in Breckland has been sufficiently investigated to provide supporting evidence for the dynamic interpretation of the plant community and also to demonstrate the usefulness in description of separating the estimate of the relative areas of the phases from the unit pattern.

Fundamentally the community consists of rings (both solid and hollow) of *Agrostis tenuis* and *A. canina*, and occasional tussocks of *Festuca ovina* set in a background of *Cladonia silvatica*. (Further reference to *Festuca* is omitted because its relations to other species in the community are not fully elucidated.) Each ring is derived by vegetative spread from one plant and epitomizes in itself the spatial and temporal changes within the community. A typical ring (Fig. 8) of medium size consists of a peripheral zone of numerous, vigorous, vegetative and flowering shoots arising from large plants with rhizomes radiating outwards, an inner zone (mature phase) with fewer and shorter vegetative and flowering shoots, or no flowering shoots at all, on smaller plants whose

12   *Pattern and process in the plant community*

rhizomes form a loose open network (cf. Bracken, p. 6). In the centre of the ring *Agrostis* as a live plant is absent, but its abundant dead remains are found below the mat of dominant *Cladonia silvatica* which is from 3 to 4 cm. thick; to this thickness it has gradually risen from a thickness of 1 cm. only in the middle of the peripheral zone.

Fuller investigation of the dynamic behaviour of the rings justifies the recognition of phases within the marginal belt and in a full cycle additional phases of degeneration and rebuilding following disruption of the lichen mat (Watt, 1938). But to keep the issue simple no more than the three phases need be further considered since they may complete a short cycle; samples from unit areas in each constitute the unit pattern.

The mutual relations between *Agrostis* and *Cladonia* have not been investigated to the degree necessary to decide whether the death of *Agrostis* is due to age, or in some measure to the influence of *Cladonia* or to both acting together. Similar problems arise at this stage of the cycle of change in all the communities examined; the problems are universal because they concern death and its causes. Their solution, although important, is not necessary to establish the fact of replacement.

The extension of the *Cladonia* phase at the expense of the mature phase (inner zone with *Agrostis*) is not, however, merely a question of time; it is primarily due to drought, which differentiates between the *Agrostis* plant of the peripheral zone (which survives) and that of the mature phase (which dies). As the result of a severe drought the mature phase passes wholesale and abruptly to the *Cladonia* phase, and the extent of the change will depend on the area of the phase capable of being affected by it. An area occupied by a large number of small patches of *Agrostis* will be relatively little affected by drought, while one which is almost entirely in the mature phase, through the spread and fusion of rings, will be almost wholly affected by it (Fig. 9). This explains the violent fluctuations met with in the populations of *Agrostis* in this community; in certain areas fluctuations have been so wide that estimates of abundance for one year are of no value as a diagnostic character. They do, however, reflect the vagaries in meteorological factors. On the other hand, data from the unit pattern would be much less affected.

Beechwood

The effect on the structure of a community of drought or other efficient cause may persist long after the cause has ceased to operate. In fact, at any given time, there may be no correspondence between structure and the current meteorological factors. The point is best illustrated by reference to communities with long-lived dominants. But first a brief note on phasic change in beechwoods.

The patchiness in some all-aged beechwoods on the Chilterns, reputedly managed on a selection system, is interpretable in terms of the temporal sequence of phases as revealed by the study of the life history of pure even-aged beechwoods of the same ecological type on the South Downs (Fig. 10) (Watt, 1925). To the three phases recognized—Bare, *Oxalis*, *Rubus*—there should be added a fourth, the gap phase, to which regeneration is confined because it is excluded from other phases. At any given place there is a cycle of change consisting of an upgrade series of phases in which there is a continual change in ecological structure, associated with changing age, rate of growth and density of the dominant trees and correlated with changes in the field layer, and a downgrade of the dying, dead and rotting stems and the vegetation of the gap.

ALEX. S. WATT   13

For the time-productivity curve for the upgrade series we have the foresters' yield tables for managed woods. Although the data take no account of the yield from the subsidiary vegetation, we may assume this to be small in relation to the yield of the trees and without appreciable effect on the course of the curve. This is of the familiar growth type, rising slowly at first, then fast, then slowly again.

The phasic phenomena of woodland are on a scale to be immediately obvious. Again because of the scale, but also because we have the data, woodlands afford an excellent illustration of the cyclic relation between tree species and of causes having long-continued

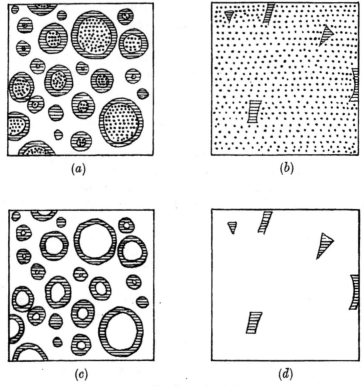

(a)   (b)

(c)   (d)

Fig. 9. Diagram to show the effect of drought on *Agrostis* in two areas of Festuco-Agrostidetum at different stages of development. In (a) there are many small (young) patches of *Agrostis*; in (b) the patches have grown and fused so that the bulk of the area is in the mature phase. In (c) and (d) drought has killed the *Agrostis* in the mature phase only of (a) and (b).

effects in structure. Both points have been dealt with recently by Jones (1945); brief reference is all that is necessary here.

Viewed against the ideal of sustained yield, aimed at most simply among foresters by allocating equal areas to each age class or group of age classes, the primeval wood by all accounts more often than not shows an unequal distribution of the areas occupied by its age classes. Aside from minor fluctuations over periods of time in, say, the number of deaths among old trees, there are exceptional factors of rare or sporadic occurrence, such as storms, fire, drought, epidemics, which create a gap phase of exceptional dimensions. If the whole gap phase is regenerated about the same time—and various circumstances, both inherent and external, like periodicity in seed years and suitable meteorological

14   *Pattern and process in the plant community*

factors for seedling survival may help in that direction—then there is initiated an age class of abnormal area. This will persist like a tidal wave moving along the age classes until at least the death of the trees; it may even influence the structure of the next generation. In other words, the relative areas under the age classes (as a super refinement of phasic subdivision) need bear no relation to current meteorological factors but be explicable in terms of some past event which happened, it may be, 200 or 300 years ago.

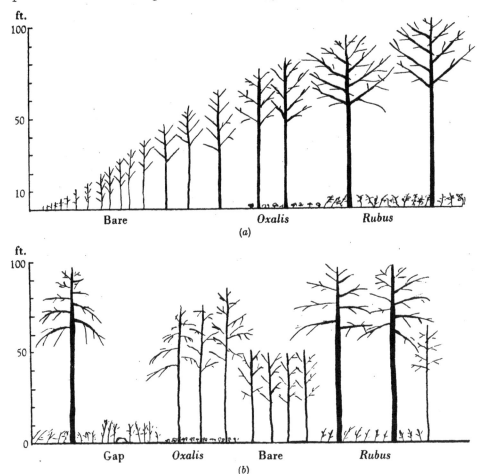

Fig. 10. Diagram to illustrate (a) the phasic change during the life history of an even-aged pure beechwood and (b) the distribution in space of the phases when the old wood is left to itself.

A series of sporadic exceptional events will obviously increase the difficulties of 'explaining' current relative areas, as also will the selective destruction of phases their juxtaposition in space. Fortuitous fluctuations of climate and other causes may thus bring about major departures from the normal or ideal wood.

The second point refers to the normal inclusion in the cycle of change of a phase dominated by another species of tree, in different kinds of beechwoods, for example, by ash, oak or birch. Such a phase (e.g. birch) may occupy small or large areas, but irrespective of the area it occupies, if the relation is cyclic and not seral (that is, one of alternating equilibrium in which successful regeneration of the beech is inhibited by itself but promoted by the conditions brought about by the other species of tree), then such a phase is part and parcel of the system. To recognize the birchwood and beechwood as separate entities may be convenient in classification, but successful practical management of such woods depends on their recognition as integral parts of a coordinated whole.

SUPPLEMENTARY EVIDENCE

Doubtless because of its universality, its familiarity and its apparent lack of significance the fact of the impermanence of the individual or group of individuals has seldom evoked comment. Butcher (1933) has, however, called attention to the changing pattern of river vegetation over a short period and suggests progressive impoverishment of the soil of the river bed as the cause of the disappearance of *Sium erectum* from a given place in it. Tutin (1938) has attributed the marked reduction in area occupied by *Zostera marina* primarily to deficiency in light intensity; this predisposed the plants to disease, which is a symptom of weakness rather than a cause of its virtual disappearance. Without contesting the accuracy of the immediate diagnosis, I would suggest that both phenomena may be more fully and satisfactorily explained against the background of cyclic change.

Phenomena similar to those described have been seen but not investigated in communities dominated respectively by *Polytrichum piliferum* (in Breckland), *Ammophila arenaria* and *Carex rigida-Rhacomitrium lanuginosum* of mountain tops. Poole (1937), from a study of mixed temperate rain forest in New Zealand, concludes that *Dacrydium cupressinum* on the one hand, and *Quintinia acutifolia* and *Weinmannia racemosa* on the other, alternate in time at one place.

The phenomena are not confined to temperate regions. In the high arctic the few species are represented by scattered individuals, and communities are distinguished by the quantitative relations of their chief components and/or by habitat factors (Griggs, 1934). Where there are few individuals competition is negligible or absent; factors of soil and climate are the primary selective agents. Griggs says there is not only change from place to place, but change in time at a given place (he refers to unpublished data from permanent quadrats as the basis for this latter statement). This distribution in place and in time, is, according to Griggs, erratic, but no details are given.

At the first remove in the low arctic a two-stage succession from open initial communities to the closed association of heath and bog shows the emergence of the organic environment into importance. Competition now is a factor limiting the distribution of species.

At the other extreme is tropical forest, the most complex of plant communities in which the biotic environment plays an outstanding part. It is not to be expected that such a complex community will be amenable to a simple dynamic interpretation. Yet presumptive evidence is given by Aubréville (1938) who has expressed views on the structure of the forest closely akin to those advocated here for a selection of communities. He finds that although there are in tropical forests very many species of tree the dominant or most abundant species are relatively few in number (about a score) their relative proportions varying from place to place, some dominating in one place, some in another.

But also at a given place there is a change in the dominant species with time, for, in the subsidiary layers the young trees (the dominants of the future) are often specifically different from the dominants of the uppermost layer. Aubréville, however, does not mention a regular cycle of change at a given place; in fact, as far as can be judged from

16   *Pattern and process in the plant community*

his account the assemblage of species in a newly formed gap depends on the fortuitous occurrence of a number of factors, including distance of parents, mobility and abundance of seed, frequency of seed production, dispersal agents and their habits, and various factors like animals eating the seed, duration of life of the seed, light intensity, root competition and the kind of undergrowth, affecting the survival of the seed and the establishment of the seedling. Although the earlier stages of gap colonization may show diversity in floristic composition the later ones, according to Dr P. W. Richards (oral communication), are much more uniform.

If we assume with Aubréville that the general climatic and soil conditions are uniform over a large area then the differentiating biological characters of the many species assume primary importance, and variation in floristic composition from place to place is the resultant of the impact of these biological characters and environmental factors, largely, but not exclusively, residing in the plant community itself; that is, the community itself largely determines the distribution, density and gregariousness of its component parts.

COMPARISON AND SYNTHESIS

The analysis of the seven communities, documented by observations over a period of time, by comparative studies and by appeal to the registration of the time change either in the plants themselves or in their subfossil remains, shows that there is an orderly sequence of phases in time at a given place and that the spatial relation between the phases can be interpreted in terms of their temporal relations. I now propose to correlate the observed phenomena and to offer a simple generalization and explanation.

The cycle of change in all the communities is divisible into two parts, an upgrade and a downgrade. The upgrade series is essentially a process or set of processes resulting in a building up of plant material and of habitat potential and the downgrade a dispersion of these mainly by fungi, bacteria, insects, etc., but also by inorganic agents. At each stage, of course, there is both a building up and a breaking down, but in the upgrade series the balance is positive, in the downgrade negative.

When bracken was being investigated the field phenomena evoked the spontaneous application of the names for the phases, pioneer, building, mature, degenerate. They undoubtedly conveyed impressions which later study hardened, for they are obviously applicable to several of the communities. The quantitative graphical expression of the change for the upgrade series for woodland and the qualitative description for other communities lead me to suggest that a curve of the growth or logistic type expresses the general course of change in total production in the upgrade series of each of the communities (Fig. 11 and appended correlation table). This is doubtless an oversimplification of the whole concrete situation. But there are data enough to show that we are dealing with something fundamental, with in fact, some biological law which operates in the plant community.

For the downgrade series a curve of similar form is tentatively suggested. There are many partial assessments of the course of the changes, but none to my knowledge dealing with the whole series in a comprehensive way.

The basis of these regular phenomena of change may be sought in the behaviour of populations and in the sequence of a plant succession on the restricted area of a patch. The community as we have seen consists of patches, each of limited area, and differentiated by floristic composition, age of dominant species and by habitat.

Patchiness can be readily enough detected in all the communities investigated and in the others referred to. In forest, the community about whose structural types more is known than about any other kind of community, patchiness is widespread. There is a form of 'irregular forest' that would appear to be an exception, although it is not known how far it is local or temporary, and until it has been examined by the use of criteria in addition to the distribution of crown classes, judgement must be suspended. A comparable structure may be found in other communities and at the moment no claim can be made that patchiness, as understood here, is universal. It certainly is widespread.

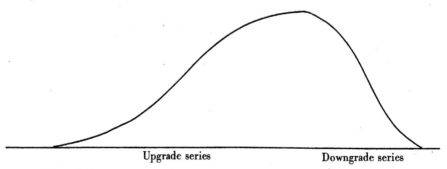

Upgrade series   Downgrade series

Fig. 11. The graphs illustrate in a general way the assumed course of change in the total productivity of the sequence of phases in the upgrade and downgrade portions of the cycle of change. The appended table correlates the phase sequences of the different communities.

Plant community				Phase	Hummock with:		
Regeneration complex	Pool	*Sphagnum cuspidatum*	*Sphagnum pulchrum*	*Sphagnum papillosum*	*Erica Scirpus*	*Calluna Cladonia*	Degenerate
Dwarf Callunetum	Bare soil	*Arctostaphylos*		*Calluna*	*Calluna*		*Cladonia*
Eroded Rhacomitrietum	Bare soil	Bryophytes *Empetrum*		*Rhacomitrium- Vaccinium*	*Rhacomitrium*		Eroding face
Pteridietum		Grass- heath	Pioneer	Building	Mature		Degenerate
Grassland A		Hollow		Building	Mature		Degenerate
Acid grass- heath		*Cladonia*	Pioneer	Building	Mature		*Cladonia*
Beechwood (sere 2)		Gap (late)	Bare stage	*Oxalis* stage	*Rubus* stage		Gap (early)

The widespread occurrence of patchiness is of interest in itself, but the persistence of the patch in time under continuing normal conditions is fundamental. Why, for example, starting with a set of patches, do we not get a general blending of the whole with the vegetation of every square foot a replica of every other?

May I first remind you that we are assuming uniformity in the fundamental soil and climatic factors of the community. Over and above this foundation we get, as we have seen, modifications of the soil and climate which are closely linked with the phases; these floating differentiae are in the first instance reactions of the phases and in the second, causes. We have, in fact, habitat variation in space as in time. (Also it should be noted that the habitat factors inherent in a phase modify and are themselves modified by the habitats of surrounding phases.) Seeds and propagules falling on a plant community therefore fall on a diversified environment (varying habitat and competitive power in the

18   *Pattern and process in the plant community*

phases), and establishment is restricted to the phase where the plants survive; for many species this phase is the gap phase, or phase corresponding to it, formed by the death of the dominant in the mature phase or end-phase of the upgrade series. For plants spreading by rhizome the same holds good; in bracken, effective invasion is limited to the grass-heath phase, although it may begin in the mature phase and continue into the pioneer phase.

But not only is establishment of many species thus restricted in space, it is also restricted in time, to the period during which the gap phase is receptive. Checks to further colonization may be made by the invaders themselves, thus setting a time limit to the period within which establishment takes place. In this way, there is a tendency towards the production of even-aged or near-aged populations. Besides this restricted period during which colonization may take place, there are many other factors which tend to produce even-aged populations, some of which have already been alluded to and will not be further discussed.

Mainly because of competition, the plants established in the gap or corresponding phase are unable to invade and suppress surrounding phases. Thus prevented from spreading laterally, the patch becomes a microcosm of limited area with continuous but restricted food supply. Hence the population of that area behaves like a population of flies or human beings under similar restrictions of space and food. It is unable to spread, but it may develop.

This development in the upgrade series may take one of two forms, either as an even-aged population of a dominant species or as a series of dominants as in a plant succession. In the former the successive phases are distinguished by the species accompanying the dominant but primarily by the age and density of the dominant itself. The justification of the recognition of age in individuals or shoots as a basis for subdivision lies in the changing needs of the individual with age, its changing form (height, spread, root distribution, area of absorptive and anchoring surface, etc.) and its changing competitive power. The need for such recognition is obvious in a plant, whose gametophyte and sporophyte are separate and have different requirements, or in an insect with complete metamorphosis, in which the structure and habits of the larva differ so widely from those of the adult; but it holds also for higher plants because their requirements change during their life history.

The second type of change is found in communities where the successive phases (or some of them) have distinct dominants, that is, the phenomena are strictly comparable with those in a sere, where new dominants come in when the environment becomes suitable for them.

Now in both these, the even-aged population and the plant succession, the course of temporal change in total production per unit area is represented by a curve of the growth type (Pearl, 1926; Braun-Blanquet & Jenny, 1926). The reason for the general equation of corresponding phases now becomes obvious. In both, the phenomena of orderly change have their basis in the universal rhythmic phenomena of population development and of plant succession. It would thus appear that certain phenomena of the plant community can be brought within the scope of mathematical inquiry. This opens up an entirely new field in the study of natural communities with the possibility of putting the practical management of communities, as the forester has done the forest, upon a sound basis of natural law.

The description of the plant community is based on the unit pattern, which is the present expression of the continuous process of development and decline in a population broadly interpreted. The qualitative and quantitative assessment, therefore, of the plant community may be expressed in terms of the temporal changes in such a population.

Now it has been shown mathematically and confirmed experimentally that certain species with different ecological requirements live together and maintain stability only in a definite proportion (Gause, 1935). In like manner we may assume (although independent confirmation is necessary) that the plant community in a constant environment will show a steady state or a definite proportion between the constituent phases. This steady state we may call the phasic equilibrium, that is, the community is in harmony with itself and with its environment. Departures from this phasic equilibrium either in space or in time could then be measured and correlated with the changed factors of the environment.

The difficulties confronting this procedure are obvious. There are doubtless some communities (e.g. Grassland A) of which it would be possible to say with some assurance that they are in phasic equilibrium. But there are others like grass-heath on acid soil, and woodland, where, on account of the lag in response or repeated and fortuitous checks to normal orderliness, or the long delay between the incidence of the disturbing factor and return to a possible normal, anything like phasic equilibrium may rarely be achieved even although the tendency is always in that direction. In such circumstances we may copy the forester whose normal forest may be defined in terms of areas as one in which equal areas are occupied by a regular gradation of age classes. If we think in terms of natural groupings (phases) of unequal duration, then the areas would be proportional to the duration of the phases, the total equalling the duration of the cycle of change. The standard of comparison or normal would then be based on the unit pattern and duration of the phases. With such a standard, departures from it could then be assessed and comparisons made, but attempts to correlate such deviations with changes in the causal environmental factors would be confronted with great if not insurmountable difficulties. But once again there is brought into prominence the need for data on the duration of life of the plant in a variety of habitats and as affected by competition.

Relative areas of the phases is only one aspect of the general problem of the texture of the pattern or of the way in which the community is put together. The other two aspects are the size of the individual sample of the phase and the arrangement of the phases. Both these are important because the conditions of establishment and growth of plants vary with the size of the sample of the phase and with its surroundings.

SOME IMPLICATIONS

Animals and micro-organisms

I have thus far alluded almost solely to the higher plants and described the communities in terms of them, but it should be clear from the insistence on process that organisms other than higher plants play an important and sometimes a fundamental role, and that in the system of which the plant community is a part they ought to be investigated. On reading *Bioecology* (Clements & Shelford, 1939) I confess to a feeling of disappointment, because the data given to support the concept of the biome fall short of demonstrating that intimacy and integration between plants and animals which we are led to expect. Perhaps this is due to the lack of data for the communities described. The degree of intimacy varies, but all organisms whether loosely attached or occupying key positions in the system should be considered part of it.

20   *Pattern and process in the plant community*

Tragårdh (1923) has given a good illustration of the interactions between plants and insects in pinewoods in northern Sweden where the nun moth is a normal inhabitant, which never becomes epidemic as it does in pure pine or spruce woods in central Europe. This he attributes to the presence in the open pinewoods of *Calluna*, *Cynoglossum*, etc., which are the food plants of species of caterpillars which provide a constant supply of hosts for the polyphagous parasitic species of *Pimpla* which also attack the nun moth and so keep it in check. In the dense conifer forest of central Europe from which *Calluna* is excluded the population of *Pimpla* rises and falls with that of the nun moth but with a distinct lag; it also dies out with the extermination of its food supply. The reappearance of the nun moth is followed by rapid multiplication, and much damage is done before *Pimpla* arrives and increases in numbers to be an effective check to it. In the Swedish forest, on the other hand, a continuity of the population of the parasite is assured here because of the presence of alternative hosts feeding on *Calluna*, etc. Here we have interrelations between components of the tree layer and the field layer and the insects peculiar to each forming a complex web which makes for stability. This stability may be assured only so long as the food supply of the alternative hosts is present in sufficient amount. Further excellent examples of the intimacy and complexity of the relations between insects and plants are given by Beeson (1941).

This address would be too long if adequate reference were made to the variety of relationship between organisms associative, co-operative, antagonistic, symbiotic, etc.—and to the change in the nature of the food supply during the period covering the death of the plants in the mature phase and the complete disintegration of the accumulated organic matter. But that the change is regular is shown by the work of Blackman & Stage (1924) on the sequence of insects on living, dying, dead and decaying walnuts, of Graham (1925) on the earlier stages in the rotting of a log, of Saveley (1939) for insects and other invertebrates on dead pine and oak logs, and of Mohr (1943) on the changes in the species population during ageing and disintegration of cow droppings.

Comparable sequences are known among fungi. Parasites, facultative parasites and saprophytes succeed each other on dying and dead stems, and Garrett (1944) writes of the succession of micro-organisms on dead and decomposing roots. Generally on organic remains the sugar fungi develop first and are succeeded later by the cellulose fungi.

The part these organisms play is essential to the maintenance of the community; no assessment of the communal life can be complete without adequate appreciation of what is taking place.

The suggestion may be made that the life histories and food habits of micro-organisms in general, and insects and fungi, in particular those with complicated life histories, may be more profitably studied against the background of phasic change and that the parallelism between the change of food and habitat requirements during the life history of an organism (e.g. some insects with soil larvae) and the spatial change in the condition and floristic composition, or in age of dominant, of juxtaposed phases may be more than fanciful speculation about origins and survival.

The nature of the plant community

Tansley (1935) has coined the term ecosystem for the basic units in nature—units in which plants, animals and habitat factors of soil and climate interact in one system. The term is primarily used in reference to the larger geographical units. The words steppe, prairie, moor, bog apply to such systems and not to the plants alone. But the use of the term ecosystem is elastic; it may be applied to the smaller units like the 'plant community'.

The first step in the shattering of this unified system is the separation of the living plants and animals (the biome) from the non-living habitat, the components of the latter being regarded as factors affecting the former. The next step, which Tansley justifies partly on the ground of practical convenience, is to associate the animals with climate and soil as factors affecting plants. Further subdivision refers to the plants alone. According to Lippmaa (1933) the plant community is composed of synusiae or strata which he regards as the fundamental units of vegetation. The final step is Gleason's; to him the plant association is 'merely a fortuitous juxtaposition of plant individuals' (1936 b).

In support of his view Gleason cites the randomicity of the component species of the association. This is by no means universal (Clapham, 1936; Blackman, 1935). The restriction of establishment in space and in time to certain phases in a diversified pattern tends towards the grouping of age classes (that is, there is overdispersed distribution in time), even although the species as a whole may be at random. Further, two species characteristic of a phase may be randomly distributed but show a high degree of association between them.

This assumed randomicity also enables Gleason to cut the community into small pieces, each of which he regards as representative of the whole. Such a proceeding ignores the significance of the diversity manifested in pattern; this pattern sets a lower limit to the representative minimum.

It is, however, important to realize that Gleason uses the word individual in the sense of an ecosystem of which the plant is the centre, that is, in which the plant and its environment form one interacting system. The individual plant (in the ordinary sense) is a part product of its environment and itself modifies the environment. Position therefore must be important.

Gleason himself cogently argues that by splitting an association into its synusiae, Lippmaa is separating synusiae which are 'united by relatively strong genetic and dynamic bonds' (1936a). The same argument is valid if it can be shown that in lateral spacing as well as in vertical there is dependence of one part on another, as is held to be shown by the evidence adduced here. His skeleton picture of a two-layered community consisting of a tree and a herb below it is divorced from its spatial moulding context and set in a moment in time. In time, both will change and their successors be influenced by their collective surroundings. In short Gleason has minimized the significance of the relations between the components of the community in horizontal space and in time. These relations constitute a primary bond in the maintenance of the integrity of the plant community; they give to it the unity of a co-ordinated system.

This co-ordinated unit may be studied as a plant community, the central object of study of plant sociology; round this unit knowledge is accumulated for use and classification. The functional relations of the community may be systematized in laws and hypotheses.

To draw a dividing line between the approach to the study of natural units through the plant community and through the ecosystem is difficult, if not impossible. But clearly it is one thing to study the plant community and assess the effect of factors which obviously and directly influence it, and another to study the interrelations of all the components of the ecosystem with an equal equipment in all branches of knowledge concerned. With a

22   *Pattern and process in the plant community*

limited objective, whether it be climate, soil, animals or plants which are elevated into the central prejudiced position, much of interest and importance to the subordinate studies and perhaps even to the central study itself is set aside. To have the ultimate even if idealistic objective of fusing the shattered fragments into the original unity is of great scientific and practical importance; practical because so many problems in nature are problems of the ecosystem rather than of soil, animals or plants, and scientific because it is our primary business to understand.

What I want to say is what T. S. Eliot said of Shakespeare's work: we must know all of it in order to know any of it.

REFERENCES

Aubréville, A. (1938). La forêt coloniale. Les forêts de l'Afrique occidentale française. *Ann. Acad. Sci. colon., Paris,* **9.**

Beeson, C. F. C. (1941). *The Ecology and Control of the Forest Insects of India and the Neighbouring Countries.* India: Dehra Dun.

Blackman, G. E. (1935). A study by statistical methods of the distribution of species in grassland associations. *Ann. Bot., Lond.,* **49.**

Blackman, M. W. & Stage, H. W. (1924). On the succession of insects living in the bark and wood of dying, dead and decaying hickory. *N.Y. St. Coll. For. Tech. Publ.* no. 17. Syracuse, N.Y.

Braun-Blanquet, J. & Jenny, H. (1926). Vegetationsentwicklung und Bodenbildung in der alpinen Stufe der Zentralalpen. *N. Denkschr. schweiz. Ges. Naturfs.* **63.**

Braun-Blanquet, J. & Pavillard, J. (1930). *Vocabulary of Plant Sociology.* Translation by F. R. Bharucha. Cambridge.

Butcher, R. W. (1933). Studies in the ecology of rivers. *J. Ecol.* **21.**

Clapham, A. R. (1936). Overdispersion in grassland communities, and the use of statistical methods in plant ecology. *J. Ecol.* **24.**

Clements, F. E. & Shelford, V. E. (1939). *Bioecology.* London.

Garrett, S. D. (1944). *Root Disease Fungi.* Waltham, Mass., U.S.A.

Gause, G. F. (1935). Vérifications expérimentales de la théorie mathématique de la lutte pour la vie. *Act. Sci. et Ind.* no. 277.

Gleason, H. A. (1936a). Is the Synusia an Association? *Ecology,* **17.**

Gleason, H. A. (1936b). In *Mem. Brooklyn Bot. Gn,* no. 4. Brooklyn, N.Y.

Godwin, H. & Conway, V. M. (1939). The ecology of a raised bog near Tregaron, Cardiganshire. *J. Ecol.* **27.**

Graham, S. A. (1925). The felled tree trunk as an ecological unit. *Ecology,* **6.**

Griggs, R. F. (1934). The problem of arctic vegetation. *J. Wash. Acad. Sci.* **24.**

Jones, E. W. (1945). The structure and reproduction of the virgin forest of the North Temperate Zone. *New Phytol.* **44.**

Lippmaa, T. (1933). Aperçu general sur la vegetation autochtone du Lautaut (Hautes Alpes). *Acta Inst. bot. tartu,* **3.**

Mohr, Carl O. (1943). Cattle droppings as ecological units. *Ecol. Monogr.* no. 13.

Osvald, H. (1923). Die Vegetation des Hochmoores Komosse. *Akad. Abh. Uppsala.*

Pearl, R. (1926). *The Biology of Population.* London.

Poole, A. L. (1937). A brief ecological survey of the Pukekura State Forest, South Westland. *N.Z. J. For.* **4.**

Saveley, H. E. (1939). Ecological relations of certain animals in dead pine and oak logs. *Ecol. Monogr.* no. 9.

Tansley, A. G. (1935). The use and abuse of vegetational concepts and terms. *Ecology,* **16.**

Tansley, A. G. (1939). *The British Islands and their Vegetation.* Cambridge.

Tragårdh, I. (1923). Ziele und Wege in der Forstentomologie. *Medd. Stat. Skog.* **20.**

Tutin, T. G. (1938). The Autecology of *Zostera marina* in relation to its wasting disease. *New Phytol.* **37.**

Walton, J. (1922). A Spitzbergen salt marsh. *J. Ecol.* **10.**

Watt, A. S. (1925). On the ecology of British beechwoods with special reference to their regeneration. Part II. The development and structure of beech communities on the Sussex Downs. *J. Ecol.* **13.**

Watt, A. S. (1937). Studies in the ecology of Breckland. II. On the origin and development of blow-outs. *J. Ecol.* **25.**

Watt, A. S. (1938). Studies in the ecology of Breckland. III. The origin and development of the Festuco-Agrostidetum on eroded sand. *J. Ecol.* **26.**

Watt, A. S. (1940). Studies in the ecology of Breckland. IV. The Grass-heath. *J. Ecol.* **28.**

Watt, A. S. (1945). Contributions to the ecology of bracken. III. Frond types and the make up of the population. *New Phytol.* **44.**

Watt, A. S. (1947). Contributions to the ecology of bracken. IV. The structure of the community. *New Phytol.* **46.**

2 The Causes and Consequences of Spatial Pattern

Introduction and Review

The beginnings of landscape ecology—the warp and weft of the tapestry—were in geography and related fields (part I). Although glimmerings of a nascent landscape perspective existed in the writings of ecologists such as Alex Watt (included in part I), the focus of most ecologists before the 1950s was on population dynamics, community structure, and physiological relations of organisms to their environments. Field studies were generally conducted in small, relatively homogeneous areas, and the ascendancy of ecological theory in the 1950s and 1960s led even more ecologists to embrace spatial homogeneity in order to match the simplifying assumptions of theory (Wiens 2000).

Changes were under way, however, and the impetus for further growth and the eventual consolidation of landscape ecology during the 1960s and 1970s came more from ecology than from geography. Changes within each of these disciplines contributed to this shift of emphasis. In geography, the themes of land-systems classification and regional synthesis that had held sway in many parts of the world earlier continued in Europe (especially Central Europe), but focus dramatically shifted in the English-speaking world, particularly in North America. This shift stemmed from a long-standing debate that had existed in geography since the 1930s. Some geographers,

chief among them Richard Hartshorne (Hartshorne 1939, 1959; Entrikin and Brunn 1989) claimed that the responsibility of geography was to view regions as spatial entities composed of discrete, spatially distributed phenomena, such as economic activities or urban features. This spatial perspective, which came to dominate geography (at least in North America), was primarily concerned with locational and distributional factors in economic and social geography, the human side of the discipline. The role of physical geography, according to Hartshorne's approach, was merely to describe the physical environment as it characterized the human environment (Butzer 1989). This approach proved unsatisfactory to physical geographers who tended to concentrate more on process-based research, often finding support for their initiatives in other, allied disciplines. Consequently, the earlier integrative themes within physical geography were often lost as geomorphologists migrated to earth science departments, soil geographers to soil science departments, and so on. Geography, therefore, lost much of its focus on the consequences of human-environment interrelationships in producing distinct landscape types, the traditions that had been leading it in the direction of landscape ecology.

At the same time, the cultural landscape theme developed by geographers such as Sauer (part I) was criticized for having failed to develop a distinct theoretical and methodological basis. Geography, it was claimed, could not develop as a truly scientific field without its own theories, models, and methods. Quantification was seen to be a key to the resolution of this issue (Burton 1963). For example, Haggett (1964) highlighted issues of both scale and quantification in problems of spatial analysis of geographic patterns. Against this backdrop, which dominated discussions in geography for several decades, there were concerns, as well, that no one really understood what *landscape* really meant or how it could be quantified. For example, Smith (1989) noted that Sauer's ideas failed because the term *landscape* was seen to be inherently confusing, especially given the various translations of the German term *Landschaft*. Furthermore, a distinction between *landscape* and the geographers' use of *area* and *region* had not been made, and *natural landscapes* not only lacked any theoretical conceptual base (as Hartshorne had also noted) but also did not exist in reality. In arguing for the retention of the traditional base of geography with a regional focus, Hart (1981) dismissed the use of *landscape* by Sauer as a misunderstanding and misinterpretation of its Germanic roots and a subsequent "corruption" in trying to form a scientific base for the discipline. So, although landscape ecology was beginning to emerge in much of the English-speaking world, its parentage in geography was headed in a different direction.

Ecology was also undergoing dramatic changes, changes that brought at least some parts of the discipline closer to the earlier focus of geographers on landscapes. This transformation was prompted by several factors: the emergence of theoretical ecology as a major force, the growing importance of systems ecology and biogeochemistry, the resolution of the Clements-Gleason debate about the cohesiveness of natural communities that had preoccupied ecologists for several decades, and increasing recognition

that models of populations based on density dependence or of communities based on interspecific competition alone were inadequate (for treatments of this history, see Kingsland 1985; Hagen 1992; Golley 1993; for some of the classic papers that fueled these changes, see Real and Brown 1991). As a result, the ecology of the late 1970s was markedly different from that of the 1950s.

One dimension of these changes was an increasing recognition of the importance of spatial pattern and heterogeneity to a variety of ecological themes. Perhaps more than any other aspect of ecology, this spatial emphasis helped dictate what landscape ecology would look like as a discipline when it did emerge during the 1980s (part III). The focus in this section is on a series of papers that helped develop this perspective on spatial pattern, what Monica Turner was later to characterize as the effect of pattern on process (part VII).

Although *spatial ecology* began to emerge in full force during the late 1960s and 1970s, it is worthwhile to step back a bit, to the mid-1950s. In 1955, a Werner-Gren symposium was held at Princeton University that focused on the human transformation of the earth. The symposium was broadly interdisciplinary, bringing together geographers, biologists, anthropologists, geologists, historians, soil scientists, economists, and others. The conference cochairs were the geographer Carl Sauer (part I), the zoologist Marston Bates, and the urban planner Lewis Mumford. The conference had several themes, but the milieu for the discussion of these themes was the landscape at various scales, from continental to local. The conference proceedings were published in a volume (Thomas 1956) that evoked considerable interest among geographers but went largely unnoticed elsewhere (for subsequent treatments, see Darling and Milton 1966; Turner et al. 1990). One contribution, however, has become a benchmark paper in both landscape ecology and conservation biology. **John Curtis,** a Wisconsin plant ecologist, was instrumental in pointing out how plant communities are made up of species responding individualistically to environmental and disturbance gradients of varying steepness. He addressed the theme of human modifications of landscapes in the upper Midwest of the United States. He documented the dramatic changes that resulted from the elimination of forests and native grasslands and drew attention to the gradual loss of species from fragmented patches of native vegetation. One of the figures from his paper, documenting the elimination of forest cover from Cadiz township in southern Wisconsin from the early 1800s to the mid-1900s, has become one of the most frequently reproduced graphics in ecology and landscape ecology. (It was the first figure in MacArthur and Wilson's classic book [1967] on island biogeography theory.)

Curtis's paper is as much a classic in conservation as in landscape ecology. This theme—of integrating conservation, ecology, and landscapes—is also evident in the paper of **H. E. Wright.** Wright integrated paleoecology (pollen and charcoal analysis) with tree-ring analysis to consider the consequences of fire management over a large area, the Boundary Waters Canoe Area of northeastern Minnesota. Wright drew from the detailed and extensive studies of Heinselman (1973), who mapped fire-origin patches over the entire landscape. Heinselman (whose paper would be included here were it not for its length) showed that a theoretical climax vegetation was never reached because of the long-term regime of fire disturbances. Wright argued that to perpetuate the very forest landscape that the wilderness area was established to preserve, fire should be used as a management tool. Given the well-established policy of fire suppression on national forests, including wilderness areas, this was, at the time, a controversial recommendation (as, indeed, it still is in some quarters). The paper also nicely illustrates a recurring theme in landscape ecology: the tradition of European landscape ecologists to focus on the dominant cultural landscapes, whereas much of North American landscape ecology has emphasized biological themes and more "natural" systems (see part III). The Wright paper shows that in North America, even when the science dealt with management issues, the target was often landscapes where human influences were less pervasive.

Although Wright's focus was on issues directly related to landscape management, the contribution from **Simon Levin and Robert Paine** illustrates quite clearly the growing perspective on spatial patterning among empirical (Paine) and theoretical (Levin) ecologists. This paper grew out of Paine's field experiments (1966, 1969; the 1966 paper is a classic in ecology, see Real and Brown 1991) on the role of predators in fostering community diversity in rocky intertidal habitats on the Pacific Coast of the United States. When Paine removed the predators (sea stars, *Pisaster*), the formerly patchy community of several species rapidly became dominated by mussels *(Mytilus)*. In this system, sea stars acted as *keystone species* (Paine's term) in maintaining the spatial heterogeneity and diversity of the intertidal systems. Levin and Paine built on these experiments to construct a mathematical patch-dynamics model that added random events and disturbance to colonization dynamics and species interactions to predict community structure. The integration of empirical work with mathematical modeling was a significant step in advancing the importance of heterogeneity in ecological work and in developing a framework for thinking about patch dynamics among landscape ecologists.

The theoretical approach to spatial pattern and its consequences is even more evident in the work of **Richard Levins.** Levins's inspiration was in the mathematical theory of island biogeography developed by MacArthur and Wilson (1963, 1967), which charted the dynamics of species extinctions and colonizations on islands in relation to island size, distance from a mainland source, clustering, and several other features of spatial configuration. The parallel between the oceanic islands emphasized by MacArthur and Wilson and patches or fragments of woodland or grassland habitat in a terrestrial matrix was evident to MacArthur and Wilson and was explicitly emphasized by later authors (e.g., Wilson and Willis 1975; Saunders et al. 1987). Levins considered the dynamics of populations occupying such a "landscape" of habitat patches and matrix in a mathematical model that considered how spatial variation might affect how a pest species might attack a crop and how a biological control agent would interact with pest, crop, and environmental patterns. The immediate focus of Levins's model was on an applied question, but his model of patch occupancy and the dynamics of populations over an array of patches—Levins coined the now-fashionable term *metapopulation*—had a far greater impact on theoretical population ecology than on pest management. Even though Levins considered patchiness in an abstract rather than a spatially explicit manner, the work set the stage for an array of spatially explicit approaches that followed a decade or two later (see Hanski and Gilpin 1997; Tilman and Kareiva 1997; Silvertown and Antonovics 2001).

By the mid-1970s, the growth of interest and literature in patchy environments among ecologists had reached a point at which some synthesis was needed. The review paper by **John Wiens** brought together a great deal of work from the 1960s and early 1970s dealing with spatial concepts and dynamics. Really more than a review, the paper used a review format to codify and advance a theory of spatial dynamics under the heterogeneous conditions of real landscapes. This was a significant advance in showing how ecology had begun to deal with space and where the needs still remained in theory and research. Wiens

also used this paper to champion what was to become known as an organism-centric perspective on landscapes: considering landscape pattern and structure from the viewpoint of the organisms being studied rather than that of the person doing the study. Perhaps to the surprise of many landscape ecologists today (some of whom consider papers of only the last five to ten years as the source of key concepts), the Wiens paper defines such current terms as *patch, edges, spatial scale, grain*, and *contrast.*

Many of the themes developed by Levin and Paine, Levins, and Wiens came together in the work of **Steward Pickett and John Thompson.** Pickett and Thompson took island biogeography theory as their starting point but immediately showed that it was inadequate in real landscapes, where the equilibrium conditions predicted by theory would seldom hold because of ecological disturbances. Their focus was on the reserve design application of island biogeography theory, and they proposed the concept of *minimum dynamic area:* the need to preserve not only habitat "islands" for conservation but an area sufficiently large to encompass all stages of disturbance and succession typical of the landscape or region. Close parallels exist with Wright's proposal for the Boundary Waters Canoe Area wilderness; indeed, Pickett and Thompson cited the papers of Wright and of Heinselman.

By the 1960s, other areas of ecology were also developing concepts that would provide part of the ecological foundation for landscape ecology. In the early 1950s, the publication of Odum's *Fundamentals of Ecology* (1953) established a role for both systems thinking and biogeochemistry in ecology. A decade later, this thinking was beginning to be implemented in large, interdisciplinary research programs. Perhaps the most obvious (and certainly the most expensive) was the International Biological Program (IBP). Because of its size and international scope, the IBP was in large part responsible for the development of systems ecology and the beginning of whole-ecosystem computer simulation modeling (Golley 1993). This thrust, in turn, provided the foundation for advances in landscape ecology to come in the 1980s and 1990s. For example, the IBP served as the base for one of the first landscape ecology studies to be funded by the U.S. National Science Foundation, culminating in a book on *Forest Island Dynamics in Man-Dominated Landscapes* (Burgess and Sharpe 1981). This theme was also the progenitor of the spatially explicit landscape simulation modeling that is in vogue today (Mladenoff and Baker 1999; Costanza and Voinov 2004; Mladenoff 2005).

Many of the programs that were part of the IBP (especially those in the United States) also emphasized the biogeochemical dynamics of systems. One of the most noteworthy studies conducted in this period (which still continues), however, was not formally part of the IBP. **Herbert Bormann, Gene Likens,** and their colleagues began a study of nutrient dynamics in forested landscapes at Hubbard Brook, New Hampshire, in 1964. This study pioneered the use of whole-watershed experimental manipulations, using forest clear-cutting and herbicide applications to test the effects on nutrient outputs via stream measurements. The Hubbard Brook studies showed surprising evidence of the magnitude of nitrogen loss from the system. The work also focused attention on the broad-scale effects of forest management and showed that these effects could be studied and measured on real landscapes in ways that integrated terrestrial with aquatic ecology. Theirs were among the first experiments conducted at the scale of complete landscape elements (watersheds).

References

Burgess, R. L., and D. M. Sharpe, eds. 1981. *Forest island dynamics in man-dominated landscapes.* New York: Springer-Verlag.

Burton, I. 1963. The quantitative revolution and theoretical geography. *Canadian Geographer* 7:151–162.

Butzer, K. 1989. Hartshorne, Hettner, and "The Nature of Geography." In *Reflections on Richard Hartshorne's "The Nature of Geography,"* eds. J. N. Entrikin and S. D. Brunn, 35–52. Washington, DC: Association of American Geographers.

Costanza, R., and A. Voinov, eds. 2004. *Landscape simulation modeling.* New York: Springer-Verlag.

Darling, F. F., and J. P. Milton. 1966. *Future environments of North America.* Garden City, NJ: Natural History Press.

Entrikin, J. N., and S. D. Brunn, eds. 1989. *Reflections on Richard Hartshorne's "The Nature of Geography."* Washington, DC: Association of American Geographers.

Golley, F. B. 1993. *A history of the ecosystem concept in ecology.* New Haven, CT: Yale University Press.

Hagen, J. B. 1992. *An entangled bank: The origins of ecosystem ecology.* New Brunswick, NJ: Rutgers University Press.

Haggett, P. 1964. Regional and local components in the distribution of forested areas in south-east Brazil: A multivariate approach. *Geographical Journal* 130: 365–380.

Hanski, I. A., and M. E. Gilpin. 1997. *Metapopulation biology: Ecology, genetics, and evolution.* San Diego, CA: Academic Press.

Hart, J. F. 1981. The highest form of the geographer's art. *Annals of the Association of American Geographers* 72:1–21.

Hartshorne, R. 1939. *The nature of geography.* Lancaster, PA: Association of American Geographers.

———. 1959. *Perspectives on the "Nature of Geography."* Chicago: Rand McNally.

Heinselman, M. L. 1973. Fire in the virgin forests of the Boundary Waters Canoe Area, Minnesota. *Quaternary Research* 3:329–382.

Kingsland, S. E. 1985. *Modeling nature: Episodes in the history of population biology.* Chicago: University of Chicago Press.

MacArthur, R. H., and E. O. Wilson. 1963. An equilibrium theory of insular zoogeography. *Evolution* 17:373–397.

———. 1967. *The theory of island biogeography.* Princeton, NJ: Princeton University Press.

Mladenoff, D. J. 2005. The promise of landscape modeling: Successes, failures, and evolution. In *Issues and Perspectives in Landscape Ecology,* eds. J. A. Wiens and M. R. Moss, 90–100. Cambridge: Cambridge University Press.

Mladenoff, D. J., and W. L. Baker, eds. 1999. *Spatial modeling of forest landscape change.* Cambridge: Cambridge University Press.

Odum, E. P. 1953. *Fundamentals of ecology.* Philadelphia: Saunders.

Paine, R. T. 1966. Food web complexity and species diversity. *American Naturalist* 100:65–75.

——. 1969. A note on trophic complexity and community stability. *American Naturalist* 103:91–93.

Real, L., and J. H. Brown, eds. 1991. *Foundations of ecology: Classic papers and commentaries.* Chicago: University of Chicago Press.

Saunders, D. A., G. W. Arnold, A. A Burbidge, and A. J. M. Hopkins, eds. 1987. *Nature conservation: The role of remnants of native vegetation.* Chipping Norton, Australia: Surrey Beatty.

Silvertown, J., and J. Antonovics, eds. 2001. *Integrating ecology and evolution in a spatial context.* Oxford, UK: Blackwell Science.

Smith, N. 1989. Geography as museum: Private history and conservative idealism. In *Reflections on Richard Hartshorne's "The Nature of Geography,"* eds. J. N. Entrikin and S. D. Brunn, 89–120. Washington, DC: Association of American Geographers.

Thomas, W. L., ed. 1956. *Man's role in changing the face of the earth.* Chicago: University of Chicago Press.

Tilman, D., and P. Kareiva, eds. 1997. *Spatial ecology: The role of space in population dynamics and interspecific interactions.* Princeton, NJ: Princeton University Press.

Turner, B. L. II., W. C. Clark, R. W. Kates, J. F. Richards, J. T. Mathews, and W. B. Meyer, eds. 1990. *The earth as transformed by human action.* Cambridge: Cambridge University Press.

Wiens, J. A. 2000. Ecological heterogeneity: An ontogeny of concepts and approaches. In *The Ecological Consequences of Heterogeneity,* eds. M. J. Hutchings, E. A. John, and A. J. A Stewart, 9–31. Oxford, U.K.: Blackwell Science.

Wilson, E. O., and E. O. Willis. 1975. Applied biogeography. In *Ecology and Evolution of Communities,* eds. M. L. Cody and J. M. Diamond, 522–534. Cambridge, MA: Harvard University Press.

The Modification of Mid-latitude Grasslands and Forests by Man

JOHN T. CURTIS[*]

Man's actions in modifying the biotic composition of mid-latitude grasslands and forests can best be studied by separating them into two groups of processes. In the first group are the effects induced by pioneer cultures in areas peripheral to main population centers. These areas may be peripheral because the main population has not had time to spread out over the entire region, as was the case during the European settlement of North America, or they may be peripheral because the severity of the environment more or less permanently prohibits the development of intensive civilization, as in rugged mountains, deserts, or taiga. In either case the exploiting peoples have economic ties with the main population in the sense that the latter furnishes both the tools for exploitation and a market for the products.

The second group of effects is composed of those that are produced by the intensive utilization of land for agricultural and urban purposes within the regions of high population. These typically follow the pioneer effects in time and are influenced by the earlier

* Dr. Curtis is Professor of Botany at the University of Wisconsin, Madison, and Research Director of the University Arboretum. He was a Guggenheim Fellow (1942) and was research director of the Société Haitiana-Americaine de Development Agricole, Gonaives, Haiti (1942–45). He is the author of papers on plant ecology, especially on the continuum concept of community relationships.

changes. Most of the available evidence on the nature of the changes induced by man is concerned with impact of European man on his environment, but it is probable that both older civilizations and aboriginal cultures exerted similar effects whenever their populations were sufficiently high.

MODIFICATIONS OF MID-LATITUDE FORESTS

The mid-latitude forests are typified by the deciduous forest formation, although several kinds of conifer forests are also to be found within the strict geographical boundaries of the mid-latitudes. In the interests of simplicity, this discussion will be concerned almost solely with the deciduous forest.

Peripheral Effects

Ordinarily, the first products of a peripheral wilderness to be exploited by an adjoining civilization are derived from the animal members of the community, especially the fur-bearers. The French *voyageurs*, the Hudson's Bay Company, and John Jacob Astor and his fur-trading competitors are familiar agents of such exploitation in America. Ecologically, the effects were not very great, since the rather minor population changes brought about in the animal species concerned were not radically different from those experienced in natural fluctuations. Of far greater significance was the utilization of the forest for its timber. The first stages in this utilization were concerned with the harvest of products of high value, such as shipmasts, spars, and naval timbers in general, followed by woods of importance in construction of houses and furniture. When the tree species suitable for these needs were common and especially when they grew in nearly pure stands, as was the case with the white pine in eastern North America, the impact of the exploitation was great. This was true both of the magnitude of the changes and of the relative size of the area affected. White pine was a favorite goal of the early American lumberman and was ruthlessly harvested far from the scenes of its ultimate use.

Some forest types were composed mostly of trees of lesser value, such as the oaks and maples. These were commonly by-passed, or their more valuable members, used as cabinet woods, were selectively logged. This emphasis on special products rather than on complete utilization was mainly a feature of young expanding cultures whose demands were small relative to the size of the resource base in the peripheral area. The phenomenon is present in current times, as exemplified by the utilization of only a few of the host of tropical species for special veneers and by the selective harvest of spruce in the conifer forests of Canada and the non-utilization of the equally abundant balsam fir and other species.

All these exploitations were and are dependent upon certain definite physical properties of the wood as it occurs in the trunks of natural trees produced under natural conditions. These properties are commonly unrelated to the ecological behavior of the species in the sense that no special growth habits or reproductive capacities are concerned. The primitive exploiter was not worried about whether or not a second or continuing crop of the species would be available. In many respects, the selec-

tive harvest resembled mining in that it was the utilization of a non-renewable resource or at least was treated as such.

As the economic demands of the main population centers grew and especially as the population centers spread out in area, the utilization of the peripheral resources became more intense. In non-industrial civilizations or non-industrial stages in the development of any culture, the forest is called on to produce a considerable share of the fuel used by that culture. This might be in the form of firewood or in the form of charcoal. Even in those countries or stages where coal was used for fuel, the forest was a source of timber for mine props and for ties or sleepers on the railroads used for hauling the coal. All these uses were more or less dependent upon a near-by source of supply. The biological significance of this lies in the fact that the harvesters now became desirous of gathering more than one crop from the same land. The ecological behavior of the species thus came to be of greater importance than the physical structure of the wood. Species were utilized regardless of behavior, but only those which possessed the ability to resprout or otherwise to reproduce themselves remained in the forest. Firewood, charcoal, and mine props all utilize small-dimension stock by preference, and thus a premium was placed on those species which could quickly return to a merchantable size. The technique of coppicing, so widely used in Eurasia, is a direct result of this situation.

In many places in the world the more intensive utilization of the trees in the subperipheral areas was accompanied by the introduction of grazing animals to the forest community. The woodchoppers, charcoal-makers, and lime-burners were more permanent inhabitants of a region than the earlier lumbermen, and they commonly broadened

Modification of Mid-latitude Grasslands and Forests   723

the base of their economy by the use of cattle, sheep, or goats. These animals were allowed to roam the woods on free range and to make use of such forage and browse as might be available there. The livestock pressure was rarely as great as that which accompanied the agricultural economies of subsequent times, but its effects cannot be overlooked (Steinbrenner, 1951). In addition to the growth behavior patterns selected by the harvesting techniques, the successful species were also those best able to withstand the effects of grazing.

Thus we find in the areas peripheral to major population centers a gradually increasing intensity of utilization, either in space or in time, from a negligible pressure with little effect on the biotic composition of the forest to a severe pressure which selectively favored species with particular behavior patterns and eliminated other species which did not conform.

Let us now inquire into the actual nature of the changes that accompanied this increase in utilization. The earlier stages in the process are best studied by examples from the United States, since these stages occurred so long ago in Asia and Europe as to have left almost no record. In the United States the fur-trapping stage was roughly a seventeenth- and eighteenth-century phenomenon, while the lumber-harvest stage was most prominent in the nineteenth century.

The later, more intensive stages began in the latter half of the 1800's in the eastern portion of the United States. In much of Europe the intensive utilization began in the 1200's and continues in marginal areas up to the present. Thus, in England, extensive forests were utilized chiefly for hunting purposes through the period of Norman domination. Later timber-harvesting resulted in such a severe depletion of suitable trees that large-scale imports

from the Baltic region were required by 1300. Widespread utilization of the forests for pasturage of swine and cattle, combined with intensive cutting of fuelwood, made further inroads on the peripheral areas. By 1544 a series of laws was passed regulating the procedures by which a coppiced woods should be managed (Tansley, 1939). In the Balkans utilization of the forests for construction timber and marine products was active in the 1200's, with many areas depleted by 1620. Intensive utilization for firewood, charcoal, and lime-burning still continues in the more mountainous regions (Turrill, 1929). In China peripheral exploitation is much older and has long since been completed in all but the most rugged and inaccessible terrain. Clearing of forests for agricultural purposes was widespread during the Shang dynasty beginning in 1600 b.c. (Needham, 1954).

The most obvious biotic change in the forest is the great shift in species composition, both qualitative and quantitative. This is best seen today in the northern states of the United States, where the original forest was a mosaic of patches of hardwoods with a few conifers and patches of conifers with a few hardwoods (Brown and Curtis, 1952). Those portions of the northern forests originally covered with hardwoods underwent a relatively slight alteration. They were composed of a mixture of species, none of which was ever in great economic demand. Large areas, therefore, were rarely cut over in anything like the intensity so common in the neighboring pine forests. In addition, fires were much less frequent and usually not so severe as those in the coniferous area. Many of the component species had the ability to resprout after cutting, like the maple and the beech, and most of them had very efficient means of reproduction, so that a stand was able to regenerate itself following partial destruction. Here and there, selective pressure reacted against one or more species. Millions of board feet of hemlock were cut in the region solely for the bark, which was used in the tanning industry. The logs were allowed to rot where they fell after they had been peeled (Goodlett, 1954). In more recent times, yellow birch has been intensively exploited because of its value as a veneer wood, and a few other species have experienced similar selective pressures. The major change resulting from all this has been an increase in the relative importance of sugar maple (*Acer saccharum*) in the remaining stands. This species is ecologically the most vigorous of all. The normal subordinate rank of the other species, accentuated by the added pressure from man, has resulted in their gradual disappearance in favor of the maple.

In contrast to this shift in relative importance of one member of the hardwood forest at the expense of others, the pine forests suffered a much more severe alteration. The march of the lumbermen from Maine in the late 1700's, to New York in 1850, to Michigan in 1870, to Wisconsin in 1880, and finally to Minnesota in 1890 was primarily a quest for white pine (*Pinus strobus*). This species, like the majority of pines, is a "fire tree" and was found in essentially pure stands in large blocks on lands subject to widespread burns. Profitable harvesting enterprises could be centered in regions where such blocks were common and where adequate facilities for transportation by river driving were available. In such regions the initial harvest obviously made great changes in forest composition by the removal of 90 per cent or more of the dominant trees. Of even greater importance were the frequent fires which broke out in the slash following the lumbering operations. These fires were allowed to burn unchecked and were often actually encouraged.

White pine is adapted to seeding-in following a fire but has no mechanism for sprouting from a burned stump. The first fire, therefore, often produced a new crop of pine seedlings, but a second fire before the trees had matured destroyed the entire population. The land became covered with weedy tree species like the aspens, birches, and oaks and with shrubs like the hazelnut, all of which could resprout following fire and thus remain in control of the ground. Excessive burning sometimes produced the so-called "barrens"—desolate tracts almost devoid of large woody plants, such as occupy extensive portions of Michigan and Wisconsin.

Accompanying the changes in species composition have been changes in the micro-environment within the forest. Selective logging or other mild harvesting practices result in an opening-up of the canopy of the forest, a breaking of the former more or less complete cover. The openings thus created possess a very different microclimate from the remaining portions of the forest. The most significant change is an increase in the rate of evaporation, and this increase is proportional to the intensity of the harvest, reaching maximum values in clear-cut and particularly in cut and burned woods.

With the exception of those lightly harvested forests containing sugar maple in which the net result is an increase in maple, practically all the environmental changes induced by the peripheral harvesting sequence are in the direction of a more xeric habitat, with greater light, more variable temperatures, more variable moisture, and much greater transpirational stress. The internal, stabilizing mechanisms of the community that lead toward homeostasis are upset or destroyed. The new environment tends to resemble that normally found in adjacent, hotter and drier regions (the "preclimax conditions" of Clements, 1936). Such condi-

Modification of Mid-latitude Grasslands and Forests   725

tions are most suitable ,for the ecologically pioneer plants of the region, which are those species that grow vigorously under the unstable climatic conditions and that possess adaptations to make use of the high light intensities. Ordinarily, such species possess highly effective means of reproduction and dispersal and, in addition, are likely to survive under severe disturbance, as by cutting or fire, through the ability to sprout from stumps or roots. The initial invaders of the disturbed areas, therefore, tend to be the pioneer species. Subsequent harvesting, as by coppicing, favors the persistence of these species and the gradual elimination of those with more climax tendencies. Thus, the original mixed forests come to be replaced by large areas of scrub oaks, aspen, box elder, sassafras, and similar species. The final selection under grazing pressure may eliminate or depress some of these, since there are very few species in the biota with the necessary combination of attributes to resist all the decimating influences. A fertile field for future investigation would be the study of the characteristics of various plants which enable them to survive under the conditions just outlined.

Agro-urban Effects

As the main centers of populations expanded into the peripheral areas, a considerable change in the land-use pattern followed. In the mid-latitudes, with their generally favorable climate during the growing season, agriculture became the dominant feature. The forests, already modified by peripheral utilization, were cleared to make room for fields with an initial selection of the best sites, followed by gradual encroachment onto less favorable land types. The actual nature of the best sites naturally varied from place to place, but the ideal appeared to be a large area of level or gently rolling land, with well-drained soils of high fertility. Frequently, the land was chosen on the basis of indicator species, black walnut being a favorite of the American settler, as it grew in rich forest stands well supplied with moisture and available nutrients. On these preferential sites the trees were killed by girdling or cutting, the logs and tops burned, the stumps pulled, blasted, or otherwise removed, and the ground plowed. Any member of the original community which persisted under the treatment was subsequently eradicated by the clean cultivation practices employed on the fields. The impact of man on these agricultural fields was thus one of total destruction with respect to the original community.

Marginal lands which were remote, difficult of access, or topographically unsuited for crop agriculture were commonly employed for intensive grazing. A continuing pressure was also exerted on them for lumber and firewood harvest. The distinction between peripheral and central activities is least clear at this stage, which is usually rather short in duration. With long-term occupancy of the land by an agro-urban culture, the remnants of the original forest come to be restricted to sites which are totally unusable for agriculture, such as cliffs, rocky ground, barren sands, ravines, or swamps. These habitats all differ markedly from the bulk of the land in their physical environment and hence also in their community composition. For various reasons, successional development is retarded in these extreme sites, and they retain a very high proportion of ecologically pioneer species. Consequently, they are less subject to drastic change by man's disturbance, since this disturbance usually leads to an increase in pioneers which are here the natural dominants.

The rate and the extent of the destruction of original cover by agricultural clearing are well demonstrated by

726   *Man's Role in Changing the Face of the Earth*

a case history covering the first century of use of a township of land in Green County, Wisconsin, along the Wisconsin-Illinois border. The vegetation in 1831 before agricultural settlement began, as derived from records of the original Government Land Survey, was mostly upland deciduous forest, dominated by basswood, slippery elm, and sugar maple except for an area of oak-hickory forest in the northwest. A small portion of prairie with surrounding oak savanna was present in the southwest corner. The extent of forest cover in 1882 and 1902 was mapped by Shriner and Copeland (1904), while that in 1935 was recorded by the Wisconsin Land Economic Survey. The present condition was determined from aerial photos taken in 1950 and from personal inspection. The changes are shown in Figure 147 and in Table 26. While the very first clearings may have been confined

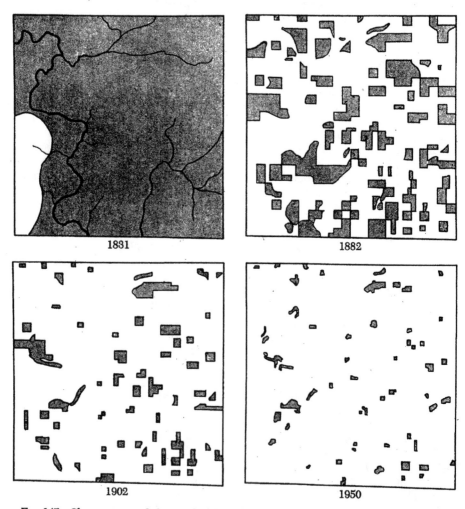

1831

1882

1902

1950

Fɪɢ. 147.—Changes in wooded area of Cadiz Township, Green County, Wisconsin (89°54′ W., 43°30′ N.), during the period of European settlement. The township is six miles on a side and is drained by the Pecatonica River. The shaded areas represent the land remaining in, or reverting to, forest in 1882, 1902, and 1950.

Modification of Mid-latitude Grasslands and Forests   727

to the best lands, by 1882 the most evident factor influencing the pattern of land clearing was the unfortunate system of land survey which resulted in square landholdings independent of terrain. Not until the forest had been reduced to less than 10 per cent of its original extent did the remaining wood lots begin to reflect the topography. Currently, the majority of the remnant forests still have one or more straight boundaries, although most of them are confined to rocky outcrops and thin-soil hilltops. The statistics for the township show a reduction in forest cover to

by 36 per cent in 1935. This was due largely to the drying-up of springs in their original headwaters, thus reflecting a decrease in subsoil water storage from the reduced infiltration on agricultural fields and pastured wood lots.

A number of important changes occur in forested regions under the impact of agriculture aside from the obvious destruction of most of the forest. Under aboriginal use or peripheral exploitation, fire was a common occurrence, with large areas involved at each burn, since both the means and the desire to stop the fires were absent. Fol-

TABLE 26

CHANGE IN WOODED AREA, CADIZ TOWNSHIP, GREEN COUNTY,
WISCONSIN (89°54′ W., 43°30′ N.), FROM 1831 TO 1950

	1831	1882	1902	1935	1950
Total acres of forest.............	21,548	6,380	2,077	1,034	786
Number of wood lots...........	1	70	61	57	55
Average size of wood lot in acres..	21,548	91.3	34.0	18.2	14.3
Total wooded area as a percentage of 1831 condition.............	100	29.6	9.6	4.8	3.6
Total periphery of wood lots in miles......................	....	99.0	61.2	47.2	39.8
Average periphery per wooded acre in feet.....................	....	82	155	241	280

29.6 per cent of the original by 1882, to 9.6 per cent by 1902, to 4.8 per cent by 1936, and to 3.6 per cent by 1954. The existing forests are used by their owners as sources of firewood and occasional saw timber. In addition, 77 per cent of the present wooded area is heavily grazed by cattle to the point where no regeneration of the trees is taking place. Thus only 0.8 per cent of the land under forest cover in 1831 is still in what might be called a seminatural state, and this tiny portion is broken up into even more minute fragments, widely scattered throughout the area.

Concomitant with the reduction in forest cover in a presumed cause-and-effect relation was a decrease in total length of the streams draining the area. The permanently flowing streams had decreased by 26 per cent in 1902 and

lowing the dissection of the landscape and the interpolation of farm land between remnant forest stands, fires were more or less automatically stopped by the bare fields or were consciously suppressed by the farmers. In consequence, the forests that escaped clearing received a degree of protection far greater than they had normally experienced. In addition, the high cost of fencing wood lots in many instances prevented their use for grazing animals. As a result of these two influences, the forest, although gradually reduced in size and contiguity, actually improved in structure, with an increased density of trees per acre and an increase in cover and hence in humidity. On many marginal sites which had been reduced to brush by recurrent presettlement fires, mature forests subsequently developed. Since the species which were able to persist through the fires were the extremely vigorous pioneers, the first forests that developed were dominated by these pioneer species. In much of the central United States, where the recovery of marginal and remnant sites has been under way for about a century, the natural processes of succession are just beginning to convert the forests to a more climax condition (Cottam, 1949).

The conversion frequently is hastened in regions where mixed forests are present by the selective utilization of the mature pioneer members of the community for farm timber and firewood. Species of oaks are most commonly involved in the process in the United States and in Europe, although other pioneer species of high economic value are sometimes important. Since these trees are removed only a few at a time as the farmer needs them, environmental conditions are not greatly altered. The major result is a liberation of the understory layer of climax species like maple, beech, and basswood and the consequent repression of regeneration of the original pioneer trees.

The early period in the agro-urban utilization of forest land, therefore, presents the anomaly of severe reduction in total amount of forest but considerable improvement in the stands that did survive. These remnants commonly suffered severe damage from grazing at a later date.

Another change accompanying agricultural occupation of forested land is in what might be called the physiognomic result. Instead of an essentially continuous forest cover, with infrequent meadow-like openings along watercourses or small grasslands where fires had been unusually severe, the landscape now presents the aspect of a savanna, with isolated trees, small clumps or clusters of trees, or small groves scattered in a matrix of artificial grassland of grains and pasture grasses, unstable and frequently devoid of plant cover as a result of regular plowing. The physical conditions of the intervening "grasslands" are such as to prevent the successful growth of practically all members of the forest biota. A few of the plant members persist along fence rows and other places of relatively infrequent disturbance, and a very few are sufficiently weedlike actually to compete in the farmer's fields, especially in the permanent pastures (bracken fern, hazelnut, etc.).

Among the animal members of the community is a group that normally made use of the original forest edges. Some birds, for example, nested within the forest but sought their food in the open places and in the tangle of vines and shrubs that commonly bordered the openings. Such animals were greatly benefited by the increased "edge" provided by the fragmented wood lots, and their relative populations increased accordingly. As shown in Table 26, the average length of the periphery per acre of forest increased from 82 feet in 1882 to 280 feet by 1950.

The artificial savanna condition provided a suitable habitat for a number of species which originally occurred in grasslands and natural savannas on the dry margin adjacent to the mid-latitude forests. A number of birds, like the prairie horned lark (Forbush, 1927, pp. 336–70) and the western meadow lark, extended their range well into the original forest country of the eastern United States. Aggressive prairie plants, like ragweed, black-eyed Susan, and big bluestem similarly advanced far beyond their original areas of prominence and became conspicuous features of the vegetation along roadsides and in agricultural fields. Similar migrations of steppe plants westward into the forests of Central Europe are known (Oltmanns, 1927, pp. 104–56). These migrations resulted in a partial blending of the components of two or more major biotic communities and served to lessen their inherent differences.

Within the remnant forest stands, a number of changes of possible importance may take place. The small size and increased isolation of the stands tend to prevent the easy exchange of members from one stand to another. Various accidental happenings in any given stand over a period of years may eliminate one or more species from the community. Such a local catastrophe under natural conditions would be quickly healed by migration of new individuals from adjacent unaffected areas (the "gap phase" concept of Watt, 1947). In the isolated stands, however, opportunities for inward migration are small or nonexistent. As a result, the stands gradually lose some of their species, and those remaining achieve unusual positions of relative abundance.

The lack of interchange of plant individuals also applies to plant pollen. Those members of the community which are regularly or usually cross-pollinated no longer have the opportunity of crossing with a wide range of individuals. It is probable, therefore, that opportunities for evolution of deviant types by random gene fixation will be increased in the future, as the isolating mechanisms have longer times in which to operate. In heavily utilized stands selection pressures engendered by the frequent disturbance, together with the shift toward pioneer conditions resulting directly from the small size of the stands, would tend to favor those ecotypes which have pioneer tendencies and to reject the more conservative climax strains. The study of this micro-evolution should be one of the most fertile fields for future investigations.

EFFECTS ON MID-LATITUDE GRASS-LANDS AND SAVANNAS

Peripheral Effects

The plant members of the mid-latitude grasslands for the most part are of no direct use to man. Their main value comes after they have been converted to high-protein foods in the form of animals. The earliest utilization of grasslands, therefore, was by hunting cultures or by peripheral exploitation for the benefit of a remote agro-urban civilization. The slaughter of the bison on the prairies of mid-continental North America is a familiar example of this process (Garretson, 1938). In the absence of adequate information as to the influence of the bison on the structure of the remainder of the grassland community, little of value can be said about the effects of their removal. In any case, such effects would have been temporary, for intensive agricultural use of much of the eastern area followed quickly afterward, while domesticated cattle were introduced in the drier ranges to the west and began to exert an effect of their own.

The nature of the changes induced by cattle were probably different from those formerly resulting from the bison. The cattle were kept on limited ranges, so that the vegetation was subject to pressure over a long season, year after year. The bison, on the other hand, may have exerted an even greater pressure for brief periods, but a recovery period of several years commonly intervened before the wandering herd revisited any particular area. The long-term effects of the two types of grazing animals thus were very different.

Cattle begin to utilize the prairie grasses as soon as growth starts in the spring. Utilization of regrowth occurs during the summer and is particularly damaging when the reproductive stems begin to elongate in the later months. The continual reduction in photosynthetic area due to leaf removal results in a decreased storage of reserves in the underground organs. At the same time, the normal control of dormant buds by growth hormones is upset by the removal of stem tips, with a resultant stimulation of new growth which further depletes the stored reserves. The bluestems (*Andropogon* sp.), Indian grass (*Sorghastrum nutans*), and switch grass (*Panicum virgatum*) are typical species which respond in this way, and the gradual weakening leads to their eventual elimination and replacement by others. The replacing forms, under normal circumstances, are grasses which are recumbent, with their stems on or near the ground surface and with a large proportion of their leaves in a similarly protected position. They accordingly escape destruction by cattle, which do not graze closely. The grama grasses (*Bouteloua* sp.) and buffalo grass (*Buchloe dactyloides*) are good examples. In the absence of competition from the former dominants, these forms rapidly increase their populations. In range parlance, they are said to be "increasers" as opposed to the species which decline under grazing, which are called "decreasers" (Dyksterhuis, 1949). Both types were present in the prairie before grazing began, but the decreasers, because of their erect habit and greater size, were dominant over the increasers.

When the carrying capacity of the grasslands is greatly exceeded, either by too many cattle per unit area or by a reduction in productivity of the grasses due to drought, then the increasers themselves begin to suffer. Their decline results in a breaking of the continuous plant cover. The bare soil thus exposed becomes available for invasion by weedy annuals, which were formerly excluded from the closed community. In the American grasslands these newcomers, termed "invaders," are frequently exotics which originated in similar situations in the grasslands of the Old World, like cheat grass, Russian thistle, and halogeton. Native invaders are typically plants indigenous to the drier shrub deserts toward the west and are frequently unpalatable as a result of spines and thorns, like prickly poppy and prickly-pear cactus. Continued

overgrazing results in the almost complete eradication of the original prairie flora and its replacement by an unstable community of annuals and thorny perennials.

The entire degradation process involves a shift from climax to pioneer plants and from mesic to xeric conditions. The upgrading effects of the original flora with respect to organic-matter accumulation in the soil, to nutrient pumping, and to water-entrance rates and other constructive activities are greatly lessened or reversed. The soil becomes compacted, less easily penetrated by rain, and much more subject to erosion, especially by wind.

Accompanying these direct effects of cattle on the plant community are a number of important secondary effects deriving from the disruption of the animal community. Misguided efforts at predator removal in the form of wolf and coyote bounties and other more direct means allow the rodent populations to get out of balance. Mice, pocket gophers, prairie dogs, and jack rabbits frequently reach epizootic levels and further add to the already excessive pressures on the plants. Control by extensive campaigns of rodent poisoning is usually temporary in effect and often serves to accentuate the unbalanced condition and to make return to stability more difficult.

In the more humid eastern portion of the grasslands in North America, the better sites were soon used for crop agriculture rather than for grazing, but marginal lands on thin soils, steep hills, or rocky areas frequently remained in use as pastures. In this region the decreasers behaved in the same manner as those farther west, and a few of the same increasers were also present. The major difference, and a very important one for the economy of large sections of the country, was the fact that the most important increaser was in reality an invader—Kentucky bluegrass (*Poa*

Modification of Mid-latitude Grasslands and Forests   731

pratensis). This species has a growth habit similar to the grama grasses and the buffalo grass of the western plains but thrives under a humid climate. It differs from the usual invaders in that it does not require bare soil or a broken cover to become established and is neither an annual nor an unpalatable perennial. The time and conditions of its origin in Eurasia are unknown. In all probability it developed under the influence of man and his pastoral habits, since it is found in the Old World in those regions where grazing has been practiced for long periods. In America it demonstrates a vigor scarcely exceeded by any other plant of similar size and has come to dominate most of the unimproved pasturages in the eastern half of the continent. This domination is virtually total in many areas, so that the original grassland species are completely lacking. The forbs and the few other grasses that do accompany bluegrass (dandelion, white clover, ox-eye daisy, quack grass, timothy, etc.) are themselves exotic and serve to replicate on this continent a man-made community that is very widespread in Europe. In fact, this great expansion of the world range of a particular community under the unintentional influence of man is one of the most powerful examples of man's role as the major biotic influence in the world today. Investigation of the origin of the component species and intensive studies of the dynamics of the assemblage in its new environment should be highly rewarding. (See the chapter by Clark below, pp. 737–62.)

Agro-urban Effects on Grasslands

The extensive utilization of the major mid-latitude grasslands for crop agriculture was restricted to those portions adjacent to the deciduous forest where the rainfall, although irregular, was usually sufficient for grain crops. The most favored places were those grass-lands which had been extended into the forest during the postglacial xerothermic period (Sears, 1942) and which had subsequently been maintained by fire when the climate again favored forest. The Corn Belt in the prairie peninsula of the United States and the European breadbasket in the steppes of the Ukraine and in the *puzta* of Hungary are outstanding examples. In the climatically suitable areas utilization of the grassland for crops instead of for pasture was dependent upon the development of the steel plow for subjugation of the tough prairie sod. The invention and widespread manufacture of such an instrument occurred in the second quarter of the nineteenth century. Hence we find that the American prairies and the European steppes were both converted to cropland at about the same time (Conard, 1951), although the inhabitants of the former were scarcely past the stage of a hunting economy, while the latter had been used for grazing for centuries or millenniums.

The conversion from grassland to cropland was far more complete than the equivalent conversion from forest. In large part this was a result of the fact that the grasslands were flat or gently rolling and presented far fewer topographical obstacles to the plow than did the forest. The destruction of the entire prairie community by clean cultivation over extensive tracts of land means that remnants of the original vegetation are very rare. The major agency preventing complete eradication in much of the American prairie is the railroad system which was extended throughout the area contemporaneously with the advance of the settler (Shimek, 1925). The railroad right of way in many instances was laid on grade and was protected by fences. The tracks themselves were placed in the middle of the right of way, thus leaving a strip of virgin grassland on either side. The only maintenance operations which affected the vegetation was an occasional burning. Since this merely continued the normal practices of aboriginal times, these linear strips of prairie have been maintained in more or less primeval condition except for the random destruction of certain species and their failure to re-enter (Curtis and Greene, 1949).

Contrary to the case of the forest remnants, these railside prairies are not necessarily on pioneer or otherwise deviant sites but rather sample the full range of environment originally present. This is indeed fortunate, since over millions of acres of middle western prairie the only prairie plants to be found are on these railroad prairies. The much-needed research on prairie ecology has been and will continue to be conducted there.

The savannas between the grasslands and the surrounding forests are physiognomically intermediate between the two major formations. In the mid-latitudes they are largely the result of repeated advances and retreats of the prairie-forest border. Along the prairie peninsula of the United States, the characteristic savanna was the oak opening, a community of widely spaced orchard-like oak trees with an understory of prairie plants and a few forest shrubs. All available evidence indicates that these savannas were created by an advance of the prairie into the forest under the driving force of fire. Their maintenance was similarly effected by recurrent fires set by the Indians. They were but little used during the peripheral period except by the early hunters. A very brief period of open-range grazing was quickly followed by agricultural settlement. The potential yields of timber were so low and the quality of the gnarled oaks so poor that no extensive lumbering was ever practiced.

Within a decade or two of settlement, the remnant oak openings that escaped

the ax and plow suddenly began to develop into dense, closed-canopy forests. In large part, this rapid increase in number of trees was due to the liberation of previously suppressed "grubs" or oak brush which had been repeatedly killed to the ground by fire and which had persisted through production of adventitious buds from underground rootstocks (Cottam, 1949). One of the major tribulations of the early settler on the savannas was the laborious hand removal of these underground growths which effectively stopped the best plow and the strongest oxen.

Agricultural occupation of the oak openings thus resulted in two very different effects. On the one hand, the majority of the land was cleared and cultivated, thus destroying the entire community. On the other, the remnant portions rapidly changed over from savanna to forest under the influence of fire protection. Those few areas which continued to be kept open by fire or other means were ordinarily thus treated so that they might be used as pasture, with consequent destruction of the understory vegetation. As a result, an oak savanna, with its full complement of original vegetation, is one of the rarest vegetation types in the United States today.

A similar release of woody vegetation by excessive fire protection has produced the mesquite stands now so common on the savannas of the southwestern range land (Humphrey, 1953).

GENERAL CONSEQUENCES OF MAN'S
UTILIZATION

Changes in Environment

The changes induced in native vegetation by man, either through peripheral, pioneer, or primitive utilization or through more intensive agro-urban occupation, range from simple modification through severe degradation to complete destruction and replacement. All these changes in plant cover are accompanied by changes in the environment within and adjacent to the affected vegetation.

Whether or not widespread deforestation can influence the amount or distribution of rainfall has been debated for decades. No satisfactory proof that such influence exists has appeared so far, but the question remains unresolved. The most convincing arguments concerning the influence of vegetation removal on over-all regional climate are those connected with the energy balance as it is influenced by the albedo or reflecting power of the earth's surface. The value of this factor is very similar for any green vegetation, whether it be forest, prairie, or corn crop. In deciduous forests it may drop during the winter, especially in regions without a permanent snow cover. In grasslands, on the other hand, it actually increases during the winter, owing to the light color of the dead and matted grasses. When the grasslands are plowed, particularly where the land is fallowed or fall-plowed, the dark prairie soils cut down reflection tremendously. The absorption of solar energy is thereby increased, with a possible appreciable change in the total energy increment and hence in the local temperature. This could be of major significance in the spring months in the northern grasslands.

The soil factor of the environment has also been altered by man's activities. Trampling and other disturbances incident to the harvesting operation, combined with activities of livestock, tend to compact the soil, destroying its loose structure and impeding the free entrance of water from rainstorms. The amount of surface runoff is thereby increased. The partial or total absence of tall trees reduces the amount of subsoil water which would normally be lost by transpiration. The excess finds its way to the stream system of the region by way of springs. The initial result of forest-cutting is an increase in the total volume of streamflow, but the complete destruction of the forest, as by fire, greatly increases the flash-flood potential of the watershed and decreases its usable water-producing abilities.

One of the major consequences of the agricultural utilization of mid-latitude forests and prairies has been the very great decrease in soil stability. The resultant soil erosion, both by water and by wind, reached terrifying proportions in many sections of the United States before concerted efforts were made to bring it under control. For the most part, this erosion is unrelated to the previous vegetation and is largely due to the misguided attempt to apply an agricultural system developed under one set of environmental and economic conditions to a totally different situation. The current severe "nutrient erosion" now accelerating in the Corn Belt under the influence of hybrid grains is another example of a faulty socioeconomic farm philosophy and is unrelated to the original prairie vegetation except in so far as the inherited soil richness, which hides the folly of the system, is a result of millenniums of prairie activity.

General Changes in Community Composition

In those cases where man's utilization has not completely destroyed the original biotic community, whether under peripheral conditions or in remnants within agricultural areas, it is possible to detect a recurrent pattern in the compositional changes that have occurred. In both forest and grassland the more conservative elements of the vegetation (the "upper middle class" and the "aristocrats" of Fernald, 1938) have tended to disappear. These are the plants that are most demanding in their requirements, with low tolerance of fluctuations in moisture, with high nutrient requirements, and with low ability to withstand frequent disturbance. They commonly have only limited powers of vegetative reproduction and usually have specialized requirements for germination. They make up the most advanced communities of a given region from the standpoints of degree of integration, stability, complexity, and efficiency of energy utilization (Sears, 1949). They are "climax" plants in the basic sense of the word.

Under the impact of man these climax plants tend to decline in numbers and importance. Their retrogression leads to decreased stability and to disorganization of the community pattern. The environmental changes accompanying the decline are in the direction of more xeric, lighter, and more variable conditions. These encourage the expansion of less conservative plants with such pioneer tendencies as the ability to withstand greater fluctuations in temperature and available moisture, the capacity for resisting disturbance through production of proliferating shoots or adventitious buds, and the possession of efficient means of rapid population increase. Particular harvesting techniques of man, either directly through logging and coppicing or indirectly through the medium of grazing animals, tend to exert a selective influence on the pioneer plants which do succeed. A premium is placed on those species which can resist the particular pressure and still maintain their populations. All others tend to decline or disappear.

This reduction in species complement increases in proportion to the intensity and duration of the utilization. In the final stages the communities completely dominated by man are composed of a small number of extremely vigorous, highly specialized weeds of cosmopolitan distribution, whose origin and distribution are in themselves man-induced phenomena. The subfinal stages are a mixture of these weeds and the most aggressive elements of the native flora. The relative proportion of indigenous and exotic elements varies with the climate, those regions most like the ancient centers of agricultural development (semidesert or Mediterranean climates) having a vegetation which is more completely exotic than that of the cool humid regions.

The highest vegetational product of evolution is the tropical rain forest. In the mid-latitudes the climax deciduous forest as found in the southern Appalachians and in the mountains of China is the ultimate in complexity, stability, and integration. Large numbers of species grow in intimate interrelationship, with maximum capture and reutilization of incident energy consistent with the seasonal nature of the climate. Many niches exist, and each has its adapted species with the necessary modifications in nutritional, growth, or photosynthetic habit to enable it to make the most of its specialized opportunities. Not only is energy capture at a maximum in this highly organized community but normal processes of peneplanation are reduced to a minimum. Indeed, there may be a decrease in randomness of the local habitat due to intake of highly dilute mineral elements from the subsoil or bedrock by tree roots and their subsequent accumulation in the humus-rich topsoil. In the sense that entropy means randomness or "mixed-up-ness" in the universe as a result of highly probable events, the climax deciduous forest may be said to possess a very low entropy, since it is an incomprehensibly improbable phenomenon existing in a dynamic steady state.

Man's actions in this community almost entirely result in a decrease in its organization and complexity and an increase in the local entropy of the system. His activity in reducing the number of major communities, climax or otherwise, and in blurring the lines of demarcation between them by increasing the range of many of their components likewise reduces the non-randomness of his surroundings. Man, as judged by his record to date, seems bent on asserting the universal validity of the second law of thermodynamics, on abetting the running-down of his portion of the universe. Perhaps the improbability of the climax biotic community was too great to be sustained, and man is the agent of readjustment. Let us hope his new powers for total entropy increase are not employed before the readjustments can be made.

REFERENCES

BROWN, R. T., and CURTIS, J. T.
1952 "The Upland Conifer-Hardwood Forests of Northern Wisconsin," *Ecological Monographs*, XXII, No. 3, 217–34.

CLEMENTS, F. E.
1936 "Nature and Structure of the Climax," *Journal of Ecology*, XXIV, No. 1, 252–84.

CONARD, H. S.
1951 *The Background of Plant Ecology*. Ames, Iowa: Iowa State College Press. 238 pp.

COTTAM, GRANT
1949 "The Phytosociology of an Oak Woods in Southern Wisconsin," *Ecology*, XXX, No. 3, 171–287.

CURTIS, J. T.
1951 "Hardwood Woodlot Cover and Its Conservation," *Wisconsin Conservation Bulletin*, XVI, No. 1, 11–15.

CURTIS, J. T., and GREENE, H. C.
1949 "A Study of Relic Wisconsin Prairies by the Species-Presence Method," *Ecology*, XXX, No. 1, 83–92.

DYKSTERHUIS, E. J.
1949 "Condition and Management of Range Land Based on Quantitative Ecology," *Journal of Range Management*, II, No. 3, 104–15.

FERNALD, M. L.
1938 "Must All Rare Plants Suffer the Fate of Franklinia?" *Journal of the Franklin Institute*, CCXXVI, No. 3, 383–97.

FORBUSH, E. H.
1927 *Birds of Massachusetts and Other New England States*. 2 vols. Boston: Massachusetts Department of Agriculture.

GARRETSON, M. S.
1938 *The American Bison*. New York: New York Zoölogical Society. 254 pp.

GOODLETT, J. C.
1954 *Vegetation Adjacent to the Border of the Wisconsin Drift in Potter County, Pennsylvania*. (Harvard Forest Bulletin No. 15.) Petersham, Mass.: Harvard Forest. 93 pp.

HUMPHREY, R. R.
1953 "The Desert Grassland, Past and Present," *Journal of Range Management*, VI, No. 3, 159–64.

KITTREDGE, JOSEPH
1948 *Forest Influences*. New York: McGraw-Hill Book Co. 394 pp.

NEEDHAM, JOSEPH
1954 *Science and Civilization in China*, Vol. I. Cambridge: Cambridge University Press. 318 pp.

OLTMANNS, F.
1927 *Das Pflanzenleben des Schwarzwaldes*. Freiburg: Badischen Schwarzwaldverein. 690 pp.

SEARS, P. B.
1942 "Xerothermic Theory," *Botanical Review*, VIII, No. 10, 708–36.
1949 "Integration at the Community Level," *American Scientist*, XXXVII, No. 2, 235–42.

SHIMEK, B.
1925 "The Persistence of the Prairies," *University of Iowa Studies in Natural History*, XI, No. 5, 3–24.

SHRINER, F. A., and COPELAND, F. B.
1904 "Deforestation and Creek Flow about Monroe, Wisconsin," *Botanical Gazette*, XXXVII, No. 2, 139–43.

STEINBRENNER, E. C.
1951 "Effect of Grazing on Floristic

736   *Man's Role in Changing the Face of the Earth*

Composition and Soil Properties of Farm Woodlands in Southern Wisconsin," *Journal of Forestry*, XL, No. 12, 906–10.

TANSLEY, H. G.
1939 *The British Islands and Their Vegetation*. Cambridge: Cambridge University Press. 930 pp.

TURRILL, W. B.
1929 *The Plant-Life of the Balkan Peninsula*. London: Oxford University Press. 487 pp.

WATT, A. S.
1947 "Pattern and Process in the Plant Community," *Journal of Ecology*, XXXV, No. 1, 1–22.

Man's Role
in
Changing the Face of the Earth

Edited by

WILLIAM L. THOMAS, Jr.
Wenner-Gren Foundation

with the collaboration of

CARL O. SAUER
University of California

MARSTON BATES
University of Michigan

LEWIS MUMFORD
University of Pennsylvania

Published for the
WENNER-GREN FOUNDATION FOR ANTHROPOLOGICAL RESEARCH
and the
NATIONAL SCIENCE FOUNDATION

by

THE UNIVERSITY OF CHICAGO PRESS
CHICAGO • ILLINOIS

The INTERNATIONAL SYMPOSIUM ON MAN'S ROLE IN CHANGING THE FACE OF THE EARTH was made possible by funds granted by the Wenner-Gren Foundation for Anthropological Research, Incorporated, a foundation endowed for scientific, educational, and charitable purposes. Publication of this volume has been aided by a grant from the National Science Foundation of the United States. Neither Foundation is the author or publisher of this volume, and is not to be understood as endorsing, by virtue of its grant, any of the statements made, or views expressed, herein.

Library of Congress Catalog Number: 56-586

THE UNIVERSITY OF CHICAGO PRESS, CHICAGO 37
Cambridge University Press, London, N.W. 1, England
The University of Toronto Press, Toronto 5, Canada

Landscape Development, Forest Fires, and Wilderness Management

Fire may provide the long-term stability needed to preserve certain conifer forest ecosystems.

H. E. Wright, Jr.

The major components of the natural landscape are landforms and vegetation. Their status at any moment is a result of a complex of interacting factors operating over various lengths of time. These dynamic systems attain quasi-equilibria lasting tens or hundreds of years but show sequential developments over longer times. Thus, the form of a hillslope represents a short-term equilibrium among processes of weathering, erosion, and vegetational stabilization under conditions of crustal and climatic stability, but in the long term the hillslope is reduced until the external controlling conditions change. Similarly, vegetation may progress in a short time through successional stages to a climax, but in the long term external factors may control the stability.

Recognition of dynamic landscape development stems from interactions between geology and biology. Lyell's (1) synthesis of paleontology and geology in 1830 helped stimulate Darwin (2) to develop the theory of organic evolution in 1859. Evolutionary trends were then sought in other sciences. Davis (3) introduced the erosion cycle in 1889, in which youth is followed by maturity and finally old age, which persists until the cycle is reinitiated by crustal uplift. In plant ecology, Cowles (4) proposed that vegetation stages parallel those of the erosion cycle. Clements (5) likened vegetation development to the life cycle of an individual.

These theories of landscape development have been severely criticized. The criticisms are based largely on quantitative analyses of the mechanisms and at least the short-term rates of geomorphic and ecologic processes. However, the schemes of Lyell, Darwin, Davis, Cowles, and Clements involve lengths of time that are difficult to evaluate. For all evolutionary processes, extrapolation from modern rates to past rates must be based on knowledge of past conditions derived from quantitative paleontology, stratigraphy, and geochronology.

The genetic mechanism of evolution is now largely understood. Rates of genetic change and speciation in organisms such as the fruit fly have been determined, but direct application of such rates to other animals is hardly justified. Particle movement on hillsides and in streams has been studied intensively, so that rates of erosion in badlands and of sediment transport in streams are known, but use of such rates to solve problems of long-term landscape development is hazardous. Short-term early plant successions on newly exposed landscapes, along with data on nutrient supply and soil stabilization, are documented, but vegetational development over subsequent decades and centuries is not easy to predict.

Landscape processes can be studied in a wide variety of situations. Even highly disturbed landscapes can provide some insights. Often experimental plots can be designed in which controlling factors can be manipulated. However, studies of long-term landscape development can be carried out only on natural landscapes, where the controlling factors have been operating for hundreds or thousands of years. Large-scale and long-term ecosystem processes can best be studied in large virgin forests.

Preservation of some natural areas may require only that exploitation be curtailed. Elsewhere, certain natural processes that have been eliminated must be restored. Specifically, in certain conifer forest areas, including most of the national parks and wilderness areas of the western and northern United States and Canada, forest fires have been effectively excluded for at least 50 years. Yet recent research shows that the maintenance of natural conditions in these forests depends on frequently occurring fire. Continued fire suppression will upset the ecosystem balance and lead to unpredictable changes.

In this article I evaluate and reconcile certain concepts of landscape development by clarifying the time scales for which the various concepts are applicable. Particular attention is devoted to problems of vegetational succession and climax in the cutover forest of the Appalachian Mountains and the fire-dependent virgin forest of the Boundary Waters Canoe Areas (BWCA) in Minnesota. Inasmuch as many areas of virgin forest, including the BWCA, are part of the national wilderness system, management must understand the effects of disrupting the natural processes that have brought the forest into being. For the BWCA a new management plan has been adopted (6) that not only ignores crucial ecological principles but also contravenes the spirit of the Wilderness Act (7, 8).

The author is Regents' Professor of Geology, Ecology, and Botany and Director of the Limnological Research Center, University of Minnesota, Minneapolis 55455.

The Erosion Cycle

In the 1880's Davis (3) developed a grand and integrated theory of landform evolution: the sequential erosion of newly exposed land surfaces through stages of youth, maturity, and old age, under conditions of stable crust and stable climate. The end form is a peneplain, which persists until crustal uplift initiates a new cycle (9).

Davis's theory of the erosion cycle was conceived for humid temperate regions, for which it received wide acceptance. He later extended the theory to arid regions (10), but tropical and polar landscapes were not easily fitted into the model.

Penck (11) was the first to offer an integrated theory of landscape development based on premises different from those of Davis. Davis assumes crustal stability during the erosion cycle, but Penck believes that crustal uplift proceeds more or less apace with erosion, and that landforms reflect the relative rates of the two processes.

A cornerstone of Davis's erosion cycle, the concept of the graded stream, was elaborated and effectively defended by Mackin (12), but Leopold and Maddock (13) showed that streams become graded very early in their development, not just at "maturity." Perhaps the most comprehensive criticism of Davis came from Hack (14), who denied the evidence for long intervals of crustal stability and thus for the development of peneplains. From studies in the Appalachian Mountains, Hack concluded that the characteristic "ridge-and-ravine" topography does not result from uplift of an older peneplain or partial peneplain followed by mature dissection, but rather from a dynamic equilibrium, in which the degradation processes of weathering, soil creep, and stream erosion are opposed by rock resistance and crustal uplift. He pointed out that differences in these processes or factors can result in slopes of differing steepness and thus in landscape of differing expression, but he implied that neither the crust nor sea level (which serves as the erosion baselevel for stream systems) is stable long enough to permit the landscape to be lowered to a peneplain.

Arguments for erosion cycles versus noncyclic dynamic equilibrium revolve around the time factor and the rates of geologic processes. The real ages of landforms are difficult to determine because of lack of radiometrically datable material. For the Appalachian Mountains, the chronology of past events and the rates of geomorphic processes are not well enough established to permit us to evaluate whether the landscape ever did or could reach the peneplain stage. The enigmatic flat crest of parts of the Blue Ridge, strikingly discordant to rock structure—a type of landform not specifically discussed by Hack—suggests the end form of a past erosion cycle, but neither it nor the basement surface beneath the coastal plain sediments is extensive enough or well enough mapped to be identified with certainty as an ancient peneplain. The Appalachian Piedmont is also a dissected plateau with discordant structure. Its identification as a peneplain by Davis was defended by Holmes (15).

The debate has reached an impasse. There is as much (or as little) historical evidence of Appalachian crustal stability with occasional uplift as there is of uplift that nearly keeps pace with erosion. Shifts in baselevel resulting from Pleistocene sea level fluctuations did not affect streams far enough inland to control erosional processes in the Appalachian Mountains. The ineffectiveness of Pleistocene baselevel controls indicates that Appalachian peneplain formation was much older than Pleistocene, perhaps as old as Cretaceous, 60 to 100 million years ago. However, the stratigraphic record is inadequate to solve the problem.

Schumm and Lichty (16) attempted to evaluate the time factor, and concluded that the controversy results from analyzing geomorphic processes on different time scales. The process geomorphologist may measure the factors in a dynamic equilibrium, but extrapolation of an erosion rate to determine the age of a landform may not be justified because modern rates may be unnatural, owing to human disturbance of the system. The historical geomorphologist may be able to evaluate such extrapolation over a short term if a datable sedimentary record of the erosional process is available. But long-term evaluations are difficult because such geomorphic factors as climate, uplift, and bedrock resistance change independently.

One can conclude with Carson and Kirby (17) that evidence for long-term crustal stability in certain regions is sufficiently strong to support the concept of the erosion cycle and the development of peneplains, although mechanisms of slope retreat and development of stream profiles may not be as simple as Davis envisioned. Much care is necessary in historical analysis, not only of the structural and hydrologic controls on the development of terraces and other landforms, but also of the time scale and the rates of geomorphic processes.

Vegetational Development

While Davis and his followers used the erosion cycle to interpret landforms worldwide, Cowles (4) and Clements (5) developed the theory that all vegetation progresses through successional stages to a stable climax, which persists until some interruption occurs. Cowles (4, 18) insisted on a precise resemblance between the climaxes of the erosional and vegetational cycles.

Clements, however, deemphasized the relation to landform evolution, arguing that vegetational succession is limited by climate. He acknowledged that local bare areas produced by vigorous erosion are more common and provide more opportunities for early stages of plant succession in youthful than in mature landscapes. But he noted that successional stages proceed to a climax much more rapidly than Davisian erosion cycles reach old age. Also, a single erosion area might transect various climatic belts with different climax vegetation. Clements used the Darwinian model to postulate that the climax formation actually is an organism, which is generated, grows, matures, reproduces, and dies.

Gleason (19) opposed climax theory by introducing the notion that a plant association represents merely an accidental juxtaposition of individuals rather than a naturally interrelated assemblage of species. This controversy continues today. Whittaker (20), in reviewing the entire climax concept literature, rejected two points: the concept of a vegetation assemblage as an organism, and the rigid connection between climax and climate. The notion of a polyclimax (21) or climax pattern is still supported. Local ecologic conditions such as high soil moisture or disturbance permit the development of a quasi-stable plant association that reproduces itself short of true climax. In Clements' view, the climax represents intermediate edaphic conditions toward which all landforms and soils eventually tend.

Virtually all evaluations of the climax

concept are based on studies of modern vegetation. Despite the use of new techniques of measurement and statistical analysis, results are inconclusive because the historical record has not been considered in the proper time framework, and the vegetational relations may have been significantly affected by timber cutting, agriculture, fire suppression, wildlife extermination, introduction of exotic insects, and other unnatural factors. These elements are taken into consideration in the following examination of the vegetation and history of Appalachian forests and fire-dependent Great Lakes forests. Especially considered are the natural forces in long-term stability, and problems in management of virgin forests.

Forests of the Appalachian Mountains

The complexity of Appalachian forest types has long been a challenge as an historical problem. Braun (*22*) found a solution by applying the climax concept to the erosion cycle. The mixed mesophytic forest type (beech, tulip, basswood, sugar maple, sweet buckeye, chestnut, red oak, white oak, and hemlock) thrives on diverse topography, so it is said to have expanded

after the uplift and dissection of a mid-Tertiary peneplain, perhaps 30 million years ago. This and other climax associations are thus thought to have persisted for millions of years. Their distribution and extent at various times were controlled by Appalachian erosion cycles. And, in Braun's view, they were essentially unaffected by Pleistocene climatic changes, except for the southward infiltration of northern forest types along coastal plains freshly exposed by sea level depression, or along youthful glacial outwash in river valleys like the Mississippi.

Deevey (*23*) discussed some untenable aspects of this picture of Appalachian vegetation history. Braun (*24*) countered vigorously, but much of her story collapses in the face of extensive paleobotanical studies of late Pleistocene lake sediments, which indicate that current forest types are only a few thousand years old. During the height of the last glacial period, 22,000 to 13,000 years ago, not only the mountains but also the piedmont and the coastal plain were characterized by a northern conifer forest (*25, 27*). Spruce was a dominant tree as far south as Virginia and North Carolina, and jack pine down to northern Georgia (Fig. 1). Alpine vegetation occurred on some

mountain crests (*28*). Components of the mixed mesophytic forest may have been present in small amounts—pollen analysis is not a sufficiently sensitive technique to establish their absence. They probably survived in favorable localities in this area of diversified topography and geology, especially in the more southerly parts. They were apparently not pushed south en masse to Florida, which featured an open vegetation with oak and herbs and xerophytic shrubs at this time (*25*). In any case, northern forest probably predominated in the Appalachians, with enclaves of southern types, rather than the reverse.

For Holocene time, significant vegetational changes are also documented. From Virginia south, oak gave way to pine about 6000 years ago (Fig. 2), when modern mesic forests became established (*29, 30*). Some of these changes may be related to climatic shifts.

In the central and northern Appalachians, the postglacial pollen sequence shows slight changes in the proportions of hemlock, beech, hickory, and chestnut, with oak dominating throughout. This was originally interpreted as representing gradual changes of climax forest types caused by a relatively warm,

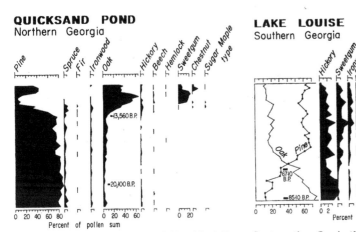

Fig. 1 (left). Selected pollen curves for Quicksand Pond, Bartow County, northern Georgia (*26*). The pine pollen type before 13,000 years before present (B.P.) is determined from pollen size and needle morphology to be jack pine (*Pinus banksiana*), which now does not grow south of New England or the Great Lakes area.   Fig. 2 (right). Selected pollen curves for Holocene sediments of Lake Louise, near Valdosta, southern Georgia (*30*). Before about 6000 years ago, dry forests of oak (*Quercus*) and hickory (*Carya*) dominated, with openings marked by prairie plants, including ragweed (*Ambrosia*), sage (*Artemisia*), and other composites and chenopods. The pine after 6000 years ago is probably longleaf pine (*Pinus palustris*). Various mesic deciduous trees, such as sweet gum (*Liquidambar*) and iron wood (*Ostrya* type), expanded about the same time, and swamp trees such as bay (*Gordonia*) and cypress (*Taxodium*) spread over the lake margin soon after. Beech (*Fagus*) apparently did not reach the area in quantity until later.

Fig. 3. Pollen (46) and charcoal (43) diagram for Lake of the Clouds, northeastern Minnesota, showing substantial uniformity of regional pollen rain since development of the pine forest 9000 years ago. The major change was immigration of *Pinus strobus* (white pine) about 6700 years ago. Increase in pollen of *Picea* (spruce) and *Cupressaceae* (probably northern white cedar) and decrease of *Pinus strobus* about 3000 years ago may reflect a trend to cooler climate. The charcoal curve shows that fires have occurred for at least 9000 years. Time scale derived from annual laminations.

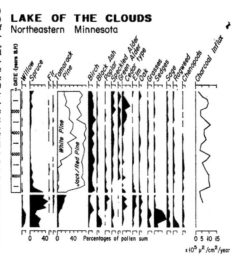

LAKE OF THE CLOUDS
Northeastern Minnesota

dry interval 4500 to 1500 years ago, supposedly indicated by an increase of hickory. However, the climatic requirements of the species are not well enough known to support such conclusions (31). The changes might represent delayed migration of particular tree dominants into already established forest, a phenomenon that Braun would not accept. Knowledge of hemlock's response to fire and human disturbances is particularly lacking. Davis (31) commented that many maps of present forests show not actual species frequencies but rather theoretical climaxes thought appropriate by some ecologists.

The pollen record thus indicates a major climatically controlled vegetation change throughout the Appalachian Mountains at the end of the Pleistocene, about 11,000 years ago, as well as significant changes in the middle of the Holocene. The present forest types in the Appalachians are closer to 10,000 years old than 10 million, indicating that they depend on the climate, which has changed in major ways during the last 15,000 years, rather than on the erosional cycle, which may not have progressed appreciably in 10 million years.

Similar changes undoubtedly occurred during earlier Pleistocene interglacial episodes. Interglacial records in southern Georgia and central Florida reveal a cyclic development similar to the incomplete Holocene cycle (25).

Vegetation changes in the Southeast were fully as pronounced as those much closer to the ice front.

Braun's failure to relate vegetation climax to an Appalachian erosion cycle does not vitiate the climax concept for that region, however. Braun simply misconceived the relative time scales of geomorphic and vegetation processes, and she lacked knowledge of the magnitudes of Pleistocene and Holocene climatic and vegetation changes.

To determine the time scale of Appalachian forest development, we must consider succession and reproduction mechanisms that perpetuate climax patterns. Hack and Goodlett (32) studied an area near the Shenandoah Valley of Virginia, where the vegetation types in the ridge-and-ravine topography form a distinctive pattern. The pattern is of a definite polyclimax. This interpretation is supported by detailed geomorphic work, which shows the sensitive response of certain tree species to moisture, slope, and soil conditions. Ridge crests and noses are characterized by pitch pine or table-mountain pine, ravines by yellow birch, basswood, and sugar maple, and intermediate slopes by the absence of these species and the presence of certain oaks. Their study shows a dynamic equilibrium between geomorphic processes and vegetation. The crucial question now is: How long after the last major disturbance—11,000 years ago— did it take the present forest to reach

climax or equilibrium? Hack (14) implied that the ridge-and-ravine topography has persisted virtually since the original Appalachian uplift many million years ago. Rock weathering and erosion proceed at a finite rate, supplying and renewing the soils that help determine forest type. The system is also always losing a finite amount of material by solution transport in streams and groundwater. Even though erosion rates have increased in many areas because man has disturbed the forest and soil cover, measurements of long-term erosion rates indicate that landscape changes controlled by geomorphic processes proceed extremely slowly compared to vegetation changes controlled by climate. In small solution depressions in the Appalachians, for example, very few meters of sediment have been deposited over tens of thousands of years, even though each drainage basin may be several times larger than its catchment area (26). These landforms have been essentially stable, therefore, while forest vegetation went through drastic transformations controlled by Quaternary climatic change.

The distribution and extent of polyclimax components are thus controlled by geomorphic processes that provide steady nutrient input, but do not change the topographic and hydrogeologic setting appreciably within the time range of climatic stability. Both the physiography and the vegetation maintain a steady state in dynamic equilibrium. How does the climax forest maintain its overall composition? What factors and mechanism determine what tree will replace one that falls?

In the mixed mesophytic Appalachian forest, windstorms are the major disruptions renewing succession. Old and infirm trees blow over, exposing the entire root mat. The opening is rapidly filled by a microsuccession of plants in which young trees take their place among the old. Goodlett's study (33) of the mound microrelief in an Allegheny Plateau forest in Pennsylvania showed that white pine is related to the incidence of windthrow of old hemlock trees. Windthrow opens the canopy, giving light for the germination and early growth of white pine in the seedbed of loose mineral material fallen from the upturned mat of roots. Windthrow is also critical in soil development. Local single-aged stands of white pine are attributed to decimation of large sections of forest by tornadoes or hurricanes that cut across a polyclimax and initiate succession (33).

Cloudbursts cause debris avalanches on slopes and washouts in ravines (*32*).

Fire is another natural disturbance that provides opportunity for regeneration in the Appalachians (*34*). The even-aged stands of white pine mentioned above have been attributed alternatively to regrowth after fire. Many pine stands are confined to sandstone ridge crests and plateaus, where dry soils and wind exposure are favorable for fire.

The mixed mesophytic forest of Appalachian hillslopes and hollows, where the forest floor is customarily moist and winds cannot easily penetrate, probably rarely burn. Many of the hardwood species resprout from roots. Here windstorms are the more significant factors in vegetation disruption. For most of the Southeast, however, the historical record of fire is not adequate for quantitative evaluations of its importance.

Although the role of fire and windthrow in Appalachian vegetation development can be generally inferred, studies of modern forests are hampered because of major disturbances by man. The extent to which features and processes are natural or artificial cannot often be determined. Lumbering and agriculture have been extensive in most of the eastern United States, and few large blocks of undisturbed forest remain. Even second-growth forests have been subject to disturbances such as chestnut blight, decimation of browsing wildlife or of predators that control their populations, and sheep and cattle grazing. Although some New England old fields have had 150 years for reforestation under fairly natural conditions (*35*), the effects of past farming on soils and forest succession are difficult to evaluate. Goodlett (*33*) showed that the frequency of windthrow was greatly reduced after clear-cutting, because no high old trees project above the regenerated forest canopy to catch the wind. The modern forest consequently contains no white pine.

Under man's watch, fires in southeastern pine and oak forests have been less frequent than in the natural regime, so even the second-growth forests after cutting are still denied a major ecological influence. Natural ecological relations are thus difficult to work out in any detail.

In a study of Catskill Mountain forests in New York, McIntosh (*36*) noted that sugar maple has gained dominance over beech and hemlock since the area was settled in about 1880, an expansion he attributed to forest dis-

turbance. Another view is that maple is the natural potential climax (*37*). McIntosh (*36*) cautioned, however, "The assessment of potentiality from primeval remnants is now largely precluded over extensive areas of the earth. . . . As the possibility of reconstructing a primeval climax from undisturbed remnants recedes and vanishes we must depend on analyses of current vegetation as it is, not as what it might become."

Forests of presettlement structure may be restored naturally to some degree, but widespread removal of seed sources through clear-cutting precludes this for hundreds of years. In areas cleared for agriculture, even root systems have been removed, so that sprouters, which usually flourish after natural disturbances, are eliminated. In any case, the situation is unsatisfactory for studies of long-term natural succession. Knowledge lost through the destruction of natural forests can never be regained, yet destruction continues. The areas preserved in the East do not include all major forest types and are too small for major studies.

Fire-Dependent Forests and the Vegetation Climax

Clements' climax concept involves successional stages leading to vegetation that reproduces in equilibrium with prevailing climatic conditions. Succession has been documented in numerous studies. Primary succession is relatively easy to describe: observations can be made over a few years, or a complex of fresh land surfaces of different known ages can be studied as a group, as in the case of abandoned river bars, landslide scars, or moraines of retreating valley glaciers. Secondary succession can be studied on old fields.

The dimensions of these situations are usually small, and seed sources for successional plants may be close-by. Starting times and conditions for succession can be determined easily, if not by observation then by techniques like tree-ring analysis. In some cases the time span can be extended by observations in second-generation forest, for example, by counting rings on downed trees on which younger trees have rooted (*38*). However, only evidence for limited spans of time can be examined.

Succession after fires is complicated. In regions where fires commonly occur, certain plants are adapted so that

their regrowth is immediate and accelerated, whereas other plants are killed and can be reintroduced only years later at the appropriate successional stage, and then only if seed sources are present nearby. Starting times for trees originating after a fire can be determined by tree-ring analysis, and the dates can be confirmed by identifying fire scars in the ring sequences of trees that survived the fire.

Tree-ring analysis and fire-scar studies were developed principally in Itasca State Park in northwestern Minnesota by Spurr (*39*) and Frissell (*40*). A comprehensive investigation has been made in the virgin forests—more than 500,000 acres (200,000 hectares)—of the Boundary Waters Canoe Area (BWCA) of northeastern Minnesota. Here a systematic survey by Heinselman (*41*) showed the fire dates for the origins of all pine stands. Fires are recorded at intervals of 1 to 8 years somewhere in the BWCA for 71 of the last 377 years. Major fires occurred in 1910, 1894, 1863–1864, 1824, 1801, 1759, 1727, 1692, and 1681. On the average, an area equivalent to the total land area was burned over during each 100-year period. This is the natural fire rotation. Reburn intervals for particular localities ranged from 30 to 300 years.

Because this BWCA study combines ecology and paleoecology and has important implications for concepts of both vegetation history and wilderness management, the ecological situation is described in detail (*41–43*).

The topography of the BWCA is diversified, with rocky hills 20 to 100 m above basins containing bogs or stream-connected lakes. The upland vegetation mosaic, however, is controlled by the incidence of fire rather than by topography (*44*).

An even-aged stand of jack pine dates to a fire that killed the previous stand but provided for regeneration, because jack pine cones persist on trees until they are opened by intense heat. Then seeds fall to the ground, where fire has burned off competitors, opened an area for sunlight, and converted nutrients to easily soluble form in ash. Stands of red and white pine also tend to be even-aged, because periodic fires clear the underbrush for seeds that fall from surviving trees. Even-aged stands of aspen and birch record vigorous sprouting after the canopy is opened by fire. Between fires, pine or hardwood stands tend to be succeeded by fir, spruce, and cedar, which are not fire-resistant and would presumably form the climax forest in the absence of fire.

Fire rotation controls the distribution of age classes of stands and the succession within stands. The resulting diversity may represent long-range stability, as implied by the paleoecological record. Suppression of fire may lead to a theoretical climax of spruce and fir, but the succession may be arrested by insect epidemics, resulting in an unpredictable forest composition. The widespread infestation of spruce budworm on balsam fir in the BWCA today may reflect the long interval without fire. The dead trees may fuel a holocaust someday—despite protection. Thus, fire suppression prevents the frequent perturbations under which the forest develops and maintains its diversity. This elimination of the perturbations may be the most profound effect of man on this natural ecosystem (45).

The 370 years covered by tree-ring studies in the BWCA represent only a short time in the total life-span of the forest. During those years explorers, fur traders, loggers, fire guards, and tourists may have modified the natural frequency of fire. To determine the natural fire frequency for prehistoric time, our knowledge must be extended backward by paleoecological studies, which depend on finding fossils in a datable stratigraphic sequence. Pollen from lake sediments records the vegetation history, and charcoal concentrations reveal the fire frequency. Radiocarbon dating is possible for the last 40,000 years. Analyses of other microfossils provide independent evidence of environmental conditions.

Such studies were made of Lake of the Clouds in the heart of the BWCA (Fig. 3). The lake is chemically stratified with organic sediments deposited in annual layers.

Recent fires known from tree-ring counts show stratigraphically as maximums in the charcoal profile in the curve for lamination thickness (possibly representing temporary increases in the inwash of soil from the drainage basin), and in pollen curves for shrubs that sprout after fires (43). The record for the last 1000 years shows major fires every 80 years on the average—about the same frequency as during the 18th and 19th centuries. Charcoal analyses of the entire depth of sediment indicate periodic fires in this forest ecosystem for more than 9000 years.

Pollen profiles generally record regional rather than local vegetation.

Consequently, even when major fires occurred in the area, the pollen rain into individual lakes was not significantly changed. The pollen diagram for Lake of the Clouds (46) shows substantial uniformity since the boreal spruce forest was transformed to pine forest about 9000 years ago (Fig. 3). Slight changes support subdivision of the profiles into pollen zones attributable to climatic change, and independent evidence for this comes from paleolimnologic studies elsewhere in Minnesota. Other long-range changes may result from other external factors, such as progressive soil leaching, or immigration of tree species controlled by such biologic factors as rate of seed dispersal.

The principal result is that the forest has been reasonably stable for 9000 years, despite severe short-term perturbations. A Clementsian climax is never reached, because the evolution of each forest stand is repeatedly interrupted by fire, but in the long range the forest mosaic as a whole maintains an equilibrium.

The long-range equilibrium is also expressed by the aquatic components of the fire-dependent forest ecosystems. Studies of hydrologic and chemical budgets of the Little Sioux fire of 1971 in the BWCA indicate that released nutrients are taken up rapidly by vegetative regrowth and soil storage, so that little reaches the lakes to nourish algal growth (47). Chemical and diatom analyses of lake sediments indicate that past fires have had a negligible effect on the quality of the lake water.

The evidence for the dependence of BWCA forests on fire is convincing because it is based on studies of a very large area of virgin forest. Where large areas of virgin forest remain in western North America, all indications point to a similar dependence (42). The western forests include most of the designated wilderness areas in North America, for which the stated goal is maintenance of natural conditions. Knowledge of successional relations in such virgin forests is essential for intelligent management. Yet where such knowledge exists, as for the BWCA, it is difficult to modify traditional fire-protection policies of management.

Management of Wilderness Areas

Most potential wilderness areas in the eastern United States are forests that were cut more than 100 years ago

and reforested with new composition. We do not know how much they differ from the primeval forests they replaced because the remaining virgin areas are too small for comparison, and the role of large-scale factors like fire cannot be evaluated. Although many second-growth forests serve some of the recreational functions of designated wilderness areas, they are man-made and cannot take the place of primeval wilderness. The wilderness experience diminishes when one encounters a cut stump or a fence in what superficially resembles a virgin forest. Further, scientific investigations of natural regimes cannot be conclusive when data have been affected by significant human disturbance.

Considering the scarcity of true wilderness, designated wilderness areas should be managed to reestablish natural regimes, so that visitors can see primeval wilderness, scientific research can progress, and complete primeval ecosystems can persist, whether for the benefit of man or not.

The 1964 Wilderness Act (7) states that wilderness areas must be managed to maintain their primitive character. Heretofore, management has suppressed fire, under the assumption that fire destroys forests. However, the ecological studies recounted above indicate that fire is a major factor in maintaining the dynamic equilibrium of many natural ecosystems. Fire-suppression policies should therefore be changed. The situation is critical for the BWCA, where evidence for fire adaptation is overwhelming, and where a new management plan (6) has been adopted. The new plan ignores ecological principles and wilderness preservation, for it authorizes timber cutting in more than a quarter of the remaining virgin forest, rejects fire as a forest-maintenance tool, and introduces instead the concept of "administrative cutting" in virgin forest to reduce fuel buildup.

Timber cutting (8) in the largest area of virgin forest east of the Rocky Mountains will reduce an area whose large size is one of its major scientific values. Large size is vital for large-scale ecological studies such as determining the distribution and extent of past fires and studying the movements and territories of mammals like moose, bear, and wolf, and for preserving the genetic variation in natural populations. Cutting also removes attractive virgin forest from the most heavily visited area of the Federal Wilderness System, at a time when use is increasing at about 10 percent per year.

Although the virgin stands of jack pine and black spruce constitute the last major source of long-fibered pulpwood in the Great Lakes region, their long-term value for wilderness recreation and scientific study far exceeds the short-term return from timber sales, especially when the cost of preparation for sale, administration, and replanting is counted. Modern logging practices of clear-cutting, rock-raking, and road-building destroy the natural landscape. Recovery takes hundreds or even thousands of years. Nutrients are removed through increased runoff and soil erosion (48), breakdown of the nitrogen cycle in the soil (49), and removal of biomass that ordinarily decomposes and releases its nutrients for vegetative regrowth (50). Streams are affected by inflow of excess nutrients (49). The stratigraphic records for lakes affected even by nonmechanized logging indicate that abrupt changes in water quality follow cutting (51).

The decision not to use fire to restore the virgin forest is based largely on fear that visitors will be endangered and that fire might spread to Canada or on to private property. It is based on the traditional view that virgin forest is destroyed by fire, and on a fear of adverse public reaction after so many years of Smokey Bear warnings. However, lakes and streams in the BWCA provide excellent firebreaks for limiting prescribed fires, and experience with free-running lightning fires at Sequoia–Kings Canyon National Park indicates that the public will accept a change in management based on sound scientific principles (52).

With rejection of fire as a management tool in fire-dependent virgin forest, the manager must either continue fire suppression, with the resulting fuel buildup and insect infestation until an uncontrollable fire inevitably occurs, or initiate "administrative cutting," which presumably means the felling of dead and diseased trees, removal of underbrush, and perhaps prescribed burning. Although administrative cutting is preferable to fire suppression, it is not a way to preserve the virgin character of the wilderness. The time has come for serious experimentation with natural and prescribed fires under controlled conditions, so that virgin forest can be renewed rather than destroyed, and so that true wilderness areas will be available for future generations of visitors.

Conclusions

Both the landforms and the vegetation of the earth develop to states that are maintained in dynamic equilibrium. Short-term equilibrium of a hillslope or river valley results from intersection between erosional and depositional tendencies, controlled by gravitational force and the efficiency of the transporting medium. Long-term equilibrium of major landforms depends on crustal uplift and the resistance of the rock to weathering. In most parts of the world landscape evolves toward a peneplain, but the reduction rate approaches zero as the cycle progresses, and the counteracting force of crustal uplift intercedes before the end form is reached.

Davis described this theoretical model in elegant terms. Leopold and Hack have provided a new and quantitative understanding of short-range geomorphic interactions that tend to discredit the Davisian model in the eyes of many. However, the substitute models of quasi-equilibrium or dynamic equilibrium merely describe short-range situations in which this or that Davisian stage is maintained despite uplift or downwasting. Given crustal stability and an unchanging climate, landforms would presumably still evolve through Davisian stages. However, the Davis model cannot be tested, for despite tremendous inventions in geochronology and impressive advances in stratigraphic knowledge, we cannot yet establish the rates or even the fact of crustal uplift in most areas. We are left with an unresolvable problem, for the sedimentary records of erosional history are largely inaccessible, undatable, and indecipherable, at least in the detail necessary to describe long-term evolution of the landscape.

We know more about the evolution and maintenance of vegetation assemblages than about landform evolution, for even long-term vegetation sequences are within the scope of radiocarbon dating, and the biostratigraphic record is detailed. Even here, however, distinctions between short-term and long-term situations must be made, so that Clements' grand scheme of vegetational climax—created soon after Davis's model of landform development—can be evaluated in terms of modern knowledge. Disillusion with the climax model paralleled disillusion with Davis's model in the 1950's, but the climax model can be tested, because the record of vegetational history is accessible, datable, and decipherable.

In the short term of a few decades, successional vegetation stages occur in a variety of situations, as confirmed by observation or by techniques such as tree-ring analysis. The successional vegetation stages are reactions to nutrients, weather, competition, and consumption. Such succession implies long-term disequilibrium, or at least unidirectional development.

The long-term controlling factor in Clements' model of vegetation development is climate. With climatic stability the succession will proceed to a climax. In the Appalachian Mountains, geomorphic, microclimatic, and edaphic conditions limit climax development, producing a polyclimax, which is generally sustained by the dominance of these factors. Death and regeneration of single forest trees is controlled mostly by windstorms. The distributional pattern may be locally transected by lightning fires, major windstorms, or washouts. However, the long-term stability of Appalachian forests is demonstrated by pollen stratigraphy.

Although we can infer the long-term stability of Appalachian forests, the trends and mechanics of short-term vegetational succession are not fully understood, because lack of sizable areas of virgin forest limits investigations of natural conditions. In this respect, the eastern United States is already much like western Europe, where climatic and disturbance factors in vegetational history cannot be disentangled.

In the Great Lakes region, a large area of virgin forest exists in the BWCA of northeastern Minnesota. Here short- and long-term studies show that for at least 9000 years the principal stabilizing factor has been the frequent occurrence of fire. Major fires occur so often that the vegetation pattern is a record of fire history. All elements in the forest mosaic are in various stages of postfire succession, with only a few approaching climax. Fire interrupts the successful sequence toward climax. Geomorphic and edaphic factors in vegetational distribution are largely submerged by the fire regime, except for bog and other lowland vegetation. Fire recycles nutrients and renews succession. Nevertheless, despite the fire regime, the resulting long-term equilibrium of the forest mosaic, characterized by severe and irregular fluctuations of individual elements, reflects regional climate.

In the BWCA and the western mountains, large virgin forests can be preserved for study and wilderness recreation. These wilderness areas must be managed to return them to the natural equilibrium which has been disturbed by 50 to 70 years of fire suppression. The goal should be to maintain virgin forests as primeval wilderness. This can be done by management that permits fire and other natural processes to determine the forest mosaic. Mechanized tree-felling and other human disturbances should be kept to an absolute minimum.

Natural landforms also should be preserved for study and for certain nondestructive recreational activities. It is somewhat late for the Colorado River and other rivers of the West, because natural balances are upset by drainage-basin disturbances. Modification of plant cover on hillslopes changes infiltration and erosion rates and thus the stream discharge and sediment load, so the stream balance is altered from primeval conditions. Scenic Rivers legislation should thus be used to restore certain river systems and their drainage basins.

Mountain meadows, badlands, desert plains, and patterned permafrost terrain are extremely fragile and sensitive. Intricate stream and weathering processes leave patterns easily obliterated by mechanized vehicles. Tire tracks can last for decades or centuries. The mineral patina or lichen cover on desert or alpine rocks are records of long stability, and slight differences in their development record the relative ages of landforms, to the year in the case of lichens. Delicate color differences in a talus slope or desert fan show long-term effects just as does the arboreal vegetation mosaic in another climatic setting.

Preservation of virgin wilderness for study is viewed by some as a selfish goal of scientists, to be achieved at the expense of commercial and recreational development. However, scientific study and nonmechanized recreational uses are compatible in wilderness areas. Furthermore, the public does appreciate intellectual stimulation from natural history, as witnessed by massive support for conservation, the Wilderness Act, and a dozen magazines like *National Geographic*. Finally, no knowledgeable American today is unaware that ecological insights are necessary to preserve the national heritage. Western dust bowls, deforested slopes, gullied fields, silted rivers, strip mine wastelands, and the like might have been avoided had long-term problems been balanced against short-term profits.

Many economic questions cannot be answered intelligently without detailed knowledge of extensive virgin ecosystems. Long-term values are enhanced by those uses of natural resources that are compatible with the preservation of natural ecosystems.

Esthetically, virgin wilderness produced by nature is comparable to an original work of art produced by man. One deserves preservation as much as the other, and a copy of nature has as little value to the scientist or discerning layman as a reproduction of a painting has to an art scholar or an art collector. Nature deserves its own display, not just in tiny refuges but in major landscapes. Man is only one of literally countless species on the earth. Man developed for a million years in a world ecosystem that he is now in danger of destroying for short-term benefits. For his long-term survival and as an expression of his rationality and morality, he should nurture natural ecosystems. Some people believe that human love of nature is self-protective. For many it is the basis of natural religion.

The opposition of many Americans to the Alaska pipeline is a manifestation of almost religious feeling; most never expect to see the Alaskan wilderness, but they are heartened to realize that it exists and is protected. The same can be said of those who contribute to save the redwoods in California. Here cost analysis fails to account for the enormous value people place on nature and on the idea of nature as contrasted to the private gain of a few developers. Americans admire European preservation of works of art. Europeans admire American foresight in setting aside national parks. However, the distribution of protected natural areas in America is uneven and inadequate, and vast areas continue to be developed or badly managed despite widespread new knowledge about long-term human interest in wilderness preservation.

Darwin turned nature study into the study of natural history. He could observe natural features in vast undisturbed areas with no thought that human interference had been a factor in their development. Today such natural landscapes have practically vanished. Those that remain should be preserved as extensively as possible, and managed with scientific knowledge of the natural processes that brought them to being. At the present accelerating rate of exploitation, massive disturbance, and unscientific management, soon no na'ural areas will be left for research

or wilderness recreation. Some say that scientific curiosity and the ability for recreation define man. This is reason enough for wilderness preservation. However, a more ominous conclusion is that the survival of man may depend on what can be learned from the study of extensive natural ecosystems.

References and Notes

1. C. Lyell, *Principles of Geology* (Murray, London, 1930).
2. C. Darwin, *On the Origin of Species* (Murray, London, 1859).
3. W. M. Davis, *Natl. Geogr. Mag.* **1**, 81 (1889); reprinted in D. W. Johnson, Ed., *Geographical Essays* (Dover, New York, 1954), pp. 485–513.
4. H. C. Cowles, *Bot. Gaz.* **31**, 73 (1901).
5. F. E. Clements, *Research Methods in Ecology* (University Publishing, Lincoln, Neb., 1905).
6. *Boundary Waters Canoe Area Land Use Plan* (U.S. Forest Service, Superior National Forest, Duluth, Minn. 1974).
7. Wilderness Act (PL 88-577, 3 September 1964).
8. The Wilderness Act, in a special section devoted to the BWCA, states that management of the BWCA shall have "the general purpose of maintaining, without unnecessary restrictions on other uses, including that of timber, the primitive character of the area." The Secretary of Agriculture thereupon authorized that the BWCA be subdivided into two zones, and that timber cutting be permitted in the Portal Zone. This was challenged in a lawsuit [Minnesota Public Interest Research Group v. Butz *et al.*, Suppl. No. 4-72 Civil 598 (District Court, Minnesota, 1973)] seeking a temporary injunction against cutting under the terms of the National Environmental Protection Act. In complying, the court stated: "The language used makes it clear that the Secretary of Agriculture is to enunciate and enforce any and all restrictions which are necessary to maintain the primitive character of the BWCA. It is only if a restriction is not necessary to fulfill this purpose that it can be challenged as 'unnecessary.' Where there is a conflict between maintaining the primitive character of the BWCA and allowing logging or other uses, the former must be supreme." Nonetheless, the new BWCA management plan retains the policy of logging in the Portal Zone.
9. W. M. Davis, *Am. Geol.* **23**, 207 (1899); reprinted in D. W. Johnson, Ed., *Geographical Essays* (Dover, New York, 1954), pp. 350–380.
10. ———, *J. Geol.* **13**, 381 (1905); reprinted in D. W. Johnson, Ed., *Geographical Essays* (Dover, New York, 1954), pp. 296–322.
11. W. Penck, *Morphological Analysis of Landforms*, H. Czech and K. C. Boswell, Transl. (St. Martin's, New York, 1953).
12. J. H. Mackin, *Geol. Soc. Am. Bull.* **59**, 463 (1948).
13. L. B. Leopold and T. Maddock, Jr., *U.S. Geol. Surv. Prof. Pap. No. 252* (1953).
14. J. T. Hack, *Am. J. Sci.* **258A**, 80 (1960).
15. C. D. Holmes, *ibid.* **262**, 436 (1964).
16. S. A. Schumm and R. W. Lichty, *ibid.* **263**, 110 (1965).
17. M. A. Carson and M. J. Kirby, *Hillslope Form and Process* (Cambridge Univ. Press, Cambridge, Mass., 1972).
18. H. C. Cowles, *Bot. Gaz.* **51**, 161 (1911).
19. H. A. Gleason, *Bull. Torrey Bot. Club* **53**, 7 (1926).
20. R. H. Whittaker, *Ecol. Monogr.* **23**, 41 (1953).
21. A. G. Tansley, *Ecology* **16**, 284 (1935).
22. E. L. Braun, *Deciduous Forests of Eastern North America* (Blakiston, Philadelphia, 1950).
23. E. S. Deevey, *Geol. Soc. Am. Bull.* **60**, 1315 (1949).
24. E. L. Braun, *Ohio J. Sci.* **51**, 1939 (1951).
25. W. A. Watts, in *Proceedings of the International Geobotany Conference* (Univ. of Tennessee Press, Knoxville, in press).
26. ———, *Ecology* **51**, 17 (1970).
27. ———, *Quat. Res.* **3**, 257 (1973); D. R. Whitehead, *Ecol. Monogr.* **42**, 301 (1972).
28. J. A. Maxwell and M. B. Davis, *Quat. Res.* **2**, 506 (1972); W. A. Watts, *Geol. Soc. Am. Bull.*, in press.
29. A. J. Craig, *Geol. Soc. Am. Spec. Pap. No.*

123 (1969), p. 283; W. A. Watts, *Geol. Soc. Am. Bull.* **80**, 631 (1969).
30. W. A. Watts, *Ecology* **52**, 676 (1971).
31. M. B. Davis, in *The Quaternary of the United States*, H. E. Wright, Jr., and D. G. Frey, Eds. (Princeton Univ. Press, Princeton, N.J., 1965), pp. 377–402.
32. J. T. Hack and J. C. Goodlett, *U.S. Geol. Surv. Prof. Pap. No. 347* (1960).
33. J. C. Goodlett, *Harv. For. Bull. No. 25* (1954).
34. K. H. Garren, *Bot. Rev.* **9**, 617 (1943).
35. S. H. Spurr, *Ecol. Monogr.* **26**, 245 (1956).
36. R. P. McIntosh, *ibid.* **42**, 143 (1972).
37. A. L. Langford and M. F. Buell, *Adv. Ecol. Res.* **6**, 84 (1969).
38. R. S. Sigafoos and E. L. Hendricks, *U.S. Geol. Surv. Prof. Pap. No. 387A* (1961); D. B. Lawrence, *Geogr. Rev.* **40**, 191 (1950).
39. S. H. Spurr, *Ecology* **35**, 21 (1954).
40. S. S. Frissell, Jr., *Quat. Res.* **3**, 397 (1973).
41. M. L. Heinselman, *ibid.*, p. 329.

42. H. E. Wright, Jr., and M. L. Heinselman, *ibid.*, p. 319.
43. A. M. Swain, *ibid.*, p. 383.
44. L. F. Ohmann and R. R. Ream, *U.S. For. Serv. Res. Pap. NC-63* (1971).
45. O. F. Loucks, *Am. Zool.* **10**, 17 (1970).
46. A. J. Craig, *Ecology* **53**, 46 (1972).
47. R. F. Wright, *Univ. Minn. Limnol. Res. Cent. Interim Rep. No. 10* (1974); J. G. McColl and D. F. Grigal, *Agron. Abstr. Am. Soc. Agron.* (1973); J. P. Bradbury, J. C. B. Waddington, S. J. Tarapchak, R. F. Wright, in *Proceedings of the 19th Congress of the International Association of Limnology* (Winnipeg, 1974).
48. L. B. Leopold, paper presented at the New Zealand Symposium on Experimental and Representative Research Basins, Wellington (1970); R. R. Curry, *Assoc. Southeast. Biol. Bull.* **18**, 117 (1971).
49. G. E. Likens *et al.*, *Ecol. Monogr.* **40**, 23 (1970).

50. G. F. Weetman and B. Webber, *Can. J. For. Res.* **2**, 351 (1972); D. W. Cole, S. P. Gessel, S. F. Dice, in *Proceedings of Symposium on Primary Productivity and Mineral Cycling in Natural Ecosystems*, H. E. Young, Ed. (Univ. of Maine Press, Orono, 1968), pp. 197–232.
51. D. M. Stark, thesis, University of Minnesota (1971); J. P. Bradbury and J. C. B. Waddington, in *Quaternary Plant Ecology*, H. J. B. Birks and R. G. West, Eds. (Cambridge Univ. Press, Cambridge, 1973), pp. 289–308.
52. B. Kilgore, *Quat. Res.* **3**, 496 (1973).
53. Discussions leading to this article started with M. L. Heinselman and E. J. Cushing around a winter campfire in the BWCA, and I thank them and numerous students for continued incentive to evaluate the time factor in landscape evolution and to demonstrate a scientific rationale for wilderness preservation. I also appreciate the editorial and the substantive critique of R. A. Watson.

Reprinted from
Proc. Nat. Acad. Sci. USA
Vol. 71, No. 7, pp. 2744–2747, July 1974

Disturbance, Patch Formation, and Community Structure

(spatial heterogeneity/intertidal zone)

SIMON A. LEVIN* AND R. T. PAINE†

* Cornell University, Ithaca, New York 14850; and † University of Washington, Seattle, Wash. 98195

Communicated by W. T. Edmondson, April 23, 1974

ABSTRACT   A model is developed to relate community structure to level of environmental disturbance in systems in which the effects of disturbance are localized in space and time. In general these disturbances create a pattern of spatio-temporal heterogeneity by renewing a limiting resource, thereby permitting utilization by species that are not dominant competitors. The proposed model predicts the frequency distribution of these renewed areas, with regard to size and age (colonization stage). The model thus allows one to relate overall system pattern to the local biology within these areas, to compare various areas with different levels of disturbance, and to predict the effects of new disturbance.

Spatial patterns within a natural community, generated by a variety of extrinsic and intrinsic factors, clearly influence apparent and emergent aspects of that assemblage. In general, however, the role of spatial heterogeneity has been ignored in most theoretical developments of population dynamics. These, spiritually tied to small-scale, closed laboratory systems, ignore the critical role of direct and indirect interactions with similar systems, and place overwhelming emphasis upon the equilibrium constitution of the closed systems (1).

An alternative approach is to view the community as a spatial and temporal mosaic of such small-scale systems, recognizing that the individual component islands or "patches" cannot be viewed as closed. Rather, they are part of an integrated "patchwork," with individual patches constantly exchanging materials directly, or indirectly through a bath. Disturbance, often in the form of extinctions due to natural catastrophe, competition, or predation-related agents, interrupts the local march to and survival of equilibrium (local climax), and the overall system patterning must be understood in terms of a balance reached between extinctions and the immigration and recolonization abilities of the various species (1–7). Disturbance operates in two ways to increase environmental heterogeneity: by providing the opportunity for local differentiation through random colonization and a kind of founder effect ensuring persistence, and by constantly interrupting the natural successional sequences (1). Such short circuits may prevent local patches from ever achieving equilibrium. The existence of such processes argues for a shift of viewpoint from the properties of the individual patch to the macroscopic statistical properties of the entire ensemble. At that level coexistence is made possible in initially homogeneous systems through the workings of local unpredictability and the creation of new opportunities for invaders. This local unpredictability is globally the most predictable aspect of the system, and may be the single most important factor in accounting for the survival of many species.

The problem of the definition of a patch depends upon the particular system under consideration. In general we shall take a patch to mean a "hole," a bounded, connected discontinuity in an homogeneous reference background which may consist of either simple or multiple components. No restrictions are imposed with regard to its size, which may be initially arbitrary and may grow or shrink; its period of persistence; its invasibility or species composition; or its geographical location which, indeed, may vary with time. In our usage, the homogeneous background mode will usually be a monoculture composed of the competitive dominant (8, 9) and associated species; but the reverse situation also occurs, with patches representing clumps of individuals against a bare background. A special case and familiar example is an oceanic island (4), fixed in size and location, which maintains its essential integrity while its constellation of species varies through time.

It is our thesis that modeling the structure of such systems must be focused initially upon those processes underlying the structure, that is, on the development of spatial heterogeneity. Our model recognizes as a first principle the uncertain or stochastic nature of local patch biology, and treats the patch as the fundamental unit of community structure. When population variation within individual patches is coupled to events generating patches, a bridge is built between population and community theory. The model allows for the consideration of divergent recolonization sequences triggered by random founder effects, and of successional transients. It is an input–output model, permitting comparisons of various geographic areas with different levels of disturbance, and direct testing of the assertion concerning the role of disturbance in accounting ultimately for patterns of diversity. The requisite variables are in most cases easily and directly measurable. Finally, although designed to deal in particular with species patterns in the marine rocky intertidal zone, the underlying model (although some specifics may change) and approach seems to hold great potential for other systems with similar characteristics: for forests faced with localized fires and fellings, for savannahs grazed by elephants, and in short for any system where space is limiting and where disturbances are localized.

Development of the model

The main aspects of the system under consideration are: (i) the distributional properties of patches, especially with regard to age and size and (ii) the biological properties, for instance, the species composition of individual patches. Any model that could predict both of these would thereby predict the essential properties of interest of the system. We attempt in this paper to relate i to the level of disturbance in the system, making the implicit assumption that i can be uncoupled from ii. If this

Proc. Nat. Acad. Sci. USA 71 (1974)

assumption is justified, then the next step would be a stochastic model for intrapatch dynamics, recognizing the unpredictabilities of recolonization episodes and environmental fluctuations. The answer to *i* we determine by consideration of the entire mosaic of patches, for which the limiting age distribution (assuming one exists) or age structure pattern (through time) is calculated. One is thus able to weight the transient stages of recolonization according to the proportions of patches in the various stages of succession, and this permits inclusion of the dynamic processes underlying *ii* rather than simply the static equilibria. Extensions of the model would allow a more general coupling of *i* and *ii*.

The procedure followed here is to consider the "population" of patches as one would a population of cells or individuals, each identified at least according to its age *a* and size ξ. As such, the model utilized here is one introduced by von Forester (10) to consider cell populations and utilized by Sinko and Streifer (11) to reconsider Frank's results (12) on the age–size structure of Daphnia populations. There is an excellent discussion of the equation in (13). We ignore the possibility of patches growing until they overlap or confront one another, a consideration of negligible importance in most of the systems of interest.

Let $n(t,a,\xi)$ = the density function for patches of age *a* and size ξ at time *t*. That is, for Δa and $\Delta \xi$ small, the number of patches that fall in the age interval *a* to $a + \Delta a$ and with size between ξ and $\xi + \Delta \xi$ is approximately given by $n(t,a,\xi)\Delta a\Delta\xi$. *n* is assumed continuously differentiable.

Let $g(t,a,\xi)$ = mean rate of growth of a patch of age *a* and size ξ at time *t*. *g* is assumed continuously differentiable, and is the average value of $(d\xi/dt)$. *g* will in many instances be negative, for example, when patches appear as full size holes in the dominant mode and eventually shrink in size. Let $\mu = \mu(t,a,\xi)$ = rate of extinction of patches of age *a*, size ξ, at time *t*. This is not to be confused with the rate of formation of patches, although the two may be related. For example, new patches may form either from disturbances that carve holes in the background or from the elimination of existing patches (through the formation of new patches engulfing them). μ (assumed continuous) measures only the latter. Patches may disappear either by such instantaneous elimination or by the more gradual process of shrinkage.

The governing equation for $n(t,a,\xi)$ then becomes (10, 11)

$$\frac{\partial n}{\partial t} + \frac{\partial n}{\partial a} + \frac{\partial}{\partial \xi}(gn) = -\mu n \qquad [1]$$

This equation has a unique solution once one specifies the initial and boundary values, $n(0,a,\xi)$ and $n(t,0,\xi)$. Specifically, $n(0,a,\xi)$ is the initial distribution and $n(t,0,\xi)$ the age-size specific birth rate of patches. Since we are interested in extrinsically generated patches, $n(t,0,\xi)$ is regarded as a given input frunction,

$$n(t,0,\xi) = b(t,\xi). \qquad [2]$$

This differs from the conventional implementation of Eq. 1 in which $n(t,0,\xi)$ is given as a feedback, dependent on the full distribution $n(t,a,\xi)$ at time *t*. Feedback is necessary if patches are not extrinsically generated but are caused by invasions from within the system.

The initial distribution is given by

$$n(0,a,\xi) = n_0(a,\xi). \qquad [3]$$

It is worth mentioning that age is just one "physiological" variable which can be incorporated into Eq. 1. Others, for example species composition, could be added, generating additional terms; and this may be necessary in dealing with the role of species whose presence seriously affects, for example, *g*.

Solution of the model

By the method of characteristics, the problem Eqs. 1–3 may be reduced to consideration of the initial value problem for $\overset{*}{\xi}(\overset{*}{a},t,\xi)$, the average size at age $\overset{*}{a}$ of a patch which at time *t* is age *a*, size ξ:

$$\frac{d\overset{*}{\xi}}{d\overset{*}{a}} = g(\overset{*}{a} + t - a,\overset{*}{a},\overset{*}{\xi}) \qquad [4]$$

$$\overset{*}{\xi}(a;t,a,\xi) = \xi$$

This always has a unique solution; and in many cases, the solution is easily obtained.

Assuming $\overset{*}{\xi}$ can be found, define

$$\phi(\overset{*}{t},\overset{*}{a},\overset{*}{\xi}) = \mu(\overset{*}{t},\overset{*}{a},\overset{*}{\xi}) + \frac{\partial}{\partial\xi} g(\overset{*}{t},\overset{*}{a},\overset{*}{\xi}),$$

for any values $\overset{*}{t},\ \overset{*}{a},\ \overset{*}{\xi}$. Then it is not difficult to show that the complete solution to Eqs. 1–3 is given by

$$n(t,a,\xi) = b[t - a,\overset{*}{\xi}(0;t,a,\xi)]$$
$$\times \exp\left[-\int_0^a \phi(\overset{*}{a} + t - a,\overset{*}{a},\overset{*}{\xi})d\overset{*}{a}\right], \quad [5]$$

provided $t \geq a$. Note that when $a = 0$, $\overset{*}{\xi}(0;t,a,\xi) = \xi$ (by Eq. 4); and so $n(t,0,\xi) = b(t,\xi)$, satisfying Eq. 2. When $t < a$, the solution can also be found using Eq. 3; but when the focus is on the asymptotic distribution, this part of the solution is not of interest. Further, since Eq. 5 does not depend on Eq. 3, it is not necessary to obtain the data $n(0,a,\xi)$.

To illustrate the workings of the model, we discuss three illustrative examples representing interesting special cases.

Example 1: No new patches being formed. This corresponds to the experimental situation where the source of patch formation is removed entirely (see *Applications* section). In this case Eq. 5 yields immediately that for $t \geq a$, $n(t,a,\xi) = 0$. In the absence of new patch formation, there are no patches younger than the elapsed time *t*. This fact is obvious without the mathematics; but it is reassuring that the model is consistent with it.

Example 2: Patches fixed in size. (Although not all of the same size). This corresponds to the usual situation in island biogeography, and our model then predicts the colonization age-structure of the population of islands. The case applies similarly to most agro-ecosystems. The Eq. 1 becomes the continuous version of the Leslie model (14). In fact, since $g = 0$, $\overset{*}{\xi} = \xi$; and so

$$n(t,a,\xi) = b(t - a,\xi) \exp\left[-\int_0^a \mu(\overset{*}{a} + t - a,\overset{*}{a},\xi)d\overset{*}{a}\right]$$

This means that since size of an individual patch is invariant, the frequency distribution with regard to ξ of patches of age *a* is simply the birth distribution *a* units earlier diminished by the accumulated deaths.

2746 Zoology: Levin and Paine

Proc. Nat. Acad. Sci. USA 71 (1974)

Example 3: g is separable $[g = g_1(a)g_2(\xi)]$. That is, the effects of age and size upon growth interact multiplicatively, and the growth rate does not depend directly on time. One important specialization of this case is when patch growth depends only on ξ, for which g_1 may be taken $\equiv 1$.

For the general case, assuming $g_2 \neq 0$

$$n(t,a,\xi) = b(t - a, \overset{*}{\xi_0}) \frac{g_2(\xi_0)}{g_2(\xi)}$$
$$\times \exp\left[-\int_0^a \mu(\hat{a} + t - a, \hat{a}, \overset{*}{\xi_0})d\hat{a}\right], \quad [6]$$

where $\overset{*}{\xi_0} = \overset{*}{\xi}(0;t,a,\xi)$ is the mean size at birth of patches of size ξ, age a at time t. Moreover, the differential equation in 4 is separable; and hence $\overset{*}{\xi_0}$ is given implicitly by the relation

$$\int_\xi^{\overset{*}{\xi_0}} \frac{d\hat{\xi}}{g_2(\hat{\xi})} = \int_a^0 g_1(\hat{a})d\hat{a}.$$

Note that $\overset{*}{\xi_0}$ is independent of t.

When the form of $g_2(\xi)$ is specified, $\overset{*}{\xi_0}$ is not usually difficult to compute. For example, if g_2 is constant,

$$\overset{*}{\xi_0} = \xi + g_2 \int_a^0 g_1(\hat{a})d\hat{a}.$$

If g_2 is proportional to ξ, say $g_2 = K\xi$, then

$$\overset{*}{\xi_0} = \xi \exp\left[K \int_a^0 g_1(a)da\right].$$

These two cases, representing arithmetic (linear) and geometric (areal) shrinking of patches, are of special interest because they represent extremes in which patch closure is strictly one- or two-dimensional.

Under some circumstances, when patches cannot be aged accurately, $n(t,a,\xi)$ cannot be measured directly. Rather, one can calculate only the distribution at time t with regard to size ξ, given by

$$N(t,\xi) = \int_0^\infty n(t,a,\xi)da.$$

For the case $g = g_2(\xi)$, assuming t is sufficiently large that $n(t,a,\xi) \cong 0$ for $a > t$, $N(t,\xi)$ can be computed, using Eq. 6. For the special case when birth rates are time-independent ($b = b_0(\xi)$) and μ is negligible ($\mu = 0$), the result is (with $g_2 < 0$)

$$N(\xi) = N(t,\xi) = \int_0^\infty b_0(\overset{*}{\xi_0})g_2(\overset{*}{\xi_0}) \frac{da}{g_2(\xi)}$$
$$= \frac{1}{|g_2(\xi)|} \int_\xi^\infty b_0(z)dz; \quad [7]$$

This provides a direct relationship between the level of disturbance, given by the function b_0, and the patch distribution with respect to size, $N(t,\xi)$. Note that in actuality, N is independent of t, and the result is a steady-state distribution achieved for t large. If $b_0(\xi)$ has the simple form of exponential decay, $b_0(\xi) = c_1 \exp(-c_2\xi)$, then

$$N(\xi) = \frac{1}{|g_2(\xi)|} \frac{c_1}{c_2} \exp(-c_2\xi). \quad [8]$$

Application

The inhabitants of many communities are clearly space-limited; and in the absence of external disturbance, such communities may tend toward a homogeneous association of the competitively dominant species and its associated fauna and flora [for instance, communities on intertidal rock platforms (8, 9), coral reefs (15), and grasslands (16)]. Under certain conditions, alternative associations may develop in a pattern reflecting historical accident.

However, local disruptions do occur, thereby making the limiting requisite available to a pool of potential invaders. In certain marine situations, patches are generated within stands of the competitively dominant mussels by the shearing force of waves (17), wave-driven logs (18), or perhaps even spontaneous decay of aged mussels. The size-specific birth distribution of patches, $b(t,\xi)$ is readily measured: at three locations on the outer coast of Washington State, patches as defined varied from <100 cm² to 38 m² with about 80% of the total $\leq$800 cm² ($N = 238$). We have experimentally produced patches of varying size and position, and have begun to measure the patch closure rate, $-g(t,a,\xi)$, as large mussels migrate into the area (9). These input functions will generate an observed age- and size-specific patch distribution, the features of which are clearly visible in the field as discontinuities in the potentially continuous mussel distribution. If the parameters of the process are slowly varying in t, an effective steady-state distribution will be reached. Model output, then, can be verified by comparing predicted to actual patch distribution with respect to size and age (Eq. 5). To a major extent, local species richness will be influenced by environmental heterogeneity, or patch structure.

We propose to test the latter relationship rigorously by sampling along disturbance gradients. Paine (8) removed a major predator (source of disturbance) and since other disturbances did not intrude, produced a monoculture of mussels. When $b = 0$, as would be the case in the absence of disturbance, the density distribution of patches, $n(t,a,\xi)$ tends to 0, and the area becomes monotonous. At the other extreme, under conditions of severe disturbance the community should be composed mainly of ephemerals. Although few data exist, Dayton (18) has clearly established that species richness is low in areas potentially dominated by mussels when the area is pounded regularly by wave-driven logs. The model predicts a reduced variety of patches under condition where the patch birth rate, b, and extinction rate, μ, are both high. (This statement is qualitative, but can be made precise.) In such conditions fugitive or transition populations should predominate in the species list. On the other hand, overall reduction in b will lead to an overall reduction in $n(t,a,\xi)$ and possible elimination of many ephemerals, perhaps at critical threshold levels of b. Maximal variety thus occurs at intermediate levels of disturbance.

Generalizations

Our model relates spatial and spatio-temporal heterogeneity or pattern to causal processes, incorporating variations in their timing and magnitude, and variations in the successional process. It specifically lends itself to interregional comparisons of these processes and their influences. The emphasis on local patterning relates immediately to theoretical explorations of heterogeneity and stability (19). Extensions of the approach would include consideration of specific dispersion patterns, for

Proc. Nat. Acad. Sci. USA 71 (1974)   Disturbance, Patch Formation, Community Structure   2747

instance nearest-neighbor relationships, and would examine the influence of such patterns on the age–size frequency distribution. Within and between habitat comparisons of species richness (20, 21) can be made following specific assumptions on the relationship between pattern (either physical structure or spatial heterogeneity) and the number of species that can coexist and degrees of dominance among them. Since the model relates disturbance to pattern, hypotheses concerning the relation between pattern and diversity translate immediately into relationships between disturbance and diversity.

The application is not confined to temperate zone rocky intertidal shores, although it is our purpose initially to apply the model to that community. Many parallels can be found for terrestrial situations. Poore (22) suggests that the integration of tropical rainforest communities is due to intrinsically or extrinsically caused alteration, combined with opportunism (a stochastic element, as in our system) and competition. Presumably, in situations of high diversity, patch size at birth tends to be small [reflected in the shape of the birth curve $b(t,\xi)$] and the successional axis is long ($|g|$ is small). Equilibrium composition will depend on disturbance frequency and the relative importance of initial opportunism.

Laws (23) describes elephants as agents of landscape patterning in East Africa. Community alteration is localized, and the disturbance may vary as a function of herd density. He suggests that a desired mix of habitat types can be maintained by controlling herd size and activity [which translates principally into $b(t,\xi)$] and natural successional relationships (local biology and the size of g). Taylor (24) demonstrates how forest fires in potentially monotonous stands of lodgepole pine enhance ecological diversity by maintaining an open canopy and sustaining those plants and animals characteristic of successional communities. Although the size distribution with regard to ξ is not explicitly given, presumably both $b(t,\xi)$ and $g(t,a,\xi)$ can be determined and our approach applied. By inverting Eq. 5 for fixed t (sufficiently large), one can then calculate the frequency and size of controlled burnings that would generate the desired species mix (for that t) in a managed forest.

Finally, as a speculation, it should be possible to devise optimal patch patterns for agriculture (25) although application will be strongly dependent on the biological relationships of both the desirable plants and their pests; and to suggest generalized responses by natural communities altered by man's activities in which the pulse rate, magnitude, and dissipation of the influences are controllable.

The development of this paper has benefitted from discussion with Drs. R. H. Whittaker, R. B. Root, E. G. Leigh, and G. Oster. We wish to acknowledge support from The National Science Foundation: GP 33031 (to S.A.L.) and GA25349 (to R.T.P.).

1. Levin, S. A. (1974) *Amer. Natur.* **108**, 207–228.
2. Skellam, J. G. (1951) *Biometrika* **38**, 196–218.
3. Huffaker, C. B. (1958) *Hilgardia* **27**, 343–383.
4. MacArthur, R. H. & Wilson, E. O. (1967) *The Theory of Island Biogeography* (Princeton University Press, Princeton, N.J.)
5. Levins, R. & Culver, D. (1971) *Proc. Nat. Acad. Sci. USA* **68**, 1246–1248.
6. Horn, H. S. & MacArthur, R. H. (1972) *Ecology* **53**, 749–752.
7. Slatkin, M. (1974) *Ecology* **55**, 128–134.
8. Paine, R. T. (1966) *Amer. Natur.* **100**, 65–75.
9. Paine, R. T. (1974) *Oecologia*, in press.
10. von Foerster, H. (1959) *The Kinetics of Cellular Proliferation*, ed. Stohlman, F., Jr. (Grune and Stratton, New York).
11. Sinko, J. & Streifer, W. (1967) *Ecology* **48**, 910–918.
12. Frank, P. W. (1960) *Amer. Natur.* **94**, 357–372.
13. Oster, G. & Takahashi, Y. (1974) *Ecology*, in press.
14. Leslie, P. H. (1945) *Biometrika* **33**, 182–212.
15. Porter, J. W. (1972) *Amer. Natur.* **106**, 487–492.
16. Harper, J. L. (1969) *Brookhaven Symp. Biol.* **22**, 48–61.
17. Harger, J. R. (1972) *Veliger* **14**, 387–410.
18. Dayton, P. K. (1971) *Ecol. Monogr.* **41**, 351–389.
19. Smith, F. E. (1972) *Trans. Conn. Acad. Arts Sci.* **44**, 309–335.
20. MacArthur, R. H. (1965) *Biol. Rev.* **40**, 510–533.
21. Whittaker, R. H. (1970) *Communities and Ecosystems* (The MacMillan Co., New York).
22. Poore, M. E. O. (1964) *J. Anim. Ecol.* **33** (supl.), 213–226.
23. Laws, R. M. (1970) *Oikos* **21**, 1–15.
24. Taylor, D. L. (1973) *Ecology*, **54**, 1394–1396.
25. Root, R. B. (1973) *Ecol. Monogr.* **43**, 95–124.

Some Demographic and Genetic Consequences of Environmental Heterogeneity for Biological Control[1]

By Richard Levins

Committee on Mathematical Biology and Biology Department
University of Chicago, Chicago, Illinois 60637

Economically important pests usually attack a crop or group of crops over a wide region in which there are geographic, local, and temporal variations in the environment. Effectiveness of any control program will therefore depend on the different responses of the crop, pest, and control organism to this pattern of environment. Usually the environmental heterogeneity is treated as an unavoidable complication in program evaluation, and attempts are made to work with "average" conditions.

The objectives of the control program do not depend only on the biology of the pest. We may want to achieve complete extirpation over part of the region, minimize average pest population over the whole region, hold total crop damage down to some acceptable level, or combine the costs of crop damage and control procedures to maximize some economic index. Therefore it is not possible to deduce an optimal strategy a priori. Rather, the purpose of this report is to show that the pattern of environmental variation in space and time can be utilized in the control of pests and to indicate the information which is needed for the selection of the most promising predator.

MIGRATION AND EXTINCTION

Since the area over which control is sought is much greater than that of the local population, the control strategy must be defined for a population or populations in which local extinctions are balanced by remigration from other populations. This situation can be described as follows:

Let N be the number of local populations at a given time; let T be the total number of sites that can support local populations; M is the migration rate (the probability that migrants from any given population reach another site) and E is the extinction probability for a local population.

Then new populations are being established at a rate which depends on the migration rate times the probability that the site reached is vacant, or

$$mN\left(1 - \frac{N}{T}\right) \quad \text{and}$$

populations are being eliminated at the rate EN. Thus the change of N with time is given by

$$\frac{dN}{dt} = mN(1 - N/T) - EN \quad (1)$$

N will reach an equilibrium when the right side of the equation is 0. This gives the equilibrium level of

$$\hat{N} = T\left(1 - \frac{E}{m}\right). \quad (2)$$

[1] This paper was presented as part of a symposium "Genetics in Biological Control" at the annual meeting of the Entomological Society of America, Dallas, Tex., Dec. 2, 1968.

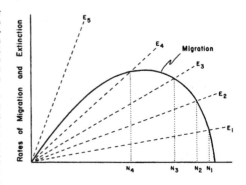

Fig. 1. Equilibrium occurs at the N for which the migration and extinction rates are equal. When N is large, a given change in E (from E_1 to E_2) produces a small change in N. But near the flat part of the migration curve the same change form E_3 to E_4 has a greater effect. If the extinction rate exceeds migration as in E_5, the population disappears.

In Fig. 1 the process is shown graphically. We see that if E is small compared with m, N is near T, and changes in E have relatively little effect. But when E is large compared with m a small change in E is more important, and when E is greater then m the pest cannot persist.

The parameter E depends on uncontrollable environmental factors which vary in time and place, and also on control procedures. If E is varying in time then N does not reach an equilibrium level but fluctuates with some probability distribution. This probability density can be approximated by the method of diffusion equations (see Appendix 1) and to the extent that the assumptions are correct approaches a steady State distribution which is readily found. This distribution can be shown to have its peak value at

$$\hat{N} = T(m - \bar{E} - \sigma_E^2)/m. \quad (3)$$

This reduces to (2) when E is constant. Since N is reduced by σ_E^2 best control is achieved for any given E when the extinction rate is most variable in time. If the factors which affect extinction are varying independently from place to place, local increases and decreases will cancel one another, and the overall variance from time to time will be small. Therefore it is to our advantage that control measures be applied simultaneously throughout the region. A parasitoid which can produce local extinction will be most effective if, for a given average effectiveness, it is least variable in space and most variable in time.

The migration parameter m is also subject to manipulation. It depends on the behavior of the insects, the average density of local populations, and the distance between sites. We do not know how m changes with distance, but for our purposes an exponential decay,

$$m = m_0 \, e^{-aD} \qquad (4)$$

is sufficient to show the qualitative effects. Here D is the distance, a is a constant that depends on the scale of measure, and m_0 is the migration rate into an adjacent plot ($D = 0$). Suppose that a field adjacent to an infested field is reinvaded on the average in 2 days. The $m_0 = \frac{1}{2}$ Suppose further that a clear plot at a given distance D is reinvaded on the average in 10 days, so that m is 0.1.

Hence e^{-aD} is 0.2. Now let us double D. Then

$$m = m_0 e^{-2aD} \qquad (5)$$

or

$$m = m_0 (e^{-aD})^2.$$

This would give

$$m = .5 (.2)^2 = .02$$

and the waiting time to reinfestation is now 50 days. We can increase the distance between available sites in several ways. One would be elimination of alternate hosts for the pest. Or it is possible that the spatial variability of the environment is such that the same predator is unevenly effective in different sites. A predator which uniformly reduces all populations by half only halves m, but one which achieves nearly complete control in half the sites can reduce m severalfold. Finally, a given amount of insecticide in an integrated program might be most effectively distributed unevenly among sites to maximize D.

Note that the recommendations we would make to achieve control by reducing m are the opposite of those for control by increasing E. In the former case we want uneven effectiveness over space, in the latter we wanted uniform control in space and maximum variation in time. Therefore the formal analysis presented here does not lead to an unambiguous optimum strategy. Rather, we have shown that regional control may be attempted by acting on either local extinction or migration rates depending on the particular situation. For this approach to be relevant at all, we must first verify that in fact there is a distance-dependent migration rate, that after local extinction there is a waiting time of at least a week for reinvasion, that we are dealing with discrete populations.

LOCAL FLUCTUATION

Consider now a local population of size x which is growing at a rate

$$\frac{dx}{dt} = rx(1 - x/k) - px \qquad (6)$$

where r is the environmentally determined intrinsic rate of increase, k is the carrying capacity of the environment, and p is the rate of predation by our control organism. Each of these parameters r, k, and p may be variables. Note that equation (6) is formally identical with equation (1) although the terms have somewhat different meanings. Therefore if the only variable is p, x will have a probability distribution with a peak at

$$\hat{x} = k(r - \bar{p} - \sigma_p^2)/r \qquad (7)$$

where p is the average predation rate and σ_p^2 its variance. For $r < \bar{p} + \sigma_p^2$ there is no peak and the population goes extinct.

We can also solve (6) directly as shown in appendix I. The solution is

$$x(t) = x_0 \bigg/ \left\{ \exp\left(-\int_0^t r - p\,dt \right) \right.$$
$$\left. + x_0 \exp\left(-\int_0^t r - p\,dt \right) \int_0^t \exp\left(\int r - p\,dt \right) \frac{r}{k}\,ds \right\}. \quad (8)$$

If the average r is greater than the average p the population persists. The average x is reduced by all factors which increase the denominator of (8). However, only $p(\tau)$ is directly controlled by us through the choice of the control organism. (Integrated control programs could also modify r and k.) Clearly a large average p (high predation rate) reduces the average pest population. In addition a large variance in p reduces the average pest population and may even result in extinction if $\sigma_p^2 > \bar{r} - \bar{p}$. Therefore, it is not desirable to seek as our control organism one which does uniformly well in all the environments it is likely to encounter. Rather for a given average effectiveness we would choose that predator which is most variable in its predation, one which is specialized to a narrower range of environments provided this is compatible with its persistence. A further examination of equation (7) shows that a negative correlation between r(t) and p(t) is beneficial. The predator should not be chosen to do best when the prey is at its best but rather at its worst. Finally, a predator will be most effective if it is maximally adapted to the conditions at the beginning of the season. For this reason, the local population of predator most likely to be the best source for introduction need not be the one living in a similar climate as the region where it will be released. Rather we should look at climates that correspond to conditions at the start of the season in the area we are concerned with.

The treatment described here is too rough to be immediately applicable. For any real pattern of environment we could calculate the terms in equation (7) and evaluate potential control agents. However, at this point there is nothing to be gained by examining hypothetical cases to obtain numbers.

So far we have treated p as an environmental variable independent of x. But the rate of predation depends on the size of the predator population as well as the physical environment. Suppose the predator population expands with the prey, so that p(t) could be replaced by $p_0(t)x$. Then equation 6 becomes

$$\frac{dx}{dt} = rx(1 - x/k) - p_0 x^2. \qquad (9)$$

Now the probability distribution of x has a peak roughly at $\dfrac{rk(1 - 2\sigma_p^2)}{r + \bar{p}k}$ when σ_p^2 is small. Thus in this situation too a variable p results in better control. However, the solution of equation (8) by the procedure in appendix I gives

$$x = x_0 \bigg/ \left\{ \exp\left(-\int r(s)ds \right) + x_0 \exp\left(-\int r(s)ds \right) \right.$$
$$\left. \times \int \exp\left(\int r(s)ds \right) \left[\frac{r}{k} + p \right] dt \right\}. \quad (10)$$

Now this is minimized by p being concentrated in the period when the uncontrolled pest population would be greatest, usually late in the season. The source of the best control population would therefore be different from the previous case. The alternative models used here assume either that the predator abundance is independent of the prey or that it is proportional to the prey. The former case corresponds to situations in which the predator has several hosts and is therefore not greatly changed by prey abundance, or to predators with much slower development or even fewer generations per year than the host so that it does not increase rapidly enough with prey density. The latter model makes the assumption that the predator is responding rapidly to its host population. But both models have the predation coefficient p depend only on physical environment. They ignore predator saturation, search difficulty, and other factors that would make p a function of x. The manner in which the model would have to be modified for particular cases will be dictated by the cases.

The treatment described considered population growth as a continuous process with overlapping generations. In many cases the pest has only one or two discrete generations per year. Then a more exact method would be a discrete model in which population size in generation two is computed from pest and predator populations in generation one. When we deal with specific cases such a representation is likely to be preferable, but at the present stage of investigation the qualitative relations are brought out by the models used.

SELECTION OF PREDATOR GENOTYPES

Once an analysis of the population dynamics and environment enables us to determine the optimum properties of the predator population, the next step is to produce such a population. We have just indicated that the optimum predator may be adapted either to the early or later parts of the season depending on the dynamics of predator-prey relations. Therefore, what is best for us is not necessarily what is best for the predator, and it would be naive to expect that the best predator is the one already found in a similar region. Nor can we throw a lot of populations together, release the mixture, and expect natural selection to mold it into an optimal control organism.

In fact, the response of the population to natural selection may be deleterious even to the predator itself. The reason for this is that when a population is being selected in a varying environment its current gene frequency depends on the previous environment, whereas its survival value depends on the present environment. Therefore fitness is increased only if there is a strong positive correlation between the environments of successive generations. And this is possible only if there are many generations per year. Where there are few generations the correlation will be negative. For example if there are two generations in the course of a summer which begins hot and dry and ends cool and wet, the first generation will be selected for survival in dry heat. Its offspring will be even better adapted to dry heat, but will have to face a wet cool environment. Its offspring, selected for the cool wet conditions, will face the dry heat of early summer. The final result is a loss of fitness if there is genetic variability for a survival in these environments. Natural populations therefore will not normally adapt to such environmental fluctuations by a genetic response to selection. Rather each population will fix those genes which are the best compromise between the conditions at different times and will develop physiological or behavioral adaptations that vitiate the need for genetic change. But the genes which are fixed will be different in different populations, so that a mixture of populations will create a gene pool subject to deleterious fluctuations.

Hence it is undesirable to release a population with too much genetic variance. Rather, we should mix lines for the purpose of selection in the laboratory and then release the selected population.

The deleterious effects of response to selection in a rapidly changing environment are a danger to the control organism but offer a means of reducing pest populations by increasing their genetic variance with respect to traits whose selective value fluctuates. This is the general principle behind Klassen's concept of conditional lethality as a control technique. I have given more quantitative treatment of this problem (Levins 1965).

CONCLUSIONS

The study of population processes in a heterogeneous environment demonstrates the fallacy of several common sense notions about pest control.

1. The best control organism is not necessarily one which is uniformly effective in all the spatial or temporal variants of the environment. On the contrary we have shown for several models that a relatively specialized predator may give the best long-term control.

2. The best predator is not necessarily adapted to the same range of environments as the pest it is to control. Depending on the particular situation we may want a predator adapted to the very early or to the late season's climate, and there may be an advantage to having its adaptive level negatively correlated with that of the pest.

3. The optimal predator is not necessarily one from the most similar climate.

4. We cannot expect natural selection in the field to produce an optimal predator provided we supply enough genetic variation. On the contrary, some kinds of genetic variance will be positively deleterious. Therefore a heterogeneous gene pool should be selected in the desired direction in the laboratory prior to introduction.

Anybody who has worked with any real insect will recognize numerous ways in which each of our model is unrealistic or inapplicable for his animal. We have chosen to use these simplified models to demonstrate the qualitative nature of the processes, indicate possibilities, show what isn't necessarily so, and suggest how the properties of environmental heterogeneity can be considered in deciding on a control strategy. The mathematical analysis itself does not prescribe that strategy. Theoretical investigation is no substitute for entomological knowledge but may be a valuable supplement.

Appendix I

Several mathematical methods may be used to study the processes defined by the equation

$$\frac{dx}{dt} = rx\left(1 - \frac{x}{k}\right) - px \qquad (11)$$

where any of the parameters r, k, or p may vary. A direct solution proceeds by the substitution

$$y = \frac{1}{x} \qquad \frac{dy}{dt} = \frac{1}{x^2}\frac{dx}{dt}.$$

Dividing (1) by $-x^2$ we obtain

$$-\frac{dx}{dt}\frac{1}{x^2} = -(r - p)\frac{r}{x} + \frac{r}{k}. \qquad (12)$$

This becomes

$$\frac{dy}{dt} = -(r - p)y + r/k. \qquad (13)$$

Adding (r-p)y to both sides and multiplying by the integrating factor

$$\exp\left(\int r - p\,dt\right)$$

We obtain

$$\exp\left(\int r - p\,dt\right)\left[\frac{dy}{dt} + \cdot(r-p)y\right]$$

$$= \exp\left(\int r - p\,dt\right)r/k. \quad (14)$$

But the left side is readily seen to be

$$\frac{d}{dt}\left\{\exp\left(\int r - p\,dt\right)y\right\}. \quad \text{Therefore}$$

we can integrate both sides and divide by $\exp\left(\int r - p\,dt\right)$

Then

$$y = y_0 \exp\left(-\int r - p\,dt\right) + \exp\left(-\int r - p\,dt\right)$$

$$\times \int \frac{r}{k}\exp\left(\int r - p\,ds\right)dt. \quad (15)$$

Finally, replacing y by $\frac{1}{x}$ and y_0 by $\frac{1}{x_0}$ we have the solution

$$x = x_0 / \left\{\exp\left(-\int r - p\,dt\right) + x_0 \exp\left(-\int r - p\,dt\right)\right.$$

$$\left. \times \int \frac{r}{k}\exp\left(\int r - p\,ds\right)dt\right\}. \quad (16)$$

In the alternative model,

$$\frac{dx}{dt} = rx(1 - x/k) - p_0 x^2 \quad (17)$$

The same substitution $y = \frac{1}{x}$ and division by x^2 gives

$$\frac{dy}{dt} = -ry + \frac{r}{k} + p_0. \quad (18)$$

Proceeding as before,

$$\frac{d}{dt}\left\{\exp\left(\int r\,dt\right)y\right\} = \exp\left(\int r\,dt\right)\left(\frac{r}{k} + p_0\right) \quad (19)$$

and

$$y = y_0 \exp\left(-\int r\,dt\right) + \left\{\exp\left(-\int r\,dt\right)\int\left[\left(\frac{r}{k} + p_0\right)\right.\right.$$

$$\left.\left. \times \exp\left(\int r\,ds\right)\right]dt\right\} \quad (20)$$

so that

$$x = x_0 / \left\{\exp\left(-\int r\,dt\right) + x_0 \exp\left(-\int r\,dt\right)\right.$$

$$\left. \times \int\left(\frac{r}{k} + p_0\right)\exp\left(\int r\,ds\right)dt\right\}. \quad (21)$$

This method gives an exact solution to the differential equation and is therefore as valid as the model. If the probability distributions of the variables r, k, and p are known we can calculate the probability distibutions of population size. But there is not always an available substitution such as we used. Kimura (1964) has introduced the method of diffusion processes to go directly to the probability density distribution of variables defined by differential equations. He defines M(x) as the mean change in x as a function of x, and V(x) as the variance of the change conditional on x. Then if the random component has not correlation between successive intervals he gives the formal solution for the probability density of x:

$$\phi(x) = \frac{c}{v(x)}\exp\left(2\int\frac{m(x)}{v(x)}dx\right). \quad (22)$$

The peak of ϕ is shown to be the solution of

$$M(x) = \frac{1}{2}\frac{dv}{dx}. \quad (23)$$

This method gives the final steady state distribution after a long time and requires that the random variations are not time-dependent. It is therefore only a crude description of the results of the process but more readily' shows the influence of the random variation of the outcome.

References Cited

Kimura, M. 1964. Diffusion models in population genetics. Methren Review Series in applied probability, vol. 2.

Klassen, W. 1969. The potential for genetic suppression of insect populations by their climatic adaptations. Ann. Entomol. Soc. Amer. (In press.)

Levins, R. 1965. Theory of fitness in a heterogeneous environment, V. Optimum genetic systems. Genetics 52: 891-904.

Ann. Rev. Ecol. Syst. 1976. 7:81–120

POPULATION RESPONSES TO PATCHY ENVIRONMENTS

♦4102

John A. Wiens
Department of Zoology, Oregon State University, Corvallis, Oregon 97331

INTRODUCTION

In the real world, environments are patchy. Factors influencing the proximate physiological or behavioral state or the ultimate fitness of individuals exhibit discontinuities on many scales in time and space. The patterns of these discontinuities produce an environmental patchwork which exerts powerful influences on the distributions of organisms, their interactions, and their adaptations.

Consideration of this environmental patch structure is critical to both the theory and management of populations. Despite the obvious heterogeneity of natural systems, most of the models that form the theoretical fabric of population biology and ecology (and that are increasingly conditioning our perception of reality) tell mathematical stories of populations existing at single points in space or time (142, 261). While some attempts have been made to relax this assumption of environmental constancy at the level of single populations (e.g. 146, 187, 245, 284), population interactions (e.g. 164, 177), or ecosystems (e.g. 72, 73, 295, 310), the task of translating patchiness into a form compatible with computer simulation of system dynamics is indeed formidable (261). In population management disciplines, the importance of environmental patch structure was realized more than a generation ago in such fundamental concepts as edge effect (156) or habitat interspersion (79, 80). More recently, resource managers have noted the influence that the size of habitat blocks (289, 319) or the presence of isolated patch types such as dead snags in forests or bodies of water may have on wildlife populations (74, 194). The maturation of both theory and practice in population biology, however, demands broader consideration of the sources of environmental heterogeneity and the dimensions of population responses to this patchiness.

Here my objective is to consider in a rather broad and general manner what I view as some important features of population responses to patchy environments. These develop from a variety of theoretical models and observations of nature in diverse areas of population biology, and my intent is more to review existing thoughts than to develop new hypotheses or theories. My approach undeniably reflects a perspective gained from experiences with consumers (primarily vertebrates) in terrestrial

81

environments. Further, I stress variation in horizontal space rather than vertical stratification or temporal variation, since responses to patchiness may perhaps be more easily discerned under these conditions. Nonetheless, these other dimensions of patchiness should be mentioned briefly.

The Dimensions of Patchiness

The patch structure of an environment is expressed in both time and space, and the effects of variation in these patch dimensions upon population attributes are often closely interwoven. Temporal variation, however, often constitutes a major source of spatial patchiness. Localized random disturbances such as fire, erosion, or tree windfalls may create patches of more immature successional status than their surroundings. The repeated but unpredictable occurrence of disturbances produces a system which is a mosaic of units of different successional development, a process that Watt (298) termed *gap phase*. Such disturbance processes substantially influence interactions between populations and thus community structure (162, 163), and may dictate the tactics adopted by fugitive species (130) or patch colonists (179, 269).

Populations of course differ in their responsiveness to temporal environmental variation. Hutchinson (130–132) noted the relations between generation length and the period of environmental variation, predicting that stability in population interactions may occur when the life cycle is either much shorter or much longer than the time between environmental changes, while instability characterizes the relations of populations whose generation length corresponds to the temporal periodicity of the environment. Populations with high turnover rates (i.e. short generations) may track environmental variation closely (180), especially when there is close temporal correlation in environmental variations (35, 166). Many adaptive responses of populations to patchiness, in fact, are influenced by the magnitude and predictability of environmental variations.

Spatial patterning is expressed in both vertical and horizontal dimensions. Vertical pattern is expressed, for example, in the layering of vegetation in in terrestrial plant communities, or variations in the attitude or display of leaves in different strata of the community (126), and such patterns may have profound effects upon the rate and efficiency of primary production processes (197, 208). Other studies (e.g. 47, 173, 253, 312) have demonstrated that vertical stratification among consumer populations may be an important mode of niche differentiation, thereby contributing to the organization and diversity of communities. The horizontal patterning of natural systems is frequently associated with variation in vertical development, and thus the two dimensions are by no means independent.

THE DEFINITION OF PATCHY ENVIRONMENTS

Part of the difficulty in conceptualizing population responses to environmental patchiness, especially in horizontal space, lies in the definition of patches. With naive directness, my dictionary defines a patch as "a surface area differing from its surroundings in nature or appearance," or "a small plot of ground." Levin & Paine

(163) considered a patch as a "hole," a "bounded, connected discontinuity in an homogeneous reference background." Whatever the definition, patches are distinguished by discontinuities in environmental character states from their surroundings; implicit are the notions that the discontinuities have biological significance, and that they matter to the organism.

Patchiness is revealed in a nearly endless spectrum of spatial *scales*, from biospheric or continental distributional patterns to patterns of clonal growth or individual responses to resource distributions (87, 316). Patterns of geographic variation within species, or of within-habitat (α) and between-habitat (β) community diversity (174, 307), or of individual activity movements within a territory or home range (3, 274, 300) exemplify differing scales of patch structure. Alternatively, patchiness may be considered in terms of the diversity of factors that produce it, as Hutchinson (130) suggested. Statisticians define patches or spatial aggregations of units (organisms, communities) by the ratio of variance to mean from randomly distributed samples, and a bewildering array of tests has been advanced to measure aggregation (e.g. 135, 171, 216, 217, 222, 258).

Operationally, of course, the patch structure of an environment is that which is recognized by or relevant to the organisms under consideration. Patchiness is thus *organism-defined*, and must be considered in terms of the perceptions of the organism rather than those of the investigator (312). In California annual grasslands, for example, vole (*Microtus californicus*) dispersion is most closely related to the distribution of food supplies (*Avena*) (20), while in other areas, *Microtus pennsylvanicus* dispersion is correlated most closely with variations in litter coverage and depth (97, 280). Obviously, such differences must be considered in evaluating environmental patchiness for these species. In a more general sense, it has been suggested that populations of small organisms with short generations are sensitive to short-term environmental fluctuations (130, 133) and have greater potential for close tracking of environmental variation than is exhibited by populations of larger organisms (181). They may thus be more responsive to physical and chemical features of their environments. Large organisms, on the other hand, may be "buffered" against small-scale variation in many abiotic features by size alone, and may be more sensitive to biotic interactions. Further, for organisms of a given body size, occupants of unstable pioneer successional stages may be more sensitive to variation in their physical environment, while biotic interactions may dominate the patch responsiveness of individuals living in the presumably more "stable" mature systems (181). In contrast, Grant & Morris (108) have proposed that in a relatively uniform habitat, intrinsic population processes may dominate the activity patterns and responses of organisms, while in a patchy environment responses to elements of the patch structure extrinsic to the population itself may assume greater importance.

Perception differences may also contribute to the generation of patchiness in systems. Visual predators, for example, may create patchiness in prey distribution by cropping individuals from areas with the least cover (261, 268), while predators relying upon olfaction for prey detection may effect decreases in prey density independent of visually defined habitat patches (108).

Such generalities are coarse, but demonstrate that patchiness must be viewed on different scales, with reference to different environmental parameters for different organisms. Recent (47, 50, 248) and older (133) studies of niche dimensionality and overlap have attempted to cope with the problem of defining meaningful scales or units for presumably relevant features by various weighting procedures or by correlations with variation in population density. Ultimately, however, the knowledge necessary to define the scale and perceptual differences in patch response is part of the old and frequently disparaged field called *natural history,* and consideration of spatial heterogeneity may well provide the arena for the integration of natural history and contemporary theoretical ecology.

The vast array of scales and perceptions of patchiness makes unambiguous definition of patch difficult. In theory, the variation in the perceptions of patch elements and patch scale by organisms should be determined by natural selection; that is, they should be attuned to the properties that produce fitness differentials. In this sense, patches may be defined as environmental units between which fitness prospects, or "quality" (91, 210), differ. Operationally, however, it is difficult to measure these fitness differentials, and patches may perhaps be more appropriately defined in terms of nonrandom distributions of activity or resource utilization among environmental units (175, 178), as recognized in the concept of *grain response.* The distinction between "coarse-grained" and "fine-grained" responses was originally framed with reference to relations between the sizes of environmental patches and individual mobility (133, 166) or to differences in the ways organisms responded to or perceived heterogeneous resource mixtures (145, 177, 179). The confusion over whether grain refers to habitat patches or to the perception of patchiness by organisms attests to the inseparability of the two in patch definition.

Given a certain environmental mosaic of resources, we may consider a fine-grained response one in which the units of the mosaic (the "grains") are utilized in direct proportion to their frequency of occurrence (i.e. in a random fashion). An individual or population exhibiting a coarse-grained response, on the other hand, distributes its utilization nonrandomly among the elements of the same mosaic, i.e. it exhibits patch "preference." A fine-grained response may be produced either if the organism does not perceptually differentiate between mosaic units that we recognize as discrete or if it discerns a difference but does not act upon that information. A coarse-grained response implies patch selection, either through active choice among recognized alternatives or passively, as a result of the actions of some external agent (e.g. a predator that removes prey from some patches but not from others may produce a coarse-grained pattern in the prey population). The suggestion that a relatively sessile individual that spends its entire life within a single patch is coarse-grained (175, 177, 179) is true but trivial, since the organism is constrained by its low mobility from responding to patch differences even if they are perceived. In such cases, grain response is best considered as a population rather than an individual attribute. Ecologists have generally not been sufficiently sensitive to the distinction between the grain response of individuals and of populations. The ambiguity of the grain concept is not diminished by the observation that it is resource- or function-specific, since an environmental mosaic may be utilized in a fine-grained fashion for some functions (e.g. feeding) and a coarse-grained manner for other (e.g. nesting).

The concept of grain is best interpreted as a behavioral response to an environmental mosaic. Organisms may thus be characterized as coarse- or fine-grained only in reference to utilization of the resources associated with a particular life-history function, and categorization of species as fine-grained or coarse-grained without specifying the context is inappropriate. If "grain" refers to the nature of a behavioral response or utilization pattern, "patch" applies to the physical environmental or resource units upon which the grain response is expressed. The grain concept may apply to either individuals or populations, but the response patterns of the two do not necessarily coincide in a given situation, and the level at which the concept is applied should always be specified.

Patches are, of course, recognized by relative differences. Gillespie (103) observed that the patchiness of an environment is a function of the relative sizes of the individual patches and the magnitude of difference between them. Environments with patches of grossly differing sizes thus have a small "effective number" of patches relative to an environment with patches of equal size, and large between-patch differences enhance heterogeneity. The potential for elaboration of coarse-grained responses is thus greatest in environments with quite different patches of equivalent size, so long as the size is of a biologically meaningful scale. The tendency toward a fine-grain response might be expected to increase as the difference between patches, patch size, and/or interpatch distance decreases, or as the temporal instability of the patch structure increases.

SPATIAL HETEROGENEITY OF VEGETATION

In terrestrial ecosystems, the most obvious framework of spatial patchiness is that created by vegetational patterns. In many respects, the population processes that produce these patterns differ from those characterizing animal populations, to a large degree because individual plants are spatially fixed, and individual responsiveness to spatial variation in the environment is thus confined to the dispersal-germination phase. Since my emphasis here is upon consumer population attributes, I do not dwell at length upon the causes of plant patterns, but because consumer responses are largely dictated by the nature of the vegetational mosaic, some review is necessary. Figure 1 depicts some of the major factors contributing to vegetational patchiness.

The most critical factors determining plant population patterns are those influencing seedling establishment, screening seeds and dictating germination and survival probabilities (117, 118, 307). Variations in edaphic and microclimatic conditions are especially important at this stage. Climatic regimes, for example, vary over several scales of spatial resolution (e.g. 121, 243), producing potential variations in seedling establishment or successful plant growth. Differences in soil texture also influence germination success, and soil texture may interact with microtopography to produce small-scale variations in microclimate and runoff patterns (335) or nutrient levels (199). Soil and microclimate variations may influence local patterns of hydrologic

86   WIENS

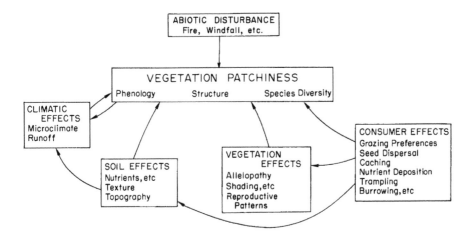

Figure 1   General relationships of some of the major factors which govern vegetational patchiness in terrestrial environments.

cycling within systems. Differences in soil concentrations of heavy metals may produce especially sharp patterns in plant distributions (12, 191).

These spatial variations may have major effects upon the species composition and structure of the vegetation in a given patch, or upon phenological relationships between patches of otherwise similar vegetation. In arid environments, small variations in soil, topography, and microclimate produce especially well-defined vegetation patterns, chiefly through their effects upon water availability (9, 21, 199, 206). Woodell et al (330), for example, suggested that the spacing pattern of creosote bush (*Larrea divaricata*) in California deserts varies as a function of rainfall, although other factors, such as allelopathy, may also be involved (10, 17).

Vegetation patchiness also reflects the influences of plant population attributes and interactions. Competitive interactions of various sorts (e.g. shading, differential rooting patterns) may lead to the suppression of some species by others or wide spacing among individuals of a single species. The production of chemical compounds that inhibit the germination or growth of other plants—allelopathy—not only influences the dominance structure of the plant community (308, 309), but the spacing patterns of vegetational complexes as well (200, 233). Further, the complex of plant secondary compounds released into the soil is superimposed upon the small-scale variations in microtopography and soil quality, enhancing the mosaic potential. In "stressful" environments, where such features are often most apparent, plants frequently reproduce vegetatively, producing clonal patches (15, 115). A similar clumping of offspring of an individual plant may occur if seed dispersal is quite limited, as in some oaks.

The effects of herbivores on plant spatial patterning may be subtle and are thus frequently unnoticed, but they are by no means as inconsequential as Poore's (224) statement likening animal effects on plant community structure to "a pattern of changing cloud shapes over a rural countryside" would suggest. In Hawaii, for example, feral goats have a controlling influence on the spatial pattern and species composition of *Acacia* parklands and lowland grasslands (270), and elsewhere domestic sheep, through selective herbivory, may influence sward structure (110). In rangelands of western United States, heavy grazing frequently leads to the replacement of native grasses by shrubs, and thus markedly influences not only the floristic composition but the spatial structure of the plant community (168). In England, depression of rabbit (*Oryctolagus cuniculus*) grazing following myxomatosis epidemics produced major changes in species abundances and vegetational patchiness in pastures (116). Elephants (*Loxodonta africana*) have profound effects on woodland vegetation in many areas in Africa (36, 154), and since bulls feed together in groups which change their feeding loci between feeding periods, grazing produces a patchy disruption of the vegetation (58). In general, herbivores which are selective in their feeding may deplete local areas of their favored foods, creating gaps in the vegetation and preventing space monopolization by single plant species, in a manner analogous to Paine's (213, 214) "keystone predators" in intertidal systems. Such grazing opens patches for colonization by other species and thereby increases the diversity of the community (116) or may alter the size or age distributions of the plant populations (208). In addition to creating patchiness in plant populations, grazing by specialist herbivores may influence the apparent habitat "preferences" of the plants. Thus in areas inhabited by the chrysomelid beetle *Chrysolina quadrigemina,* Klamath weed (*Hypericum perforatum*) attains its greatest densities in partially shaded habitats, but the most suitable habitat is actually on good soil in open sun—areas in which the beetle effectively controls the plant populations (129). In fact, specialist herbivores should be expected to play a greater role in the generation of patchiness than generalist grazers, and the magnitude of this effect is related to the degree to which food supply normally limits the herbivore populations (116). Even generalist herbivores, however, can influence vegetational heterogeneity by concentrating grazing for short periods of time in given locations and then moving elsewhere, or by responding to variations in plant community composition by feeding in a frequency-dependent fashion (116).

Given these grazer impacts on vegetation it is not surprising to find a diversity of "defense mechanisms" against herbivores represented in plant populations—dense hairs (160), spines, production of tannins (83) or alkaloids (69), or various allelochemics (88, 309). If the herbivore-plant associations persist, a "coevolutionary race" may ensue, in which increasing herbivore specialization to circumvent the plant defenses generates increased specificity in the plant defenses, and so on (75, 116). Root (240) and Feeny (84) have offered an especially intriguing suggestion, that early successional plant populations which are intermingled on a local scale may possess an "associational resistance" to certain types of herbivores, as the herbivores are thwarted not only by the defensive adaptations of individual plant species but also by the combination of defenses of the vegetational mosaic as a whole. This "associational resistance" is an emergent property of the vegetational system. It may represent an additive effect of the defensive adaptations of individual organisms, although Feeny (84) has referred to it as a "bizarre kind of mutualism."

Animals may have a variety of less apparent effects upon vegetation structure. Activities such as trampling or burrowing can alter the texture or topography of the soil, and thus soil "quality." Localized deposition of dung may markedly affect soil nutrient levels and influence the floral composition as well as the growth patterns of vegetation in such patches (116). Granivorous consumers exert predation pressures on the seeds produced by individual plants. Coupled with the patterns of seed dispersal, this produces spatially defined germination success patterns (136, 137). Many birds and mammals gather seeds from wide areas and thereby may play a major role in seed dispersal (13, 153, 285); by storing the gathered seeds in localized caches this dispersal becomes nonrandom, and, to the degree that buried caches are not later recovered, germination and growth may occur in tight, widely separated patches (1, 93). Finally, pollinators such as insects, bats, or birds, by virtue of characteristic flight distances, affect the genetic structure of plant populations (161) and, by influencing the probability of fertilization of isolated plant individuals, probably affect the spatial structure of the population as well.

The multitude of factors governing spatial heterogeneity of vegetation usually acts in concert. West (304), for example, described several levels of pattern in ponderosa pine (*Pinus ponderosa*) forests in central Oregon. Stands differed in density, dominance, and interspecific competitive relationships with respect to gross moisture gradients. But within stands there was a mosaic of relatively small clusters of differing age structures, produced by periodic fires. The trees within individual clusters were even-aged, but clusters of similar age structure differed in density, presumably as a result of chance factors during establishment. Finally, individuals within a recently established cluster were regularly dispersed, probably as a result of competition for water, minerals, and light during the sapling stage. West did not investigate animal influences on this array of pine patterns, but several small mammals, including golden-mantled ground squirrels (*Spermophilus lateralis*) and yellow-pine chipmunks (*Eutamias amoenus*), harvest ponderosa seeds and bury them in concentrated caches (personal observations). Not all of these caches are recovered by the animals, and the small clusters of seedlings which germinate in these caches may represent an important component of forest regeneration.

It is tempting to believe that generalities exist regarding the patchiness of various broadly defined vegetation types. Unfortunately, the diversity of spatial scales of patch structure reduces the clarity of these relations, if indeed they exist. Whittaker (306, 307), for example, was unable to detect any clear correlation of primary production with plant community diversity over a wide range of vegetation types, although Margalef (181) depicted an inverse relationship.

In my own work with bird communities in western North America (318), I measured vegetation patchiness using a scale appropriate for small birds. The work was restricted to grassland and shrubsteppe systems, but within that range of environments there was a clear relationship between increasing structural heterogeneity of the vegetation, decreasing coverage by grasses, and increasing coverage of woody vegetation and of bare unvegetated areas. These changes paralleled the series from tallgrass prairie to shortgrass prairie to palouse prairie to shrubsteppe, substantiating the broader generalization that more arid systems are generally more heterogeneous, at least on this particular spatial scale. This probably is an effect of the greater role of water stress than of competition for sunlight in governing plant interactions. However, the life form or reproductive pattern of the vegetation dominants may also be important. The woody shrub vegetation which dominates many arid and semiarid habitats (e.g. *Artemisia, Atriplex, Larrea*) realizes annual net production as incremental growth to existing structure, and thus short-term variations in climate and production are not associated with large variations in structure and horizontal pattern. Grassland vegetation, in contrast, dies back and regrows each year, and annual variations in climate and production may be accompanied by spectacular variations in plant growth and pattern (317). The patch structure of woody vegetation, and perhaps of perennial plants in general (116), is thus more stable than that of herbaceous or annual vegetation.

All of this may suggest that arid systems are more sensitive to the various factors promoting vegetation patchiness (Figure 1), but the more arid systems are also temporally less predictable (206, 317), and this undoubtedly influences patch structure. Further, because of water limitations, the repair process following disturbance is notoriously slow in arid environments, so that gap phases may persist for decades or even centuries, while the relative swiftness of repair in more mesic systems may contribute to their lower apparent patchiness.

Can we discern changes in vegetation patchiness during succession? As Drury & Nesbit (71) have noted, unambiguous generalizations of any sort regarding succession are not easy to come by. The simple dogma that species diversity (and thus, perhaps, spatial diversity) increases during succession (e.g. 207) is insecure (14, 205), and the presumably stable, self-replacing "climax" may actually be a mosaic of patches of different structure and ages following gap disturbance (172). Certainly vertical layering or patchiness of vegetation exhibits a general increase during forest successions (126), but changes in the horizontal patch structure of vegetation are less apparent. Whittaker (307) suggested that the "internal patterning" of forests may decrease in late successional stages, and Morrison & Yarranton (198) discerned increases in spatial heterogeneity during transitional phases of sand dune succession, due to the fact that different patches underwent successional transition at different times. The longer the persistence of a given successional stage, the greater the probability, of course, that some large- or small-scale disturbance (e.g. fire or a falling tree) will create patch openings which revert to earlier successional status, creating gap phase. But as succession proceeds toward closed-canopy stages, microvariations in environment may become "buffered" by the community structure, and thus of less import in generating large-scale patchiness. Herbivore influences may also change during succession. Root (240; see also 84) suggested that plants of early successional stages may produce allelochemics and depend upon "associational resistance" as defense against herbivores, while the dominants of late successional communities, by virtue of their size and/or abundance, form vegetational patches that persist for long periods. These patches support large herbivore faunas, but the herbivores are less specialized to consume certain plants, and are thus unable to "override" the generalized plant defense mechanisms. As a result, herbivores may play a relatively minor role in the generation of vegetation patchiness in late successional stages. Root, however, was concerned only with insect herbivores, and the same speculations may not apply to larger vertebrate herbivores, which influence plant patterns on a different scale. In mature woodlands in some areas of Europe, for example, wild boar (*Sus scrofa*) may have important influences on the age structure and spatial dispersion of individuals in plant populations through their rooting activities, excavating areas of several square meters and destroying saplings and small trees (personal observation).

CONSUMER POPULATION ATTRIBUTES IN PATCHY ENVIRONMENTS

The habitat patterns produced by these varied effects provide an environmental mosaic of patches. A variety of attributes or adaptive responses of populations develop within the selective regimes associated with this heterogeneity. I consider them separately here for convenience, although in nature they interact to form what Pitelka and his colleagues (223) have called an "exploitation system."

A Cautionary Note on Optimization and Group Selection

Nearly all of the theory of population biology and ecology, including that relating populations to patchy environments, is built upon optimization principles, at least in the informal sense that natural selection, in maximizing individual fitness, produces optimal adaptive responses (46). Selection is viewed as incessantly trimming all but the best-adapted or optimal phenotypes from a population, and competition is usually presumed to be the major driving force of this selection. Theories predicting the conditions of optimum response, however, usually contain several limiting assumptions: the population-environment system is considered to be in equilibrium; at least some environmental components (resources) are constantly limiting, and there is therefore no temporary relaxation of selection pressures favoring the optimum; the suitable environment (or niche space) is fully occupied or saturated; the optimum is stable, that is, what is optimal at one point in time is equally adaptive at another; and the theoretically predicted optimum is in fact attainable, i.e. the past history of the population (e.g. founder effects) does not place some optimal states "off limits." Obviously, these assumptions do not unfailingly hold in the real world. For example, in arid or semiarid habitats with highly unpredictable climates (206, 317) or areas with relatively high production concentrated into a short growing season, such as tundra, selection may often be relaxed and competition reduced or absent. Under such conditions, we must ask what penalties (in relative fitness) are paid by the individual which behaves in a non-optimal fashion; the answer may frequently be "none." In other situations, predators may act to prevent prey populations from reaching competitive levels (51), and the "optimum" may be determined by factors quite unrelated to competition. We must thus exercise some restraint in embracing optimality based theories as anything other than hypotheses of population response under idealized conditions, and should not consider Nature to be remiss if the theoretical optimum is not the normal state of affairs.

Discussions of the evolution of population attributes also frequently develop arguments on the adaptiveness of some feature in terms of the population as a whole rather than individual organisms. Such approaches at times portray selection acting upon traits "for the good of the group," without regard for their effects upon individual fitnesses. While mechanisms by which selection may operate through differential success of groups of individuals are known (170, 326), there are severe difficulties with such group selection explanations (e.g. 311, 324, 328). It seems unlikely that group selection has normally played an important role in the evolutionary development of various population attributes, or that our understanding of these features is aided much by such explanations.

With these cautions in mind, we may now consider a spectrum of population responses to patchiness. A good deal of the theory carries connotations of optimization, and at times group selection arguments are implicit.

Habitat Selection

If patches differ in "quality," individuals should be expected to exhibit some degree of patch or habitat selection. This response amplifies the effects of habitat structure upon population attributes by increasing the discontinuity or nonrandomness of dispersion of individuals within the population, and the magnitude of this effect is closely related to the grain-response pattern of individuals. Bryant's (34) simulation analysis demonstrates that a coarse-grained strategy is optimal if the environmental patches differ in quality but are relatively stable or predictable through time. As environmental variation increases, habitat selection tends increasingly toward fine-grainness (i.e. individuals distributed randomly among habitats, in Bryan's simulation), and this is intensified if the patch variations are relatively predictable. This results in an "opportunistic" mode of habitat response, with relatively close tracking of variations in patch structure or frequency. Levins (166) drew essentially similar conclusions from fitness-set analyses, although his emphasis was upon temporal rather than spatial environmental variations.

Habitat "selection" may result from individuals exercising a choice among compared alternatives (true habitat *selection*), or from some external agent imposing a differential distribution of individuals among patch types upon the population (habitat *correlation*). Thus, although *Microtus* density is correlated with vegetation cover in Wisconsin marshes, these density variations may be related to differential predation in areas of differing cover rather than to active selection of areas by individual voles (97). The distribution of morphs of *Cepaea* land snails in some areas is largely a result of differential predation under different habitat conditions (39, 86), and the habitat associations of melanic and "normal" moths in England and North America apparently involve habitat-dependent predation as well as some degree of habitat preference by the moths (27, 144, 250, 251). Unfortunately, most studies of habitat selection actually measure correlations of individual or population distributions with environmental variables (e.g. 44, 98, 120, 242, 312), and thus cannot distinguish between habitat selection and habitat correlation. Attempts to unravel these processes experimentally (e.g. 148, 209, 215, 249, 299, 313, 314) have proven difficult. Further, the selection of habitat types may involve an interplay of genetic ("innate") and experiential influences on preferences (134). Tadpoles of different *Rana* species exhibit different "receptivity" to learning substrate pattern characteristics during development. The learned preferences may influence habitat orientation in the

92   WIENS

patchy environments characteristically occupied by these species (313, 314), while the determination of what may or may not be learned may be largely genetic. In other species (e.g. *Peromyscus,* 299; *Salmo,* 227, 228, 234, 235), responses to habitat stimuli may be more directly determined by inheritance. In some bird populations, learning of habitat preferences has apparently produced ecological variants within the species, individuals occupying different habitat types in a mosaic fashion as a function of their early experience (120, 220).

Under field conditions, the expression of individual habitat preferences is strongly influenced by population densities in the various patch types. In an "ideal" (i.e. "other things being equal") situation, individuals should select habitat types on the basis of fitness prospects (166), which may be a function of the intrinsic "quality" of the habitat type and the population density (intensity of competition) in the habitat (90, 91). At low total population sizes, only the optimal or highest quality patch types should be occupied, but as density increases, the quality of that habitat decreases, so that a point is reached at which some other habitat type has equal potential quality (Figure 2). At this point, patch occupancy should expand to include the additional habitat of now equal quality, given the assumption of perfect choice capabilities of the organisms. With further increases in population density,

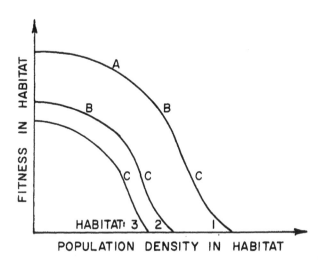

Figure 2   The Fretwell-Lucas model of habitat selection and distribution. Fitness prospects for a given habitat are greatest at low densities, and as density increases, habitat suitability decreases following the functions shown for habitats 1, 2, and 3. In this example, habitat 1 has the greatest intrinsic quality; when total population size is low (*A*), all individuals should settle there. As density increases, the quality of habitat 1 decreases to a point at which fitness prospects in habitats 1 and 2 are equal, and individuals should now begin to select both types (*B*). At higher densities (*C*), additional habitat types may be incorporated into the range of equivalent options. Modified from Fretwell & Lucas (91).

the range of habitat types which are of equal quality continues to increase. A major consequence of this ideal habitat distribution is relative equality of individual success in the various habitats. Fretwell & Lucas (91) suggested that departures from this expectation may reflect the influences of other factors, such as territorial behavior, upon population densities, although this view has been challenged by Klomp (147). It is nonetheless apparent that density does influence patterns of habitat occupancy. In a Wisconsin grassland, for example, Grasshopper Sparrows (*Ammodramus savannarum*) upon their arrival in the spring initially occupied breeding habitats with relatively low grass cover and litter depth and tall emergent herbaceous vegetation, while Savannah Sparrows (*Passerculus sandwichensis*) occupied habitat patches with characteristics at the opposite extremes. Individuals arriving later, however, settled in the remaining habitat not already occupied by their species, so that as breeding populations increased in size, the two species converged in patterns of overall habitat occupancy (315). At the peak of the breeding season, habitat occupancy by the two species was virtually identical (312). In a similar vein, the habitat patch occupancy of *Microtus* is apparently density related in many areas. Anderson (11) distinguished between "survival habitat" and "colonization habitats"; within survival habitat, population densities remain relatively constant from year to year, while colonization habitat is normally occupied by "excess" individuals emigrating from the survival habitat. Only in an exceptional year are conditions in the colonization habitats suitable to permit overwinter survival and reproduction during the following year. The studies of Semeonoff & Robertson (256) and Tamarin & Krebs (277) have indicated differences in the genetic structure of populations occupying survival and colonization habitats.

These observations and the Fretwell-Lucas model (91) suggest that measures of niche breadth based upon the range of environmental conditions occupied by a species (e.g. 47, 50, 166, 190) are in fact density dependent, and may be used in comparing species only if the populations are in equilibrium, or if adjustments for density inequities between the species relative to their "carrying capacities" are made.

The manner in which density changes affect the habitat distribution of a population is influenced by the grain-response pattern. Individuals exhibiting a fine-grained response may spread randomly over patch types as density increases (Figure 3); tendencies toward aggregation may appear only if social effects intervene. On the other hand, the dispersion of individuals in a coarse-grained population, such as the sparrows I studied (315), may initially be highly aggregated, and become less so at high densities as a broader range of habitat types becomes equally suitable (as predicted by the Fretwell-Lucas model). At extremely high densities, a coarse-grained population may converge toward a fine-grained population in dispersion pattern. Grant & Morris (108) have presented data on *Microtus* populations that demonstrate that dispersion tends toward uniformity as density increases, and that this trend is more clearly defined in uniform than in patchy habitats.

The selection of habitat patches among an environmental mosaic has population effects beyond those on dispersion patterns. Segregation of individuals into different

94   WIENS

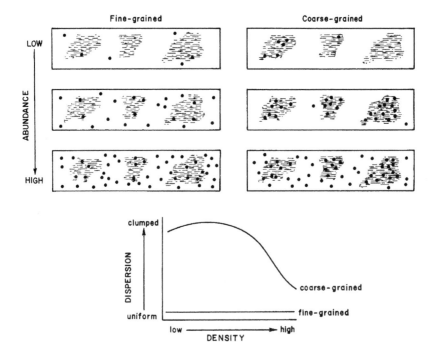

Figure 3   Habitat or patch occupancy patterns for a fine-grained (*left*) and a coarse-grained (*right*) population as density increases from low to high levels. The fine-grained population may retain a uniform dispersion pattern over a wide range of densities, while the dispersion of the coarse-grained population is initially clumped (as individuals settle in the preferred patches) but tends toward greater uniformity as density increases (and individuals are forced to settle in less preferred patch types), as shown in the bottom diagram. Modified from Rotenberry & Wiens (244).

patch types may lead to assortative mating, and thus affect the levels and patterns of genetic variation of the population as a whole (159, 167, 187). To the degree that genetic morphs are habitat-specific (or fitness is patch-related), increased refinement in habitat selection should be favored by selection (241), and under stable conditions this may initiate population divergence (186).

Dispersal and Homing

Some sort of dispersal phase occurs during the life cycle of most organisms, and the broad variety of dispersal patterns and distances (140, 329) has provided abundant fuel for theoretical considerations of the adaptiveness or evolution of dispersal behavior (48, 92, 165, 176), some of which rely directly upon group selection arguments (287, 332). We should expect dispersal behavior to be relatively well developed in coarse-grained populations occupying patchy environments, since it may lead to colonization of unoccupied but favorable habitat patches (146, 179). For example, many insects that are coarse-grained in their environmental responses have well-developed dispersal mechanisms (140, 269), and in some butterfly species, the dispersal characteristics of populations are attuned to local or regional habitat patchiness (101, 102). Temporal variability in the patch structure of an environment adds additional emphasis to the adaptiveness of dispersal, as individuals or their offspring must frequently change location as patches become unsuitable or newly available through disturbance or successional change. The complex of attributes characterizing fugitive species (130) represents responses to rapid patch turnover (see also 169, 269).

Dispersal patterns are frequently bimodal, with many individuals undertaking very short movements, a few moving intermediate distances, and somewhat more dispersing longer distances. In the *Perognathus* populations studied by French and his colleagues (89), 25–30% of the population dispersed each year, with a sharply bimodal pattern. Many insects undertake long-distance "migrations" of more or less set distances, during which they are apparently insensitive to stimuli such as potential mates, habitats, etc. These movements contrast with shorter "trivial" within-habitat movements (52, 140, 269), and the bimodal pattern of individual movements may represent a combination of these two dispersal types. From work on small mammals, Howard (127) suggested that individuals might undertake two sorts of emigratory movements, fixed-distance (usually long-range) "innate" dispersal and usually shorter "environmental" movements representing adjustments to local population density. His argument embraced group selection explanations, however, and Murray (201) proposed instead that the dispersal patterns result from age structure-dominance relationships and have no inherent adaptive function of their own. Others (e.g. 41, 301–303) have attributed the pattern of dispersal to behavioral polymorphism, the occurrence in a population of both relatively sedentary and relatively vagile individuals.

The movement of individuals has fundamental effects upon the genetic and social composition of populations. Dispersal represents potential gene flow between groups, but the realization of this potential requires that the dispersing individuals be successful in discovering suitable habitat conditions, be sexually active, be able to merge into the social or reproductive structure of an existing population, and have no disadvantage in mating, none of which is at all certain. To the extent that dispersal is accompanied by genetic interchange, it may act to thwart the development of local adaptation or differentiation of populations (e.g. 188, 271), although not to the degree once thought (76, 81, 119, 259). Berndt & Sternberg, (22), for example, demonstrated that dispersal distances were negatively correlated with the number of described geographic races in three European bird species, and attributed the relationship to the magnitude of gene flow. Much of the research on dispersal patterns in *Drosophila* (what some population geneticists mistakenly call "dispersion") is undertaken to document gene flow patterns and magnitudes (e.g. 59, 68, 294). Populations located in environmentally or spatially marginal situations may be maintained by a more or less continuing influx of emigrants from central populations (11, 272), and thus differ in age structure and perhaps genetic structure from the "donor" populations. Colonizing episodes often involve a small nucleus of dispersing individuals, and many of the attributes of the newly established population may reflect founder effects (179).

Wilson (326) has recently presented a different conception of the relation between dispersal and genetic processes in populations. His model considers small "trait-groups" within demes; while dispersal from these groups leads to genetic mixing which determines the deme boundaries, most of the ecological interactions of individuals (e.g. competition, feeding, mating, predation) are carried out in the more spatially restricted "trait-groups." Wilson proposed that if this sort of population structure occurs, it can promote the selection of "altruistic" traits that favor the fitness of the group over that of the individual, a process of group selection. This may be most likely, Wilson claimed, where a spatially heterogeneous habitat partitions the deme into many small trait-groups. Unfortunately, Wilson offered no evidence in support of his model, and demonstrating that a model shows that such a mechanism can work is no assurance that it is commonplace in nature.

The relationships between homing or site tenacity and environmental patchiness on population attributes are generally opposite those of dispersal. Thus individual patch specialists should be expected to exhibit strong orientation to their "home" patch, especially if the favored patch types are widely dispersed (i.e. are habitat "islands") and temporally fairly stable. On the other hand, if the spatial location of favorable habitat units changes during the reproductive life of an individual, close site tenacity may be a poor strategy, and individual fitness may be increased by a flexible habitat search pattern covering a larger area. If homing is tied closely to reproductive sites, it may act to reduce the effective population size by increasing inbreeding, thereby eroding genetic variability in the population as a whole and promoting population differentiation (42). Unfortunately, most of our knowledge of homing comes from studies of experimentally displaced individuals, and we know little of the patterns of individual allegiances to locations for populations as a whole.

Social Organization

A complex web of factors influences the pattern of social and space-related behavior exhibited by a population. Thus sociality may be related to mobility, the availability of specialized sites (e.g. roosts, dens), climate, the sources and intensity of predation, population density and age structure, food abundance and dispersion, and habitat configuration, among other factors (49, 54, 55, 77, 85, 143, 223). This diversity of underlying determinants has led to seemingly endless discussions over "the" function of a given social pattern (e.g. 332), but it is obvious that such attempts at simplicity are doomed to failure, as the review by Brown & Orians (32) amply demonstrates. Social patterns have no unitary adaptive function, but are the creations of multiple selective pressures, and are thus likely to confer multiple adaptive advantages to individuals. Attempts to construct general theories of the evolution of social patterns (5, 30, 54, 85) have incorporated the diversity of proximate determinants of sociality into a common framework linking aggression to the feasibility of site-specific defense of important resources, with predation probabilities influencing the outcome in various ways. It is certainly not my intent here to review social behavior (see 32, 55, 82, 328), but only to indicate how one circumstance—environmental patchiness—can influence its expression. Further, there are variations of social organization not considered here: arena or lek display systems (37,

157, 321), communal breeding (334), family groupings (31, 331), and the rich diversity of highly specialized organizations in social insects (327, 328).

The patch structure of resources in space and/or their transiency in time governs the form of social organization expressed within a population, largely through its effects upon the expense or efficiency of resource procurement within a specified location (30, 66, 114). Thus, as the spatial or temporal clumping of resources (e.g. food, breeding space, mates, nest or shelter sites) increases, an increasingly larger area is required to ensure an adequate supply, and the task of maintaining individual rights to those resources becomes increasingly difficult, inefficient, and demanding of energy. The expense of defending a given quantity of resources thus increases as some function of decreasing resource predictability. At different points along this expense function, different forms of social organization may be most appropriate ("optimal"), as visualized in a simplistic manner in Figure 4.

When resources are plentiful and evenly distributed or accessible and predictable, a rigid form of territorial behavior may be optimal, since there is a high probability that an individual territory will contain sufficient resources for an individual and its offspring. Staunch defense of the territory will then be worth the effort. Under

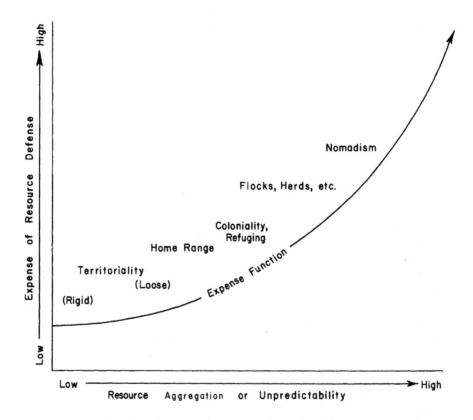

Figure 4   The expression of spatially-defined social organization as function of the increasing expense of defense of resources as their unpredictability or aggregation in space increases. Modified extensively from Orians (212).

uniform environmental conditions, traveling time and energy to gather food or defend boundaries will be minimal if the territory is circular in shape (211), and as population density increases the ideal territory shape may be expected to tend toward polyhedral, permitting closer packing without sacrificing area (107). Clear demonstrations of such optimal shaping are scarce, and are likely to occur only if the territories are small enough that the boundaries can be continuously and completely defended, as in the cichlid fish studied by Barlow (18). Territory shape should thus be expected to become increasingly irregular as the distribution of resources deviates from uniformity, and the amoeboid changes in boundary configurations through time noted for some species (312) may reflect temporal changes in the small-scale patch structure of the resources as well as redefinitions of dominance relationships between neighbors. Territory size may increase with changes in the abundance or dispersion of resources, maintaining a supply adequate to meet individual needs [(260, 275, 291) but see (189)]. However, this entails an inevitable increase in the expense of defense of the area, and at some point the rigidity of boundary defense must loosen and permit spatial overlaps. The antbirds studied by Willis (325), for example, depend upon food supplies related to the activities of army ant (*Eciton*) swarms, which are locally unpredictable, and the birds thus occupy overlapping "loose territories" of localized dominance. In small or nocturnal mammals, close visual monitoring of area boundaries may be impossible, and individuals occupy overlapping home ranges regardless of resource distribution.

As resources become increasingly concentrated into separated and spatially unpredictable patches, close individual attachment to even a semi-exclusive area becomes overly expensive, and social organization may shift to a more aggregated grouping of individuals. These aggregations may be localized in some "central place," a breeding colony, roosting site, or den, from which individuals or groups (flocks, herds, etc) disperse in daily movements to obtain resources (food) which occur in high abundance within scattered patches, as visualized in Hamilton & Watt's (114) "refuging" concept. [Similar conceptual or analytic models may be applied to consideration of the economic systems of urban areas, suggesting optimal patterns for human urban social organization and group (town) size in relation to the patchiness of their surrounding "resource support" areas (6–8, 60).] Social aggregations such as colonies, herds, flocks, or schools have frequently been interpreted in terms of reduced individual probabilities of suffering predation (e.g. 28, 38, 111, 218, 219, 226, 292, 293, 323), since a group may have a higher probability of detecting danger than an individual, can mount group defense against a predator, may be more difficult for a predator to locate, or since an individual in a group may be more difficult for a predator to capture than when solitary, especially if it is in the center of the group. Others (e.g. 138, 202, 281, 297, 333) have argued that such groupings find their primary adaptive function in the exploitation of patchily distributed food. Lazarus (155) has suggested that the demands of feeding efficiency may account for qualitative variations in individual dispersion (solitariness, flocking, etc), while predation may effect quantitative differences (e.g. flock density, territory size). The model developed by Horn (125) demonstrates that if food is clumped and unpredictable, travel time to exploit these resources is minimized by individuals nesting colonially in the geometric center of all food patches, while an even dispersion of food favors individual territorialism. Ward & Zahavi (296, 297, 333) have further proposed that central gathering places of group-foraging organisms may function as "information-centers," enabling individuals that lack the knowledge of the locations of patchily distributed but abundant food concentrations to profit from the searching abilities of other individuals in the social group. Individuals that have been unsuccessful in finding resource concentrations may follow successful individuals out of the roost or colony on foraging trips to good feeding areas, and successful individuals may "advertise" their success by behavioral differences from unsuccessful individuals, and thus attract followers. Studies of starling (*Sturnus vulgaris*) roost dynamics by Hamilton and his co-workers (112, 113) indicated that some individuals did follow others in morning foraging flights, and "in-transit" stops along flight paths were sugggested to be a mechanism to permit continuing evaluation of resource conditions closer to the roost, potentially reducing the distance flown and thus reducing energy expeditures. Krebs (150) observed Great Blue Heron (*Ardea herodias*) flock movements in British Columbia and suggested that his results supported the information-center idea. But while the birds did follow one another, it was not clear that "successful" individuals were followed by "unsuccessful" ones, or that any sort of "information transfer" occurred at the breeding colony. The efficiency of individuals feeding in flocks was greater than that of solitary birds, but this was likely a manifestation of the occurrence of groups of individuals in the more favorable food patches, as Goss-Custard (106) also found for estuarine-feeding shorebirds. Studies in Oregon indicate that heron social organization may involve a complex interplay of individual feeding territories, group foraging areas, and changing flight patterns (R. Bayer, personal communication), and the information-center concept seems inappropriate there.

Social grouping in relation to irregularly dispersed food supplies may confer other individual benefits. Close association of foraging individuals may facilitate observational learning of feeding techniques or profitable prey types (4, 151, 202). Individuals feeding in a group may make more efficient use of the resources than solitary individuals, since the feeding group more thoroughly gleans an area before moving to other feeding locations, and thus maximizes the disparity and recognizability of food abundance between that area and the sites not yet visited by the group (45, 47). Also, social groups may employ various cooperative hunting tactics not available to individuals (e.g. 184, 193, 252). If spatial aggregation of individuals is related to food exploitation, we should expect groups to increase in size as the food supply decreases and/or becomes concentrated into more widely separated patches (e.g. 45, 112). Under such conditions, upper limits on group size may be determined by the food types and abundance (138) or by the balance between increased efficiency of energy capture by groups of increasing size and the increased energy demands of the group (56, 328).

These social responses to the spatial distribution of food resources are not without applied implications. Many granivorous bird species, for example, probably evolved in environments in which food production was locally concentrated but highly unpredictable, and occupy large breeding colonies or nonbreeding roosts from which they radiate in wide-ranging foraging flock movements. As human civilizations developed, the grassland and savannah habitats of such species were among the first to be converted to agricultural croplands, and the birds were, in a sense, "pre-adapted" to assume a pest role. Species such as *Quelea* in sub-Saharan Africa (142, 296) or various blackbirds (Icteridae) in North America (192) are outstanding examples of such "pre-adapted" pest populations. Recently, Dyer & Wiens (320) completed a simulation study of the effects of one such "refuging" species, Red-winged Blackbirds (*Agelaius phoeniceus*), on corn crops in northern Ohio. By coupling the distributional patterns of the foraging birds with information on dietary habits and energy demands, they predicted the spatial patterns of corn consumption, and on this basis proposed several management tactics.

In addition to patterns of social grouping, other elements of social organization, such as mating systems, also may vary in relation to environmental heterogeneity. Since males and females usually have unequal "investments" in individual offspring, selection should favor polygamy as the fundamental mating system (283, 328). A superabundance of food during the breeding season may enhance the development of polygamy (especially polygyny), since it enables a single female to raise the offspring while the male partakes of additional matings (53). Orians, Verner & Willson (210, 288, 290) have extended this argument to consider the effects of local variations in food supply or habitat "quality." Thus, if male territories differ markedly in quality, a female may realize greater success by pairing with a mated male on a prime territory and rearing the young alone than by forming a monogamous bond with a male on an inferior territory, even if the male assists. Environments that exhibit considerable patchiness in productivity, such as marshes or early successional stages, should provide the best setting for such a process, and indeed most cases of polygyny among birds do occur in such habitats (53, 54, 290). Several careful field studies (124, 182, 183, 336) have strengthened the hypothesis. There are, of course, other factors that may contribute to the evolution of polygyny (e.g. 19, 70, 109).

There is no clear separation of mating system and social grouping responses to environmental circumstances in real populations, as the analyses of Pitelka et al (223) and Crook (54) persuasively demonstrate. Among the Arctic sandpipers studied by Pitelka's group, most of the species follow a "conservative" strategy of exploiting the variable resource conditions of the tundra: individuals disperse on large territories, which provide adequate food, or on small territories used in conjunction with communal feeding grounds in "rich" patches, and the monogamously mated pair maintains residence on the territory through the breeding season, concentrating in breeding areas where and when environmental conditions are particularly favorable. Several other social patterns are associated with more "opportunistic" exploitation of the patchy resource distribution. Some species exhibit serial polygamy, in which the female is freed from incubating the first clutch and may mate again with the same or a different male to produce a second clutch of eggs. Other species are polygynous, males exhibiting a mating success in accordance with the quality of their territory. Finally, the promiscuous species have variable mating relationships and fluid spacing patterns, in which territories function chiefly as mating stations. While environmental patchiness is the chief factor promoting these alternative exploitation systems, other factors, such as predation, are also involved.

Predator-Prey Relationships

There is a rich body of theory narrating how optimal predators should forage (see 40, 78, 247), but most work assumes a random distribution of prey, and thus does not relate to the task at hand. Predation may relate to environmental patchiness in its effects on the spatial distribution of prey individuals, in the determination of predator tactics, and in the potential stabilization of population interactions.

Predators may act on prey populations in a spatially discontinuous manner and thereby prevent the monopolization of resources by single prey populations. Such "keystone predators" (213, 214) may play a major role in creating localized patches or gaps in prey distributions (61, 116, 164). Predation may also contribute to the adaptiveness of clumped social dispersions among prey, as noted above. Aggregation may decrease the chances of a predator finding the prey (29, 279, 293), especially if such prey are more responsive to the approach of a predator and react earlier, as is the case with European hare (*Lepus europaeus*) (139). R. Taylor (279) modeled the susceptibility of individual prey to ambush predators, concluding that the prey almost always benefit from clumping, largely because of limitations on the number of prey such a predator may consume upon encountering a clump. But clumps of prey are inevitably more conspicuous than solitary individuals, and the adaptiveness of aggregation must involve some compromise between the reduced probability of detection of a clump and the increased probability of being captured once a clump is discerned. When prey groups are small, the benefits of aggregation may not outweigh the disadvantages of increased conspicuousness, and there should therefore be lower limits on group size, at least in relation to predation selective forces. Under some conditions, then, selection may operate against increased clumping of prey, and instead favor "spacing-out" of prey, so that the distance between individuals exceeds the distance at which a predator can detect an individual prey. The predator must then search for each prey individually, and realizes no "bonus" from its discovery of one prey individual. Tinbergen (282) and Croze (57) have advanced this hypothesis with particular reference to camouflaged or cryptic prey, which might "blow their cover" by close spatial association. J. Taylor (278) has developed a geometric model which predicts that spacing-out may be more important for small or cryptic prey than for large, easily perceived ones.

Differences in the dispersion of prey, from whatever cause, should be expected to influence the effectiveness of the various foraging tactics potentially employed by a predator. The "spacing-out" hypothesis of Tinbergen et al (282), for example, assumes that predators practice "area restricted search," concentrating their foraging activities for some time in the vicinity of their last successful capture. Such behavior should be expected to be well defined in predators feeding upon clumped prey, and reduced if the prey are uniformly or randomly dispersed, and the field experiments of Croze (57) and Smith (262, 263) have confirmed such patterns. Prey dispersion may also influence the foraging pathway followed by a predator. Smith's elegant experiments (262, 263) showed that thrushes (*Turdus*) restricted large turns and alternated right and left turns when foraging, and that this continually took the birds into previously unsearched ground. With increased prey density the turn rate increased, maintaining the search paths within the high density areas. Cody (45) considered the foraging pathways of entire flocks, and went so far as to suggest that the foraging path structure was adjusted to the renewal rate of the prey populations, so that the flock would return to previously gleaned areas as a fresh supply of prey became available. His observational substantiation of this suggestion is not convincing, however.

Some predators may exploit patchily distributed prey by responding to habitat patches rather than individual prey types. Royama (246) has formalized this idea, submitting that individuals may survey (by some means) many patch types within their environment, select those in which foraging is most profitable (by some measure, such as energy yield/energy expended in foraging), and forage more or less opportunistically within the selected patch types. A predator foraging on the basis of patch profitability must, however, make the decision when to "give up" on one patch and move to another. Gibb (99) hypothesized that individuals might hunt by "expectation," forming an expectation of patch yield on the basis of experience and leaving one patch for another after realizing that expectation. This view has been challenged (149, 257), and Charnov (40, 152) developed an alternative model. Charnov assumed that capture rate within a patch decreases with searching time because prey resources become depleted. When the capture rate drops below the average rate for the habitat as a whole, the predator should leave the patch for another. The model, given some simplifying assumptions, predicts that a predator should have a constant "giving up time" for all patch types within a habitat, but that giving up time should be shorter in better habitats, where the average capture rate is higher. Experimental studies by Smith & Dawkins (264) and Simons & Alcock (257) have lent strength to the patch profitability idea, and Krebs et al (152) have obtained findings consistent with Charnov's model. Smith & Sweatman (265) have undertaken one of the most thorough analyses of these hypotheses, supplementing experiments with field observations of tits (*Parus*). On the average, the birds did select the most profitable feeding patches, and exhibited some ability to respond to temporal changes in patch quality. Upon finding a profitable patch, they tended to stick with it, shifting patch utilization more or less in accordance with Charnov's predictions. Their evidence suggested that tits forage efficiently, although somewhat less than predicted by optimization theory.

The problem of how many patch types to include in a foraging itinerary was addressed by MacArthur & Pianka (178) in terms of the reduction in interpatch travel time and the increase in hunting time per prey capture accompanying the addition of another patch type to the repertoire. Their model produced several interesting predictions: predators which spend considerable time in pursuit of prey should be expected to show greater patch restriction when food is abundant than predators which spend most of their foraging time searching for prey, and as food density in the patches is reduced (by competitors), the optimal predator should reduce the number of patch types it visits but not the range of prey taken within the patches [this has come to be known as the "compression hypothesis" (176)]. Unfortunately, no convincing tests of these ideas have yet appeared.

Such patch-dependent foraging may have fundamental implications in consideration of the trophic webs about which most analyses and models of ecosystems are built. If prey densities are initially unequal among patches, for example, patch profitability foraging may lead to greater uniformity of prey densities (265). And if a predator population has a patchy distribution relative to that of its prey (i.e. its dispersion is determined by factors other than food supply), the effects of predation (and the consequent energy and nutrient flows) will be concentrated on certain segments of the prey population rather than evenly distributed, as is usually assumed.

Population Stability

Spatial patterning may also contribute to the stabilization of predator-prey population dynamics. Early bottle experiments (e.g. 94) and more recent experiments (e.g. 128) as well as many theoretical arguments (e.g, 26, 176, 185, 261, 286) generally point to the conclusion that spatial patchiness, which provides elements of the prey population temporary refuges from the predator and thus profoundly alters the mean number of prey caught, contributes to the stabilization of otherwise unstable predator-prey interactions. Much depends, however, on the prey having a more rapid dispersal or interpatch migration rate, or a greater reproductive rate, than the predator. In an opposing vein, Steele (273) has suggested that small-scale spatial variability in a system may displace prey and predator populations from their stable limit cycle and drive the system to extinction regardless of the rates of dispersal.

The most famous "natural" example of the importance of prey refuges is probably the relation between prickly pear cactus (*Opuntia*) and the cactus-feeding moth (*Cactoblastis cactorum*). *Opuntia* was introduced to Australia in the early nineteenth century and escaped from cultivation to eventually cover some 60 million acres of rangeland. The moth was introduced to control the cactus, and was eminently successful, destroying the cactus over large areas and leaving only small widely distributed enclaves of cactus. The decline of the cactus of course resulted in the decline of the moth. At the present time, the system exists in an unsteady equilibrium: patches of *Opuntia* in which moths occur are eventually destroyed by the moths, but, in the meantime, seed scattered in new areas may produce new clusters of cactus that will grow until they are eventually discovered by dispersing moths. The patchy distribution of *Opuntia* in time and space prevents its total eradication by *Cactoblastis.* Monro (196) has investigated the *Cactoblastis-Opuntia* relationship in some detail.

Spatial heterogeneity may have broader population effects in contributing to stability than those related to predator-prey interactions alone. den Boer & Reddingius (63, 230, 232) have suggested that populations can achieve increased stability by "spreading of risk" or "gambling for existence." They argue that a population occupying a range of habitat patches is subject to different events and controls in each patch type, and thus the probability of local perturbations leading to population extinction decreases as some function of the number of units (patch types) over which this risk is spread. The theory has group selection overtones, and has recently been criticized on mathematical grounds (158, 236), but there is some simulation and empirical support (64, 65, 231, 266).

Genetic Structure

I have previously mentioned some ways in which population responses to the patch structure of environments can affect the genetic composition of the population, and other reviews in this volume consider genetic correlates of environmental heterogeneity in greater detail. Here I touch briefly upon two aspects of this relationship.

The theoretical treatments of Levene (159) and Levins (165, 166) predict that genetic heterozygosity can increase with spatial and/or temporal environmental variation. In particular, coarse-grained species occupying a patchy environment may be exposed to different intensities and directions of selection in different patch types, and may occupy small patches in which gene frequencies may be subject to founder effects or drift; further, the patch structure, and thus the selection pressures, are likely to change through time (67). Under such conditions a wide array of theoretical and observational studies (e.g. 12, 42, 103, 104, 167, 187, 204, 276) indicate that genetic polymorphism should be great. For fine-grained organisms, however, the expectations are different. Here individuals encounter a variety of environmental conditions, but different individuals within a population probably encounter roughly the same average set of conditions. Thus individual flexibility in responsiveness rather than fixed polymorphism should be favored by selection. This might be achieved either through a general "all-purpose" genetic structure (relatively low heterozygosity with alleles subject to a wide range of environmental canalization) or through possessing alternative enzyme forms to contend with widely different reaction conditions (high heterozygosity). Both strategies may in fact be followed, and population geneticists have yet to agree on predictions. Selander & Kaufman (255) submitted that vertebrates, by virtue of their larger body size, greater individual mobility, and greater degree of homeostatic control, may be considered fine-grained in comparison to invertebrates, which may exhibit a greater degree of patch specialization. Their review of levels of enzyme variation demonstrated that vertebrates (unfortunately, primarily small mammals in their sample) displayed less genic heterozygosity than invertebrates (primarily *Drosophila* species). Johnson (141) and Bryant (34), on the other hand, presented evidence that fine-grained individuals respond to the variability of environmental conditions they encounter through diversity of enzyme forms. In a simulation model, Bryant (34) considered the effects of both environmental variance and temporal predictability (autocorrelation) on heterozygosity for an environment containing two habitat patch types (Figure 5). Low environmental variance relative to the difference between patches may favor a coarse-grained response by the population (i.e. the patches are stable, and patch specialization is feasible), favoring homozygosity. As environmental variance increases, habitat distribution tends toward fine-grainness, and heterozygosity should increase. The tendency toward fine-grainness is ac-

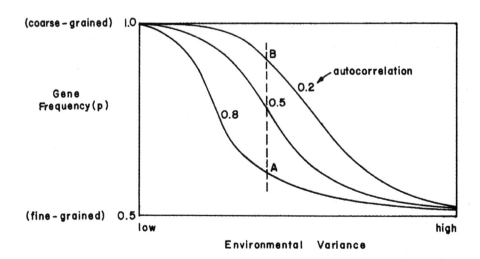

Figure 5 An hypothesized relationship between environmental variance in space (or time) and the extent of heterozygosity in a population. As environmental variation increases, gene frequencies shift toward greater heterozygosity, and the speed of this shift is a function of the "predictability" or autocorrelation of environmental variations. To the extent that different alleles are associated with different environmental situations, a homozygous condition is equivalent to coarse-grainness, heterozygosity to fine-grainness. Modified from the simulations of Bryant (34).

celerated if there is high environmental autocorrelation, resulting in an "opportunistic" mode of habitat response with relatively close tracking of environmental variations. Or, as Johnson (141, p. 36) has independently noted, "when a predictably variable environment is rendered less predictable, the effect may be to increase the probability that an individual will encounter constant conditions, thus reducing the advantage of the heterozygote." In Figure 5 this is represented by a shift from point *A* to point *B*. More complex arguments have been developed by Bryant (35) and Gillespie (103, 104). Considerations of the genetic structure of central and marginally located populations (e.g. 16, 225, 267) express such ideas upon a larger spatial scale. We may note in passing that one of the consequences of genetic polymorphism may be a "spreading of risk" over a variety of phenotypes, which may contribute to population stability (63, 123, 232).

Social organization, interacting with environmental patchiness, may also influence the breeding system of populations, and thus their genetic structure. Such interactions are perhaps most evident in small mammal populations, which have received extensive study. Populations of house mice (*Mus musculus*) in Alberta granaries (11) or barns in Texas (254) or Poland (2) are subdivided into small social family groups of perhaps 4–7 reproductively active individuals. Males are dominant within these groups and mate polygynously. The animals defend territories and exercise some degree of habitat selection. Young individuals disperse from these social groups, and are the chief means of colonization of newly opened habitat, but interchange between social groups is apparently slight. According to Anderson (11,

p. 309) the life expectancy and survival rates of these dispersing individuals are low, and their contribution to the ongoing gene pool minor; nonetheless, "their physiological and behavioural commitment to ecological and evolutionary exploration is crucial to the occupancy of niches in which nature appears particularly capricious." Selander (254) proposed that such a social structure promotes efficient resource utilization, but there are obvious and documented effects upon the genetic features of the populations as well. Since most groups are initiated by small numbers of dispersing individuals, founder effects may be commonplace. Further, the dominance-defined breeding system depresses the effective population size (62), enhancing drift and inbreeding. All of this contributes to a rather substantial genetic heterogeneity between the small "social demes." Similar breeding structures have been reported for *Peromyscus* (229) and *Apodemus* (33), although the genetic consequences are less well documented.

Recent studies in more natural settings have cast doubt upon the generality of these patterns among small mammals, although not upon their existence in semi-enclosed "artificial" situations. Dispersal and intergroup mixing seem far more commonplace in "natural" habitats (203); up to one-quarter of the individuals may breed in areas other than their birthplace (25). Group territories are thus only partial barriers to gene flow, and these "natural" populations appear to lack the genetic heterogeneity that characterizes barn or granary groups. Unfortunately, both field situations were located on islands, so their generality may perhaps also be questioned.

AN EXAMPLE: HARVESTER ANT COLONIES

Harvester ants of the genus *Pogonomyrmex* may play a major role in creating habitat patchiness where they occur, and an interconnected web of patch-related population features is founded upon this patch structure. The ants occupy a broad array of arid or early successional habitats, specializing to a degree upon exploitation of the rich (but erratic) seed production of these habitats; many of the features of their behavior and ecology may be interpreted in terms of adaptation to utilization of this seed resource (327). The ants establish underground colonies of up to several thousand individuals, and clear the vegetation from a well-defined disc about the surface mound, thus creating well-defined patches. Colonies tend to be evenly spaced (Figure 6), and are actively defended against small or large intruders, especially if the colony is large and well established (328). Individuals forage in a typical "refuging" pattern in the area surrounding the colony, and their foraging activities are closely related to proximate environmental conditions, including variations in food density (23, 24, 122, 305). Species of *Pogonomyrmex* differ in important fashions, some of which have been related to competition (24, 122, 305). In Figure 6, I have suggested some of the general relationships associated with the generation of such crisply defined patchiness by the disc-clearing activities of harvester ants. Obviously, not all of these effects occur at a given colony, or even in a given species.

The area of the discs cleared by the ants varies with habitat conditions and the density of colonies. In Colorado shortgrass prairies, Rogers (239) recorded up to 31

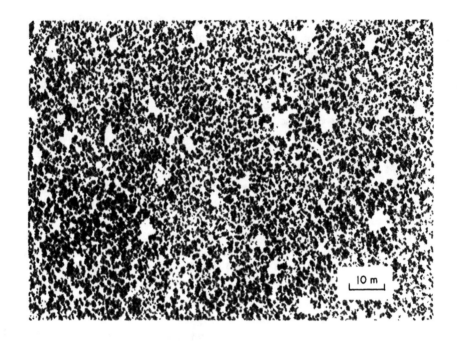

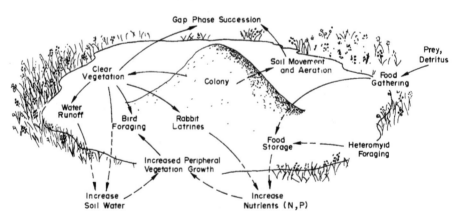

Figure 6 (*Top*) Aerial photograph of the spatial pattern of *Pogonomyrmex* colonies in *Chrysothamnus-Artemisia* shrub desert in central Oregon. (*Bottom*) A generalized conception of the interactions which are related to the colonial activities of *Pogonomyrmex* harvester ants.

colonies/ha, and the cleared discs covered 1.9–27.3 m²/ha, up to 0.3% of the total ground area. In semiarid rangeland in central Oregon, colonies may number 8–12/ha. The ants clear an average area of 22.5 m² about the colony, so 11–17% of the ground surface may be denuded (322). Clearing of the vegetation influences soil water characteristics below the disc. Soil water is increased in this zone, largely because of the elimination of transpiring plants in the disc area, but also because of water runoff from the colony mound into the surrounding disc (238). The increased soil water as well as the increased nutrient availability in the vicinity of the colony (see below) results in an increase in vegetation growth and standing crop around the disc periphery (96, 105, 238). This zone of stimulated plant growth appears to be attractive to some grassland bird species for foraging (personal observation), perhaps by providing areas of insect concentrations or because windblown seeds become trapped in the denser vegetation about the disc edge. Sage grouse (*Centrocercus urophasianus*) cocks may enhance their visibility during lek courtship dances by strutting on the elevated mounds (100).

The ant colony also has important influences on soil characteristics. Rogers (238) calculated that harvester ants in shortgrass prairies may move as much as 2.8 kg of soil per colony per year. This soil movement affects the redistribution and aeration of the soil, and may increase the "quality" of the colony area for seed germination of a number of plant species, although in active colonies the seedlings are rapidly destroyed by the ants (322). Upon abandonment of the colony, typical succession may ensue (105), and the ants may thus contribute markedly to the vegetation mosaic of their habitats. While there may be considerable movement of colonies, once established they appear to persist for a considerable time. Gentry (95), for example, observed a series of 100 colonies for 15 years, during which time only five colonies "died." Michener (195) reported one colony at least 40 years old!

Perhaps the major influences of colonies are through their prey gathering and concentration of nutrients in the colony area. Harvester ants forage primarily upon seeds, although litter, insects, bird droppings, and other detritus may comprise a significant portion of their gatherings. Material is gathered over an area up to 15–50 m from the colony, depending on the species and food density (24, 105, 122, 237). In *Pogonomyrmex,* individuals apparently forage independently of one another, some species employing "trunk trails" leading from the colony to feeding areas (122). Another desert harvester ant (*Veromessor pergandei*) employs group foraging tactics, all foragers from a colony searching for food in the same direction, forming a long continuous column from the nest which in time rotates in a circular fashion about the nest (24). Annual energy flow into a single *Pogonomyrmex* colony in shortgrass prairies may be as much as 1.45 $kcal/m^2/yr$ (239), which is equal to the entire energy consumption by the bird populations of such areas during a normal breeding season (316). A good deal of the food is stored underground, concentrating nutrients such as P, K, or NO_3 from the surrounding area into a small patch at the colony location (96, 238). Since the colonies have considerable longevity, this may introduce a substantial time delay into nutrient cycling in the system. The increase in nutrient levels may be accentuated by the behavior of rabbits, which in at least some areas apparently prefer to use the cleared disc areas as latrines (116). The underground food stores may also play an important role in the population dynamics of heteromyid rodents such as kangaroo rats (*Dipodomys*), which may excavate the food stores in early spring, when the seed production from the prior growing season is depleted and the current season's production has not yet begun (43).

Harvester ants are eaten by a variety of organisms (95), and some exhibit responses or adaptations to the patchy distribution of this prey. Horned lizards (*Phrynosoma*), for example, specialize upon ants, and have been seen gleaning individuals from harvester ant colonies (195). They possess an integrated constellation of anatomical, behavioral, physiological, and ecological adaptations centered about exploitation of ants as spatially concentrated prey. These have been documented by Pianka & Parker (221), who, following optimal foraging theory (178), proposed that *Phrynosoma* can "afford" to be ant specialists only because the ants occur in clumped concentrations.

CONCLUDING COMMENTS

These varied considerations of environmental and population heterogeneity have fundamental implications for attempts to model the dynamics of total ecosystems. Thus, a patchy distribution of predator activities creates uneven effects upon prey populations, and thereby upon the trophic pathways which bind model ecosystems together. If predator and prey populations respond to different scales or perceptual dimensions of patch structure, the effects may be especially obscure. The creation of local concentrations or time lags in energy or nutrient flows (e.g. at harvester ant colonies) may introduce important discontinuities into model simulations or ecosystem behavior. Nor can genetic polymorphisms within populations be ignored, especially if they characterize the dominant or "important" populations in the system. There are additional questions which real-world heterogeneity forces us to address if we are to unravel the biology of total ecosystems. Thus:

1. What are the effects of a patchy distribution or concentration of consumption on the trophic structure of an ecosystem? If consumers forage by "patch profitability," does this tend to loosen the interconnectedness of food webs, or to reduce their potential for "functional" control roles in the system?

2. Patchiness, by providing local "refugia," may enhance spreading of risk, and thus system persistence or resiliency. Can this be taken to imply that patchy systems have a higher "fitness" than homogeneous systems?

3. Does a polymorphic population structure, driven by coarse-grained responses, tend to reduce the effect of temporal variations on systems properties, as Gillespie (103) has suggested?

4. Grant & Morris (108) have predicted that patchy systems may be more subject to extrinsic (abiotic) controls than uniform systems, where intrinsic (biotic or populational) controls are more important. How does this influence our consideration of abiotic driving variables in ecosystem analyses?

5. Does a patchy environment allow both coarse-grained and fine-grained strategies to be pursued by different populations within a functional group or "guild," and thus provide a mixture of close tracking of environmental variations with relative immunity to short-term variations?

6. Gap phase processes create patches of different successional "maturity." Margalef (180) suggested that more mature systems should exploit or draw energy and nutrients from less mature systems. Do these relationships hold between patch types within a system? Does the development of patchiness tend to accelerate the discontinuity of nutrient or energy flow patterns in a system, and does this affect system "stability"?

7. Are patchy systems more or less efficient in energy and/or nutrient flow relations than more homogeneous systems?

110   WIENS

Much of what I have reviewed in this paper resides more in the realm of theory than fact, and, indeed, there are many instances in which the theory is ambiguous, superficial, or contradictory. I think this to a large degree reflects the general paucity of good information on how patchiness is expressed in nature, how individual organisms respond to these discontinuities in their environments, and how these responses collectively produce attributes characterizing entire populations, communities, or ecosystems. Obtaining the necessary information will require intensive field studies undertaken with a conceptual foundation of environmental heterogeneity, with patches as the organizing framework. Two generalities emerge from this review which bear upon the conduct of such studies. First, it is essential that the fabric of spatial scales on which patchiness is expressed be unraveled, and the structure of spatial heterogeneity be related to the variations in environmental states on diverse time scales. The key to achieving this is in shedding our own conceptions of environmental scale and instead concentrating on the perceptions of the organisms, attempting to view environmental structure through their senses. This is what old natural historians used to do, and in the rush to become scientifically accepted (that is, quantitative and abstract), we have often left the plants and animals behind. Unfortunately, the proper approach requires long-term studies, and this does not conveniently fit a thesis project or a two-year grant period.

The proper study of patchiness is also hindered by the multiplicity of disciplines involved. Theory tends to develop within somewhat discrete, conventionally defined areas of science, but natural populations care little about how scientists and theoreticians define their specialities. Population responses to patchiness form an integrated whole, and these in turn have feedback effects on environmental dispersion patterns over a wide range of spatial and temporal scales. Further, we have frequently studied these systems as if they were in equilibrium, but our intuitions of Nature should tell us that they are likely not. We must therefore give greater emphasis to variability, not just as a statistical measure, but as a fundamental attribute of biological systems.

ACKNOWLEDGMENTS

The idea for attempting this review developed from discussions with Bill Hunt, Jim Ellis, Charles Rodell, and Mel Dyer during a sabbatical leave at the Natural Resource Ecology Laboratory, Colorado State University. I am indebted to George Van Dyne and Jim Gibson for affording me the facilities of the laboratory during that time. Many of my colleagues in the Behavioral Ecology Laboratory at Oregon State University argued theory with me, and N. H. Anderson, Kenton Chambers, H. H. Crowell, Peter Dawson, Wayne Hoffman, Russell Riddle, John Rotenberry, and Jim Taylor commented on the manuscript. My own interest in the relations between populations and their environments was kindled by my father and reinforced by George Sutton, Charles Carpenter, and John Emlen. My work has been supported by the National Science Foundation, most recently through Grants GB-42741 and BMS-75-11898. This is contribution number 51 of the Behavioral Ecology Laboratory, Oregon State University.

Literature Cited

1. Abbott, H. G., Quink, T. F. 1970. Ecology of eastern white pine seed caches made by small forest mammals. *Ecology* 51:271–78
2. Adamczyk, K., Petrusewicz, K. 1966. Dynamics, diversity and intrapopulation differentiation of a free-living population of house mouse. *Ekol. Polska Ser. A* 14:725–40
3. Adams, L., Davis, S. D. 1967. The internal anatomy of home range. *J. Mammal.* 48:529–36
4. Alcock, J. 1969. Observational learning in three species of birds. *Ibis* 111:308–21
5. Alexander, R. D. 1974. The evolution of social behavior. *Ann. Rev. Ecol. Syst.* 5:325–83
6. Amson, J. C. 1972. The dependence of population distribution on location costs. *Environ. Plan.* 4:163–81
7. Amson, J. C. 1972. Equilibrium models of cities: 1. An axiomatic theory. *Environ. Plan.* 4:429–44
8. Amson, J. C. 1973. Equilibrium models of cities: 2. Single-species cities. *Environ. Plan.* 5:295–338
9. Anderson, D. J. 1971. Spatial patterns in some Australian dryland plant communities. See Ref. 216, pp. 271–86
10. Anderson, D. J. 1971. Pattern in desert perennials. *J. Ecol.* 59:555–60
11. Anderson, P. K. 1970. Ecological structure and gene flow in small mammals. *Symp. Zool. Soc. London* 26:299–325
12. Antonovics, J. 1971. The effects of a heterogeneous environment on the genetics of natural populations. *Am. Sci.* 59:593–99
13. Ashby, K. R. 1967. Studies on the ecology of field mice and voles (*Apodemus sylvaticus, Clethrionomys glareolus* and *Microtus agrestis*) in Houghall Wood, Durham. *J. Zool. London* 152:389–513
14. Auclair, A. N., Goff, F. G. 1971. Diversity relations of upland forests in the western Great Lakes area. *Am. Nat.* 105:499–528
15. Austin, M. P. 1968. Pattern in a *Zerna erecta* dominated community. *J. Ecol.* 56:197–218
16. Ayala, F. J., Powell, J. R., Dobzhansky, Th. 1971. Polymorphism in continental and island populations of *Drosophila willistoni. Proc. Natl. Acad. Sci. USA* 68:2480–83
17. Barbour, M. G. 1973. Desert dogma reexamined: Root/shoot productivity and plant spacing. *Am. Midl. Nat.* 89:41–57
18. Barlow, G. W. 1974. Hexagonal territories. *Anim. Behav.* 22:876–78
19. Bartholomew, G. A. 1970. A model for the evolution of pinniped polygyny. *Evolution* 24:546–59
20. Batzli, G. O. 1968. Dispersion patterns of mice in California annual grassland. *J. Mammal.* 49:239–50
21. Beals, E. W. 1968. Spatial pattern of shrubs on a desert plain in Ethiopia. *Ecology* 49:744–46
22. Berndt, R., Sternberg, H. 1968. Terms, studies and experiments on the problems of bird dispersion. *Ibis* 110:256–69
23. Bernstein, R. A. 1974. Seasonal food abundance and foraging activity in some desert ants. *Am. Nat.* 108:490–98
24. Bernstein, R. A. 1975. Foraging strategies of ants in response to variable food density. *Ecology* 56:213–19
25. Berry, R. J., Jakobson, M. E. 1974. Vagility in an island population of the house mouse. *J. Zool. London* 173:341–54
26. Birch, L. C. 1971. The role of environmental heterogeneity and genetical heterogeneity in determining distribution and abundance. *Proc. Adv. Study. Inst. Dyn. Numbers Popul. Oosterbeek, 1970* pp. 109–28
27. Boardman, M., Askew, R. R., Cook, L. M. 1974. Experiments on resting site selection by nocturnal moths. *J. Zool. London* 172:343–55
28. Breder, C. M. 1967. On the survival value of fish schools. *Zoologica* 52:25–40
29. Brock, V. E., Riffenburgh, R. H. 1963. Fish schooling: A possible factor in reducing predation. *J. Cons. Cons. Int. Explor. Mer.* 25:307–17
30. Brown, J. L. 1964. The evolution of diversity in avian territorial systems. *Wilson Bull.* 76:160–69
31. Brown, J. L. 1974. Alternate routes to sociality in jays—with a theory for the evolution of altruism and communal breeding. *Am. Zool.* 14:63–80
32. Brown, J. L., Orians, G. H. 1970. Spacing patterns in mobile animals. *Ann. Rev. Ecol. Syst.* 1:239–62
33. Brown, L. E. 1969. Field experiments on the movements of *Apodemus sylvaticus* L. using trapping and tracking techniques. *Oecologia* 2:198–222
34. Bryant, E. H. 1973. Habitat selection in a variable environment. *J. Theor. Biol.* 41:421–29

112   WIENS

35. Bryant, E. H. 1974. On the adaptive significance of enzyme polymorphisms in relation to environmental variability. *Am. Nat.* 108:1–28
36. Buechner, H. K., Dawkins, H. C. 1961. Vegetation change induced by elephant and fire in the Murchison Falls National Park, Uganda. *Ecology* 42:752–66
37. Buechner, H. K., Roth, H. D. 1974. The lek system in Uganda Kob antelope. *Am. Zool.* 14:145–62
38. Carl, E. A. 1971. Population control in Arctic Ground Squirrels. *Ecology* 52:395–413
39. Carter, M. A. 1968. Thrush predation of an experimental population of the snail *Cepaea nemoralis* (L.). *Proc. Linn. Soc. London* 179:241–49
40. Charnov, E. L. 1973. *Optimal foraging: some theoretical explorations.* PhD thesis. Univ. Washington, Seattle, Wash.
41. Chitty, D. 1960. Population processes in the vole and their relevance to general theory. *Can. J. Zool.* 38:99–113
42. Christiansen, F. B. 1974. Sufficient conditions for protected polymorphism in a subdivided population. *Am. Nat.* 108:157–66
43. Clark, W. H., Comanor, P. L. 1973. The use of Western Harvester Ant, *Pogonomyrmex occidentalis* (Cresson), seed stores by heteromyid rodents. *Biol. Soc. Nevada Occas. Pap.* No. 34, 6 pp.
44. Cody, M. L. 1968. On the methods of resource division in grassland bird communities. *Am. Nat.* 102:107–47
45. Cody, M. L. 1971. Finch flocks in the Mohave Desert. *Theor. Popul. Biol.* 2:142–58
46. Cody, M. L. 1974. Optimization in ecology. *Science* 183:1156–64
47. Cody, M. L. 1974. *Competition and the Structure of Bird Communities.* Princeton, New Jersey: Princeton Univ. Press. 318 pp.
48. Cohen, D. 1967. Optimization of seasonal migratory behavior. *Am. Nat.* 101:1–17
49. Cohen, J. E. 1971. *Causal Groups of Monkeys and Men: Stochastic Models of Elemental Social Systems.* Cambridge: Harvard Univ. Press. 175 pp.
50. Colwell, R. K., Futuyma, D. J. 1971. On the measurement of niche breadth and overlap. *Ecology* 52:567–76
51. Connell, J. H. 1971. On the role of natural enemies in preventing competitive exclusion in some marine animals and in rain forest trees. *Proc. Adv. Study*

Inst. Dyn. Numbers Popul. Oosterbeek, 1970 pp. 298–312
52. Corbet, P. S. 1962. *A Biology of Dragonflies.* London: Witherby. 247 pp.
53. Crook, J. H. 1962. The adaptive significance of pair formation types in weaver birds. *Symp. Zool. Soc. London* 8:59–70
54. Crook, J. H. 1965. The adaptive significance of avian social organizations. *Symp. Zool. Soc. London* 14:181–218
55. Crook, J. H. ed. 1970. *Social Behaviour in Birds and Mammals.* New York: Academic. 492 pp.
56. Crook, J. H. 1972. Sexual selection, dimorphism, and social organization in the primates. In *Sexual Selection and the Descent of Man,* ed. B. G. Campbell, pp. 231–81. Chicago: Aldine
57. Croze, H. 1970. Searching image in Carrion Crows. *Z. Tierpsychol. Beih.* 5: 85 pp.
58. Croze, H. 1974. The Seronera bull problem. II. The trees. *E. Afr. Wildl. J.* 12:29–47
59. Crumpacker, D. W., Williams, J. S. 1973. Density, dispersion, and population structure in *Drosophila pseudoobscura. Ecol. Monogr.* 43:499–538
60. Dacey, M. F. 1971. Regularity in spatial distributions: A stochastic model of the imperfect central place plane. See Ref. 216, pp. 287–309
61. Dayton, P. K. 1971. Competition, disturbance, and community organization: The provision and subsequent utilization of space in a rocky intertidal community. *Ecol. Monogr.* 41:351–89
62. DeFries, J. C., McClearn, G. E. 1972. Behavioral genetics and the fine structure of mouse populations—a study in microevolution. In *Evolutionary Biology,* ed. T. Dobzhansky, M. R. Hecht, W. C. Steere, 5:279–91. New York: Appleton-Century-Crofts
63. den Boer, P. J. 1968. Spreading of risk and stabilization of animal numbers. *Acta Biotheor.* 18:165–94
64. den Boer, P. J. 1971. Stabilization of animal numbers and the heterogeneity of the environment: The problem of the persistence of sparce populations. *Proc. Adv. Study Inst. Dyn. Numbers Popul. Oosterbeek, 1970* pp. 77–97
65. den Boer, P. J. 1974. An answer to the comment of Roff. *Am. Nat.* 108:394–96
66. Denham, P. J. Energy relations and some basic properties of primate social organization. *Am. Anthropol.* 73:77–95
67. Dickinson, H., Antonovics, J. 1973. The effects of environmental hetero-

geneity on the genetics of finite populations. *Genetics* 73:713–35

68. Dobzhansky, Th., Wright, S. 1943. Genetics of natural populations. X. Dispersion rates in *Drosophila pseudoobscura. Genetics* 28:304–40

69. Dolinger, P. M., Ehrlich, P. R., Fitch, W. L., Breedlove, D. E. 1973. Alkaloid and predation patterns in Colorado lupine populations. *Oecologia* 13:191–204

70. Downhower, J. F., Armitage, K. B. 1971. The yellow-bellied marmot and the evolution of polygamy. *Am. Nat.* 105:355–70

71. Drury, W. H., Nisbet, I.C.T. 1973. Succession. *J. Arnold Arbor. Harv. Univ.* 54:331–68

72. Dubois, D. M. 1975. A model of patchiness for prey-predator plankton populations. *Ecol. Model.* 1:67–80

73. Dubois, D. M. 1975. Simulation of the spatial structuration of a patch of prey-predator populations in the southern bight of the North Sea. *Mem. Soc. Ro. Sci. Liége Ser. 6* 7:75–82

74. Dwyer, P. D. 1972. Feature, patch and refuge area: Some influences on diversity of bird species. *Emu* 72:149–56

75. Ehrlich, P. R., Raven, P. H. 1965. Butterflies and plants: A study in coevolution. *Evolution* 18:586–608

76. Ehrlich, P. R., Raven, P. H. 1969. Differentiation of populations. *Science* 165:1228–32

77. Eisenberg, J. F. 1966. The social organization of mammals. *Handb. Zool.* 10(7):1–92

78. Ellis, J. E., Wiens, J. A., Rodell, C. F., Anway, J. C. 1976. A conceptual model of diet selection as an ecosystem process. *J. Theor. Biol.* 43:In press

79. Elton, C. S. 1949. Population interspersion: An essay on animal community patterns. *J. Ecol.* 37:1–23

80. Elton, C. S. 1966. *The Pattern of Animal Communities.* London: Methuen. 432 pp.

81. Endler, J. A. 1973. Gene flow and population differentiation. *Science* 179:243–50

82. Etkin, W. ed. 1964. *Social Behavior and Organization among Vertebrates.* Chicago: Chicago Univ. Press. 307 pp.

83. Feeny, P. 1970. Seasonal changes in oak leaf tannins and nutrients as a cause of spring feeding by winter moth caterpillars. *Ecology* 51:565–81

84. Feeny, P. 1975. Biochemical coevolution between plants and their insect herbivores. In *Coevolution of Animals and Plants,* ed. L. E. Gilbert, P. H. Raven, pp. 3–19. Austin: Univ. Texas Press

85. Fisler, G. F. 1969. Mammalian organizational systems. *Los Angeles County Mus. Contrib. Sci. No. 167* 32 pp.

86. Ford, E. B. 1975. *Ecological Genetics.* London: Chapman & Hall. 442 pp. 4th ed.

87. Forman, R. T. T. 1964. Growth under controlled conditions to explain the hierarchial distribution of a moss, *Tetraphis pellucida. Ecol. Monogr.* 34:1–25

88. Freeland, W. J., Janzen, D. H. 1974. Strategies in herbivory by mammals: The role of plant secondary compounds. *Am. Nat.* 108:269–89

89. French, N. R., Tagami, T. Y., Hayden, P. 1968. Dispersal in a population of desert rodents. *J. Mammal.* 49:272–80

90. Fretwell, S. D. 1972. *Populations in a Seasonal Environment.* Princeton, New Jersey: Princeton Univ. Press. 217 pp.

91. Fretwell, S. D., Lucas, H. L. Jr. 1969. On territorial behavior and other factors influencing habitat distribution in birds. I. Theoretical development. *Acta Biotheor.* 19:16–36

92. Gadgil, M. 1971. Dispersal: Population consequences and evolution. *Ecology* 52:253–61

93. Gashwiler, J. S. 1970. Further study of conifer seed survival in a western Oregon clearcut. *Ecology* 51:849–54

94. Gause, G. F. 1934. *The Struggle for Existence.* Baltimore: Williams & Wilkins

95. Gentry, J. B. 1974. Response to predation by colonies of the Florida Harvester Ant, *Pogonomyrmex badius. Ecology* 55:1328–38

96. Gentry, J. B., Stiritz, K. L. 1972. The role of the Florida Harvester Ant, *Pogonomyrmex badius,* in old field mineral nutrient relationships. *Environ. Entomol.* 1:39–41

97. Getz, L. L. 1970. Influence of vegetation on the local distribution of the meadow vole in southern Wisconsin. *Univ. Conn. Occas. Pap. Biol. Sci. Ser.* 1:213–41

98. Ghiselin, J. 1970. Edaphic control of habitat selection by kangaroo mice (*Microdipodops*) in three Nevadan populations. *Oecologia* 4:248–61

99. Gibb, J. A. 1962. L. Tinbergen's hypothesis of the role of specific search images. *Ibis* 104:106–11

100. Giezentanner, K. I., Clark, W. H. 1974. The use of western Harvester Ant mounds as strutting locations by Sage Grouse. *Condor* 76:218–19

114 WIENS

101. Gilbert, L. E., Singer, M. C. 1973. Dispersal and gene flow in a butterfly species. *Am. Nat.* 107:58–72
102. Gilbert L. E., Singer, M. C. 1975. Butterfly ecology. *Ann. Rev. Ecol. Syst.* 6:365–97
103. Gillespie, J. 1974. Polymorphism in patchy environments. *Am. Nat.* 108:145–51
104. Gillespie, J. 1974. The role of environmental grain in the maintenance of genetic variation. *Am. Nat.* 108:831–36
105. Golley, F. B. Gentry, J. B. 1964. Bioenergetics of the southern Harvester Ant, *Pogonomyrmex badius. Ecology* 45:217–25
106. Goss-Custard, J. D. 1970. Feeding dispersion in some over-wintering wading birds. See Ref. 55, pp. 3–35
107. Grant, P. R. 1968. Polyhedral territories of animals. *Am. Nat.* 102:75–80
108. Grant, P. R., Morris, R. D. 1971. The distribution of *Microtus pennsylvanicus* within grassland habitat. *Can. J. Zool.* 49:1043–52
109. Haartmann, L. von 1969. Nest-site and evolution of polygamy in European passerine birds. *Ornis Fenn.* 46:1–12
110. Hamilton, B. A., Hutchinson, K. J., Annis, P. C., Donnelly, J. B. 1973. Relationships between the diet selected by grazing sheep and the herbage on offer. *Aust. J. Agric. Res.* 24:271–77
111. Hamilton, W. D. 1971. Geometry for the selfish herd. *J. Theor. Biol.* 31:295–311
112. Hamilton, W. J. III, Gilbert, W. M. 1969. Starling dispersal from a winter roost. *Ecology* 51:886–98
113. Hamilton, W. J. III, Gilbert, W. M., Heppner, F. H., Planck, R. J. 1967. Starling roost dispersal and a hypothetical mechanism regulating rhythmical animal movement to and from dispersal centers. *Ecology* 48:825–33
114. Hamilton, W. J. III, Watt, K. E. F. 1970. Refuging. *Ann. Rev. Ecol. Syst.* 1:263–87
115. Harberd, D. J. 1961. Observations on population structure and longevity of *Festuca rubra* L. *New Phytol.* 60:184–206
116. Harper, J. L. 1969. The role of predation in vegetational diversity. *Brookhaven Symp. Biol.* 22:48–62
117. Harper, J. L., White, J. 1971. The dynamics of plant populations. *Proc. Adv. Study Inst. Dyn. Numbers Popul. Oosterbeek, 1970* pp. 41–63
118. Harper, J. L., White, J. 1974. The demography of plants. *Ann. Rev. Ecol. Syst.* 5:419–63
119. Hastings, A., Rohlf, F. J. 1974. Gene flow: Effect in stochastic models of differentiation. *Am. Nat.* 108:701–5
120. Hildén, O. 1965. Habitat selection in birds. *Ann. Zool. Fenn.* 2:53–75
121. Hinds, W. T. 1975. Energy and carbon balances in cheatgrass: An essay in autecology. *Ecol. Monogr.* 45:367–88
122. Hölldobler, B. 1974. Home range, orientation and territoriality in harvesting ants. *Proc. Natl. Acad. Sci. USA* 71:3274–77
123. Holling, C. S. 1973. Resilience and stability of ecological systems. *Ann. Rev. Ecol. Syst.* 4:1–23
124. Holm, C. H. 1973. Breeding sex ratios, territoriality, and reproductive success in the Red-winged Blackbird (*Agelaius phoeniceus*). *Ecology* 54:356–65
125. Horn, H. S. 1968. The adaptive significance of colonial nesting in the Brewer's Blackbird (*Euphagus cyanocephalus*). *Ecology* 49:682–94
126. Horn, H. S. 1971. *The Adaptive Geometry of Trees.* Princeton, New Jersey: Princeton Univ. Press. 144 pp.
127. Howard, W. E. 1960. Innate and environmental dispersal of individual vertebrates. *Am. Midl. Nat.* 63:152–61
128. Huffaker, C. B. 1958. Experimental studies on Predation: Dispersion factors and predator-prey oscillations. *Hilgardia* 27:343–83
129. Huffaker, C. B. 1962. Some concepts on the ecological basis of biological control of weeds. *Can. Entomol.* 94:507–14
130. Hutchinson, G. E. 1953. The concept of pattern in ecology. *Proc. Acad. Nat. Sci. Philadelphia* 105:1–12
131. Hutchinson, G. E. 1957. Concluding remarks. *Cold Spring Harbor Symp. Quant. Biol.* 22:415–27
132. Hutchinson, G. E. 1965. *The Ecological Theater and the Evolutionary Play.* New Haven, Conn: Yale Univ. Press. 139 pp.
133. Hutchinson, G. E., MacArthur, R. H. 1959. A theoretical ecological model of size distributions among species of animals. *Am. Nat.* 93:117–25
134. Immelmann, K. 1975. Ecological significance of imprinting and early learning. *Ann. Rev. Ecol. Syst.* 6:15–37
135. Iwao, S. 1972. Application of the $\overset{*}{m}$-m method to the analysis of spatial patterns by changing the quadrat size. *Res. Popul. Ecol.* 14:97–128
136. Janzen, D. H. 1970. Herbivores and the

number of tree species in tropical forests. *Am. Nat.* 104:501–28

137. Janzen, D. H. 1971.Seed predation by animals. *Ann. Rev. Ecol. Syst.* 2:465–92

138. Jarman, P. J. 1974. The social organization of antelope in relation to their ecology. *Behaviour* 58:215–67

139. Jezierski, W. 1972. Elements of the space structure of European hare (*Lepus europaeus* Pallas, 1778) population. *Ekol. Polska* 20:593–607

140. Johnson, C. G. 1969. *Migration and Dispersal of Insects by Flight.* London: Methuen. 766 pp.

141. Johnson, G. B. 1974. Enzyme polymorphism and metabolism. *Science* 184:28–37

142. Katz, P. L. 1974. A long-term approach to foraging optimization. *Am. Nat.* 108:758–82

143. Keast, A. 1968. Seasonal movements in the Australian honeyeaters (Meliphagidae) and their ecological significance. *Emu* 67:159–209

144. Kettlewell, H. B. D. 1961. The phenomenon of industrial melanism in Lepidoptera. *Ann. Rev. Entomol.* 6:245–62

145. King, C. E. 1971. Resource specialization and equilibrium population size in patchy environments. *Proc. Natl. Acad. Sci. USA* 68:2634–37

146. Kitching, R. 1971. A simple simulation model of dispersal of animals among units of discrete habitats. *Oecologia* 7:95–116

147. Klomp, H. 1972. Regulation of the size of bird populations by means of territorial behaviour. *Neth. J. Zool.* 22:456–88

148. Klopfer, P. 1963. Behavioral aspects of habitat selection: The role of early experience. *Wilson Bull.* 75:15–22

149. Krebs, J. R. 1973. Behavioural aspects of predation. In *Perspectives in Ethology,* ed. P. P. G. Bateson, P. H. Klopfer, Chap. 3. New York: Plenum

150. Krebs, J. R. 1974. Colonial nesting and social feeding as strategies for exploiting food resources in the Great Blue Heron (*Ardea herodias*). *Behaviour* 51:99–134

151. Krebs, J. R., MacRoberts, M. H., Cullen, J. M. 1972. Flocking and feeding in the Great Tit *Parus major*—an experimental study. *Ibis* 114:507–30

152. Krebs, J. R., Ryan, J. C., Charnov, E. L. 1974. Hunting by expectation or optimal foraging? A study of patch use by chickadees. *Anim. Behav.* 22:953–64

153. LaTourrette, J. E., Young, J. A., Evans, R. A. 1971. Seed dispersal in relation to rodent activities in seral big sagebrush

communities. *J. Range Manage.* 24:118–20

154. Laws, R. M., Parker, I. S. C., Johnstone, R. C. B. 1975. *Elephants and Their Habitats.* London: Clarendon. 376 pp.

155. Lazarus, J. 1972. Natural selection and the functions of flocking in birds: A reply to Murton. *Ibis* 114:556–58

156. Leopold, A. 1933. *Game Management.* New York: Scribner

157. Leuthold, W. 1966. Variations in territorial behavior of Uganda Kob *Adenota kob thomasi* (Neumann 1896). *Behaviour* 27:215–58

158. Levandowsky, M. 1974. A further comment on the model of Reddingius and den Boer. *Am. Nat.* 108:118–20

159. Levene, H. 1953. Genetic equilibrium when more than one ecological niche is available. *Am. Nat.* 87:331–33

160. Levin, D. A. 1973. The role of trichomes in plant defense. *Q. Rev. Biol.* 48:3–15

161. Levin, D. A., Kerster, H. 1974. Gene flow in seed plants. *Evol. Biol.* 7:139–220

162. Levin, S. A. 1974. Dispersion and population interactions. *Am. Nat.* 108:207–28

163. Levin, S. A., Paine, R. T. 1974. Disturbance, patch formation, and community structure. *Proc. Natl. Acad. Sci. USA* 71:2744–47

164. Levin, S. A., Paine, R. T. 1975. The role of disturbance in models of community structure. In *Ecosystem Analysis and Prediction,* ed. S. A. Levin, pp. 56–67. Philadelphia: Soc. Ind. Appl. Math. 337 pp.

165. Levins, R. 1965. The theory of fitness in a heterogeneous environment: IV. The adaptive significance of gene flow. *Evolution* 18:635–38

166. Levins, R. 1968. *Evolution in Changing Environments.* Princeton, New Jersey: Princeton Univ. Press. 120 pp.

167. Levins, R., MacArthur, R. H. 1966. The maintenance of genetic polymorphism in a spatially heterogeneous environment: Variations on a theme by Howard Levene. *Am. Nat.* 100:585–89

168. Lewis, J. K. 1969. Range management viewed in the ecosystem framework. In *The Ecosystem Concept in Natural Resource Management,* ed. G. M. Van Dyne. pp. 97–187. New York: Academic

169. Lewontin, R. C. 1965. Selection for colonizing ability. In *The Genetics of Colonizing Species.* ed. H. G. Baker,

116   WIENS

G. L. Stebbins, pp. 77–94. New York: Academic

170. Lewontin, R. C. 1970. The units of selection. *Ann. Rev. Ecol. Syst.* 1:1–18

171. Lloyd, M. 1967. "Mean crowding." *J. Anim. Ecol.* 36:1–30

172. Loucks, O. L. 1970. Evolution of diversity, efficiency, and community stability. *Am. Zool.* 10:17–25

173. MacArthur, R. H. 1958. Population ecology of some warblers of the northeastern coniferous forest. *Ecology* 39:599–619

174. MacArthur, R. H. 1965. Patterns of species diversity. *Biol. Rev.* 40:510–33

175. MacArthur, R. H. 1968. The theory of the niche. In *Population Biology and Evolution,* ed. R. Lewontin. Syracuse, New York: Syracuse Univ. Press

176. MacArthur, R. H. 1972. *Geographical Ecology.* New York: Harper & Row. 269 pp.

177. MacArthur, R. H., Levins, R. 1964. Competition, habitat selection, and character displacement in a patchy environment. *Proc. Natl. Acad. Sci. USA* 51:1207–10

178. MacArthur, R. H., Pianka, E. R. 1966. On optimal use of a patchy environment. *Am. Nat.* 100:603–9

179. MacArthur, R. H., Wilson, E. O. 1967. *The Theory of Island Biogeography.* Princeton, New Jersey: Princeton Univ. Press. 203 pp.

180. Margalef, R. 1968. *Perspectives in Ecological Theory.* Chicago: Univ. Chicago Press. 111 pp.

181. Margalef, R. 1969. Diversity and stability: A practical proposal and a model of interdependence. *Brookhaven Symp. Biol.* 22:25–37

182. Martin, S. G. 1971. *Polygyny in the Bobolink: Habitat quality and the adaptive complex.* PhD thesis. Oregon State Univ., Corvallis, Oregon. 181 pp.

183. Martin, S. G. 1974. Adaptations for polygynous breeding in the Bobolink, *Dolichonyx oryzivorus. Am. Zool.* 14:108–19

184. Martinez, D. R., Klinghammer, E. 1970. The behavior of the whale *Orcinus orca*: A review of the literature. *Z. Tierpsychol.* 27:828–39

185. May, R. M. 1973. *Stability and Complexity in Model Ecosystems.* Princeton, New Jersey: Princeton Univ. Press. 235 pp.

186. Maynard Smith, J. 1966. Sympatric speciation. *Am. Nat.* 100:637–50

187. Maynard Smith, J. 1970. Genetic polymorphism in a varied environment. *Am. Nat.* 104:487–90

188. Mayr, E. 1963. *Animal Species and Evolution.* Cambridge: Harvard Univ. Press. 797 pp.

189. McLaren, I. A. 1972. Polygyny as the adaptive function of breeding territory in birds. *Trans. Conn. Acad. Arts Sci.* 44:191–210

190. McNaughton, S. J., Wolf, L. L. 1970. Dominance and the niche in ecological systems. *Science* 167:131–39

191. McNeilly, T., Bradshaw, A. D. 1968. Evolutionary processes in populations of copper tolerant *Agrostis tenuis* Sibth. *Evolution* 22:108–18

192. Meanley, B. 1971. Blackbirds and the southern rice crop. *US Dep. Inter. Fish Wildl. Serv. Res. Publ. No. 100* 64 pp.

193. Mech, L. D. 1970. *The Wolf: The Ecology and Behavior of an Endangered Species.* Garden City, New York: Nat. Hist. 384 pp.

194. Meslow, E. C., Wight, H. M. 1975. Avifauna and succession in Douglas-fir forests of the Pacific Northwest. *Proc. Symp. Manage. Forest Range Habitats for Nongame Birds, US Dep. Agric. Forest Serv. Gen. Tech. Rep. WO–1* pp. 266–71

195. Michener, C. D. 1942. The history and behavior of a colony of harvester ants. *NY Entomol. Soc.* 56:248–58

196. Monro, J. 1967. The exploitation and conservation of resources by populations of insects. *J. Anim. Ecol.* 36:531–47

197. Monsi, M., Uchijima, Z., Oikawa, T. 1973. Structure of foliage canopies and photosynthesis. *Ann. Rev. Ecol. Syst.* 4:301–27

198. Morrison, R. G., Yarranton, G. A. 1974. Vegetational heterogeneity during a primary sand dune succession. *Can. J. Bot.* 52:397–410

199. Mott, J. J., McComb, A. J. 1974. Patterns in annual vegetation and soil microrelief in an arid region of western Australia. *J. Ecol.* 62:115–26

200. Muller, C. H. 1966. The role of chemical inhibition (allelopathy) in vegetational composition. *Bull. Torrey Bot. Club* 93:332–51

201. Murray, B. G. 1967. Dispersal in vertebrates. *Ecology* 48:975–78

202. Murton, R. K. 1971. Why do some bird species feed in flocks? *Ibis* 113:534–36

203. Myers, J. H. 1974. Genetic and social structure of feral house mouse populations on Grizzly Island, California. *Ecology* 55:747–59

204. Nevo, E., Dessauer, H. C., Chuang, K. C. 1975. Genetic variation as a test of natural selection. *Proc. Natl. Acad. Sci. USA* 72:2145–49

205. Nicholson, S. A., Monk, C. D. 1974. Plant species diversity in old-field succession on the Georgia Piedmont. *Ecology* 55:1075–85

206. Noy-Meir, I. 1973. Desert ecosystems: Environment and producers. *Ann. Rev. Ecol. Syst.* 4:25–51

207. Odum, E. P. 1969. The strategy of ecosystem development. *Science* 164:262–70

208. Ogden, J. 1970. Plant population structure and productivity. *Proc. NZ Ecol. Soc.* 17:1–9

209. O'Hara, R. K. 1974. *Effects of developmental state and prior experience on habitat selection in three species of anuran larvae.* MSc thesis. Mich. State Univ., East Lansing, Michigan. 71 pp.

210. Orians, G. H. 1969. On the evolution of mating systems in birds and mammals. *Am. Nat.* 103:589–603

211. Orians, G. H. 1971. Ecological aspects of behavior. In *Avian Biology,* eds. D. S. Farner, J. R. King. 1:513–46. New York: Academic

212. Orians, G. H. 1973. *The Study of Life.* Boston: Allyn & Bacon. 656 pp. 2nd ed.

213. Paine, R. T. 1966. Food web complexity and species diversity. *Am. Nat.* 100:65–75

214. Paine, R. T. 1969. The *Pisaster-Tegula* interaction: Prey patches, predator food preference, and intertidal community structure. *Ecology* 50:950–61

215. Partridge, L. 1974. Habitat selection in titmice. *Nature* 247:573–74

216. Patil, G. P., Pielou, E. C., Waters, W. E. ed. 1971. *Statistical Ecology. Vol. 1. Spatial Patterns and Statistical Distributions.* University Park, Pa: Penn. State Univ. Press. 582 pp.

217. Patil, G. P., Stiteler, W. M. 1974. Concepts of aggregation and their quantification: A critical review with some new results and applications. *Res. Popul. Biol.* 15:238–54

218. Patterson, I. J. 1965. Timing and spacing of broods in the Black-headed Gull *Larus ridibundus. Ibis* 107:433–59

219. Patterson, I. J., Dunnet, G. M., Fordham, R. A. 1971. Ecological studies of the Rook, *Corvus frugilegus* L., in north-east Scotland. Dispersion. *J. Appl. Ecol.* 8:815–33

220. Peitzmeier, J. 1951. Zum oekologischen Verhalten der Misteldrossel (*Turdus v.*

viscivorus L.) in Nordwesteuropa. *Bonn. Zool. Beitr.* 2:217–24

221. Pianka, E. R., Parker, W. S. 1975. Ecology of horned lizards: A review with special reference to *Phrynosoma platyrhinos. Copeia* 1975:141–62

222. Pielou, E. C. 1969. *An Introduction to Mathematical Ecology.* New York: Wiley. 286 pp.

223. Pitelka, F. A., Holmes, R. T., MacLean, S. F. Jr. 1974. Ecology and evolution of social organization in arctic sandpipers. *Am. Zool.* 14:185–204

224. Poore, M. E. D. 1964. Integration in the plant community. *J. Ecol.* 52 (Suppl.):213–26

225. Prakash, S. 1973. Patterns of gene variation in central and marginal populations of *Drosophila robusta. Genetics* 75:347–69

226. Pulliam, H. R. 1973. On the advantages of flocking. *J. Theor. Biol.* 38:419–22

227. Raleigh, R. F. 1967. Genetic control in the lakeward migration of sockeye salmon (*Oncorhynchus nerka*) fry. *J. Fish. Res. Board Can.* 24:2613–22

228. Raleigh, R. F. 1971. Innate control of migrations of salmon and trout fry from natal gravels to rearing areas. *Ecology* 52:291–97

229. Rasmussen, D. I. 1970. Biochemical polymorphisms and genetic structure in populations of *Peromyscus. Symp. Zool. Soc. London* 26:335–49

230. Reddingius, J. 1971. Gambling for existence. *Acta Biotheor. Suppl.* 1:1–208

231. Reddingius, J. 1974. Models in biology: A comment. *Am. Nat.* 108:393–94

232. Reddingius, J., den Boer, P. J. 1970. Simulation experiments illustrating stabilization of animal numbers by spreading of risk. *Oecologia* 5:240–84

233. Rice, E. L. 1974. *Allelopathy.* New York: Academic. 353 pp.

234. Ritter, J. A., MacCrimmon, H. R. 1973. Effects of illumination on behavior of wild Brown (*Salmo trutta*) and Rainbow Trout (*Salmo gairdneri*) exposed to black and white backgrounds. *J. Fish. Res. Board Can.* 30:1875–80

235. Ritter, J. A., MacCrimmon, H. R. 1973. Influence of environmental experience on response of yearling Rainbow Trout (*Salmo gairdneri*) to a black and white substrate. *J. Fish. Res. Board Can.* 30:1740–42

236. Roff, D. A. 1974. A comment on the number-of-factors model of Reddingius and den Boer. *Am. Nat.* 108:391–93

237. Rogers, L. E. 1974. Foraging activity of the Western Harvester Ant in the short-

118 WIENS

grass plains ecosystem. *Environ. Entomol.* 3:420–24

238. Rogers, L. E., Lavigne, R. J. 1974. Environmental effects of Western Harvester Ants on the shortgrass plains ecosystem. *Environ. Entomol.* 3:994–97

239. Rogers, L. E., Lavigne, R., Miller, J. L. 1972. Bioenergetics of the Western Harvester Ant in the shortgrass plains ecosystem. *Environ. Entomol.* 1:763–68

240. Root, R. B. 1975. Some consequences of ecosystem texture. See Ref. 164, pp. 83–97

241. Rosenzweig, M. L. 1974. On the evolution of habitat selection. *Proc. 1st Int. Congr. Ecol.* 1:401–4

242. Rosenzweig, M. L., Winakur, J. 1969. Population ecology of desert rodent communities: Habitats and environmental complexity. *Ecology* 50:558–72

243. Rotenberry, J. T., Hinds, W. T., Thorp, J. M. 1976. Microclimatic patterns on the Arid Lands Ecology Reserve. *Northw. Sci.* In press

244. Rotenberry, J. T., Wiens, J. A. 1976. A method for estimating species dispersion from transect data. *Am. Midl. Nat.* 94:64–78

245. Roughgarden, J. 1974. Population dynamics in a spatially varying environment: how population size "tracks" spatial variation in carrying capacity. *Am. Nat.* 108:649–64

246. Royama, T. 1970. Factors governing the hunting behaviour and selection of food by the Great Tit (*Parus major* L.). *J. Anim. Ecol.* 39:619–68

247. Royama, T. 1971. A comparative study of models for predation and parasitism. *Res. Popul. Biol. Suppl.* 1:1–91

248. Sabath, M. D., Jones, J. M. 1973. Measurement of niche breadth and overlap: the Colwell-Futuyma method. *Ecology* 54:1143–47

249. Sale, P. F. 1969. Pertinent stimuli for habitat selection by the juvenile manini, *Acanthurus triostegus sandvicensis. Ecology* 50:616–23

250. Sargent, T. D. 1966. Background selection of Geometrid and Noctuid moths. *Science* 154:1674–75

251. Sargent, T. D. 1969. Behavioral adaptations of cryptic moths. III. Resting attitudes of two bark-like species, *Melanolophia canadaria* and *Catacala ultronia. Anim. Behav.* 17:670–72

252. Schaller, G. B. 1972. *The Serengeti Lion: A Study of Predator-Prey Relations.* Chicago: Univ. Chicago Press. 480 pp.

253. Schoener, T. W. 1974. Resource partitioning in ecological communities. *Science* 185:27–39

254. Selander, R. K. 1970. Behavior and genetic variation in natural populations. *Am. Zool.* 10:53–66

255. Selander, R. K., Kaufman, D. W. 1973. Genic variability and strategies of adaptation in aminals. *Proc. Natl. Acad. Sci. USA* 70:1875–77

256. Semeonoff, R., Robertson, F. W. 1968. A biochemical and ecological study of plasma esterase polymorphism in natural populations of the field vole, *Microtus agrestis* L. *Biochem. Genet.* 1:205–27

257. Simons, S., Alcock, J. 1971. Learning and the foraging persistence of White-crowned Sparrows, *Zonotrichia leucophrys. Ibis* 113:477–82

258. Skellam, J. G. 1951. Random dispersal in theoretical populations. *Biometrika* 38:196–218

259. Slatkin, M. 1973. Gene flow and selection in a cline. *Genetics* 75:733–56

260. Smith, C. C. 1968. The adaptive nature of social organization in the genus of tree squirrels *Tamiasciurus. Ecol. Monogr.* 38:31–63

261. Smith, F. E. 1972. Spatial heterogeneity, stability, and diversity in ecosystems. *Trans. Conn. Acad. Arts Sci.* 44:309–35

262. Smith, J. N. M. 1974. The food searching behaviour of two European thrushes. I. Description and analysis of search paths. *Behaviour* 48:276–302

263. Smith, J. N. M. 1974. The food searching behaviour of two European thrushes. II. The adaptiveness of the search patterns. *Behaviour* 49:1–61

264. Smith, J. N. M., Dawkins, R. 1971. The hunting behaviour of individual Great Tits in relation to spatial variations in their food density. *Anim. Behav.* 19:695–706

265. Smith, J. N. M., Sweatman, H. P. A. 1974. Food-searching behavior of titmice in patchy environments. *Ecology* 55:1216–32

266. Solomon, M. E. 1971. Elements in the development of population dynamics, *Proc. Adv. Study Inst. Dyn. Numbers Popul. Oosterbeek, 1970* pp. 29–40

267. Soulé, M. 1973. The epistasis cycle: A theory of marginal populations. *Ann. Rev. Ecol. Syst.* 4:165–87

268. Southern, H. N., Lowe, V. P. W. 1968. The pattern of distribution of prey and predation in Tawny Owl territories. *J. Anim. Ecol.* 37:75–97

269. Southwood, T. R. E. 1962. Migration of terrestrial arthropods in relation to habitat. *Biol. Rev.* 37:171–214

270. Spatz, G., Mueller-Dombois, D. 1973. The influence of goats on koa tree reproduction in Hawaii Volcanoes National Park. *Ecology* 54:870–76

271. Spieth, P. T. 1974. Gene flow and genetic differentiation. *Genetics* 78:961–65

272. States, J. B. 1976. Local adaptations in chipmunk (*Eutamias amoenus*) populations and evolutionary potential at species' borders. *Ecol. Monogr.* In press

273. Steele, J. H. 1974. Spatial heterogeneity and population stability. *Nature* 248:83

274. Stefanski, R. A. 1967. Utilization of the breeding territory in the Black-capped chickadee. *Condor* 69:259–67

275. Stenger, J. 1958. Food habits and available food of ovenbirds in relation to territory size. *Auk* 75:335–46

276. Strobeck, C. 1974. Sufficient conditions for polymorphism with N niches and M mating groups. *Am. Nat.* 108:152–56

277. Tamarin, R. H., Krebs, C. J. 1969. Microtus population biology. II. Genetic changes at the transferrin locus in fluctuating populations of two vole species. *Evolution* 23:24–49

278. Taylor, J. 1976. The advantage of spacing-out. *J. Theor. Biol.* In press

279. Taylor, R. J. 1976. Value of clumping to prey and the evolutionary response of ambush predators. *Am. Nat.* 110:13–29

280. Tester, J. R., Marshall, W. H. 1961. A study of certain plant and animal interrelations on a native prairie in northwestern Minnesota. *Minn. Mus. Nat. Hist. Occas. Pap.* 8:51 pp.

281. Thompson, W. A., Vertinsky, I., Krebs, J. R. 1974. The survival value of flocking in birds: A simulation model. *J. Anim. Ecol.* 43:785–820

282. Tinbergen, N., Impekoven, M., Franck, D. 1967. An experiment on spacing-out as a defence against predation. *Behaviour* 28:307–21

283. Trivers, R. L. 1972. Parental investment and sexual selection. See Ref. 56, pp. 136–79

284. Usher, M. B., Williamson, M. H. 1970. A deterministic matrix model for handling the birth, death, and migration processes of spatially distributed populations. *Biometrics* 26:1–12

285. van der Pijl, L. 1969. *Principles of Dispersal in Higher Plants.* New York: Springer

286. Vandermeer, J. H. 1973. On the regional stabilization of a locally unstable predator-prey relationship. *J. Theor. Biol.* 41:131–70

287. Van Valen, L. 1971. Group selection and the evolution of dispersal. *Evolution.* 25:591–98

288. Verner, J. 1965. Breeding biology of the Long-billed Marsh Wren. *Condor* 67:6–30

289. Verner, J. 1975. Avian behavior and habitat management. See Ref. 194, pp. 39–58

290. Verner, J., Willson, M. F. 1966. The influence on habitats of mating systems of North American passerine birds. *Ecology* 47:143–47

291. Verner, J., Engelsen, G. H. 1970. Territories, multiple nest building, and polygyny in the Long-billed Marsh Wren. *Auk* 87:557–67

292. Vine, I. 1971. Risk of visual detection and pursuit by a predator and the selective advantage of flocking behavior. *J. Theor. Biol.* 30:405–22

293. Vine, I. 1973. Detection of prey flocks by predators. *J. Theor. Biol.* 40:207–10

294. Wallace, B. 1968. *Topics in Population Genetics.* New York: Norton. 481 pp.

295. Walsh, J. J., Dugdale, R. C. 1971. A simulation model of the nitrogen flow in the Peruvian upwelling system. *Invest. Pesq.* 35:309–30

296. Ward, P. 1965. Feeding ecology of the Black-faced Dioch *Quelea quelea* in Nigeria. *Ibis* 107:173–214

297. Ward, P., Zahavi, A. 1973. The importance of certain assemblages of birds as "information-centres" for food-finding. *Ibis* 115:517–34

298. Watt, A. S. 1947. Pattern and process in the plant community. *J. Ecol.* 35:1–22

299. Wecker, S. C. 1963. The role of early experience in habitat selection by the Prairie Deer Mouse, *Peromyscus maniculatus bairdi. Ecol. Monogr.* 33:307–25

300. Weeden, J. S. 1965. Territorial behavior of the Tree Sparrow. *Condor* 67:193–209

301. Wellington, W. G. 1957. Individual differences as a factor in population dynamics: The development of a problem. *Can. J. Zool.* 35:293–323

302. Wellington, W. G. 1960. Qualitative changes in natural populations during changes in abundance. *Can. J. Zool.* 38:289–314

303. Wellington, W. G. 1964. Qualitative changes in populations in unstable environments. *Can. Entomol.* 96:436–51

304. West, N. E. 1969. Tree patterns in cen-

120   WIENS

tral Oregon Ponderosa Pine forests. *Am. Midl. Nat.* 81:584–90

305. Whitford, W. G., Ettershank, G. 1975. Factors affecting foraging activity in Chihuahuan Desert harvester ants. *Environ. Entomol.* 4:689–96

306. Whittaker, R. H. 1965. Dominance and diversity in land plant communities. *Science* 147:250–60

307. Whittaker, R. H. 1969. Evolution of diversity in plant communities, *Brookhaven Symp. Biol.* 22:178–96

308. Whittaker, R. H. 1970. Biochemical ecology of higher plants. In *Chemical Ecology*, ed. E. Sondheimer, J. B. Simeone, pp. 43–70. New York: Academic

309. Whittaker, R. H., Feeny, P. P. 1971. Allelochemics: Chemical interactions between species. *Science* 171:757–70

310. Wiegert, R. G. 1975. Simulation modeling of the algal-fly components of a thermal ecosystem: Effects of spatial heterogeneity, time delays, and model condensation. In *Systems Analysis and Simulation in Ecology*, ed. B. C. Patten, 3:157–81. New York: Academic

311. Wiens, J. A. 1966. On group selection and Wynne-Edwards' hypothesis. *Am. Sci.* 54:273–87

312. Wiens, J. A. 1969. An approach to the study of ecological relationships among grassland birds. *Ornithol. Monogr.* 8:1–93

313. Wiens, J. A. 1970. Effects of early experience on substrate pattern selection in *Rana aurora* tadpoles. *Copeia* 1970:543–48

314. Wiens, J. A. 1972. Anuran habitat selection: Early experience and substrate selection in *Rana cascadae* tadpoles. *Anim. Behav.* 20:218–20

315. Wiens, J. A. 1973. Interterritorial habitat variation in Grasshopper and Savannah sparrows. *Ecology* 54:877–84

316. Wiens, J. A. 1973. Pattern and process in grassland bird communities. *Ecol. Monogr.* 43:237–70

317. Wiens, J. A. 1974. Climatic instability and the "ecological saturation" of bird communities in North American grasslands. *Condor* 76:385–400

318. Wiens, J. A. 1974. Habitat heterogeneity and avian community structure in North American grasslands. *Am. Midl. Nat.* 91:195–213

319. Wiens, J. A. 1975. Avian communities, energetics, and functions in coniferous forest habitats. See Ref. 194, pp. 226–65

320. Wiens, J. A., Dyer, M. I. 1975. Simulation modelling of Red-winged Blackbird impact on grain crops. *J. Appl. Ecol.* 12:63–82

321. Wiley, R. H. 1974. Evolution of social organization and life-history pattern among grouse. *Q. Rev. Biol.* 49:201–27

322. Willard, J. R., Crowell, H. H. 1965. Biological activities of the harvester ant, *Pogonomyrmex owyheei,* in central Oregon. *J. Econ. Entomol.* 58:484–89

323. Williams, G. C. 1964. Measurement of consociation among fishes and comments on the evolution of schooling. *Publ. Mus. Mich. State Univ. Biol. Ser.* 2:349–84

324. Williams, G. C. 1966. *Adaptation and Natural Selection.* Princeton, New Jersey: Princeton Univ. Press. 307 pp.

325. Willis, E. O. 1966. The role of migrant birds at swarms of army ants. *Living Bird* 5:187–231

326. Wilson, D. S. 1975. A theory of group selection. *Proc. Natl. Acad. Sci. USA* 72:143–46

327. Wilson, E. O. 1971. *The Insect Societies.* Cambridge: Harvard Univ. Press. 548 pp.

328. Wilson, E. O. 1975. *Sociobiology.* Cambridge: Harvard Univ. Press. 697 pp.

329. Wolfenbarger, D. O. 1959. Dispersion of small organisms. *Lloydia* 22:1–105

330. Woodell, S. R. J., Mooney, H. A., Hill, J. A. 1969. The behaviour of *Larrea divaricata* (Creosote Bush) in response to rainfall in California. *J. Ecol.* 57:37–44

331. Woolfenden, G. E. 1975. Florida scrub jay helpers at the nest. *Auk* 92:1–15

332. Wynne-Edwards, V. C. 1962. *Animal Dispersion in Relation to Social Behaviour.* Edinburgh: Oliver & Boyd. 653 pp.

333. Zahavi, A. 1971. The function of preroost gatherings and communal roosts. *Ibis.* 113:106–9

334. Zahavi, A. 1974. Communal nesting by the Arabian Babbler. A case of individual selection. *Ibis* 116:84–87

335. Zedler, J. B., Zedler, P. H. 1969. Association of species and their relationship to microtopography within old fields. *Ecology* 50:432–42

336. Zimmerman, J. L. 1971. The territory and its density dependent effect in *Spiza americana. Auk* 88:591–612

PATCH DYNAMICS AND THE DESIGN OF NATURE RESERVES

S. T. A. Pickett†

Plant Ecology Laboratory, Department of Botany, 289 Morrill Hall, University of Illinois, Urbana, Illinois 61801, USA

John N. Thompson

Ecology Program, 320 Morrill Hall, University of Illinois, Urbana, Illinois 61801, USA

ABSTRACT

Island biogeographic theory has been applied to the design of nature reserves. However, immigration, which is important in maintaining species equilibrium on true islands, will not contribute significantly to the maintenance of equilibrium on reserves in the future because of the disappearance of recolonisation sources. Consequently, extinction becomes the dominant population process, and the internal disturbance dynamics become the critical design feature of reserves. The design of reserves should be based on 'minimum dynamic area', the smallest area with a natural disturbance regime which maintains internal recolonisation sources and hence minimises extinctions. Determination of minimum dynamic area must be based on knowledge of disturbance-generated patch size, frequency, and longevity, and the mobilities of the preserved species. These features have not all been explicitly considered in the previous island biogeographic design recommendations.

INTRODUCTION

Nature reserves are habitat islands. They are areas of natural landscape surrounded by expanses of culturally modified habitat. Because of this, the theory of island biogeography has been applied to the design of nature reserves (see review by Diamond & May, 1976). The degree of isolation from sources of colonisers determines the rate of immigration, while the island area determines the population sizes and thus the rate of extinction. The immigration and extinction rates together determine the number of species that will inhabit the island at equilibrium (MacArthur & Wilson, 1967). The purpose of this paper is to stress the importance

---

† *Present address:* Department of Botany, Rutgers University, New Brunswick, New Jersey 08903, USA.

27

Biol. Conserv. (**13**) (1978)—© Applied Science Publishers Ltd, England, 1978
Printed in Great Britain

of the internal dynamics of habitats in the design of reserves as colonisation sources become increasingly less available.

COMPONENTS OF EQUILIBRIUM

The components of equilibrium on both true and habitat islands are the same, but the relative importance of these differs between the two types of system. True islands, even very isolated ones, have very large continental species pools from which colonisers are drawn. The island characteristic which largely determines the magnitude of the flux of colonisers is distance from the source. More distant islands will receive fewer colonists. The second island character important in determining its equilibrium species number is area. Increased area permits large population sizes and thereby reduces the probability of extinction. Thus, larger islands have lower extinction rates. The number of species that an island will support at equilibrium is that number where extinction equals immigration. Most true islands will approach equilibrium from below (Wilson, 1969). The only non-island character important in determining equilibrium is the continued existence of the species pool supplying colonists.

Nature reserves as habitat islands are influenced by the same components: the island characters of area and isolation, and the population processes of immigration and extinction. However, for reserves the rich, extensive source areas do not, or will not, continue to exist (Meijer, 1973; Diamond, 1976; Terborgh, 1976). The reserves themselves must perform most if not all of this function as increasingly more land in all biomes is developed or disrupted. Because of this, the immigration rate, which is so important in maintaining equilibrium on true islands, is likely to decline significantly (Terborgh, 1974a; Willis, 1974; Diamond, 1976). Extinction will then become the dominant population process affecting equilibrium in reserves and species numbers will decline to a new level. There will still be immigrants, but such species will be the widespread, fugitive types which do not need reserves for their survival (Willis, 1974; Terborgh, 1976; Whitcomb et al., 1976). Species requiring continuous habitat (Peterken, 1974) or specialised habitat types (Geist, 1971) for their survival or dispersal will not be able to disperse to sites where they have become extinct (Terborgh, 1974b).

The equilibrium process on habitat islands will somewhat mimic that on land bridge islands where species number is determined by extinction. These land bridge islands, isolated since the end of the Pleistocene, have more species than oceanic islands of similar area and isolation (Diamond, 1972), but have fewer species than the continental areas with which they were once continuous. Their species numbers are decreasing, or relaxing, to new lower equilibria, due to the current dominance of extinction over immigration (Diamond, 1972; Willis, 1974). It is the species characteristic of closed, undisrupted habitats to become extinct first.

Because extinction is the dominant process in ecological time on land bridge and newly-created habitat islands, the factors influencing it deserve closer examination. As for true islands, area affects extinction, with smaller areas increasing the probability of extinction due principally to reduced population sizes (Simberloff, 1974; Terborgh, 1974a; Elton, 1975). Small populations are particularly subject to fluctuations and random events. However, in addition, small populations, or populations necessarily confined to small areas, will be more subject to extinction due to the internal dynamics of the island landscape.

PATCH DYNAMICS

Internal dynamics are generated by patterns of disturbance and subsequent patterns of succession, and may be called patch dynamics. Three sets of phenomena and processes define the dynamics of patches on a landscape scale. The disturbance regime determines the size, density and temporal frequency of patches. The internal structure of patches is determined by species composition, population densities, population dispersions and organism geometry. As a result of their internal features, patches will have characteristic longevities. Many attributes of patches have been considered by Levin (1976) for the landscape scale, and by Root (1975) and Thompson (in press) for the within-community scale.

On continental landscapes, patch dynamics are complex and of profound importance. Patches are generated by many agents whose direct effect is on vegetation, for example, fire, which is one of the most common. It is important, for example in temperate deciduous forest (Loucks, 1970), boreal and montane coniferous forest (Heinselman, 1971; Wright, 1974), grassland (Old, 1969), chaparral (Vogl & Schorr, 1972), and tropical deciduous forest (Walter, 1971). Winds and windstorms are important disturbance agents in deciduous forest (Jones, 1945; Lindsay, 1972; Wright, 1974), coniferous forest (Sprugel, 1976) and the tropics (Webb, 1958). Ice and snow (Siccama et al., 1976), floods and landslides are important in some systems (Hack & Goodlett, 1960). Biotic factors also contribute to the disturbance regime. Natural death of old individuals is an important gap generator in all systems (Bray, 1956; Forcier, 1975). Disease and herbivory (Mattson & Addy, 1975) are also very common. Animals disturb some systems (Bratton, 1975; Platt, 1975). The contribution of these various agents to the disturbance regime and hence patch dynamics depends on the climate, topography and biota of the site. Disturbance may occur on scales ranging from the removal of one individual (Horn, 1976) to the opening of many km^2 (Wright, 1974) as a result of different disturbance agents or the change in severity of one type.

Disturbance and succession are ubiquitous landscape features. As a result, the natural landscape is a mosaic of successional patches of various sizes (Whittaker, 1953) whose diversity depends partially on patch dynamics (Loucks, 1970) and which provides opportunities for evolutionary divergence of species (Pickett, 1976). Even climax species must have a fugitive aspect (Harper, 1965) to exploit gap disturbance in the mature community (Jones, 1945; Forcier, 1975). Preservation of landscapes must allow for the continued operation of these processes (Stone, 1965; Houston, 1971; Wright, 1974; McLean, 1976).

Patch dynamics must occur with equal importance on true islands. As a mechanism for causing local extinction of species they are an unfactored component of the species-area relation. Disturbance of a given size and frequency is much more likely to cause extinction on a small island by reducing population sizes below threshold levels. Patch dynamics have remained unfactored because the comparative and empirical approach to island biogeography could rely on area to predict equilibrium species number, and because of the importance of the immigration component on true islands. Processes which we would identify as patch dynamics have been recognised by students of oceanic island biogeography (Willis, 1974; Diamond, 1975b, 1976; Wilson & Willis, 1975). The predictive role of patch dynamics in species equilibria will increase from oceanic through land bridge to habitat islands because of the increased importance of extinction and consequent relaxation.

The use of patch dynamics in the design of nature reserves then requires knowledge of the disturbance regime and the associated patterns of succession. The type of disturbance will influence all other patch characteristics. Thus disturbance type is fundamental background information. Additionally, species are differentially sensitive to different types and degrees of disturbance so that this factor interacts with species biology in determining successions. The critical characters of disturbed patches are their size and frequency of occurrence in time. Different kinds of disturbance occur with varying severity in a region and may act jointly to determine the size and configuration of patches.

Two cautions concerning the disturbance regime are in order. First, it is important to know the size and timing of rare, extensive disturbance events as well as normal patches. The effect of rare patches may not be obvious in contemporary time, but it may be critical in the structure and function of natural landscapes (Henry & Swan, 1974). Rare and extreme events can be extremely influential in geomorphic processes (Hack & Goodlett, 1960; Leopold et al., 1964). Second, disturbance regimes will differ among regions as a function of climate, geomorphology and biota. For example, Lorimer (1977) reports that in Maine large-scale disturbances occur at a site in the order of every 800 years for fire and 1150 years for winds. Small-scale disturbances occur every 250–300 years (Lorimer, 1977). In contrast, a given site in Minnesota boreal forests burns every 5–50 years (Heinselman, 1971). These differences must be taken into account when planning for different areas.

Table 1 presents some sizes, causes and frequency of disturbance in North American landscapes. Only references giving measurements or estimates of area of disturbed patches have been selected from the extensive literature on disturbance

TABLE 1

SOME SIZES, CAUSES AND FREQUENCIES OF DISTURBANCE IN NORTH AMERICAN LANDSCAPES

System	Type of disturbance	Mean size	Maximum size	Frequency	Reference
Beech/maple	Windfall	0·010 ha	0·025 ha		Williamson (1975)
Beech/maple	Tornado track		400 m[b]		Lindsay (1972)
Northern deciduous	Wind		15533 ha		Irving in Curtis (1959)
Deciduous	Hurricane	1·61–2·4 km[b]			Beebe in Goodlet (1954)
Grassland/forest	Tornado track	77 ha	217 ha	8/yr	Eshelman & Stanford (1977)
Northern deciduous	Hurricane			1635, 1815, 1938[c]	Spurr (1956)
Deciduous	Flood		60·96 m[b]		Hack & Goodlett (1960)
Deciduous	Landslide	15/24 m[b]	304 m[b]		Hack & Goodlett (1960)
Boreal	Fire		809716 ha		Schmidt (1970)
Pond pine	Fire	11860 ha[a]			Wade & Ward (1973)
Sand hills	Fire	2995 ha[a]			DeCoste et al. (1968)
Boreal	Fire	5922 ha[a]			Sando & Haines (1972)
Montane conifer	Fire	20242 ha[a]			Anderson (1968)
Interior Alaska	Fire (total)	1226 ha		218/yr	Barney (1969)
Interior Alaska	Fire (lightning)	3095 ha		83/yr	Barney (1969)
Boreal	Fire	1739 ha		33/yr	Johnson & Rowe (1975)
Lodgepole/Ponderosa	Fire			0·4/yr	Franklin et al. (1972)
Boreal	Fire			4·8/152700 ha/yr	Martinka (1976)
Montane conifer	Fire	99071 ha[a]			Neiland (1958)

32

S. T. A. PICKETT, JOHN N. THOMPSON

TABLE 1—*contd.*

SOME SIZES, CAUSES AND FREQUENCIES OF DISTURBANCE IN NORTH AMERICAN LANDSCAPES

System	Type of disturbance	Mean size	Maximum size	Frequency	Reference
Pine/spruce/fir	Fire	up to 97% of site showed fire evidence			Patten (1963)
Southern Appalachians	Fire (lightning)	3·4 ha	33 ha	6/400000 ha/yr	Barden & Woods (1973)
Southern Appalachians	Fire (man-made)	5·4 ha			
Mixed conifer	Fire			0–13/yr	Biswell (1967)
Deciduous	Fire	6·6 ha		10/198138 ha/yr	Haines et al. (1975)
Deciduous	Fire	4·4 ha		10/349635 ha/yr	Haines et al. (1975)
Deciduous	Fire	10·4 ha		39/504655 ha/yr	Haines et al. (1975)
Deciduous	Fire	12·6 ha		212/622307 ha/yr	Haines et al. (1975)
Deciduous/fir	Fire	2·0 ha		7/153441 ha/yr	Haines et al. (1975)
Deciduous	Fire	11·1 ha		140/407165 ha/yr	Haines et al. (1975)
Deciduous	Fire	8·1 ha		22/482348 ha/yr	Haines et al. (1975)
Deciduous/conifer	Fire	6·4 ha		148/381255 ha/yr	Haines et al. (1975)
Deciduous/conifer	Fire	3·9 ha		16/495303 ha/yr	Haines et al. (1975)
Deciduous	Fire	12·8 ha		41/198623 ha/yr	Haines et al. (1975)
Deciduous/boreal	Fire	5·0 ha		60/1102024 ha/yr	Haines et al. (1975)
Deciduous/boreal	Fire	1·9 ha		7/319068 ha/yr	Haines et al. (1975)

[a] Single event.
[b] Width.
[c] Years of occurrence.

effects. Disturbances and systems of economic importance, e.g. fires in coniferous forests, are probably over-represented. Some organisms within disturbed patches may survive or rapidly regenerate, but the data give an indication of the kinds and extents of disturbances that may be important for reserve design. Disturbances of various sizes are indicated as important components of system dynamics in many regions (Spurr & Barnes, 1973) even though quantitative estimates of single disturbance events are lacking. Wind in temperate (Goodlett, 1954; Henry & Swan, 1974), and especially tropical, systems (Longman & Jenik, 1974) is important, but disturbance patch sizes are unavailable. Similarly earthquake and landslides are important but areas are unavailable (Dyrness, 1967; DuMontelle *et al.*, 1971; O'Loughlin, 1972).

The biology of the species in a reserve will interact with the disturbance regime to determine extinction probabilities (Simberloff & Abele, 1976*a,b*). Taxa may be differentially sensitive to a given disturbance regime. The mobility of the taxon is an important consideration (Whitcomb *et al.*, 1976). The sedentary nature of many climax (Peterken, 1974) and tropical taxa (Ashton, 1969; Terborgh, 1974*b*) must be taken into account. Whether close competitors require separate patches to avoid extinction should also be determined (Simberloff & Abele, 1976*a*). The requirement of a large resource base by populations of large consumers will affect their minimum area requirements. Resource requirements may be expressed as broad, continuous areas or as areas of concentration ('hot spots', Diamond, 1975*b*), and either may alter in time and space (Willis, 1974). The low population densities of some taxa, particularly in diverse communities, increase local extinction probabilities (Elton, 1975).

The succession within patches after disturbance, i.e. patch longevity, may affect the extinction rates of some taxa. If some species require patches of a given age which support a particular kind of resource (Wright, 1974; Eisenberg & Seidensticker, 1976; Seidensticker, 1976) or do not harbour superior competitors (Pickett, 1976), then sufficient area of such patches must exist in a reserve. Many wildlife species can maintain marginal populations in late successional forests but reach highest population levels only in a mixture of successional habitats. Examples include snowshoe hare ((*Lepus americanus*) Grange, 1965), deer (Byelich *et al.*, 1972) and some Asian ungulates (Eisenberg & Seidensticker, 1976). Gullian & Svoboda (1972) suggest that to maintain highest population levels of ruffed grouse (*Bonsas umbellelus*), all age classes of aspen must be available to each wintering and breeding grouse within a normal range of about 4 ha. Knowledge of patch generation and graduation through resource and successional stages is therefore required to preserve such species.

DESIGN CONSIDERATIONS

Island biogeography theory and work on land bridge islands has been used to generate the following suggestions for nature reserve design. Reserves should be (1) large, (2) circular, (3) undivided, and if divided, (4) connected by dispersal corridors or (5) close to one another (Willis, 1974; Terborgh, 1974a,b; Diamond, 1975a; Wilson & Willis, 1975; Diamond & May, 1976). Through the additional consideration of patch dynamics which are important in the organisation of natural landscapes, we suggest that the design of nature reserves be based on analysis of 'minimum dynamic area', the smallest area with a natural disturbance regime, which maintains internal recolonisation sources, and hence minimises extinction. The minimum dynamic area of various taxa will undoubtedly differ, however, the minimum dynamic area of a reserve will be defined by the most extinction-prone taxon. Nature reserves should:

(1)   be considerably larger than the largest disturbance patch size, including rare patches,
(2)   include internal recolonisation sources,
(3)   include different ages of disturbance-generated patches,
(4)   encompass areas sufficient to support large consumer populations in habitats not made unsuitable by disturbance, and
(5)   contain separate minimum dynamic areas of each included habitat type.

Direct determination of minimum dynamic area could be made by surveys, similar to those of land bridge islands, for fully relaxed habitat islands, i.e. those which are at equilibrium with respect to their normal disturbance regime. Surveys of habitat islands that may be less than fully relaxed, i.e. those in which the normal disturbance regime has not yet established the new equilibrium (e.g. Galli *et al.*, 1976; Forman *et al.*, 1976) provide minimum area in the absence of disturbance and must be combined with study of patch dynamics to make recommendations on effective reserve size.

ACKNOWLEDGEMENTS

We thank G. Brown for references and comments on the ecology of rare events, F. A. Bazzaz, G. O. Batzli, J. R. Karr, T. D. Lee, M. N. Melampy, J. A. D. Parrish, P. W. Price and A. R. Zangerl for comments, criticism and discussion.

REFERENCES

ANDERSON, H. E. (1968). *Sundance fire: An analysis of fire phenomena.* Ogden, Utah, USDA For. Serv. Intermountain Forest and Range Expt. Station.
ASHTON, P. S. (1969). Speciation among tropical forest trees. Some deductions in the light of recent evidence. *Biol. J. Linn. Soc.*, **1**, 155–96.
BARDEN, L. S. & WOODS, F. W. (1973). Characteristics of lightning fires in southern Appalachian forests. *Proc. Ann. Tall Timber Fire Ecol. Conf.*, **13**, 345–61.

BARNEY, R. J. (1969). *Interior Alaska wildfires* 1956–1965. Pacific Northwest Forest & Range Experiment Station, USDA For. Serv. Institute of Northern Forestry, Juneau, Alaska.

BISWELL, H. H. (1967). Forest fire in perspective. *Proc. Ann. Tall Timbers Fire Ecol. Conf.*, **7**, 43–64.

BRATTON, S. P. (1975). The effect of the European wild boar (*Sus scrofa*) on gray beech forest in Great Smoky Mountains National Park. *Ecology*, **56**, 1356–66.

BRAY, J. R. (1956). Gap phase replacement in maple-basswood forest. *Ecology*, **37**, 598–600.

BYELICH, J. D., COOK, J. L. & BLOOCH, R. I. (1972). Management for deer. *Aspen: Symp. Proc. USDA, For. Serv. Gen Tech. Report* NC-1, 120–5. North Central Forest Experimental Station.

CURTIS, J. T. (1959). *The vegetation of Wisconsin.* Madison, Wisconsin, University of Wisconsin Press.

DeCOSTE, J. H., WADE, D. D. & DEEMING, J. E. (1968). The Gaston fire. *Southeast. For. Expt. Sta. USDA For. Serv. Res. Pap.* SE-43, 36 pp.

DIAMOND, J. M. (1972). Biogeographic kinetics: Estimation of relaxation times for avifaunas of southwest Pacific Islands. *Proc. natn. Acad. Sci. USA*, **69**, 3199–203.

DIAMOND, J. M. (1975a). The island dilemma: Lessons of modern biogeographic studies for the design of natural reserves. *Biol. Conserv.*, **7**, 129–46.

DIAMOND, J. M. (1975b). Assembly of species communities. In *Ecology and evolution of communities*, ed. by M. L. Cody & J. M. Diamond, 342–444. Cambridge, Belknap Press of Harvard University.

DIAMOND, J. M. (1976). Island biogeography and conservation: Strategy and limitations. *Science, N.Y.*, **193**, 1027–9.

DIAMOND, J. M. & MAY, R. M. (1976). Island biogeography and the design of natural reserves. In *Theoretical ecology*, ed. by R. M. May, 163–86. Philadelphia, Saunders.

DuMONTELLE, P. B., HESTER, N. C. & COLE, R. E. (1971). Landslides along the Illinois River valley south and west of La Salle and Peru, Illinois. *Environmental Geology Notes*, No. 48. Illinois State Geological Survey, Urbana, Ill.

DYRNESS, C. T. (1967). Mass soil movement in the H. J. Andrews Experimental Forest. *U.S. For. Serv. Res. Pap* PNW-4L, 12 pp.

EISENBERG, J. F. & SEIDENSTICKER, J. (1976). Ungulates in southern Asia: A consideration of biomass estimates for selected habitats. *Biol. Conserv.*, **10**, 293–308.

ELTON, C. S. (1975). Conservation and the low population density of invertebrates inside neotropical rain forest. *Biol. Conserv.*, **7**, 3–15.

ESHELMAN, S. & STANFORD, J. L. (1977). Tornadoes, funnel clouds and thunderstorm damage in Iowa during 1974. *Iowa State J. Res.*, **51**, 327–61.

FORCIER, L. K. (1975). Reproductive strategies and the co-occurrence of climax tree species. *Science, N.Y.*, **189**, 808–10.

FORMAN, R. T. T., GALLI, A. E. & LECK, C. F. (1976). Forest size and avian diversity in New Jersey woodlots with some land use implications. *Oecologia*, **26**, 1–8.

FRANKLIN, J. F., HALL, F. C., DYRNESS, C. T. & MASER, C. (1972). *Federal research natural areas in Oregon and Washington. A guidebook for scientists and educators.* Portland, Oregon, Pacific N.W. Forest and Range Experimental Station.

GALLI, ANNE E., LECK, C. F. & FORMAN, R. T. T. (1976). Avian distribution patterns in forest islands of different sizes in central New Jersey. *Auk*, **93**, 356–64.

GEIST, V. (1971). *Mountain sheep: A study in behavior and evolution.* Chicago, University of Chicago Press.

GOODLETT, J. C. (1954). Vegetation adjacent to the border of the Wisconsin drift in Potter County, Pennsylvania. *Harv. For. Bull.*, **25**.

GRANGE, W. (1965). Fire and tree growth relationships to snowshoe rabbits. *Proc. Ann. Tall Timber Fire Ecol. Conf.*, **4**, 111–25.

GULLION, G. W. & SVOBODA, F. J. (1972). The basic habitat resource for ruffed grouse. *Aspen: Symp. Proc. USDA, For. Ser. Gen. Tech. Report* NC-1. 113–19. North Central Forestry Experiment Station.

HACK, J. T. & GOODLETT, J. C. (1960). Geomorphology and forest ecology of a mountain region in the central Appalachians. *Prof. Pap.*, *U.S. Geol. Surv.* No. 347.

HAINES, D. A., JOHNSON, V. J. & MAIN, W. A. (1975). Wildfire atlas of the north central states. *USDA For. Serv. Gen. Tech. Pap.* NC-16.

HARPER, J. L. (1965). Establishment, aggression and cohabitation in weedy species. In *The genetics of colonizing species*, ed. by H. G. Baker and G. L. Stebbins, 243–65. New York, Academic Press.

HEINSELMAN, M. L. (1971). The natural role of fire in northern conifer forests. In *Fire in the northern environment*, 61–72. Portland, Oregon, Pacific Northwest Forest and Range Experiment Station USDA Forest Service.

HENRY, J. D. & SWAN, J. M. A. (1974). Reconstructing forest history from live and dead plant material—An approach to the study of forest succession in southwest New Hampshire. *Ecology*, **55**, 772–83.

HORN, H. S. (1976). Succession. In *Theoretical ecology principles and applications*, ed. by R. M. May, 187–204. Philadelphia, Saunders.

HOUSTON, D. B. (1971). Ecosystems of national parks. *Science, N.Y.*, **172**, 648–51.

JOHNSON, E. A. & ROWE, J. S. (1975). Fire in the subarctic wintering ground of the Beverley caribou herd. *Am. Midl. Nat.*, **94**, 1–14.

JONES, E. W. (1945). The structure and reproduction of the virgin forests of the north temperate zone. *New Phytol.*, **44**, 130–48.

LEOPOLD, L. B., WOLMAN, M. G. & MILLER, J. P. (1964). *Fluvial processes in geomorphology*. San Francisco, Freeman.

LEVIN, S. A. (1976). Population dynamic models in heterogeneous environments. *Ann. Rev. Ecol. Syst.*, **7**, 287–310.

LINDSAY, A. A. (1972). Tornado tracks in the presettlement forests of Indiana. *Proc. Indiana Acad. Sci.*, **82**, 181 (Abstr.).

LONGMAN, K. A. & JENIK, J. (1974). *Tropical forest and its environment*. London, Longman.

LORIMER, C. G. (1977). The presettlement forest and natural disturbance cycle of northeastern Maine. *Ecology*, **58**, 139–48.

LOUCKS, O. L. (1970). Evolution of diversity, efficiency and community stability. *Am. Zool.*, **10**, 17–25.

MACARTHUR, R. H. & WILSON, E. O. (1967). *The theory of island biogeography*. Princeton University Press, Princeton, New Jersey.

MARTINKA, C. J. (1976). Fire and elk in Glacier National Park. *Proc. Tall Timber Fire Ecol. Conf.*, **14**, 377–89.

MATTSON, W. J. & ADDY, N. D. (1975). Phytophagous insect as regulators of forest primary production. *Science, N.Y.*, **190**, 515–22.

MCLEAN, A. (1976). Protection of vegetation in ecological reserves in Canada. *Can. Fld Nat.*, **90**, 144–8.

MEIJER, W. (1973). Devastation and regeneration of lowland dipterocarp forests in south-east Asia. *BioScience*, **23**, 528–33.

NEILAND, B. J. (1958). Forest and adjacent burn in the Tillamook burn area of northwestern Oregon. *Ecology*, **39**, 660–71.

OLD, S. M. (1969). Microclimates, fire, and plant production in an Illinois prairie. *Ecol. Monogr.*, **39**, 355–84.

O'LOUGHLIN, C. L. (1972). A preliminary study of landslides in the coast mountains of southwestern British Columbia. In *Mountain geomorphology: Geomorphological processes in the Canadian Cordillera*, ed. by H. O. Ataymayker and H. J. McPherson, 101–11. *B. C. Geographical Series*, No. 14. Vancouver, Tantalus Research Ltd.

PATTEN, D. T. (1963). Vegetational pattern in relation to environments in the Madison Range, Montana. *Ecol. Monogr.*, **33**, 375–406.

PETERKEN, G. F. (1974). A method of assessing woodland flora for conservation using indicator species. *Biol. Conserv.*, **6**, 239–45.

PICKETT, S. T. A. (1976). Succession: An evolutionary interpretation. *Am. Nat.*, **110**, 107–19.

PLATT, W. J. (1975). The colonization and formation of equilibrium plant species associations on badger disturbances in a tall-grass prairie. *Ecol. Monogr.*, **45**, 285–305.

ROOT, R. B. (1975). Some consequences of ecosystem texture. In *Ecosystem analysis and prediction*, ed. by S. A. Levin, 83–92. Philadelphia, Society for Industrial and Applied Mathematics.

SANDO, R. W. & HAINES, D. A. (1972). Fire weather and behavior of the Little Sioux fire. *USDA For. Serv. Res. Pap.* NC-76, 6 p. North Central Forest Experiment Station, St Paul, Minn.

SCHMIDT, R. L. (1970). A history of pre-settlement fires on Vancouver Island as determined from Douglas fir ages. In *Tree ring analysis with special reference to northwest America. Univ. B. C. Fac. For. Bull.*, **7**, 107–8.

SEIDENSTICKER, J. (1976). Ungulate populations in Chitawan Valley, Nepal. *Biol. Conserv.*, **10**, 183–210.

SICCAMA, T. G., WEIR, G. & WALLACE, K. (1976). Ice damage in a mixed hardwood forest in Connecticut in relation to *Vitis* infestation. *Bull. Torrey bot. Club*, **103**, 180–3.

SIMBERLOFF, D. S. (1974). Permo-Triassic extinctions: The effects of area on biotic equilibrium. *J. Geol.*, **82**, 267–74.

SIMBERLOFF, D. S. & ABELE, L. G. (1976a). Island biogeography theory and conservation practice. *Science, N.Y.*, **191**, 285–6.

SIMBERLOFF, D. S. & ABELE, L. G. (1976b). Island biogeography and conservation: strategy and limitations. *Science, N.Y.*, **193**, 1032.

SPRUGEL, D. G. (1976). Dynamic structure of wave-generated *Abies balsamea* forests in the northeastern United States. *J. Ecol.*, **64**, 889–911.

SPURR, S. H. (1956). Natural restocking of forests following the 1938 hurricane in central New England. *Ecology*, **37**, 443–51.

SPURR, S. H. & BARNES, B. V. (1973). *Forest ecology*, 2nd edition. New York, Ronald Press.

STONE, E. C. (1965). Preserving vegetation in parks and wilderness. *Science, N.Y.*, **150**, 1261–7.

TERBORGH, J. (1974a). Preservation of natural diversity: The problem of extinction prone species. *BioScience*, **12**, 715–22.

TERBORGH, J. (1974b). Faunal equilibria and the design of wildlife preserves. In *Tropical ecological systems. Trends in terrestrial and aquatic research*, ed. by F. B. Golley and E. Medina, 369–80. New York, Springer Verlag.

TERBORGH, J. (1976). Island biogeography and conservation: Strategy and limitations. *Science, N.Y.*, **193**, 1029–30.

THOMPSON, J. N. (in press). Patch dynamics in *Pastinaca sativa* and resource availability to a specialized herbivore.

VOGL, R. J. & SCHORR, P. K. (1972). Fire and manzanita chaparral in the San Jacinto Mountains, California. *Ecology*, **53**, 1179–88.

WADE, D. D. & WARD, D. E. (1973). An analysis of the Air Force bomb range fire. *Southeast. For. Exp. Sta., USDA For. Serv. Res. Pap.* SE-105. 38 pp.

WALTER, H. (1971). *Vegetation of the earth*. New York, Springer-Verlag.

WEBB, L. J. (1958). Cyclones as an ecological factor in tropical lowland rain forest, North Queensland. *Austr. J. Bot.*, **6**, 220–8.

WHITCOMB, R. F., LYNCH, J. F., OPLER, P. A. & ROBBINS, C. S. (1976). Island biogeography and conservation: Strategy and limitations. *Science, N.Y.*, **193**, 1030–2.

WHITTAKER, R. H. (1953). A consideration of climax theory: The climax as a population and pattern. *Ecol. Monogr.*, **23**, 41–78.

WILLIAMSON, G. B. (1975). Pattern and seral composition in an old-growth beech-maple forest. *Ecology*, **56**, 727–37.

WILLIS, E. O. (1974). Populations and local extinctions of birds on Barro Colorado Island, Panama. *Ecol. Monogr.*, **44**, 153–69.

WILSON, E. O. (1969). The species equilibrium. *Brookhaven Symp. Biol.*, **22**, 38–47.

WILSON, E. O. & WILLIS, E. O. (1975). Applied biogeography. In *Ecology and evolution of communities*, ed. by M. L. Cody and J. M. Diamond, 522–34. Cambridge, Belknap Press of Harvard University.

WRIGHT, H. E. (1974). Landscape development, forest fires, and wilderness management. *Science, N.Y.*, **186**, 487–95.

WRIGHT, H. M. (1974). Non-game wildlife and forest management. In *Wildlife and forest management in the Pacific northwest*, ed. by H. C. Black, 27–38. Corvallis, Forest Research Laboratory School of Forestry, Oregon State University.

Nutrient Loss Accelerated by Clear-Cutting of a Forest Ecosystem

Science, 159:882–884 (1968)

F. H. Bormann, G. E. Likens, D. W. Fisher, and R. S. Pierce

Nutrient Loss Accelerated by Clear-Cutting of a Forest Ecosystem

Abstract. *The forest of a small watershed-ecosystem was cut in order to determine the effects of removal of vegetation on nutrient cycles. Relative to undisturbed ecosystems, the cut ecosystem exhibited accelerated loss of nutrients: nitrogen lost during the first year after cutting was equivalent to the amount annually turned over in an undisturbed system, and losses of cations were 3 to 20 times greater than from comparable undisturbed systems. Possible causes of the pattern of nutrient loss from the cut ecosystem are discussed.*

One-third of the land surface of the United States supports forest, and much of it is occasionally harvested. Yet, apart from nutrient losses calculated on the basis of extracted timber products, we have little quantitative data on the effects of harvesting on either the nutrient status of forest ecosystems or the chemistry of stream water draining from them—which is closely related to the increasingly important problem of eutrophication of stream and river water (*1*). This paucity of information partly reflects the difficulties of measuring characteristics of massive forest ecosystems, and the fact that nutrient cycles are closely tied to hard-to-measure hydrologic parameters (*2*).

The input and output of chemicals can be measured and nutrient budgets constructed for forest ecosystems by use of the small-watershed approach (*2*). For several years we have measured these parameters on six small undisturbed watersheds in the Hubbard Brook Experimental Forest in central New Hampshire (*3*); here the bedrock is practically impermeable (*4*), and all liquid water leaves the watersheds by way of first- or second-order streams; the runoff pattern is typical of northern regions having deep snow packs (*5*). Additional information on topography, geology, climate, and biology is given by Likens *et al.* (*4*).

Chemical relations for these undisturbed forest ecosystems (watersheds) are being established by weekly measurements of dissolved cations and anions entering the ecosystem in all forms of precipitation and leaving the system in stream water. These data, combined with measurements of precipitation and stream flow, enable computation of the input and output of these various elements, as well as annual budgets (*2*). These results have been reported (*4, 6, 7*).

In 1965 the forest of one watershed (W-2) was clear-cut in an experiment designed to: (i) determine the effect of clear-cutting on stream flow, (ii) examine some of the fundamental chemical relations of the ecosystem, and (iii) evaluate the effects of forest manipulation on nutrient relations and on eutrophication of stream water. This is a preliminary report of chemical effects observed during the subsequent year.

The experiment began during the winter of 1965–66 when the beech-maple-birch forest (15.6 hectares) was leveled by the U.S. Forest Service. All trees, saplings, and shrubs were cut, dropped in place, and limbed so that no slash was more than 1.5 m above ground. No products were removed from the forest, and great care was taken to minimize erosion of the surface. On 23 June 1966, regrowth of vegetation was inhibited by application of the herbicide Bromacil ($C_9H_{13}BrN_2$-

O_2) at 28 kg/hectare; approximately 80 percent of the mixture applied was Bromacil; 20 percent was largely inert carrier (*8*).

Samples of stream water were collected and analyzed weekly, as they had been for 2 years before the cutting; the loss of ions was calculated in terms of kilograms per hectare. Similar measurements on adjacent undisturbed watersheds provided comparative information.

The cutting had a pronounced effect on runoff, which began to increase in May 1966; the cumulative runoff value for 1966 exceeded the expected value by 40 percent. The greatest difference occurred during June through September, when runoff values were 418 percent greater than expected. The difference is directly attributable to the removal of transpiring surface and probably reflects wetter conditions within the soil profile.

The striking loss of nitrate nitrogen in stream water (Fig. 1) suggests that alteration of normal patterns of nitrogen flow played a major role in loss of nutrients from the cutover ecosystem. This loss is best understood by consideration of nitrogen patterns in the undisturbed ecosystem. Runoff data from such systems (*7*) (Fig. 1) indicate a strong and reproducible seasonal cycle of concentration of nitrate in stream water. High concentrations are associated with the winter period from November through April, while low concentrations persist from April through November.

The decline of nitrate concentrations during May and the low concentrations throughout the summer correlate with heavy nutrient demands by the vegetation and increased heterotrophic activity associated with warming of the soil. The winter concentration pattern of NO_3^- may be explained in strictly physical terms, since the input of nitrate in precipitation from November through May largely accounts for nitrate lost in stream water during this period. Evaporation from the snow pack may account for some increase in concentration of NO_3^- in the stream water in the spring. Also, since yearly input of nitrate in precipitation exceeds losses in stream water (Table 1), concentrations in stream water provide little conclusive evidence of the occurrence of nitrification in these undisturbed acid soils.

Results from the cut watershed demonstrate nitrogen relations of such an

Table 1. Partial nitrogen budgets for watersheds 6 (13.2 hectares) and 2 (15.6 hectares); all data are in kilograms of elemental nitrogen per hectare. Gains by biological fixation and losses by volatilization are not included. Watershed 2 was cut in the winter of 1965–1966.

Year	Input in precipitation		Output in stream water		Net gains (+) losses (−)	
	NH_4–N	NO_3–N	NH_4–N	NO_3–N	NH_4–N	NO_3–N
Watershed 6 (undisturbed)						
1965	2.1	2.8	0.6	1.0	+1.5	+1.8
1966	2.0	4.5	0.5	1.5	+1.5	+3.0
Watershed 2 (cutover)						
1965	2.1	2.8	0.4	1.3	+1.7	+1.5
1966	2.0	4.5	1.2	58.1	+0.8	−53.6

ecosystem and indirectly those of the undisturbed ecosystem. Comparison of nitrate concentrations in stream water from watersheds W-6 (undisturbed) and W-2 (cutover) indicates a similar pattern of concentrations throughout 1964 and 1965, prior to cutting, and through May of 1966 (Fig. 1). Beginning on 7 June 1966, 16 days before application of the herbicide, nitrate concentrations in W-2 show a precipitous rise, while the undisturbed ecosystem shows the normal late-spring decline. Allison (9) has documented similar losses of nitrate from uncropped fields or fields carrying poorly established crops. The increase in nitrate concentrations is a clear indication of the occurrence of nitrification in the cutover ecosystem. Since an NH_4^+ substrate is required, the occurrence of nitrification also indicates that soil C : N ratios were favorable for the production of NH_4^+, in excess of heterotrophic needs, sometime before 7 June.

Some of these conclusions must hold for the undisturbed ecosystem; that is to say, sometime prior to 7 June C : N ratios were favorable for the flow of ammonium either to higher plants or to the nitrification process. The low levels of NH_4^+ and NO_3^- in the drainage water of the undisturbed ecosystem (W-6) may attest to the efficiency of the oxidation of NH_4^+ to NO_3^-, and to the efficiency of the vegetation in utilizing NO_3^-. However, Nye and Greenland (10) state that growing, acidifying vegetation represses nitrification; thus the vegetation may draw directly on the NH_4^+ pool, and little nitrate may be produced within the undisturbed ecosystem. In this case, one must assume that cutting drastically altered conditions controlling the nitrification process.

The action of the herbicide in the cutover watershed seems to be one of reinforcing the already well-established trend, of loss of NO_3^-, induced by cutting alone. This action is probably effected through herbicidal destruction of the remaining vegetation—herbaceous plants and root sprouts. In the event of rapid transformation of all nitrogen in the Bromacil, this source could at best contribute only 5 percent of the nitrogen lost as nitrate.

During 1966 the cutover area showed a net loss of N of 52.8 kg/hectare, compared to a net gain of 4.5 kg/hectare for the undisturbed system (Table 1). If one assumes that the cutover system would have normally

gained 4.5 kg/hectare, the adjusted net loss from the cutover system is about 57 kg/hectare. The annual turnover of nitrogen in undisturbed systems is approximately 60 kg/hectare on the basis of an equilibrium system in which annual leaf fall is about 3200 kg/hectare (11) and annual losses of roots are about 800 kg/hectare. Consequently an amount of elemental nitrogen, equivalent to the annual turnover, was lost during the first year following cutting.

Nitrogen losses from W-2 do not take into account volatilization, which accounted for about 12 percent of the total losses from 106 uncropped soils (9). Moreover, denitrification, an an-

aerobic process, requires a nitrate substrate generated aerobically (12); consequently, for substantial denitrification to occur in fields, aerobic and anaerobic conditions must exist in close proximity. The large increases in subsurface flow of water from the cutover watershed suggests that such conditions may have been more common throughout the watershed.

A high level of nitrate ion in the soil solution implies a corresponding concentration of cations and ready leaching (10); precisely this situation prevailed in W-2. Simultaneously with the rise of nitrate, concentrations of Ca^{++}, Mg^{++}, Na^+, and K^+ rose ultimately severalfold. These increases, in

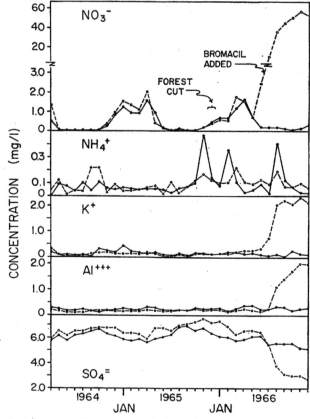

Fig. 1. Average monthly concentrations of selected cations and anions in stream water draining from forest ecosystems undisturbed (solid lines) and clear-cut during the winter of 1965–66 (dashed lines).

combination with the increase in drainage water, led to net losses 9, 8, 3, and 20 times greater, respectively, than similar losses from five undisturbed ecosystems between June 1966 and June 1967. Concentrations of Al^{+++} rose about 1 month later than the initial rise in nitrate, while sulfate showed a sharp drop in concentration, coincident with the rise in nitrate (Fig. 1).

These results indicate that this ecosystem has limited capacity to retain nutrients when the bulk of the vegetation is removed. The accelerated rate of loss of nutrients is related to the cessation of uptake of nutrients by plants and to the larger quantities of drainage water passing through the system. Accelerated losses may also relate to increased mineralization resulting from changes in the physical environment, such as change in temperature or increase in available substrate.

However, the effect of the vegetation on the process of nitrification cannot be overlooked. In the cutover ecosystem the increased loss of cations correlates with the increased loss of nitrate; consequently, if the intact vegetation inhibits the process of nitrification [13] and if removal of the vegetation promotes nitrification, release from inhibition may account for major losses of nutrients from the cutover ecosystem.

These results suggest several conclusions important for environmental management:

1) Clear-cutting tends to deplete the nutrients of a forest ecosystem by (i) reducing transpiration and so increasing the amount of water passing through the system; (ii) simultaneously reducing root surfaces able to remove nutrients from the leaching waters; (iii) removal of nutrients in forest products; (iv) adding to the organic substrate available for immediate mineralization; and (v), in some instances, producing a microclimate more favorable to rapid mineralization. These effects may be important to other types of forest harvesting, depending on the proportion of the forest cut and removed. Loss of nutrients may be greatly accelerated in cutover forests where the soil microbiology leads to an increase of dissolved nitrate in leaching waters [10].

2) Management of forest ecosystems can significantly contribute to eutrophication of stream water. Nitrate concentrations in the small stream from the cutover ecosystem have exceeded established pollution levels (10 parts per million) [14] for more than 1 year, and algal blooms have appeared during the summer.

F. H. BORMANN
*School of Forestry, Yale University,
New Haven, Connecticut 06520*
G. E. LIKENS
*Department of Biological Sciences,
Dartmouth College, Hanover,
New Hampshire 03755*
D. W. FISHER
*U.S. Geological Survey,
Washington, D.C. 20025*
R. S. PIERCE
*Northeastern Forest Experimental
Station, U.S. Forest Service,
Durham, New Hampshire*

References and Notes

1. W. H. Carmean, *Science* **153**, 695 (1966); E. S. Deevey, *ibid.* **154**, 68 (1966).
2. F. H. Bormann and G. E. Likens, *ibid.* **155**, 424 (1967).
3. Hubbard Brook Experimental Forest is part of the Northeastern Forest Experiment Station.
4. G. E. Likens, F. H. Bormann, N. M. Johnson, R. S. Pierce, *Ecology* **48**, 772 (1967).
5. W. E. Sopper and H. W. Lull, *Water Resources Res.* **1**, 115 (1965).
6. F. H. T. Juang and N. M. Johnson, *J. Geophys. Res.* **72**, 5641 (1967).
7. D. Fisher, A. Gambell, G. E. Likens, F. H. Bormann, in preparation.
8. H. J. Thome, personal communication, 1967.
9. F. E. Allison, *Advan. Agron.* **3**, 213 (1955).
10. R. H. Nye and D. J. Greenland, *Tech. Commun. 51* (Commonwealth Bureau of Soils, Harpenden, England, 1960).
11. G. Hart, R. E. Leonard, R. S. Pierce, *Forest Res. Note 131* (Northeast. Forest Exp. Sta., Upper Darby, Pa., 1962).
12. S. L. Jansson, *Kgl. Lantbrukshogskol. Ann.* **24**, 101 (1958).
13. E. L. Rice, *Ecology* **45**, 824 (1964).
14. F. H. Rainwater and L. L. Thatcher, *U.S. Geol. Surv. Water-Supply Paper 1454* (U.S. Government Printing Office, Washington, D.C., 1960).
15. Supported by NSF grants GB 1144, GB 4169, GB 6757, and GB 6742. We thank Noye Johnson, David Pramer, William Smith, and Garth Voigt for suggestions and for reviewing the manuscript. Contribution 4 of the Hubbard Brook Ecosystem Study; a contribution to the U.S. program for the International Hydrological Decade, and to the International Biological Program.

28 December 1967

Antigen Binding to Cells: Determination by Enzymic Fluorogenic Group Hydrolysis

Abstract. A sensitive method for detecting cells containing antibody to β-galactosidase has been devised. The enzyme attached to the cells containing antibody can hydrolyze a fluorogenic substrate and yield fluorescent products which are measured microphotofluorometrically. This method of detecting a few molecules of antibody is applicable to other enzyme antigen systems.

A method for measuring cellular antibody in very small amounts is necessary when only a few molecules of antibody may be present in a cell, such as very early in the antibody response, during states of immunologic unresponsiveness, or possibly at receptor sites on the membranes of cells bearing immunologic memory [1]. Our method was designed to detect only a few molecules of cellular antibody.

Using the substrate fluorescein-di-β-galactopyranoside (FDGal), Rotman et al. measured the β-galactopyranosidic activity of single molecules [2], individual bacterial cells [3], and viruses [1]. The substrate is nonfluorescent and upon hydrolysis of its glycosidic bonds yields fluorescent products (fluorescein-mono-β-galactoside or fluorescein, depending on whether one or both glycosidic bonds are cleaved, respectively). Antibody to β-galactosidase does not impair the activity of the enzyme [5], and so the enzyme-antienzyme complex can hydrolyze the substrate also.

For the assay, the spleens of immunized or normal A/Jax mice were removed and macerated; the fragments were screened through a 60-gauge and then a 250-gauge stainless steel gauze in Hanks's balanced salt solution, with 50 μg of streptomycin, 60 units of penicillin, and 20 units of heparin per milliliter. The resultant single-cell suspension was washed once in the same solution and centrifuged. The cell pellet was treated with 95 percent ethanol for 10 minutes at room temperature, washed again, and suspended in medium to yield a 20-percent suspension of cells.

A portion of the suspension was removed for measurement of "intrinsic" β-galactosidase, including the activity of endogenous mouse-spleen β-galactosidase and activity of any possible residual antigen from the immunization. The rest of the cells were incubated with 15 μg of β-galactosidase per milliliter of cell suspension for 1 hour at 37°C and then overnight at 4°C [6]. The cells were then washed four times with 40 ml of medium to remove the free enzyme, and suspend-

3 The Emergence of Multiple Concepts of What Landscape Ecology Is About

Introduction and Review

Modern landscape ecology began to develop its own identity in the early 1980s, building on the spatial approaches of geography, soil science, and plant ecology and the increasing awareness among both basic and applied ecologists of the importance of spatial patterns. The International Association for Landscape Ecology (IALE) was founded in 1982, one year after a congress in Veldhoven, The Netherlands, brought together landscape ecologists from around the world. This meeting was particularly significant in exposing scientists from both sides of the Iron Curtain to perspectives that previously had not been widely recognized (Tjallingii and de Veer 1982). The first volume of the journal *Landscape Ecology* appeared in 1987. By outward appearances, landscape ecology had arrived.

But appearances can be deceiving. Beneath the facade of a single professional association and a flagship journal, there were clear differences in concepts about what the subject of landscape ecology really was, how such studies should be approached, and, at the core, how landscapes should be viewed. Such varied views might be expected of a science that had its roots in diverse disciplines (part I) with diverse cultures and histories of land use. Few disciplines emerge fully formed and conceptually unified, of course, so the conceptual immaturity of landscape ecology during the 1980s (and perhaps much later; Wiens 1999) is understandable. Nonetheless, for some time landscape ecology remained, in Richard Hobbs's words, "a science in search of itself" (1994:170).

The development of these multiple views of landscape ecology has been discussed by Schreiber (1990), Forman (1990), Naveh and Lieberman (1994), Zonneveld (1995), Wiens (1999), and Turner et al. (2001). The five papers included here represent part, but by no means all, of the varied perspectives about landscape ecology as it began to coalesce and emerge.

One view, prevalent in Central Europe, derived from a foundation in the ideas of Berg and Solnetsev (part I), which represented a geographic perspective on landscape ecology. **Ernst Neef** outlined the state of thinking and progress made from these earlier foundations. In the translated excerpts from one of his standard texts, Neef outlines his concept of landscape, arguing for a much firmer definition and conceptual base. Although he begins with a global/planetary perspective, he raises several problems that were to become significant issues in the later developments in landscape ecology in the 1980s and 1990s (part IV) as it developed from its ecosystem base. Issues of both scale and hierarchy are raised together with critical questions concerning the abstraction and the limitations of spatial data and the chorological (i.e., spatial) perspective in landscape study. The capacity of individual subdisciplines to contribute to the solving of landscape problems as an interdisciplinary exercise without a firm conceptual understanding of landscape structure and function is deemed an issue requiring immediate attention.

Beginning in the mid-1950s, Neef led a school of landscape ecologists in the German Democratic Republic whose influence, together with that of workers in Poland and Czechoslovakia, consolidated and defined a geoecological approach to landscape ecology. Neef's approach was much more scientific than the earlier work of fellow geographers. The objective was to investigate the nature of the landscape and the interactions of its component parts through field investigations of processes, model development, and laboratory analyses. Neef and his coworkers developed many versions of general models of homogeneous geocomplexes and more structured models of heterogeneous geocomplexes. The overall goal was to enhance the understanding of landscape functioning and its natural potential in ways that could be used in state and regional planning. Regrettably, much of Neef's work and that of his colleagues was unknown in the English language until the 1980s, when IALE meetings and related publications provided a broader outlet for their work. (A useful and accessible source for gaining an overview of this work can be found in a special issue of *GeoJournal* [1983, volume 7, number 2] devoted to the theme of landscape synthesis.)

One feature of an emerging discipline is a classification phase, in which a taxonomy of the objects of study is developed. Ideally, this enables everyone to use the same terms to describe the same things. **Richard Forman and Michel Godron** developed such a structural typology of patch types in a landscape. Their approach followed directly from the adoption of two simplifying premises about landscapes: that they

could be considered as a series of patches surrounded by a homogeneous matrix and that landscapes should be viewed on a kilometers-wide scale. The perspective was explicitly structural rather than functional; patch boundaries, for example, were considered in terms of various structural configurations rather than their functional properties (for a more functional perspective, see Wiens et al. 1985). Nonetheless, the description and categorization of patch types represented an attempt to formalize what had previously been vague references to "patches." A similar distinction between structure and function typified some of the landscape work of geographers. This was expressed both in their approaches to landscape classification (part I) and in the contrast between the work of Neef and his Central European and Soviet colleagues and the general neglect of integrated landscape processes by physical geographers in North America (part II).

Although Forman and Godron's focus was on landscape structure, a more functionally slanted perspective on landscapes was beginning to emerge in North America at about the same time. To some degree, this development grew out of a structure-function polarity that had emerged in mainstream ecology in which community ecologists (and, to some extent, population ecologists) focused on structure, whereas ecosystem ecologists largely emphasized processes or functions. Drawing from their shared interests in spatial relationships and inspired by developments in landscape ecology in Europe, U.S. ecologists **Paul Risser, James Karr, and Richard Forman** convened a workshop in Allerton Park, Illinois, in early 1983 to develop a framework for landscape ecology. The attendees represented a broad range of disciplines and included both basic and applied perspectives; many, however, had emphasized ecological processes in their previous work, and perhaps one third had been involved in the ecosystem studies conducted as part of the International Biological Program (IBP). Not surprisingly, then, the workshop conclusions emphasized the importance of landscape structure in controlling the redistribution of organisms, materials, and energy among landscape components. Landscape dynamics were also linked to feedback controls, perhaps reflecting the cybernetic approach to ecosystems prevalent at that time. Finally, the workshop conclusions emphasized the overarching role of spatial heterogeneity, both as an essential feature of landscapes and as an important determinant of movements and interactions across landscapes, and, therefore, as a focus for resource management. This emphasis also set landscape ecology apart from mainstream ecology, which was still largely bound to notions of homogeneity and equilibrium.

The Allerton Park workshop was a watershed event in the development of landscape ecology, at least in North America. Within two years, a U.S. Chapter of IALE had been formed. Somewhat earlier, in 1984, the Canadian Society for Landscape Ecology and Management had been established as a chapter within IALE. Its initial foundations, however, were more aligned to the applied land resource perspective outlined in part I (Rubec et al. 1988) than to the biologically based approach.

Landscape was becoming recognized (if not yet respected) by the granting agencies, and publications dealing with landscape structure and function were beginning to appear in ecological and management journals with increasing regularity. By 1987, the discipline had developed to the point at which a synthetic paper of sorts, by **Dean Urban, Robert O'Neill, and Hank Shugart,** appeared in a general science journal, *BioScience.* This paper was remarkable in several respects. It represented a clear evolution from the earlier patch-matrix perspective of Forman and Godron, instead viewing a landscape as a mosaic of patches of varying sizes, origins, and dynamics. Although Forman and Godron thought of landscape as a single, broad scale, Urban et al. emphasized that landscapes were influenced by events at multiple scales of space and time. Indeed, rather than thinking of landscape as a scale, Urban et al. considered it as a level in a hierarchy and couched most of their thinking about landscapes in the context of hierarchy theory (O'Neill et al. 1986). In this sense, the approach reflected the continuing influence of ecosystem or systems-based thinking, tracing back to the work of Gene Odum and H. T. Odum and of Bernard Patten during the IBP (see Hagen 1992; Golley 1993). This approach was later reinforced by a group based at Oak Ridge National Laboratory (Robert O'Neill, Robert Gardner, Monica Turner, Anthony King, and Virginia Dale) that came to have a powerful influence over the development of North American landscape ecology.

The influence of systems thinking evident in the hierarchical approach of Urban et al. is reflected with even greater force in the writings of **Zev Naveh.** Perhaps more than anyone else, Naveh has championed a view of landscapes not only as hierarchical but also as self-organizing systems in which humans are not only occupants of the landscape but are also participants in the systems. In Naveh's view, landscapes have a natural homeostasis that results from multiscale cybernetic feedback controls. Because human activities tend to disrupt this homeostasis, one cannot consider contemporary landscapes separately from their human participants. Interesting parallels may be drawn between the holistic view of landscapes espoused by Naveh and recent developments in the panarchy theory developed by C. S. Holling, Lance Gunderson, and others (Gunderson and Holling 2002).

References

Forman, R. T. T. 1990. *Land mosaics.* Cambridge: Cambridge University Press.

Golley, F. B. 1993. *A history of the ecosystem concept in ecology.* New Haven, CT: Yale University Press.

Gunderson, L. H., and C. S. Holling, eds. 2002. *Panarchy: Understanding transformations in human and natural systems.* Washington, DC: Island Press.

Hagan, J. B. 1992. *An entangled bank: The origins of ecosystem ecology.* New Brunswick, NJ: Rutgers University Press.

Hobbs, R. J. 1994. Landscape ecology and conservation: Moving from description to application. *Pacific Conservation Biology* 1:170–176.

Naveh, Z., and A. Lieberman. 1994. *Landscape ecology: Theory and application.* New York: Springer-Verlag.

O'Neill, R. V., D. L. DeAngelis, J. B. Waide, and T. F. H. Allen. 1986. *A hierarchical concept of ecosystems.* Princeton, NJ: Princeton University Press.

Rubec, C. D. A., E. B. Wicken, J. Thie, and G. R. Ironside. 1988. Ecological land classification and landscape ecology in Canada: The role of the CCELC and the formation of the CSLEM. In *Landscape Ecology and Management*, ed. M. R. Moss, 51–54. Montreal: Polyscience.

Schreiber, K.-F. 1990. The history of landscape ecology in Europe. In *Changing Landscapes: An Ecological Perspective*, eds. I. S. Zonneveld and R. T. T. Forman, 21–33. New York: Springer-Verlag.

Tjallingii, S. P., and A. A. de Veer, eds. 1982. *Perspectives in landscape ecology: Proceedings of the International Congress arranged by the Netherlands Society for Landscape Ecology, Veldhoven, The Netherlands, 1981.* Wageningen, The Netherlands: Pudoc.

Turner, M. G., R. H. Gardner, and R. V. O'Neill. 2001. *Landscape ecology in theory and practice: Pattern and process.* New York: Springer-Verlag.

Wiens, J. A. 1999. The science and practice of landscape ecology. In *Landscape Ecological Analysis: Issues and Applications*, ed. J. M. Klopatek and R. H. Gardner, 371–383. New York: Springer-Verlag.

Wiens, J. A., C. S. Crawford, and J. R. Gosz. 1985. Boundary dynamics: A conceptual framework for studying landscape ecosystems. *Oikos* 45:421–427.

Zonneveld, I. S. 1995. *Land ecology.* Amsterdam: SPB Academic.

The Theoretical Foundations of Landscape Study

Ernst Neef

The Axiomatic Foundations

Generalizations

Reality provides the basic concepts about an object, concepts that are both derived from experience and confirmed by evidence. These basic concepts are an image of the nature of a given object. They cannot be proven because there are no arguments beyond these basic concepts to support this evidence. Nor can they be defined because there are no fundamental dimensions available for their definition. These basic concepts, however, present the beginning of all scientific effort because they provide the fundamental divisions of a science for which proof, defined terms, and conclusions can be established. Wallot[1] notes that, in the technical realm, basic dimensions, such as the concept of length, often cause more difficulty than all spatially derived notions and that errors are often made because basic dimensions and basic concepts have remained unclear.

It is necessary for geography to clarify its basic concepts and establish a firm foundation for future geographic work. Because these basic concepts will arise from evidence and experience, they have an axiomatic character. The geographer, therefore, must clarify these axiomatic foundations because axioms form the basis of work in the scientific domain and contribute to providing science with an appropriate foundation. There is no more urgent need for this than in geography, the objectives of which cause many difficulties. Such difficulties are a consequence of its variety, complex character, and its temporal and spatial concepts.

Axioms cannot be defined, merely formulated. All geographic concepts of an axiomatic character—among them the concept of landscape—elude definition. How much energy has been wasted in trying to define basic geographic concepts without arriving at any recognizable results! One might recall Schmithüsen's much-quoted remark that, to this day, there is no generally recognized definition of landscape! Conclusions can be derived from basic axiomatic concepts. These, in turn, allow terms to be defined and proofs to be substantiated. This is of fundamental importance and is the requirement for an unambiguous technical language. This is difficult to manage successfully, however, given the characteristics of geography and the methodological route that this science must follow in order to clarify successfully the organization and structure of its subject matter and to and make any research available. These questions play a major role in geography because geographic reality is of a complex character, one that is not immediately open to scientific analysis. Such questions are also of importance to geography because, by this type of analysis, its inherent interconnections must be broken apart, at least in part. In so doing, any investigation will deviate from geographic reality.

Only those characteristics that have a unifying character and that are derived directly from the geosphere and from landscape structure will be considered axioms. Like the basic fundamentals of physics, these may only make a minimal number of assertions, and conclusions cannot be deduced from the basic facts. Rather, these axioms are to be formulated as special theorems or as fundamental propositions. Axiomatic theory, therefore, consists of the formulation of axioms proper and the subsequent elaboration of fundamental theorems.

It appears to be possible to lay down three axioms for geography. They are:

1. The planetary axiom.
2. The landscape axiom.
3. The chorological axiom.

The three major theorems that follow are derived from these axioms:

1. The theorem of the geographic continuum.
2. The theorem of geographic limits.
3. The theorem of geographic matter.

Ultimately, a number of other basic concepts like zonality, azonality, chore, physiotope, and others can be derived from these axioms and major theorems.

The Planetary Axiom

The initial, basic fact is that the geosphere is part of the earth as a planet. It follows that its form closely confers a number of features or geographic phenomena that result from their shape, size, and astronomical position on the planet. Through various connections and with different degrees of geographic content, this interdependence is ultimately shared by every point on the earth's surface because of the earth's geospheric form. The climatic situation at any point on the earth is, therefore, determined by the overall terrestrial interrelationships of the atmosphere and also by the fact that every point is integrated with the planetary circulation. Certain basic features result from this. For example, if a weather station is to be taken as a representative location, then *representative* must refer to the general geophysical arrangement and to the exclusion of local features.

The planetary axiom can, therefore, be formulated in quite a general way without reference to any particular phenomena:

All geographic facts are in some way related to the *earth* as a planet and are endowed by this with certain geographic features.

This applies to isolated material facts as well as to every section of the earth's surface.

The forms in which the dependence upon the planet earth is expressed vary a great deal, however, and some commonly known geographic phenomena lack important geographic content. In the first instance, this applies to geographic coordinates that give the exact geographic location of a place on the surface of the earth. In fact, these are definitions of position. Geographic content that results from geospheric interrelationships cannot be expressed by coordinates. This cannot, however, alter the fact that coordinates can allow people with the appropriate knowledge to point out the geographic character of an area as defined by coordinates. Likewise, there is little significance of a geographic nature in the fact that both the former geographic mile and the meter, the international unit of length, are derived from dimensions of the earth. Much more important are the distances and areas that can be derived from the size of the planet and that provide the basic concepts for the terrestrial relationships of position, spatial connection, and perception of distance. Such perceptions may result from polar or equatorial distance or from the size of the space between meridians and parallels at the different latitudes. The total area of the earth, with its 510 million square kilometers, sets the limits to the extent of the earth's surface. The dimensions of the earth, therefore, provide a measure of the terrestrial portion of the earth in length and area. These properties run through all relationships.

The same applies to the basic measures of time, such as day and night, summer and winter half year, and length of day and night, all of which result from the rotation and revolution of the earth. They have become accepted as matters of course, but assuming these basic units of time were different, then many basic facts would change. If day and night, or even the summer or winter half year, lasted longer, radiation and outward radiation would alter significantly and thus produce serious geographic effects that would be seen in greater extremes and considerably greater amplitudes of thermal conditions. All conditions of life, therefore, would be profoundly influenced. This shows that planetary facts also cause a measure of the temporal conditions, which are themselves characteristics of the earth.

Attention is drawn to these trivial matters to show that planetary facts reveal themselves in many basic phenomena that are not yet objectively linked to basic geographic facts; that is, they present a kind of pregeographic, general basis for terrestrial phenomena. When they link up with concrete facts, however, they become immediately effective in geographic relationships and, therefore, must not be ignored. Even within the context of the planetary axiom, with the possibility of involving quantitative data like the abstraction of special geographic features, basic planetary facts are turned into important ordering principles such as the notion of zonality. The qualities of a spherical surface, which lead to the basic concepts of the geographic continuum, produce decisive starting points for problems of perception in geography. Because they are linked with concrete facts, they no longer maintain the character of independent axioms but can be derived from them as primary axioms.

The Landscape Axiom

Every part of the earth's surface has a particular form, resulting both from the combination of forces similar to those that produced the geosphere and from the combined effects of the biosphere and the three inorganic spheres.

What could only be formulated in general terms in the case of the geosphere can, however, be put in more concrete terms in this instance. The lithosphere, atmosphere, and hydrosphere, like the biosphere, are individually characterized by an extraordinary variety of phenomena. Unified in one place, the different substances react upon one another: Physical and chemical processes take place, some fast and some gradually, and physical conditions vary from season to season so that there are degrees of intensity. The components, which have a role in these processes, act as conditional factors. Their influence results not only in material variety but also in varieties of interplay between the forces operating within these material systems. This reciprocal dependence of landscape elements is known as *interdependence*, and the overall phenomenon of the interlacing of all the components of this complicated play of forces is known as *interrelationship*. This expression is eminently suited to the characterization of the state of the material system as it is presented at every point on the earth's surface. It includes the nexus of structure as well as activity, and the term *connections within the landscape* is widely used. In German, therefore, one may refer to this axiom as the *landscape axiom*. The following definition is suggested:

At every point of the earth's surface, elements, components, and factors of geographic substance find themselves in varied relationships and correlations in accordance with the laws of nature.

The existence of geography as a science results from the validity of this landscape axiom, but every isolation of an element of the landscape from these interrelationships within the landscape deviates from geographic reality. The development of other special subjects such as hydrology, climatology, geomorphology, and so on, which apply themselves to one particular geofactor, does not remove anything from the primary objective of the science of geography: the landscape. Consideration of the landscape axiom demands the investigation of the earth's surface and every one of its components to its fullest complexity. Only then will geography make a significant contribution to the investigation of reality.

Once the landscape axiom has been recognized as being of the utmost importance for geography, it becomes obvious that the investigation of its vertical interconnections, the methods it employs, and the results obtained will acquire a decisive significance both for the scientific character and for the strength of statements made by geography. The complexity of the material and the dynamic relationships in the geographic complex are admittedly difficult to comprehend. Specialist sciences tend to tackle only one of the elements of the landscape and are often forced to be abstractions from the essential nature of this concrete, interconnected complex. By no means can the sum of their findings result in a reliable picture of geographic reality. Therefore, geography cannot adopt the findings of these disciplines without examining whether they retain their relevance to this total complex. In other words, the geographer must be able to be a good judge of the geographic complex, as it is established, at any place on the earth. No alternative is open to geography but to engage in the investigation of this geographic complex.

Many geographers are, however, of the opinion that such research lies outside the field of geography. The validity of the landscape axiom, however, leads to the conclusion that geography must develop the study of those phenomena, which are basic to the geographic complex, because any other approach would result in the neglect of the most important foundations of its potential for comprehensive understanding.

Here is an additional fact. A scientifically valid interpretation of the vertical interconnections assumes that quantitative methods are also employed. Quantitative work, however, permits incontestable assertions only when all data used have been obtained under the same conditions. Customarily, one may refer to these same conditions as *homogeneity* and base the comparison of areas only on geographically homogenous areal units. This concept of geographic homogeneity, therefore, requires thorough clarification, as does the need for clarification of the analysis of facts in the context of the science of relationships, which tends to isolate objects from the geographic complex. This will then permit individual phenomena to be appreciated within the geographic context given; that is, the forms of complex analysis. In this way, the landscape axiom would lead to insights that are fundamental to the development of geography and its methods and to the application of geographic points of view.

The Chorological Axiom

The components of the geosphere are also unified in a geographic complex at every point across the earth's surface by interrelationships in the same manner as individual places; that is, in the horizontal dimension. They are connected in many different ways. The chorological axiom can be referred to as the spatial interconnection and formulated as:

All geographic facts have a geographic location that is marked by its site but especially by site relationships to adjacent places and areas.

It follows that it is impossible to imagine any portion of the geographic realm that is not woven into these spatial relationships. Because all phenomena of a spatial order are included in the term *chorological*, it is expedient to refer to this axiom as the chorological axiom.

Partly as a defense against the philosophical tendencies of his era, of which Rickert was the major proponent, Alfred Hettner, influenced by Ritter and with a biased overemphasis, described geography as a chorological science. This description, as a chorological or spatial science, has been heavily contested. Chorology alone cannot constitute the essence of geography. Nor can the chorological connection be denied or bypassed either in the case of geography or among the different forms of matter on the surface of the earth. Denial of the chorological aspect would entirely do away with geography. In addition to the other axioms mentioned previously, the principle of interconnection—the spatial weft and warp—characterizes the nature of the geographic surface of every part of the earth.

The chorological axiom is valid for every geographic fact. Geographic areas as well as separate components of the geographic connection are subject to it. For particular areas, territories, or other regions, this means that a scientifically valid understanding is possible only if one does not consider and present them in isolation—one could say as islands—but takes into consideration how both short- and long-distance relationships affect them. The same applies to single objects. Only in their individual manifestation, which includes site, site relationships, and these effects, are they real in the sense of geographic objects.

These three axioms characterize the nature of the objective of geography. Apart from this, all axiomatic bases of a general epistemological nature also apply to geography. It is self evident that geographic matter is also subject to development and can therefore only be imagined as existing in time. Against this, of course, it could be said that an area is also a valid condition of matter apart from it being part of the geographic realm. Geographic axioms are derived not from mathematical or abstract space but from actual space, bound to the facts of terrestrial and planetary existence. The arrangement of things on the surface of the earth, in the geosphere, is what binds geography.

The three axioms are closely interrelated, and geographic reality presupposes all three of them. Therefore, they cannot be applied separately, even if it is possible by certain methods to leave one or the other temporarily out of consideration.

The Theorem of the Geographic Continuum

Geography is indebted to H. Lautensach[2] for having drawn renewed attention to the fact that geographic phenomena are embedded in the continuum of the earth's surface. This is of significance in the history of ideas insofar as it is the one-time effort to present geographic spatial units as a wholes in the same sense as in psychology and biology in which this has been given a basically negative slant. The spherical surface of the earth is complete in itself. Its geometric form does not, therefore, impose limits on the earth's surface. Connections are possible in all directions. Preferred directions do not arise from the spherical shape but from the position of the planet, which results in zonality. The resulting differences, however, do not cancel out connections in all directions. In part, they even have the effect of leading toward a greater diversity of connections created by this connectedness.

The spherical surface permits complete movements around the globe and thus the inclusion of all points on earth into an all-embracing continuity of the planetary, atmospheric circulation. These mass movements (of air masses in the atmosphere, of water masses in ocean circulation) transport not only the media but also their qualities. The latter are transformed from place to place, where they then take effect through this geographic connection that exercises influence in a number of ways. Thanks to this unlimited movement, no part of the surface of the earth remains isolated or without impact or influence from outside. Likewise, influences from any place on earth can also be brought to bear upon parts of the earth's surface that are either close at hand or far away.

The theorem of the geographic continuum can be formulated in the following way:

All geographic phenomena and all areas of the earth's surface are ultimately part of an overall terrestrial connection and only fully understandable within it.

In these diverse connections, attention must be paid also to the intermediate links, which vary from case to case. What is near at hand will obviously tend to have a greater influence than what is remote. Moreover, in time, the network of spatial relationships will undergo rhythmic or periodic change as well as alteration as a consequence of historic developments. The theorem, however, remains:

There is no place on earth that is not tied in with the overall terrestrial-geographic relationships.

No one can therefore experience a scientifically correct assessment if connections are neglected with both adjacent and with remote areas of the earth. Even if this is not constantly repeated, the theorem of the geographic continuum will remain in force as an unstated precondition.

The geographic continuum and the incorporation of all geographic phenomena and places into this overall terrestrial connection meet the requirement of a hologaic approach. The unity of the earth's surface not only forces many movements into complete circulation systems but also provides the framework for bringing order to the diversity that appears in both chorological differentiation and in quantitative definitions. The ultimate measure of this is the overall terrestrial balance.

The Theorem of Geographic Boundaries

The planetary axiom and the theorem of the geographic continuum are the basis for further fundamental conclusions that concern the character of geographic boundaries. All geographic boundaries exist within a continuum. All areas that are confined by geographic boundaries nonetheless remain part of the geographic continuum. They are open boundaries with effects from near and far overlapping in both directions. This makes the geographic boundary lose its absolute character. It not only confines but also permits communication at the same time. Therefore, questions arise as to what scientific justification there is in talking of geographic boundaries or of boundaries at all. Often, so-called geographic boundaries are nothing more than lines of particular features. Areal limits for the distribution of a particular feature can, in reality, often be based on demonstrable facts. Such frontiers can be defined more precisely, for instance as a boundary line that includes the extreme limits of a phenomenon, or as a borderline that includes continuous areas of distribution, as in the differentiation of the terms *tree line* and *forest line*. The fact that some features have no pronounced frontiers but form broad and ill-defined border areas, with gradual transitions and degrees of intensity, necessitates the introduction of certain defined conditions in the place of the actual, but elusive, transition to which the frontier is then attached. Such a frontier is, of course, nothing but a mental construct, which presents a substitute for an otherwise elusive aid. It does not represent a real picture of an objective fact. If mere areal divisions of particular features present such difficulties, then geographic complexes with a number of characteristics will render it even more difficult to establish an objective boundary. Because particular features do not coincide in their spatial areas but overlap to a greater or lesser degree, an objective delineation of the entire complex is difficult to establish. Here, too, for reasons of expediency, a boundary line commonly must be defined, but it is only as an approximation that it can be referred to as a natural boundary.

There is no case in geographic reality where a boundary appears as a sharp dividing line. Whenever a boundary is mentioned in scientific geography, it cannot be in the sense of a boundary of changing forms.[3] This then implies that, as a rule, the boundary cannot be a feature of geographic spatial units. The recurring assertion of landscape theory—that the landscape is a product of our thought—applies unconditionally to the boundaries of a landscape but not to the landscape itself. When seen from the perspective of this statement, the question of what a landscape actually is appears in an interesting light. The overall result is:

All geographic boundaries are boundaries in a continuum. They do not separate mutually interdependent parts of the earth's surface but are lines or boundaries of changing forms. Boundaries of specific landscape elements are spatial limits and must be characterized as such. If they are used as characteristics for geographic boundary areas, the fixing of which is difficult, they are merely substitutes for geographic boundaries (Sölch[4]); that is, they are a methodological aid. This applies even more so if the fixing of boundary areas is achieved by the use of index values or other indicator values chosen because they are expedient. If boundaries in a continuum are transitions, this characteristic of geographic boundaries must be indicated.

The Theorem of Geographic Matter

The chorological axiom is valid for every set of geographic circumstances. Geographic areas as well as specific components of the geographic complex are subject to it. For particular areas, territories, or other geographic regions, this means that a scientifically valid understanding is possible only if they are seen by their spatial linkages of short- and long-distance relationships and are neither considered nor presented in isolation. The same applies to single objects in their sense as geographic objects based on site, site relationships, and their effects.

The theorem of geographic matter can be formulated as follows:

There is nothing in geographic reality which is not endowed with the geographic qualities of site and site relationships.

If in the course of the geographic investigation of an area or object any of these geographic site qualities is omitted, the phenomenon in question will be robbed of its geographic individuality. Abstraction can take place at several levels. Even the first level makes the individual geographic object a component of a complex, transforming it into a type. If site relationships are in part maintained, site types are the result. These are applied predominantly in settlement geography; terms like *sea-port town, bridging point,* and *border town* serve as examples. In physical geography, such terms occur less often, and site description tends to be in the foreground in expressions like *weather-side, slope position, summit position,* and *shady side.* If site descriptions are also eliminated, the complex is defined by its content: terms like *university town* and *fair town,* which are used to indicate their most important function; *small town* or *large town* when it is a matter of indicating size; or *tidal coast, drumlin ledge,* or *oak copse* when the physiogeographic characteristics of the complex is intended. All these complexes are types that geography needs to employ in order to have at its disposal well-ordered and manageable groups of defined units instead of vast quantities of individual formations.

The axioms regard the formation of types as a method that considers the chorological axiom only in part, if at all. If only a little is abstracted from reality, the result will be a narrowly defined type, whereas widely defined types, which include a great deal, result from substantial abstracting. If too much is abstracted, perfectly empty and meaningless complexes will be obtained that are of no value as a geographic tool. The neglect of the chorological axiom implies, however, a departure from geographic reality in every case. It is important to keep an eye on this departure, which necessarily arises from the analytical method, and employ suitable methods that restore the connection to geographic reality.

The theorem of geographic matter appears to be significant because geography employs statistical methods a great deal. The statistical method, however, needs complexes that are obtained by stripping identical objects of their individual, geographic site quality and arranging them in mathematically definable units. Geographic assertions do, of course, get lost in this process, and careful examination is needed as to whether this implies the neglect of essential geographic features. Statistically derived values are often reemployed for geographic comparisons; in this, the resulting numerical values may imply a particularly high degree of accuracy.

Other considerations, however, may also be brought to bear on statistical methods and their value in geographic research. In order to achieve the most accurate regional differentiation possible, geography must derive small statistical quantities that provide values for a narrowly defined area. In a statistical sense, these quantities are too small to make use of the laws of statistics. In unfavorable instances, they result in the abandoning of some geographic reality, even if there is no question as to the veracity of the statistical method applied. The question arises, therefore, whether there are methods that are not accompanied by such disadvantages. These appear to be given by representative methods that can be referred to as model methods if the quantitative side is strictly adhered to. In this case, an exact and actual example is worked on (even measured as far as possible) so that it can stand as a representative method for other cases. Other examples can then be compared to the model, deviations can be established, and, therefore a way can be found for quantitative treatment that comes closer to geographic reality and that can be more easily reexamined at any given time. Therefore, discussion of the axiomatic foundations can lead to an elucidating evaluation or a clearer judgment of geographic working methods.

Abstract Spatial Theory

If the chorological axiom is allowed to be considered on its own while the concrete context of geographic reality is ignored, an abstraction of geographic space and of the site relationships operating within it will take place. Such methods, when stripped of geographic content, are known in nautical science and spherical trigonometry and are employed for the calculation of distances on the surface of a sphere. If the planetary surface is not accounted for, abstract site patterns then result. Such purely theoretical abstractions are to be found in two experiments in geographic science: in Thünen's and Christaller's systems. In more recent times this direction has been the subject of special attention in American geography.[5] Generally, one tries to simplify distributions and connections of different geographic phenomena by abstraction so that they are suited to mathematical treatment. The point of all these efforts lies in the fact that every traversing of space requires energy, even if, in this case, the approach is a purely scientific one. Therefore, it is possible to calculate, for every case, the optimal pattern of a particular constellation in space in the form of quantitative statements. Subsequently, this must be translated into a concrete form of expenditure involved (costs, time, fuel, etc.) for economic-geographic purposes. Apart from such an abstraction, providing potentially valuable fundamental knowledge and measures for the cybernetic treatment of economic-geographic problems, an abstract basic theory of site relationships offers a large number of possibilities for reification. This field of the methodological investigation of principles has hitherto received very little attention, but it will be important in this approach that the orientation toward geographic reality is not lost.

Axioms as Conditions of Reality

What now requires analysis is how to place into one operation the complicated structure for the scientific investigation of all material geographic systems according to their structures and linking effects as well as managing this multilateral and complex reality; that is, the division of the total complex of reality into its components. This must be constructed in such a simple way that it permits the recognition and definition of causal relationships. To begin with, geographic reality is, of course, lost with this. A danger exists that the results are remote from reality and that dummy results are obtained. The history of science shows that such seemingly plausible dummy solutions may well be used with intent to deceive, and one must find arguments that allow a result to be controlled and its concurrence with geographic reality to be established. Axioms presenting nothing but formulas that express a given reality, as well as fundamental theorems, may therefore serve as conditions of reality for such a control. They make it possible to indicate how close to reality the result comes—a highly important argument because, when all is said and done, geographic research cannot yield more than an approximation of reality.

The theorem of the geographic object allows for deviation from reality to be checked at every stage of abstraction. This must be compensated by suitable methods for this deviation, but such corrections often remain undone.

Even the evaluation of available data requires the application of conditions of reality. For example, if the amalgamation of previously independent, rural agricultural parishes into larger administrative units combines places of different character into one statistical unit, statements are made that suppress differentiation within the new parish. Because, in this case, the parishes thus affected tend to be purely agricultural communities, their numbers will be reduced by amalgamation. The new statistical values will indicate that the parishes with a large quota of farmers—that is, the true farming communities—will show a percentage decrease. In fact, however, there may not have been any change except by the classification criteria used, which derived new, but less-differentiated statements. The new statement is therefore only valid for the scale at which basic materials are obtained. All derived statements are therefore found to be of an appropriately limited validity. Results from the old statistics can no longer be compared with those newly obtained. Because there is no evident decline in the number of parishes with a high agricultural quota in many countries, it is doubly regrettable that administrative changes are no longer capable of accurately reflecting regionally the processes that are evident in the employment statistics. This is the reason that the basic data for statistical investigations should always refer to the same census units. It also becomes obvious that controlling the conditions of reality must be applied not only to the result but also to the foundations. In other words, in geographic research both the scale and the strength of evidence of the bases employed become part of the result. It is therefore necessary to validate the strength of evidence of the resulting assertions upon the initial data. It then becomes clear also that the value of geographic comparison, as a method of great importance to geography, depends on the extent to which comparable objects are included.

If the climatic geomorphology of an area states that a climate, like the one prevalent in southern China today, must be assumed for a certain phase in the genesis of its surface forms, then the conditions drawn upon, in the examination of a result, will show that the climate of southern China is tied to the eastern flanks of the continent within the context of the planetary order of the atmospheric circulation. A climatic type of the eastern sides of the continents in the subtropics, however, cannot possibly occur on the western sides of the continents. The *tertium comparationis* have evidently been wrongly selected here. The conclusion might perhaps have been that climatic elements, like those influencing the morphological processes now in southern China, must have been at work during the phases of development under investigation. It is not the climatic type, but rather certain soil temperatures and certain humidity levels—possibly together with the distribution of humidity throughout the seasons—that direct the chemical processes of weathering and soil formation as well as the conditions of drainage. These are decisive factors for surface erosion and did contribute to the morphological development of the area under investigation. If, however, this climatic type was deduced from the occurrence of particular indicator plants, the error in this conclusion (quite apart from the botanical question of the conformity of the ecological conditions of the species) is about as great as if one had postulated a southern Chinese climatic type for the European Mediterranean area because tropical fruits with an East Asian origin were cultivated there. Climatic types obey a planetary and chorological order. A transfer of type to another situation of the planetary or chorological order—even if only for comparative purposes for presentation—contradicts both the axioms.

In landscape ecology, the soil type is the most important feature for the characterization of a location and its ecological qualities. Indeed, it has frequently been assumed that the justification for a soil survey is a substitute for the more complex analysis of landscape ecology that could be derived from it. A communication from A. Zworikin, to the effect that more than five hundred soil types and variants have been recorded by soil geography in the Ukraine, in contrast to a mere fraction of ecological types in the same area, does, however, show that these two lines of research do not fully overlap. To achieve congruence, it is necessary to analyze further the effects that soil types exert on ecological connections so that the same effects are treated as a whole, unless other ordering and classification principles exclude from the very outset a parallelism of the soil type with ecotype phenomena.

The number of examples could be enlarged upon at will. Time and again, this shows that the axioms are a valuable aid as conditions of reality in:

1. The securing of comparability.
2. The establishing of limits of validity for assertions.
3. The definition of the degree of approximation for a knowledge of complete reality.
4. The choice of the correct terms and definitions in presentation.

Basic Methodological Problems

The Extent of Knowledge of Landscape Theory

The preceding discussion of the geospheric and the regional order of things on the earth's surface, together with the formulation of axioms and the derivation of major assertions from them, produces a portrait of a distinct and specialist realm of reality. Nonetheless, it is clear that this field is not investigated by any other branch of science. The intention here is to widen the horizons of geography from their current state into a specialist discipline. This geography must therefore include more than either the contemporary discipline of geography now does or defines as its content and purpose. Missing from this characterization are many of the current special aspects of geography found in geomorphology, climatology, settlement, or transport geography. Nothing is said about the particular parallelism of subjects in general geography or with the independent geosciences. A very broad framework is set up and a very wide horizon marked out for this scientific task. The term *geography* for this field, therefore, should be avoided. Because the characteristics of the proposed field have been derived from the phenotypes of the landscape, it would be better to relate it to the landscape or to the doctrine of the landscape, for which the existing forms of regional units, regional associations, and regional linkages appear to be decisive. These are the very features of a reference system of spatial properties that do not arise in any other science or that are to be found reduced to simple data on distributions. Furthermore, one ought to speak of research in regional science, first because the word *Landschaft* does not easily translate into other languages, and second because other fields of knowledge also use the principles of regional ordering and consequently could also be described as regional sciences. For instance, at the University of Waterloo in Canada, geography and urban and regional planning are brought together in one unit that could justifiably be called regional science.

In addition to the doctrine of landscape, with landscape ecology as its core, there are many more sciences for which regional problems are a primary consideration. Nowhere in the literature of the theory of science, however, has the regional aspect been adequately presented and methodically penetrated in such a way that any guiding principles could be discerned for this task. The division of the arts and sciences came about in the last century at a time when the sciences were organizing themselves for work at universities and academies that had not yet recognized regional problems as thematic fields of central importance. This organization of scholarship, however, with its two main branches in the arts and the sciences, does not provide a starting point for addressing major regional problems. A complementary one is therefore required. A trinity of the arts, natural sciences, and regional sciences would meet these contemporary requirements better. It would then become an important objective to establish the research principles and the methodology for the regional sciences under these general guidelines for scientific research. The starting point would have to be landscape research in geography because it is broadest in terms of the scope of its subject matter. That is not all, however; for decades this approach has been under pressure from other scientifically orientated colleagues, who claim it to be unscientific in its methods because geography alone, overwhelmed as it is by the reality of regional diversity, cannot make specific judgments on any scientific relationships by way of causal analysis.

Even if the author were convinced that a number of regional sciences would base their scientific work on sound theory, the intention here is to provide an adequate clarification of the problems of the idea of landscape. Therefore, before discussing pure and applied landscape research, it would be useful to examine some questions of earlier landscape studies because this would reveal the narrowness of professional thinking in geography under its existing disciplinary direction. By contrast, we can see what is to be offered by the proposed breadth and scope of a regional perspective in landscape research.

If the concept of the landscape is based on the axiomatic characteristics outlined, and the term *landscape* is derived from the basic concept of the geographic entity, then this must generate detailed and varied questions. These questions have remained unanswered because the existing philosophical approach to geography does not address these types of problems. Because a narrow disciplinary evaluation exposes some of these deficiencies, it is worthwhile to discuss some of these because they may contribute to a better understanding of this field of knowledge.

Landscape as a Geographic Individuum or as a Type

Every section of the earth's surface, as a part of the geosphere, is a unique, individual, concrete form. Because investigation of such concrete phenomena is the task of geography, inquiry into individual phenomena ultimately must also be the aim of geographic research. If, as some distinguished geographers have suggested, the term *landscape* is to be understood as a type derived from reality, then individual phenomenon must be accorded another term; the term *land* has been suggested for this. Thus *land* would be the individuum, and *landscape* would be the type into which such a land unit would be classified. Problems remain, however, for understanding the new term *land*. If this is so, then the role of geography as the science of landscapes, which is so frequently heard in German geography, cannot then be tenable. Also, can the recognition of a type be declared the subject of a science? Types are scientifically devised for ordering material according to different characteristics. Therefore, it cannot be the task of geography proper to arrange highly varied geographic reality according to types, but this necessary, and indeed indispensable, methodical step cannot be elevated to be the object of a science, just as the elucidation of relationships cannot be the sole basis for the definition of a science. This has happened in the case of the chorological science, no matter how significant the chorological aspect may be. If landscape is to be of fundamental importance for geography, it can only be in the sense of landscape being the concrete phenomenon of the geosphere.

Landscape as a Physiognomic Unit, as an Image of a Landscape

A wide spectrum of opinion exists based on geographic work, which ranges from the evaluation of what is visible to the total rejection of physiognomy as an unscientific approach. There can be no doubt, however, that every geographically concrete formation on the earth's surface is characterized not only by a particular material structure and a definite linkage of effects but also that this is obviously expressed by an outward appearance. Therefore, the general principle states that content and physiognomy are indivisible. If, however, such a division is to be made in the methodological literature, it is to be taken as a transfer of certain methodological research steps in the perception of the object itself. The description of the appearance—that is, landscape physiognomy—without elucidation of the content, which, after all, determines the outer appearance, is undoubtedly unscientific and implies a renunciation of geographic knowledge. The denial of this physiognomic aspect in geography leads only to volumes of illustrations without any geographic clarification, thereby promoting an unscientific approach.

Aesthetic and artistic perceptions of geographic images start from this point, with one, like the other, denying scientific principles, no matter whether it is because science is believed to be inadequate to comprehend reality or because the scientific foundation is being refuted altogether. It is absurd to maintain that the physiognomic approach is unscientific and therefore not entitled to a place in science, because geography is the very subject in which outward appearance tends to be the starting point of analysis. Time and time again, some not unimportant geographers have drawn attention to the fact that observation (evidently meaning what is visible; i.e., the physiognomy of the landscape) is the basis of geography and, in particular, of landscape research. Every landscape has an outward appearance, a physiognomy, and this image of the landscape is the embodiment of the total geographic reality. When attention is drawn to the physiognomic side only, however, reference is made to the image of the landscape instead of to the landscape itself. A dispute over either the physiognomy or the content of the landscape will, however, neither promote scientific knowledge nor advance geography by a single step.

Scale of Landscape

If, as O. Granö has postulated, the landscape may be regarded as that sector of the earth's surface that is visible from any one point, the landscape of that place can be defined in many different ways according to the direction of observation, weather conditions, and altitude. It is obvious that this does not allow for a precise description of the landscape as a concrete phenomenon of the geosphere. If its individual character, which distinguishes it from other sections of the earth's surface, can be determined by the structure and interrelated effects of its geographic content, a landscape will extend as far as that same content is present. This, of course, leads to the question of the "same" content and its discernibleness. The response to this question depends upon the scale of inquiry, which also determines comparability, similarity, and identity. An inquiry at a large scale reveals some differences that one at a small scale cannot embrace. Is size to be regarded as a characteristic feature of the landscape or even included in its definition? Such a definition would include a quality that does not depend on concrete content but on the scale of inquiry. This must imply that, in the appearance of several scale-dependent landscapes, the existence in the literature of terms like *miniature landscape, landscape cell,* or *large-scale landscape* proves that statements on scale are understood to be complementary. Even a narrower interpretation of geographic reality reveals considerable differences, however. Surely the smallest homogeneous areal units are not landscapes that, after all, always represent a mosaic of such homogeneous, basic units. In addition, it must be remembered that it was human beings who served as the standard for the popular term *landscape;* in other words, that the concept of a landscape gives credibility to the human frame of reference. Nonetheless, even in the popular concept of landscape, content, together with its outward appearance, is the prime consideration. The boundary of the landscape is often scarcely noticed, apart from special cases with particularly well-marked borders. The size of the landscape plays no role whatsoever in this. In the vast steppes, homogeneous landscape areas extend over considerable distances, whereas changes occur very frequently in mountains, and small divisions in regionalization can be discerned. The question of the normal size of a landscape thus lacks a basis in reality. The scale is set by the science itself according to the case being studied. A term for landscape, in which a definition stipulates a characteristic size, or largeness, therefore cannot be objectively correct. At this point, the question of the size of a landscape also touches on that of its boundaries.

The Problem of the Boundaries of a Landscape

As long as the landscape was only treated as a research object in physical geography, its definition tended to contain the formulation of natural boundaries in the landscape. This asserts that an indispensable feature of the landscape is its separation by boundaries, though this brief definition fails to mention whether these are complex boundaries or lines of particular dominant elements. This, however, is the salient point. Lines of particular landscape elements can generally be given precisely. If it is a matter of dominant qualities—that is, features prevailing in a nexus with others—such a boundary can make itself felt in other geographic elements so that they, too, experience a noticeable change in their strength or as phenomena along this boundary. A bundling together of feature boundaries can take place, and the boundary of such landscape complexes, as a rule, will consolidate as a narrow seam. In that case, it is impossible to define a boundary for the entire landscape complex because this will tend to omit some, not inconsiderable, elements that do not fit easily into this bundling. More often than not, these are climatic features. The boundaries of landscape complexes, so sharp in Central Europe between Pleistocene sand cover and loess cover, are not reflected in the climatic scene. A comparison of different characteristic values of natural qualities at this boundary shows very noticeable differences in the properties of the hydrological cycle (e.g., soil moisture), for chemical and physical reactions of the soils, as well as for many other factors—in fact, a division into two groups. Climatic measurements, however, would not produce a separation into two distinct groups at this line because neither the values for temperature nor those for precipitation experience rapid change here. From a climatic point of view, there is no landscape boundary. In this case, macroclimatic values, obtained from geophysical—not geographic—investigations, cannot be used.

The position is quite different, however, if those climatic sequences that are codetermined by the geographic complex are also considered from a geographic point of view; for example, in the question of the storage of precipitation water in the soil. Such climatic deviations in the field are also present as a result of the exposure or the delimitation of the soil horizons. These are rarely clearly defined boundary lines, however. In the vast majority of cases, boundaries are imprecise and characterized by the interdigitating and overlapping of particular elements and by the interdigitation of areas on both sides of the boundary. Though the content of the landscape can be assessed for every element as to its structure, linkage, and development, a uniform boundary, for the most part, cannot be discerned. This shows that even boundary features are unable to determine the term *landscape.* The continuum of geographic reality does not permit such a definition. As a rule, landscape does not exist as a naturally delineated object. Boundaries tend to be projected into geographic reality by the human mind. Though certainly indispensable for practical purposes and as an aid to ordering, it nevertheless is less than a helpful feature when it comes to defining the landscape.

Much of the discussion of boundaries has been contributed by those scientists whose research objects show clear and unambiguous boundaries or delineation. Their norms, however, do not apply to geographic boundaries. Boundaries, in landscape interrelationships, have much more important functions as geographic boundaries in the continuum and tend to have the effect of membranes and act as starting points for special investigations.

Landscape Connection

Time and time again in the foregoing sections, it has been demonstrated that the singling out of landscape elements leads to all sorts of contradictions and that knowledge of isolated facts hinders the perception of the landscape as a whole. Special importance is therefore accorded to the knowledge of connections among the elements that are united in one landscape. United in one geographic place, components and elements can enter into a relationship in different ways, depending on the way in which they are included within the system by the interplay of its forces. It can happen that two or more components may occur in the same space without causal connections being found among them. Such a coming together (coincidence) in the same place does not result in geographic relationships. This is referred to as *synchor*. These components are independent of one another; they have no influence on dynamics or development, although both of them may be visible phenomena. They are a kind of accessory component. Other components are joined in a linking of effects, sometimes closely, sometimes less so, or they may vary in their relationship according to the seasons. The closer the relationship of such components in the dynamics of the system, the more regularly—to the point of being a firmly established link—are these components connected to one another, the more characteristic they are for a particular geographic location. Following a suggestion by J. Schmithüsen, components thus connected in such a linkage are known as being *synergetic*. This distinction is important for the topological approach because intensely synergetic components are best suited to the definition of features or types. Synchoric components, on the other hand, are not suitable for the definition of types. They only play a role in individual description. The majority of the components contributing to the geographic form of a locality can be classified according to their degree of synergetic relationships. Because the components with the strongest relationships are also most instrumental in determining the character of the landscape, there is a possibility of employing these features—which are referred to as the dominants—primarily for characterization. This characterization would be included in the compilation of a terminology. The more components are introduced into the interrelationship of features of a particular type of landscape, or, in other words, the more attention is paid to the geographic context, the closer to reality will be the type.

Conclusions

Taking the landscape to be a special kind of material structure on the earth's surface, any discussion of landscape, even citing examples from previous decades, clearly does not do justice to the phenomenon of landscape. This demonstrates a considerable difference in scientific thinking because the general concept of landscape corresponds to the axiomatic perception of geographic objects. This places it at least one level above questions of the size or the boundaries of a landscape. These are the topics that, in the past, have tended to be the focus of discussion in the discipline but have failed to provide any genuinely new knowledge as to the character of the landscape. It is a long way from the landscape considered in this way as being the equivalent of geographic reality to that proposed here, in which new elements and components are included and give both a concrete expression and direction as well as providing a foundation for geography as a discipline.

No matter what the combination of words, the use of the term *landscape* indicates that the integrative linking of phenomena into complex units is being considered. Used as adjectives, *geographic* and *landscape* may seem to imply the same thing, but according to the axiomatic characteristics outlined, an orientation by basic concept is, in fact, more focused. The spatial ordering of objects in geographic space may be reduced merely to a distribution without any further connection, but in a landscape spatial linkages can hardly be re-formed as a simple distribution. In part, this is the fault of the long and rather mixed history of geography. The word *geographic* is used for so many purposes that one is tempted to ask, in every case, what is actually meant by its use. Geographic coordinates, for instance, are geodetic definitions of position but are devoid of content in the sense of the axioms of the landscape doctrine. Any information on the vertical and horizontal linkages of that location is missing. Often geographic is understood to mean "topographical." These are misdemeanors from former times that have survived to the present. The perception of landscape, as proposed here, offers strong resistance to slipping into general-purpose prescription. Topographical knowledge, though very useful to the geographer, does not necessarily embrace geographic understanding. In the vernacular (as well as in other sciences!), the word *geography* is used predominantly in its primitive sense and is also implied in statements on distributions. This has led to the *geography of dialects,* a *geography of houses,* and to other special aspects that employ the venerable word *geography* in the pursuit of particular regional studies. If regional studies of this kind get hold of the word *landscape* and begin to speak of *linguistic landscapes,* and so on, such use must be put down to the follies of fashion. This issue can flourish only because geography has failed to provide a clear definition of the foundations of its knowledge base.

The concept of landscape, as a reflection of the form of matter existing on the earth's surface, thus implies that a material system is being considered and that this involves all the components and their interlinkage effects. The concept of the landscape (one should avoid using the term *landscape* as a short form at this level) can now be formulated as follows:

A landscape is a concrete part of the earth's surface that is characterized by a uniform structure with interrelated links and effects.

This general concept of the landscape allows the derivation of numerous terms. Because the term *landscape* seems to safeguard essential geographic connections, additional characterizations may support a great variety of meanings and that may be used so long as they do not result in contradictions to this concept of the landscape. Thus, by stressing the dominant elements, one can obviously speak of moraine landscapes, coastal landscapes, steppe landscapes, and industrial landscapes, to name only a few. These are inevitably sections of the earth's surface that, being subject to natural ordering, receive their physiognomy from their structure and their interrelationships.

Additional terms chiefly serve to achieve a concise and vivid description when referring to the essence of the character of the landscape. Much more important, however, is the fact that the broad scope of this field can be divided for research purposes. In fact, it needs to be because of the range of problems confronting research. The body of knowledge entitled *landscape* is investigated by a number of sciences with different aims and methods. This interdisciplinary character for research into these realities should definitely be preserved; indeed, it should be extended in accordance with the historical development of those subjects and disciplines that include the landscape as a part of their work.

Because the distinction between the arts and sciences came about and because space and time are decisive for all geographic development, several schools of thought have emerged in landscape research. They differ in respect of their aims and methods, and they operate with different terminologies. Briefly, they are circumscribed by the terms *natural landscape* and *cultural landscape*. The natural landscape has been the research focus for natural scientists, with geologists and biologists joining geographers and, more recently, specialized geoscientists. A totally different orientation, and therefore one with little mutual interest to the research carried out by the natural scientists, is that found among those investigating the cultural landscape. Primarily, this examines the historical development of culture in the landscape as well as its formation and conscious transformation through cultivation and the development of settlements, the results of which are now seen before us in the contemporary expression of the landscape, with its relicts from different periods of development. Economic geographic aspects, however, tend to accord priority to economic evaluation and give less value to any cross-connections with cultural developments. Though more appropriate to the variety of problems tackled by this field, this aspect has only developed into a strong and important subdiscipline at a relatively late date. Cultural geography is a subdiscipline of geography in its own right, but it is not a part of the study of the cultural landscape, in which a particular region is investigated in terms of all its interrelationships. Historical development has led to ever-greater specialization in various specialist branches.

The consistent application of causal analysis, and its further perfection in the sciences, has led to an unexpectedly rich yield of new scientific findings. The specialization required for this, however, has increasingly found itself set against the requirements of the research necessitated by the environmental problems within the complicated and multifaceted nature of the landscape. Some of the specialist terms quoted previously, which belong to a period some decades back, are no longer applicable because they no longer meet current requirements. The upshot of this must be a methodologically based division of the landscape.

Setting out from a consideration that there is only one situation in which natural objects are joined by others, and which human beings have either transformed or created entirely anew, outward appearance or physiognomy is always determined by the sum of its varied material phenomena. The relationships in which the particular phenomena find themselves are of varying character and are the result of different natural and societal processes. Only those that are subject to a similar combining of effects can be regarded, and investigated, as systems with an established order and can be the object of scientific endeavor. Starting from the present-day landscape, the extent of its transformation will be readily recognized from its original form, as initially represented by nature, to that of a cultural landscape. The original vegetation cover will have been replaced by a covering of cultivated plants; rivers will have been greatly altered by technical measures and construction, and much more evidence reflects the recent cultural overlay.

Nonetheless, soil fertility, with its related physical processes, continues to be effective in crop production, just as precipitation, evaporation, and runoff obey the laws of nature in regulated rivers, the only difference being that some of the original conditions have been replaced by secondary ones. The aims of resource utilization must be supported by additional economic investigation. As such, an actual landscape is objectively visible, and all its elements are identifiable. There can be no doubt as to its reality, but there must be no doubt that a scientific investigation of such complicated and multiformed landscape units can be tackled in a simple way. Any investigation must be carried out on different aspects, many of which are also often determined by historical processes. The dominant, and often biased, work and emphasis on results, which appeals to particular specialties, together with the simultaneous neglect of the multilateral structure of landscape, are a result that is a characteristic of the reduction of landscape research into the scientific subdisciplines. The *potential vegetation*, to which biology applies itself and is often used as an overall characterization of the landscape, is a striking example of this. *Potential vegetation* is defined as the vegetation that would be growing if all human measures of cultivation stopped and gave way to purely natural processes of growth. A realistic grasp of actual landscape variability is therefore substituted by the narrow focus of one particular discipline. The same occurs in geology, in which actual geomorphological processes, which transform the landscape, tend to be regarded as mere geomorphological specialties, not as processes that transform the entire landscape. The formation of special disciplines in the organization of science and the neglect of interrelated events in the actual landscape generally become the starting point for a very narrow view of problems that pay no attention to landscape diversity. The structure of landscape-related problems deserves greater attention:

One uniform scientific approach cannot do justice to the full geographic reality but tends to give way to an investigation of the diverse relationships that, in reality, overlap.

Physical geographic investigations of the landscape are now clearly oriented toward the view that the nature of a country is, in fact, only a subsystem of the whole. These subsystems are referred to as natural units or, in short, the natural units of landscape but no longer the natural landscapes. Most contemporary and most anthropogenically altered natural features tend to allow structure and dynamics to be determined by measurements. The previously mentioned changes in landscapes, measured by the increasing numbers of technical sciences involved (including chemistry) and that are becoming more and more focused, are applied mainly to nature's response to human effects on the landscape. Natural properties are considerably reduced so that the ability of landscapes to function is questioned when technical and economic revision of human supply and utilization systems are raised. The influence of the landscape on human beings thus extends to many aspects of their lives; the field of knowledge concerning landscape continually increases in importance for gaining an understanding of the relationships between nature and society. Landscape research and knowledge of the dynamics of landscape thus acquire a topicality, even for fields that seem to be remote when they are brought into context with the landscape. The field of knowledge dealing with landscape thus becomes an interdisciplinary field of interest for all problems in regional development, and numerous working relationships are bound to grow out of it. To start with, however, it is necessary to clarify the theoretical problems of landscape research to the extent that they are able to provide the principles for practical application.

Notes

1. J. Wallot, *Grössungleichungen, Einheiten und Dimensionen* (Leipzig, East Germany, 1953).
2. H. Lautensach, "Der geographische Formenwandel. Studien zur Landschaftssystematik." *Colloquium Geographicum* 3 (1952).
3. H. Lautensach and R. Bögel, "Der Jahresgang des mittleren geographischen Höhengradienten der Lufttemperatur in den verscheidenen Klimagebeiten der Erde. Erdkunde," *Archiv für wissenschaftliche Geographie* 10 (1956): 270–282.
4. J. Sölch, *Die Auffassung der "naturlichen Grenzen" in der wissenschaftlichen Geographie* (Innsbruck, Austria, 1924).
5. Compare the overview of literature in Marc Anderson, "Discussion Paper No. 2: A Working Bibliography of Mathematical Geography," *Michigan Inter-University Community of Mathematical Geographers* (1963). See also W. Bunge *Theoretical Geography*, Lund Studies in Geography Series C, General and Mathematical Geography, No. 1 (Lund, Sweden, 1962). Attention should also be directed to the comprehensive publications with the term *regional science*, especially in the United States.

Patches and Structural Components
For A Landscape Ecology

Richard T. T. Forman and Michel Godron

Landscapes as ecological units with structure and function are composed primarily of patches in a matrix. Patches differ fundamentally in origin and dynamics, while size, shape, and spatial configuration are also important. Line corridors, strip corridors, stream corridors, networks, and habitations are major integrative structural characteristics of landscapes. (Accepted for publication 29 May 1981)

Landscapes surround us, yet curiously it is hard to find people with the same concept of a landscape. Artists and humanists commonly portray the landscape as what the eye can perceive and sometimes limit the idea to natural landforms or communities. Such a landscape generally includes a high degree of spatial heterogeneity. In geographical literature, the landscape plays a central role, with most definitions focusing on the dynamic relationship between two characteristics—natural landforms or physiographic regions and human cultural groups (Grossman 1977, Mikesell 1968, Sauer 1963). In this article we ask whether the landscape is a recognizable and useful unit in ecology, with a distinctive structure and function that can be analyzed, as is done for organisms or ecosystems. What are the structural components of a landscape and their characteristics? Are there interesting, indeed critical, ecological questions facing us that may be solved using a landscape approach?

Walking in a small area of an agricultural landscape, one might encounter a corn field, a bean field, an abandoned old field, an upland oak stand, and a lowland elm-ash-sycamore woods adjacent to one another. If one studied this cluster of five specific communities or stands, one would find fluxes of energy, mineral nutrients, and species between adjacent stands, indicating considerable interaction among stands of the cluster. If one moved several kilometers away within the landscape, one would find a similar cluster of stands with similar interactions. Moving on, one would find this cluster repeated until entering a different geomorphological area, or an area subjected to different natural or human disturbances. Here, a different cluster of interacting stands would be evident as one entered, for example, a landscape of ridges and valleys, a suburban landscape, or a sandy forested landscape. Such observations are at the heart of the landscape concept, which we describe as follows.

A landscape is a kilometers-wide area where a cluster of interacting stands or ecosystems is repeated in similar form. The landscape is formed by two mechanisms operating together within its boundary—specific geomorphological processes and specific disturbances of the component stands.

Landscapes vary considerably in areal extent, and a localized area of a few meters or hundreds of meters across is at a finer level of scale than a landscape. Because of the area's geomorphology, the complex of landforms and parent materials present is relatively constant over a landscape. Each stand has a characteristic disturbance regime (the sum of the frequencies, intensities, and types of individual disturbances). A cluster of stands or a "stand cluster," therefore, has a disturbance regime cluster, which in turn is fairly constant throughout the landscape. Disturbances include both natural events and human activities such as fire, hurricanes, agricultural practices, or forest cutting.

Between the stands of a cluster are transition zones or ecotones, which may vary from being abrupt to gradual and wide. In less disturbed landscapes, gradual community gradients may be common or uncommon, depending upon how sharp environmental changes are with distance. However, with greater disturbance, especially by human activity, a landscape mosaic of ecosystem patches with distinct boundaries comes into sharper focus.

We suggest that landscape[1] is a distinct, measurable unit with several interesting ecological characteristics. Within the landscape is a recognizable and repeated cluster of ecosystems and disturbance regimes. The boundary between landscapes (which differ in geomorphology and disturbance) is relatively distinct, particularly in vegetation structure. Ecologically, landscape structure is measured by the distribution of energy, mineral nutrients, and species in relation to the numbers, kinds, and configurations of the component ecosystems. Landscape dynamics is the flux of energy, mineral nutrients, and species among the component ecosystems, and consequent changes in those systems.

The key landscape structure questions today center on the importance of numbers, kinds, and configurations of ecosystems. Randomness is rare within a landscape. The overwhelming number of

Forman is with the Department of Botany, Rutgers University, New Brunswick, NJ 08903. Godron is with the Centre d'Etudes Phytosociologiques et Ecologiques, CNRS, B. P. 5051, Route de Mende, 34033 Montpellier, France.

[1]Related concepts: A region is bounded by a complex of physiographic, economic, social and cultural characteristics (Dickinson 1970, Isard 1975). A stand (or a localized community) is the group of organisms at a specific locality, and is homogeneous enough to be considered a unit (Greig-Smith 1964, Daubenmire 1968). The ecosystem concept—organisms and their encompassing abiotic environment—may be applied at any level of spatial scale (Odum 1971). However, in practice one looks for relative homogeneity so as to characterize an ecosystem with a limited number of measurements (Woodwell and Whittaker 1968, Forman 1979a, Bormann and Likens 1980). Though one may apply the ecosystem concept to a heterogeneous region or landscape, in this article we limit its use to stands within a landscape.

species exhibits an aggregated or clustered distribution of individuals (Chessel 1978, Greig-Smith 1964, Kershaw 1973), and even in random and regular species distributions, some aggregations of individuals normally are present in a surrounding area of lower density (Godron 1966, 1971). This basic aggregation pattern of individuals of a species underlies the patchiness of vegetation and animal communities so commonly seen in nature.

In simplest terms, patches are communities or species assemblages surrounded by a matrix with a dissimilar community structure or composition. The matrix exhibits several characteristics itself, such as the degree of heterogeneity and connectivity, but in this article we focus on patches and the other structural components, corridors, networks and habitations. We further limit the analysis to patches at a single level of scale, the landscape, though most of the resulting patterns appear to apply to all levels of scale.

PATCH ORIGINS

Causal Mechanisms

Five causal mechanisms predominate and the five types of patches produced differ strikingly in their dynamics and stability (Forman 1979b). A *spot disturbance patch* results from disturbance of a small area in the matrix (Figure 1). For example, patches are produced by a small fire in a grassland, a large blowdown in a forest, overgrazing by a local exploding population of rodents, or local spraying of a generalized insecticide. Other examples are given by Heinselman (1973), Levin and Paine (1974), Pickett and Thompson (1978), and Forman and Boerner (1981). Following the disturbance, succession proceeds until the patch disappears by becoming like the matrix; that is, population changes and immigrations and extinctions of species take place until the relative abundances of the species are similar to those of the surrounding matrix. In unusual cases, especially where the intensity of disturbance is severe or the matrix is undergoing rapid change, succession may lead to a semi-stable patch that differs significantly from the matrix. The spot disturbance patch typically has high population changes and species immigration causes its ultimate disappearance.

A *remnant patch* is caused by widespread disturbance surrounding a small area, the inverse of the spot disturbance

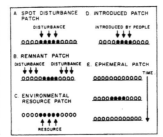

Figure 1. Patch origins. Species dynamics within a patch and turnover of the patch itself differ substantially according to the mechanisms causing a patch. ○ ○ ○ = matrix; ● ● ● = patch; disturbance = a sudden severe environmental change.

mechanism. This patch is a remnant of the previous community embedded in a matrix that has been disturbed. Examples of remnant patches are woodlots in an agricultural area, a shrub-covered island produced by flooding a valley, a breeding warbler community on a southfacing slope that survived a rare freeze, or a pocket of herbivores that escaped the invasion of an aggressive non-native species. Further examples are described by Galli et al. (1976), Gottfried (1979), Helliwell (1976), Pollard et al. (1974), Seignobos (1978), and Southwood (1961). If the disturbance in the matrix is temporary, succession will proceed until the matrix converges with the patch in species similarity. Here again the patch disappears. If this convergence is rapid, the patch may change relatively little in species composition. However, if the disturbance of the matrix is chronic, inhibiting the normal successional rate and direction, the patch will remain. In this case, a net loss of species may take place (Diamond 1972, Willis 1974). This hypothesized net loss would be rapid at first, finally dropping to zero, a response referred to as a relaxation period. The species lost are those requiring a habitat larger than the remnant patch or those sensitive to a modified microenvironment within a patch.

Hence, remnant patches vary from short-lived, as produced by a single natural or human disturbance, to long-lived, resulting from chronic human disturbance. In the same way, spot disturbance patches may be short- or long-lived. In remnant patches caused by chronic disturbance, the net loss of species during a relaxation period results in the patch remaining, but with a species composition differing from the original patch.

An *environmental resource patch* reflects the normal heterogeneous distribution of resources in the environments and results from the environmental resources of a relatively permanent and discrete area which differ from the surrounding area. Unlike the other patches, the environmental resource patch is not dependent on disturbance. Concentrations of amphibians and reptiles in a desert oasis, patches of heaths on an exposed mountain ridge, acid-tolerant mosses in a glacier-caused bog, and pollinators in a moist alpine gully are examples. Other examples are described by Brown (1971), MacArthur and Wilson (1967), Simberloff (1976), Smith (1974), Stiles (1979), and Willis (1974). Since the cause of the environmental resource patch is relatively permanent, the patch is permanent, and species changes simply reflect those normal in the interaction between a small community in dynamic equilibrium with a surrounding matrix community.

An *introduced patch* is dominated by an aggregation of individuals introduced into a matrix by people. Pine and eucalyptus plantations (*Pinus, Eucalyptus*), golf courses, fields of wheat and corn (*Triticum, Zea*), or a large feeding station that attracts vertebrate species to a small area are examples. Introduced patches remain as long as the human disturbance regime maintains them. Thereafter, species from the matrix colonize, and like the spot disturbance patch, the patch disappears as it converges with the matrix.

An *ephemeral patch* is a transient aggregation of species caused by normal short-lived fluctuations in resource levels, that is, levels of biotic or abiotic environmental change that are frequent enough and of a low enough intensity that species have adapted to them. Especially rare or severe environmental changes are considered disturbances, which in turn cause spot disturbance and remnant patches. Examples of ephemeral patches are mammals feeding at dawn around a large savanna mudhole, a localized bloom of annuals in the desert, or a large shrubby area in field-to-forest succession. However, the rapid-turnover ephemeral patch appears more prominent at finer levels of scale than the landscape.

In summary, patch is a spatial concept focused on a small area. Both the causal mechanisms of patches and the resulting dynamics of patches differ greatly. As with most biological patterns, some overlaps exist among the five basic patch origins. For example, a severe chronic disturbance in the matrix might so change the matrix that convergence with a remnant patch community is prevented, and the remnant patch develops into an environmental resource patch.

Other Characteristics

A multitude of possible community types, named primarily by physiognomy or predominant species, may characterize patches in a landscape. The number of such different community types is a key structural characteristic of landscapes, not only for mapping, but to provide an index of the range of biotic richness, productivity, and nutrient and water fluxes in a landscape.

In addition, each ecosystem component is normally patchy in the landscape. For example, superimposing maps of soil types, tree communities, and herbivorous mammal communities for a landscape may show several places where boundaries coincide and many places where they do not. The degree of congruity in space among the units of different components is useful in mapping, land use planning, and analyses of landscape structure (Forman 1979b, McHarg 1969).

PATCH SIZE

Productivity, nutrient and water flux, and species dynamics are all affected by the size of landscape patches. Island biogeographic theory developed from studying archipelagoes in a matrix of water lends insight into the relationship between species and area. The number of species, S, (= species diversity) on an island was related directly to three factors in order: the island area, its isolation, and its age (MacArthur and Wilson 1967). The basic island area effect, though, is mainly due to habitat diversity; in most cases, larger islands simply have more habitats which, therefore, support more species. However, there is also an area effect: When the habitat diversity of large and small islands or patches does not differ, more species are typically found on the former (Forman et al. 1976, Simberloff 1976). Finally, one of the major factors determining diversity on an island or patch is the history and present regime of disturbance (Carlquist 1974, Pickett and Thompson 1978). Summarizing the patterns for islands, species diversity is a function of certain island characteristics listed in the suggested order of overall importance:[2]

$$S = f(\text{habitat diversity} \pm \text{disturbance} + \text{area} - \text{isolation} + \text{age})$$

Patches in the landscape, however, differ significantly from islands surrounded by water (Forman 1979b). Above, we analyzed patch origins and here note that average turnover rates (the appearance and disappearance) of landscape patches may be high, whereas islands are essentially permanent. Similarly, the sharpness of the patch boundary varies greatly in the landscape (Whittaker 1973), and gradual gradients may be more conducive to the movement of species between patch and matrix. The heterogeneity of the landscape matrix is often high, which implies a large source of species in the matrix and strong directional effects of the matrix on different sides of the patch. The landscape matrix may be used as a rest stop for many species moving between patches, particularly in the limited area of a landscape compared with extensive oceanic archipelagoes. Here the importance of isolation, a central characteristic of island biogeographic theory, is lessened.

Within a landscape, the "species rain" appears to be high, that is, most species reach most patches within their life cycle. Therefore, when species extinctions take place in patches, rapid recolonization is facilitated and the effect of isolation minimized. While this species rain is suggested to be high for a community, as measured by species diversity, a relatively small proportion of individual species has a limited dispersability within the landscape. Isolation in the landscape may be critical for these individual species, many of which are uncommon. Thus, in conservation not only must the basic community pattern be considered, but also the specific populations (Diamond and May 1976, Forman et al. 1976, Pickett and Thompson 1978, Simberloff 1976, Terborgh 1976). Summarizing the species diversity patterns for landscape patches we suggest:

$$S = f(\text{habitat diversity} \pm \text{disturbance} + \text{area} + \text{age} + \text{matrix heterogeneity} - \text{isolation} - \text{boundary discreteness})$$

Landscape patch area has been shown to correlate strongly with species diversity (Galli et al. 1976, Gottfried 1979, Moore and Hooper 1975, Peterken 1974, Robbins 1980, Whitcomb 1977), but rarely has area been considered separate

[2] + = positively related to diversity; − = negatively related; ± = usually negatively, but sometimes positively, related. Units are not considered in this encapsulation.

from habitat diversity. When patch area alone is evaluated, we find it to be an important determinant of species diversity, and that species groups (such as trees, seed-eating birds and insectivorous birds) respond differently to patch area (Elfstrom 1976, Forman et al. 1976).

PATCH EDGE AND SHAPE

The microenvironment in the center of a tiny patch of woods differs strikingly from the center of an extensive woods. This results largely from penetration of air from the surrounding matrix throughout the tiny woods, whereas this air penetrates only a limited distance into the edge of the extensive forest. The outer band of a patch, which has an environment significantly different from the interior of the patch, is known as the *patch edge*. This produces an *edge effect*, that is, a difference in species composition and abundance in the edge. For example, differences between the edge and interior of deciduous forests in North America and Europe have been documented for a host of meteorological factors, vegetational characteristics (Jakucs 1972, Wales 1967, 1972), and animal communities (Galli et al. 1976, Johnston 1947, Leopold 1933, Patton 1975). Soil and fire characteristics probably also differ.

Several factors affect the width of the patch edge. The angle of the sun plays a major role, with edges facing equatorward typically wider than those facing poleward (Wales 1972), and those in temperate areas wider than in tropical areas. Wind also exerts a major influence, with the prevailing wind direction during the active or growth period having a wider edge than other sides. The degree of species difference between the patch and matrix is significant, too.

The patch edge appears to vary in width from a few meters to a few tens of meters in patches at the landscape level. Different groups of organisms respond differently to the environmentally determined edge width. For example, in woodlots, avian and tree communities appear to differ from the interior only in the outer portion of a forest edge, while herbs and mosses appear sensitive to essentially the entire edge width.

Patch shape as a variable is important in several ways, such as a target for dispersal or home range suitability; here we consider patch shape in the context of the edge concept. A large isodiametric patch is mostly interior, with a band of edge in the outer portion of the patch. A rectangular patch of the same size has proportionally less patch interior and more patch edge. Finally, a narrow strip patch of the same size may be all edge. Since community and population characteristics differ between the interior and the edge, comparing these characteristics with the interior to edge ratio of patches may be useful in evaluating the importance of patch shape in a landscape.

Whitmore (1975) noted that plant species composition and community structure varied according to the shape of openings in tropical rain forests. Stiles (1979) found sharp differences in wasp nesting density in the New Jersey Pine Barrens according to the width of the habitat. In Idaho rockslides, small mammal density correlated best with the length of the rockslide perimeter (Bunnel and Johnson 1974). Unpublished data (Forman and Clay) on mushroom diversity in old New Jersey two-hectare oak woodlots indicate a halving of species diversity and a threshold response in proceeding from isodiametric through rectangular to strip patches. Patch width or shape, therefore, appears to be a major ecological variable in the landscape.

Several special cases of shape bear mention. Ring zones are belts of vegetation, commonly within a particular altitudinal range, which extend around a mountain, and contain a "hole" with different vegetation at a different altitude (Hedberg 1955, MacArthur and Wilson 1967). The interior to edge ratio indicates that ring zones are more similar to strip patches than isodiametric patches. Linear patches and dendritic patterns contain special characteristics and are considered below.

The peninsula, where a narrow portion projects from a large patch, is a common shape, and species diversity commonly decreases progressively toward the tip. The reason for this pattern in major continental peninsulas of North America is hypothesized to be species extinction on the peninsula during the Pleistocene and subsequent gradual recolonization from the continent (Simpson 1964, Taylor and Regal 1978). An alternative explanation based on the edge effect, that the peninsular edge has a climate strongly modified by the surrounding water leaving little if any interior environment, is well known to farmers who must grow different crops on peninsulas (e.g., Squier 1877). Apparently the peninsular effect has not been studied at the landscape patch level.

PATCH NUMBERS AND CONFIGURATION

So far we have focused on the characteristics of individual patches. Patches, however, generally do not exist singly, but vary in numbers and in their configuration and juxtaposition to one another. Patches exhibiting any of the above described patch characteristics may, of course, vary from zero to many in a landscape. In understanding a landscape, determining the number of patches in each of at least four categories appears essential. How many patches are there of each of the patch origins? How many of each community type are there for each of the patch origins? In each category thus formed, what is the size distribution of the patches? And what is the distribution of patch shapes in each of these?

Determining the numbers in each of these four categories is not difficult in some landscapes. A subsample can then be selected for measurement of the species, energy, or nutrient component of interest, and by simple multiplication the status of the component in the patches of a landscape can be estimated with a measure of variability. However, this estimation is inadequate, because the spatial configuration among the patches has been ignored. For example, a landscape with ten evenly-distributed large patches differs fundamentally in most ecological fluxes from a landscape with the ten patches clustered at one end.

Various spatial configurations (Figure 2) can be examined using standard statistical techniques (Chessel 1978, Daget 1979, Godron 1971, Kershaw 1973) applied to the distribution of patches in each of the categories just described. The patches of a category may be random, regular, or aggregated; or positive or negative associations among patches of different categories may be present. This provides insight into both the cause of the patches and the potential for interpatch interaction. For example, common nonrandom patterns of patches are seen in limestone karst topography, in dendritic stream basins, along roads and property lines, or encircling towns. Finally, the actual distance between patches is an important measure of potential patch interactions.

CORRIDORS

There are four types of corridors in landscapes: *Line corridors,* such as paths, roads, hedgerows, property

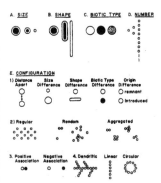

Figure 2. Patch characteristics in a landscape.

boundaries, drainage ditches, and irrigation channels, are narrow and typically have only species characteristic of patch edges. *Strip corridors* are wider bands containing a patch interior environment in which interior species may migrate or live. *Stream corridors,* which border water courses and vary in width according to the size of the stream, control water and mineral nutrient runoff, minimizing flooding, siltation, and soil fertility loss. *Networks* are formed by intersecting or anastomosing corridors and therefore contain loops. Some overlap among the four basic types exists, such as edge species moving in all four, or a wide stream corridor also functioning as a strip corridor for movement of patch interior species.

Line corridors are particularly characteristic of landscapes dominated by human disturbance. They originate in the same ways as patches, e.g., remnant tree lines left between fields from an earlier forest, paths as spot disturbance lines, and introduced lines as shrub and tree plantings for defense, enclosing livestock or decreasing wind (Kellogg 1934, Rotzien 1963, Seignobos 1978, Van Eimern et al. 1964).

The plant and animal species of line corridors generally also characterize patch edges (Pollard et al. 1974). These corridors provide habitat and breeding sites for species requiring the surrounding matrix environment for protection or feeding. Introduced nonnative species are common in line corridors, especially the disturbance-caused corridors.

The microstructure of the line provides insight into its potential functions (Les Bocages 1976, Lewis 1969, Pollard and Relton 1970, Pollard et al. 1974,

Southwood 1961). Hence, a path line contains mainly disturbance-resistant species and has compacted soil, often with attendant erosion along the line. In contrast, the hedgerow line of shrubs or trees, which is higher than the matrix, cuts wind velocity, shades the adjacent matrix, and has a high evapotranspiration rate. Irrigation channels, and often roads and hedgerows, include adjoining ditches and embankments with considerable microhabitat diversity where amphibians, reptiles, and moisture-tolerant plants are often favored. Changes in line corridors through time are little known. M. D. Hooper, however, found a linear correlation between hedgerow age and shrub species diversity in managed British hedgerows, with an average one species gained per century (Les Bocages 1976, Pollard et al. 1974).

General characteristics of the wider strip corridors are reasonably well known, despite a paucity of direct studies. The corridor must provide protective cover for species from natural predators, domestic animals, and human effects lining each side of the corridor. The outer portions of the strip corridor have the edge effect, while the central portion contains the interior environment required for many patch interior species (Anderson et al. 1977, Johnson et al. 1979). For this reason, the width of a strip corridor is critical, since the interior environment must be present and sufficiently wide itself to be used by interior species.

In contrast to the line and strip corridors, the stream corridor is normally a dendritic pattern formed by intersecting narrow fingers upstream which gradually widen downstream. The stream corridor is the most widespread corridor type, and the concept has developed from considerations of water and mineral nutrient flows. This corridor strongly affects the erosion rate of the stream banks and adjoining upland and the absorption rate of water from precipitation and runoff. These, in turn, control siltation and flood levels in downstream ecosystems. The stream corridor is optimum when it doubles as a strip corridor for the migration of interior species. Since many species cannot survive the occasional floods of the stream lowland or the wet soils of the lowland and adjoining banks, the corridor must include a strip of interior environment on well-drained soil atop the stream bank.

A corridor should be continuous for maximum effectiveness (Getz et al. 1978, Schreiber and Graves 1977). In land-

scapes with ample human activity, one type of corridor, such as a road, commonly crosses another type, such as a hedgerow. The degree to which such crossings are effective barriers to the migration of different species needs study.

The corridor may exist as an isolated unit or it may interconnect patches in the landscape. In patches, species become extirpated for many reasons. Following loss of a species in a patch, a connected corridor facilitates rapid reestablishment of certain species in the patch. A strip corridor that links small patches may enrich those patches with species that otherwise could not survive in small isolated patches, because many species have minimum patch size requirements for survival (Galli et al. 1976, Robbins 1980, Terborgh 1976). Additionally, corridors facilitate gene flow across the landscape.

Networks are particularly widespread in landscapes bearing the heavy imprint of human activity. Anastomosing line corridors generally form networks, though occasionally networks may be composed of strip corridors. Familiar examples are the interconnected hedgerows or "bocage" and the networks of roads and railroads. A few networks reflect natural conditions, such as the polygon soils of arctic tundra areas or the reticulate trails of large mammals in east African savannas.

As isolated units, single corridors are considered to enhance the movement of species. However, as a series of interconnected links and loops, a network provides a more efficient migratory system, since alternative pathways are present. This structure is important for animal foraging efficiency, predator avoidance, and minimizing the barrier or isolating effect of a local disturbance or break in a corridor link. The frequency of intersections of corridors and the degree to which such intersections are expanded nodes or patches may play an important role in migration efficiency. Some networks, such as paths and roads, are especially effective for movement of people and domestic animals. We hypothesize networks to be important migration routes for native species, but as yet, the evidence is meager (Pollard et al. 1974).

In short, networks are prominent features of most landscapes today. Their functional importance rests not only in movement along the links, but in their impact on the matrix and patches in the surrounding landscape.

HABITATIONS

A final major structural characteristic of many landscapes is human habitation, including the house with its associated yard, courtyard, farm buildings and immediate surroundings. Habitations, of course, are disturbance-caused, partially or totally eliminating the natural ecosystem at that spot. The continued existence of the habitation depends on maintaining a chronic disturbance level.

The primary ecological structure of habitations is based on the types of organisms that have replaced the naturally occurring ones. Foremost are people, who provide the continued disturbance regime to maintain the habitation area. Most of the plants, in turn, are introduced by people. Some may be native species, but humans exhibit a propensity for surrounding themselves with a diverse and exotic species assemblage. People also generally introduce domestic, rather than native animals into homes, and both animal and plant pests are inadvertently introduced. Native species from the surrounding matrix or patches immigrate into habitation areas, but their success depends upon the level of disturbance maintained.

Distance between habitations in effect defines urban, suburban, town, village, and various rural areas. The greatest density and diversity of introduced species appears, on the average, in suburban areas, and indeed, in all areas with contiguous homes, the ecosystem is dominated by humans and introduced species (Schmid 1975). In rural areas with isolated homes, the entire border of the habitation interfaces with patches, networks, corridors or the matrix, so that interaction with other landscape elements is at a maximum. This interaction is the primary ecological importance of habitations in rural landscapes.

DYNAMICS OF THE LANDSCAPE

Our primary objective in this article is to lend insight into the ecological structure of landscapes, particularly patches (Figure 3). Yet, the structure is ultimately of importance as it relates to function. We have touched on the dynamics of patches themselves. Here we briefly suggest some examples of fluxes between structural components of the landscape (Forman 1981), that is, interactions between patch and matrix, patch and patch of the same type, patches of different types, network and matrix, and the like.

Interactions between patch and matrix are important in both directions. Heat energy carried by wind from one to the other accelerates the evapotranspiration rate and desiccates the microenvironment for certain species. Similarly, wind carries moisture, ash, dust, and propagules back and forth. Fire and other disturbances start in one and enter the other, and many types of animals forage from one to another.

Corridors and networks facilitate movement of species from patch to patch in the landscape, but also play a major role in inhibiting migration of matrix species by subdividing the matrix into isolated units. Networks and stream corridors retard surface water and nutrient runoff, and subsequent siltation and floods in downstream ecosystems, and in a similar fashion, a network modifies the flow of air and heat energy over the landscape, which in turn alters evapotranspiration and the moisture patterns of the soil.

Finally, habitations, as species sources, provide people and nonnative plants and animals. They, in turn, harvest species, form corridors and networks, produce various disturbances, and colonize both the surrounding patches and matrix.

CONCLUSION

The structure of a landscape is primarily a series of patches surrounded by a matrix. The origins of patches differ according to the disturbance regime in the patch, disturbance in the matrix, natural distribution of environmental resources, species introductions by people, and time. These differences in patch origin determine the species dynamics and the stability and turnover of patches themselves.

Patch area, and secondarily isolation, have traditionally been considered the major variables indicating the species diversity of a patch. We hypothesize that species diversity in a landscape patch is a function of the following patch variables in order of overall importance: habitat diversity ± disturbance + area + age + matrix heterogeneity − isolation − boundary discreteness. Patch shape as a modifier of area is important to species diversity and is mediated through the patch edge or edge effect.

The numbers of patches of each patch origin, biotic patch type, size, and shape determine in part the landscape structure. However, the spatial configuration among the patches present may be just as important as the numbers.

Corridors vary in width and function. Line corridors, particularly those resulting from human activities, are very narrow and used primarily for movement of edge species or people. Strip corridors, for effective movement of species characteristic of the interior of a patch, are wide enough to include an interior microenvironment as well as edge effect on both sides. A special case is the stream corridor, which also controls water and nutrient flows across the landscape.

Networks composed of intersecting corridors are prominent features of most landscapes. Networks provide an efficient migratory route as well as alter the flow of nutrients, water, and air across the landscape.

The concept of repetitive patterns in the structure of landscapes opens up a host of ecological questions related to both structure and function, and provides a relatively simple framework for testing them. It also provides a land management tool for helping to determine priorities in the land use. Finally, it emphasizes that no patch stands alone.

ACKNOWLEDGMENTS

We thank Steward T. A. Pickett and Mark J. McDonnell for significantly improving this manuscript, and the Nation-

Figure 3. Portion of an agricultural landscape in New Jersey. Farming practices for corn and beans since 1701 have molded this landscape. Limited suburbanization effects are recent. The geomorphology is a level Triassic red shale, on which a well-drained silt loam of the Penn series predominates. The biotic patch types present are dominated by white, red and black oak (*Quercus alba, Q. borealis, Q. velutina*), except in stream corridors and wet spots where pin oak, red maple, ash and elm (*Q. palustris, Acer rubrum, Fraxinus, Ulmus*) predominate. Photograph taken May 29, 1970. **A.** Spot disturbance patch (small opening in forest). **B.** Strip corridor (powerline crossing stream corridor). **C.** Narrow patch with no forest interior. **D.** Strip corridor (wooded). **E.** Tiny patches with no forest interior. **F.** Peninsula. **G.** Tiny remnant patch affected by proximity to larger patch. **H.** Introduced patch (golf course). **I.** Introduced line corridor (*Platanus* planted along road). **J.** Large remnant patch (well-developed forest interior; patch edge about twice as wide to south as north). **K.** Road network. **L.** Dwellings clustered (village). **M.** Introduced patch (cemetery conifers and grass). **N.** Environmental resource patch (lowland tree species on wet spot). **O.** Temporal patch (area of shrubs and successional trees undergoing rapid change). **P.** Wide stream corridor (containing both river and canal). **Q.** Narrow stream corridor. **R.** Matrix (corn and bean fields). **S.** Line corridor (road). **T.** Habitation (area of farm buildings). **U.** Hedgerow network (connecting woods patches). **V.** Small remnant patch (contains limited area of forest interior).

al Science Foundation for grant DEB-80-04653 in support of a portion of this work.

REFERENCES CITED

Anderson, S. H., K. Mann, and H. H. Shugart, Jr. 1977. The effect of transmission line corridors on bird populations. *Am. Midl. Nat.* 97: 216–221.

Bormann, F. H., and G. E. Likens. 1980. *Pattern and Process in a Forested Ecosystem.* Springer-Verlag, New York.

Brown, J. H. 1971. Mammals on mountaintops: non-equilibrium insular biogeography. *Am. Nat.* 105: 467–478

Bunnell, S. D., and D. R. Johnson. 1974. Physical factors affecting pika density and dispersal. *J. Mammal.* 55: 866–869.

Carlquist, S. J. 1974. *Island Biology.* Columbia Univ. Press, New York.

Chessel, D. 1978. Description non paramétrique de la dispersion spatiale des individus d'une espèce. *Biometrie et Ecologie* 1: 45–135.

Daget, P. 1979. La nombre d'espèces par unité d'échantillonnage de taille croissante. *La Terre et la Vie* 32: 461–470.

Daubenmire, R. 1968. *Plant Communities: A Textbook of Plant Synecology.* Harper and Row, New York.

Diamond, J. M. 1972. Biogeographical kinetics: estimation of relaxation times for avifaunas of southwest Pacific islands. *Proc. Nat. Acad. Sci. USA* 69: 3199–3203.

Diamond, J. M. and R. M. May. 1976. Island biogeography and the design of natural reserves. Pages 163–186 in R. M. May, ed. *Theoretical Ecology.* Saunders, Philadelphia, PA.

Dickinson, R. E. 1970. *Regional Ecology: The Study of Man's Environment.* John Wiley & Sons, Inc., New York.

Elfstrom, B. A. 1976. Tree species diversity and forest island size on the Piedmont of New Jersey. M.S. thesis, Rutgers Univ., New Brunswick, NJ.

Forman, R. T. T., ed. 1979a. *Pine Barrens: Ecosystem and Landscape.* Academic Press, New York.

———. 1979b. The Pine Barrens of New Jersey: An ecological mosaic. Pages 569–585 in R. T. T. Forman, ed. *Pine Barrens: Ecosystem and Landscape.* Academic Press, New York.

———. 1981. Interactions among landscape elements: a core of landscape ecology. In *Perspectives in Landscape Ecology.* Proc. Int. Congr. Landscape Ecol., 1981, Veldhoven. Pudoc Publ., Wageningen, The Netherlands, in press.

Forman, R. T. T., A. E. Galli, and C. F. Leck. 1976. Forest size and avian diversity in New Jersey woodlots with some land use implications. *Oecologia (Berl.)* 26: 1–8.

Forman, R. T. T., and R. E. J. Boerner. 1981. Fire frequency and the Pine Barrens of New Jersey. *Bull. Torrey Bot. Club* 108: 34–50.

Galli, A. E., C. F. Leck, and R. T. T. Forman. 1976. Avian distribution patterns

within different sized forest islands in central New Jersey. *Auk* 93: 356–364.

Getz, L. L., F. R. Cole, and D. L. Gates. 1978. Interstate roadsides as dispersal routes for *Microtus pennsylvanicus. J. Mammal.* 59: 208–212.

Godron, M. 1966. Application de la théorie de l'information à l'étude de l'homogénéité et de la structure de la végétation. *Oecol. Plant.* 2: 187–197.

———. 1971. *Essai sur une approche probabiliste de l'écologie des végétaux.* Thèse d'Etat, Univ. Sci. Tech. Languedoc, Montpellier, France.

Gottfried, B. M. 1979. Small mammal populations in woodlot islands. *Am. Midl. Nat.* 102: 105–112.

Greig-Smith, P. 1964. *Quantitative Plant Ecology.* Butterworths, London.

Grossman, L. 1977. Man-environment relationships in anthropology and geography. *Assoc. Am. Geogr. Ann.* 67: 126–144.

Hedberg, O. 1955. Vegetation belts of the East African mountains. *Sven. Bot. Tidskr.* 45: 140–202.

Heinselman, M. L. 1973. Fire in the virgin forests of the Boundary Waters Canoe Area, Minnesota. *Quat. Res. (NY)* 3: 329–382.

Helliwell, D. R. 1976. The effects of size and isolation on the conservation value of wooded sites in Britain. *J. Biogeogr.* 3: 407–416.

Isard, W. 1975. *Introduction to Regional Science.* Prentice-Hall, New York.

Jakucs, P. 1972. *Dynamische Verbindung der Walder und Rasen.* Akad. Kiado, Verlag Ungarischen Akad., Wissenschaften, Budapest.

Johnson, W. C., R. K. Schreiber, and R. L. Burgess. 1979. Diversity of small mammals in a powerline right-of-way and adjacent forest in east Tennessee. *Am. Midl. Nat.* 101: 231–235.

Johnston, V. R. 1947. Breeding birds of the forest edge in Illinois. *Condor* 49: 45–53.

Kellogg, R. S. 1934. The shelterbelt scheme. *J. Forestry* 32: 947–977.

Kershaw, K. A. 1973. *Quantitative and Dynamic Plant Ecology.* American Elsevier Publishers, Inc., New York.

Leopold, A. 1933. *Game Management.* Charles Scribner's Sons, New York.

Les Bocages: Histoire, Ecologie, Economie. 1976. Université Rennes, Rennes, France.

Levin, S. A., and R. T. Paine. 1974. Disturbance, patch formation, and community structure. *Proc. Nat. Acad. Sci. USA* 71: 2744–2747.

Lewis, T. 1969. The distribution of insects near a low hedgerow. *J. Appl. Ecol.* 6: 443–452.

MacArthur, R. H., and E. O. Wilson. 1967. *The Theory of Island Biogeography.* Princeton University Press, Princeton, NJ.

McHarg, I. L. 1969. *Design with Nature.* Natural History Press, Garden City, New York.

Mikesell, M. W. 1968. Landscape. *Int. Encyclo. Soc. Sci.* 8: 575–580.

Moore, N. W., and M. D. Hooper. 1975. On the number of bird species in British woods. *Biol. Conserv.* 8: 239–250.

Odum, E. P. 1971. *Fundamentals of Ecology.* W. B. Saunders Co., Philadelphia, PA.

Patton, D. R. 1975. A diversity index for quantifying habitat edge. *Wildl. Soc. Bull.* 394: 171–173.

Peterken, G. F. 1974. A method of assessing woodland flora for conservation using indicator species. *Biol. Conserv.* 6: 239–245.

Pickett, S. T. A., and J. N. Thompson. 1978. Patch dynamics and the design of nature reserves. *Biol. Conserv.* 13: 27–37.

Pollard, E., M. D. Hooper, and N. W. Moore. 1974. *Hedges.* W. Collins Ltd., London.

Pollard, E., and J. Relton. 1970. Hedges V. A study of small mammals in hedges and cultivated fields. *J. Appl. Ecol.* 7: 549–557.

Robbins, C. S. 1980. Effect of forest fragmentation on bird populations. Pages 198–212 in R. M. DeGraaf and K. E. Evans, compilers, Management of North Central and Northeastern Forests for Nongame Birds. *U.S. Dep. Agric. For. Serv., Gen. Tech. Rept.* NC-51.

Rotzien, C. L. 1963. A cumulative report on winter bird population studies in eight deciduous shelterbelts of the Red River Valley, North Dakota. *Proc. ND Acad. Sci.* 17: 19–23.

Sauer, C. O. 1963. The morphology of landscape. Pages 315–350 in J. Leighly, ed. *Land and Life: A Selection from the Writings of Carl Ortura Sauer.* University of California Press, Berkeley.

Schmid, J. A. 1975. *Urban Vegetation: A Review and Chicago Case Study.* Department of Geography, University of Chicago, Chicago, IL.

Schreiber, R. K., and J. H. Graves. 1977. Powerline corridors as possible barriers to the movement of small mammals. *Am. Midl. Nat.* 97: 504–508.

Seignobos, C. 1978. Les systèmes de défense végétaux pré-coloniaux. *Annales de l'Université du Tchad. Série Lettres, Langues Vivantes et Sciences Humaines.* Numéro Spécial. Tchad.

Simberloff, D. S. 1976. Experimental zoogeography of islands: Effects of island size. *Ecology* 57: 629–648.

Simpson, G. G. 1964. Species density of North American recent mammals. *Syst. Zool.* 13: 57–73.

Smith, A. T. 1974. The distribution and dispersal of pikas: Consequences of insular population structure. *Ecology* 55: 1112–1119.

Southwood, T. R. E. 1961. The number of species of insect associated with various trees. *J. Anim. Ecol.* 30: 1–8.

Squier, E. G. 1877. *Peru: Incidents of Travel and Exploration in the Land of the Incas.* Holt, Rinehart & Winston, New York.

Stiles, E. W. 1979. Animal communities of the New Jersey Pine Barrens. Pp. 541–553 in R.T.T. Forman, ed. *Pine Barrens: Ecosystem and Landscape.* Acad. Press, NY.

Taylor, R. J., and P. J. Regal. 1978. The peninsular effect on species diversity and the biogeography of Baja California. *Am. Nat.* 112: 583–593.

Terborgh, J. 1976. Island biogeography and conservation: Strategy and limitations. *Science* 193: 1029–1030.

Van Eimern, J., R. Karshon, L. A. Razumova, and G. W. Robertson. 1964. Windbreaks and shelterbelts. *World Meteorol. Organiz. Tech. Note* No. 59. 188 pages.

Wales, B. A. 1967. Climate, microclimate and vegetation relationships on northern and southern forest boundaries in New Jersey. *William L. Hutcheson Mem. For. Bull.* 2: 1–60.

———. 1972. Vegetation analysis of northern and southern edges in a mature oak-hickory forest. *Ecol. Monogr.* 42: 451–471.

Whitcomb, R. F. 1977. Island biogeography and "habitat islands" of eastern forest. *Am. Birds* 31: 3–5.

Whitmore, T. C. 1975. *Tropical Rain Forests of the Far East*. Oxford University Press, New York.

Whittaker, R. H., ed. 1973. *Ordination and Classification of Communities*. Junk Publ., The Hague.

Willis, E. O. 1974. Populations and local extinctions of birds on Barro Colorado Island, Panama. *Ecol. Monogr.* 44: 153–169.

Woodwell, G. M., and R. H. Whittaker. 1968. Primary production in terrestrial ecosystems. *Am. Zool.* 8: 19–30.

PAPER 16

LANDSCAPE ECOLOGY
Directions and Approaches

Paul G. Risser
Illinois Natural History Survey

James R. Karr
University of Illinois

Richard T. T. Forman
Rutgers University

A workshop held at
Allerton Park
Piatt County, Illinois
April 1983

Illinois Natural History Survey Special Publication Number 2

Published March 1984 by the Illinois Natural History Survey
Natural Resources Building
607 East Peabody Drive
Champaign, Illinois 61820

A Division of the Illinois Department of Energy and Natural Resources

INTRODUCTION

In recent years, several attempts have been made to define a field of science entitled "regional ecology" or "landscape ecology." These initiatives have originated from a number of scientific points of view (Watt 1947; Whittaker and Levin 1977), yet no clear set of general principles has emerged.

Current ideas about landscape ecology (e.g., Burgess and Sharpe 1981; Forman 1981; Forman and Godron 1981; Luder 1981; Minnich 1983; Naveh 1982; Romme and Knight 1982; Sharpe et al. 1981) are influenced by (a) a preoccupation with the extension of island biogeography theory to continental landscape patches, (b) the presumption that ecosystem-level characteristics are adequate to address landscape-level characteristics, (c) a recognition of the need to address landscape issues in land and resource management, (d) a belief that map-overlay methodology is sufficient to capture the essential attributes of multiunit landscapes, (e) the realization that human activities are an integral part of any meaningful concept of landscape ecology, and (f) the recognition that the inclusion of many appropriate scientific disciplines results in an exceedingly complex field. Collectively, these influences appear to have stalled the crystallization and communication of current understanding of "landscape ecology," especially as the concept might facilitate basic and applied research on natural resources.

A landscape perspective in ecology is not new (Neef 1967; Troll 1968); indeed, this is the perspective embodied in *A Sand County Almanac* (Leopold 1949) and in many early writings in ecology, natural history, and wildlife biology. Similarly, this landscape perspective is represented in related disciplines, such as landscape planning, economic geography, and cultural anthropology. However, these ideas have never been coalesced, organized, and confronted rigorously to produce a theoretically sound basis for understanding landscape-scale interactions. Further, the ecological base of this disciplinary integration is especially weak, and so developing definitive and ecologically based methods and models for managing natural resources is essential.

In spite of this conceptual bottleneck, ideas and concepts are developing (albeit slowly), research is being designed, and resource managers are grasping at even fragments of generalizations about the ecology of landscapes that can focus research efforts and guide resource management decisions (Forman 1979; Hansson 1977; Isard 1975; Klopatek et al. 1983; Naveh and Lieberman 1984; Samson and Knopf 1982). A mechanism for speeding the integration of a landscape ecology approach was to gather together experienced individuals with different viewpoints but with a strong desire to examine landscapes through the ideas of ecology and related disciplines. This report summarizes the deliberations of the 25 individuals (see List of Participants) who spent three days attempting to outline the disciplinary area of landscape ecology, to evaluate the potential of such a discipline, and to describe its application to basic and applied natural-resource issues. Although the group represented diverse points of view, an ecological perspective prevailed. Ideas contained in this report represent the collective efforts of the group, and no attempt has been made to identify specific thoughts with any particular individual.

DEFINITION AND CONCEPT OF LANDSCAPE ECOLOGY

Ecology deals with the understanding of fundamental processes and consequences of management of spatially and temporally homogeneous and heterogeneous geomorphic and living systems.

Landscape ecology differs from subdisciplines of ecology, such as population, community, and ecosystem ecology, in matters of primary emphasis. Landscape ecology focuses explicitly upon spatial pattern. Specifically, *landscape ecology considers the development and dynamics of spatial heterogeneity, spatial and temporal interactions and exchanges across heterogeneous landscapes, influences of spatial heterogeneity on biotic and abiotic processes, and management of spatial heterogeneity.* Thus, the primary focus of landscape ecology is on (a) spatially heterogeneous geographic areas, e.g., pine barrens, regions of row crop agriculture, Mediterranean woodland landscapes, and areas of urban and suburban development; (b) fluxes or redistribution among landscape elements; and (c) human actions as responses to, and their reciprocal influences on, ecological processes. Principles of landscape ecology help to provide theoretical and empirical underpinnings for a variety of applied sciences, e.g., regional planning, landscape architecture, and natural-resource management.

The relationship between spatial pattern and ecological processes is not restricted to a particular scale. One's understanding of landscape ecology issues focused at one scale may profit from experiments and observations on the effects of pattern at both finer and broader spatial scales. In turn, results from landscape studies may find application in understanding the way organisms interact with patterned environments at other levels of scale (Wiens in press).

Ecological processes vary in their effects or importance at different scales. Thus, biogeographic processes may be relatively unimportant in determining local patterns but may have major effects upon regional patterns. Processes leading to population decline may produce extinction at a local scale, but may only appear as spatial redistributions or alterations in age structure at broader levels.

Different species and groups of organisms (e.g., plants, herbivores, predators, parasitoids) may operate at different spatial scales, and thus, investigations undertaken at a given scale may not treat such components with equivalent resolution. Operationally,

8   ILLINOIS NATURAL HISTORY SURVEY SPECIAL PUBLICATION   No. 2

scales of landscape elements are defined arbitrarily, using spatial "filters" of specified sizes determined by the specific objectives of the investigation. At the landscape level, there may be spatial units or elements, such as fields, woodlots, clearings, or hedgerows. With some well-exercised caution, patterns and processes studied at other levels of scale, e.g., within a simple, relatively homogeneous landscape element, such as an old field, may provide concepts useful in understanding the landscape level, and vice versa. Processes and patterns occurring at much finer scales are not always perceived because of filtering or averaging effects, whereas those occurring at much broader scales may be missed simply because the investigation is focused within a single landscape unit. Although hierarchical approaches may offer some thoughtful insights (Allen and Starr 1982), there are few simple answers to ecological questions regardless of scale, and there will be no substitute for carefully constructed experiments and tests of ideas drawn from many disciplines.

Because of the spatial patterning of landscapes, flows and transfers among spatial components assume special importance. The process of the redistribution of materials, energy, and/or individuals among landscape elements is an essential feature of landscape ecology. Redistribution among landscape elements thus represents a dynamic in ecological systems that has largely been ignored by ecologists; until very recently dynamics of heterogeneous environments were largely ignored by the ecological sciences.

Traditionally, ecological studies have operated with the assumption that systems are more or less homogeneous and, in some cases, at approximate equilibrium. These considerations, primarily made for convenience, emphasized relatively undisturbed, "natural" habitats for study. Adoption of a perspective that admits the importance of spatial and temporal heterogeneity creates a major argument for the merger of the more or less independently developing European school of landscape geography and the growing body of ecological theory resulting from the study of heterogeneity and instability in ecological systems. Thus, the need to consider spatial pattern in ecological systems and the consequences of pattern on the dynamics and persistence of landscapes is clear.

At present, the theoretical treatment of these patterns and even the measurement of pattern is possible, but too little basic empirical information exists to document the phenomena and their consequences for landscapes. Thus, a clear need exists to focus basic ecological investigation on the patterns and interrelationships among the elements of landscapes to provide the empirical base from which to test and refine current models and societal policies. Ultimately, this approach will contribute to the development of general principles of landscape ecology.

A major forcing function of landscapes is the activity of mankind, especially associated cultural, economic, and political phenomena. Given the history of ecology, it may be tempting to draw a sharp line between landscape ecology and the applied management of landscapes. This distinction is not sharp, however, particularly since management practices of the past had much to do with the structure of landscapes that have developed today. Understanding landscapes requires that we deal with human impacts contributing to the landscape phenomenon, without attempting to draw the traditional distinction between basic and applied ecological science or ignoring the social sciences.

Landscape ecology could be viewed as the intersection of many disciplines, as a separate discipline, or as one branch of ecology. The first option is intellectually and practically the most persuasive. Arguments can be made for the inclusion of concepts and methods from such disciplines as terrestrial and aquatic ecology, geography, history, agricultural economics, civil engineering, landscape architecture, and wildlife management. To be recognized as a distinct scientific discipline, an area of activity must have a set of general, guiding principles, a conceptual framework of its own. Landscape ecology cannot now be viewed as a discipline because principles developed by practitioners (landscape ecologists) are few; most applicable principles have been developed in other established disciplines. In time, new principles should emerge and landscape ecology will develop its own body of theory. The third option, viewing landscape ecology as a branch of ecology, would emphasize natural spatial processes and patterns and, at least by tradition, would tend to exclude the formal analysis of human cultural processes that form landscapes. Viewing landscape ecology as an interdisciplinary area of research avoids the issue of which discipline "owns" landscape ecology. In human ecology (Young 1974), several disciplines (i.e., geography, sociology, anthropology) have claimed ownership, which has been a futile exercise, because human ecology, like landscape ecology, is also inherently interdisciplinary.

REPRESENTATIVE QUESTIONS ADDRESSED BY LANDSCAPE ECOLOGY

What does landscape ecology, so defined, have to offer? To test this potential, we must be able to frame fundamental questions concerning the development, maintenance, and effects of temporal and spatial heterogeneity of the landscape. Four examples follow.

How are fluxes of organisms, of material, and of energy related to landscape heterogeneity?

Thoughtful observation and experimentation indicate that units of landscape are not ecologically independent. Patches in a landscape mosaic are coupled by fluxes of organisms, biotic and abiotic energy, and nutrients. Understanding the fundamental behavior of such operationally defined units requires specific study of the fluxes at the landscape scale and a recognition that anthropocentrically defined landscape patches are differentially significant to different species. Fluxes of organisms among landscape patches include an array of organisms along a wide spectrum of size, mobility, developmental rates, and resource requirements. Each species views the landscape differently, and what appears as a homogeneous patch to one species may comprise a very heterogeneous, patchy environment to another. At each species-specific scale of reference, species survival often depends on interpatch fluxes. For example, some populations of small vertebrates survive in agricultural landscapes of the north-central United States and central Canada only because they can move between wooded patches along fencerow corridors through intervening farmland. Both white-footed mice (*Peromyscus leucopus*) and chipmunk (*Tamias* sp.) suffer frequent local extinctions and, lacking a durable source area, depend for survival at the landscape scale on fluxes of colonists among patches (Fahrig, Lefkovitch, and Merriam 1983; Henderson, Merriam, and Wegner submitted; Middleton and Merriam 1981; Wegner and Merriam 1979). In a comparison of avian distribution in Australia and Wisconsin, Howe (in press) concluded that the regional mosaic in Australia with a larger proportion of its landscape in forest as compared with Wisconsin, may account for the differing species-area relations for birds in isolated woodlots.

These examples demonstrate the dynamics of populations in the farmland mosaic. There is also evidence that even within continuous forests, bird species may require a mosaic of habitat patches (Karr 1982a, b; Karr and Freemark 1983; Middleton and Merriam 1983). This dependency of many terrestrial and aquatic vertebrates on habitat mosaics seems to be a general phenomenon (Karr and Freemark in press). Comprehension of this dependency may yield new insights into species survival and local population phenomena as they relate to patch dynamics in general as elements in the ecology of regions.

Grazing systems offer an opportunity to evaluate the transport of nutrients across the landscape pattern (Woodmansee 1979). Nutrients in rangelands are transported among patches by four principal agents: large animals, water, wind, and man (via supplemental feeding). Large animals are important transport mechanisms because they typically graze from patches (remove material) that contain relatively large amounts of high-quality forage. Usually these grazing areas (patches) are separated spatially from areas (patches) where the animals water, rest, bed down, and ruminate. Overall, material is removed from the grazed areas and accumulates in resting areas via defecation and urination. Most behaviors of animals in western rangelands (except watering, which may be partially determined by the manager) are determined primarily by soil type and fertility, plant community relationships, and topography. Though these studies have been conducted with cattle, the results should be generalizable to many large and small herbivores and are demonstrated from the Serengeti (McNaughton 1979). Historically, spatial and temporal heterogeneity in species attributes and population characteristics have been a problem for ecologists as they searched for communities that were homogeneous in space and time. Recognition of the role of spatial and temporal dynamics on the integrity and continuity of ecosystem processes may be essential to our understanding of basic ecology and of the problems that derive from landscape ecology (McNaughton 1983; Karr and Freemark in press).

Many species of insects require resources in two or more landscape units (patches) to complete their complex life cycles. For example, many herbivorous species feeding within crop units must move to wooded areas for overwintering, and many predaceous insects, such as vespid wasps, colonize hedgerows but forage in cultivated fields. Thus, crop mix and phenology, as well as natural events, have major influences on the flow of insect herbivores. Similar patterns also occur in unmanaged landscapes, such as those of the western tent caterpillar in Canada (Wellington et al. 1975; Thompson, Vertinsky, and Wellington 1979) and *Heliothis zea* in North Carolina (Stinner, Rabb, and Bradley 1977). Indeed, managing elk herds and waterfowl habitats also represent classic examples which are implicitly concerned with patch dynamics.

Plant breeders and pest control scientists are finding it desirable to use a landscape approach in choosing strategies for developing and using the different types of resistance. For example, if antibiosis is introduced into a crop variety that represents the chief food biomass of an insect herbivore and this variety is used uniformly over the area representing the habitat of an isolated population of the herbivore species, the population of insects develops tolerance to the antibiotic properties of the crop variety. However, if the crop variety represents but a small fraction of the food biomass of the ambient herbivore population, insect resistance does not decrease so readily (Kennedy 1983).

These paragraphs introduced three topics of significance at the landscape level, all involving redistribution of populations, energy, and materials: habitat selection by small mammals and birds, grazing system transfer of materials, and management of insect pests. Other examples could have been chosen, e.g., acid precipitation (Krug and Fink 1983). The point, however, is importance of the common theme of spatial heterogeneity and of redistribution processes.

10 ILLINOIS NATURAL HISTORY SURVEY SPECIAL PUBLICATION No. 2

What formative processes, both historical and present, are responsible for the existing pattern in a landscape?

Formative processes in the landscape can be organized in various ways. At the broadest level, virtually all processes of climate, geology, vegetation, animals, microorganisms, and human culture could be included. Both natural and man-created processes are important. For example, the natural movement of water is an important formative process in a landscape, but as man alters water movement through such means as dam construction or extensive irrigation of cropland, a new landscape may be created.

No clear method has emerged as a conceptual framework to organize the study of formative processes. Such a framework must transcend a complexity of dynamics in physical and biological processes at a variety of spatial and temporal scales and, for biology, at individual, population, community, and ecosystem levels.

One framework proposed during the workshop places organized "processes" into several general categories. Conserving processes tend to restrict change, while expanding processes promote growth and development of a landscape attribute. In conserving processes, for example, competitively superior organisms resist colonization by other species. Expanding processes occur when organisms or groups of organisms significantly expand the geographic area that they occupy within a landscape. Resisting processes protect the landscape from outside forces. In a biological context, these are analogous to resilience (Holling 1973). To a major extent, these types of "processes" are not really synthetic system-level processes. Rather, each seems to be a passive result of an accumulation of dynamics of physical and biological components of landscape systems.

These and other dynamics interact to create pattern on the landscape. Consider a landscape composed of two types of land systems. Assume that both types are exceptional in their ability to maintain themselves at a site (conservative processes) and that each type is able to hold the space against the other. A landscape composed of such subsystems would be a fixed mosaic of unchanging patches. Change made in the number or areal extent of such patches would not equilibrate to the initial state. Such a landscape might have a self-sustaining natural system (such as a forest) and an equivalently self-sustaining man-made system (cropland). This landscape dominated by conservative processes could be quite static.

As a second example, consider system types that are extreme in their expansive processes. If the two types are exactly at parity with respect to their ability to hold space, the landscape could appear quite unchanging like the "conservative-conservative" landscape just described.

Levin (1978) discusses the range of landscape system behavior for landscapes composed of expansive systems (*diffusion* is the word used for expanding processes). With regular disturbance, these "expansive-expansive" land systems can behave as a shifting dynamic mosaic. At a finer scale, the dynamic sorting due to erosion and silt deposition in stream channels is an example; the locations of pools and riffles, sand bars, and woody debris shift. At a broader scale, the natural vegetation of the northern hardwood forest with waves of disturbance (Sprugel 1976; Sprugel and Bormann 1981) or the role of fire in the regional pattern of vegetation of southern California (Minnich 1983) are examples.

Landscapes with interacting processes, perhaps with one type of land system expansive-process dominated and the other conservative-process dominated, would respond in the manner of classic succession theory. With regular disturbance, such a landscape would tend to have a replacement sequence and would require regular disturbance to maintain both land-system types in the mosaic.

The strong need for an integrative framework to organize the disparate set of processes and dynamics that influence landscape pattern is clear from these few examples. Other examples that might be cited include the feedback between landscape elements (such as natural and man-disturbed), dynamics due to disturbances that are either acute or chronic in intensity, the distinction between landscape processes that operate at evolutionary and ecological time scales, and processes mediated by global control, such as climatic pattern.

How does landscape heterogeneity affect the spread of disturbances?

The complexity of types of disturbances, their spatial and temporal scales, and their differential effects on biological processes (due to variations among taxa and levels, e.g., individual, population) defy simple comprehension (Karr and Freemark in press). As a generality, homogeneity often enhances the spread of disturbance. Examples include the spread of pests in agroecosystems, wildfire perpetuation, the spread of Dutch elm disease, and erosional patterns. Indeed, the effects of disturbance may increase the heterogeneity of the environment and, thus, alter the impact of a later disturbance of the same magnitude.

Heterogeneity also may enhance the spread of disturbance, as, for example, is the case where small woodlots harbor white-tailed deer populations that disturb surrounding crops. Heterogeneity may act as a stabilizing factor (e.g., by the spreading of risk) as well as fostering disturbance.

Many species require two or more landscape elements to complete their life cycle, and the impact of a specific disturbance (e.g., a severe dry period) may

March 1984 RISSER, KARR, AND FORMAN: LANDSCAPE ECOLOGY 11

be a function of the spatial pattern of landscape elements. Of course, temporal and spatial heterogeneity may also affect or limit the perpetuation of disturbance. Finally, landscape heterogeneity may impact the rate of recovery of landscapes from disturbance by providing refuges for recolonizing organisms.

Again, we emphasize the complexity of responses of organisms (and the biota) at the landscape level to perturbations, both large (= disturbance) and small. Indeed, a given perturbation may be a disturbance to one organism occupying a landscape, yet it may not be perceived by another organism in the same landscape (Karr and Freemark in press). Similarly, the same organism in two different landscapes may be affected differently by the same disturbance.

How can natural resource management be enhanced by a landscape ecology approach?

Natural resource managers, by necessity, often implement landscape ecology from an observational rather than an empirical or theoretical approach. For example, silviculturists have traditionally been concerned with the size, shape, distribution, and timing of timber harvests for the regeneration of forest stands. Agricultural grassland, shrub land, and aquatic ecosystems have been managed with similar single-resource objectives and multiresource consequences.

Ruffed grouse (*Bonasa umbellus*) use timber stands of various ages for feeding (mature forest), breeding (pole size), and brooding (regeneration stands). As a result, habitat managers manage forest stands to include these age classes in close proximity (Gullion 1977). Similarly, pheasants (*Phasianus colchicus*) use agricultural cropland for feeding, but populations are enhanced if 10–15 percent of the habitat is in hay crops to provide nesting and roosting cover (Warner 1981). The same principles are employed when managing habitat for wildlife that migrates over long distances, e.g., elk and waterfowl. Both breeding and overwintering requirements must be met, but in addition, a habitat mosaic may be important in each season. For example, migrating and overwintering geese (*Branta canadensis*) and mallard ducks (*Anas platyrhynchos*) require agricultural crop residue and forage, while canvasback (*Aythya valisineria*) and redhead ducks (*A. americana*) require aquatic vegetation and invertebrates, such as the fingernail clam (*Musculium transversum*), for food (Bellrose et al. 1979). The management of recreation resources also requires a mosaic of landscape classes that support different types or combinations of dispersed recreation (Driver and Brown 1978).

No body of ecological theory exists by which to evaluate or alter current management programs that result in various production levels of wildlife, timber, or recreation. As a result, decisions about the size, shape, distribution, and timing of resource management actions are based on observations. Landscape ecology, by defining responses to management apparent at the landscape scale, together with appropriate characterizations of heterogeneity, can provide a unifying framework for developing consistent predictive models of utility in resource management.

The recognition that informed resource management decisions often cannot be made exclusively at the site level is essential. A shift to regional and national decisions will place greater emphasis on landscape ecology concepts (Joyce et al. 1983). The realization of this fact is expressed in regional consortia, such as the Ohio River Basin Commission in water resource management and the Flyway Council concept in migratory waterfowl management. Multiple or joint resource production interactions inherent in landscape management must consider: (a) optimizing trade-offs in production, since not all natural values can be maximized simultaneously, (b) imposing economic values in deciding among trade-offs between alternative landscape management actions, and (c) considering the socio-economic impacts, as well as the production of natural values, in choices of alternative landscape management actions.

Clearly, the evaluation of the trade-offs, when accomplished comprehensively, demands a set of principles based on landscape ecology issues discussed in this paper.

METHODOLOGIES APPLICABLE TO LANDSCAPE ECOLOGY

In earlier sections of this report we have discussed the concept of landscape ecology, and have presented examples of fundamental ecological questions that require the application of landscape ecology concepts. In developing a framework for landscape ecology, it is necessary to consider techniques and methods that are particularly appropriate and how these techniques will provide for the needs of landscape ecology. The following paragraphs address four methodological questions which may demonstrate the value of a variety of appropriate techniques.

Methods for Measuring Landscape Heterogeneity

Many issues in landscape ecology require quantification of spatial heterogeneity or landscape pattern. Currently, several descriptive techniques exist for quantifying the spatial relationships among entities, such as individuals in populations, habitats, and land-use types. In addition, there are schemes or terms for describing landscape pattern, e.g., species-dependent patch, matrix, corridor, and mosaic.

However, some problems require modifications or extensions of these methods to include (a) human cultural and demographic characteristics, (b) capture of the dynamics of changes in patterns over time,

12 ILLINOIS NATURAL HISTORY SURVEY SPECIAL PUBLICATION No. 2

and (c) incorporation of value judgments with empirical data associated with landscape pattern. The following paragraphs discuss useful techniques and describe additional methods from other fields of science that might be effective in addressing the required modifications.

Topological Measure of Spatial Heterogeneity. A variety of graph theoretical methods, as well as statistical, metric, and topological methods, could be applied to landscape ecology (Haggett, Cliff, and Frey 1977; Lowe and Moryadas 1975; Mandelbrot 1977; Sugihara 1983). Movement of migrants or propagules among elements of the mosaic can be expressed in graph theoretic measures of connectivity. Experimental manipulation (e.g., introducing corridors or barriers) can be used to test hypotheses about the importance of connectance in landscape dynamics (Fahrig, Lefkovitch, and Merriam 1983). Statistical methods employ a basic distribution model for the whole population of spatially distributed characters (e.g., Poisson model) and test the observed distribution against the model. In probabilistic methods, heterogeneity is measured as the probability of obtaining the observed spatial distribution by a random process. Macroheterogeneity and microheterogeneity can also be distinguished as functions of scale, and results are often expressed by information theory terms or concepts to emphasize the progression of knowledge given by each successive step in the analysis (e.g., Batty and Sikdar 1982).

Spectral Analysis. Techniques used in spectral analysis have most often been applied to time series, generally with the hope of associating underlying processes with pattern. That is, spectral analysis is a technique to elucidate the autocorrelation structure of the underlying process. These techniques have also been used to quantify patterns in space. For example, Denman and Platt (1976) used spectral analysis to study the patch structure of marine phytoplankton fields.

Spectral analysis (Shugart 1978) techniques often require treatment of the data (e.g., detrending, filtering to reduce noise) and require considerable amounts of data collected at regular intervals. Analysis is in the form of a graph of variance accounted for plotted against the log of the frequency of a sine function. The power of the method is in quantification of pattern in a precise mathematical way coupled with a tradition of interpreting patterns in terms of functional characteristics.

Artificial Intelligence Methods. The problem of recognizing and classifying patterns embedded in heterogeneous, spatial, and temporal arrays has become a prime focus of the emerging field of artificial intelligence (Michalski and Chilausky 1980). The term "pattern" is closely allied with "concept." This is not a single technique, but the application of several related techniques to the synthetic problem of pattern analysis. Landscape ecology may benefit from and ultimately contribute to this effort.

Data Acquisition Approaches

Data bases with spatial components useful in landscape ecology are available from several sources. Many of these sources are conventional, e.g., aerial photography, multispectral scanner imagery, and biological sampling schemes as well as various statistical measures of demography. Several problems require particular attention in acquisition, management, and display of spatial data in the context of landscape ecology: merging data from multiple sources with various levels of precision, resolution, and timing; introduction of distortions when reporting and displaying mapped information; decisions about the appropriate resolution for a particular problem; and choice of display formats for particular uses and users.

Field samples may result in statistical replications in a universe that is assumed to be homogeneous. If the universe is heterogeneous, use of systematic sampling or stratified sampling, which takes into account the heterogeneity, becomes necessary. Also there is the need to determine the "grain" of heterogeneity to know the mean radius of validity of any physical or biological measure. A traditional approach has been to collect a cumulative series of measures such that the magnitude of the variances in the sampled parameters corresponds to the grain size.

Useful Modelling Approaches

Purposes and Applications. Modelling techniques are useful for the static description of spatial heterogeneity and for the elaboration of the dynamics of pattern. In the former category are such well-established methodologies as direct and indirect gradient analysis and other multivariate statistical methodologies as well as more recent approaches to pattern analysis, noted above. The focus of this section is on the dynamics of pattern, primarily in ecological time, and on their consequences for the species and systems on the landscape.

Models serve many purposes: as predictive tools, allowing estimation of future consequences of past, present, and future events; as vehicles for the design of management schemes in which an adaptive or feedback component may be central; as descriptors and explainers of historic patterns; and as general frameworks for arranging ideas, defining research priorities, and understanding natural systems. In an applied context, a spectrum of models ranging from generic ones, which serve as preliminary screens, to site-specific ones, which serve as the basis for solving more detailed problems, is needed.

When using models, one must also recognize the multiplicity of interrelated scales—hierarchical, spatial, and temporal—on which processes take place.

March 1984   RISSER, KARR, AND FORMAN: LANDSCAPE ECOLOGY   13

Such models as the JABOWA-FORET class of forest growth descriptors (Shugart and West 1980) and patch models of intertidal zones (Levin and Paine 1974; Paine and Levin 1981) treat a spatial scale small enough that recruitment may be regarded as independent of internal dynamics. Such models exhibit the importance of an explicit consideration of spatial scale.

Individual and Population-Level Models. Spatial effects often can be appraised from individual or population-level models (Levin 1974, 1978). Considerable attention has been devoted to patch effects upon the foraging behavior of individuals, generally in the context of optimization. Such models consider the effects of patchy distributions of resources upon proximate features of behavior (capture rates, patch residency times, etc.) but generally do not consider (but see Stinner, Rabb, and Bradley 1977) the dynamics of the patches themselves (e.g., resource depletion, disturbance, succession) or patch interrelationships (e.g., patch position effects). A variation on this theme addresses central-place foraging behavior (Orians and Pearson 1978), in which the key landscapelike components are a central place from which individuals forage (e.g., den, nest site, colony) and the surrounding area, which is often assumed to be homogeneous. However, one application of such central-place ideas has considered patchiness in resource distributions in the foraging zone. Ford et al. (1982) modeled the distribution of foraging seabirds in the oceanic areas about a colony, assigning various areas different resource production levels. In this model the landscape components had internal dynamics (depletion and renewal of resources) that had effects on population structure and persistence, but patch interactions or position effects were not included in the model. Another model by Fahrig, Lefkovitch, and Merriam (1983) considering the rate of exchange between patches showed that interpatch connectiveness was capable of stabilizing population levels in patches.

Other population-level attributes may also be affected in the spatial configuration of the landscape, and these attributes are candidates for spatial modelling approaches (Schluter 1981). The probability of the persistence of a single-species population, for example, may be affected by spatial heterogeneity (Chesson 1981). At a basic level, the concept has been envisioned as "spreading of risk" (den Boer 1968, 1981) in which a population distributed among patches in a landscape has patch-specific probabilities for growth, extinction, and recolonization, and the population as a whole persists because of the spatial subdivision of these probabilities. These processes have been modelled by applying conventional population-dynamics models to individual patches and then aggregating the responses over the population (landscape) as a whole.

Elements of island biogeography theory predict that as patch size decreases and isolation increases, the probability of local extinction should increase and that of recolonization should decrease. Such theoretical expectations have not been integrated into a landscape perspective that would consider the effects of patch dynamics (including fragmentation and patch regeneration) in a mosaic landscape rather than an island setting.

Simple extensions of single-species population systems to two-species systems permit incorporation of the effects of processes, such as predation and competition, and consideration of their relationship to environmental patchiness (Noy-Meir 1981). However, a modelling approach that simply aggregates or sums within-patch population dynamics over the various patch types displayed in a landscape will be inadequate (Wiens 1984). Spatial relationships and patch configuration must be considered in such models, perhaps capturing the redistribution process by variations of diffusion-reaction approaches that are cast in terms of the dynamics of various populations.

JABOWA-FORET Models. One class of models that has been useful in some aspects of the landscape are the forest vegetation models based on the reproduction behavior of species populations. These temporally dynamic models are based on sets of individual patches. The patches are not connected to each other, but each patch is connected to a universal pool of seeds. These models can reproduce temporal changes in a heterogeneous environment, because each patch is characterized by a specific soil, climate, elevation, slope, aspect, and disturbance regime (e.g., Weinstein 1982; Harwell and Weinstein 1982).

Other Patch Models. Other models have been developed based on a paradigm similar to the JABOWA-FORET models—that patchiness is important but that specific spatial position is not, because a common seed or larval source, or both, is available. These models include the simple Markov transition matrices describing forest secondary succession (Horn 1976) and the patch dynamics models of Levin and Paine (1974), which allow patches to change in size and which consider explicitly changes in the age-size distribution of patches in relation to disturbance. Position is not explicitly considered, except for a causal relation of successional patterns to physical characteristics. Nonequilibrium island biogeographical models (Faaborg 1979; Williamson 1981) represent an intermediate class in which some aspects of position are considered, e.g., distance to source.

Interaction-Redistribution Models. The most generically important class of models is those which examine growth, interaction, and spatial redistribution on a common time scale. These are extensions of the prototypical diffusion-advection-reaction models, which, though they have limitations (Chesson 1981), have played a fundamental role in a wide spectrum of fields (e.g., Levin 1974). In these models (possibly stochastic), local dynamics are coupled with movement either within and among a mosaic of cells

14   ILLINOIS NATURAL HISTORY SURVEY SPECIAL PUBLICATION No. 2

or patches or in a continuum. More recent work has developed the mathematical detail to sophisticated levels, including density-dependent and space-dependent movement patterns, aggregation, the effects of geometry, and interacting populations with diffusion and long-distance transport (e.g., Haggett, Cliff, and Frey 1977). Much current biological work examines the applicability of the generally used descriptors of movement, including especially insect mark-and-recapture experiments, radio tracking of vertebrates, and comparison of model predictions with data on seed, pollen, and particulate dispersal. In general, this approach has increasingly proved to be an adaptable, realistic, and accurate basis for describing movement of many organisms (e.g., Gunn and Rainey 1979; Rabb and Kennedy 1979).

Optimization Models. Optimization models that consider landscape features have been employed in integrated pest management programs operative at the regional level, controlled burning of forests, optimizing clear-cut settings, land allocation strategies, resource management, and optimal design of reserves. Optimization techniques build on a combination of the diffusion-reaction approach described earlier, dynamic programming, and control theory, which includes stochastic and econometric methods. Frequently used approaches blend ecology and economics; consider costs and benefits, such as yield; and present net worth.

Other Approaches. Not explicitly covered in this report, but of recognized potential importance and highly developed, are:

the influence of spatial heterogeneity upon the generation and maintenance of genetic diversity and the implications for the development of resistance to pesticides, heavy metals, and other anthropogenic stresses

the development of physical models (e.g., microcosms) and experimental field studies coupled with mathematical models (Physical models are extremely important when site-specific information is necessary, as contrasted with the generic approaches emphasized in this report.)

relationships to techniques in geography (It would be profitable to develop interactions among ecologists, epidemiologists, and geographers interested in similar landscape problems and analyses to develop common methodologies.)

Landscape Ecology Procedures Useful in the Solutions of Pest Management Problems

It has been more traditional to follow landscape approaches to forest pest problems (e.g., spruce budworm in North America) and to pests of man and animals (e.g., tsetse flies in Africa) than it has been in attacking agricultural problems, although there are

exceptions (e.g., locusts in Africa and Asia and beet leafhoppers in North America). In contrast, most agricultural pest management studies in the past have been site specific (i.e., they relate to one crop in one habitat). In recent years, however, emphasis on wide-area population studies has increased and greater recognition has been given the need for manipulating landscape heterogeneity (e.g., see many papers in Proceedings of the Tall Timbers Conference on Ecology and Animal Control by Habitat Management, No. 1-6, 1969-1974, Tallahassee, Florida).

Ecotones have been recognized as sources of agricultural pests (e.g., weeds, plant pathogens, invertebrate and vertebrate herbivores) and their natural enemies for many years (van den Bosch and Telford 1964; Southwood and Way 1970; and Rabb et al. 1976). The potential for enhancing natural control of agricultural insect pests through managing field borders (ditch banks, fence rows, and wood edges) has been noted by many researchers; however, little overt management has resulted. Price (1976) drew fresh attention to this potential in his attempt to apply island biogeography theory in studying fluxes of insect pests and their natural enemies in soybean fields. Stinner et al. (1974) modelled the role of spatial and temporal patterns of several crops in the population dynamics of *Heliothis zea*. They computed the movement of moths among field types as a function of an attraction index of hosts for adults for each crop-maturity combination and the average distance between fields of different types (immature stages do not disperse among fields). Agricultural landscapes can be designed to reduce *H. zea* density in the most susceptible crops.

As yet, no studies have been designed to measure and specifically evaluate for pest management purposes a significant portion of the landscape (i.e., the interaction of several habitat types). Such studies are strongly recommended. These long-term investigations would aid our understanding of long-range dispersal patterns of pest species, how the landscape mosaic impacts their life histories, and how the landscape can be managed for purposes of biological control.

BIBLIOGRAPHY

ALLEN, T. F. H., and T. B. STARR. 1982. Hierarchy: perspectives for ecological complexity. University of Chicago Press.

BATTY, D. J., and P. N. SIKDAR. 1982. Spatial aggregation in gravity models, an information-theoretic framework. Environmental Planning 14:377-407.

BELLROSE, F.C., F. L. PAVEGLIO, JR., and D. W. STEFFECK. 1979. Waterfowl populations and the changing environment of the Illinois River Valley. Illinois Natural History Survey Bulletin 32:1-54.

BURGESS, R. L., and D. M. SHARPE, eds. 1981. Forest island dynamics in man-dominated landscapes. Springer-Verlag, New York.

CHESSON, P. L. 1981. Models for spatially distributed populations: the effect of within-patch variability. Theoretical Population Biology 19:288-325.

DEN BOER, P. J. 1968. Spreading of risk and stabilization of animal numbers. Acta Biotheoretica (Leiden) 18:165-194.

DEN BOER, P. J. 1981. On the survival of populations in a heterogeneous and variable environment. Oecologia 50:39-53.

DENMAN, K. L., and T. PLATT. 1976. The variance spectrum of phytoplankton in a turbulent ocean. Journal of Marine Research 34:593-601.

DRIVER, B. L., and P. J. BROWN. 1978. The opportunity spectrum concept and behavioral information in outdoor recreation resource supply inventories: a rationale. Pages 24-31 in H. G. Lund, V. J. LaBau, P. F. Folliott, and D. Robbins, technical coordinators, Integrated inventories of renewable natural resources: proceedings of the workshop. U. S. Department of Agriculture Forest Service General Technical Report RM-55. Rocky Mountain Forest and Range Experiment Station, Fort Collins, CO.

FAABORG, J. 1979. Qualitative patterns of avian extinction on neotropical land bridge islands: lessons for conservation. Journal of Applied Ecology 16:99-107.

FAHRIG L., L. LEFKOVITCH, and G. MERRIAM. 1983. Population stability in a patchy environment. Pages 61-67 in W. K. Lauenroth, G. V. Skogerboe, and M. Flug, eds., Analysis of ecological systems: state-of-the-art in ecological modelling. Elsevier, New York.

FORD, R. G., J. A. WIENS, D. HEINEMANN, and G. L. HUNT. 1982. Modelling the sensitivity of colonially breeding marine birds to oil spills: guillemot and kittiwake populations on the Pribilof Islands, Bering Sea. Journal of Applied Ecology 19: 1-31.

FORMAN, R. T. T., ed. 1979. Pine barrens: ecosystem and landscape. Academic Press, New York.

FORMAN, R. T. T. 1981. Interaction among landscape elements: a core of landscape ecology. Pages 35-48 in Perspectives in landscape ecology. Proceedings of the International Congress, Society of Landscape Ecology, Veldhoven, Pudoc, Wageningen, Netherlands.

FORMAN, R. T. T., and M. GODRON. 1981. Patches and structural components for a landscape ecology. BioScience 31:733-740.

GULLION, G. W. 1977. Forest manipulation for ruffed grouse. North American Wildlife and Natural Resources Conference Transactions 42:449-458.

GUNN, D. L., and R. C. RAINEY, eds. 1979. Strategy and tactics of control of migrant pests. Philosophical Transactions of the Royal Society of London B Biological Sciences 287: 249-488.

HAGGETT, P., A. D. CLIFF, and A. FREY. 1977. Locational analysis in human geography. John Wiley and Sons, New York.

HANSSON, L. 1977. Landscape ecology and stability of populations. Landscape Planning 4:85-93.

HENDERSON M., G. MERRIAM, and J. WEGNER. Patchy environments and species survival: chipmunks in an agricultural mosaic. Biological Conservation (submitted).

HARWELL, M. A., and D. A. WEINSTEIN. 1982. Modelling the effects of air pollution on forested ecosystems. Ecosystems Research Center Report 6. Cornell University, Ithaca, NY.

HOLLING, C. S. 1973. Resilience and stability of ecological systems. Annual Review of Ecology and Systematics 4:1-24.

HORN, H. S. 1976. Markovian properties of forest succession. Pages 196-211 in M. L. Cody and J. M. Diamond, eds., Ecology and evolution of communities. Harvard University Press, Cambridge, MA.

HOWE, R. W. In press. Local dynamics of bird assemblages in small forest habitat islands in Australia and North America. Ecology.

ISARD, W. 1975. Introduction to regional science. Prentice-Hall, New York.

JOSELYN, G. B., J. E. WARNOCK, and S. L. ETTER. 1968. Manipulation of roadside cover for nesting pheasants—a preliminary report. Journal of Wildlife Management 32:217-233.

JOYCE, L. A., B. MCKINNON, J. G. HOF, and T. W. HOEKSTRA. 1983. Analysis of multiresource production for national assessments and appraisals. U.S. Department of Agriculture Forest Service General Technical Report RM-101. Rocky Mountain Forest and Range Experiment Station, Fort Collins, CO.

KARR, J. R. 1982a. Population variability and extinction in the avifauna of a tropical land bridge island. Ecology 63:1975-1978.

KARR, J. R. 1982b. Avian extinction on Barro Colorado Island, Panama: a reassessment. American Naturalist 119:220-239.

KARR, J. R., and K. E. FREEMARK. 1983. Habitat selection and environmental gradients: dynamics in the "stable" tropics. Ecology 64:1481-1494.

KARR, J. R., and K. E. FREEMARK. In press. Disturbance, perturbation, and vertebrates: an integrative pespective. In S. T. A. Pickett and P.S. White, eds., Natural disturbance: an evolutionary perspective. Academic Press, New York.

KLOPATEK, J. M., J. R. KRUMMEL, J. B. MANKIN, and R. V. O'NEILL. 1983. A theoretical approach to regional environmental conflicts. Journal of Environmental Management 16: 1-15.

KRUG, E. C., and C. R. FINK. 1983. Acid rain on acid soil: a new perspective. Science 221:520-525.

LEOPOLD, A. 1949. A Sand County almanac. Oxford University Press, New York.

LEVIN, S. A. 1974. Dispersion and population interactions. American Naturalist 108:207-228.

LEVIN, S. A. 1978. Pattern formation in ecological communities. Pages 433-465 in J. H. Steele, ed., Spatial pattern in plankton communities. Plenum Publishing Corporation, New York.

LEVIN, S. A., and R. T. PAINE. 1974. Disturbance, patch formation and community structure. National Academy of Sciences (USA) Proceedings 71:2744-2747.

LOWE, J. C., and S. MORYADAS. 1975. The geography of movement. Houghton Mifflin, Boston, MA.

LUDER, P. 1981. The diversity of landscape ecology. Definition and attempt at empirical identification. Angewandte Botanik 55:321-329.

MANDELBROT, B. B. 1977. Form, chance, and dimension. W. H. Freeman Press, San Francisco, CA.

MCNAUGHTON, S. J. 1979. Grassland-herbivore dynamics. Pages 46-81 in A. R. E. Sinclair and M. Norton-Griffiths, eds., Serengeti. Dynamics of an ecosystem. University of Chicago Press.

MCNAUGHTON, S. J. 1983. Serengeti grassland ecology: the role of composite environmental factors and contingency in community organization. Ecological Monographs 53:291-320.

MICHALSKI, R. S., and R. L. CHILAUNSKY. 1980. Knowledge acquisition by encoding expert rules versus computer induction from examples: a case study involving soybean pathology. International Journal of Man-Machine Studies 12:63-87.

MIDDLETON, J. D., and G. MERRIAM. 1981. Woodland mice in a farmland mosaic. Journal of Applied Ecology 18:703-710.

MIDDLETON, J. D., and G. MERRIAM. 1983. Distribution of woodland species in farmland. Journal of Applied Ecology 20: 625-644.

MINNICH, R. A. 1983. Fire mosaics in southern California and northern Baja California. Science 219:1287-1294.

NAVEH, Z. 1982. Landscape ecology as an emerging branch of human ecosystem science. Advances in Ecological Research 12:189-237.

NAVEH, Z., and A. S. LIEBERMAN. 1984. Landscape ecology. Theory and application. Springer-Verlag, New York.

NEEF, E. 1967. Die theoretischen grundlagen landschaftslehre. Verlag VEB Herman Harck, Geogrophisch-Kartogrophische Anstalt Gotha, Leipzig.

NOY-MEIR, I. 1981. Spatial effects in modelling of arid ecosystems. Pages 411-432 in D. W. Goodall, R. A. Perry, and K. M. W. Howes, eds., Arid-land ecosystems: structure, functioning and management. Cambridge University Press, Cambridge, England.

ORIANS, G. H., and N. E. PEARSON. 1978. On the theory of central place foraging. Pages 155-177 in D. J. Horn, G. R. Stairs, and R. D. Mitchell, eds., Analysis of ecological systems. The Ohio State University Press, Columbus.

PAINE, R. T., and S. A. LEVIN. 1981. Intertidal landscapes: disturbance and the dynamics of pattern. Ecological Monographs 51:145-178.

PICKETT, S. T. A., and P. S. WHITE. In press. Natural disturbance: an evolutionary perspective. Academic Press, New York.

16   Illinois Natural History Survey Special Publication No. 2

Price, P. W. 1976. Colonization of crops by arthropods: non-equilibrium communities in soybean fields. Environmental Entomology 5:605-611.

Rabb, R. L., R. E. Stinner, and R. Van den Bosch. 1976. Conservation and augmentation of natural enemies. Pages 233-254 in D. B. Huffaker and P.S. Messenger, eds., Theory and practice of biological control. Academic Press, New York.

Rabb, R. L., and G. G. Kennedy, eds. 1979. Movement of highly mobile insects: concepts and methodology in research. Proceedings of a conference, Movement of selected species of Lepidoptera in the southeastern United States. North Carolina State University, Raleigh.

Romme, W. H., and D. H. Knight. 1982. Landscape diversity: the concept applied to Yellowstone Park. BioScience 32: 664-670.

Samson, F. B., and F. L. Knopf. 1982. In search of a diversity ethic for wildlife management. North American Wildlife and Natural Resources Conference Transactions 47:421-431.

Schluter, D. 1981. Does the theory of optimal diets apply in complex environments? American Naturalist 118:139-147.

Sharpe, D. M., F. W. Stearns, R. L. Burgess, and W. C. Johnson. 1981. Spatio-temporal patterns of forest ecosystems in man-dominated landscapes of the eastern United States. Proceedings of the International Congress, Society of Landscape Ecology, Veldhoven, Pudoc, Wageningen, Netherlands.

Shugart, H. H., ed. 1978. Time series and ecological processes. SIAM. Institute of Mathematics and Society, Philadelphia, PA.

Shugart, H. H., and D. C. West. 1980. Forest succession models. BioScience 30:308-313.

Southwood, T. R. E., and M. J. Way. 1970. Ecological background to pest management. Pages 6-29 in R. L. Rabb and F. E. Guthrie, eds., Concepts in pest management. North Carolina State University, Raleigh.

Sprugel, D. G. 1976. Dynamic structure of wave-regenerated Abies balsamea forests in the north-eastern United States. Journal of Ecology 64:889-911.

Sprugel, D. G., and F. H. Bormann. 1981. Natural disturbance and the steady state in high-altitude balsam fir forests. Science 211:390-393.

Stinner, R. R., R. L. Rabb, and J. R. Bradley, Jr. 1974. Population dynamics of Heliothis zea (Boddie) and H. virescens (F.) in North Carolina: a simulation model. Environmental Entomology 3:163-168.

Stinner, R. E., R. L. Rabb, and J. R. Bradley, Jr. 1977. Natural factors operating in the population dynamics of Heliothis zea in North Carolina. Pages 622-642 in Proceedings of the XV International Congress on Entomology, August 19-27, 1976. Washington, D.C.

Sugihara, G. 1983. Peeling apart nature. Nature 304:94.

Thompson, W. A., I. B. Vertinsky, and W. G. Wellington. 1979. The dynamics of outbreaks: further simulation experiments with western tent caterpillar. Research in Population Ecology 20:188-200.

Troll, C. 1968. Landschaftsokogie. Pages 1-21 in R. Tuxen, ed., Pfanzensoziologic und landschaftsokogie. Verlag Dr. W. Junk, Der Haag.

Van den Bosch, R., and A. D. Telford. 1964. Environmental modification and biological control. Pages 459-488 in P. DeBach, ed., Biological control of insect pests and weeds. Reinhold Publishing Corporation, New York.

Van den Bosch, R., O. Bleingolea, M. Hafez, and L. A. Falcon. 1976. Biological control of insect pests of row crops. Pages 443-456 in C. B. Huffaker and P. S. Messenger, eds., Theory and practice of biological control. Academic Press, New York.

Warner, R. E. 1981. Illinois pheasants: population, ecology, distribution, and abundance, 1900-1978. Illinois Natural History Survey Biological Notes 115.

Watt, A. S. 1947. Pattern and process in the plant community. Journal of Ecology 35:1-22.

Wegner, J., and G. Merriam. 1979. Movements by birds and small mammals between a wood and adjoining farmland habitats. Journal of Applied Ecology 16:349-357.

Weinstein, D. A. 1982. A user's guide to FORNUT: a simulation model of forest dynamics and nutrient cycling during succession. Ecosystems Research Center, Cornell University, Ithaca, NY. (draft).

Wellington, W. G., P. J. Cameron, W. A. Thompson, J. B. Vertinsky, and A. S. Landsberg. 1975. A stochastic model for assessing the effects of external and internal heterogeneity on an insect population. Research in Population Ecology 17: 1-28.

Whittaker, R. H., and S. A. Levin. 1977. The role of mosaic phenomena in natural communities. Theoretical Population Biology 12:117-139.

Wiens, J. A. In press. Vertebrate responses to environmental patchiness in arid and semi-arid ecosystems. In S. T. A. Pickett and P. S. White, eds., Natural disturbance: an evolutionary perspective. Academic Press, New York.

Williamson, M. 1981. Island populations. Oxford University Press, New York.

Woodmansee, R. G. 1979. Factors influencing input and output of nitrogen in grasslands. Pages 117-134 in N. R. French, ed., Perspectives in grassland ecology. Springer-Verlag, New York.

Young, G. L. 1974. Human ecology as an interdisciplinary concept: a critical inquiry. Pages 1-105 in A. MacFadyen, ed., Advances in ecological research. Vol. 8. Academic Press, New York.

Landscape Ecology

A hierarchical perspective can help scientists understand spatial patterns

Dean L. Urban, Robert V. O'Neill, and Herman H. Shugart, Jr.

A terrestrial landscape is a mosaic of heterogeneous land forms, vegetation types, and land uses. The study of landscape—its spatial patterns and how they develop—is presently emerging as a new discipline in the field of ecology (Forman 1981, 1983, Forman and Godron 1981, 1986, Naveh and Lieberman 1984, Noss 1983, Risser et al. 1984). Landscape ecology is motivated by a need to understand the development and dynamics of pattern in ecological phenomena (Clark et al. 1978, Levin 1976a,b, 1978, Whittaker and Levin 1977, Wiens 1976), the role of disturbance in ecosystems (Mooney and Godron 1983, Pickett and White 1985, Sousa 1984, White 1979), and characteristic spatial and temporal scales of ecological events (Allen and Starr 1982, O'Neill et al. 1986).

Pattern, generated by processes at various scales, is the hallmark of a landscape. In this paper we outline an approach to landscape study that employs a hierarchical paradigm of pattern and behavior. Although our em-

Dean L. Urban is a research associate and Herman H. Shugart, Jr. is the Corcoran Professor of Environmental Sciences in the Department of Environmental Sciences, University of Virginia, Charlottesville, VA 22903. Robert V. O'Neill is a senior research ecologist in the Environmental Sciences Division, Oak Ridge National Laboratory, Oak Ridge, TN 37831. They study the development and dynamics of scaled pattern in ecosystems. © 1987 American Institute of Biological Sciences.

A landscape is a mosaic of patches, the components of pattern

phasis is on forested landscapes, we can generalize a theory of landscape ecology.

Landscape pattern and process

We will first focus on the wide range of phenomena in a natural terrestrial landscape by considering the apparent complexity of landscape dynamics and illustrating how a hierarchical paradigm lends itself to simplifying such complexity. Our perspective also affords insights into the management of man-dominated landscapes.

Development of landscape pattern. A landscape is a mosaic of patches, components of pattern. The agents of pattern formation on natural landscapes can be categorized as disturbances, biotic processes (especially the demographic processes of birth, death, and dispersal), and environmental constraints (Levin 1978). Each of these agents can be considered across a spectrum of spatial and temporal scales. For example, disturbances that affect terrestrial landscapes vary in spatial extent, recurrence interval, and intensity (Pickett and White 1985, Sousa 1984, White 1979). Disturbances range from the localized effects of an individual

death to the large-scale effects of wildfires, drought, and epidemic disease. Biotic, or regenerative, processes also vary in scale from the regrowth of an individual to the reorganization of species assemblages. Environmental constraints include microclimatic and fine-scale soil conditions governing seed germination, and also subcontinental climatic regimes that delineate biomes, such as the Eastern Deciduous Forest.

The agents of pattern formation are interwoven in landscape development. This interaction allows some sites to be especially prone to, or sheltered from, disturbances. For example, topographic position interacts with fire frequency; dry ridges burn more frequently than moist (mesic) coves. Regenerative processes are influenced by site quality, and also vary with the age and life-history attributes of the regenerating individuals (Odum 1969, Shugart and Hett 1973). Moreover, both disturbances and regeneration may be constrained by the existing spatial pattern (Curtis 1956, Forman 1981, Watt 1947). Finally, new patches are continually superimposed on existing patches (Reiners and Lang 1979). The emergent scenario is a mosaic of patches of various size, of various origins, in various stages of regeneration, approaching microenvironmental equilibria at various rates. Such complexity would seem overwhelming at first and any attempt to fully understand landscapes would appear futile.

Organization of landscape pattern. But importantly, the complexity of landscape pattern is organized in a special way: the component events and patches occur at characteristic scales that are positively correlated in time and space. Disturbances, for example, occur over a wide range of space and time scales, but those affecting a particular landscape typically can be divided into events occurring on characteristic scales. For example, Romme and Knight (1982) were able to break the fire regime in Yellowstone Park into two components: frequent, small fires (every few decades, affecting areas of less than 100 ha); and larger, less-frequent fires (every few centuries, affecting areas of several hundred ha). Similarly, Heinselman (1973) documented a two-scale fire regime in the Boundary Wa-

ters Canoe Area of Minnesota. The concept of disturbance is itself implicitly scaled; if disturbances are events that kill trees prematurely, then "disturbance" is confined to a relatively narrow window of time corresponding to the lifespan of trees (Figure 1a). Events outside this domain, while they may affect trees, are not considered disturbances. Small-scale disturbances tend to occur more frequently while larger events tend to be less frequent.

Demographic processes are also scaled. Individuals have characteristic sizes and lifespans that dictate correspondingly scaled patterns (Watt 1947). Species dispersal distances and rates define the scale of larger and longer-term patterns (Figure 1b). At

still larger scales and over longer spans of time, genetic fluxes define the domains of adaptation and evolutionary processes.

Environmental constraints are scaled by the manner in which we witness them. Atmospheric conditions can be observed at virtually any scale, but "microclimate," "weather," and "climate" connote phenomena witnessed over increasingly larger areas and longer timespans. Likewise, we could also measure soil processes at virtually any scale, but we would not measure the chemical weathering of minerals over the entire Canadian Shield (we would use a small, representative surface), nor would we observe the evolution of landforms on a single hillslope (where we would

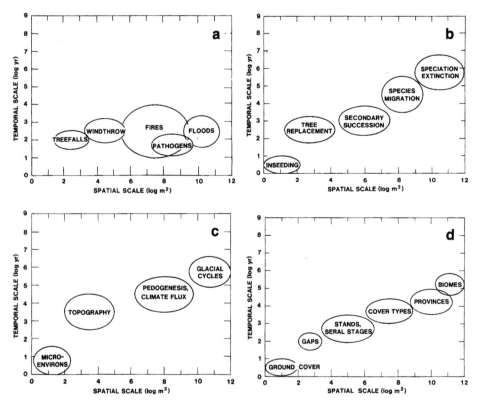

Figure 1. (a) Disturbance regimes, (b) forest processes, (c) environmental constraints, and (d) vegetation patterns, viewed in the context of space-time domains. Modified from Delcourt et al. (1983).

study erosional processes). The phenomena we study tend to be positively correlated in time and space (Figure 1c).

Vegetation patterns can be resolved on different scales (Figure 1d). Whittaker's (1953) notion of climax pattern was a two-level description of vegetation pattern that reconciled the broad-scale idea of a single stable community of plants dictated by climate (Clements 1916) with fine-scale observations of individual plant communities (Gleason 1939). Moreover, the large-scale generalization is only loosely coupled to the fine-scale details of a phenomenon. The successional schemes described by Clements may apply in general to large areas, but they do not predict with certainty what one will actually find on a particular site at a given time. This loosely coupled, multileveled organization of landscapes requires a new conceptual model. We suggest a paradigm that comes from hierarchy theory (Allen and Starr 1982, Allen et al. 1984, O'Neill et al. 1986, Pattee 1973, Whyte et al. 1969).

The paradigm

Hierarchy theory is concerned with systems that have a certain type of organized complexity. Hierarchically organized systems can be divided, or decomposed, into discrete functional components operating at different scales (Simon 1962). As applied to landscape ecology, the hierarchical paradigm provides guidelines for defining the functional components of a system, and defines ways components at different scales are related to one another (e.g., lower-level units interact to generate higher-level behaviors and higher-level units control those at lower levels). This paradigm can aid the design of studies in landscape ecology and the prediction of how external factors will alter an ecosystem.

Natural phenomena often are not perfectly decomposable: spatial boundaries may be difficult to define precisely and components may interact. Yet many complex, natural phenomena are nearly decomposable (Allen and Starr 1982, O'Neill et al. 1986, Simon 1973) and thus can be conceptualized usefully as hierarchical systems.

Hierarchical structure. Components of a hierarchical system are organized into levels according to functional scale (Figure 2). Events at a given level have a characteristic natural frequency and, typically, a corresponding spatial scale. In general, low-level events are comparatively small and fast; higher-level behaviors are larger and slower. More strictly, components of a hierarchical system may be ordered into levels according to a number of criteria (Allen et al. 1984). Higher levels may be larger than, slower than, constrain (control), or contain lower levels. In many of the hierarchies we will consider, all criteria will apply. The rules structuring hierarchies can be conveniently illustrated through an extended example.

Let us develop a hierarchy to study the species-composition dynamics of a deciduous forest system in the eastern United States. The forest gap has long been recognized as a functional unit in forest systems (Watt 1925, 1947; see also Bormann and Likens 1979, Shugart 1984). When a large tree dies, it creates a gap where subordinate trees may thrive under a regime of greater resources now available to them. A transitional stage of scramble competition ensues, until another tree grows large enough to become dominant. When this tree dies, the cycle repeats. The spatial unit of gap dynamics is equal to the area affected by the death of a canopy-dominant tree; its natural frequency reflects the lifespan of the dominant species (Shugart 1984, Shugart et al. 1981).

Trees within a forest gap interact much more among themselves, by virtue of their shared regime of available resources (especially light), than they do with trees beyond the gap. These interactions define the boundaries of a gap. By extension, a larger forest area can be decomposed into a mosaic of gap-sized patches, in which each gap undergoes its own dynamics. But the gaps are neither identical nor completely independent. The gaps comprising a mesic cove share similar species under similar growing conditions, and they exchange seeds and nutrients more often within the cove than with gaps on a nearby ridge. Again, these similarities allow us to delineate an area of characteristic size within which gaps interact at a characteristic frequency, and allow us to define stands as higher-level compo-

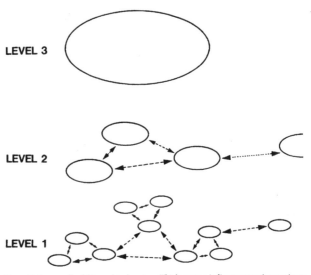

LEVEL 3

LEVEL 2

LEVEL 1

Figure 2. A generalized hierarchical system. Thick arrows indicate strong interactions; broken arrows, weak interactions.

FOUR LEVELS OF A FOREST HIERARCHY

LEVEL	BOUNDARY DEFINITION	SCALE
LANDSCAPE	PHYSIOGRAPHIC PROVINCES; CHANGES IN LAND USE OR DISTURBANCE REGIME	10000s ha
WATERSHED	LOCAL DRAINAGE BASINS; TOPOGRAPHIC DIVIDES	100s–1000s ha
STAND	TOPOGRAPHIC POSITIONS; DISTURBANCES PATCHES	1s–10s ha
GAP	LARGE TREE'S INFLUENCE	0.01–0.1 ha

Figure 3. A forested landscape as a hierarchy of gaps, stands, and watersheds.

nents of a forest hierarchy.

Moving upscale once again, we might define watersheds at the next higher level, because stands within a watershed share a similar resource base and interact more among themselves than they do with stands in other watersheds. At a still higher level, we might define landscapes as units of similar, interacting watersheds. At the landscape scale, boundaries might be coincident with large-scale physiographic features (e.g., mountain ranges) that govern weather patterns and limit frequencies of species movement. Of course, such landscapes interact as well, giving definition to still higher levels, for example, regional forest provinces, such as a spruce-fir forest.

We have constructed a four-level hierarchy to represent a forest (Figure 3). At each level, similar and interacting components become the functional aggregates at the next higher level. This is a rate-structured hierarchy, because components of one aggregate interact more frequently and intensively among themselves than with components of other aggregates. This rule defines the horizontal structure (within levels) as well as the vertical structure (between levels) of a hierarchical system. Interactions among components at one level generate the behaviors of a component at the next higher level. A gap has its own internal dynamics, but it also contributes to the behavior of a stand. In turn, a stand's behaviors are not only its own, but also are a part of watershed function. Each patch, at any level, is at once an integral whole and a part

of a higher-level component (Koestler 1967).

Landscapes have a special kind of vertical structure: they are nested spatially. Each level of the hierarchy contains the levels below it. This property of containment provides for a number of special features (Allen and Hoekstra 1984). Levels in a nested hierarchy may coincide when defined by a variety of ordering criteria or measured attributes. Forest gaps might be recognizable in terms of species composition, biomass, ambient sunlight (insolation), or other attributes. They might also be recognizable in terms of the limited basic resources that constrain tree growth. Thus, the hierarchy is also ordered on the criterion of constraint.

Indeed, it seems that constraint and interaction can be mutually reinforcing as ordering criteria: patches delineated by a spatially distributed constraint (e.g., topographic moisture) may interact to generate higher-level aggregates. Forest stands defined on topographic moisture may be joined by seed dispersal to generate an interacting landscape mosaic. As a further consequence, note that because landscapes contain stands, higher levels of this hierarchy are larger than lower levels; also, because stands interact more within themselves than among themselves, higher levels are slower than lower levels. This special relationship among ordering criteria in a landscape hierarchy makes its structure very robust. The hierarchical levels are often evident as patches in nature.

Each hierarchy is constructed in

relation to a specified phenomenon of interest. Different phenomena may call for differences in the hierarchies we use to study them. While the hierarchy, gap-stand-watershed-landscape, might be appropriate for a study of either species composition or nutrient cycling in forests, these levels need not be relevant for other purposes. For a landscape dominated by recurrent fires, hierarchical levels corresponding to the scales of individual fires and to the larger fire-mosaic might be a more appropriate conceptualization. For a study of forest birds, an obvious focal scale would be territories. The higher levels, for example, the forest stand, may or may not be suitable to studies of fires and birds. Thus, the species-composition hierarchy for a forest system is not the only one possible.

Mechanistic explanation. The behavior of a forest gap through time depends on the individual trees in the gap. Their growth rates, lifespans, shade tolerance, response to moisture and nutrient levels, and initial sizes collectively determine which trees come into dominance and which trees are suppressed. Gaps, in turn, interact to generate stand dynamics. Whether sugar maple (*Acer saccharum*) will become more important than white oak (*Quercus alba*) in a particular stand may depend on the interaction of nearby gaps. If oak-dominated gaps tend to become maple-dominated gaps as a result of seed rain from nearby maples, stand composition will shift toward maple. This general rule carries upscale: stands, then watersheds, then landscapes interact to generate successively higher-level behaviors.

We can go a long way toward understanding a complex phenomenon when we explain its behavior in terms of interactions among its parts. But understanding a hierarchical phenomenon requires more than mechanisms. Understanding requires that the mechanisms be considered in context.

Constraint and higher-level context. Specific conditions within the gap influence the processes, such as tree growth and longevity, that generate gap dynamics. The pattern of available light in the gap, for example, provides a context that constrains individual tree growth, usually limiting replacement of a dominant tree to shade-tolerant species. This knowledge may not, however, allow us to predict with certainty which shade-tolerant species will come into dominance: it might be sugar maple, or beech *(Fagus grandifolia)*, or basswood *(Tilia americana)*, or some other equally tolerant species. The greater the functional redundancy among trees, the less predictable is case-specific behavior. The constraint in effect judges all trees by a single standard.

Some other constraints on tree growth are soil moisture levels, nutrient availability, and disturbance regimes such as fires or wind. All these constraints affect the trees within the gap, favoring some over others. The constraints sort individually since trees differ individually with respect to such characteristics as fire tolerance, drought tolerance, and ability to withstand wind.

The factors producing these constraints are themselves patterned on some spatial scale. Topographic pattern governs soil factors, insolation, and moisture and defines the characteristic spatial scale of these constraints. Each of these constraints provides a context for the behaviors of the lower levels of the hierarchy.

Nested hierarchical organization further specifies the context for lower-level behaviors. Returning to the example of tree species replacement in gap dynamics, we might not be able to specify precisely which tree will come into dominance, but we can predict that it will be a species that is well-represented in the local seed pool and that has not been excluded by the other constraints acting on that particular site. Collectively, the various constraints provide a context that allows us to make sense of what we observe in the forest at a given time, a contextual explanation. In general, the more constraints we consider at one level that are relevant to one criterion, the greater our predictive power.

Thus, to understand a complex, hierarchically organized system we must consider multiple levels. The reference level is the scale on which the phenomenon is witnessed as an interesting event. Once specified, the event has its mechanistic explanation at the next lower level, and its significance in the context of higher-level constraints (O'Neill et al. 1986).

Hierarchy as an analytic tool. Simultaneously considering a complex ecological system on many scales is an intimidating prospect that is made simpler by the manner in which patterns translate across hierarchical levels. Hierarchy isolates the phenomenon of interest. In a forest, a tree integrates fine-scale variation in its physical environment, experiencing only a blend. For example, a tree's growth ring is an integration of growing conditions during the year, and it is an instant cross-section of any longer-term trends. In addition, a tree experiences large patterns as a relatively constant context. Each tree within a gap is not subject to the full range of soil patterns in a watershed, but only to the small sample of such patterns that constitute the local condition.

A component at a given level of a hierarchy experiences as variable only those patterns that are similarly scaled in rate, as well as in size. By comparison lower-level dynamics are so fast that they are experienced as average values; higher-level dynamics are too slow to be experienced as variable. Thus, the complexity of dynamics and spatial patterns at several scales is resolved into a few variables and a set of constants, defined relative to the reference level.

A powerful consequence of a hierarchical paradigm is that it allows one to focus on an event at a particular scale, while recognizing that there are other scales relevant to that event. When describing an event, its characteristic scale dictates an appropriate sampling scale and frequency. The scale of an event defines the observational level through which the rest of the hierarchy is accessed. The next lower level provides the components of the event and its mechanistic explanation. Higher levels provide the context that gives the event greater significance. These higher-level factors can be treated as constants when viewed from the reference level, though they may be quite variable at larger scales.

Levels of the hierarchy further removed from the reference level are not of immediate concern, in that they contribute very little toward understanding the event of interest. For example, in describing gap dynamics, we know that safe sites for germination are important to seedling establishment. However, we need not consider these details at the level of the gap, because within gaps safe sites occur with predictable frequency. We also know that the specific forest exists because the post-Pleistocene climate favors trees, but this knowledge has little bearing on the event at hand.

To recap, a hierarchical perspective would emphasize three strategic concerns in an analysis of landscape pattern: (1) to detect pattern and define its spatial and temporal scale, which is to define functional patches at a specific level; (2) to infer which factors generate the pattern, whether they be demographic processes, environmental constraints, disturbances, or a combination of these; and (3) to relate this pattern to adjacent levels. We have emphasized that the notions of mechanistic interaction and constraining context explicitly involve multiple levels of reference. In the next section we pursue a variation on this theme.

Consequences of landscape pattern

We have seen that the apparent complexity of landscapes can be partially resolved by decomposing them into a hierarchical framework. It is when one considers landscape phenomena at different levels that the consequences of pattern at characteristic scales emerge. These consequences are far reaching; reviews by Levin (1976a) and Wiens (1976) illustrate the scope of the subject. One fundamental consequence of hierarchical structure is that events causing pattern on one scale can be incorporated into higher-level behavior (O'Neill et al. 1986). Through this incorporation, effectively nonequilibrium dynamics or spatial heterogeneity at one scale can be translated to equilibrium or constancy at a higher level.

There are two aspects of incorporation that are especially pertinent to landscape ecology. The first concern is whether the biological mechanisms necessary to a pattern disturbance at a given scale are available in the system; that is, do the mechanisms exist? Obviously a logical consequence of patchiness would be the evolution of mechanisms to deal with it. Both plants and animals have evolved diverse tactics to utilize patchily distributed resources. One strategy for dealing with patchiness is to employ superior competitive ability to persist within a given patch, maybe even exerting some control over the environment. An alternative strategy is to concede competitive advantage by playing the role of fugitive on a larger scale.

The cherry (*Prunus* spp.), which requires open sites for regeneration, is a fugitive in both time and space (Auclair and Cottam 1971, Marks 1974). Cherry seeds may be dispersed great distances by birds, and at the site of deposition, the seeds may remain viable for several decades. Thus, cherry can take advantage of an appropriate regeneration site elsewhere, or it can await the recurrence of a disturbance that brings an appropriate site to it. At the level of gap dynamics, cherry is a loser: it is eventually replaced by longer-lived, shade-tolerant species. But the cherry maintains a nearly constant abundance

over the landscape, although not at specific local sites. Cherry wins the war, though it loses every battle. Thus the dynamics of regional patchiness are incorporated at a higher level.

That species assemblages in nature can be categorized into types (constant assemblages) at a variety of scales (e.g., Braun 1950, Holdridge 1967, von Humboldt 1807) suggests that many natural patterns are incorporated by ecological systems. In particular, a disturbance regime that can be incorporated is not disturbing at all (Allen and Starr 1982). At a higher level, disturbance frequency can be viewed as a constraint that governs an equilibrium species assemblage. An analysis of prairie plant species composition in response to fire illustrates this two-level approach. Allen and Wyleto (1983) used the "time since fire" variable to emphasize the successional dynamics of plant species composition following a burn. Using the "fire frequency" as a variable, they demonstrated that different equilibrium species assemblages were maintained under different fire regimes. In effect, data transformation

changed the level of resolution in the data, and accessed a hierarchical system at two different levels (Allen et al. 1984).

Incorporation can be passive; a disturbance is incorporated simply by increasing the scale of reference. In the example of cherry abundance, the cherry does not affect the regional disturbance regime. In other cases, adaptive mechanisms may evolve such that the disturbance regime is actually modified. Many fire-adapted systems evolve such that the member species produce volatile substances and serotinous seeds, which require the heat of a fire to germinate. These species both exert some control over fire frequency and capitalize on fires for episodic regeneration (Mutch 1970). This incorporation with evolutionary integration has repercussions that we will discuss later.

A second aspect of incorporation pertinent to landscape ecology concerns whether incorporation can be realized within a particular bounded system. That is, given a geographically defined region (e.g., a park) and the perturbation affecting it, is the region of sufficient scale to incorporate the disturbance? Shugart and West (1981) have addressed this idea, using a forest simulation model. In their simulations, individual trees suffered stochastic, age-related mortality, and each model plot (0.08-ha gap) was independent of other plots. They found that 50 model plots were necessary to stabilize the statistical variance in biomass associated with gap dynamics, that is, to incorporate gap dynamics using biomass as a standard.

Shugart and West then compared several bounded landscapes to the scale of their disturbance regimes. They defined a quasi-equilibrating landscape as one in which the area ratio of bounded landscape to disturbance regime was at least 50:1. Smaller landscapes were called effectively nonequilibrating (Figure 4).

A bounded landscape that is large enough to incorporate the factors that disturb its component patches has a constant frequency distribution of patches of all types at all times, and is considered to be a strictly equilibrating landscape. A smaller landscape that is unable to incorporate a disturbance has a transient frequency distribution of patch types, which changes

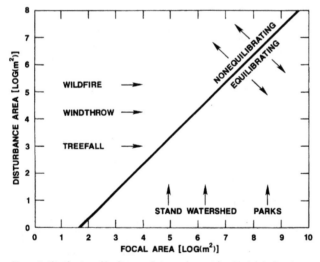

Figure 4. Classification of landscapes relative to the spatial scale of their disturbance regimes, according to a 50:1 ratio as calculated for forest biomass dynamics (Shugart and West 1981). A forest stand may incorporate single treefalls but not larger windthrows or fires; a small watershed can incorporate treefalls but perhaps not larger windthrows, and not wildfire; a larger park may incorporate all of these disturbances.

in response to each disturbance event. These are called nonequilibrating.

The implication is that the relative abundance of each patch type can be predicted readily at the higher level, if the lower-level dynamics can be incorporated. In a national park that is large enough to incorporate wildfire, a constant and predictable proportion of fire-successional vegetation (perhaps critical habitat for some species) is maintained. Conversely, a park that is too small to incorporate a natural wildfire regime does not lend itself to straightforward predictions of the relative abundance of patch types.

The phenomenon of incorporation is a natural extension of pattern at characteristic scales (O'Neill et al. 1986). This perspective is especially illuminating when applied to man-dominated landscapes.

Man-dominated landscapes

There is a tendency to view man-dominated landscapes as being different from natural landscapes. Excellent discussions of human impacts on landscapes are available (Burgess and Sharpe 1981, Forman 1981, Forman and Godron 1981 and 1986, Mooney and Godron 1983). We focus here on man's influence on the characteristic scales of landscape phenomena.

Effects of anthropogenic scaling. A primary influence of man is to rescale patterns in time and space (Figure 5). Human control of forest fires illustrates several ramifications of this rescaling. Fire suppression retards the natural frequency of burns in systems that have incorporated fire. When wildfires do occur as a result of fuel buildup, they may escape to burn over a larger area and at a greater intensity than they would otherwise. A thick-barked tree that could survive a low-intensity fire might succumb to a hotter fire. A species with seeds that require episodic fires in order to regenerate might decrease in regional abundance because fire suppression removes opportunities for germination. In each case, an incorporating mechanism has been short-circuited.

Under the constraint of periodic burns, the role of a fugitive, like the cherry, might have a winning strategy at the landscape scale. With fire suppression, the constraining rules

ANTHROPOGENIC EFFECTS ON LANDSCAPES

HUMAN ACTIVITY	CONSEQUENCES
RESCALE PATCH DYNAMICS	RENDER ADAPTIVE MECHANISMS LESS EFFECTIVE
	CHANGE CONSTRAINING RULES
	ALTER PATCH INTERACTIONS
RESCALE BOUNDED REGIONS	REDEFINE FROM EQILIBRATING TO NONEQUILIBRATING STATE
INTRODUCE NOVEL PATCHES AND DYNAMICS	RENDER ADAPTIVE MECHANISMS LESS EFFECTIVE
	REDUCE POTENTIAL FOR SPECIES TO EVOLVE ADAPTIVE MECHANISMS
HOMOGENIZE PATTERNS THROUGH LAND USE	REDUCE TREE SPECIES DIVERSITY
	REDUCE HABITAT DIVERSITY FOR FOREST WILDLIFE

Figure 5. Summary of the effects of anthropogenic rescaling of natural landscape patterns and processes.

change and that role may no longer be advantageous. Finally, small burns in a fire mosaic might depend on an immediate seed rain from adjacent unburned forest in order to regenerate. Rescaling the burns to cover larger areas might decrease the influx of seeds, slowing the regenerative dynamics of the mosaic. Though these effects are not independent, the specific actions of rescaling are (1) to render natural incorporating mechanisms less effective; (2) to change the set of constraints (including disturbance frequencies) governing lower-level biotic processes; and (3) to change the degree of interaction among patches, thus altering behaviors that influence higher levels.

Man also rescales natural regions by establishing new boundaries. Pipelines, drainage canals, and roads all set new bounds if they are effective barriers to patch interactions, especially species dispersal. This is critical when the scale is redefined relative to the scale at which disturbances can be incorporated. In such cases, an equilibrating system may be rescaled to a nonequilibrating state. Forest fragments in the eastern United States illustrate this effect. Some fragments are large enough to incorporate disturbances; most are not (Pickett and Thompson 1978).

Man also introduces novel perturbations that might differ in spatial or temporal scale from natural regimes. For example, the spatial scale and dynamics of human land use may be different from any natural forest process. Many changes in land use cover large areas but are frequent or chronic, contrary to the natural rule of large/slow or small/fast. Man-dominated landscapes may change according to such nonecological factors as price of commodities or transfers of land ownership. One would expect that such anthropogenic regimes would disrupt the natural system, leaving only behaviorally plastic species. This is consistent with the frequent association of generalists and weeds with man-dominated regions.

Man's activities at some scales may homogenize a forest stand's fine-scale patterns that result from gap dynamics. Chronic use of woodlots for grazing or as a fuelwood source can obliterate natural patterns in regeneration, so that the entire woodlot assumes a high degree of similarity. Such an effect can be indirectly imposed by natural edge effects in very small woodlots. In small woodlots, increased insolation and convection can alter the physical environment to such an extent that natural gap-phase replacement mediated by shade and moisture is not expressed. Very small woodlots do not develop an interior of mesic species; they remain essentially all edge (Levenson 1981). This has an obvious effect on local and regional forest diversity (Noss 1983) but may have further consequences. The regional abundance of many forest birds and small mammals may depend on a continual availability of specific forest microhabitats (Seagle et al. 1984, Whitcomb et al. 1981). Man's homogenizing effect may thus contribute to a regional decline in forest microhabitat specialists.

Man's various effects on landscape pattern are neither exclusive nor independent but are typically interactive and cumulative. A forest fragment has an imposed size. It may have its component events rescaled and its internal patterns altered. It may be operating under a new or rescaled set of higher-level constraints. Each of these factors contributes to the confounded and confounding behavior of woodlots (Burgess and Sharpe 1981, Curtis 1956, 1959).

Prescriptive scaling in land management. A knowledge of the characteristic spatial scale and natural frequency of patch dynamics on a landscape lends itself to prescriptive applications in natural resource management. In general, resource management should be scaled to mimic natural patch dynamics, so as to take advantage of preselected adaptive mechanisms in the local species pool. Foresters use prescriptive scaling when they mimic natural disturbances with clearcuts, a practice that represents the collective wisdom of generations of foresters, who have found a successful clearcutting strategy through trial and error. A hierarchical approach dictates the same strategy deductively. It seems likely that the mimicry approach to resource management should prove useful in other land use applications.

Another method of prescriptive scaling might capitalize on the notion that lower-level interactions are propagated upward in a hierarchy to generate higher-level behaviors. This suggests that management could be tailored such that a minimal amount of management, at the proper time and place, could be amplified at higher levels to have maximal effect. One might envision steering long-term, large-scale forest dynamics by

thoughtfully cutting or planting just a few trees.

A third mode of prescriptive scaling proceeds from our discussion of incorporation. We suggest that an inherently nonequilibrium landscape that cannot incorporate its internal dynamics can be equilibrated by rescaling its internal dynamics to effect smaller patches. Thus, prescribed burning of small patches could reinstate fire into a park that could otherwise not incorporate wildfire. In rangelands, controlled rotation of grazing pressure in small paddocks mimicks but rescales the natural regime of far-ranging herbivores, effectively creating rangeland microcosms. It seems fruitful to attempt to generalize these familiar examples to other systems in managed landscapes.

Experimental landscapes. Man-dominated landscapes can provide natural experiments from which we can learn a great deal about ecological scaling in natural systems. Human land-use patterns may be more variable than many natural environmental patterns, because human land use reflects not only natural constraints (Bowen and Burgess 1981) but also the financial resources and personal whims of private landowners. Thus, these landscapes often provide a spectrum of anthropogenic patches of various sizes within the same area (e.g., Ambuel and Temple 1983, Burgess and Sharpe 1981, Forman et al. 1976). Such landscapes can provide the necessary empirical observations from which to infer critical thresholds of interaction among components of hierarchical phenomena. Specifically, studies of man-modified landscapes may indicate how inter-patch distance, connectivity, and spatial configuration modify patch interactions to generate higher-level behaviors in mosaics; what the minimum difference in scale might be for a perturbation to be incorporated at the next higher level; and whether these rules are constant or vary systematically for different kinds of phenomena. We suggest that general answers to such questions will form the basis of a theory of landscape ecology.

Conclusion

The purpose of a paradigm is to serve

as a conceptual and analytic model to be exercised as long as it is useful. That hierarchy theory is conceptually appropriate to landscape ecology should be apparent from our discussion. We have stated very little that is actually novel: we have merely rephrased familiar notions in terms of patterns at characteristic scales. Indeed, several early classics (e.g., Tansley 1935, Watt 1947, Whittaker 1953) and a host of more recent efforts (Allen and Starr 1982) embrace ideas that are implicitly hierarchical. Hierarchy theory takes this conceptually good fit and pushes it deductively toward new insights into complex, natural phenomena. As a nascent interdisciplinary endeavor, landscape ecology can benefit from a hierarchy theory as a unifying conceptual and analytic framework.

Acknowledgments

This research was sponsored by the National Science Foundation's Ecosystem Study Program, under Interagency Agreement No. BSR-8103181-A02 with the US Department of Energy, under Contract No. DE-AC05-840R21400 with Martin Marietta Energy Systems, Inc. Publication No. 2807, Environmental Sciences Division, ORNL. We appreciate the helpful comments of T. F. H. Allen, B. Milne, D. Miller, P. Delcourt, D. DeAngelis, and A. King.

References cited

Allen, T. F. H., and T. W. Hoekstra. 1984. Nested and non-nested hierarchies: a significant distinction for ecological systems. Pages 175–180 in A. W. Smith, ed. *Proceedings of the Society for General Systems Research. I. Systems Methodologies and Isomorphies.* Intersystems Publ., Coutts Lib. Serv., Lewiston, NY.

Allen, T. F. H., T. W. Hoekstra, and R. V. O'Neill. 1984. Interlevel relations in ecological research and management: some working principles from hierarchy theory. *USDA Forest Service Gen. Tech. Rep. RM-110.* Rocky Mountain Forest and Range Experiment Station, Fort Collins, CO.

Allen, T. F. H., and T. B. Starr. 1982. *Hierarchy: Perspectives for Ecological Complexity.* University of Chicago Press, Chicago.

Allen, T. F. H., and E. P. Wyleto. 1983. A hierarchical model for the complexity of plant communities. *J. Theor. Biol.* 101: 529–540.

Ambuel, B., and S. A. Temple. 1983. Area-dependent changes in the bird communities and vegetation of southern Wisconsin woodlots. *Ecology* 64: 1057–1068.

Auclair, A. N., and G. Cottam. 1971. Dynamics of black cherry (*Prunus serotina* Erhr.) in southern Wisconsin oak forests. *Ecol. Monogr.* 41: 153–177.

Bormann, F. H., and G. E. Likens. 1979. *Pattern and Process in a Forested Ecosystem.* Springer-Verlag, New York.

Bowen, G. W., and R. L. Burgess. 1981. A quantitative analysis of forest island pattern in selected Ohio landscapes. ORNL/TM-7759. Oak Ridge National Laboratory, Oak Ridge, TN.

Braun, E. L. 1950. *Deciduous Forests of Eastern North America* (1974 facsimile edition). Hafner Press, New York.

Burgess, R. L., and D. M. Sharpe, eds. 1981. *Forest Island Dynamics in Man-dominated Landscapes.* Springer-Verlag, New York.

Clark, W. C., D. D. Jones, and C. S. Holling. 1978. Patches, movements, and population dynamics in ecological systems: a terrestrial perspective. Pages 385–432 in J. S. Steele, ed. *Spatial Pattern in Plankton Communities.* Plenum Press, New York.

Clements, F. E. 1916. Plant succession: an analysis of the development of vegetation. *Carnegie Institute Publ.* 242, Washington, DC.

Curtis, J. T. 1956. The modification of mid-latitude grasslands and forests by man. Pages 721–736 in W. L. Thomas ed. *Man's Role in Changing the Face of the Earth.* University of Chicago Press, Chicago.

———. 1959. *The Vegetation of Wisconsin: An Ordination of Plant Communities.* University of Wisconsin Press, Madison.

Delcourt, H. R., P. A. Delcourt, and T. Webb III. 1983. Dynamic plant ecology: the spectrum of vegetation change in space and time. *Quat. Sci. Rev.* 1: 153–175.

Forman, R. T. T. 1981. Interactions among landscape elements: a core of landscape ecology. Pages 35–48 in S. P. Tjallingii and A. A. deVeer, eds. *Perspectives in Landscape Ecology.* Pudoc, Wageningen, the Netherlands.

———. 1983. An ecology of the landscape. *BioScience* 33: 535.

Forman, R. T. T., A. E. Galli, and C. F. Leck. 1976. Forest size and avian diversity in New Jersey woodlots with some land use implications. *Oecologia* 26: 1–8.

Forman, R. T. T., and M. Godron. 1981. Patches and structural components for a landscape ecology. *BioScience* 31: 733–740.

———. 1986. *Landscape Ecology.* John Wiley & Sons, New York.

Gleason, H. A. 1939. The individualistic concept of the plant association. *Am. Midl. Nat.* 21: 92–110.

Heinselman, M. L. 1973. Fire in the virgin forests of the Boundary Waters Canoe Area, Minnesota. *Quat. Res.* 3: 329–382.

Holdridge, L. R. 1967. *Life Zone Ecology.* Tropical Science Center, San Jose, Costa Rica.

Koestler, A. 1967. *The Ghost in the Machine.* Macmillan Publ., New York.

Levenson, J. B. 1981. Woodlots as biogeo-
graphic islands in southeastern Wisconsin. Pages 13–39 in R. L. Burgess and D. M. Sharpe, eds. *Forest Island Dynamics in Man-dominated Landscapes.* Springer-Verlag, New York.

Levin, S. A. 1976a. Population dynamic models in heterogeneous environments. *Ann. Rev. Ecol. Syst.* 7: 287–310.

———. 1976b. Spatial patterning and the structure of ecological communities. Pages 1–36 in S. A. Levin, ed. *Some Mathematical Questions in Biology. VII. Lectures on Mathematics in the Life Sciences,* vol. 8. American Mathematical Society, Providence, RI.

———. 1978. Pattern formation in ecological communities. Pages 433–465 in J. S. Steele, ed. *Spatial Pattern in Plankton Communities.* Plenum Press, New York.

Marks, P. L. 1974. The role of pin cherry (*Prunus pensylvanica*) in the maintenance of stability in northern hardwood ecosystems. *Ecol. Monogr.* 44: 73–88.

Mooney, H. A., and M. Godron. 1983. *Disturbance and Ecosystems.* Springer-Verlag, New York.

Mutch, R. W. 1970. Wildland fires and ecosystems: a hypothesis. *Ecology* 51: 1046–1051.

Naveh, Z., and A. S. Lieberman. 1984. *Landscape Ecology: Theory and Application.* Springer-Verlag, New York.

Noss, R. F. 1983. A regional landscape approach to maintain diversity. *BioScience* 33: 700–706.

Odum, E. P. 1969. The strategy of ecosystem development. *Science* 164: 262–270.

O'Neill, R. V., D. L. DeAngelis, J. B. Waide, and T. F. H. Allen. 1986. *A Hierarchical Concept of the Ecosystem.* Princeton University Press, Princeton, NJ.

Pattee, H. 1973. *Hierarchy Theory.* George Braziller, New York.

Pickett, S. T. A., and J. N. Thompson. 1978. Patch dynamics and the design of nature reserves. *Biol. Conserv.* 13: 27–37.

Pickett, S. T. A., and P. S. White, eds. 1985. *The Ecology of Natural Disturbance and Patch Dynamics.* Academic Press, Orlando, FL.

Reiners, W. A., and G. E. Lang. 1979. Vegetation patterns and processes in the balsam fir zone, White Mountains, New Hampshire. *Ecology* 60: 403–417.

Risser, P. G., J. R. Karr, and R. T. T. Forman. 1984. Landscape ecology: directions and approaches. *Illinois Natural History Survey Special Publ. No. 2.* Illinois Natural History Survey, Champaign.

Romme, W. H., and D. H. Knight. 1982. Landscape diversity: the concept applied to Yellowstone Park. *BioScience* 32: 664–670.

Seagle, S. W., H. H. Shugart, and D. C. West. 1984. Habitat availability and animal community characteristics. ORNL/TM-8864, Oak Ridge National Laboratory, Oak Ridge, TN.
Shugart, H. H., Jr. 1984. *A Theory of Forest Dynamics.* Springer-Verlag, New York.

Shugart, H. H., Jr., and J. M. Hett. 1973. Succession: similarities of species turnover rates. *Science* 180: 1379–1381.

Shugart, H. H., Jr., and D. C. West. 1977. Development of an Appalachian deciduous forest succession model and its application to the assessment of the impact of the chestnut blight. *Environ. Manage.* 5: 161–179.

———. 1981. Long-term dynamics of forest ecosystems. *Am. Sci.* 69: 647–652.

Shugart, H. H., Jr., D. C. West, and W. R. Emanuel. 1981. Patterns and dynamics of forests: an application of simulation models. Pages 74–94 in D. C. West, H. H. Shugart, Jr., and D. B. Botkin, eds. *Forest Succession: Concepts and Application.* Springer-Verlag, New York.

Simon, H. A. 1962. The architecture of complexity. *Proc. Am. Philos. Soc.* 106: 467–482.

———. 1973. The organization of complex systems. Pages 3–27 in H. H. Pattee, ed. *Hierarchy Theory.* George Braziller, New York.

Sousa, W. P. 1984. The role of disturbance in natural communities. *Annu. Rev. Ecol. Syst.* 15: 353–391.

Tansley, A. G. 1935. The use and abuse of vegetational concepts and terms. *Ecology* 16: 284–307.

von Humboldt, A. 1807. *Ideen zu einer Geographic der Pflanzen nebst einem Naturgemälde der Tropenländer.* Bey F. G. Cotta, Tübingen, FRG.

Watt, A. S. 1925. On the ecology of British beech woods with special reference to their regeneration. II. The development and structure of beech communities on the Sussex Downs. *J. Ecol.* 13: 27–73.

———. 1947. Pattern and process in the plant community. *J. Ecol.* 35: 1–22.

Whitcomb, R. F., C. S. Robbins, V. F. Lynch, B. L. Whitcomb, M. K. Klimkiewicz, and D. Bystrak. 1981. Effects of forest fragmentation on avifauna of the eastern deciduous forest. Pages 125–205 in R. L. Burgess and D. M. Sharpe, eds. *Forest Island Dynamics in Man-dominated Landscapes.* Springer-Verlag, New York.

White, P. S. 1979. Pattern, process, and natural disturbance in vegetation. *Bot. Rev.* 45: 230–299.

Whittaker, R. H. 1953. A consideration of climax theory: the climax as a population and pattern. *Ecol. Monogr.* 23: 41–78.

Whittaker, R. H., and S. A. Levin. 1977. The role of mosaic phenomena in natural communities. *Theor. Popul. Biol.* 12: 117–139.

Whyte, L. L., A. G. Wilson, and D. Wilson, eds. 1969. *Hierarchical Structures.* Elsevier Science Publ., New York.

Wiens, J. A. 1976. Population responses to patchy environments. *Annu. Rev. Ecol. Syst.* 7: 81–120.

3. Biocybernetic Perspectives Of Landscape Ecology And Management

Zev Naveh

Technion, Israel Institute of Technology
Haifa 32 000
Israel.

INTRODUCTION - SOME RELEVANT TRANS-DISCIPLINARY PREMISES OF GENERAL SYSTEMS THEORY

One of the major challenges for landscape ecology is to fuse its different emerging approaches into one broad stream, to form a unified theoretical and methodological framework for a trans-disciplinary science that is oriented to both problem-inquiry and problem-solving. But as a first prerequisite, landscape ecologists must beaware of being infected by what Boulding has called the most contagious disease of Academia, namely specialized deafness. Of great importance in this respect is general systems theory, a trans-disciplinary metatheory by which we may be able to overcome academic and professional barriers and link quantitative and formal approaches with qualitative and descriptive approaches. These ideas have been developed extensively in Naveh and Lieberman (1984).

The basic paradigm of general systems theory is a view of the hierarchical organization of nature as ordered wholes of multi-levelled, stratified open systems, with each higher level having additional emerging qualities (Laszlo, 1972). In landscape ecology we can recognize the total human ecosystem as the highest ecological level, above the population, community and ecosystem level, integrating in space and time the biosphere and the man-made technosphere with the geosphere in concrete, voluminous patches of nature or landscape units.

An important development in the epistemology of this systems hierarchy is the holon concept (from the Greek *holos* = whole + *proton* = part), coined by Koestler (1969) in order to overcome the contradiction between holistic and atomistic perceptions of systems. Within this hierarchy each intermediate level has a "Janus faced" dichotomic nature: it functions as a self-contained whole toward its sub-system components, but at the same time also as a dependent part toward its super-system. It has, therefore, both an assertive and an integrative tendency. Such a hierarchically organized holon whole (or "holarchy") cannot be reduced to its elementary parts, but only dissected into the constituent branches of this holarchy. Taking into consideration these concepts, we can now define landscapes by substituting for the narrow and rigid, one-dimensional, spatial definitions of the hierarchy a more flexible, holistic and multi-dimensional holarchy, as follows: landscapes in their totality are physical, ecological and geographical ordered wholes, integrating all natural and cultural patterns and processes along multi-dimensional spatial, temporal and conceptual scales. As such they are holons of the total human ecosystem with increasing complexity from the ecotope as the smallest and simplest landscape holon to the ecosphere as the largest and most complex global one.

Another important epistemological development in systems theory relevant to this discussion is the recognition of natural ordered wholes, and therefore also of landscapes, as self-transcendent natural Gestalt systems, and their description by natural and for-

24

mal languages. They have the capability to represent not only objects but themselves, and their uniqueness can be described only by homology with the help of other Gestalt systems such as natural (everyday) language. Their formal openness to the flow of energy/matter and information can be described by analogy and with the help of the formal "scientific" language of logical, graphical, or mathematical symbols. But the self-transcendent openness of such natural Gestalt systems cannot be fully grasped by their formal openness. In other words, only if the means of representation are themselves self-transcendent can we represent other self-transcendent objects. Our direct and ultimate means of representation is our use of language. This is the "organ" of consciousness (the perception of perception) and the most direct and accessible means of both cognitive and affective representation.

Pankow (1976), the German philosopher, has described in detail these concepts in an important book edited by Jantsch and Waddington (1976). He used self-transcendence as the conceptual bridge for inter- and trans-disciplinarity. He stated:

"Self-transcendence becomes the common interdisciplinary beginning and end for the humanistic, as well as the natural sciences. Interdisciplinarity through self-transcendence does not require the formalization of disciplines, but unifies the disciplines while preserving the variety of their way for thinking and speaking (points or angles of view)."

In fact, landscape ecology has to deal not only with formal openness in the study, appraisal, planning and management of our total human ecosystem landscapes, but also with self-transcendent openness in the perception and evaluation of their scenic, historic, educational and other intrinsic cultural values. For this reason it has to become a unified, natural and humanistic science. Presently such a unification has been made much easier to achieve with the help of mathematical fuzzy set theory as applied in advanced computer programs of expert systems, based on knowledge engineering and artificial intelligence. Here are their most salient features, as lucidly defined by Negoita (1985):

"Knowledge engineering is a discipline devoted to integrating human knowledge as "knowledge-based" computer systems. Its processes are not hard-coded but state-driven knowledge, telling how the data of a problem can be manipulated to solve it. By internalising this procedural knowledge of the world, the machine becomes intelligent. An expert system is an information system that can pose and answer questions relating to information borrowed from human experts and stored in the systems knowledge base."

Usually, knowledge is derived in linguistic form from experts. The theory of fuzzy sets offers a way to incorporate subjective evaluations in knowledge bases. A linguistic variable differs from a numeric variable in that its values are not numbers but words. Fuzzy set theory is therefore a tool to achieve the goal of having precisely manipulable natural language expressions. The key idea in fuzzy set theory is that an element need not be simply true or false as in an ordinary set with a binary choice, but may be partly true to any degree. It has a degree of membership in a fuzzy set function as a real number in the interval [0,1]. Such a function allows a continuum of possible choices and can be used to describe imprecise terms, such as "old". For example, the degree of membership of an age of seventy years or greater is 1.0 and that of a "partially" old age of sixty is 0.7. In this way the vagueness of the term "old" can be captured mathematically and dealt with in algorithmic fashion.

In fuzzy linear programming verbal models are used for the automatic transfer of knowledge from such linguistic values for optimization. This is, of course, of special relevance for decision making in land use planning and management options.

THE APPLICATION OF BIOCYBERNETIC PRINCIPLES TO LANDSCAPE ECOLOGY

One of the best examples of this trans-disciplinary approach is the application of biocybernetic principles and their use as isomorphic models (*sensu* von Bertalanffy, 1968) for problems of landscape ecological planning, management and conservation. Biocybernetics is a recent ramification of cybernetics which can be defined as the theory of regulation in biological and ecological systems, enabling their self-organization by positive-deviation amplifying - feedback loops and their self-stabilization by negative-deviation reducing - feedback loops. In both cases the output values are fed back into the input values.

Vester (1976) was probably the first to apply biocybernetic principles to problems of land use and landscape planning and management. He suggested eight basic biocybernetic rules which reinforce each other and ensure the self-regulation and long-term viability of the biosphere. These rules, based chiefly on negative feedback couplings, have been applied by Vester and von Hesler (1980) as heuristic tools in the computerized preparation of a simulated sensitivity model of the congested Untermain Region in West Germany. In this model, cybernetic criteria for the following seven most relevant system compartments were evaluated: economy, population, land use, human ecology, natural resources, infrastructure, and the communal and public sectors. These were related to energy, material and information flow, as well as to other important systems criteria, such as degree of couplings, input-output relations, diversity and irreversibility. This dynamic simulation model was treated as an open system with a stepwise hierarchical structure enabling constant rechecking by the operator, who acts as an internal feedback regulator and becomes an integral part of the investigated interaction system (Naveh and Liberman, 1984). Dynamic biocybernetic models were applied also in the study of forest die-off in West Germany by Haber et al. (1984) as part of a three-level hierarchical approach integrating geographical information systems with sensitivity feedback models (Figure 3.1), and in scenarios of resource management strategies and land use intensities aimed at minimizing urban-industrial technospheric impacts on the open landscape.

The advantage of using such cybernetic models lies in their capacity to express, in an explicit way, the complex inter- and cross-linked physico-chemical, biological and socio-economical connectivity which is governed by mutually and causally related positive and negative feedback loops. Thus, in Figure 3.1, the deposition of air pollutants is increased by a larger foliage area and thereby air pollution is reduced. But if these pollutants are causing damage to sensitive trees, thereby reducing the effective leaf or needle surface area, the concentration of air pollutants will increase even if emission is kept constant.

This biocybernetic approach is of great relevance for the inter- and cross-linked relations between biotic and abiotic landscape state factors and landscape functions as affected by land use patterns. These are mostly not governed by single cause-effect reactions but by much more complex positive and negative feedback loops. From the point of view of human land uses, the most important landscape functions are their capacities for biological and geophysical production and protection, and for carrying, filtering, buffering, storing and regulating the flows of energy, matter, organisms and information. When landscapes carry out all these functions to their fullest ecological potentials on a long-term basis, we can consider them healthy, stable and attractive.

However, in recent times and at a rapidly increasing speed and extent land uses have been devoted to the short-term exploitation of their economic production, as carrier functions and to the conversion of open biosphere landscapes into human habitation and the industrial production of the technosphere. At the same time, all other functions and intrinsic landscape values have been taken for granted as "free ecological services" (Westman, 1977). These are therefore neglected and are very often not included in land use decisions and cost/benefit calculations of environmental impact statements. By such biocybernetic models and their quantification, we may gain deeper insights and arrive at some general law with predictive values, which may help us to develop biocybernetic indices for landscape health and stability as practical guides in land use decisions.

THE BIOCYBERNETIC REGULATION OF NATURAL AND SEMI-NATURAL LANDSCAPES

From a biocybernetic point of view we can distinguish two major groups of landscape functions with opposing feedback regulations. These are:

- biological production and self-organization functions coupled by positive feedback loops between photosynthetic and assimilative growth processes of the living sponge of the soil/plant/animal complex and the energy/material and information fluxes, and
- ecological protection and self-stabilization functions coupled by negative feedback loops between the moderating and attenuating processes of the living sponge and these fluxes.

In natural and sub-natural landscapes, such as the virgin tropical forests and the remote Arctic tundras, human disturbances are negligible and have not changed substantially. The landscape structure and its functions, their inherent capacities for self-or-

26

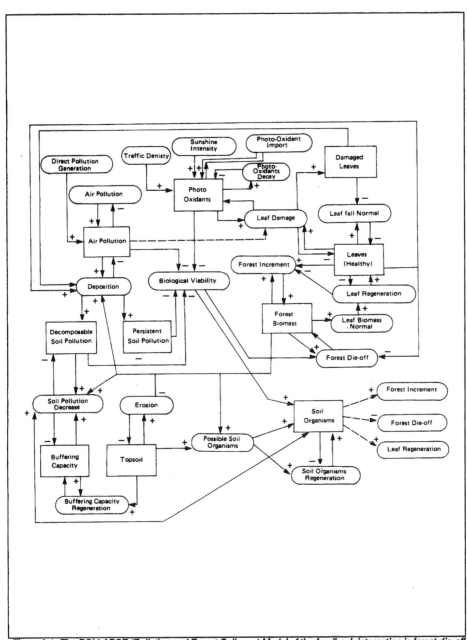

Figure 3.1 *The POLLAPSE (Pollution and Forest Collapse) Model of the feedback interaction in forest die-off (Haber et al., 1984). The model shows clearly the central role of photo-chemical oxidants and air pollutants from point-generated sources, enhancing this process, and that of soil organisms ensuring the biological viability and resilience of the living sponge.*

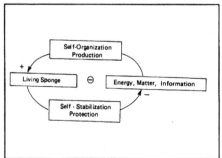

Figure 3.2 *Homeostatic regulation in stable natural landscapes. The greater the fluxes of solar energy, nutrients, water and organisms, the greater the photosynthetic activity, growth and organization processes in the food chain and therefore the capacity of the living sponge to absorb and utilize these fluxes, but also to moderate and attenuate their impacts and thereby to function as a major site-protecting and stabilizing feedback.*

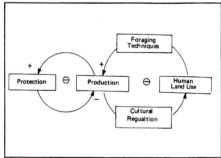

Figure 3.3 *Homeorhetic regulation in metastable semi-natural landscapes. By pre-technological land uses the homeostatic regulation is shifted into homeorhetic regulation if the foraging techniques for exploitation of the production functions are constrained by traditional ecological wisdom, as well as by spiritual, ritual and other cultural regulative feedback.*

ganization and self-stabilization, and their production and protecting functions remain intact. Therefore, the mutual balancing between positive and negative feedback loops has probably helped to maintain a kind of homeostatic steady state between the production and protection functions, as illustrated by a simple cybernetic model in Figure 3.2

In semi-natural landscapes, shaped by long-lasting pre-agricultural and pastoral land uses which have disturbed the living sponge in the more recent or distant past, landscape structure and functions have been changed by varying degrees. But if they are not exposed to extreme natural or human-caused perturbations to which they have not become adapted throughout their long evolutionary history, and as long as their biotic communities are reproducing themselves spontaneously, we can assume that they have also retained their capacities for self-organization and self-regulation and also their production and protection functions. Such landscapes presently make up the majority of our yet unspoiled "natural" and "wilderness" areas, nature reserves and parks. Many of these are studied by ecologists as "natural ecosystems." It is therefore important to clarify further their biocybernetic status.

It may be assumed, because of long-lasting, periodically repeated human perturbations of selective

foraging and intentional burning, the chief management tool in these pre-technological landscapes, that they are no longer maintained in a stationary state of homeostasis, but in a state of evolutionary metastability by a dynamic flow equilibrium or homeorhesis. In this case the dynamic interplay of both natural and cultural regulative feedback couplings (Figure 3.3) ensures that the system continues to alter in the same way that it has in the past (Waddington, 1975). Good illustrations of this were to be found in the semi-arid pastoral landscapes of the subtropical and tropical regions. In these traditional, pre-technological management systems prolonged drought periods, unreliable rainfall, lack of water and nutritious food in critical periods controlled the number of livestock and their movements. Together with grazing and burning practices, which were well adapted to local ecological conditions, they acted as combined natural and cultural protection and regulation functions. In closed negative feedback loops with human populations, this ensured also the long-term productivity of these pastoral total human ecosystems (Figure 3.4).

However, in the process of modernization and by the so-called "economic pasture and livestock development" methods, most natural, as well as traditional, cultural negative feedbacks have been removed by the provision of food and water supple-

28

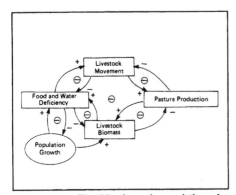

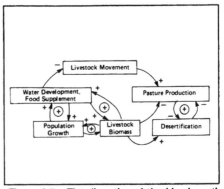

Figure 3.4 *The biocybernetic regulation of a semi-arid landscape by water development and food supplement. The greater the seasonal or annual food and water deficiency, the more time and effort has to be spent on livestock movement to preserve the pastures close to permanent water sources for critical periods' and prevent drought-starvation of livestock and people.*

Figure 3.5 *The disruption of the biocybernetic regulation of a semi-arid landscape by water development and food supplement. All negative feedback loops of Figure 3.4 have been replaced by positive runaway feedbacks and a new one has been added between pasture production and desertification: As pastures with permanent water supply are depleted, especially in drought years, by uncontrolled livestock pressure, desertification is enhanced and, in turn, the greater the desertification, the lower the pasture productivity.*

ments in critical periods, in order to increase biological production by cultural positive technological feedbacks. At the same time, most attempts to replace these negative feedbacks by the introduction of new cultural regulative feedbacks have failed to counteract the destabilizing feedbacks of the modern management practices of livestock and grazing control based on carrying capacity and range conditions. Thereby, well-meant but ill-advised technological improvements have turned into vicious runaway feedbacks of pasture deterioration and population explosion (Figure 3.5). These have then been followed, in general, by more and more severe drought-starvation cycles, leading to desertification (Naveh, 1966; Farvar and Milton, 1972; Le Houerou, 1985; Western and Finch, 1986).

In the Mediterranean region traditional agro-pastoral land uses modified most of the original semi-natural mountain and hill landscapes into very heterogeneous and attractive landscape mosaics with great patch dynamics of regeneration patterns after burning, grazing and coppicing, which ensured high structural, floristic and faunistic diversity and metastability. During these long lasting, and periodically repeated disturbances, a new dynamic homeorhetic flow equilibrium was established between the woody and herbaceous vegetation com-

ponents, which ensured also continued production of very species rich herbaceous pastures (Naveh, 1982a; Naveh, 1984; Naveh and Dan, 1973; Naveh and Whittaker, 1979; Naveh and Lieberman, 1984). The biocybernetic feedback regulation of a Mediterranean annual grassland can be illustrated by a "demostat" (population stability) model (Figure 3.6). In this, the dynamic homeorhetic flow equilibrium, ensuring the set point of highest spatio-temporal diversity and metastability (or global stability), is maintained by annual and seasonal climatic fluctuations in pasture production and composition, and by alternation of heavy grazing with rest periods. In traditional grazing systems, higher grazing pressure in the early growth period, when green forage was scarce, favored legumes and forb dominance, but its reduction during the main spring flush, when forage was abundant, favored the grasses and ensured their continued seed production. Therefore, the number of livestock that could survive the critical dry summer and early winter periods, could never overgraze these pastures in the spring period.

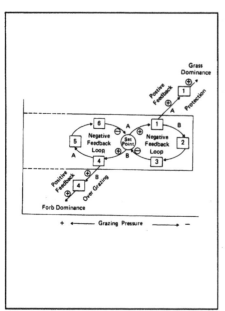

Figure 3.6 *Demostat model of Mediterranean grassland, maintained by cybernetic feedback control of grazing pressure.*

All such semi-natural landscapes, as well as those pastoral landscapes in which such a homeorhetic flow equilibrium has been ensured by periodic natural and cultural perturbations of fire, grazing, cutting, etc., should be considered metastable "perturbation-dependent" systems (Vogl, 1980). Here it would be futile to attempt the restoration of the self-stabilizing and organizing processes of virgin, natural and subnatural landscapes by simply stopping all human disturbances. On the contrary, we have to conserve and to re-establish the homeorhetic flow equilibrium by continuing or simulating all ecological processes, including defoliation pressures, to which these systems have been adapted throughout their long cultural history (Ricklefs et al. 1984; Naveh, 1987a).

This dynamic flow equilibrium is now apparently disrupted, either by complete cessation of all defoliation pressures from human disturbances, or the synergistic impacts of intensified traditional and neo-technological pressures, which include mass recreation and tourism. Using the decline in struc-tural and floristic diversity of the vegetation, followed closely also by faunistic impoverishment as a measure of the distortion of this homeorhetic flow equilibrium, it has been found that these parameters were low under intensified land pressures, and also under complete, prolonged protection, and lowest under mono-species pine afforestation.

It should be realized also that in cultivated agricultural landscapes the modernization and intensification of traditional crop farming systems, aimed at greater short-term production and profits, has been achieved by the introduction of new and much more powerful, and therefore also much more destabilizing, positive feedback loops. These exist between the scientific and agro-technological information and production functions and are driven by huge, energy demanding and costly machinery and by great inputs of chemicals such as fertilizers, herbicides and pesticides. Although these agro-industrial landscapes, are part of the open landscape they are not only incapable of fulfilling any of the vital protection, filtration carrying and buffering landscape functions, but also add additional detrimental ecological burdens to the ecosphere. As indicated by the studies of Phipps (1984) in Canada, and explained in more detail elsewhere (Naveh, 1987b) this conversion of the heterogeneous, fine-grained and metastable agro-ecological landscape patterns into homogeneous, coarse-grained and unstable monocultural steppes of cotton, wheat or corn reduces the rate of negentropic order and information in the landscape. The high inputs of fossil energy and chemicals result in high entropy which produces degradation, destructive kinetic energy and toxic waste products. This process is further enhanced by the above-described loss of natural landscape protection and regulation functions from the remaining open degraded and denuded landscapes.

A BIOCYBERNETIC APPROACH TO MULTIPLE-BENEFIT LANDSCAPE MANAGEMENT AND RESTORATION

Studies by Naveh (1979) and Naveh and Lieberman (1984) of multi-beneficial and multi-purpose restoration and afforestation programs in degraded Mediterranean landscapes were aimed at both the tangible landscape functions, such as livestock and forest production, which are transferable into market goods and can be expressed in monetary

30

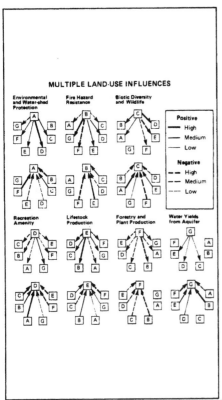

Figure 3.7 *Cybernetic model of land use and environmental variables.*

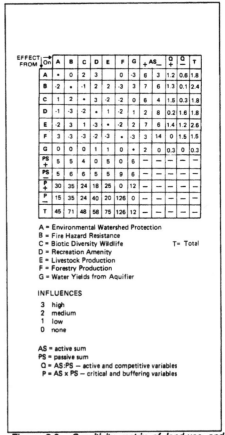

EFFECT→ On FROM ↓	A	B	C	D	E	F	G	+AS_	Q/+	Q/-	T
A	•	0	2	3		0	-3	6	3	1.2	0.6	1.8
B	-2	•	-1	2	2	-3	3	7	6	1.3	0.1	2.4
C	1	2	•	3	-2	-2	0	6	4	1.5	0.3	1.8
D	-1	-3	-2	•	1	-2	1	2	8	0.2	1.6	1.8
E	-2	3	1	-3	•	-2	2	7	6	1.4	1.2	2.6
F	3	-3	-3	-2	-3	•	-3	3	14	0	1.5	1.5
G	0	0	0	1	1	0	•	2	0	0.3	0	0.3
PS+	5	5	4	0	5	0	6	—	—	—	—	—
PS-	5	6	6	5	5	9	6	—	—	—	—	—
P+	30	35	24	18	25	0	12	—	—	—	—	—
P-	15	35	24	40	20	126	0	—	—	—	—	—
T	45	71	48	58	75	126	12	—	—	—	—	—

A = Environmental Watershed Protection
B = Fire Hazard Resistance
C = Biotic Diversity Wildlife   T= Total
D = Recreation Amenity
E = Livestock Production
F = Forestry Production
G = Water Yields from Aquifer

INFLUENCES

3 high
2 medium
1 low
0 none

AS = active sum
PS = passive sum
Q = AS:PS — active and competitive variables
P = AS x PS — critical and buffering variables

Figure 3.8 *Sensitivity matrix of land-use and environmental variables.*

terms, and also intangible cultural, aesthetic, psycho-hygienic and scientific functions, producing non-economic richnesses. The objective was to ensure that both of these sets of functions were recognized, and studies revealed that this could be achieved by the optimization of all intrinsic landscape values through manipulation of the plant-soil-animal complex. For this purpose the biocybernetic method developed by Vester (1976) was applied in the preparation of a sensitivity matrix which included the major protection and production functions and their bio-ecological, socio-ecological and socio-economic benefits. As shown in Figures 3.7 and 3.8, their mutual, relative positive and negative effects on each other in actual land uses were evaluated. By summarizing these values from left

to right for the active sums (AS), and from top to bottom for the passive sum (PS), a quantitative assessment of these interactions was achieved. The quotient (Q = AD/PS) defined the most active and competitive function, which will most affect all others and be least affected (in this case C in a positive way and D in a negative way), and the least compatible and influential with the lowest Q values (F without positive influence and C without negative influence). The product AS×PS defined the most critical functions exerting the greatest effect and which were most affected. In this case this is B and E in a positive way and F in a negative way. E and

F have the highest over-all Q and P values and those with the lowest P values are the "buffering elements" F and G. (For more details on this methodology see Naveh (1979) and Naveh and Lieberman (1984).)

The main conclusion of this work was that the highest overall bio-ecological, socio-ecological and socio-economical benefit can be derived from the establishment of closely interwoven, flexible single and multipurpose land use patterns. As far as possible these should be interlaced with multi-layered conservation and recreation and/or fodder forests and parklands. However, recreation, because of its potentially high negative effect, should be managed with great caution. It will conflict not only with forestry and livestock production but also with protection functions to which highest priority should be given. Grazing, on the other hand, can serve as a very effective regulative feedback device, but if not controlled effectively it will exert adverse effects on all protection functions. It also became obvious that fire hazard resistance, because of its important active role, should be taken into account in any rational Mediterranean landuse policy. On the other hand, water yields, as the most passive element, cannot be considered a major landuse goal, but only a desirable by-product. Thus, even with the crude biocybernetic models available at this time, we can already indicate major directions for improved land use options. However, now, with the help of computerized expert systems using fuzzy set linear programming, much more robust integrative models can be developed and applied in actual land use decision making.

BIOCYBERNETICS ON A GLOBAL LANDSCAPE SCALE

The realization of the far-reaching global implications of human induced changes to the living sponge and the vegetation, even in restricted terrestrial landscapes, by causing complex mutual-causal feedback relations with the geosphere, is broadening the biocybernetic perspectives into what could be called "geo-biocybernetics." This is well reflected in the following statement in a comprehensive report on remote sensing of the biosphere:

"Unlike the perspective we gain when we view life as a collection of organisms, the global perspective reveals the biosphere as a complex, hierarchical system composed of many positive feedback loops with constraints formed by interrelated biogeochemical cycles. This perspective introduces new and exciting theoretical problems that must be solved before we will understand life on this planet." (National Research Council 1986, page 53)

The importance of positive feedback loops between the removal of vegetation canopy and climate, driven by changes in albedo, evaporation and surface roughness and leading to desertification and to long-term global climatological changes, has also been emphasized in a recent important workshop (Rosenzweig and Dickinson, 1986). The same is true also for the exponential rise in atmospheric carbon dioxide, acid rain and other destabilizing changes in atmospheric chemistry driven by positive feedbacks caused by industrialization, land clearing and agricultural intensification.

Probably the most advanced geo-biocybernetic approach has been applied in the "Gaia" hypothesis by Lovelock (1979). This views the earth as a homeostatic self-regulating and therefore also self-cleansing system that can even switch over to homeorhetic regulation. But if the human-induced, neo-technological changes go beyond the limits of homeorhetic regulation capacities it may fail completely and break down. This holistic view is based on the assumption that the evolution of the biota is not independent of the evolution of the environment but that both evolutionary processes are in fact tightly coupled and have evolved together as a single system. Accordingly, life and environment are two parts of a closely coupled geo- biocybernetic system with mutual, amplified active feedback loops and not, as viewed conventionally by climatologists

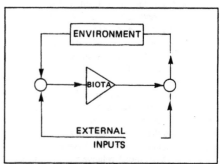

Figure 3.9 *A diagram drawn from control theory to illustrate as a single system the active feedback between the biota and the environment.*

32

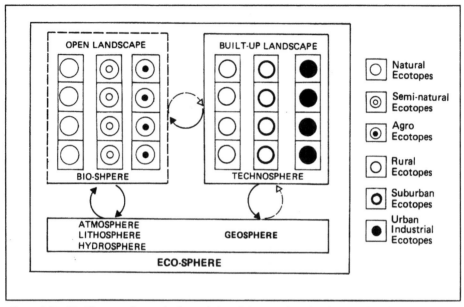

Figure 3.10 *The threats to homeorhetic regulation of the ecosphere by the overwhelming impact of the technosphere on the biosphere and geosphere. This regulation and thereby the full geo-biocybernetic integration of the biosphere and technosphere with the geosphere can be ensured only if the destabilizing impacts of the technosphere on the biosphere and geosphere are counteracted by cultural regulation.*

and biochemists, passive feedbacks between the biosphere and the geosphere (Figure 3.9).

This geo-biocybernetic approach, called by Lovelock (1985) "Geophysiology", has much in common with our biocybernetic landscape approach, viewing landscapes as the concrete entities of closely coupled life and environment systems. These evolved together in a closely interconnected landscape holarchy, and not, as regarded in the traditional ecosystem approach, as isolated forest, grassland or wetland and other natural ecosystems. But we should also realize that from the Pleistocene onwards, *Homo erectus* and *Homo sapiens* have played such an increasingly important role in this evolutionary process that we have only very few truly "natural" ecosystems left on earth. We have now apparently reached a critical stage in which *Homo industrialis*, by creating the exponentially expanding technosphere with his great technological power and skill (far exceeding his ecological wis-

dom and knowledge), has become a geo-biocybernetic driving force. This is threatening to decouple the negative, stabilizing feedback loops between the biosphere and geosphere, and thereby the homeorhetic regulation of the ecosphere (Figure 3.10).

The death of remote lakes, and the rapidly expanding decline of forests in temperate zones by acid rain and by photochemical oxidant pollutants are some of the alarming syndromes not only of the connectivity between open and built-up landscape holons on multidimensional spatio-temporal scales, but also between them and their atmospheric, hydrospheric and lithospheric mantle.

Unfortunately we cannot yet answer the crucial question of how much more of the living sponge of open landscapes can be converted into wasteland, asphalt and concrete before causing irreversible damage to the ecosphere. We are also still very far from being able to predict at what stage human-caused changes in the physics and chemistry of the atmosphere and in the over-all thermodynamic balance of the ecosphere may cause such a breakdown of the entire system. As Lovelock points out, this would be comparable to trying to predict, on theoretical grounds, how much of a victim's skin can be burnt before death is inevitable. But while we can learn from burn cases we cannot afford to be empirical in our approach to tropical forest clearance and other open land conversions and must be, therefore, ultra-cautious.

CONCLUSIONS

Most of us are not dealing in our day-by day studies and research projects with these continental and global spatio-temporal landscape scales. But nevertheless, we have to broaden the conceptual and cognitive scales of our work to these dimensions. We should not be content with the provision of the semantic written or verbal information on the results of our work in scientific and professional publications and reports; in most cases this might be filed away for good. According to von Weizsaecker (1974), semantic information becomes meaningful only as pragmatic information through a feedback reaction by the receiver, and when expressed in his actions. In our case this should be the regulative feedback information for land use decisions which we must provide for landscape planning, management, conservation and restoration. The recognition of the vital importance of this pragmatic information transcends the paradigms of prevailing "normal scientific knowledge." This distinguishes between theoretical and practical knowledge and denies ethical and aesthetic value-oriented cognition (Wojiechowski, 1975). If landscape ecology is viewed as a science, contributing to the global survival of life on earth, it cannot be such a value-free cognitive science. As an order seeking holistic science of man and nature in their totality, it must include the evaluation of the long-term, global and cumulative effects of the lower landscape holons on the upper ones and upon their biocybernetic interconnectiveness. It must then apply these judgements in practice for the protection of our threatened open landscapes from biological, ecological and cultural decline and impoverishment.

In the context of conceptual connectivity, the recognition of human society in its relation to open landscapes and the nature of this holon is of greatest importance. We are both autonomous wholes of the technosphere and its built-up landscapes, and also dependent parts of the biosphere and its life-supporting landscapes. The latter has a self-organizing and self-stabilizing feedback relationship with the geosphere. These relationships are threatened by the waste products and the stresses of our technosphere. We are, therefore, both affecting and affected links of the total human ecosystem. Our global survival and the survival of the technosphere will depend on the creation of a new geobiocybernetic symbiosis, and by a reconciliation between these contradicting holon tendencies of self-assertive wholes and integrative parts (Naveh and Lieberman, 1984). This should not be interpreted as theoretical and philosophical speculation but as a practical guideline in our work. Each landscape ecologist can contribute to this symbiosis by supplying the missing regulative feedback loops, with the help of scientific, technological, educational and pragmatic information derived from his own and his colleagues' scientific and professional work.

REFERENCES

Bertalanffy, L. von, 1968, *General System Theory: Foundations, Development and Applications.* George Braziller, New York.

Farvar, T. and J.P. Milton (eds.), 1972, *The Careless Technology.* Natural History Press, New York.

Haber, W., Grossmann, W.D. and J. Schaffer, 1984, Integrated evaluation and synthesis of data by connection of dynamic feedback models with a geographical information system. In: J. Brandt and P. Agger (eds.) *Methodology in Landscape Ecological Research and Planning.* Proceedings First International Seminar IALE, Roskilde University Centre, Roskilde, Denmark; pp. 147-162.

Jantsch, E. and C.H. Wadington (eds.), 1976, *Evolution and Consciousness: Human Systems in Transition.* Addison-Wesley, Reading, Massachusetts.

Koestler, A., 1969, Beyond atomism and holism - the concept of the holon. In: A. Koestler and J.R. Smithies (eds.), *Beyond Reductionism: New Perspectives in the Life Sciences.* Hutchinson, London; pp. 192-216.

Laszlo, E., 1972, *Introduction to Systems Philosophy: Toward a New Paradigm of Contemporary Thought.* Harper Torchbooks, New York.

Le Houerou, H.N., 1985, Pastoralism. In: R.W. Kates, J.H. Ausubel and M. Berberian (eds.), *Climatic Impact Assessment.* SCOPE, John Wiley and Sons, Ltd., New York; pp. 155-185.

Lovelock, J.E., 1979, *Gaia: A New look at Life on Earth.* Oxford University Press, Oxford.

Lovelock, J.E., 1985, Are we destabilizing world climate? The lessons of geophysiology. *The Ecologist,* 15: 52-55.

34

National Research Council, 1986, *Remote Sensing of the Biosphere*. Joint report by Committee on Planetary Biology, Space Science Board, Commission on Physical Sciences, Mathematics and Resources, and the National Research Council. National Academy Press, Washington, D.C.

Naveh, Z., 1966, The development of Tanzania Masailand- a sociological and ecological challenge. *African Soils*, 10/13: 449-539.

Naveh, Z., 1979, A model of multiple-use strategies of marginal and untillable Mediterranean upland ecosystems. In: J. Cairns, G.P. Patil, and W.E. Waters (eds.), *Environmental Biomonitoring, Assessment, Prediction and Management: Certain Case Studies and Related Quantitative Issues*. International Cooperative Publishing House, Fairland, Maryland; pp. 269-286.

Naveh, Z., 1982a, The dependence of the productivity of a semi-arid Mediterranean hill pasture ecosystem on climatic fluctuations. *Agriculture and Environment*, 7: 47-61.

Naveh, Z., 1982b, Mediterranean landscape evolution and degradation as multivariate biofunctions: Theoretical and practical implications. *Landscape Planning*, 9: 125-146.

Naveh, Z., 1984, The vegetation of the Carmel and Nahal Sefunim and the evolution of the cultural landscape. In: A Ronen (ed.), *Sefunim Prehistoric Sites Mount Carmel, Israel*. BAR International Series 230; pp. 23-63.

Naveh, Z., 1987a, *Multifactorial Restoration of Semi-arid Mediterranean Landscapes for Multi-purpose Land Uses*. Westview Press, Boulder, Colorado (In press).

Naveh, Z., 1987b, Biocybernetic and thermodynamic perspectives of landscape functions and land use patterns. *Landscape Ecology*, 1(2): 75-83.

Naveh, Z. and J. Dan, 1973, The human degradadtion of Mediterranean landscapes in Israel. In: F. di Castri and H.A. Mooney (eds.), *Mediterranean - Type Ecosystems: Origin and Structure. Ecological Studies: Analysis and Synthesis*, Vol. 7, Springer Verlag, New York. pp. 373-390.

Naveh, Z. and R.H. Whittaker, 1979, Structural and floristic diversity of shrublands and woodlands in northern Israel and other Mediterranean areas. *Vegetatio*, 41: 171-190.

Naveh, Z. and A.S. Lieberman, 1984, *Landscape Ecology, Theory and Application*. Springer Verlag, New York.

Negotia, C.V., 1985, *Expert Systems and Fuzzy Systems*. The Benjamin/Cummings Publishing Company, Menlo Park, California.

Pankow, W., 1976, Openness as self-transcendence. In: E. Jantsch and C.H. Waddington (eds.), *Evolution and Consciousness: Human Systems in Transition*. Addison-Wesley, Reading, Massachusetts; pp. 16-36.

Phipps, M., 1984, Rural landscape dynamic: The illustration of some key concepts. In: J. Brandt and P. Agger (eds.), *Methodology in Landscape Ecological Research and Planning*. Proceedings First International Seminar IALE, Roskilde University Centre, Roskilde, Denmark; pp. 46-54.

Ricklefs, R.E., Z. Naveh and R.E. Turner, 1984, Conservation of ecological processes. *The Environmentalist* 4. IUCN Commission on Ecology, Paper 8.

Rosenzweig, C. and R. Dickson, 1986, *Climate-Vegetation Interaction. Workshop proceedings*, NASA/Goddard Space Flight Centre, Report 01ES-2.

Vester, F., 1976, *Urban Systems in Crisis*, Deutsche Verlags-Anstalt GmbH, Stuttgart.

Vester, F. and A. von Hesler, 1980, *Sensitivity Model: Ecology and Planning in Metropolitan Areas*. Regionale Planungsgemeinschaft Untermain, Frankfurt am Main.

Vogl, R.J., 1980, The ecological factors that produce perturbation-dependent ecosystems. In: J. Cairns (ed.), *The Recovery Process in Damaged Ecosystems*. Ann Arbor Science Publishers Inc., Ann Arbor, Michigan; pp. 63-94.

Von Weizsaecker, E., 1974, *Offene System Beitraege zu Zeitstrukturen, Information, Entropie und Evolution*, Klett Verlag, Stuttgart.

Waddington, C.H., 1975, *The Evolution of an Evolutionist*. Cornell University Press, Ithaca, New York.

Western, D. and V. Finch, 1986, Cattle and pastoralism: survival and production in arid lands, *Human Ecology*, 14: 77-94.

Westman, W.E., 1977, How much are nature's services worth? *Science*, 197: 960-964.

Wojciechowski, J.A., 1975, The ecology of knowledge. In: N.H. Steneck (ed.), *Science and Society, Past, Present and Future*. University of Michigan Press, Ann Arbor; pp. 258-302.

4 The Central Role of Scale

Introduction and Review

Today, nearly all ecologists recognize the ubiquity of scale influences and the importance of defining scale explicitly. Although ecologists had long appreciated the effects of quadrat size and had devoted considerable attention to species-area relationships (which are really scaling functions), general awareness of the importance of scale in ecology dawned slowly. Several key publications in the 1980s (e.g., Allen and Starr 1982; Delcourt et al. 1983; O'Neill et al. 1986) prompted more rigorous consideration of how patterns and processes varied with scale, initiating a search for general scaling rules (e.g., Meentemeyer and Box 1987; Turner et al. 1989). Only in the 1990s, however, did issues of scale surge to the forefront of thinking and (even later) practice in ecology.

In large part, the heightened awareness of scale issues in ecology was a result of the attention it received from landscape ecologists. Issues of scale were a focus of landscape ecology from its beginnings, perhaps because of its importance in geography, where scale is central to everything. Consideration of scale effects—on conceptual frameworks, models, empirical studies, and spatially explicit formulations of both pattern and process—was the norm in landscape ecology well before it became prevalent in ecology (for recent treatments,

see Peterson and Parker 1998; Gardner et al. 2001). Uncertainty continued, however, about whether landscape should be thought of as a spatial scale or a level in an organizational hierarchy (see part III). Despite valiant attempts to put the issue to rest (e.g., O'Neill and King 1998; Allen 1998; Wiens 2001; King 2005), this uncertainty continues, some workers using the terms *landscape* and *level* interchangeably, others clearly equating *landscape* with *scale*. The seven papers included here exemplify this central role of scale in landscape ecology; the first four emphasize spatial scale, whereas the last three reveal the importance of expanding the temporal scale on which one views landscapes.

John Wiens approached scaling issues from an ecological perspective. Wiens had been considering the effects of spatial patterns on populations well before it was in vogue (part II), considering scale from the perspective of an organism. Looking across taxa, it became apparent that the scale of human perception (or management) was inappropriate for many organisms and that functional scales needed to be defined differently for various types of organisms. Moreover, Wiens proposed that the effects of changing scale on the structural or functional properties of systems were likely to be discontinuous, following predictable relationships within certain ranges of a scaling spectrum (what Wiens called *domains of scale*) that might suddenly dissolve with further changes in scale, as a scaling threshold was crossed. Although Wiens made only fleeting mention of hierarchy theory, the relations between scaling domains and levels in a hierarchy are clear. This paper laid out an approach to an organism-based definition of scale that strongly influenced the direction of landscape ecology both in North America and Australia.

The organismal perspective to scale advanced by Wiens was not entirely new, of course. Wiens had hinted at such ideas in earlier writings (part II; Wiens 1981), but **John Addicott** and several of his students formalized this thinking by reviewing concepts of environmental heterogeneity and then casting the responses of organisms to this heterogeneity in terms of scale-dependent ecological neighborhoods. Despite the emphasis on heterogeneity (i.e., spatial pattern), the real contribution of Addicott et al. was to shift attention to processes, by considering how the activity and behavioral traits of organisms would determine the scales in space and time on which they would respond to heterogeneity. The appropriate scales for studying these phenomena, Addicott et al. suggested, must therefore be defined by properties of the organisms rather than by arbitrary decisions of the investigator; studies conducted at inappropriate scales could produce irrelevant or incorrect conclusions, affecting comparisons among studies or between theories and empirical results. Addicott et al. went beyond developing a fresh way to consider scaling, however, to present concrete ways in which the spatial and temporal dimensions of an organism's ecological neighborhood could be estimated by frequency distributions of activity or influence.

Robert O'Neill took an entirely different, and much more theoretical, approach to scaling issues. Like Urban et al. and Naveh (part III), O'Neill cast his approach in the framework of hierarchy theory. One of the most enduring contributions of hierarchy theory was the explicit recognition of the relevance, for any question, of three scales: the scale at which the process of interest actually operates, the broader scale that constrains the process and provides context, and the finer scale that encompasses the mechanisms underlying the process. Although O'Neill developed a formal framework more with reference to hierarchical levels than to scale per se and did not explicitly relate his approach to landscapes, the real value of the work was in drawing attention to what O'Neill termed *transmutations*. Transmutations occur when a process determining patterns at one level (or scale) is changed to a different controlling process when one moves to another hierarchical level (or scale). In other words, it is a formalized expression of the scale domains and thresholds that Wiens was talking about.

All three of these papers approached scale from an ecological perspective. Scale, of course, was nothing new to geographers, who had long recognized well-accepted rules about scale, such as cartographic scale or relative scale. **Vernon Meentemeyer** introduced the diversity of geographic scale concepts to landscape ecologists. Among other things, Meentemeyer drew attention to several fallacies that may trap one who is not attentive to or explicit about scale, fallacies that potentially follow as well from neglecting Wiens's scale domains, Addicott et al.'s ecological neighborhoods, or O'Neill's transmutations. Although Meentemeyer's paper provided only an overview of these rich geographic concepts, it opened the door to the primary literature of that discipline to the current generation of landscape ecologists, especially those in North America, who were farther removed from the close linkage with geography that characterized landscape ecology in Europe. This article is from a series of papers published in a special issue of *Landscape Ecology* that emerged from a workshop that focused on predicting across scales; other papers in that issue provide useful perspectives from a variety of disciplines.

These four papers focused primarily on spatial scaling. This emphasis is understandable, given the strong emphasis on spatial patterns in landscapes. After all, landscapes are usually defined or described in terms of spatial patterns and relationships. Landscapes are also dynamic, however, and these changes may occur over a wide range of temporal scales. For example, **William Romme and Dennis Knight** used data on fire-scar histories and patterns of forest succession to reconstruct changes in landscape diversity of Yellowstone National Park in Wyoming over a period of 240 years. By extending the time scale well beyond that customary for ecological or landscape studies, Romme and Knight were able to consider large, infrequent fire disturbances as well as smaller, short-term disturbances. This showed that the proportion of the landscape occupied by different forest successional stages was not stable over time. Thus, although the Yellowstone landscape was indeed a dynamic mosaic, it was not at steady state. This was one of the early studies that suggested that nonequilibrium dynamics could be the norm in some systems. Another notable contribution of this paper and the more detailed monograph by Romme (1982) was the application of indexes of diversity and evenness, commonly used to describe ecological communities, to the description of landscapes and their changes through time.

The landscape disturbances and dynamics considered by Romme and Knight were largely natural. In contrast, the temporal dynamics of the landscape patterns in The Netherlands considered by **Bas Pedroli and Guus Borger** were determined by human land uses and the sociopolitical factors that influenced land uses. Pedroli and Borger used a variety of sources of information, from before the Middle Ages onward, to reconstruct human influences on groundwater hydrology and landscape patterns. The results of considering landscapes on such an expanded temporal scale clearly show that, although landscape patterns proximately reflect land-use practices, these practices are in turn influenced by groundwater hydrology, and the interaction is dynamic over multiple time scales. Interpreting contemporary landscape patterns solely in terms of contemporary processes would be incomplete and misleading. Obviously, also, considering the landscape without human influence would be impossible. This is clear enough in heavily modified landscapes such as those in Europe (and, increasingly, in much of the world); the imprint of human activities on supposedly natural landscapes, such as those of Yellowstone, are less evident, but that does not mean that they can be forgotten.

Both the importance of understanding the past in order to understand the present and the importance of incorporating both historic and prehistoric influences of humans on landscapes were given impetus in landscape ecology by the work of **Hazel and Paul Delcourt**. The Delcourts adopted a paleoecological perspective, charting the environmental and cultural changes that have influenced landscapes in the southern Appalachian Mountains of the United States during the Quaternary period, spanning the past 1.8 million years. During this time period, multiple thresholds of changing climate affected landscapes over broad spatial scales, whereas over the past five thousand years, human cultural evolution and changing land use have shifted the controls of landscape patterns from natural to anthropogenic factors (a shift documented in a different way by Diamond 1999). Thus, the ecological communities and landscapes seen now, and often interpreted as if they were stable, are in fact very recent. (An even broader temporal perspective on human modification of landscapes is provided by the historical and prehistoric uses of fire by aboriginal cultures, such as in Australia [Pyne 1991; Mulvaney and Kamminga 1999; Bowman 2000; Bradstock et al. 2002].)

Another message of the Delcourts' study resonates with the interdisciplinary origins of landscape ecology: As the temporal scale of analysis is lengthened, it becomes necessary to draw on a broader range of disciplines to assess the multiple forces influencing landscape dynamics.

References

Allen, T. F. H. 1998. The landscape "level" is dead: Persuading the family to take it off the respirator. In *Ecological Scale: Theory and Applications*, eds. D. L. Peterson and V. T. Parker, 35–54. New York: Columbia University Press.

Allen, T. F. H., and T. B. Starr. 1982. *Hierarchy: Perspectives for ecological complexity*. Chicago: University of Chicago Press.

Bowman, D. M. J. S. 2000. *Australian rainforests: Islands of green in a land of fire*. Cambridge: Cambridge University Press.

Bradstock, R. A., J. E. Williams, and A. M. Gill, eds. 2002. *Flammable Australia: The fire regimes and biodiversity of a continent*. Cambridge: Cambridge University Press.

Delcourt, H. R., P. A. Delcourt, and T. Webb. 1983. Dynamic plant ecology: The spectrum of vegetational change in space and time. *Quaternary Science Review* 1:153–175.

Diamond, J. 1999. *Guns, germs, and steel*. New York: W.W. Norton.

Gardner, R. H., W. M. Kemp, V. S. Kennedy, and J. E. Petersen, eds. 2001. *Scaling relations in experimental ecology*. New York: Columbia University Press.

King, A. W. 2005. Hierarchy theory and the landscape . . . level? Or, words do matter. In *Issues and Perspectives in Landscape Ecology*, eds. J. A. Wiens and M. R. Moss, 29–35. Cambridge: Cambridge University Press.

Meentemeyer, V., and E. O. Box. 1987. Scale effects in landscape studies. In *Landscape Heterogeneity and Disturbance*, ed. M. G. Turner, 15–34. New York: Springer-Verlag.

Mulvaney, J., and J. Kamminga. 1999. *Prehistory of Australia*. Washington, DC: Smithsonian Institution Press.

O'Neill, R. V., D. L. DeAngelis, J. B. Waide, and T. F. H. Allen. 1986. *A hierarchical concept of ecosystems*. Princeton, NJ: Princeton University Press.

O'Neill, R. V., and A. W. King. 1998. Homage to St. Michael; or, Why are there so many books on scale? In *Ecological Scale: Theory and Application*, eds. D. L. Peterson and V. T. Parker, 3–15. New York: Columbia University Press.

Peterson, D. L., and V. T. Parker. 1998. *Ecological scale: Theory and applications*. New York: Columbia University Press.

Pyne, S. J. 1991. *Burning bush: A fire history of Australia*. Seattle: University of Washington Press.

Romme, W. H. 1982. Fire and landscape diversity in subalpine forests of Yellowstone National Park. *Ecological Monographs* 52:199–221.

Turner, M. G., R. V. O'Neill, R. H. Gardner, and B. T. Milne. 1989. Effects of changing spatial scale on the analysis of landscape pattern. *Landscape Ecology* 3:153–162.

Wiens, J. A. 1981. Scale problems in avian censusing. *Studies in Avian Biology* 6:513–521.

Wiens, J. A. 2001. Understanding the problem of scale in experimental ecology. In *Scaling Relationships in Experimental Ecology*, eds. R. H. Gardner, M. Kemp, V. Kennedy, and J. Petersen, 61–88. New York: Columbia University Press.

Functional
Ecology 1989,
3, 385-397

ESSAY REVIEW
Spatial scaling in ecology[1]

J. A. WIENS
*Department of Biology and Natural Resource
Ecology Laboratory, Colorado State University,
Fort Collins, Colorado 80523, USA*

'The only things that can be universal, in a sense,
are scaling things'

(Mitchell Feigenbaum[2])

Introduction

Acts in what Hutchinson (1965) has called the
'ecological theatre' are played out on various
scales of space and time. To understand the drama,
we must view it on the appropriate scale. Plant
ecologists long ago recognized the importance of
sampling scale in their descriptions of the dispersion or distribution of species (e.g. Greig-Smith,
1952). However, many ecologists have behaved as
if patterns and the processes that produce them are
insensitive to differences in scale and have
designed their studies with little explicit attention
to scale. Kareiva & Andersen (1988) surveyed
nearly 100 field experiments in community
ecology and found that half were conducted on
plots no larger than 1 m in diameter, despite
considerable differences in the sizes and types of
organisms studied.

Investigators addressing the same questions
have often conducted their studies on quite
different scales. Not surprisingly, their findings
have not always matched, and arguments have
ensued. The disagreements among conservation
biologists over the optimal design of nature
reserves (see Simberloff, 1988) are at least partly
due to a failure to appreciate scaling differences
among organisms. Controversies about the role of
competition in structuring animal communities
(Schoener, 1982; Wiens, 1983, 1989) or about
the degree of coevolution in communities (Connell, 1980; Roughgarden, 1983) may reflect the
imposition of a single scale on all of the species in
the community. Current ecological theories do
little to resolve such debates, because most of these
theories are mute on scale – they can be applied at
any scale on which the relevant parameters can be
measured.

Recently, however, ecologist studying a wide
range of topics have expressed concern about
scaling effects (see Dayton & Tegner, 1984; Wiens
et al., 1986a; Giller & Gee, 1987; Meetenmeyer &
Box, 1987; Frost *et al.*, 1988; Rosswall, Woodmansee & Risser, 1988). 'Scale' is rapidly becoming a
new ecological buzzword.

Scientists in other disciplines have recognized
scaling issues for some time. The very foundation
of geography is scaling. In the atmospheric and
earth sciences, the physical processes that
determine local and global patterns are clearly
linked (e.g. Schumm & Lichty, 1965; Clark, 1985;
Dagan, 1986; Ahnert, 1987) and their importance
is acknowledged in hierarchies of scale that guide
research and define subdisciplines within these
sciences. Physical and biological oceanographers
often relate their findings to the spectrum of
physical processes from circulation patterns in
oceanic basins or large gyres to fine-scale eddies or
rips (e.g. Haury, McGowan & Wiebe, 1978; Steele,
1978; Legrende & Demers, 1984; Hunt &
Schneider, 1987; Platt & Sathyendranath, 1988).
Physicists and mathematicians studying fractal
geometry, strange attractors, percolation theory,
and chaos address scaling as a primary focus of
their investigations (Nittman, Daccord & Stanley,
1985; Orbach, 1986; Grebogi, Ott & Yorke, 1987;
Gleick, 1987).

Why have ecologists been so slow to recognize
scaling? Ecologists deal with phenomena that are
intuitively familiar, and we are therefore more
likely to perceive and study such phenomena on
anthropocentric scales that accord with our own
experiences. We have also been somewhat tradi-
tion-bound, using quadrats or study plots of a
particular size simply because previous workers
did. Unlike the physical and earth sciences (and
many laboratory disciplines of biology), where our
perceptual range has been extended by technology, few tools have been available to expand

[1] Adapted from the first Katharine P. Douglass Distinguished Lecutre at the Rocky Mountain Biological
Laboratory, Gothic, Colorado, 23 July 1987.
[2] Quoted in Gleick, 1987, p. 186.

our view of ecological phenomena (but see Platt & Sathyendranath, 1988; Gosz, Dahm & Risser, 1988).

My thesis in this paper is that scaling issues are fundamental to all ecological investigations, as they are in other sciences. My comments are focused on spatial scaling, but similar arguments may be made about scaling in time.

The effects of scale

Some examples

The scale of an investigation may have profound effects on the patterns one finds. Consider some examples:

● In hardwood forests of the north-eastern United States, Least Flycatchers (*Empidonax minimus* Baird & Baird) negatively influence the distribution of American Redstart (*Setophaga ruticilla* L.) territories at the scale of 4-ha plots. Regionally, however, these species are positively associated (Sherry & Holmes, 1988). Apparently the broad-scale influences of habitat selection override the local effects of interspecific competition. Similar scale-dependency has been found in the habitat relationships of shrubsteppe birds (Wiens, Rotenberry & Van Horne, 1986b), interspecific associations among plant species (e.g. Beals, 1973) or phytoplankton and zooplankton (Carpenter & Kitchell, 1987), and the patterns of coexistence of mosses colonizing moose dung (Marino, 1988) or of ants on mangrove islands (Cole, 1983).

● In the Great Barrier Reef of Australia, the distribution of fish species among coral heads at the scale of patch reefs or a single atoll may be strongly influenced by chance events during recruitment and the species composition of local communities of fish may be unpredictable (Sale, 1988; Clarke, 1988). At the broader scales of atolls or reef systems, community composition is more predictable, perhaps because of habitat selection, niche diversification, or spatial replacement of species within trophic guilds (Ogden & Ebersole, 1981; Anderson *et al.*, 1981; Green, Bradbury & Reichelt, 1987; Galzin, 1987).

● On the basis of experiments conducted at the scale of individual leaf surfaces, plant physiologists have concluded that stomatal mechanisms regulate transpiration, whereas meterologists working at the broader scale of vegetation have concluded that climate is the principal control (Jarvis & McNaughton, 1986; Woodward, 1987). In

a similar manner, most of the variation in litter decomposition rates among different species at a local scale is explained by properties of the litter and the decomposers, but at broader regional scales climatic variables account for most of the variation in decomposition rates (Meentemeyer, 1984).

● Domestic cattle grazing in shortgrass prairie use elements of local plant communities quite nonrandomly on the basis of short-term foraging decisions, but use of vegetation types is proportional to their coverage at the broader scale of landscape mosaics (Senft *et al.*, 1987).

● The distribution of phytoplankton in marine systems is dominated by horizontal turbulent diffusion at scales up to roughly 1 km (Platt, 1972; Denman & Platt, 1975). At somewhat broader scales, phytoplankton growth, zooplankton grazing, and vertical mixing override these local effects (Denman & Platt, 1975; Lekan & Wilson, 1978; Therriault & Platt, 1981). At scales of >5 km, phytoplankton patchiness is controlled largely by advection, eddies, and local upwelling occurring over areas of 1–100 km (Gower, Denman & Holyer, 1980; Legrende & Demers, 1984). The same controls operate in lakes, although the transitions occur at finer scales (Powell *et al.*, 1975).

These examples could easily be extended. The salient point is that different patterns emerge at different scales of investigation of virtually any aspect of any ecological system.

Linkages between physical and biological scales

In the marine phytoplankton and other aquatic systems, physical features may be primary determinants of adaptations of organisms, and physical and biological phenomena may scale in much the same way. However, in many terrestrial environments, atmospheric and geological influences may often be obscured by biological interactions (Clark, 1985). The relationships between climate and vegetation that are evident at broad scales, for example, may disappear at finer scales, overridden by the effects of competition and other biological processes (Greig-Smith, 1979; Woodward, 1987). Local biological interactions have the effect of decoupling systems from direct physical determination of patterns by introducing temporal or spatial lags in system dynamics or creating webs of indirect effects. However, at broader scales, physical processes may dominate or dissipate these biological effects (Levin, 1989). There are exceptions: plant distributions on fine scales may be controlled by edaphic or microtopographic factors, and vegetation may influence climate at regional scales.

System openness and the scale of constraints

Ecological systems become closed when transfer rates among adjacent systems approach zero or when the differences in process rates between adjacent elements are so large that the dynamics of the elements are effectively decoupled from one another. In open systems, transfer rates among elements are relatively high, and the dynamics of patterns at a given scale are influenced by factors at broader scales. However, 'openness' is a matter of scale and of the phenomena considered. At the scale of individual habitat patches in a landscape mosaic, for example, population dynamics may be influenced by between-patch dispersal, but at the broader scale of an island containing that landscape, emigration may be nil and the populations closed. The same island, however, may be open with regard to atmospheric flows or broad-scale climatic influences.

The likelihood that measurements made on a system at a particular scale will reveal something about ecological mechanisms is a function of the openness of the system. The species diversity of a local community, for example, is influenced by speciation and extinction, and by range dynamics at regional or biogeographic scales (Ricklefs, 1987). Changes in population size at a location may reflect regional habitat alterations, events elsewhere in a species' range, or regional abundance and distribution rather than local conditions (May, 1981; Väisänen, Järvinen & Rauhala, 1986; Roughgarden, Gaines & Pacala, 1987; Wiens, 1989). Habitat selection by individuals may be determined not only by characteristics of a given site but by the densities of populations in other habitats over a larger area (O'Connor & Fuller, 1985). den Boer (1981) suggested that small local populations may frequently suffer extinction, only to be reconstituted by emigrants from other areas. The fine-scale demographic instability translates into long-term persistence and stability at the scale of the larger metapopulation (Morrison & Barbosa, 1987; DeAngelis & Waterhouse, 1987; Taylor, 1988).

Ecologists generally consider system openness in the context of how broad-scale processes constrain finer-scale phenomena. This is one of the primary messages of hierarchy theory (Allen & Starr, 1982) and of 'supply-side' ecology (Roughgarden *et al.*, 1987) and it is supported by

studies of the temporal dynamics of food webs as well (Carpenter, 1988). However, the ways in which fine-scale patterns propagate to larger scales may impose constraints on the broad-scale patterns as well (Huston, DeAngelis & Post, 1988; Milne, 1988). Ecologists dealing with the temporal development of systems (e.g. forest insect epidemics: Barbosa & Schultz, 1987; Rykiel *et al.*, 1988) recognize this sensitivity to small differences in fine-scale initial conditions as the effects of historical events on the subsequent state of the system.

Extent and grain

Our ability to detect patterns is a function of both the *extent* and the *grain*[1] of an investigation (O'Neill *et al.*, 1986). Extent is the overall area encompassed by a study, what we often think of (imprecisely) as its scale[2] or the population we wish to describe by sampling. Grain is the size of the individual units of observation, the quadrats of a field ecologist or the sample units of a statistician (Fig. 1). Extent and grain define the upper and lower limits of resolution of a study; they are analogous to the overall size of a sieve and its mesh size, respectively. Any inferences about scale-dependency in a system are constrained by the extent and grain of investigation — we cannot generalize beyond the extent without accepting the assumption of scale-independent uniformitarianism of patterns and processes (which we know to be false), and we cannot detect any elements of patterns below the grain. For logistical reasons, expanding the extent of a study usually also entails enlarging the grain. The enhanced ability to detect broad-scale patterns carries the cost of a loss of resolution of fine-scale details.

Variance, equilibrium and predictability

When the scale of measurement of a variable is changed, the variance of that variable changes. How this happens depends on whether grain or extent is altered. Holding extent constant, an increase in the grain of measurement generally decreases spatial variance. In a perfectly homogeneous area (i.e. no spatial autocorrelation among

---

[1] This use of 'grain' differs from that of MacArthur & Levins (1964), who considered grain to be a function of how animals exploit resource patchiness in environments.

[2] Note that what is a fine scale to an ecologist is a large scale to a geographer or cartographer, who express scale as a ratio (e.g. 1:250 000 is a smaller scale than 1:50 000).

388
J. A. Wiens

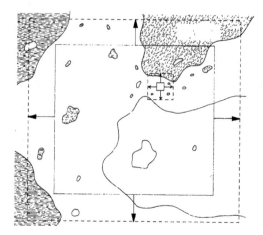

Fig. 1. The effects of changing the grain and extent of a study in a patchy landscape. As the extent of the study is increased (large squares), landscape elements that were not present in the original study area are encountered. As the grain of samples is correspondingly increased (small squares), small patches that initially could be differentiated are now included within samples and the differences among them are averaged out.

sample locations), the log-log plot of variance versus grain (or N) has a slope of −1 (Fig. 2a). In a heterogeneous area, this slope will generally be between −1 and 0 (O'Neill *et al.*, unpublished), although the relationship may be curvilinear (Fig. 2a; Levin, 1989). As grain increases, a greater proportion of the spatial heterogeneity of the system is contained within a sample or grain and is lost to our resolution, while between-grain heterogeneity (= variance) decreases (Fig. 2b). If the occurrence of species in quadrats is recorded based on a minimal coverage criterion, rare species will be less likely to be recorded as grain size increases; this effect is more pronounced if the species are widely scattered in small patches than if they are highly aggregated (Turner *et al.*, unpublished). If the measurement criterion is simply the presence or absence of species in quadrats, however, more rare species will be recorded as grain increases, and diversity will increase rather than decrease with increasing grain. Exactly how variance changes with grain scale thus depends on the magnitude and form of the heterogeneity of an area (Milne, 1988, unpublished; Palmer, 1988) and on the type of measurement taken.

Spatial variance is also dependent on the extent of an investigation. Holding grain constant, an increase in extent will incorporate greater spatial heterogeneity, as a greater variety of patch types or landscape elements is included in the area being studied (Fig. 1). Between-grain variance increases with a broadening of scale (extent) (Fig. 2b).

These considerations also relate to the patterns of temporal variation or equilibrium of ecological systems. Ecologists have often disagreed about whether or not ecological systems are equilibrial (e.g. Wiens, 1984, in press; Chesson & Case, 1986; DeAngelis & Waterhouse, 1987; Sale, 1988). Whether apparent 'equilibrium' or 'nonequilibrium' is perceived in a system clearly depends on the scale of observation. Unfortunately, current theories provide little guidance as to what we might expect: models in population biology (e.g. May & Oster, 1976; Schaffer, 1984; May, 1989) and physics (Gleick, 1987) show that order and stability may be derived at broad scales from finer-scale chaos *or* that fine-scale determinism may produce broad-scale chaos, depending on circumstances. Perhaps ecological systems follow principles of universality, their final states at broad scales depending on general system properties rather than fine-scale details (cf. Feigenbaum, 1979). Brown (1984) has championed this view, but we still know far too little

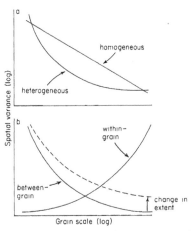

Fig. 2. (a) As the grain of samples becomes larger, spatial variance in the study system as a whole decreases, albeit differently for homogeneous and heterogeneous areas. This is related to the within- and between-grain (sample) components of variation. (b) With increasing grain scale, less of the variance is due to differences between samples and more of the overall variation is included within samples (and therefore averaged away). An increase in the extent of the investigation may increase the between-grain component of variance by adding new patch types to the landscape surveyed (Fig. 1), but within-grain variance is not noticeably affected.

temporal scales of variation (Fig. 3). With increased spatial scale, the time scale of important processes also increases because processes operate at slower rates, time lags increase, and indirect effects become increasingly important (Delcourt, Delcourt & Webb, 1983; Clark, 1985). The dynamics of different ecological phenomena in different systems, however, follow different trajectories in space and time. An area of a few square metres of grassland may be exposed to ungulate grazing for only a few seconds or minutes, whereas the temporal scale of microtines in the same area may be minutes to hours and that of soil arthropods days to months or years. There are no standard functions that define the appropriate units for such space-time comparisons in ecology. Moreover, the continuous linear scales we use to measure space and time may not be appropriate for organisms or processes whose dynamics or rates vary discontinuously (e.g. 'physiological time' associated with diapause in insects; Taylor, 1981).

Any predictions of the dynamics of spatially broad-scale systems that do not expand the temporal scale are *pseudopredictions*. The predictions may seem to be quite robust because they are made on a fine time scale relative to the actual dynamics of the system (Fig. 3), but the mechanistic linkages will not be seen because the temporal extent of the study is too short. It is as if we were to take two

about the scaling behaviour of ecological systems to consider universality as anything other than an intriguing hypothesis.

Predictability and space-time scaling

Because the effects of local heterogeneity are averaged out at broader scales, ecological patterns often appear to be more predictable there. Whether or not the predictions are mechanistically sound depends on the importance of the fine-scale details. The Lotka-Volterra competition equations may predict competitive exclusion of species that in fact are able to coexist because of fine-scale spatial heterogeneity that is averaged away (e.g. Moloney, 1988). These predictions are not really scale-independent but are instead insensitive to important scale-dependent changes.

Our ability to predict ecological phenomena depends on the relationships between spatial and

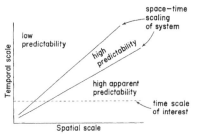

Fig. 3. As the spatial scaling of a system increases, so also does its temporal scaling, although these space-time scalings differ for different systems. Studies conducted over a long time at fine spatial scales have low predictive capacity. Investigations located near to the space-time scaling functions have high predictive power. Short-term studies conducted at broad spatial scales generally have high apparent predictability (pseudopredictability) because the natural dynamics of the system are so much longer than the period of study. Often, ecologists and resource managers have been most interested in making and testing predictions on relatively short time scales, regardless of the spatial scale of the investigation.

390

J. A. Wiens

snapshots of a forest a few moments apart and use the first to predict the second. This problem may be particularly severe in resource management disciplines, where the application of policies to large areas is often based on very short-term studies.

Detecting patterns and inferring processes

The characteristics of ecological systems at relatively fine scales differ from those at relatively broad scales (Table 1), and these differences influence the ways ecologists can study the systems. The possibilities for conducting replicated experiments vary inversely with the scale of investigation. The potential for sampling errors of several kinds are greater at finer scales, although the intensity of sampling is generally lower at broader scales. Fine-scale studies may reveal greater detail about the biological mechanisms underlying patterns, but generalizations are more

likely to emerge at broader scales. Because the time-frame of ecological processes tends to be longer at broader scales (Fig. 3), long-term investigations are more often necessary to reveal the dynamics of the system. The scale of investigation thus determines the range of patterns and processes that can be detected. If we study a system at an inappropriate scale, we may not detect its actual dynamics and patterns but may instead identify patterns that are artifacts of scale. Because we are clever at devising explanations of what we see, we may think we understand the system when we have not even observed it correctly.

Dealing with scale

Scale arbitrariness

The most common approach to dealing with scale is to compare patterns among several arbitrarily selected points on a scale spectrum. In his analysis

Table 1. General characteristics of various attributes of ecological systems and investigations at fine and broad scales of study. 'Fine' and 'broad' are defined relative to the focus of a particular investigation, and will vary between studies.

Attribute	Scale	
	Fine	Broad
Number of variables important in correlations	many	few
Rate of processes or system change	fast	slow
Capacity of system to track short-term environmental variations	high	low
Potential for system openness	high	low
Effects of individual movements on patterns	large	small
Type of heterogeneity	patch	landscape mosaic
Factors influencing species' distribution	resource/habitat distribution, physiological tolerances	barriers, dispersal
Resolution of detail	high	low
Sampling adequacy (intensity)	good	poor
Effects of sampling error	large	small
Experimental manipulations	possible	difficult
Replication	possible	difficult
Empirical rigor	high	low
Potential for deriving generalizations	low	high
Form of models	mechanistic	correlative
Testability of hypotheses	high	low
Surveys	quantitative	qualitative
Appropriate duration of study	short	long

of reef-fish communities, for example, Galzin (1987) compared distributions within a single transect, among several transects on the same island, and among five islands. Roughgarden et al. (1987) compared the dynamics of rocky intertidal barnacle communities and assemblages of *Anolis* lizards on islands at 'small', 'medium', and 'large' spatial scales. Senft et al. (1987) examined herbivore foraging in relation to vegetation patterns at the scales of the local plant community, the landscape, and the region. Multiscale studies of birds have considered patterns at three to five scales, and Wiens et al. (1986a) recognized four scales of general utility in ecological investigations.

In these examples, the definition of the different scales makes intuitive sense and the analyses reveal the scale-dependency of patterns. Casting the relationships in the context of hierarchy theory (Allen & Starr, 1982; O'Neill et al., 1986) may further sharpen our focus on scaling by emphasizing logical and functional linkages among scales. The scales chosen for analysis are still arbitrary, however: they tend to reflect hierarchies of spatial scales that are based on our own perceptions of nature. Just because these particular scales seem 'right' to us is no assurance that they are appropriate to reef fish, barnacles, anoles, cattle, or birds. We need nonarbitrary, operational ways of defining and detecting scales.

Dependence on objectives and organisms

What is an 'appropriate' scale depends in part on the questions one asks. Behavioural ecologists, population ecologists, and ecosystem ecologists, for example, all probe the relationship between resources and consumers, but differences in their objectives lead them to focus their investigations at different scales (Pulliam & Dunning, 1987). Conservation of key species or habitats may target particular patches or landscape fragments for management, whereas programmes emphasizing species richness or complexes of communities may concentrate on preserving broader-scale landscape mosaics (Noss, 1987; Scott et al., 1987).

Differences among organisms also affect the scale of investigation. A staphylinid beetle does not relate to its environment on the same scales as a vulture, even though they are both scavengers. What is a resource patch to one is not to the other. The scale on which an oak tree 'perceives' its environment differs from that of an understorey bluebell or a seedling oak (Harper, 1977). Local

populations of vagile organisms may be linked together into larger metapopulations and their dynamics may be less sensitive to the spatial configuration of local habitat patches than more sedentary species (Morrison & Barbosa, 1987; Fahrig & Paloheimo, 1988; Taylor, 1988). Chronically rare species may follow different dispersal and scaling functions than persistently common species. Consumers that use sparse or clumped resources are likely to operate at larger spatial scales than those using abundant or uniformly distributed resources, especially if the resources are critical rather than substitutable (Tilman, 1982; O'Neill et al., 1988).

Such scaling differences among organisms may be viewed in terms of 'ecological neighbourhoods' (Addicott et al., 1987) or 'ambits' (Hutchinson, 1953; Haury et al., 1978); areas that are scaled to a particular ecological process, a time period, and an organism's mobility or activity. The ecological neighbourhood of an individual's daily foraging may be quite different from that of its annual reproductive activities. The ecological neighbourhood of the lifetime movements of a tit in a British woodland may comprise an area of a few square kilometres whereas a raptor may move over an area of hundreds or thousands of square kilometres; a nomadic teal of ephemeral desert ponds in Australia may range over the entire continent. Incidence functions (Diamond, 1975) or fragmentation response curves (Opdam, Rijsdijk & Hustings, 1985) depict the ecological neighbourhoods of species with respect to colonization and persistence of populations in areas of different sizes (scales).

To some extent, differences in ecological neighbourhoods among taxa parallel differences in body mass. This raises the possibility of using allometric relationships (e.g. Calder, 1984) to predict scaling functions for organisms of different sizes. On this basis, for example, one might expect the scale of the home range of a 20-g lizard to be approximately 0·3 ha, whereas that of a 20-g bird would be in the order of 4 ha; the parallel scale for a 200-g bird would be 92 ha. Although such an approach ignores variation in allometric relationships associated with diet, age, season, phylogeny, and a host of other factors, it may still provide an approximation of organism-dependent scaling that is less arbitrary than those we usually use.

Because species differ in the scales of their ecological neighbourhoods, studies of interactions among species may be particularly sensitive to scaling. The population dynamics of predators and of their prey, for example, may be influenced

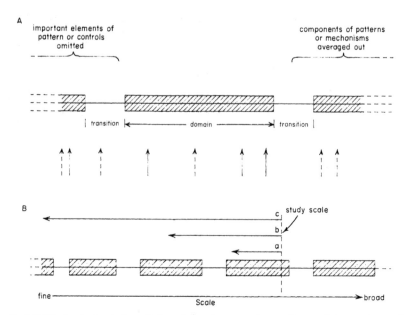

Fig. 4. (A) The domain of scale of a particular ecological phenomenon (i.e. a combination of elements of a natural system, the questions we ask of it, and the way we gather observations) defines a portion of the scale spectrum within which process-pattern relationships are consistent regardless of scale. Adjacent domains are separated by transitions in which system dynamics may appear chaotic. If the focus is on phenomena at a particular scale domain, studies conducted at finer scales will fail to include important features of pattern or causal controls; studies restricted to broader scales will fail to reveal the pattern or mechanistic relationships because such linkages are averaged out or are characteristic only of the particular domain. Comparative investigations based on sampling the scale spectrum at different points in relation to the distribution of scale domains and transitions (solid and dashed vertical arrows) will exhibit different patterns. (B) If a reductionist approach is adopted to examine patterns found at a particular scale of study, the findings (and inferences about causal mechanisms) will differ depending on how far the reductionism is extended toward finer scales and how many domains are crossed (compare a, b, and c).

by factors operating on different scales (Hengeveld, 1987), and attempts to link these dynamics directly without recognizing the scale differences may lead to greater confusion than enlightenment. The competitive interactions among species scaling the environment in similar ways may be more direct or symmetrical than those between organisms that share resources but operate on quite different scales. If we arbitrarily impose a particular scale (e.g. quadrat size) on a community of organisms that operate on different scales, we truncate the interactions to different degrees for different species.

Domains of scale

Scale-dependency in ecological systems may be continuous, every change in scale bringing with it changes in patterns and processes. If this is so, generalizations will be hard to find, for the range of extrapolation of studies at a given scale will be severely limited. If the scale spectrum is not continuous, however, there may be *domains* of scale (Fig. 4a), regions of the spectrum over which, for a particular phenomenon in a particular ecological system, patterns either do not change or change monotonically with changes in scale. Domains are separated by relatively sharp transitions from dominance by one set of factors to dominance by other sets, like phase transitions in physical systems. Normally well-behaved deterministic systems may exhibit unpredictable behaviour at such transitions (Kitchell *et al.*, 1988), and nonlinear relations may become unstable (O'Neill, personal communication). The resulting chaos makes translation between domains difficult (Heaps & Adam, 1975; May, 1989). The argument over the relative merits of linear versus nonlinear models in ecology (e.g. Patten, 1983) may reflect a failure to recognize the differences in system dynamics within versus between domains.

How may we recognize domains of scale in a way that avoids the arbitrary imposition of pre-conceived scales or hierarchical levels on natural variation? Several statistical approaches are based on the observation that variance increases as transitions are approached in hierarchical systems (O'Neill et al., 1986). If quadrats in which plant species abundances have been recorded are aggregated into larger and larger groupings, the variance of differences in abundance between pairs of contiguous groups fluctuates as a function of group size (scale). Peaks of unusually high variance indicate scales at which the between-group differences are especially large, suggesting that this may represent the scale of natural aggregation or patchiness of vegetation in the communities (Greig-Smith, 1952, 1979), the boundary of a scale domain. Similar techniques may be used to analyse data gathered on continuous linear transects (Ludwig & Cornelius, 1987). Coincidence in the variance peaks of different features of the system (e.g. plants and soil nutrients, seabirds and their prey) may indicate common spatial scalings and the possibility of direct linkages (Greig-Smith, 1979; Schneider & Piatt, 1986). For a series of point samples, the average squared difference (semi-variance) or the spatial autocorrelation between two points may be expressed as a function of the distance between them to estimate the scale of patchiness in a system (Sokal & Oden, 1978; Burrough, 1983). Other investigators have used spectral analysis (Legrende & Demers, 1984) or dimensional analysis (Lewis & Platt, 1982). Obviously, the degree to which any of these methods can reveal scales of spatial patterning is sensitive to grain and extent.

Another approach involves the application of fractal geometry (Mandelbrot, 1983; Peitgen & Saupe, 1988) to ecological patterns. In many physical systems, such as snow crystals, clouds, or flowing fluids, the configuration of patterns differs in detail at different scales but remains statistically 'self-similar' if the changes in pattern measurements are adjusted to the changes in measurement scale (Burrough, 1983; Hentschel & Procaccia, 1984; Nittman et al., 1985). The way in which detail differs with scale is characterized by a fractal dimension, D, which indexes the scale-dependency of the pattern. Statistical self-simila-

rity of patterns (constant D) occurs when processes at fine scales propagate the patterns to broader scales, although self-similar patterns may also arise from the operation of different but complementary processes (Milne, 1988). A change in the fractal dimension of a pattern, on the other hand, is an indication that different processes or constraints are dominant. Regions of fractal self-similarity of pattern may therefore represent domains of scale, whereas rapid changes in fractal dimension with small changes in measurement scale (e.g. the landscape patterns analysed by Krummel et al., 1987 or Palmer, 1988) may indicate transitions between domains. There is a relationship between the sizes and movement patterns of organisms and the fractal dimensions of their habitats (Morse et al., 1985; Weiss & Murphy, 1988, Wiens & Milne, in press), so it may be possible to define ecological neighbourhoods or domains using functions that combine allometry and fractals.

Domains of scale for particular pattern-process combinations define the boundaries of generalizations. Findings at a particular scale may be extrapolated to other scales within a domain, but extension across the transition between domains is difficult because of the instability and chaotic dynamics of the transition zone. Measurements made in different scale domains may therefore not be comparable, and correlations among variables that are evident within a domain may disappear or change sign when the scale is extended beyond the domain (as in the examples of species associations given on p. ?). Explanations of a pattern in terms of lower-level mechanisms will differ depending on whether we have reduced to a scale within the same domain, between adjacent domains, or across several domains (Fig. 4b). The peril of reductionism in ecology is not so much the prospect that we will be overcome by excessive detail or distracted by local idiosyncrasies but that we will fail to comprehend the extent of our reduction in relation to the arrangement of domains on a scale spectrum.

Of course, not all phenomena in all systems will exhibit the sort of discontinuities in scale-dependence necessary to define domains. Some phenomena may change continuously across broad ranges of scale. The boundaries of even well-defined domains may not be fixed but may vary in time, perhaps in relation to variations in resource levels. The notion of domains, like other conceptual constructs in ecology, may help us to understand nature a bit better, but it should not become axiomatic.

394

J. A. Wiens

Developing a 'science of scale' in ecology

Recently, Meentemeyer & Box (1987) have called for the development of a 'science of scale' in ecology, in which scale is included as an explicitly stated variable in the analysis. I think that we must go further, to consider scaling issues as a *primary focus* of research efforts. Instead of asking how our results vary as a function of scale, we should begin to search for consistent patterns in these scaling effects. How does heterogeneity affect the size of scale domains? Are the ecological neighbour-hoods of organisms in high-contrast landscapes scaled differently from those in areas where the patch boundaries are more gradual? Are there regularities in the transitions between orderly and seemingly chaotic states of ecological systems with changes in scale, in a manner akin to tur-bulence in fluid flows? Do selective forces influ-ence how organisms scale their environments, so that particular life-history traits are related to responses to particular scales of environmental patchiness? If one adjusts for the size differences between organisms such as a beetle and an ante-lope that occur in the same prairie, can they then be seen to respond to the patch or fractal structure of the 'landscapes' they occupy in the same way? Are differences between them interpretable in terms of differences in their physiology, reproduc-tive biology, or social organization? Does the spatial heterogeneity of soil patterns at different scales have different effects on how forest eco-systems respond to climatic changes? Is the spread of disturbances a function of the fractal structure of landscapes? Does nutrient redistribution among patches at fine scales lead to instability or stability of nutrient dynamics at broader scales?

To address such questions, we must expand and sharpen the ways we think about scaling. Our ability to detect environmental heterogeneity, for example, depends on the scale of our measurements, whereas the ability of organisms to respond to such patchiness depends on how they scale the environment. Proper analysis requires that the scale of our measurements and that of the organisms' responses fall within the same domain. Because of this, however, the 'proper' scale of investigation is usually not immediately apparent. Moreover, the ecological dynamics within a domain are not closed to the influences of factors at finer or broader domains; they are just different. Ecologists therefore need to adopt a multiscale perspective (Legrende & Demers, 1984; Clark, 1985; Wiens *et al.*, 1986a; Blondel, 1987; Addicott *et al.*, 1987). Studies conducted at several scales or

in which grain and extent are systematically varied independently of one another will provide a better resolution of domains, of patterns and their determinants, and of the interrelationships among scales.

We must also develop scaling theory that generates testable hypotheses. One particular focus of such theory must be on the linkages between domains of scale. Our ability to arrange scales in hierarchies does not mean that we understand how to translate pattern-process rela-tionships across the nonlinear spaces between domains of scale, yet we recognize such linkages when we speak of the constraining effects of hierarchical levels on one another or comment on the openness of ecological systems. Perhaps there is a small set of algorithms that can serve to translate across scales. Discovering them requires that we first recognize that ecological patterns and processes are scale-dependent, that this scale-dependency differs for different ecological systems and for different questions that we ask of them, that ecological dynamics and relationships may be well-behaved and orderly within domains of scale but differ from one domain to another and become seemingly chaotic at the boundaries of these domains, and that an arbitrary choice of scales of investigation will do relatively little to define these scaling relationships.

Acknowledgments

A Katharine P. Douglass Distinguished Lectureship at Rocky Mountain Biological Labora-tory provided the atmosphere and the impetus to focus my thinking about scaling, and I thank the staff of the laboratory for their hospitality and Mrs Douglass for endowing the lectureship. Paul Dayton, Kim Hammond, Victor Jaramillo, Natasha Kotliar, Bruce Milne, Bob O'Neill, Eric Peters, LeRoy Poff, Rick Redak, Bill Reiners, David Schimel, Dick Tracy, Monica Turner, James Ward, and Greg Zimmerman commented on the manu-script and agreed on almost nothing. Si Levin, Bob May, Bruce Milne, and Bob O'Neill shared un-published manuscripts. Ward Watt provided useful editorial suggestions. My investigations of ecological scaling are supported by the United States National Science Foundation (BSR-8805829) and Department of Energy (DE-FG02-88ER60715).

References

Addicott, J.F., Aho, J.M., Antolin, M.F., Padilla, M.F.,

395
Scale in ecology

Richardson, J.S. & Soluk, D.A. (1987) Ecological neighborhoods: scaling environmental patterns. *Oikos,* **49,** 340–346.

Ahnert, F. (1987) Process-response models of denudation at different spatial scales. *Catena Supplement,* **10,** 31–50.

Allen, T.F.H. & Starr, T.B. (1982) *Hierarchy: Perspectives for Ecological Complexity.* University of Chicago Press, Chicago.

Anderson, G.R.V., Ehrlich, A.H., Ehrlich, P.R., Roughgarden, J.D., Russell, B.C. & Talbot, F.H. (1981) The community structure of coral reef fishes. *American Naturalist,* **117,** 476–495.

Barbosa, P. & Schultz, J.C. (1987) *Insect Outbreaks.* Academic Press, New York.

Beals, E.W. (1973) Ordination: mathematical elegance and ecological naivete. *Journal of Ecology,* **61,** 23–36.

Blondel, J. (1987) From biogeography to life history theory: a multithematic approach illustrated by the biogeography of vertebrates. *Journal of Biogeography,* **14,** 405–422.

Brown, J.H. (1984) On the relationship between abundance and distribution of species. *American Naturalist,* **124,** 255–279.

Burrough, P.A. (1983) Multiscale sources of spatial variation in soil. I. The application of fractal concepts to nested levels of soil variation. *Journal of Soil Science,* **34,** 577–597.

Calder, W.A. III (1984) *Size, Function, and Life History.* Harvard University Press, Cambridge, Massachusetts.

Carpenter, S.R. (1988) Transmission of variance through lake food webs. In *Complex Interactions in Lake Communities* (ed. S.R. Carpenter), pp. 119–135. Springer-Verlag, New York.

Carpenter, S.R. & Kitchell, J.F. (1987) The temporal scale of variance in limnetic primary production. *American Naturalist,* **129,** 417–433.

Chesson, P.L. & Case, T.J. (1986) Overview: nonequilibrium community theories: chance, variability, history, and coexistence. In *Community Ecology* (ed. J. Diamond & T.J. Case), pp. 229–239. Harper & Row, New York.

Clark, W.C. (1985) Scales of climate impacts. *Climatic Change,* **7,** 5–27.

Clarke, R.D. (1988) Chance and order in determining fish-species composition on small coral patches. *Journal of Experimental Marine Biology and Ecology,* **115,** 197–212.

Cole, B.J. (1983) Assembly of mangrove ant communities: patterns of geographic distribution. *Journal of Animal Ecology,* **52,** 339–348.

Connell, J.H. (1980) Diversity and coevolution of competitors, or the ghost of competition past. *Oikos,* **35,** 131–138.

Dagan, G. (1986) Statistical theory of groundwater flow and transport: pore to laboratory, laboratory to formation, and formation to regional scale. *Water Resources Research,* **22,** 120S–134S.

Dayton, P.K. & Tegner, M.J. (1984) The importance of scale in community ecology: a kelp forest example with terrestrial analogs. In *A New Ecology. Novel Approaches to Interactive Systems* (ed. P.W. Price, C.N. Slobodchikoff & W.S. Gaud), pp. 457–481. John Wiley & Sons, New York.

DeAngelis, D.L. & Waterhouse, J.C. (1987) Equilibrium and nonequilibrium concepts in ecological models. *Ecological Monographs,* **57,** 1–21.

Delcourt, H.R., Delcourt, P.A. & Webb, T. (1983) Dynamic plant ecology: the spectrum of vegetation change in space and time. *Quaternary Science Review,* **1,** 153–175.

den Boer, P.J. (1981) On the survival of populations in a heterogeneous and variable environment. *Oecologia,* **50,** 39–53.

Denman, K.L. & Platt, T. (1975) Coherences in the horizontal distributions of phytoplankton and temperature in the upper ocean. *Memoirs Societie Royal Science Liege* (Series 6), **7,** 19–36.

Diamond, J.M. (1975) Assembly of species communities. In *Ecology and Evolution of Communities* (ed. M.L. Cody & J.M. Diamond), pp. 342–444. Harvard University Press, Cambridge, Massachusetts.

Fahrig, L. & Paloheimo, J. (1988) Effect of spatial arrangement of habitat patches on local population size. *Ecology,* **69,** 468–475.

Feigenbaum, M. (1979) The universal metric properties of nonlinear transformations. *Journal of Statistical Physics,* **21,** 669–706.

Frost, T.M., DeAngelis, D.L., Bartell, S.M., Hall, D.J. & Hurlbert, S.H. (1988) Scale in the design and interpretation of aquatic community research. In *Complex Interactions in Lake Communities* (ed. S.R. Carpenter), pp. 229–258. Springer-Verlag, New York.

Galzin, R. (1987) Structure of fish communities of French Polynesian coral reefs. I. Spatial scale. *Marine Ecology – Progress Series,* **41,** 129–136.

Giller, P.S. & Gee, J.H.R. (1987) The analysis of community organization: the influence of equilibrium, scale and terminology. In *Organization of Communities Past and Present* (ed. J.H.R. Gee & P.S. Giller), pp. 519–542. Blackwell Scientific Publications, Oxford.

Gleick, J. (1987) *Chaos: Making a New Science.* Viking, New York.

Gosz, J.R., Dahm, C.N. & Risser, P.G. (1988) Long-path FTIR measurement of atmospheric trace gas concentrations. *Ecology,* **69,** 1326–1330.

Gower, J.F.R., Denman, K.L. & Holyer, R.J. (1980) Phytoplankton patchiness indicates the fluctuation spectrum of mesoscale oceanic structure. *Nature,* **288,** 157–159.

Grebogi, C., Ott, E. & Yorke, J.A. (1987) Chaos, strange attractors, and fractal basin boundaries in nonlinear dynamics. *Science,* **238,** 632–638.

Green, D.G., Bradbury, R.H. & Reichelt, R.E. (1987) Patterns of predictability in coral reef community structure. *Coral Reefs,* **6,** 27–34.

Greig-Smith, P. (1952) The use of random and contiguous quadrats in the study of the structure of plant communities. *Annals of Botany,* New Series, **16,** 293–316.

Greig-Smith, P. (1979) Pattern in vegetation. *Journal of Ecology,* **67,** 755–779.

Harper, J.L. (1977) *Population Biology of Plants.* Academic Press, New York.

Haury, L.R., McGowan, J.A. & Wiebe, P.H. (1978) Patterns and processes in the time-space scales of plankton distribution. In *Spatial Pattern in Plankton Communities* (ed. J.H. Steele), pp. 277–327. Plenum, New York.

Heaps, N.S. & Adam, Y.A. (1975) Non-linearities associated with physical and biochemical processes in the sea. In *Modelling of Marine Ecosystems* (ed. J.C.P. Nihoul), pp. 113–126. Elsevier, Amsterdam.

396

J. A. Wiens

Hengeveld, R. (1987) Scales of variation: their distribution and ecological importance. *Annales Zoologici Fennici*, **24**, 195–202.

Hentschel, H.G.E. & Procaccia, I. (1984) Relative diffusion in turbulent media: the fractal dimension of clouds. *Physics Review A*, **29**, 1461–1470.

Hunt, G.L. Jr & Schneider, D.C. (1987) Scale-dependent processes in the physical and biological environment of marine birds. In *Seabirds: Feeding Ecology and Role in Marine Ecosystems* (ed. J.P. Croxall), pp. 7–41. Cambridge University Press, Cambridge.

Huston, M., DeAngelis, D. & Post, W. (1988) New computer models unify ecological theory. *BioScience*, **38**, 682–691.

Hutchinson, G.E. (1953) The concept of pattern in ecology. *Proceedings of the National Academy of Science of the USA*, **105**, 1–12.

Hutchinson, G.E. (1965) *The Ecological Theater and the Evolutionary Play*. Yale University Press, New Haven, Connecticut.

Jarvis, P.G. & McNaughton, K.G. (1986) Stomatal control of transpiration: scaling up from leaf to region. *Advances in Ecological Research*, **15**, 1–49.

Kareiva, P. & Andersen, M. (1988) Spatial aspects of species interactions: the wedding of models and experiments. In *Community Ecology* (ed. A. Hastings), pp. 38–54. Springer-Verlag, New York.

Kitchell, J.F., Bartell, S.M., Carpenter, S.R., Hall, D.J., McQueen, D.J., Neill, W.E., Scavia, D. & Werner, E.E. (1988) Epistemology, experiments, and pragmatism. In *Complex Interactions in Lake Communities* (ed. S.R. Carpenter), pp.263–280. Springer-Verlag, New York.

Krummel, J.R., Gardner, R.H., Sugihara, G., O'Neill, R.V. & Coleman, P.R. (1987) Landscape patterns in a disturbed environment. *Oikos*, **48**, 321–324.

Legrende, L. & Demers, S. (1984) Towards dynamic biological oceanography and limnology. *Canadian Journal of Fishery and Aquatic Science*, **41**, 2–19.

Lekan, J.F. & Wilson, R.E. (1978) Spatial variability of phytoplankton biomass in the surface waters of Long Island. *Estuarine and Coastal Marine Science*, **6**, 239–251.

Levin, S.A. (1989) Challenges in the development of a theory of ecosystem structure and function. In *Perspectives in Ecological Theory* (ed. J. Roughgarden, R.M. May & S.A. Levin), pp. 242–255. Princeton University Press, Princeton, N.J.

Lewis, M.R. & Platt, T. (1982) Scales of variation in estuarine ecosystems. In *Estuarine Comparisons* (ed. V.S. Kennedy), pp. 3–20. Academic Press, New York.

Ludwig, J.A. & Cornelius, J.M. (1987) Locating discontinuities along ecological gradients. *Ecology*, **68**, 448–450.

MacArthur, R.H. & Levins, R. (1964) Competition, habitat selection, and character displacement in a patchy environment. *Proceedings of the National Academy of Science of the USA*, **51**, 1207–1210.

Mandelbrot, B. (1983) *The Fractal Geometry of Nature*. W.H. Freeman & Company, San Francisco.

Marino, P.C. (1988) Coexistence on divided habitats: mosses in the family Splachnaceae. *Annales Zoologici Fennici*, **25**, 89–98.

May, R.M. (1989) Levels of organization in ecology. In *Ecological Concepts. The Contribution of Ecology to an Understanding of the Natural World*. BES Symposium, No. 29, pp. 339–361. Blackwell Scientific Publications, Oxford.

May, R.M. (1981) Modeling recolonization by neotropical migrants in habitats with changing patch structure, with notes on the age structure of populations. In *Forest Island Dynamics in Man-dominated Landscapes* (ed. R.L. Burgess & D.M. Sharpe), pp. 207–213. Springer-Verlag, New York.

May, R.M. & Oster, G.F. (1976) Bifurcations and dynamic complexity in simple ecological models. *American Naturalist*, **110**, 573–599.

Meentemeyer, V. (1984) The geography of organic decomposition rates. *Annals of the Association of American Geographers*, **74**, 551–560.

Meentemeyer, V. & Box, E.O. (1987) Scale effects in landscape studies. In *Landscape Heterogeneity and Disturbance* (ed. M.G. Turner), pp. 15–34. Springer-Verlag, New York.

Milne, B.T. (1988) Measuring the fractal geometry of landscapes. *Applied Mathematics and Computation*, **27**, 67–79.

Moloney, K.A. (1988) Fine-scale spatial and temporal variation in the demography of a perennial bunchgrass. *Ecology*, **69**, 1588–1598.

Morrison, G. & Barbosa, P. (1987) Spatial heterogeneity, population 'regulation' and local extinction in simulated host-parasitoid interactions. *Oecologia*, **73**, 609–614.

Morse, D.R., Lawton, J.H., Dodson, M.M. & Williamson, M.H. (1985) Fractal dimension of vegetation and the distribution of arthropod body lengths. *Nature*, **314**, 731–733.

Nittman, J., Daccord, G. & Stanley, H.E. (1985) Fractal growth of viscous fingers: quantitative characterization of a fluid instability phenomenon. *Nature*, **314**, 141–144.

Noss, R.F. (1987) From plant communities to landscapes in conservation inventories: a look at the Nature Conservancy (USA). *Biological Conservation*, **41**, 11–37.

O'Connor, R.J. & Fuller, R.J. (1985) Bird population responses to habitat. In *Bird Census and Atlas Studies: Proceedings of the VII International Conference on Bird Census Work* (ed. K. Taylor, R.J. Fuller, & P.C. Lack), pp. 197–211. British Trust for Ornithology, Tring, England.

Ogden, J.C. & Ebersole, J.P. (1981) Scale and community structure of coral reef fishes: a long term study of a large artificial reef. *Marine Ecology – Progress Series*, **4**, 97–104.

O'Neill, R.V., DeAngelis, D.L., Waide, J.B. & Allen, T.F.H. (1986) *A Hierarchical Concept of Ecosystems*. Princeton University Press, Princeton, N.J.

O'Neill, R.V., Milne, B.T., Turner, M.G. & Gardner, R.H. (1988) Resource utilization scales and landscape patterns. *Landscape Ecology*, **2**, 63–69.

Opdam, P., Rijsdijk, G. & Hustings, F. (1985) Bird communities in small woods in an agricultural landscape: effects of area and isolation. *Biological Conservation*, **34**, 333–352.

Orbach, R. (1986) Dynamics of fractal networks. *Science*, **231**, 814–819.

Palmer, M.W. (1988) Fractal geometry: a tool for describing spatial patterns of plant communities. *Vegetation*, **75**, 91–102.

Patten, B.C. (1983) Linearity enigmas in ecology. *Ecolo-*

397

Scale in ecology

gical Modelling, **18**, 155–170.

Peitgen, H.-O. & Saupe, D., eds. (1988) *The Science of Fractal Images*. Springer-Verlag, New York.

Platt, T. (1972) Local phytoplankton abundance and turbulence. *Deep-Sea Research*, **19**, 183–187.

Platt, T. & Sathyendranath, S. (1988) Oceanic primary production: estimation by remote sensing at local and regional scales. *Science*, **241**, 1613–1620.

Powell, T.M., Richardson, P.J., Dillon, T.M., Agee, B.A., Dozier, B.J., Godden, D.A. & Myrup, L.O. (1975) Spatial scales of current speed and phytoplankton biomass fluctuations in Lake Tahoe. *Science*, **189**, 1088–1090.

Pulliam, H.R. & Dunning, J.B. (1987) The influence of food supply on local density and diversity of sparrows. *Ecology*, **68**, 1009–1014.

Ricklefs R.E. (1987) Community diversity: relative roles of local and regional processes. *Science*, **235**, 167–171.

Rosswall, T., Woodmansee, R.G. & Risser, P.G., eds. (1988) *Scales and Global Change*. John Wiley & Sons, New York.

Roughgarden, J. (1983) Competition and theory in community ecology. *American Naturalist*, **122**, 583–601.

Roughgarden, J., Gaines, S.D. & Pacala, S.W. (1987) Supply side ecology: the role of physical transport processes. In *Organization of Communities Past and Present* (ed. J.H.R. Gee & P.S. Giller), pp. 491–518. Blackwell Scientific Publications, Oxford.

Rykiel, E.J. Jr, Coulson, R.N., Sharpe, P.J.H., Allen, T.F.H. & Flamm, R.O. (1988) Disturbance propagation by bark beetles as an episodic landscape phenomenon. *Landscape Ecology*, **1**, 129–139.

Sale, P.F. (1988) Perception, pattern, chance and the structure of reef fish communities. *Environmental Biology of Fishes*, **21**, 3–15.

Schaffer, W.M. (1984) Stretching and folding in lynx fur returns: evidence for a strange attractor in nature? *American Naturalist*, **123**, 798–820.

Schneider, D.C. & Piatt, J.F. (1986) Scale-dependent correlation of seabirds with schooling fish in a coastal ecosystem. *Marine Ecology – Progress Series*, **32**, 237–246.

Schoener, T.W. (1982) The controversy over interspecific competition. *The American Scientist*, **70**, 586–595.

Schumm, S.A. & Lichty, R.W. (1965) Time, space, and causality in geomorphology. *American Journal of Science*, **263**, 110–119.

Scott, J.M., Csuti, B., Jacobi, J.D. & Estes, J.E. (1987) Species richness. *BioScience*, **37**, 782–788.

Senft, R.L., Coughenour, M.B., Bailey, D.W., Rittenhouse, L.R., Sala, O.E. & Swift, D.M. (1987) Large herbivore foraging and ecological hierarchies. *BioScience*, **37**, 789–799.

Sherry, T.W. & Holmes, R.T. (1988) Habitat selection by breeding American Redstarts in response to a dominant competitor, the Least Flycatcher. *The Auk*, **105**, 350–364.

Simberloff, D. (1988) The contribution of population and community biology to conservation science. *Annual Review of Ecology and Systematics*, **19**, 473–511.

Sokal, R.R. & Oden, N.L. (1978) Spatial autocorrelation in biology. 2. Some biological implications and four applications of evolutionary and ecological interest. *Biological Journal of the Linnean Society*, **10**, 229–249.

Steele, J., ed. (1978) Spatial pattern in plankton communities. NATO Conference Series, Series IV: Marine Sciences, vol. 3. Plenum Press, New York.

Taylor, A.D. (1988) Large-scale spatial structure and population dynamics in arthropod predator-prey systems. *Annales Zoologici Fennici*, **25**, 63–74.

Taylor, F. (1981) Ecology and evolution of physiological time in insects. *The American Naturalist*, **117**, 1–23.

Therriault, J.-C. & Platt, T. (1981) Environmental control of phytoplankton patchiness. *Canadian Journal of Fishery and Aquatic Science*, **38**, 638–641.

Tilman, D. (1982) *Resource Competition and Community Structure*. Princeton University Press, Princeton, New Jersey.

Väisänen, R.A., Järvinen, O. & Rauhala, P. (1986) How are extensive, human-caused habitat alterations expressed on the scale of local bird populations in boreal forests? *Ornis Scandinavica*, **17**, 282–292.

Weiss, S.B. & Murphy, D.D. (1988) Fractal geometry and caterpillar dispersal: or how many inches can inchworms inch? *Functional Ecology*, **2**, 116–118.

Wiens, J.A. (1983) Avian community ecology: an iconoclastic view. In *Perspectives in Ornithology* (ed. A.H. Brush & G.A. Clark Jr), pp. 355–403. Cambridge University Press, Cambridge.

Wiens, J.A. (1984) On understanding a non-equilibrium world: myth and reality in community patterns and processes. In *Ecological Communities: Conceptual Issues and the Evidence* (ed. D.R. Strong Jr, D. Simberloff, L.G. Abele & A.B. Thistle), pp. 439–457. Princeton University Press, Princeton, N.J.

Wiens, J.A. (1986) Spatial scale and temporal variation in studies of shrubsteppe birds. In *Community Ecology* (ed. J. Diamond & T.J. Case), pp. 154–172. Harper & Row, New York.

Wiens, J.A. (1989) *The Ecology of Bird Communities*, Vol. 2. Processes and Variations. Cambridge University Press, Cambridge.

Wiens, J.A., & Milne, B.T. (1989) Scaling of 'landscapes' in landscape ecology from a beetle's perspective. *Landscape Ecology*, **3**, in press.

Wiens, J.A., Addicott, J.F., Case, T.J. & Diamond, J. (1986a) Overview: The importance of spatial and temporal scale in ecological investigations. In *Community Ecology* (ed. J. Diamond & T.J. Case), pp. 145–153. Harper & Row, New York.

Wiens, J.A., Rotenberry, J.T. & Van Horne, B. (1986b) A lesson in the limitations of field experiments: shrubsteppe birds and habitat alteration. *Ecology*, **67**, 365–376.

Woodward, F.I. (1987) *Climate and Plant Distribution*. Cambridge University Press, Cambridge.

Received 9 December 1988; revised 2 February 1989; accepted 7 February 1989

OIKOS 49: 340–346. Copenhagen 1987

Ecological neighborhoods: scaling environmental patterns

John F. Addicott, John M. Aho, Michael F. Antolin, Dianna K. Padilla, John S. Richardson and Daniel A. Soluk

Addicott, J. F., Aho, J. M., Antolin, M. F., Padilla, D. K., Richardson, J. S. and Soluk, D. A. 1987. Ecological neighborhoods: scaling environmental patterns. – Oikos 49: 340–346.

In this paper we review, develop, and differentiate among concepts associated with environmental patterning (patch, division, and heterogeneity), spatial and temporal scales of ecological processes (ecological neighborhoods), and responses of organisms to environmental patterning (relative patch size, relative patch duration, relative patch isolation, and grain response). We generalize the concept of ecological neighborhoods to represent regions of activity or influence during periods of time appropriate to particular ecological processes. Therefore, there is no single ecological neighborhood for any given organism, but rather a number of neighborhoods, each appropriate to different processes. Neighborhood sizes can be estimated by examining the cumulative distribution of activity or influence of an organism as a function of increasingly large spatial units. The spatial and temporal dimensions of neighborhoods provide the scales necessary for assessing environmental patterning relative to particular ecological processes for a given species. Consistent application of the neighborhood concept will assist in the choice of appropriate study units, comparisons among different studies, and comparisons between empirical studies and theoretical postulates.

J. F. Addicott and D. K. Padilla, Dept of Zoology, Univ. of Alberta, Edmonton, Alberta, Canada T6G 2E9. J. M. Aho, Savannah River Ecology Lab., Aiken, SC 29801, USA. M. F. Antolin, Dept of Biol. Sci., Florida State Univ., Tallahassee, FL 32306, USA. J. S. Richardson, Inst. of Animal Resource Ecology, Univ. of British Columbia, Vancouver, B. C., Canada V6T 1W5. D. A. Soluk, Dept of Zoology, Erindale College, Mississauga, Ontario, Canada L5L 1C6.

Introduction

Environmental patterning refers to the non-uniform, spatial and temporal distribution of resources and abiotic conditions that influence species or species interactions. Such patterning is pervasive in nature and is known or hypothesized to affect many ecological processes and phenomena, including population dynamics, life histories, dispersal, foraging behavior, patterns of natural selection, coexistence of species, predation, and species diversity (e.g. Huffaker 1958, Southwood 1962, Levin and Paine 1974, May 1974, Roff 1974, Wilbur et al. 1974, Levin 1976, Chesson 1978, Pleasants and Zimmerman 1979, Shorrocks et al. 1979, Denno et al. 1980, Hassell 1980, Wilson 1980, den Boer 1981, Paine and Levin 1981, Kareiva 1982, Spence 1983). Therefore, the study of environmental patterning and how organisms respond to it has become a central focus of current ecological research.

However, there are no general procedures and criteria for determining how organisms respond to or are affected by environmental patterning. Convenient but arbitrary spatial and temporal study units may be inappropriate for the processes being studied (see discussion by: Brown and Kodric-Brown 1977, Connell and Sousa 1983, Connor et al. 1983, Cunningham 1986), and conclusions appropriate to one scale of environmental or population patterning may be inappropriately transferred to another scale (e.g. Price 1980). Without a reasonable means of scaling, it is difficult to compare results from the same species in different environments, from different species in the same environment, or be-

Accepted 12 February 1987
© OIKOS

Fig. 1. Examples of combinations of divided and heterogeneous environments. Different patch types are denoted by different patterns of shading. Notice the variation in patch size, and the existence of patches of patches.

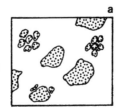

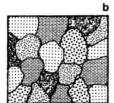

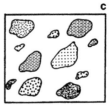

a

b

c

DIVIDED
HOMOGENEOUS

UNDIVIDED
HETEROGENEOUS

DIVIDED
HETEROGENEOUS

tween theoretical postulates and empirical results. Therefore, a precise and unified approach to the study of patterning and responses to patterning is needed.

In this paper we first discuss concepts associated with environmental patterning: patch and environmental division/heterogeneity (Wiens 1976, Shorrocks et al. 1979, Gould and Stinner 1984). Second, we develop a generalized concept of ecological neighborhoods (Wright 1943, 1946, Southwood 1977, Antonovics and Levin 1980). Ecological neighborhoods then provide a mechanism for scaling both spatial and temporal patterns of the environment relative to organisms. Next, we develop explicit procedures and criteria for implementing the concept of ecological neighborhoods. Finally, we discuss why scaling is necessary in ecological studies. Our intent is to encourage the use of more explicit criteria for defining and reporting on environmental patterning and the responses of organisms to it.

Environmental patterning

The study of environmental patterning involves descriptions of the spatial distributions of both resources and abiotic conditions. When environmental variables are discontinuous among arbitrarily chosen sampling units, environments are said to consist of patches. A patch is defined as a discontinuity in environmental character states where the discontinuity matters to the organism (Wiens 1976), as a "bounded, connected discontinuity in a homogeneous reference background" (Levin and Paine 1974), or as any place in the environment where the abundance of either resources or organisms is high or low relative to its surroundings (Roughgarden 1977). These definitions can be implemented by using any of a variety of statistical procedures (see Pielou 1969, Ripley 1981).

We distinguish two qualitatively different kinds of patterning: division and heterogeneity (Shorrocks et al. 1979, Gould and Stinner 1984). Division involves the separation of patches by regions of relatively unsuitable environmental conditions (Fig. 1a,c), where suitability is defined in terms of the fitness (or components of fit-

ness) that an organism would experience within a patch (Southwood 1977). Heterogeneity involves the existence of two or more qualitatively different patch types (Fig. 1b,c) that may or may not differ in suitability. Environments could be divided and homogeneous (Fig. 1a), heterogeneous but not divided (Fig. 1b), or both divided and heterogeneous (Fig. 1c). In a divided environment, the interpatch region could represent environments where the fitness of an organism would be zero (Southwood 1977), such as marine environments for terrestrial insects or all plants except flowering yuccas for yucca moths. Alternatively, the interpatch region could represent environments in which fitness is relatively low, but non-zero. Each type of patterning exists as a continuum, with variation from continuous to divided environments, and from homogeneous to heterogeneous environments.

Patchy environments increase the difficulty of making appropriate choices of temporal and spatial scale for study units. The distribution of organisms and patterns of resources may not reflect an organism's use of the environment. Simply identifying the existence of patchiness does not mean that the patchiness is important for a particular process. Next, environments can simultaneously exhibit patchiness at a number of different spatial scales, from millimeters to kilometers (Krebs 1978). However, a simple description of patch structure of either an organism or its environment would not indicate which spatial scale is appropriate for the study of a given process for a given organism (Connor et al. 1983, Heads and Lawton 1983). Finally, different species may treat patch sizes differently. A simple description of environmental or population patch structure provides no biologically meaningful way of comparing patches among different organisms (Southwood 1977). Comparisons must be based on the responses of organisms.

A general concept of ecological neighborhoods

Therefore, we need a mechanism for determining the scale at which experiments or observations should be made, and this scale will in turn measure patch size, patch duration, patch isolation and use of heterogeneous patches relative to organisms. The concept of ecological neighborhoods is appropriate for this task. Here we expand and generalize neighborhood concepts developed by Wright (1943, 1946), Southwood (1977) and Antonovics and Levin (1980) for scaling genetic or ecological processes. Our generalized version of the concept of ecological neighborhoods refines earlier uses to make it applicable to any ecological process and to both mobile and sessile organisms.

We define ecological neighborhoods by three properties: an ecological process, a time scale appropriate to that process, and an organism's activity or influence during that time period. The ecological neighborhood of an organism for a given ecological process is the region within which that organism is active or has some influence during the appropriate period of time. This concept of neighborhood could apply to any ecological process, such as population growth, predator-prey interactions, competition, foraging behavior, or territorial defense. The choice of an ecological process will define an appropriate time scale over which to measure neighborhood size. As with genetic neighborhoods, it may be convenient to refer to either the physical size of the neighborhood or the number of individuals within the neighborhood.

For relatively mobile organisms, the movement of individuals will usually define the neighborhood (Dobzhansky and Wright 1943, Lloyd 1967, Koeppl et al. 1975, Jumars 1976, Ford and Krumme 1979). For sessile organisms, neighborhoods will frequently be restricted to the immediate vicinity of a single individual, depending upon the precise mechanism of resource utilization or interference with other individuals (Connell 1961, Mack and Harper 1977, Antonovics and Levin 1980). However, in some cases regions of influence can extend well beyond individuals whether they are mobile or sessile. This will occur when 1) consumable resources are mobile with respect to an organism such as for sit-and-wait predators or filter feeding invertebrates (e.g. Buss and Jackson 1981), 2) an organism produces some potentially deleterious by-product that is carried in the air or water (e.g. allelopathy) (Pratt 1966, Lewis 1986), and 3) an organism attracts its consumers or its mutualistic foragers from a wide region (e.g. Manasse and Howe 1983).

Responses to environmental patterning

Relative patch size, isolation, and duration

Ecological neighborhoods provide appropriate scales for measuring the relative size, isolation, and duration

of patches in a patterned environment. Relative patch size is simply the ratio of patch size to neighborhood size. Patches of a given size are relatively small if a neighborhood encompasses a number of them, and relatively large if a neighborhood encompasses only part of one patch (see Southwood 1977). In a divided environment, patches are separated from each other. The size of the region separating patches may be just as important as the size of patches themselves (Southwood 1977). Therefore, it is also appropriate to scale interpatch distances by neighborhood sizes, leading to the concept of relative patch isolation. For example, Southwood's (1977) definition of an isolated habitat includes the idea that interpatch distance must be large relative to an organism's migratory range, "the area over which it can move when it is not reproducing".

Temporal patterning of environments is also important, because patch duration (= temporal persistence) may range from seconds to millennia (Levin and Paine 1974, Paine and Levin 1981). The basis for comparing patch duration is the time period appropriate to a given neighborhood (see Connell and Sousa 1983). Relative patch duration provides an assessment of whether patches are permanent or ephemeral with respect to a given process and organism (Southwood 1977, Connell and Sousa 1983). As with patch size, any given patch could be relatively permanent or ephemeral simultaneously, depending upon the neighborhoods and organisms being considered.

The interaction between relative patch size and relative patch duration may be complex. Temporal environmental patterning may affect the mobility or life history of organisms, and these will in turn affect neighborhood size. If resources in a given patch are insufficient for the completion of an entire life history stage, then individual organisms must disperse from one patch to another as patches are depleted (see Chew 1977). An alternative response to ephemeral resources is the evolutionary modification of either dispersal behavior or duration of life history stages to match the temporal availability of resources (e.g. Taylor 1980a,b).

Grain: responses to environmental heterogeneity

For heterogeneous patches, utilization of the different patch types can be random, in proportion to their availability, or non-random, with some patch types used more frequently than their relative availability and others less frequently. These represent fine and coarse grain responses, respectively. Although the term grain has been used in other contexts (MacArthur and Levins 1964, Levins 1968), it is most useful as a concept for the response of an organism to environmental heterogeneity. As with relative patch size and duration, grain is relative to a given process and period of time. With this definition, grain is not a property of an environment; there are not fine and coarse grained environments,

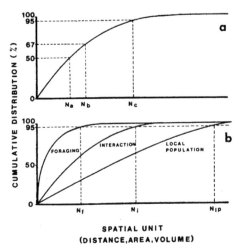

SPATIAL UNIT
(DISTANCE, AREA, VOLUME)

Fig. 2. Hypothetical examples of the relationship between cumulative distribution of movement or influence against spatial unit. Horizontal dotted lines at arbitrary points on the cumulative distribution indicate different neighborhood sizes for different decision criteria (Fig. 2a). Each curve in Fig. 2b represents a cumulative distribution with an associated neighborhood size for a different ecological process, using a decision criterion of 95%.

only fine and coarse grained responses by organisms to environmental heterogeneity.

Neighborhoods, relative patch sizes, and grain are also potentially interrelated. If patches are large relative to a given neighborhood, grain response to those patches must be coarse. If patches are relatively small, however, grain response could be either fine or coarse. More complex situations will arise when patches of different quality also differ in size or duration.

Estimating neighborhood size

There are two problems involved in measuring neighborhoods. The first is to develop procedures for measuring the distribution of activity or influence among sampling units. The second is to develop criteria for determining the boundaries of a neighborhood. These criteria will be arbitrary, but they should be explicit. The procedures needed to measure the distribution of activity or influence will depend on the kinds of neighborhoods and organisms being studied.

For neighborhoods such as local reproductive population neighborhoods that involve relatively long time periods, the most appropriate indicator of activity may be a measure of net movement of individuals (or their gametes or progeny) among sampling units during the period of time appropriate to the neighborhood. There are two general sets of procedures for measuring net movements. One is the direct measurement of dispersal distances, using a mark-release-recapture technique at a number of sampling sites to determine the cumulative frequency distribution of dispersal distances (Dobzhansky and Wright 1943). The other is to measure the distribution of locations where an organism occurs throughout an appropriate period of time (Koeppl et al. 1975, Ford and Krumme 1979). For some kinds of organisms and neighborhoods, assessment of influence will not be a function of the movement of that organism. For example, for interaction neighborhoods in sessile organisms, the regions from which resources are obtained or depleted, or the regions from which predators or mutualistic foragers are attracted by neighboring individuals must be measured (e.g. Buss and Jackson 1981, Manasse and Howe 1983).

Given a distribution of activity or influence, neighborhood size is then defined by the distance, area, or volume representing some arbitrary but explicit point on the distribution (Fig. 2a). If the concept of neighborhood is to be useful, the decision criterion must be consistent, explicit and reasonable. For population dynamics we want to identify a neighborhood where most of the population growth is due to in situ reproduction rather than due to immigration (Connor et al. 1983). For interaction groups we would want to identify the neighborhood from which most potentially limiting resources are obtained (Antonovics and Levin 1980). It is not clear a priori whether a criterion of 50%, 67% or 95% is reasonable (Koeppl et al. 1975). Moreover, the arbitrary nature of the decision criterion is clearly shown by comparing the importance of dispersal in a genetic and ecological context. The movement of a very small number of individuals can maintain genetic uniformity among populations (Roughgarden 1979:203), but the same amount of dispersal may have little effect on population dynamics. Until there is an accumulation of studies of neighborhoods, it is not so important that the criterion be common to all workers, but it is imperative that the criterion be explicit. Fig. 2b provides an example of how a decision criterion of 95% would be applied to define three different neighborhoods for a single organism.

Why scaling is necessary

Three major problems arise if the size of study units is arbitrary rather than based upon a scale appropriate to organisms and questions. First, it will be difficult or impossible to make comparisons of experimental or observational results among different species in either the same or different systems, because arbitrarily chosen study units may represent different scales (= neighborhood types) for each species. For example, in a study of the establishment, extinction, and dynamics of aphids on fireweed, Addicott (1978a) assumed that aphids on individual ramets represented local populations for each of four species of aphids. This arbitrary assumption seemed reasonable based upon the mobility of apterous aphids, population densities of aphids on individual ramets, and the work of others (Hughes 1963, Sanders and Knight 1968). However, we now know that this assumption was appropriate for *Aphis varians* but not for *Macrosiphum valerianae*; there is little or no movement of apterous *A. varians* from ramet to ramet within a clone of fireweed, whereas *M. valerianae* are relatively mobile (Antolin and Addicott, unpubl.). Therefore, analysis of data from individual ramets of fireweed relates to different levels of population structure in the two species. Comparisons of species with distinctly different life histories will be even more difficult. For example, it will be difficult to compare the responses to patterned environments for organisms as different as aphids (e.g. Addicott 1979) and birds (e.g. Rotenberry and Wiens 1980) unless we can tell whether or not each has been studied at the same relative scale.

The second problem arising from an arbitrary choice of spatial units is that a given patch size need not necessarily correspond to the particular ecological neighborhood appropriate for examining a given ecological model (see Connor et al. 1983, Heads and Lawton 1983). Therefore, it will be difficult to relate field experiments or observations to particular theoretical models or conceptual postulates, and conclusions about those models could be flawed. This would clearly be the case in the fireweed system, where individual ramets do not represent local population neighborhoods for *M. valerianae*. Data on *M. valerianae* on individual ramets should not be related to models of the dynamics of patch occupation that are based upon local populations as the unit of observation. This does not mean that it is necessarily inappropriate to study systems at arbitrary and convenient spatial scales. However, it may be inappropriate to compare data sets among different systems or with conceptual models. Since such comparisons are an essential part of our science (Connell 1983, Schoener 1983), it is important that study units be based upon appropriate and clearly identified scales.

The third problem is that different processes in the same system may occur at different scales, and therefore it may not be sufficient to examine a particular system at only one spatial or temporal scale. For example, in the aphid-fireweed system there are a number of processes that affect the dynamics of local populations of aphids, but these processes occur at a variety of scales. Competition neighborhoods can be as small as the immediate vicinity of an individual aphid for behavioral interactions (e.g. Whitham 1979), individual ramets for competition based upon plant resources, or clones of fireweed for competition for the services of ants (Addi-

cott 1978b). Predation neighborhoods can be as small as single ramets for the predaceous mites of aphids, one or more clones of fireweed for insect predators or parasitoids, or all fireweed ramets in whole drainages for avian predators of aphids. The variety of neighborhood sizes complicates the study of organisms in patterned environments, because it is extremely difficult to examine processes simultaneously that may be occurring at two or more spatial or temporal scales. However, an adequate understanding of the dynamics or behavior of a system may demand just that.

Discussion

In this paper we have made three main points. First, we believe that it is essential to distinguish between three conceptually distinct but biologically interrelated concepts: 1) the distribution of environmental variables as represented by temporal and spatial patterns; 2) the area of activity or influence of organisms during appropriate periods of time that define what we call ecological neighborhoods; 3) the use of ecological neighborhoods to scale environmental patterning, thereby allowing assessment of relative patch size, relative patch isolation, relative patch duration, and relative utilization of heterogeneous patches (grain response).

Second, the concept of ecological neighborhoods needs to be general; there should not be just one kind of process or a limited number of time or dispersal distances that should be used to scale ecological processes. Instead, each process and interaction should be examined to determine an appropriate neighborhood, defined by movement or influence during an appropriate period of time. Thus, a single, general concept of ecological neighborhoods can be applied at a variety of levels, rather than having a series of apparently unrelated concepts and terms such as ambit (Lloyd 1967, Jumars 1976), home range (Koeppl et al. 1975), territory, local population (Wright 1946), etc. Therefore, we have generalized the concepts of scaling and ecological neighborhoods, as developed by Southwood (1977) and Antonovics and Levin (1980).

Third, scaling of ecological observations is extremely important. Ecological neighborhoods provide a basis for scaling the patterning of environments relative to organisms, which in turn provides the basis for comparing species in the same system, comparing processes among organisms in different systems, making appropriate comparisons between field studies and theoretical models, and for designing field studies of population interactions at one or more appropriate scales.

Implementing the neighborhood concept by measuring neighborhood sizes and using them to scale environmental patterning will be hindered by at least five major problems. First, there will continue to be conflict between the ease of choosing study units to correspond to patch sizes and the fact that ecological neighborhoods do not necessarily correspond to patches. Similarly, the

OIKOS 49:3 (1987)

scale at which a study can be conveniently conducted may not correspond very closely to a neighborhood that should be ultimately of interest to the investigator. Third, as has been the case with genetic neighborhoods, it will be difficult to actually obtain the data required to assess neighborhood sizes. Fourth, the criteria for establishing neighborhood size are necessarily arbitrary, and there is as yet no consensus on rationale for choosing one decision criterion over another. Finally, organisms are affected by a variety of processes, each of which may have a different neighborhood size associated with it.

Despite these practical difficulties, it is imperative for ecologists to begin to report the scale(s) at which their observations and experiments are made. This could involve a detailed analysis of neighborhood sizes relative to the sizes of study units and patches, or it could simply involve reporting qualitative impressions of the natural history of the study organism to justify the use of a particular study unit. However, the future development of ecology will be hindered not by the absence of appropriate concepts and techniques for scaling environmental patterning, but rather by the failure to recognize the overriding importance of scaling for facilitating comparisons between studies and comparisons between theoretical postulates and empirical work.

Acknowledgements - This paper was conceived in 1982 during a Population Ecology Seminar on the role of heterogeneity in ecology. We would like to thank the following persons for reading and commenting on the manuscript: J. H. Brown, J. C. Holmes, P. Kareiva, P. C. Marino, M. A. Mort, P. W. Price, J. R. Spence, and J. A. Wiens.

References

Addicott, J. F. 1978a. The population dynamics of aphids on fireweed: a comparison of local and metapopulations. – Can. J. Zool. 56: 2554–2564.
– 1978b. Competition for mutualists: Aphids and ants. – Can. J. Zool. 56: 2093–2096.
– 1979. A multispecies aphid-ant association: density dependence and species-specific effects. – Can. J. Zool. 57: 558–569.
Antonovics, J. and Levin, D. A. 1980. The ecological and genetic consequences of density-dependent regulation in plants. – Ann. Rev. Ecol. Syst. 11: 411–452.
Brown, J. H. and Kodric-Brown, A. 1977. Turnover rates in insular biogeography: effect of immigration on extinction. – Ecology 58: 445–449.
den Boer, P. J. 1981. On the survival of populations in a heterogeneous and variable environment. – Oecologia (Berl.) 50: 39–53.
Buss, L. W. and Jackson, J. B. C. 1981. Planktonic food availability and suspension-feeder abundance: evidence for in situ depletion. – J. Exp. Mar. Biol. Ecol. 49: 151–161.
Chesson, P. L. 1978. Predator-prey theory and variability. – Ann. Rev. Ecol. Syst. 9: 323–347.
Chew, F. S. 1977. Coevolution of pierid butterflies and their cruciferous foodplants. II. The distribution of eggs on potential foodplants. – Evolution 31: 568–579.
Connell, J. H. 1961. The influence of interspecific competition and other factors on the distribution of the barnacle *Chthamalus stellatus*. – Ecology 42: 133–146.
– 1983. On the prevalence and relative importance of interspecific competition: evidence from field experiments. – Am. Nat. 122: 661–696.
– and Sousa, W. P. 1983. On the evidence needed to judge ecological stability or persistence. – Am. Nat. 121: 789–824.
Connor, E. F., Faeth, S. H. and Simberloff, D. 1983. Leafminers on oak: the role of immigration and in situ reproductive recruitment. – Ecology 64: 191–204.
Cunningham, M. A. 1986. Dispersal in white-crowned sparrows: A computer simulation of the effect of study-area size on estimates of local recruitment. – Auk 103: 79–85.
Denno, R. F., Raupp, M. J., Tallamy, D. W. and Reichelderfer, C. F. 1980. Migration in heterogeneous environments: Differences in habitat selection between the wing forms of the dimorphic planthopper, *Prokelisia marginata* (Homoptera: Delphacidae). – Ecology 61: 859–867.
Dobzhansky, T. and Wright, S. 1943. Genetics of natural populations. X. Dispersion rates in *Drosophila pseudoobscura*. – Genetics 28: 304–340.
Ford, R. G. and Krumme, D. W. 1979. The analysis of space use patterns. – J. Theor. Biol. 76: 125–155.
Gould, F. and Stinner, R. E. 1984. Insects in heterogeneous habitats. – In: Huffaker, C. B. and Rabb, R. L. (eds), Ecological entomology. Wiley, New York, pp. 427–449.
Hassell, M. P. 1980. Some consequences of habitat heterogeneity for population dynamics. – Oikos 35: 150–160.
Heads, P. A. and Lawton, J. H. 1983. Studies on the natural enemy complex of the holly leaf-miner: the effects of scale on the detection of aggregative responses and the implications for biological control. – Oikos 40: 267–276.
Huffaker, C. B. 1958. Experimental studies on predation: dispersion factors and predator-prey oscillations. – Hilgardia 27: 343–383.
Hughes, R. D. 1963. Population dynamics of the cabbage aphid *Brevicoryne brassicae* (L.). – J. Anim. Ecol. 32: 393–424.
Jumars, P. A. 1976. Deep-sea species diversity: does it have a characteristic scale? – J. Mar. Res. 34: 217–246.
Kareiva, P. 1982. Experimental and mathematical analyses of herbivore movement: quantifying the influence of plant spacing and quality on foraging discrimination. – Ecol. Monogr. 52: 261–282.
Koeppl, J. W., Slade, N. A. and Hoffmann, R. S. 1975. A bivariate home range model with possible application to ethological data analysis. – J. Mammal. 56: 81–90.
Krebs, C. J. 1978. Ecology: the experimental analysis of distribution and abundance. 2nd Ed. – Harper & Row, New York.
Levin, S. A. 1976. Population dynamic models in heterogeneous environments. – Ann. Rev. Ecol. Syst. 7: 287–310.
– and Paine, R. T. 1974. Disturbance, patch formation and community structure. – Proc. Nat. Acad. Sci. U.S.A. 71: 2744–2747.
Levins, R. 1968. Evolution in changing environments: some theoretical explorations. – Princeton Univ. Press, Princeton, NJ.
Lewis, W. M., Jr. 1986. Evolutionary interpretations of allelochemical interactions in phytoplankton algae. – Am. Nat. 127: 184–194.
Lloyd, M. 1967. "Mean crowding." – J. Anim. Ecol. 36: 1–30.
MacArthur, R. H. and Levins, R. 1964. Competition, habitat selection, and character displacement in a patchy environment. – Proc. Nat. Acad. Sci. U.S.A. 51: 1207–1210.
Mack, R. N. and Harper, J. L. 1977. Interference in dune annuals: spatial pattern and neighborhood effects. – J. Ecol. 65: 345–363.
Manasse, R. S. and Howe, H. F. 1983. Competition for dispersal agents among tropical trees: influences of neighbors. – Oecologia (Berl.) 59: 185–190.
May, R. M. 1974. Ecosystem patterns in randomly fluctuating environments. – In: Rose, R. and Snell, F. (eds), Progress

in theoretical biology. Academic Press, New York, pp. 1–50.

Paine, R. T. and Levin, S. A. 1981. Intertidal landscapes: disturbance and the dynamics of pattern. – Ecol. Monogr. 51: 145–178.

Pielou, E. C. 1969. An introduction to mathematical ecology. – Wiley, New York.

Pleasants, J. M. and Zimmerman, M. 1979. Patchiness in the dispersion of nectar resources: Evidence for hot and cold spots. – Oecologia (Berl.) 41: 283–288.

Pratt, D. M. 1966. Competition between *Skeletonema costatum* and *Olisthodiscus luteus* in Narragansett Bay and in culture. – Limnol. Oceanogr. 11: 447–455.

Price, P. W. 1980. Evolutionary biology of parasites. – Princeton Univ. Press, Princeton, NJ.

Ripley, B. D. 1981. Spatial statistics. – Wiley, New York.

Roff, D. A. 1974. Spatial heterogeneity and the persistence of populations. – Oecologia (Berl.) 15: 245–258.

Rotenberry, J. T. and Wiens, J. A. 1980. Temporal variation in habitat structure and shrub steppe bird dynamics. – Oecologia (Berl.) 47: 1–9.

Roughgarden, J. D. 1977. Patchiness in the spatial distribution of a population caused by stochastic fluctuations in resources. – Oikos 29: 52–59.

– 1979. Theory of population genetics and evolutionary ecology: an introduction. – Macmillan, New York.

Sanders, C. J. and Knight, F. B. 1968. Natural regulation of the aphid *Pterocomma populifoliae* on bigtooth aspen in northern Lower Michigan. – Ecology 49: 234–244.

Schoener, T. W. 1983. Field experiments on interspecific competition. – Am. Nat. 122: 240–285.

Shorrocks, B., Atkinson, W. D. and Charlesworth, P. 1979. Competition on a divided and ephemeral resource. – J. Anim. Ecol. 48: 899–908.

Southwood, T. R. E. 1962. Migration of terrestrial arthropods in relation to habitat. – Biol. Rev. 37: 171–214.

– 1977. Habitat, the templet for ecological strategies? – J. Anim. Ecol. 46: 337–365.

Spence, J. R. 1983. Pattern and process in co-existence of water-striders (Heteroptera: Gerridae). – J. Anim. Ecol. 52: 497–512.

Taylor, F. 1980a. Optimal switching to diapause in relation to the onset of winter. – Theor. Pop. Biol. 18: 125–133.

– 1980b. Timing in the life histories of insects. – Theor. Pop. Biol. 18: 112–124.

Whitham, T. G. 1979. Territorial behaviour of *Pemphigus* gall aphids. – Nature (Lond.) 279: 324–325.

Wiens, J. A. 1976. Population responses to patchy environments. – Ann. Rev. Ecol. Syst. 7: 81–120.

Wilbur, H. M., Tinkle, D. W. and Collins, J. B. 1974. Environmental certainty, trophic level, and resource availability in life history evolution. – Am. Nat. 108: 805–817.

Wilson, D. S. 1980. The natural selection of populations and communities. – The Benjamin/Cummings Publishing Company, Inc., Menlo Park, California.

Wright, S. 1943. Isolation by distance. – Genetics 28: 114–138.

– 1946. Isolation by distance under diverse systems of mating. – Genetics 31: 39–59.

G. S. Innis and R. V. O'Neill, (eds.),
Systems Analysis of Ecosystems, pp. 59-78. All rights reserved.
International Co-operative Publishing House, Fairland, Maryland.

TRANSMUTATIONS ACROSS HIERARCHICAL LEVELS

ROBERT V. O'NEILL

Environmental Sciences Division
Oak Ridge National Laboratory
Oak Ridge, Tennessee 37830 USA

SUMMARY. Hierarchy theory (Weiss, 1971) plays an implicit role in
many ecological models. A set of phenomena, here called Transmutations, occurs when a process at one level of the hierarchy is
represented by a process occurring at another level. The phenomena
can result in erroneous representations unless they are carefully
considered. The present paper discusses transmutations which
occur when population processes are represented by physiological
responses of individual organisms, when communities are represented
by the behavior of the dominant population and when ecosystem
processes are represented simplistically.

KEY WORDS. model, hierarchy theory, ecosystem, error, function.

1. INTRODUCTION

The development of large-scale ecological models depends
implicitly on a concept known as hierarchy theory (Weiss, 1971),
which views biological systems in a series of hierarchical levels
(i.e., organism, population, trophic level, ecosystem). The theory
states that an explanation of a biological phenomenon is provided
when it is shown to be the consequence of the activities of the
system's components, which are themselves systems in the next lower
level of the hierarchy. Thus, the behavior of a population is to
be 'explained' by the behavior of the organisms in the population.
The initial step in any modeling project is, therefore, to identify
the system components and the interactions between them. Thus, the
modeler attempts to predict or explain the behavior of his system by
dealing with the known behavior of its components.

60   R. V. O'NEILL

Since hierarchy theory plays an important role in model conceptualization and design, it is helpful to bring together a diverse
group of phenomena under the common title of transmutation. Transmutation occurs when a process (or the mathematical function describing the process) is changed as one moves from one hierarchical
level to the next. A number of different types of change occur,
and these transmutations must be considered carefully to avoid
errors in the application of hierarchy theory to ecological problems.

A series of examples of transmutation will be presented to show
how and why these changes occur. The types of changes will be
summarized and possible implications of transmutation for hierarchy
theory, for the modeler and for the ecological theoretician will
be developed.

2.   TRANSMUTATIONS BETWEEN THE ORGANISM AND
      POPULATION LEVELS

One type of transmutation occurs when total population processes are 'explained' by physiological responses of organisms
within the population. I will try to show that the manner in
which a population rate (e.g., respiration) varies with an environmental variable (e.g., temperature) may not be the same as that
determined for individuals.

A common assumption is that the population responds like the
mean individual. If an organism's rate is an exponential function
of temperature, the population response is assumed to be exponential with rate constants equal to the mean of the rate constants
associated with the individuals in the population. This assumption follows logically, but naively, from hierarchy theory since
the population is to be 'explained' by the behavior of its components.

The examples below will attempt to show that the organism-
level process can be transmuted on the population level. One of
two phenomena may occur. In some cases, the mathematical form
of the dependency may change. In other cases, the form of the
dependency will remain but the rate constants will not be the
mean constants determined by sampling individuals in the populations.

2.1 Transmutation of a Threshold Function. Consider a population
in which each individual organism shows the dependence of some
arbitrary rate, R , on temperature, T , which is illustrated
in Figure 1a. Below a critical temperature, Tc , the rate is
0.1. At Tc , a threshold is reached, and the rate becomes 1.0.
Thus,

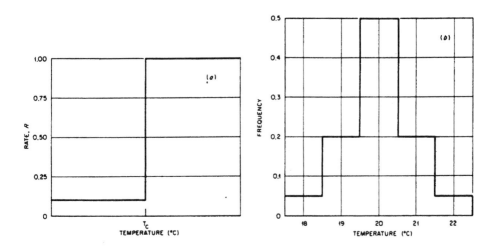

FIG. 1: (a) Rate R as a function of temperature (equation 1). It is assumed that each individual in the population responds in this manner, but the critical temperature, Tc, is distributed across the population due to genetic variability. (b) Frequency distribution of Tc across the hypothetical population.

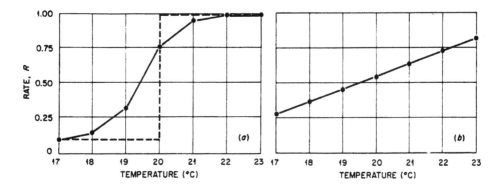

FIG. 2: (a) The transmuted dependency of rate R on temperature. Dotted line shows the dependency exhibited by each member of the population. Solid line shows the population rate if the critical temperature is distributed as shown in Figure 1b. (b) The limit of the transmutation of rate R on temperature as the genetic variability in the population reaches a maximum.

62   R. V. O'NEILL

$$R = \begin{cases} 1.0 & , \quad T \geq Tc \quad , \\ 0.1 & , \quad T < Tc \quad , \end{cases} \qquad (1)$$

In this hypothetical population, each individual responds as equation 1, but the value of Tc differs due to genetic variability. The mean value of Tc is 20.0°C. The null hypothesis is that the population responds as equation 1 with Tc = 20. Figure 1b shows the frequency distribution of Tc for individuals in the population. The distribution is represented as discrete to simplify calculations. The distribution shows that 50% of the population reaches the threshold at 20.0°C; the other half switches either above or below this temperature. The population rate is the sum, weighted by fraction, of the rates of these five subgroups. At 18°C, for example, 5% of the population has reached the threshold and R = 1.0. Thus, the population rate is (0.05)(1.0) + (0.95)(0.1) = 0.145.

The transmuted function is shown in Figure 2a with the original threshold function superimposed as a dotted line. The dependency is no longer a sharp discontinuity; rather, there is a smoothing of the curve. The inflection point of the curve also has shifted downward to approximately 19.5°C. The form of the original function has been transmuted due to the genetic variability of the population. It should be intuitively obvious that if we replaced the discrete distribution (Figure 1b) with a continuous distribution, Figure 2a would be a smooth curve. This smooth curve has been described in some modeling studies as an arctangent function (Kohlemainen and Nelson, 1969; Bledsoe et al., 1971). Note, however, that it is difficult to choose the proper parameters for the arctangent function, since the inflection point does not correspond to the mean Tc of the individuals in the population.

Let us assume that genetic variability increases such that the distribution in Figure 1b approaches a uniform distribution. To simplify calculations, assume that Tc becomes uniformly distributed between 16° and 25°C, and 10% of the population reaches a threshold at each intermediate temperature. Figure 2b shows that the transmuted function approaches a straight line:

$$R = 0.09 \, T - 1.25 \quad . \qquad (2)$$

Thus, the transmutation of a threshold step function (equation 1) approaches a straight line as the genetic variability of the population approaches a uniform distribution.

If a wider range of temperatures is considered, e.g., from 0° to 40°C, then the transmuted function becomes a piece-wise linear function of the form:

$$R = \begin{cases} 1.0 & , \quad T > 25 , \\ 0.09T-1.25 & , \quad 16 \leq T \leq 25 , \\ 0.1 & , \quad T < 16 . \end{cases} \qquad (3)$$

While functions of the form of equation 3 have been utilized in ecosystem models (e.g., Bledsoe *et al.*, 1971; Patten *et al.*, 1975), they have been presented as approximations to some more complex function. In the present analysis equation 3 is shown to describe the actual population dependency for the conditions stated above. If each individual in a population shows a threshold response, and the critical threshold value is distributed due to genetic variability, the transmuted function is an arctangent function (Figure 2a) which will tend toward a linear (equation 2) or piece-wise linear (equation 3) function as genetic variability tends toward a uniform distribution.

2.2 Transmutation of a More Complex Discontinuous Function. Transmutation is not limited to simple threshold functions. Consider the rate, S , of some processes to be a function of temperature, T , as shown in Figure 3a. In this case, the process (e.g., egg-laying or flowering) is activated at a critical temperature, Tc , and deactivated at a temperature 1°C higher, T_{c+1} . The process is only slightly more complex than the dependency in Figure 1a and can be described as the product or sum of step functions.

Assume that the value of Tc is distributed across the population as shown in Figure 1b. The transmutation is due to the same genetic variability as in the first example. For example, at 19°C 25% of the population is activated and 75% is deactivated. The population rate becomes: (0.25)(1.0) + (0.75)(0.0) = 0.25.

The transmuted function is shown in Figure 3b. The original function (Figure 3a) has been smoothed and spread over a wider range of temperatures. In addition, the maximum rate has been reduced from 1.0 to 0.7. Unlike the first example, the point of inflection remains unaltered between 20° and 21°C. Once again the form of the function has been transmuted.

Let us now consider that genetic variability has increased so that Figure 1b becomes a uniform distribution between 16° and 25°C. As the population approaches this uniform distribution, the rate, S , approaches a constant, 0.2, between 17° and 24°C. The importance of this particular transmutation must be carefully considered. As the genetic variability of the population increases, the rate, S , approaches a constant. This means that the population, composed of temperature-dependent organisms, tends toward homeothermy. The effect of the transmutation is a tendency toward a more constant rate over a wider range of temperatures.

64   R. V. O'NEILL

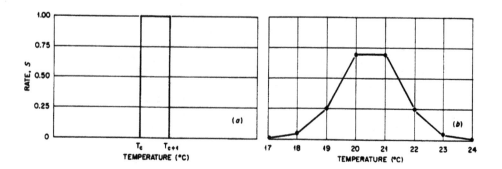

FIG. 3: (a) Rate S as a function of temperature. It is assumed
that each individual in the population responds in this manner,
but the critical temperature, Tc, is distributed across the population as shown in Figure 1b. (b) The transmuted dependency of S
on temperature, calculated as the actual population rate.

2.3 Discussion. In developing examples of the transmutation of
physiological functions at the population level, discontinuous
functions have been deliberately chosen. This is simply because the
transmutations tend to be more dramatic and, therefore, are better
pedagogical tools. Later examples will show that transmutations
occur with other classes of functions as well.

Those with a broad familiarity with probability theory may be
inclined to dismiss the concept of transmutation. It is well
known that for a function, g(T) , with a parameter, x , that is
a random variable distributed according to the probability density,
f(x) , the expected value for the function at T is given by

$$E \ g(T) = \int_{-\infty}^{\infty} g(T,x)f(x) \ dx \ . \tag{4}$$

The first two examples follow from equation 4 and careful
examination will show that the computations are discretized forms
of equation 4. However the theoretical result given by equation 4
is of little value since the integral cannot be explicitly determined
for the probability distributions of interest in ecology, such as

the Gaussian or log-normal distributions. The error of assuming that a population responds like the average individual also can be found in almost every large scale ecological model. Therefore, it remains of some interest and importance to point out that transmutation does occur.

In summary, then, as a result of genetic variability, population rates will tend to be maintained over a wider range of an environmental variable than any individual organism. The rates at the extremes will be higher than the individual rates and the rates at the optimum will be lower.

3. TRANSMUTATION BETWEEN THE POPULATION AND COMMUNITY (OR TROPHIC) LEVEL

By considering the transmutations that occur across the population/community boundary, three additional points can be made. First, transmutation is not limited to discontinuous functions but occurs with more complex continuous functions as well. Second, phenomena other than genetic variability will cause transmutations. Third, transmutation can occur between any two hierarchical levels, not only between organisms and populations.

3.1 Transmutation of a Complex, Continuous Function. Consider a community of sympatric populations with a growth rate, U, which responds to temperature as

$$U = Um \ V^X \ e^{X(1-V)} \ , \tag{5}$$

where $V = \dfrac{Tm - T}{Tm - To}$, $X = \dfrac{W^2}{400} (1 + \sqrt{1 + 40/W})^2$, and
$W = \ln Q \ (Tm - To)$, and where Tm is the upper lethal temperature, To is the optimal temperature at which growth reaches a maximum rate, Um , and Q represents the Q_{10} factor. The Q_{10} factor is the number of times a rate increases with a 10°C increase in temperature, over intermediate temperatures (Prosser and Brown, 1961).

Equation 5 has been applied in a number of ecological models (Shugart *et al.*, 1974; Kitchell *et al.*, 1974; Park *et al.*, 1974; Titus *et al.*, 1975) and a closely related function has been found useful in additional studies (e.g., Bledsoe *et al.*, 1971). The equation is merely an empirical function which shows the same numerical properties as numerous poikilothermic phenomena. The complexity results from a reparameterization into Tm , To , and Q which are biologically meaningful and measurable.

66   R. V. O'NEILL

*TABLE 1: Parameters values for five populations in a hypothetical
community. Parameters are inserted in equation 5 and used to
generate Figure 4.*

Population	T_0	Q	T_m	U_m	Fraction of total community
1	10	2.4	22	1.4	0.05
2	15	2.2	26	1.2	C.20
3	20	2.0	30	1.0	0.50
4	25	1.8	34	0.8	0.20
5	30	1.6	38	0.6	0.05

Table 1 presents parameter values for the five populations in
a hypothetical community. Since the discussion focuses on populations, rather than individual organisms, the parameters are not
only distributed among populations but also interrelated within a
population. Values within each population are interrelated because
of adaptive strategies to minimize competition. Thus, the first
population reproduces rapidly at lower temperatures and can tolerate a wider range. Conversely, population five is adapted to
a higher and narrower temperature range and reproduces less rapidly.

Unlike genetic variability, the rate constants in a community
are not distributed around a mean according to any probability
distribution, but are determined by evolutionary shifts in gene
frequencies. As a result, the parameters will differ among
populations and will (1) be as widely separated as possible to
minimize competition and (2) be interrelated within a population
to optimize a particular adaptive strategy.

The populations in Table 1 are not intended to represent any
existing community, but have been chosen to show the type of parameter distribution and relationship which would be expected. These
hypothetical populations will suffice to show that, as a result of
competition and evolution equation 5 will be transmuted at the
community level.

If the five populations are equal in size, the mean parameter
values are equal to those of population #3. The dotted line in
Figure 4 was generated by inserting these average population values
into equation 5. The actual growth rates (solid line) were generated by calculating the growth rate separately for each population,
summing, and dividing by five since each population represents 1/5
of the community.

The response of the community to temperature has been transmuted. However, the transmutation is less radical than the discon-

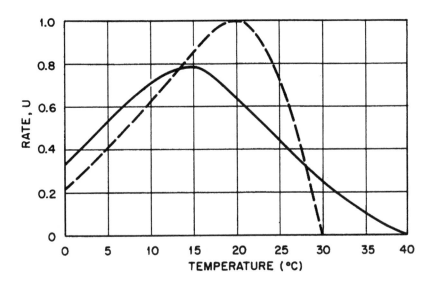

FIG. 4: The rate U as a function of temperature (equation 5). Dotted line is calculated from equation 5 assuming that the community responds like the average population. Solid line is the actual community rate calculated separately and summed for the populations.

tinuous functions in previous examples. Comparison of the two curves in Figure 4 illustrates the general features of the transmutation. As noted in previous examples, the function is spread over a wider temperature range. The maximum rate has lowered and the upper lethal temperature for the community is not an average but is equal to the upper lethal temperature of the most thermophilic population. The shift to a lowered optimal temperature results from the correlation of lower optimal temperatures and higher growth rates (Table 1). The general form of equation 5 has been retained, but the parameters have changed significantly. Inspection of the solid line in Figure 4 suggests: $T_o = 15$, $T_m = 40$, $U_m = 0.79$, and $Q = 2.0$. A reasonable fit to the solid line can be obtained by inserting these parameters into equation 5.

3.2 Transmutation Due to Differential Response. To this point, rate processes have been considered as a function of some environmental variable. In the context of model-building, equations 1-5 would be individual terms in the difference or differential equations describing the system. For the remainder of the paper, transmutations will be discussed in the context of the number and form of *equations* in the model, rather than transmutations on individual *terms* within the equations.

68   R. V. O'NEILL

Consider a community composed of two populations. The growth rate of the populations is given by:

$$\frac{dx_i}{dt} = \begin{cases} a_i x_i & , \quad x_1 + x_2 < 100 \\ \\ 0 & , \quad x_1 + x_2 \geq 100 \end{cases} \qquad (6)$$

where x_i is the number of individuals in the ith population ($i=1,2$) and a_i is the intrinsic rate of increase. In this simple system, the populations grow exponentially until the total community numbers 100 individuals, at which point some spatial (e.g., territorial) limitation stops all growth.

If $a_1 = 2.0$ and $a_2 = 0.1$, then the first population would be dominant, composing about 99% of the individuals in the community ($t=0$, $x_1(0)=x_2(0)=0.0$). One might assume that community dynamics could be adequately described by the response of the dominant population. Often, this assumption is adequate and transmutation is not important. However, a simple example can be constructed in which a transmutation will dominate the dynamics of the community.

Let us assume that one population is sensitive to some stress which we apply to the system. The stress causes a_1 to change from 2.0 to -1.0. However, the other population is immune to this stress and a_2 remains 0.1. If the community responds like the dominant population, x_1, the system goes rapidly to zero following the application of stress (Figure 5, solid line). The stress eliminates the community. However, considering the minor population, x_2, as well, (Figure 5, dashed line) the community recovers, since the second population is not sensitive to stress. Indeed, the number of individuals in the community has reached 100 by 46 time intervals. As a result of stress, the community response is transmuted across the hierarchical level. Instead of eliminating the community (as predicted by considering only the dominant population), stress merely changes the species composition and the perturbation has absolutely no effect on the number of individuals in the community when viewed 50 time intervals later.

It should be obvious that if the community included additional species with varying sensitivities, the response would differ still more from the response of the individual populations. The depression away from 100 in Figure 5 would be less and the recovery would be more rapid. The dotted line in Figure 5 was generated

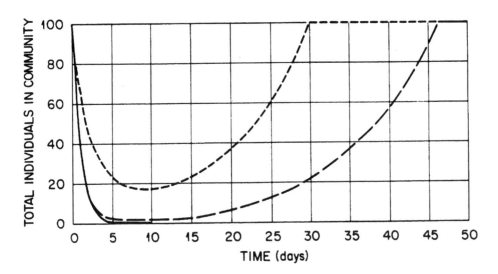

FIG. 5: The number of individuals in a hypothetical community following the introduction of a stress. Solid line is generated from equation 6 assuming that the community responds like the dominant population. Dashed line is generated from equation 6 assuming a second minor population is considered which is immune to the stress. Dotted line considers a community of five populations with three additional populations having sensitivities intermediate between the other two.

from a community of five populations with post-stress values for a_1 of -1.0, -0.725, -0.45, -0.175, and +0.1 and relative frequencies in the original community of 5%, 20%, 50%, 20%, and 5%. As the number of species becomes large the community would tend to remain at a constant 100 individuals. Assuming that the sensitivities of the additional populations continued to vary widely, the depression away from 100 would continue to diminish and the recovery rate would increase. Therefore, no response of the community to the stress would occur. The community would tend to respond like the least sensitive population, rather than like the average population. One could expect shifts in species dominance, but little overall change in the total number of individuals in the community.

4. TRANSMUTATIONS ON THE ECOSYSTEM LEVEL

At the ecosystem level of resolution, it is difficult to construct simple examples that lend insight to transmutation

70   R. V. O'NEILL

processes, unless, we restrict ourselves to simple linear models. Although the resulting models are simple, their simplicity makes it possible to present complete analyses of the phenomena.

4.1 Transmutation of 'Shunt.' Consider the two component system shown in Figure 6. The first component, x_1 , (e.g., conductile tissues of plants) passes material through while the second component x_2 , (e.g., bole wood in a tree) represents a 'shunt,' receiving material from and returning it to x_1 . It would seem natural to aggregate these components and consider a single, combined system (e.g., tree boles). The time solutions of the two component system and the aggregated, single variable representation can now be compared.

Assume that the system in Figure 6 can be represented by

$$\frac{dx_1}{dt} = 1 - (a + b)x_1 + cx_2 \ ,$$

$$\frac{dx_2}{dt} = bx_1 - cx_2 \ , \tag{7}$$

where $I = 10.0$, $a = b = 1.0$, $c = 0.01$. The corresponding single variable system would be

$$\frac{dy}{dt} = I - ky \ . \tag{8}$$

The value of k is established by setting $y(\infty) = x_1(\infty) + x_2(\infty)$. Therefore,

$$\frac{I}{k} = \frac{I}{a} + \frac{bI}{ac} \ , \quad \text{so that}$$

$$k = 0.0099 \ .$$

Also let $x_1(0) = x_2(0) = y(0) = 0.0$ at $t = 0$. Because of the simplicity of equations 7 and 8, it is possible to arrive at explicit time solutions. The methods of solution as well known and are discussed in an ecological context by Funderlic and Heath (1971). Rather than showing the solution methods in detail, we will simply state that

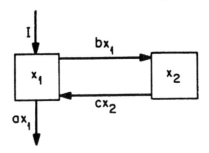

FIG. 6: Flow diagram for a simple two component system. The direct flow of material is through x_1 (e.g., conductile tissue in a tree). The second component, x_2, is a kind of 'shunt' which receives material from and returns material to x_1 (e.g., bolewood).

$$x_1(t) = 10 - 5\ e^{-2.005t} - 5\ e^{-0.005t} \ ,$$

$$x_2(t) = 1000 - 7.5\ e^{-2.005t} - 992.5\ e^{-0.005t} \ , \quad (9)$$

$$y(t) = 1010 - 1010\ e^{-0.0099t} \ .$$

The time behavior of the two component system is determined by the sum of two exponentials, while the single component system is determined by a single exponential. The sum of two exponentials cannot be represented by a single exponential unless the coefficients are equal. Thus, an error, Z , has been introduced. The magnitude of the error is the difference between $y(t)$ and $x_1(t) + x_2(t)$:

$$Z = 1010\ e^{-0.0099t} - 12.5\ e^{-2.005t} - 997.5\ e^{-0.005t} \ . \quad (10)$$

By setting the first time derivative of equation 10 equal to zero and solving iteratively for t , it can be shown that the maximum error occurs at t = 142. By inserting this value of t into equations 9 and 10, the maximum error is shown to be 46% of the true value. Thus, for the parameter values presented, the error is quite significant.

By repeating the calculations with other parameter values, that the error can be shown to increase as the difference between a + b and c increases. The rror also increases as the difference between $x_1(\infty)$ and $x_2(\infty)$ increases. A general analysis of the aggregation error introduced in linear models, as well as important classes of nonlinear models, is presented in a forthcoming paper (O'Neill and Rust, manuscript).

72   R. V. O'NEILL

In the terminology developed earlier, equation 10 represents
an error of transmutation. The error, in the present case, is
not due to genetic variability of competitive interactions but due
to the direct interactions (i.e., material exchange) between components of the system. System components (i.e., x_1 and x_2)
respond like single exponentials when isolated from each other.
But in combination, the components respond like the sum of two
exponentials. As a result of the interactions between the components, the system cannot be described by a single exponential,
even though its components respond like single exponents in isolation. Therefore, a transmutation results since the total system
cannot be described by the same function which models the components in isolation.

4.2 Transmutation of a Rapid Bypass. From the modeler's viewpoint
all of the examples above have led to rather undesirable results.
In each case, the naive application of hierarchy theory has led
to errors. Therefore, it seems only just to present an example
in which the transmutation leads to a simplification.

Consider the two component system shown in Figure 7. In this
system, the major flow of material is through x_1 while x_2 represents a component 'feeding' on x_1 . This system can be
represented by

$$\frac{dx}{dt} = I - (a + b) x_1 \; ,$$

$$\frac{dx}{dt} = bx_1 - dx_2 \; , \qquad (11)$$

where $I = 10$, $a = b = 0.001$, $d = 1.0$. For the case $I = 0.0$,
$t = 0$, $x_1(0) = 5000$, and $x_2(0) = 5$, a solution is given by

$$x_1(t) = 5000 \; e^{-0.002t} \; , \qquad (12a)$$

$$x_2(t) = 4.98 \; e^{-0.002t} + 0.02 \; e^{-1.0t} \; . \qquad (12b)$$

This example could be considered to represent a large mass of
decomposing organic matter (x_1) and a population of small decomposers (x_2), perhaps soil arthropods. Equations 12 reveal that the
organic matter (x_1) will slowly decompose and its dynamics will be
governed by a single exponential. The behavior of x_2 through
time is determined by the sum of two exponentials. However, the

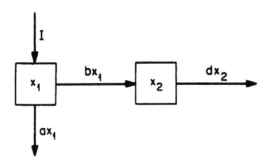

FIG. 7: Flow diagram for a simple two component system. The majority of the material flows directly through x_1 *(e.g., decomposing organic matter). The second component,* x_2, *represents a population consuming* x_1 *as a food supply (e.g., soil arthropods).*

second term of the series, 0.02 exp(-1.0 t), goes to zero very rapidly. For example, after ten time intervals, this term equals 9.08×10^{-7} . Therefore, as time increases, the dynamics of x_2 will be almost totally dominated by the first term, 4.98 exp(-0.002 t).

Let us examine what error would be introduced by dropping the second term and assuming that

$$x_2(t) = 5 \, e^{-0.002t} \quad . \tag{13}$$

The error, Z , which is introduced is equal to the difference between equation 13 and equation 12b:

$$Z = 0.02(e^{-t} - e^{-0.002t}) \quad . \tag{14}$$

By setting the first derivative of equation 14 to zero, it can be shown that the maximum error occurs at $t = 6$. At that point, the error is 0.0197. The true value of x_2 at t=6 is 4.9206 and, thus, the maximum error is about 0.4%. Such a small error would be acceptable for most purposes. Therefore, a system (Figure 7) with a small fast component feeding on a large, slow component, can be represented as a single-exponential system:

$$\frac{dx}{dt} = I - (a + b)x_1 \quad ,$$
$$x_2(t) = 0.001 \, x_1(t) \quad . \tag{15}$$

74   R. V. O'NEILL

The transmutation from equations 11 to equations 15 is fortunate
since it reduces the complexity (i.e., order) of the system.
Transmutation does not always make the system more complex than
the behavior of its components. In this and the following example,
the system behaves as though it were simpler than the sum of its
components.

4.3 Transmutations of Complex System Output. Consider an unspeci-
fied, linear, system, and assume we are interested in the flow of
some material out of this system. To study system behavior, a constant daily supply of radiotracer is introduced for 30 days. Subsequently, the tracer remaining in the system is monitored over the
following 30 days. This is a common type of experiment, since the
rate at which the material leaves the system is related to the
turnover of that substance by the components within the system.

Figure 8 shows the data points produced from this experiment.
If the system can be represented by a system of linear differential
equations, the time behavior for such a system can be represented
by a sum of exponential terms. Therefore, we approach the analysis
of the data in Figure 8 by fitting it to an exponential series.
We arrive at

$$x_t = 499.3 \; e^{-0.0092t} + 56.1 \; e^{-0.215t} \; , \tag{16}$$

which is the equation of the solid line in Figure 8. The coefficients of equation 16 were fitted by a graphical process commonly
applied to radiotracer data. The procedure is discussed in ecological context in O'Neill (1971). The procedure permits a rapid
approximation, but does *not* result in a least squares fit to the
data. Thus, it should be possible to arrive at a closer fit.
Nevertheless, equation 16 yields an R^2 of 0.984. The mean residual is 2 plus or minus a standard deviation of 6. Therefore, the
conclusion seems well justified that the system is a two component
system.

The data in Figure 8 were actually generated from the function

$$x_t = 148 \; e^{-0.001t} + 139 \; e^{-0.005t} + 130 \; e^{-0.01t} + 78 \; e^{-0.05t}$$

$$+ \; 48 \; e^{-0.1t} + 10 \; e^{-0.5t} + 5 \; e^{-t} + 1 \; e^{-5.0t} + 0.5 \; e^{-10t}$$

$$+ \; 0.05 \; e^{-100t} \; . \tag{17}$$

A random error of $\pm$ 2% was subsequently added to each data point
to produce Figure 8. When equation 17 is compared to the data, the
R^2 is 0.985. Thus, the two component model is within 0.1% of
explaining all of the variance in the data that was not purely random.

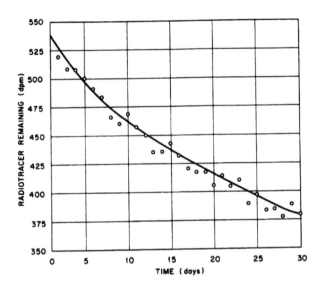

FIG. 8: Data on the rate of elimination of a radiotracer from a hypothetical ecological system. Solid line represents an empirical fit to the data (equation 16).

The transmutation of equation 17 into equation 16 can be explained in part by an examination of the coefficients in equation 17. The first data point was taken one day after the last injection of isotope. By that time, the sum of the last four terms in the series was less than 0.04 isotope units out of the total 540 units remaining in the system. The data contain almost no information on the contribution of these four terms to the dynamics of the system. On the other hand, the first two terms of the series approach zero very slowly so that 30 days of monitoring were insufficient to characterize these exponentials. Therefore, the result is partially due to the inadequacies of the data. Nevertheless, data of this type seldom require more than three exponentials, no matter how many components are operating in the system.

The coefficients that were fitted in equation 16 do not correspond simply to the coefficients of the generating equation (equation 17). Thus, the fitted coefficients cannot be legitimately assigned any ecological significance, and do not correspond to the rate constants of any component in the system. Nevertheless, we can argue that equation 17 has been transmuted to equation 16 in the sense that system behavior in this example can be described by the simpler equation. While it is possible to explain the data by applying the original equation (equation 17) with its ten components, the data can also be explained by a simpler model without any direct reference to system components.

5. DISCUSSION AND IMPLICATIONS

On the basis of a series of example, it is obvious that transmutation is caused by a number of well-known and previously described
phenomena. For example, transmutation can be caused by the trivial
observation that the sum of two exponentials cannot be exactly
described by a single exponential. Nevertheless, the examples
serve to demonstrate that changes do occur across hierarchical
levels, and that the changes can be counterintuitive.

Nothing in this paper contradicts the basic tenet of hierarchy
theory that the behavior of a system can be 'explained' by the
behavior of its components. However, the examples demonstrate
the dangers of a naive application of this approach. To summarize
the observations: (1) the relevant 'components' of a system may
not be intuitively obvious. Thus, Figure 6 and equation 10 illustrated that 'tree boles' is not a relevant component of material
passage through a tree or forest. (2) A system may not show the
same dependence on the environment that is shown by its components. Thus, a poikilothermic population with the temperature
dependence shown in Figure 3a tended toward homiothermy as genetic
variability tends toward a uniform distribution. (3) Although a
system may be explained in terms of its components, another explanation, simpler and equally valid, may exist. Thus equation 15 and
16 show system behavior explained by functions simpler than the
sum of the components. These observations indicate the dangers of
a naive application of hierarchy theory. They also suggest that
further investigation of transmutation would help refine and
clarify the principles of the theory.

The implications of transmutation for modeling seem painfully
obvious. Many simplifying assumptions, such as, the population
responds, like the mean individual, must be opened to serious
question. The possibility of transmutation must be considered
throughout the modeling process. However, by direct consideration
of transmutation, it may also be possible to introduce new
approaches to simplification (e.g., equations 15 and 16).

Perhaps the most intriguing implications of transmutation are
directed toward the theoretician. Through the diversity of examples
presented in this paper, a few patterns seem to emerge. The strongest pattern (and the most interesting) is that transmutation appears
to increase the homeostasis of the system relative to its components. In the first three examples, transmutation permitted the
system to operate over a wider range of conditions than was possible
for any single component. The fourth example (Figure 5) showed that
an increased resistence to stress and more rapid recovery could be
attributed to transmutation. Further investigations of transmutation would certainly be fruitful if it could be demonstrated that
certain classes of transmutation always lead to, and therefore
explain, increased homeostasis on the next hierarchical level.

TRANSMUTATIONS 77

ACKNOWLEDGEMENTS

Research sponsored by the Eastern Deciduous Forest Biome, US-IBP, funded by the National Science Foundation under Interagency Agreement AG-199, DEB76-00761 with the U. S. Department of Energy, Oak Ridge National Laboratory (operated by Union Carbide Corporation under contract W-7405-eng-26 with the U. S. Department of Energy). Contribution No. 323, Eastern Deciduous Forest Biome. Publication No. 1248, Environmental Sciences Division, ORNL.

REFERENCES

Bledsoe, L. J., Francis, R. C., Swartzman, G. L., and Gustafson, J. D. (1971). *PWNEE: A Grassland ecosystem model.* Technical Report 64, Grassland Biome, Colorado State University, Fort Collins Colorado.

Funderlic, R. E., and Heath, M. T. (1971). *Linear compartmental analysis of ecosystems.* ORNL-IBP-71-4. Oak Ridge National Laboratory, Oak Ridge, Tennessee.

Kitchell, J. F., Koonce, J. F., O'Neill, R. V., Shugart, H. H., Magnuson, J. J., and Booth, R. S. (1974). A model of fish biomass dynamics. *Transactions of the American Fisheries Society,* 103, 786-798.

Kolehmainen, S. E. and Nelson, D. J. (1969). *The balances of ^{137}Cs, stable cesium and the feeding rates of bluegill (Lepomis macrochirus Raf.) in White Oak Lake.* ORNL-4445, Oak Ridge National Laboratory, Oak Ridge, Tennessee.

O'Neill, R. V. (1971). Systems approaches to the study of forest floor arthropods. In *Systems Analysis and Simulation in Ecology, Vol. I,* B. C. Patten, ed. Academic Press, New York. 441-477.

Park, R. A. *et al.* (1974). A generalized model for simulating lake ecosystems. *Simulation,* 23, 33-50.

Patten, B. C., Egloff, D. A., and Richardson, T. H. (1975). Total ecosystem model for a cove in Lake Texoma. In *Systems Analysis and Simulation in Ecology, Vol. III,* B. C. Patten, ed. Academic Press, New York. 205-421.

Prosser, C. L. and Brown, F. A. (1961). *Comparative Animal Physiology.* Saunders, Philadelphia, Pennsylvania.

Shugart, H. H., Goldstein, R. A., O'Neill, R. V., and Mankin, J. B. (1974). TEEM: a terrestrial ecosystem energy model for forests. *Oecologia Plantarum,* 9, 231-264.

78   R. V. O'NEILL

Titus, J., Goldstein, R. A., Adams, M. S., Mankin, J. B., O'Neill, R. V., Weiler, P. R., Shugart, H. H., Booth, R. S. (1975). A production model for *Myriophyllum spicatum*. *Ecology*, 56, 1129–1138.

Weiss, P. A., ed. (1971). *Hierarchically Organized Systems in Theory and Practice*. Hafner, New York.

[*Received June* 1977. *Revised April* 1979]

Landscape Ecology vol. 3 nos. 3/4 pp 163-173 (1989)
SPB Academic Publishing bv, The Hague

Geographical perspectives of space, time, and scale

Vernon Meentemeyer
Department of Geography, University of Georgia, Athens, Georgia 30602

'... We usually opt for one level of analysis exclusively, without considering the range of other alternatives. To judge from the literature this choice is a private act of faith, not to be reported publically.'

Mary Watson 1978

1. Introduction

In the discipline of geography, scale has always been a major issue; however, geographers do not seem to explicitly state their scales of analysis any more fully than scientists in other disciplines. Nevertheless, the geographic literature is rich in philosophical discussions of spatial scales and methodological solutions for dealing with scale (*e.g.*, Harvey 1969). These solutions should not need to be reinvented as the ecologic and biological sciences attempt to more fully incorporate the spatial dimension into their work and to move to ever broader-scales of spatial analysis.

This chapter reviews the major scale issues in geography and the manner in which spatial scale problems have been manipulated and resolved. In particular I discuss examples of the nature of the variables used in spatial/regional models at various scales, the methodological dilemmas and inferential fallacies encountered in spatial analyses, and some common solutions. In addition I examine the basis for selection of scales (including time scales) and some of the trade-offs or concessions needed to move to analyses at broader continental and global scales. Finally, as part of my conclusions, a case is made for a fuller incorporation of space and spatial scales into hierarchy theory.

First I must admit that I am a geographer, specifically a physical geographer, and that this essay fully reflects my biases. I review the literature in human and economic geography, climatology, geomorphology, and remote sensing, which are often a part of the discipline as well as literature in related disciplines of landscape ecology, ecology, and meteorology.

Geography has often been criticized for its breadth of topics and divergent points of view (Hart 1982). The discipline spans human, biological, and physical environmental arenas and includes spatial scales from a single point to the entire globe. It follows then, that geography has abundant literature on methodologies and the merits of various research agendas. Problems in the search for causality and the predictions of spatial patterns are often discussed (Harvey 1969). It is interesting, however, that the question of whether one is working at a 'fundamental' level is never discussed in geography.

Remarkably, the common bond of the spatial point of view seems to cement the discipline (Clark *et al.* 1987). This finding leads to the question: what ingredients differentiate a study as geographic or spatial? It seems that when one or more spatial variables are explicit, distinct variables in an analysis, the study becomes a spatial analysis (Meentemeyer and Box 1987). Examples of spatial variables include area, range, distance, direction, spatial geometries and patterns, spatial connectivity, isolation, diffusion, spatial associations, and scale (Abler *et al.* 1971). These variables may be considered 'geographic primitives' (Mitchelson unpublished).

Watson (1978) maintains that ' . . . scale is a 'geographic' variable almost as sacred as distance.' Perhaps cartography is the geographic subdiscipline that is most adept at handling spatial scale. Well-developed rules have been developed to balance the scale versus resolution-information content of a map (Board 1967). One of the first decisions is selection of a map scale; indeed creative selection of map scales may be part of the art in cartography. Very likely it is the geographer's affinity with mapmaking that makes scale 'sacred,' but that does not mean that scales are always stated explicitly. Nor is scale for most researchers simply a question of balancing the size (extent) of a region with desired levels of resolution. One's purpose and philosophical viewpoint toward space has much to do with the nature of research designs and results.

2. Absolute versus relative space

It is necessary in my view to recognize *a priori* whether a study involves absolute or relative space. Harvey (1969) presents an excellent review of the evolution of these two points of view. He points out that Kant had a great influence on geography but that Kant expressed in his latter works an absolute view of space, (*i.e.*, space may exist for its own sake independent of matter). Accordingly, space just 'is,' and it may therefore be viewed as a 'container' for elements of the earth's surface (Table 1). In other words, the job of geography should consist mainly of filling the 'container' with information. Absolute scale involves primarily an Euclidian point of view usually based on a defined grid system. The location of elements within the grid of the

Table 1. Philosophical views of space: the difference between absolute and relative space.

Absolute space	Relative space
Space can exist independent of any matter	Space exists only with reference to things and processes
Space as a 'container'	Space is defined by things and processes
Associated primarily with inventory and mapping	Associated primarily with studies of forms, patterns, functions, rates, diffusion
Euclidean space	May involve non-Euclidean (transformed) space

region under consideration is critical, as is the size (scale) of the region. This is the point of view of conventional cartography, remote sensing, and the mapping sciences. It is the appropriate approach in inventory, planning, and most mapping and descriptive studies. Moreover, it is quite easy to view 'subcontainers' within a 'container' and to devise appropriate classification schemes. A city may be viewed as having several districts, areas, or neighborhoods, all of which may show ever-smaller areal units. Depending on the classification scheme and skills of discrimination, the creation of spatial hierarchies is quite straightforward, albeit in absolute space.

The relativistic point of view involves two considerations. First of all, space is defined by the spatial elements and processes under consideration. The 'relevant' space is defined by the spatial processes, *e.g.*, migration and commuting patterns, watersheds, dispersion of pollutants, and even the diffusion of ideas and information. In studies of the relationship between (among) spatial patterns/forms and functions, processes and rates often define the scales and regions. Secondly this approach may result in space being defined in non-Euclidean terms. Even distance may be relative (Harvey 1969). Two areas separated by a barrier may be close in absolute space and very distant in relative space when time, rates, and interactions are considered. Thus a functional (spatial process) region may be difficult to map in terms of absolute space.

The need for more broad-scale studies generated

by the International Geosphere-Biosphere Program (IGBP) has often produced calls for the use of advanced techniques in remote sensing and applications of geographic information systems (GISs) (*e.g.*, NASA 1986; Kotlyakov *et al.* 1988). Some problems can realistically be solved only by these techniques; however, they involve absolute space almost exclusively.

Most modern work in geography involves a relative view of space (Harvey 1969; Abler *et al.* 1971) because much of this work involves spatial processes and mechanisms. Both absolute and relative space involve scale, but each approach tends to produce distinctly different research results. Moreover the nature of the resulting models is influenced by scale, especially for spatial models produced from the relativistic point of view. However, this fact leads to the additional complication that spatial scales need not be viewed only in absolute terms. Scale is also relative when scales change across a map. It is instructive to examine changes in model structures and relevant variables in the geographic (broadly defined) literature caused by changes in spatial scales.

3. Variables changing with scale

As in many other disciplines, geography has debated the appropriate scale of analysis for various processes (Nir 1987). There is, however, widespread agreement that changes in scale change the important, relevant variables. Moreover the value of a phenomenon at a particular place is usually driven by causal processes which operate at differing scales (Mitchelson unpublished). In studies of human migration, the models for predicting the spatial patterns of intrastate movement usually involve regionally aggregated data for groups. Often included are variables related to labor demand, investment and business climate, and income, *i.e.*, group and 'structural-contextual' variables. Intra-urban migration models often involve the age, education, and income of individuals, as well as kinship and other affinity measures. Distance and status may also be useful measures, but at this scale most variables delimit the individual (Pandit, personal communication.

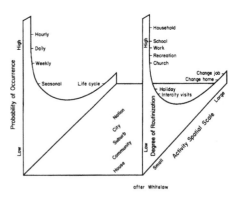

Fig. 1. The activity space of individuals as it relates to time involved, distance travelled, degree of routinization, and probability of occurrence.

In planning and modeling of water supply networks for third-world countries, studies at a national scale often involve urban and regional water demands. At a village scale, walking time and distance to a spigot may be preeminent concerns (Logan, personal communication). In other words, group and regional aggregation variables are replaced by measures of the individual person or family.

Behavioral geography is a subdiscipline in geography which examines the use of space by individuals and the timing of this use. Portions of this discipline have been termed activity space and time-space geography (Carlstein and Thrift 1978). The approaches taken have shown that human activities which are the most routine involve the smallest spaces and are correlated with the shortest periods of time (Fig. 1). Rare, unroutine activities often involve movement over large spaces or distances and can be so rare as to recur only a few times (or once) in a lifetime (*e.g.*, changing careers). The most frequent movements are of the shortest distance and may also display effort-minimization principles (Zipf 1949, Holley 1978). Thus different spatial activities have radically different time and space scales. Perhaps it is now time to incorporate spatial activities of nonhuman entities into this framework.

166

Table 2. The correspondence among time scales, scales of atmospheric variables and topographic variables most frequently used in studies of orographic precipitation.

Time	Variables	
	Atmospheric	Topographic
Minute	Local convection Dew point depression	Slope %
Hour	Feeder cloud Potential instability Wind speed	Orientation Elevation
Day	Synoptic events Vorticity Short-wave patterns	Exposure?
Year	Precipitable H_2O Upper-level divergence Baroclinic zones SST and ENSO	Elevation Exposure Slope % Orientation
Normals	Baroclinic zones Long wave patterns Wind persistence and direction Wind speed	Exposure Orientation

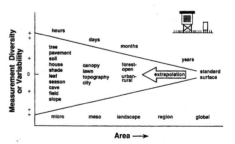

Fig. 2. Time and spatial area relationships with measurement diversity in physical climatology.

3.1. Studies in orographic precipitation

It is also possible to review the research literature on a particular phenomenon with a view to the time and space scales which have been used. Presumably a sufficient number of studies across a sufficient diversity of spatial scales have been conducted, and some indication of the changes in relevant variables should be evident. I have done this for studies of precipitation patterns in mountainous areas (Table 2). For precipitation events at a point (weather station) lasting for minutes to perhaps an hour, the studies are highly process oriented and often involve atmospheric variables defining local convection and dew point depression. The actual topography itself does not seem especially important, although percentage slope is sometimes considered.

At the time scale of an hour or more, the formation of feeder clouds at low levels, potential instability, and current windspeeds are often examined. As the time scale lengthens, the appropriate atmospheric variables involve even broader scales

(Table 2). The topographic variables used show less-well-defined relationships with scale. Spatial scales are poorly defined in many studies, and studies at contintental and global scales are nonexistent (Basist 1989). Probably the interactions of elevation with exposure and of slope with orientation are more appropriate at broad regional scales; surprisingly, elevation alone is a less-useful predictor than other measures of topography (Basist 1989).

3.2. Studies in physical climatology

It has been argued elsewhere (Meentemeyer and Box 1987) that processes and phenomena which involve broad spatial scales appear to be changing so slowly that very long time scales are needed to observe and model these entities. The literature of the physical climatology of the earth's surface is also illustrative of the pragmatic problem of matching time and space scales (Flohn 1981), as well as the nature of the variables which appear important. At the scales of micrometeorology, measurements are rarely conducted for more than a few hours or days; however, the variety of situations and the number of environmental variables studied have been exceptionally large (Fig. 2).

Nearly every conceivable location and environmental variable (including factors such as temperature, moisture, radiation, wind, and heat flux) has been monitored. As measurement time scales are lengthened, the areas become more aggregated and

Table 3. Some observations and speculations on spatial scale.

1. Broad-scale patterns (aggregate scales) generate hypotheses; fine scales (individual level) determine cause and effect (Watson 1978).
2. Sciences dealing mostly with processes, *e.g.*, meteorology are better able to switch scales (Steyn *et al.* 1981).
3. Sciences dealing mostly with phenomenon have more difficulty with time and space scales (*e.g.*, geography, climatology, landscape ecology) because the size of the phenomenon decides the scale.
4. As the spatial scale becomes finer (smaller spatial units) the vertical three-dimensional aspect becomes more important.
5. Meso scales are usually the most difficult to define and model.

defined by terms such as slope, city versus rural, and land versus water. The variety of earth surface 'classes' becomes more restricted, as well as the number of environmental variables measured continuously. In addition, the three-dimensional spaces of microclimatology, in which the vertical dimension defined by the boundary layer is significant, are progressively collapsed to a standard two-dimensional surface at broader scales (Table 3). The vast majority of long-term measurements are made only at standard weather stations, where by international agreement measurements are made in the same manner over most of the globe.

Such standard or reference stations at which long-term measurements are made are so expensive to maintain that only governments have sufficient resources. It should be mentioned, however, that these stations were not designed to answer questions about biotic-abiotic interactions or even about climate change but for meteorologists producing weather forecasts (Mather 1974), justifying the expense of their operation. Yet it is these point measurements which have been used to interpolate, extrapolate, and describe abiotic environments across the diverse elements of landscapes, regions, and the entire globe. In fact, there is a clear trend in the literature of physical climatology to extrapolate from coarser to finer spatial and time scales – rarely in the opposite direction. Unfortunately, weather stations are biased toward low-elevation areas, regions near higher-population densities, and land masses. Even this very rich data source is not 'global.'

3.3. Scale thresholds?

Reviews of research literature on narrowly defined phenomena should be conducted in a systematic way to find additional order in the effects of changes in scale. Some phenomena show distinct scale thresholds. In geomorphology/hydrology, small watersheds in temperate zones display a very peaked discharge response. At about 300 km^2, the peak flattens because at this size many watersheds support floodplains (Klein 1976, in Beven *et al.* 1988).

The search for changes in model structures and even thresholds in spatial systems can fruitfully be started now. Moreover, it is likely that research in broad-scale spatial phenomena and processes will proliferate. Unfortunately spatial analyses and varying scales of time-space resolution can produce some difficult methodological problems.

4. Methodological dilemmas in spatial analyses

Tobler (1969) stated the problem of spatial autocorrelation succinctly in his first law of geography: near things are more related than distant things. Thus every spatial element may be correlated, *i.e.*, it is similar to its neighboring element. Without spatial autocorrelation, however, the surface of the earth would appear entirely random. Spatial autocorrelation is, in fact, the basis for the recognition of spatial variability, of land versus water, field versus forest, high density versus low density, etc. Often it is useful to search for the level of resolution which maximizes the spatial variability of a phenomenon (Harvey 1969). This is then the level at which spatial patterns may be most easily recognized and studied. The underpinnings of spatial autocorrelation are treated elsewhere (*e.g.*, Cliff and Ord 1973).

Although spatial autocorrelation has received much recent attention, especially by soil scientists (*e.g.*, Kachanoski 1988), it is one of the more esoteric methodological problems in spatial analyses. Perhaps the two most important problems in geographic research are the lack of experimental control and the size of the observational unit. It is safe

168

to say that nearly every geographic primitive (area, shape, distance, scale, etc.) needs to be controlled for if the results (models?) are to be general and transferable to other settings. The dependency of results on the size of the spatial unit in an analysis provides ample examples of potentially erroneous inferences.

4.1. Erroneous inferences

Generalizations across spatial scales and units of aggregation have generated three types of erroneous inference (Mitchelson, unpublished): (a) individualistic fallacy – imputing macrolevel (aggregate) relationships from microlevel (individualistic) relationships, (b) cross-level fallacies – making inferences from one subpopulation to another at the same level of analysis, and (c) ecological fallacy – making inferences from higher levels of aggregation to a lower level.

In geography, when spatial units (patches, districts, areas, regions, gaps) are the elements of a correlation-regression analysis, the results are termed 'ecological correlation' (Robinson 1950). Generally, when the size of the observational unit, is large, the estimate of variation for the phenomenon is low because the means vary less than the values upon which they are based. This can lead to an erroneous inference termed the 'ecological fallacy' in economic geography, *i.e.*, making inferences about the individual or lower levels from the higher levels of aggregation. Robinson (1950) demonstrated 'ecological fallacy' in the correlation between race and measured IQ. When the United States was divided into nine regions, the correlation coefficient was 0.946 ($r^2 = 0.89$), and a value of 0.733 ($r^2 = 0.537$) at the level of 48 states (regions). However, at the individual level, the coefficients were only r = 0.203 ($r^2 = 0.04$). Good statistical designs can, however, overcome this potential fallacy.

Johnston (1976) provides an extremely simple example of the problem of unit size, autocorrelation, and 'ecological fallacy' (Fig. 3). The diagram on the left represents a plot of the relationship between people aged 65 and over and the percentage of

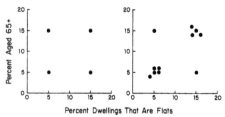

Fig. 3. An influence of the size of a spatial observation on spatial correlations.

dwellings that are flats in each of four areas in London. The correlation coefficient is zero. Two regions are large, however, and when broken down into equal-sized census tracts (right-hand figure), the correlation coefficient is about 0.6 (Johnston 1976).

A particularly demanding methodological problem in geography has been inference of spatial process from spatial form. Indeed, it is from spatial form that most processes are 'discovered.' Unfortunately empirical results are usually scale specific. Patterns which appear to be ordered at one scale may appear random at other scales (Miller 1978). Moreover, different spatial processes can generate exactly the same spatial patterns. Often fine-scale processes can cause clumping patterns, but the clumps show the results of processes leading to as much dispersion as possible. For example, shoe stores tend to clump to increase comparative shopping, but each clump desires to be as far as possible from another clump of shoe stores.

The size of the observational unit may also influence statistical distributions. Generally Poisson distributions are generated from small sampling quadrats, and large quadrats generate negative binomial distributions (Watson 1978). This can influence inference (process from form) as well as spatial correlation.

The rules for optimal spatial sampling and data grouping to reduce the loss of information on individuals have been developed (*e.g.*, Clark and Avery 1976), and these rules can reduce some common fallacies in inference. Missing spatial data can, however, produce special problems. If the goal is a

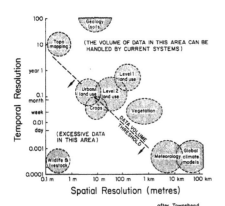

Fig. 4. The influence of levels of temporal and spatial resolution on data-handling thresholds for various phenomena.

map of a process or phenomenon, or if a model is the goal, then missing data are a serious problem. Does one interpolate, extrapolate, or produce other estimates of values for missing spaces? Certainly spatial averaging is possible, and it is also possible to fit trend surfaces of varying complexity. Unfortunately these approaches are also scale dependent and therefore scales must be considered in estimating missing data.

4.2. Coupling hierarchical levels

One solution to poor spatial data coverage is the development of a model of spatial relationships that couples two hierarchical levels. Watson (1978) notes, however, that few studies in geography have combined macrospatial and microspatial levels of analysis because of the incredibly large amounts of data needed. Indeed many scale problems seem in actuality to involve thresholds in data-handling abilities. Figure 4, based on Townsend (1987), demonstrates the boundary between scales which produce 'excessive' data and data volumes which can be handled by current systems in remote sensing, GISs, and atmospheric circulation models. The constraints may be caused by any combination of hardware, software, or model structure. Naturally, the detection of spatial processes and phenomena

which display great temporal variability and require high levels of resolution produce excessive volumes of data. Thus for some fine-scaled phenomenon, simple extrapolation might be acceptably accurate but meet data-handling thresholds. Furthermore, multiple time or space scales would push the data volume threshold to the upper right-hand portion of Fig. 4.

Hierarchical coupling is common in applied climatology as is extrapolation from broad to fine scales. Climatology was in its infancy a weak stepsister of meteorology (Mather 1974). The data collected by meteorologists in their attempts to forecast the weather were the basis of the discipline. At various spatial scales, it seems natural then to correlate these temperature and precipitation records with ecosystem processes. Unfortunately, temperature and precipitation are measures of the state of the atmosphere and may not represent well that part of climate which is actually entering into an environmental process. It is necessary to conceive an 'effective climate' (term coined by D.B. Carter): that climate or abiotic environment most intimately involved in an environmental process. For some processes, for example, soil temperature may be more 'effective' than air temperature and soil moisture more so than precipitation. In the terminology of hierarchy theory (O'Neill et al. 1986; O'Neill 1988; Salthe 1985), this appropriate climatic (abiotic) environment could be considered the 'constraints' on lower levels. I suggest that these constraints must be the effective climate, the environment closest to the actual processes, and not just weather records.

4.3. The data-rich to data-poor solution

Sometimes available data determine research designs and space-time scales. This may be especially true for broad-scale geographical-ecological problems of the type proposed as part of IGBP. Geographers and climatologists have coupled hierarchical levels with success when the higher level (constraints) have been data-rich. In fact, many spatial models are based on the concept of predicting the spatial patterns of data-poor (especially in-

170

volving poor spatial coverage and missing data) phenomena and processes on the basis of data-rich constraint variables. My own work has involved predicting the geography of litter decomposition rates at continental scales on the basis of abiotic (climatic) constraints (Meentemeyer 1984). In most of these spatial models, the climatic variables have not been precipitation and air temperature, but instead have been evapotranspiration and measures of seasonality, which apparently enter more effectively into decomposer systems. These models have been criticized for not including fully the organismic, chemical and physical variables well known to control decay rates. However, adding such information produces exceedingly complex models, which when coupled with the driving variables of climate, do very little to improve the prediction of broad-scale geographic patterns. Apparently a threshold is reached at which the 'costs' of additional causal or mechanistic information is not balanced by improved predictions of spatial patterns. At this point information on the lower levels cannot simply be moved upscale.

Fortunately we already have many of the data-rich variables at near global scales which can be used as the driving variables in predicting spatial patterns at the broader scales. Information on climate, soil, topography, vegetation, and land use comes readily to mind. Remote sensing has produced spatial coverage for additional variables, especially for the oceans (Walsh and Dieterle 1988). Perhaps the innovative spatial-environmental models of the future will involve higher- to lower-level couplings to produce new geographies of processes and their rates which cannot now be mapped.

As shown above, in Fig. 2, and in the hierarchy literature, extrapolation from higher to lower levels has been successful, with much less success for fine-to-broad extrapolations. The challenge for the global climate change program then is exceedingly difficult because it involves analysis of the levels and constraints which are above that of some of our most useful and data-rich constraints (e.g., weather records). To improve the spatial modeling component in landscape ecology, it may be helpful to find the appropriate constraints for the spatial hierarchical level of concern.

4.4. Loss of detail in spatial analysis

The selection of spatial scales involves much more than selection of levels of spatial resolution. Nevertheless it should be clear that the addition of the spatial dimension in the study of nearly any process or phenomenon may involve a variety of trade-offs. Models for broader-scale patterns result in less predictive accuracy at specific points or places. Since geography is primarily an empirical science, the generalizations (models) are only as good as the finest-grain spatial data available. Often it is necessary to sample for just one or two variables at many points (regions, places, etc.) in order to develop good spatial data sets. The details of entities and processes at places often cannot be used. Thus the model may appear to be 'superficial,' but, without the sacrifice in detail, a spatial model and/or a predictive map would not have been possible.

The incorporation of the spatial dimension in landscape ecology and projects under the global change programs require the substitution of more samples geographically but sampling of less detail at each site or place. It seems that the addition of the spatial dimension forces attention to higher hierarchical levels: the broader the scale, the higher the level.

The history of spatial modeling has shown the success of modeling on the basis of higher-level constraints. Lower levels provide data for testing of hypotheses and the search for causality (Table 3). Therefore it is apparent that much of the cherished detail of the reductionist sciences may not be needed, and indeed cannot be used, in broad-scale spatial modeling.

5. How spatial scales are selected (apparently)

Steyn et al. (1981) make the interesting point that disciplines concerned primarily with processes, such as meteorology, are able to switch scales with relative ease (e.g., Gedgelman 1985). On the other hand, disciplines dealing with phenomenon are often restricted by the size of the phenomenon (Table 4). Many phenomena come in characteristic

Table 4. The selection of spatial scales: some apparent determinants and constraints.

1. The size and 'speed' of a spatial phenomenon or process
2. Existing maps and map scales
3. Scales of aerial photography and remote sensing images
4. Size of the spatial units (e.g., quadrat, tract, patch, area, gap)
5. Mathematical-statistical constraints (e.g., spatial-temporal autocorrelation, centrality bias, missing data)
6. Within-site versus across-site variability
7. Data handling thresholds
   A. Time
   B. Technology
   C. Money
8. Practical-empirical considerations
9. Philosophical propensities (e.g., micro versus macro, absolute space versus relative space)
10. Arbitrary

size classes. Moreover phenomena associated with ephemeral processes or fast relaxation times may need to be studied at fine time-space scales (Table 4).

The tremendous burden of sampling spatial variables adequately often means that existing data sources and map scales (e.g., 1:24,000, 1:50,000) must be used. Thus it is common to define the spatial scale of a study by the approximate corresponding map scales (e.g., Krummel et al. 1987). Similarly the scales of aerial photography and remote sensing images may constrain the spatial scales chosen. The size of quadrats, census tracts, patches, and even pixel size may fix the limits of suitable scales.

Mathematical and statistical considerations may affect the selection of scales. Spatial and temporal autocorrelation for phenomena and processes may vary with scale, depending on the degree of spatial and temporal heterogeneity. In essence the scales need to match the heterogeneity; i.e., the phenomenon dictates the scale (e.g., White 1987). Some techniques, such as those based on nearest-neighbor analyses, have a centrality bias which changes with scale. Studies of spatial interaction are especially sensitive to scale. Larger regions tend to incorporate more potential interactions and have a larger centrality bias, depending on the nature of

the interactions. Similarly, scale may be determined by the degree of within-site versus across-site variability. Generally the scale selected is the one which maximizes across-site variability (Table 4).

Data-handling thresholds are intertwined with time and space scales. This data-handling threshold has been moved to higher time-space resolution by technology. However, time and money constraints often seem to limit spatial scales, the number of variables considered, and the number of hierarchical level used.

The abundant arguments in geography regarding the merits of microscale versus macroscale analyses and of all scales in between point to basic differences in philosophical stances on scale. Researchers with similar propensities select similar scales and seem therefore to group together. Perhaps this is caused by dominant paradigms, data sources, and other realities. Is it thus possible to categorize disciplines, subdisciplines, and groups on the basis of their 'favorite' time and space scales? In the end it seems that scales are unconsciously selected and therefore may seem to be entirely *arbitrary*.

6. Summary and conclusions

This article reviews space and time scales from a geographer's point of view. Because spatial phenomena come in incredibly different size classes, geographers have conducted analyses across many orders of spatial magnitude. Geographers seem adept at moving from one scale to another, but they are not prone to explicitly state these scales a priori. Moreover, in spite of many appeals for multiscaler research (e.g., Abler 1987; Miller 1970; Stone 1968; Kirkby 1985), this is seldom done, although higher-level information is often used to predict lower levels. Good multiscale work apparently meets data-handling thresholds rather quickly.

Most geographic research is now conducted with a relativistic view of space rather than a view of space as a 'container.' Spatial scales for relative space are more difficult to define, however, than those for the absolute space of cartography and remote sensing.

The relevant, important, and useful variables

172

from a modeling standpoint change with spatial scale. By reviewing the literature on a topic in a systematic way, as was done here for physical climatology and orographic precipitation, this scale change in variables can be seen. We do not as yet have models of the changes in models caused by changes in scale.

Spatial data violate nearly every requirement for parametric statistical analysis (Meentemeyer and Box 1987), which is partially responsible for fallacies and erroneous inference. Many of these problems are scale dependent. Based on the work of Harvey (1969), we see that there are three primary methodological problems in spatial analyses. There are first of all the differences in inference and relevant variables caused by different scales or hierarchical levels. This has been called the 'scale problem' in geographic literature. Secondly, the description and modeling of spatial patterns, as noted above, may defy easy solutions, and finally the relationships between spatial patterns and process remain a challenge.

The geographic literature contains many examples of extrapolations to lower levels from higher levels. Often the higher levels have been more widely sampled geographically (e.g., weather and climate, topography) and may be data rich. Models which predict spatial patterns and process often use the data-rich higher levels as driving variables for lower levels. Young (1978) argues that central place theory in geography should be a component of hierarchy theory. Indeed it can be argued here that space is inherently hierarchical and needs to be more fully incorporated into hierarchy theory.

As the various disciplines under the umbrella of the environmental sciences more fully incorporate the spatial dimension into their research agendas, problems associated with spatial scale will be encountered. Many of these problems have in varying degrees been recognized if not solved. Nevertheless it is worth noting Clark's (1985) warning, 'No simple rules can automatically select the "proper" scale for attention.'

Good geographic models require good geographic coverage, but this may mean that lower-level details are simply not needed. As mentioned earlier, the question of whether one is working at a 'fundamental' level is never discussed in geography. The Long-Term Ecological Reserve (LTER) sites are a step in the right direction, but a geographer would prefer much more intensive spatial sampling, even if that means a sacrifice in accuracy or detail. Otherwise a spatial analysis may not be possible. It remains to be seen to what degree the reductionist sciences can contribute to IGBP. More work with explicitly stated scales is needed, as well as across-scales research. Scale has been treated philosophically in this essay. But I am reminded of Couclelis's caution, 'Philosophizing in an empirical discipline is a sure sign of trouble' (cited in Abler 1987).

7. Acknowledgements

I wish to thank my colleagues in the Department of Geography, University of Georgia for their assistance. Ronald Mitchelson provided me with his unpublished paper on scale and read an early draft of this paper. Kavita Pandit and Bernard Logan provided specific examples of scale problems, and James Wheeler provided me with new references. Alan Basist shared with me his knowledge of orographic precipitation. I thank my colleagues at Oak Ridge National Laboratory for the invitation to attend the workshop which helped initiate this paper. Elgene Box was instrumental in getting me to give serious thought to scale issues. Audrey Hawkins and Dorothy Osborn prepared all drafts of this paper.

References

Abler, R.F. 1987. What shall we say? To whom shall we speak? Ann. Assoc. Am. Geogr. 77: 511–524.

Abler, R.F., Adams, J. and Gould, P. 1971. Spatial organization: the geographer's view of the world. Prentice-Hall, Inc., Englewood Cliffs, New Jersey.

Basist, A.N. 1989. A comparison of orographic effects on precipitation at 10 regions worldwide. Masters Thesis, University of Georgia.

Beven, K.J., Wood, E.F. and Sivapalan, M. 1988. On hydrological heterogeneity – catchment morphology and catchment response. J. Hydrol. 100: 333–375.

Board, C. 1967. Maps as models. In Models in Geography. pp. 671–726. Edited by R.J. Chorley and P. Haggett. Methuen and Co., Ltd., London.

Carlstein, T. and Thrift, N. 1978. Afterword: towards a time-space structured approach to society and environment. *In* Human Activity and Time Geography. pp. 225–263. Edited by T. Carlstein, D. Parkes and N. Thrift. John Wiley and Sons, New York.

Clark, M.J., Gregory, K.J. and Gurnell, A.M. 1987. Introduction: change and continuity in physical geography. *In* Horizons in Physical Geography. pp. 1–5. Edited by M.J. Clark, K.J. Gregory and A.M. Gurnell. Barnes and Noble Books, Totowa, New Jersey.

Clark, W.A.V. and Avery, K.L. 1976. The effects of data aggregation in statistical analysis. Geogr. Anal. 8: 428–438.

Clark, W.C. 1985. Scales of climatic impacts. Clim. Change 7: 5–27.

Cliff, A.D. and Ord, J.K. 1973. Spatial autocorrelation. Pion, London.

Flohn, H. 1981. Climatic variability and coherence in time and space. *In* Food-climate Interactions. pp. 423–441. Edited by W. Bach, J. Pankrath and S.H. Schneider. D. Reidel Publishing Company, Dordrecht, The Netherlands.

Gedgelman, S.D. 1985. Atmospheric circulation systems. *In* Handbook of Applied Meteorology. pp. 3–61. Edited by D.D. Houghton. John Wiley and Sons, New York.

Hart, J.F. 1982. The highest form of the geographer's art. Ann. Assoc. Am. Geogr. 72: 1–29.

Harvey, D. 1969. Explanation in Geography. St. Martin's Press, New York.

Holley, B.P. 1978. The problem of scale in time-space research. *In* Time and Regional Dynamics. pp. 5–18. Edited by T. Carlstein, D. Parkes and N. Thrift. John Wiley and Sons, New York.

Johnston, R.J. 1976. Residential area characteristics: research methods for identifying urban and sub-areas – social area analysis and factorial ecology. *In* Social Areas in Cities, Volume I: Spatial Processes and Form. pp. 193–235. Edited by D.T. Herbert and R.J. Johnston. John Wiley and Sons, London.

Kachanoski, R.G. 1988. Processes in soils – from pedon to landscape. *In* Scales and Global Change: Spatial and Temporal Variability in Biospheric and Geospheric Processes. SCOPE, Vol. 35. pp. 153–177. Edited by T. Rosswall, R.G. Woodmansee and P.G. Risser. John Wiley and Sons, Chichester.

Kirkby, A. 1985. Pseudo-random thoughts on space, scale and ideology in political geography. Polit. Geogr. Q. 4(1): 5–18.

Kotlyakov, V.M., Mather, J.R., Sdasyuk, G.V. and White, G.F. 1988. Global change: geographical approaches (a review). Proc. Nat. Acad. Sci. U.S.A. 85: 5986–5991.

Krummel, J.R., Gardner, R.H., Sugihara, G., O'Neill, R.V. and Coleman, P.R. 1987. Landscape patterns in a disturbed environment. Oikos 48: 321–324.

Mather, J.R. 1974. Climatology: fundamentals and applications. McGraw-Hill, New York.

Meentemeyer, V. 1984. The geography of organic decomposition rates. Ann. Assoc. Am. Geogr. 74: 551–560.

Meentemeyer, V. and Box, E.O. 1987. Scale effects in landscape studies. *In* Landscape Heterogeneity and Disturbance. Ecological Studies, Vol. 64. pp. 15–34. Edited by M.G. Turner. Springer-Verlag, New York.

Miller, D.H. 1970. Geographic scales of field research on the energy-mass budget in the Soviet Union. Prof. Geogr. 22(2): 97–98.

Miller, D.H. 1978. The factor of scale: ecosystem, landscape mosaic and region. *In* Sourcebook on the Environment. pp. 63–88. Edited by K.A. Hammond. University of Chicago Press, Chicago.

Mitchelson, R.L. Concerns About Scale, unpublished.

NASA. 1986. Earth system science overview: a program for global change. Earth System Science Committee, NASA Advisory Council. National Aeronautics and Space Administration, Washington, D.C.

Nir, D. 1987. Regional geography considered from the systems approach. Geoforum 18(2): 187–202.

O'Neill, R.V. 1988. Hierarchy theory and global change. *In* Scales and Global Change: Spatial and Temporal Variability in Biospheric and Geospheric Processes. SCOPE, Vol. 35. pp. 29–46. Edited by T. Rosswall, R.G. Woodmansee and P.G. Risser. John Wiley and Sons, Chichester.

O'Neill, R.V., DeAngelis, D.L., Waide, J.B. and Allen, T.F. 1986. A hierarchical concept of ecosystems. Monographs in Population Biology, Vol. 23. Edited by R.M. May. Princeton University Press, Princeton, New Jersey.

Robinson, W.S. 1950. Ecological correlations and the behavior of individuals. Am. Soc. Rev. 15: 351–357.

Salthe, S.N. 1985. Evolving Hierarchical Systems: Their Structure and representation. Columbia University Press, New York.

Steyn, D.G., Oke, T.R., Hay, J.E. and Knox, J.L. 1981. On scales in meteorology and climatology. Clim. Bull. 39: 1–8.

Stone, K.H. 1968. Scale, Scale, Scale. Econ. Geogr. 44: 94.

Tobler, W.R. 1969. Geographical filters and their inverses. Geogr. Anal. 1: 234–253.

Townsend, J.R.G. 1987. Remote sensing – Global and local views. *In* Horizons in Physical Geography. pp. 63–85. Edited by M.J. Clark, K.J. Gregory and A.M. Gurnell. Barnes and Noble Books, Totowa, New Jersey.

Walsh, J.J. and Dieterle, D.A. 1988. Use of satellite ocean colony observations to refine understanding of global geochemical cycles. *In* Scales and Global Change: Spatial and Temporal Variability in Biospheric and Geosperic Processes. SCOPE, Vol. 35. pp. 287–318. Edited by T. Rosswall, R.G. Woodmansee and P.G. Risser. John Wiley and Sons, Chichester.

Watson, M.K. 1978. The scale problem in human geography. Geogr. Ann. 60B: 36–47.

White, P.S. 1987. Natural disturbance, patch dynamics, and landscape patterns in natural areas. Nat. Areas J. 7(1): 14–22.

Whitelaw, J.S. 1972. Scale and urban migrant behavior. Aust. Geogr. Stud. 10: 101–106.

Young, G.L. 1978. Hierarchy and central place: some questions of more general theory. Geogr. Ann. 60B: 71–78.

Zipf, G.K. 1949. Human behavior and the principle of least effort. Addison-Wesley Press, Cambridge.

Reprinted from *BioScience*

Landscape Diversity: The Concept Applied to Yellowstone Park

William H. Romme and Dennis H. Knight

Changes in landscape patterns may influence a variety of natural features including wildlife abundance, nutrient flow, and lake productivity. Data suggest that cyclic changes in landscape diversity occur on areas of 100 km² in Yellowstone National Park. When properly managed, large wilderness areas provide the best and probably the only locale for studying the kind of landscape changes that occurred for millenia in presettlement times. *(Accepted for publication 12 May 1982)*

Each successive level of biological organization has properties that cannot be predicted from those of less complex levels (Odum 1971). Thus, populations have certain attributes distinct from the characteristics of the individuals of which they are composed, and communities have unique properties beyond the attributes of their component populations. An important level of organization that is now receiving more attention is the landscape, or mosaic of communities that covers a large land unit such as a watershed or a physiographic region (Forman and Godron 1981).

The importance of large-scale landscape patterns has been widely recognized (e.g., Bormann and Likens 1979, Forman 1979, 1982, Forman and Boerner 1981, Forman and Godron 1981, Habeck 1976, Habeck and Mutch 1973, Hansson 1977, Heinselman 1973, Loucks 1970, Luder 1981, Pickett 1976, Reiners and Lang 1979, Rowe 1961, Shugart and West 1981, Sprugel 1976, Sprugel and Bormann 1981, Swain 1980, White 1979, Wright 1974, Zachrisson 1977, and others). A few studies have quantitatively treated changes in landscape patterns (e.g., Hett 1971, Johnson 1977, Johnson and Sharpe 1976, Shugart et al. 1973). We recently made a detailed analysis of landscape composition and diversity in a pristine watershed in Yellowstone National Park in relation to fire and forest regrowth following fire (Romme 1982). In this paper we describe the natural changes that have occurred in landscape

pattern over a period of 240 years and the possible consequences of these changes for certain aspects of ecosystem structure and function. Although we focus on Yellowstone in this analysis, the concepts are applicable to other ecosystems as well.

The term *landscape diversity* refers to the diversity of plant communities making up the vegetational mosaic of a land unit. Landscape diversity results from two superimposed vegetation patterns: the distribution of species along gradients of limiting factors, and patterns of disturbance and recovery within the communities at each point along the environmental gradients (Forman and Godron 1981, Reiners and Lang 1979). Both of these patterns contribute to the vegetational diversity of the Yellowstone landscape.

Over the park's 9000 km², elevation ranges from about 1800 m along the Yellowstone River in the northern portion to over 3000 m on the high peaks of the east and northwest. As a result, there are pronounced gradients of temperature and moisture, with related patterns in species distribution. The areas at lower elevations in the north support open sagebrush *(Artemisia tridentata)* parks on drier sites and aspen *(Populus tremuloides)* woodlands and Douglas fir *(Pseudotsuga menziesii)* forests in more mesic locations (Despain 1973). On the cooler subalpine plateaus one finds extensive upland coniferous forests of lodgepole pine *(Pinus contorta* var. *latifolia)*, subalpine fir *(Abies lasiocarpa)*, Engelmann spruce *(Picea engelmannii)*, and whitebark pine *(P. albicaulis)*, broken by occasional meadows and sagebrush parks on alluvial and lacustrine soils. The high peaks are covered by forests of spruce,

fir, and whitebark pine on sheltered slopes, with alpine or subalpine meadows and boulder fields on the more exposed sites. Pollen analysis of pond sediments indicates that these basic patterns of species distribution have been relatively stable during the last 5000 years (Baker 1970).

However, vegetational patterns related to the second source of landscape diversity—perturbation—have undergone changes during this time. Most of the changes have been natural, as described below, but some aspen and sagebrush communities in northern Yellowstone appear to have been altered somewhat by fire suppression during the last century. Comparisons of 100-year-old photographs with recent photographs of the same sites show that forests today are generally more dense, with an increase in conifers and a decrease in aspen, and that many sagebrush parks now contain more shrubs and fewer grasses and forbs. Streamside thickets of willow *(Salix* spp.) and alder *(Alnus* spp.) also appear less extensive and robust than formerly (Houston 1973).[1] Some have attributed these changes to excessive browsing by elk *(Cervus elaphus)* (Beetle 1974, Peek et al. 1967).

A more common explanation appears to be the virtual elimination of fire in this area from 1886 to 1975. Houston (1973) found that fires formerly recurred at average intervals of 20–25 years in northern Yellowstone, a disturbance frequency that probably was essential for the persistence of plant species and communities representing early stages of secondary succession (notably aspen and herbaceous plants). In the absence of fire, succession has proceeded unchecked and other species such as Douglas fir and sagebrush have become increasingly predominant. Thus fire

Romme is with the Department of Natural Science, Eastern Kentucky University, Richmond. KY 40475. Knight is with the Department of Botany, University of Wyoming, Laramie, WY 82071.

[1]Houston, D. B. 1976. The Northern Yellowstone Elk. Parts III and IV. Vegetation and habitat relations. Unpublished report, Yellowstone National Park, Wyoming.

prevention appears to have modified the overall composition of the northern Yellowstone landscape, reducing landscape diversity by increasing the area covered by late successional plant communities at the expense of early successional communities. The magnitude of this change is relatively small in the context of the entire northern Yellowstone landscape, however, since early and herbaceous communities comprised a small fraction of the landscape even in presettlement times (Despain 1973). Similar changes have also been described in several other western parks and wilderness areas following effective fire control (Habeck 1976, Habeck and Mutch 1973, Kilgore and Taylor 1979, Loope and Gruell 1973, Lunan and Habeck 1973). Because a major management goal in the large national parks is to preserve ecosystems in their primeval state (Houston 1971), Yellowstone recently instituted a new fire management policy that allows lightning-caused fires to burn without interference if they do not threaten human life, property, or other values (US National Park Service 1975).

The situation seems to be different on the high subalpine plateaus that dominate most of the central, western, and southern areas of the park. Because of inaccessibility, effective fire control was not accomplished here until about 1950 when fire-fighting equipment and techniques were greatly improved (US National Park Service 1975). Moreover, our research indicated that fire occurs naturally at very long intervals because of very slow forest regrowth and fuel accumulation after fire (Romme 1982). On an average site, 200 years or more are required for a fuel complex to develop that is capable of supporting another destructive fire. On dry or infertile sites, 300–400 years may be necessary. Fires ignited prior to that time are likely to burn a very small area and have a minimal impact on the vegetation (Despain and Sellers 1977, Romme 1982). Recent uncontrolled fires in the park that burned intensely in 300-year-old forests have been observed to stop when they reached a 100-year-old stand, even though weather conditions remained favorable for fire (Despain[2], Despain and Sellers 1977). Thus, in an ecosystem where fire historically occurred at intervals of 200+ years on any particular site, suppression during the last 20–30 years probably has had very little effect on overall landscape pattern. Any major

[2] D. G. Despain, personal communication.

changes that have occurred are largely the result of natural processes that would have taken place even in man's absence.

Although the subalpine landscape apparently has not been substantially altered by man's activities (excluding, of course, those areas of intensive development for visitor use), it has by no means been static during the last 100 years. We found evidence that major fires occur cyclically, i.e., thousands of hectares may burn at intervals of 300–400 years with relatively few major fires in the same area during the intervening periods (Romme 1982). Such a fire cycle can occur because: geologic substrate, soils, and vegetation are very similar over much of the plateau region; forests over large contiguous areas grow and develop a fuel complex at approximately the same rates; and the plateau topography has low relief and few natural barriers to fire spread. Thus one extensive fire tends to be followed by another fire in the same area some 300–400 years later. In other parts of the Rocky Mountains where topographic barriers are more numerous, where succession occurs more rapidly, or where fuel characteristics are different, this particular type of fire cycle may not occur.

LITTLE FIREHOLE RIVER WATERSHED

We conducted our study in the Little Firehole River watershed, which covers 73 km^2 on the Madison Plateau, a large rhyolite lava flow in west-central Yellowstone. Coniferous forests predominate, with lodgepole pine occurring throughout and subalpine fir, Engelmann spruce, and whitebark pine being found on more mesic sites. Alluvial deposits in the central and northern parts of the watershed support subalpine meadows or open coniferous forests with rich shrub and herbaceous understories. The topography is generally flat or gently sloping, with an average elevation of about 2450 m.

Fire history during the last 350 years was determined using the fire-scar methods developed by Heinselman (1973) and Arno and Sneck (1977). Major fires occurred in 1739, 1755, and 1795 (± 5 years), collectively burning over half of the upland area. Of the forested areas that did not burn at that time, nearly all were located either on topographically protected sites (ravines, lower northeast-facing slopes) that burn rarely (Romme and Knight 1981, Zachrisson 1977), or in places that had been burned by a moder-

ately large fire in 1630, less than 200 years earlier, and were covered by young forests. Since 1795 only three fires >4 ha have occurred, and all three were relatively small (<100 ha). The absence of recent large fires is almost certainly due to a lack of suitable fuel conditions over most of the watershed, not to fire suppression by man. In fact, park records show that only one fire has been controlled in this area, a 90-ha burn in 1949. The fire probably would not have covered a much larger area even without suppression, since it was surrounded by young forests and topographically sheltered sites. Today the areas burned in the 1700s support lodgepole pine forests that are all developing more-or-less synchronously; in another 100–150 years extensive portions of the watershed will again have fuel conditions suitable for a large destructive fire.

Three stages of forest regrowth following fire (early, middle, and late successional) can be recognized on upland sites. Early successional stages are usually present for about the first 40 years and are characterized by an abundant growth of herbs and small shrubs. The large dead stems of the former forest remain standing throughout most of this period, and an even-aged cohort of lodgepole pine becomes established. Middle successional stages are marked by the maturation and dominance of the even-aged pine cohort, beginning with canopy closure around 40 years and lasting until senescence around 250–300 years. Herbaceous biomass and species diversity are lowest during this period (Taylor 1973). During late successional stages (250–300+ years) the even-aged pine canopy deteriorates with heavy mortality and is replaced by trees from the developing understory to produce an all-aged, usually mixed-species stand, which then persists until the next destructive fire.

We used our data on fire history and on the rates and patterns of forest succession after fire to reconstruct the sequence of vegetation mosaics that must have existed in the Little Firehole River watershed during the last 240 years. Past landscape patterns were reproduced by first making a map showing the age (time since the last destructive fire) of all homogeneous forest units in 1978, based on extensive field sampling and aerial photography. Then, to reconstruct the landscape of 1738, for example, we subtracted 240 years from the age of each stand in 1978 and determined in which successional stage a stand of that age would have been. Where a fire had occurred more recently than the date of interest (e.g., areas that burned in 1739 in the reconstruction for 1738) we assumed that the stand was in a fire-susceptible late successional stage (Romme 1982).

Figure 1 shows the proportions of the Little Firehole River watershed covered by early, middle, and late successional stages at different times since 1738. In 1738 most of the area was covered by late successional forests, but fires in 1739, 1755, and 1795 greatly reduced the old-growth forests and replaced them with early successional stages. Middle successional stages became most abundant around 1800 and have dominated the watershed since. The early successional stages that were common in the late 1700s and early 1800s have been very uncommon since the mid-1800s. A decrease in middle successional stages after 1938 and an associated increase in late successional stages reflect forest maturation on areas burned in 1739.

To further describe historic patterns in landscape diversity, we calculated three diversity indices (similar to those used for measuring species diversity) and applied them to our landscape reconstructions for 1778–1978. We computed a richness index, based on the number of community types present; an evenness index, reflecting the relative amount of the landscape occupied by each community type; and a patchiness index, indicating the size and interspersion of individual community units as well as the structural contrast between adjacent communities (Romme 1982). Figure 2 shows the results of plotting a weighted average of all three indices, as a measure of overall landscape diversity, and the Shannon index (Pielou 1975), which we calculated by using the proportion of the watershed covered by a community type as a measure of abundance. Both indices reveal a similar pattern: Landscape diversity was high in the late 1700s and early 1800s following the extensive fires of 1739, 1755, and 1795; it fell to a low point in the late 1800s during a 70-year period with no major fires; and it increased again during this century as a result of two small fires plus some variation in the rate of forest maturation in areas burned in 1739 and 1795. This variation in rates of succession is attributable to several factors including localized high densities of the mountain pine beetle *(Dendroctonus ponderosae)* (Romme 1982).

The dramatic changes in landscape composition and diversity in the Little Firehole River watershed during the last 240 years (Figures 1 and 2) must have been associated with significant changes in ecosystem structure and function, including net primary productivity, nutrient cycling, total biomass, species diversity, and population dynamics of individual species. These relationships cannot be fully quantified at this time, but speculation based on existing knowledge is useful.

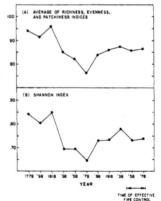

Figure 2. Changes in two measures of landscape diversity in the Little Firehole River watershed from 1778–1978.

IMPLICATIONS FOR WILDLIFE

Taylor and Barmore (1980) censused breeding birds in a series of lodgepole pine stands representing a gradient from the earliest successional stages after fire through late successional stages in the park. Their data show the pattern of avifaunal succession in a single homogeneous stand. In attempting to answer the question of how breeding bird species and populations change with time in an entire subalpine watershed, we used Taylor and Barmore's (1980) census data to estimate the number of breeding pairs in each stand within our reconstructed vegetation mosaics, summing the estimates for all to arrive at an estimate of breeding pairs in the entire watershed.

Figure 3 shows the results for three representative species and for the total number of breeding pairs of all species. Mountain bluebirds *(Sialia currucoides)* require open habitats with dead trees for nesting. Such habitat was most abundant in the Little Firehole River watershed during the late 1700s and early 1800s when 25–50% of the area was covered by early forest successional stages following the large fires of the 1700s (Figure 1). Consequently, bluebirds may have been very numerous at that time. However, as forests matured bluebird populations

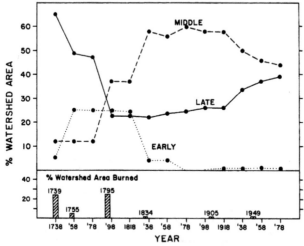

Figure 1. Percent of watershed area covered by early, middle, and late stages of forest succession from 1738–1978 in the 73-km² Little Firehole River watershed, Yellowstone National Park.

probably dropped dramatically (Figure 3). Today bluebirds are uncommon in the watershed except in the 90-ha area that burned in 1949. Note that this probable population decline was a perfectly natural event, occurring at a time when European man had not yet entered the area.

In contrast to the bluebird, ruby-crowned kinglets *(Regulus calendula)* prefer mature forests. Thus, kinglets were less common when bluebirds were most abundant (Figure 3). The yellow-rumped warbler *(Dendroica auduboni)* breeds successfully in a variety of habitats and as a result the population of this species probably has fluctuated little during the last 240 years despite the major landscape changes that have occurred (Figure 3). Figure 3 also shows that the total number of breeding pairs of all species has probably fluctuated greatly in the last few centuries. The highest numbers apparently were in the late 1700s and early 1800s when landscape diversity was also greatest (Figure 2).

The population estimates shown in Figure 3 can be challenged easily on the basis that they were derived solely from habitat availability, i.e., the number of hectares of forest present in each age class. Of necessity we have ignored other critical determinants of population

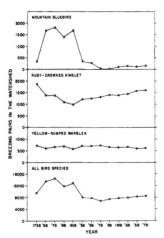

Figure 3. Estimated population sizes of breeding birds in upland forests of the Little Firehole River watershed, based on data from Taylor and Barmore (1980) and the trends shown in Figure 1. Populations in meadows and riparian forests, which cover approximately 16% of the watershed, are not included because appropriate population density data are not available for these habitats.

Table 1. Relative values of plant communities and successional stages for elk habitat in the Little Firehole River watershed.[*]

Plant community type	Potential forage value	DISTANCE COEFFICIENT			
		DISTANCE (m) TO COVER OR WATER			
		0–320	320–800	800–1500	1500+
Alluvial woodland adjacent to moist meadow	10	1.0	0.9	0.7	0.5
Meadow	9	1.0	0.9	0.7	0.5
Upland forest; early successional stages	7	1.0	0.9	0.7	0.5
Upland forest; late successional stages	3	1.0	0.9	0.7	0.5
Upland forest; middle successional stages	2	1.0	0.9	0.7	0.5

[*]Based on models and discussions by Asherin 1973, Basile and Jensen 1971, Black et al. 1976, Hershey and Leege 1976, Lonner 1976, Lyon 1971, Marcum 1975, 1976, Pengelly 1963, Reynolds 1966, Stelfox et al. 1976, Thomas et al. 1976, and Winn 1976.

density. Nevertheless, the overall patterns are valid to the extent that they show the constraints of habitat on potential populations.

We were also able to consider the effect of landscape change on elk. Using a model much like that developed by Thomas et al. (1976), we examined changes in three critical habitat features during the last 200 years, namely, forage quantity and palatability, shelter (or cover), and water. The forage and shelter provided by an individual forest may change greatly during postfire succession. Early successional stages usually have the best forage whereas middle and late successional stages provide the best shelter. However, because elk use several different kinds of habitat, the distribution and interspersion of plant communities and successional stages is critical. Thus the center of a large meadow or recently burned area may receive little elk use, despite abundant forage, if it is too distant from shelter or water, and the potential shelter of very extensive tracts of mature forest may be largely ignored if little forage is available (Black et al. 1976, Hershey and Leege 1976, Marcum 1975, Reynolds 1966, Stelfox et al. 1976, Thomas et al. 1976, Winn 1976).

We developed a relative ranking system by which every type of plant community and successional stage in the Little Firehole River watershed was assigned a value from 0–10 to indicate potential forage value (Table 1). These values were subjective, based on published literature and our own observations in the study area. We then divided the watershed into 1429 units of 5 ha each, identified the dominant vegetation type within each unit, and assigned appropriate values to each. Every value was multiplied by a distance coefficient

reflecting the distance to the nearest shelter or water if those features were not present within the unit itself (Table 1), the product being our elk habitat index. In this manner we analyzed our reconstructed vegetation mosaics for 1778, 1878, and 1978.

Figure 4 (a, b, and c) shows the results for three 5-ha units having different histories of fire and forest regrowth. As a result of changes in stand structure, the quality of elk habitat has varied greatly. However, when we averaged the values for all 1429 individual 5-ha units to obtain an estimate of elk habitat quality for the watershed as a whole, we found much less difference among the landscapes of 1778, 1878, and 1978 (Figure 4d). There are probably two main reasons for this result. First, temporary increases in habitat quality in one part of the watershed (due primarily to the great improvement in forage after fire) have been balanced by decreases resulting from forest maturation on other areas burned earlier. Second, and probably more important, the best habitat is in and around moist meadows where forage, shelter, and water all occur in close proximity. In fact, our model may underestimate the habitat quality of subalpine meadows in the park, since we reduced our elk ha index in the centers of large meadows to reflect the distance to shelter. However, the shelter requirement apparently is much less critical for elk populations that are not hunted by man, and elk in the park are frequently observed feeding in the centers of large meadows.[3] We were unable to determine whether the large fires in the surrounding uplands had burned the meadows and adjacent allu-

[3]L. Irwin, personal communication.

vial woodlands. We assumed that the fires in these areas were of low intensity and produced little change in community structure or elk habitat. Although our results suggest that fires may not greatly influence the overall quality of elk summer range on the high plateaus of Yellowstone, elk are attracted to recently burned areas (Davis 1977), and over much of the subalpine zone, moist meadows are less common than in the Little Firehole River watershed. Where meadows are less common, summer elk populations may fluctuate in response to changes in the upland landscape.

IMPLICATIONS FOR AQUATIC ECOSYSTEMS

One of the most interesting and attractive features of the park is Yellowstone Lake. This virtually unpolluted subalpine lake covers 354 km^2 and contains populations of the native cutthroat trout (*Salmo clarkii*). The trout support a complex food chain including pelicans, ospreys, otters, and bears. Some evidence indicates that the lake's net primary productivity has declined during the last century, as has its carrying capacity for trout and associated top predators (Shero 1977, US National Park Service 1975). Because the period of apparent decline coincides with attempts at fire control, some have suggested that the cause is reduced nutrient input to the lake due to biotic immobilization by forests. As noted earlier, however, our research indicates that the natural fire regime has not been greatly altered by

man's activities in the Yellowstone subalpine zone, particularly in the very remote areas that drain into Yellowstone Lake.

Rather than attribute the cause to fire suppression, we favor the hypothesis that lake productivity is to some extent synchronized with the long-term fire cycle that seems to prevail in the watershed of Yellowstone Lake. A variety of evidence supports this hypothesis. For example, experiments in the Rocky Mountains have shown that removal of mature forest from 40% of a subalpine watershed results in an increase in total water discharge of 25% or more (Leaf 1975). The increase is due to several factors related to the distribution and melting of the winter snowpack. Albin (1979) compared two small tributary streams of Yellowstone Lake; about 20% of one watershed was burned by fires 36 and 45 years previously, whereas the other watershed was unburned. The burned watershed had greater seasonal variation in streamflow and greater total water discharge per hectare. If a large portion of a subalpine watershed burns at intervals of approximately 300 years, as seems to occur in the Little Firehole River watershed, then streamflow also may exhibit a long-term cycle over and above yearly and seasonal fluctuations. During the high-discharge portion of the cycle, especially in years of high snowfall, debris is washed out of stream channels, new channels are cut, and new alluvial deposits are created. Such events influence habitat for fish as well as for floodplain species like willow and alder, which in

turn are important browse species for elk and other terrestrial animals (Houston 1973).

But more important to the question of Yellowstone Lake is the nutrient content of stream water. Immediately after deforestation by fire or cutting there often is an increase in dissolved minerals due to erosion, reduced plant uptake, increasd microbial activity, increased leaching, and the release of elements from organic matter by fire (Bormann and Likens 1979, McColl and Grigal 1975, Wright 1976). The increase is usually short-lived, lasting several years at most (Albin 1979, Bormann and Likens 1979), but it may be important as a periodic nutrient subsidy (Odum et al. 1979) to oligotrophic aquatic ecosystems. As young forests become established, biotic immobilization is so effective that nutrient concentrations in stream water fall to very low levels (Bormann and Likens 1979, Marks and Bormann 1972, Vitousek and Reiners 1975). Thus a watershed dominated by early and middle forest successional stages (e.g., the Little Firehole River watershed during the 1800s) would produce relatively nutrient-poor water. As forests reach late successional stages, tree growth and net primary productivity decrease, nutrient uptake is less, and consequently the leachate is richer in dissolved minerals (Bormann and Likens 1979, Vitousek and Reiners 1975).[4]

Thus, although the possible connection between fire suppression and reduced productivity in Yellowstone Lake is plausible, an equally attractive alternative hypothesis is that extensive fires in the watershed about 100 years ago replaced many late successional forests with early successional stages. As young forests over much of the watershed began utilizing soil nutrients more efficiently, the total amount leached into stream water feeding the lake was reduced accordingly. If this is true, any recent decline in lake productivity may be a natural phenomenon that has occurred many times in the past and will be alleviated as forests in the watershed mature. Of course, the Yellowstone Lake watershed is very large (ca. 2600 km^2), and landscape patterns over this large area may be in a state of dynamic equilibrium, or what Bormann and Likens (1979) have referred to as a shifting mosaic steady

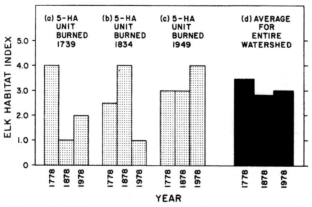

Figure 4. Elk habitat index (see text) for three representative 5-ha units and for the entire Little Firehole River watershed (d) in 1778, 1878, and 1978.

[4]Pearson, J. A., D. H. Knight, and T. J. Fahey. Unpublished ms. Net ecosystem production and nutrient accumulation during stand development in lodgepole pine forest, Wyoming.

state. If this is found to be true for the Yellowstone Lake watershed, then total nutrient input to the lake should be about the same from year to year (though the source would vary), and some other explanation for the decline in lake productivity will be required.

CONCLUSIONS

After a century of ecological research that focused largely on species or individual communities or ecosystems, there now is a growing interest in still higher levels of organization such as the landscape and biosphere. Changes in landscape patterns influence a variety of natural features including wildlife, water and nutrient flow, and the probability of different kinds of natural disturbances. Given a sufficiently large area and a natural disturbance regime, various measures of landscape pattern may remain fairly constant over time despite dramatic cyclic changes in localized areas such as a small watershed. Such "steady states" have been demonstrated or hypothesized for a Swedish boreal forest (Zachrisson 1977), high-elevation fir forests in New England and elsewhere (Sprugel 1976, Sprugel and Bormann 1981), primeval northern hardwood forests of North America (Bormann and Likens 1979), and mesic deciduous forests of the southern Appalachians (Shugart and West 1981). Our results suggest that strong cyclic changes occur on areas of at least 100 km^2 in Yellowstone National Park, but more research is needed to determine if the landscape patterns in the park as a whole are in a state of equilibrium. Large wilderness areas, when protected from pollutants and managed so that natural perturbations can continue, provide the best and probably the only locale for studying the kind of landscape changes that occurred for millenia in presettlement times.

ACKNOWLEDGMENTS

This research was supported by grants from the University of Wyoming—National Park Service Research Center. We thank D. G. Despain and D. B. Houston for sharing their observations and records of natural fires in Yellowstone; L. Irwin for advice on our elk habitat model; N. Stanton and M. Boyce for guidance in estimating diversity; R. Levinson and R. Marrs for assistance with aerial photograph interpretations; M. Cook for computer programming assistance; R. Levinson, L. van Dusen, K. White, and P. White for field assistance; D. G. Despain, L. Irwin, W. H. Martin, W. G. Van der Kloot, and two anonymous reviewers for helpful comments on the manuscript; and K. Diem, M. Meagher, J. Donaldson, and the staff of the Old Faithful Ranger Station for administrative and logistical support.

REFERENCES CITED

Albin, D. P. 1979. Fire and stream ecology in some Yellowstone Lake tributaries. *Calif. Fish Game* 65: 216–238.

Arno, S. F. and K. M. Sneck. 1977. A method for determining fire history in coniferous forests of the mountain west. *US For. Serv. Gen. Tech. Rep.* INT-42.

Asherin, D. A. 1973. Prescribed burning effects on nutrition, production and big game use of key northern Idaho browse species. Doctoral Thesis, University of Idaho, Moscow, ID.

Baker, R. G. 1970. Pollen sequence from late quaternary sediments in Yellowstone Park. *Science* 168: 1449–1450.

Basile, J. V. and C. E. Jensen. 1971. Grazing potential on lodgepole pine clearcuts in Montana. *US For. Serv. Res. Pap.* INT-98.

Beetle, A. A. 1974. Range survey in Teton County, Wyoming. Part IV: quaking aspen. *Univ. of Wyom. Agri. Exp. Station Publ.* SM 27.

Black, H., R. J. Scherzinger, and J. W. Thomas. 1976. Relationships of Rocky Mountain elk and Rocky Mountain mule deer habitat to timber management in the Blue Mountains of Oregon and Washington. Pages 11–31 in *Proceedings of the Elk-Logging-Roads Symposium.* Forest, Wildlife, and Range Experiment Station, University of Idaho, Moscow, ID.

Bormann, F. H. and G. E. Likens. 1979. *Pattern and Process in a Forested Ecosystem.* Springer-Verlag, New York.

Davis, P. R. 1977. Cervid response to forest fire and clearcutting in southeastern Wyoming. *J. Wildl. Manage.* 41: 785–788.

Despain, D. G. 1973. Major vegetation zones of Yellowstone National Park. *Information Paper Number 19,* Yellowstone National Park, WY.

Despain, D. G. and R. E. Sellers. 1977. Natural fire in Yellowstone National Park. *West. Wildlands* 4: 20–24.

Forman, R. T. T., ed. 1979. *Pine Barrens: Ecosystem and Landscape.* Academic Press, New York.

Forman, R. T. T. 1982. Interaction among landscape elements: a core of landscape ecology. In Perspectives in landscape ecology. *Proceedings of the 1981 Symposium of the Netherlands Society for Landscape Ecology,* Veldhoven, Pudoc, Wageningen, the Netherlands, in press.

Forman, R. T. T. and R. E. Boerner. 1981. Fire frequency and the Pine Barrens of New Jersey. *Bull. Torrey Bot. Club* 108: 34–50.

Forman, R. T. T. and M. Godron. 1981. Patches and structural components for a landscape ecology. *BioScience* 31: 733–740.

Habeck, J. R. 1976. Forests, fuels and fire in the Selway-Bitterroot Wilderness, Idaho. Pages 305–352, in E. V. Komarek, general chairman *Proceedings of the Montana Tall Timbers Fire Ecology Conference and Fire and Land Management Symposium Number 14, 1974.* Tall Timbers Research Station, Tallahassee, FL.

Habeck, J. R. and R. W. Mutch. 1973. Fire-dependent forests in the northern Rocky Mountains. *Quat. Res. (NY)* 3: 408–424.

Hansson, L. 1977. Landscape ecology and stability of populations. *Landscape Planning* 4: 85–93.

Heinselman, M. L. 1973. Fire in the virgin forests of the Boundary Waters Canoe Area, Minnesota. *Quat. Res. (NY)* 3: 329–382.

Hershey, T. J., and T. A. Leege. 1976. Influences of logging on elk on summer range in north-central Idaho. Pages 73–80 in *Proceedings of the Elk-Logging-Roads Symposium;* Forest, Wildlife, and Range Experiment Station, University of Idaho, Moscow, ID.

Hett, J. M. 1971. Land-use changes in east Tennessee and a simulation model which describes these changes for three counties. ORNL-IBP-71-8. Oak Ridge National Laboratory, Ecological Sciences Division, Oak Ridge, TN.

Houston, D. B. 1971. Ecosystems of National Parks. *Science* 172: 648–651.

———. 1973. Wildfires in northern Yellowstone National Park. *Ecology* 54: 1111–1117.

Johnson, W. C. 1977. A mathematical model of forest succession and land use for the North Carolina Piedmont. *Bull. Torrey Bot. Club* 104: 334–346.

Johnson, W. C. and D. M. Sharpe. 1976. An analysis of forest dynamics in the northern Georgia Piedmont. *Forest Science* 22: 307–322.

Kilgore, B. M. and D. Taylor. 1979. Fire history of a sequoia-mixed conifer forest. *Ecology* 60: 129–142.

Leaf, C. F. 1975. Watershed management in the central and southern Rocky Mountains: A summary of the status of our knowledge by vegetation types. *US For. Serv. Res. Pap.* RM-142.

Lonner, T. N. 1976. Elk use-habitat type relationships on summer and fall range in Long Tom Creek, southwestern Montana. Pages 101–109 in *Proceedings of the Elk-Logging-Roads Symposium.* Forest, Wildlife, and Range Experiment Station, University of Idaho, Moscow, ID.

Loope, L. L. and G. E. Gruell. 1973. The ecological role of fire in the Jackson Hole area, northwestern Wyoming. *Quat. Res. (NY)* 3: 425–443.

Loucks, O. L. 1970. Evolution of diversity, efficiency, and community stability. *Am. Zool.* 10: 17–25.

Luder, P. 1981. The diversity of landscape ecology. Definition and attempt at empirical identification. *Angew. Botanik* 55: 321–329.

Lunan, J. S. and J. R. Habeck. 1973. The effects of fire exclusion on ponderosa pine communities in Glacier National Park, Montana. *Can. J. For. Res.* 3: 574–579.

Lyon, L. J. 1971. Vegetal development following prescribed burning of Douglas fir in south-central Idaho. *US For. Serv. Res. Pap.* INT-105.

Marcum, C. L. 1975. Summer-fall habitat selection and use by a western Montana elk herd. Doctoral Thesis, University of Montana, Missoula, MT.

———. 1976. Habitat selection and use during summer and fall months by a western Montana elk herd. Pages 91–96 in *Proceedings of the Elk-Logging-Roads Symposium*. Forest, Wildlife, and Range Experiment Station, University of Idaho, Moscow, ID.

Marks, P. L. and F. H. Bormann. 1972. Revegetation following forest cutting: Mechanisms for return to steady-state nutrient cycling. *Science* 176: 914–915.

McColl, J. G. and D. F. Grigal. 1975. Forest fire: effects on phosphorus movement to lakes. *Science* 188: 1109–1111.

Odum, E. P. 1971. *Fundamentals of Ecology*. Third edition. W. B. Saunders Co. Philadelphia, PA.

Odum, E. P., J. T. Finn, and E. H. Franz. 1979. Perturbation theory and the subsidy-stress gradient. *BioScience* 29: 349–352.

Peek, J. M., A. L. Lovaas, and R. A. Rouse. 1967. Population changes within the Gallatin elk herd, 1932–65. *J. Wildl. Manage.* 31: 304–316.

Pickett, S. T. A. 1976. Succession: an evolutionary interpretation. *Am. Nat.* 110: 107–119.

Pengelly, W. L. 1963. Timberlands and deer in the Northern Rockies. *J. For.* 61: 734–740.

Pielou, E. C. 1975. *Ecological Diversity*. Wiley-Interscience, New York.

Reiners, W. A. and G. E. Lang. 1979. Vegetational patterns and processes in the Balsam fir zone, White Mountains, New Hampshire. *Ecology* 60: 403–417.

Reynolds, H. G. 1966. Use of openings in spruce-fir forests of Arizona by elk, deer, and cattle. *US For. Serv. Res. Note* RM-66.

Romme, W. H. 1982. Fire and landscape diversity in subalpine forests of Yellowstone National Park. *Ecol. Monogr.* 52: 199–221.

Romme, W. H. and D. H. Knight. 1981. Fire frequency and subalpine forest succession along a topographic gradient in Wyoming. *Ecology* 62: 319–326.

Rowe, J. S. 1961. Critique of some vegetational concepts as applied to forests of northwestern Alberta. *Can. J. Bot.* 39: 1007–1017.

Shero, B. R. 1977. An interpretation of temporal and spatial variations in the abundance of diatom taxa in sediments from Yellowstone Lake, Wyoming. Doctoral Dissertation, Univ. of Wyoming, Laramie.

Shugart, H. H., Jr., and D. C. West. 1981. Long-term dynamics of forest ecosystems. *Am. Sci.* 69: 647–652.

Shugart, H. H., Jr., T. R. Crow, and J. M. Hett. 1973. Forest succession models: a rationale and methodology for modeling forest succession over large regions. *For. Sci.* 19: 203–212.

Sprugel, D. G. 1976. Dynamic structure of wave-regenerated *Abies balsamea* forests in the north-eastern United States. *J. Ecol.* 64: 889–911.

Sprugel, D. G. and F. H. Bormann. 1981. Natural disturbance and the steady state in high-altitude balsam fir forests. *Science* 211: 390–393.

Stelfox, J. G., G. M. Lynch, and J. R. McGillis. 1976. Effects of clearcut logging on wild ungulates in the central Albertan foothills. *For. Chron.* 52: 65–70.

Swain, A. M. 1980. Landscape patterns and forest history in the Boundary Waters Canoe Area, Minnesota: A pollen study from Hug Lake. *Ecology* 61: 747–754.

Taylor, D. L. 1973. Some ecological implications of forest fire control in Yellowstone National Park, Wyoming. *Ecology* 54: 1394–1396.

Taylor, D. L. and W. J. Barmore. 1980. Post-fire succession of avifauna in coniferous forests of Yellowstone and Grand Teton National Parks, Wyoming. Pages 130–145 in R. M. DeGraff, Technical Coordinator, *Proceedings of the Workshop on Management of Western Forests and Grasslands for Nongame Birds. US For. Serv. Gen. Tech. Rep.* INT-86.

Thomas, J. W., R. J. Miller, H. Black, J. E. Rodiek, and C. Maser. 1976. Guidelines for maintaining and enhancing wildlife habitat in forest management in the Blue Mountains of Oregon and Washington. Pages 452–476 in *Transactions of the Forty-first North American Wildlife and Natural Resources Conference*, Washington, DC.

US National Park Service. 1975. The Natural Role of Fire: A Fire Management Plan for Yellowstone National Park. Unpublished report, Yellowstone National Park, WY.

Vitousek, P. M. and W. A. Reiners. 1975. Ecosystem succession and nutrient retention: a hypothesis. *BioScience* 25: 376–381.

White, P. S. 1979. Pattern, process, and natural disturbance in vegetation. *Bot. Rev.* 45: 229–299.

Winn, D. S. 1976. Terrestrial vertebrate fauna and selected coniferous forest habitat types on the north slope of the Uinta Mountains. United States Department of Agriculture, Wasatch National Forest, Region 4.

Wright, H. E., Jr. 1974. Landscape development, forest fires, and wilderness management. *Science* 186: 487–495.

Wright, R. F. 1976. The impact of forest fire on the nutrient influxes to small lakes in northeastern Minnesota. *Ecology* 57: 649–663.

Zachrisson, O. 1977. Influence of forest fires on the north Swedish boreal forest. *Oikos* 29: 22–32.

Landscape Ecology vol. 4 no. 4 pp 237-248 (1990)
SPB Academic Publishing bv, The Hague

Historical land use and hydrology. A case study from eastern Noord-Brabant

G. Bas M. Pedroli[1*] and Guus J. Borger[2]
[1]*Landscape and Environmental Research Group, University of Amsterdam, Dapperstraat 115, NL-1093
BS Amsterdam;* [2]*Seminar of Historical Geography, University of Amsterdam, Jodenbreestraat 23,
NL-1011 NH Amsterdam**

Keywords: landscape ecology, historical geography, land use, groundwater, Kempen, Netherlands

Abstract

The historical geography of the landscape of a lowland brook valley in the sandy Kempen area (eastern Brabant, Netherlands) shows the interaction of ecological processes and land use, and helps to understand processes in the present landscape. In this location the human influence, especially on the groundwater hydrology, played a major role in the development of the landscape. Levels and flow of different types of groundwater interacted with vegetation development and human interference, to produce landscape patterns.

Four main stages have been identified. In the prehistoric period, a natural deciduous forest covered the higher grounds, ombrotrophic peat was formed in the valley, and groundwater was relatively deep. In the medieval stage man settled on the edge of the valley, cleared parts of the forest and dug part of the peat. Groundwater levels were raised, which increased the rate of groundwater discharge and increased the amount of associated lowland peat formation in the valley. This tendency continued in modern times, when the area was completely deforested. Groundwater levels increased further due to decreased evapotranspiration, which gave rise to the use of ponds for fish and for water mills. Finally, in the most recent period the groundwater level has been lowered by extensive artificial drainage, partly on a regional scale. It was concluded that evaluation of historical changes in the landscape help provide landscape planners with a sound idea of the nature of the landscape.

Introduction

The influence of man on the development of landscape is one of the main themes of landscape ecology (Golley 1987). It might even be stated that landscape, as such, is studied only in those situations where human involvement is evident (Neef 1982b, p. 21). Where this is not the case, scientists are more likely to be involved with ecosystem analysis (*e.g.*, Bormann and Likens 1979; Ellenberg *et al.* 1986). However, the relationship between human activi-

ties and soil and landscape attributes is not clear.

Iverson (1988), in a study of land use change over 160 years in Illinois, showed that most of the present land use patches are poorly correlated with natural characteristics of the landscape or with the original cover. He concluded that land use is often situated in the wrong place from a landscape ecological point of view. The comprehensive land use study of Kopp *et al.* (1982) emphasizes the importance of topological relationships, as a natural basis for land use.

* *Present affiliation and correspondence address:* Ministry of Agriculture, Nature Management and Fisheries, Directie Bos- en Landschapsbouw, P.O. Box 20023, NL-3502 LA Utrecht.

238

Most of these land use studies stress pattern analysis (Gardner *et al.* 1987; Schaller and Haber 1988). Processes, linking the patches in these patterns, are mainly studied in animal ecology (Opdam *et al.* 1985; Van Dorp and Opdam 1987; Van Winden and Opdam 1987; O'Neill *et al.* 1988). In such studies the historical development of landscape is rarely included. It has even been suggested that the traditional analytical approach is not appropriate to the study of the historical landscape (Billinge 1977; Gregory 1978; Neef 1982a; Williams 1989). Our question was whether historical geography, through the study of historical sources and maps, could provide information on the spatial processes playing a role in landscape development. Thus, we attempted to follow a more synthetical approach in this paper.

The aim of the paper is to evaluate the available evidence of former land use with regard to landscape development, and eventually landscape management in a small area in the Netherlands, the Strijper Aa landscape in the southeastern part of Noord-Brabant. Special attention is paid to the historical patterns of groundwater dynamics.

The area of study

The Strijper Aa is a small lowland brook, draining a flat area of about 30 km² in the Dutch Kempen area, adjacent to the Dutch-Belgian border (Fig. 1). The Strijper Aa landscape is a typical representative of the cultural landscapes in the Low Countries. It is characterized by a variety of fields, forest patches and open heath between small settlements. Present land use in the area is composed of strongly fertilized cropland and meadows, and coniferous plantations, all on nutrient-poor sandy soils. In the central valley, some wetland still occurs on peaty soils. Elevations in the area range from 24 to 29 m a.s.l., and the groundwater table varies over the area between 0 and 5 m below the soil surface. Hydrology plays a crucial role in the interrelationship between the higher groundwater recharge sites and the discharge sites in the valley. See Pedroli (1990a) for an evaluation of groundwater distribution and dynamics in the area.

In this region there is an increasing discrepancy between the market-directed objectives of the agri-

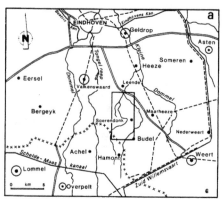

Fig. 1. a. Location of the study area studied, with the area depicted in Figs. 2 and 3 indicated by a rectangle. b. The approximate distribution range of plaggen soils after Pape (1970).

culture in the cultivated areas, and nature management objectives aimed at the uncultivated areas. In the 1950s, when plans to reclaim parts of the wetlands in the area became known, many objections to the plans were expressed, because the wetlands were so unique. The direct connections of dry heathland, through wet heathland and *Myrica gale* shrubland, with marshland forming the source of a brook (Schimmel 1952), were especially valued. Van Leeuwen (1966) mentioned that this area belonged to the region with the highest species diversity in the country, around the turn of the centuries. In 1968 very small parts of this connection remained, representing a complete gradient at three places only (Londo 1968). The originally mesotrophic vegetation had been changed through eutrophication almost everywhere.

Development of land use in the Strijper Aa area

Prehistory to the Middle Ages

Presettlement vegetation in the area consisted of deciduous forests. Palynological records from a heath moorland pool in the area show the presence of a rich Oak-Birch forest (*Quercetum mixtum*) from about 7000 yr BP at dry upland locations, whereas the wet valleys must have been covered with Alder-Ash-Limetree forest (Van Leeuwaarden 1982). Locally, patches of ombrotrophic bog may have been present on top of the marshland peat. Since this area is on the southernmost border of potential plateau-bog formation (Joosten and Bakker 1987), and the discharge of mineral-rich water in the brookvalley does not allow a bog-cover to form, patches of bog must have been located on the margins of the peatland (Ivanov 1981, p. 23).

Late-Paleolithic to Mesolithic finds indicate that non-sedentary people visited the area regularly in the period 12,000–6000 yr BP (Arts 1987). From Neolithic time on, farmers settled in the area and probably cultivated part of the zones adjacent to the lowland brookvalleys. During the late Bronze and early Iron Age (about 1000–500 yr BC) extensive urnfields in neighbouring areas (Slofstra 1982; Slofstra *et al.* 1985; Van Mourik 1988) indicate an expansion of the population may have taken place. Later, up to the early 3rd century AD, human habitation associated with the brookvalleys seems to have been widespread. Archeological finds demonstrate the presence of 2nd century Roman settlements in the neighbourhood (Slofstra 1982). Little is known, however, of the Early Middle Ages (Theuws 1988).

Middle Ages

Expanding occupation and reclamation of forest for agriculture took place between AD 800 and 1100 (Janssen 1972; Hoppenbrouwers *et al.* 1986). The existing local patches of peat bog must have been reclaimed during the Middle Ages, since comparable situations have been described for western Brabant (Leenders 1986; Renes 1987). However, there is no reason to suppose that a large mire existed in this area since some of it would have been preserved, like the *Peel*-bog (Joosten and Bakker 1987). Also, there are no remains of canals for transport of peat turves like in western Brabant (Leenders 1986).

Growth of the human population during the twelfth and thirteenth centuries made it profitable for the Dukes of Brabant to assert their control over the unimproved wastes (Mertens 1982; Coopmans 1987). They encouraged the colonizing movement to increase their own profits and to assure larger areas of political loyalty and safe military movements (Van Uytven 1958). Extension of the farmland around pre-existing hamlets could now come about only by agreement between individual reclaimers and the territorial lord. Meanwhile, peasant squatters reclaimed tiny stretches of arable land from the heath, their settlements forming isolated pockets of cultivation in the extensive wastes (Lambert 1985).

Locally, oak forests remained at least until the 16th century. For example, in 1527 extensive forests of large oaks were still present to the north of Leende and were used as grazing land for pigs (Arch, Leende, inv. no. 7*). The wetter parts of the valleys were in use as hayland and coppice, and were at times dug out for peat. In periods of increasing population pressure, parts of the commons were enclosed and given out for private use. In 1527 for example, the meadows in the Strijper Aa Valley were largely in private use (Arch. Leende, inv. no. 7), whereas in the early 14th century no distinction was made between heathland or marshland waste (Martens 1943, pp. 123–124).

* The municipal historical archives are referenced by their inventory-numbers and appear before the list of published references.

240

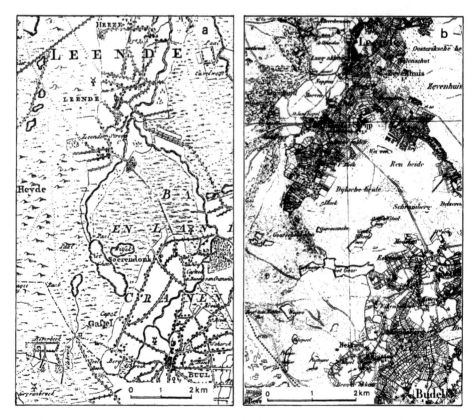

Fig. 2. Topographic maps. a. Verhees' 1794 map. Enlarged; originally of scale 1:180,000. b. Dutch Topographic Service, edition 1863, surveyed 1863; original scale 1:50,000.

The most obvious remains of the land use system, which came into being in the High Middle Ages and lasted locally up to the 1950's, are the thick humous plaggen soils (Pape 1970). These have been formed over a large area in northwestern Europe (Fig. 1) by the almost yearly fertilization of the originally humus-poor podzolic soils with a mixture of manure, sods, litter or sand. For centuries, the main food crop on the plaggen soils was rye. Crop growing was only possible by the use of extensive areas of common waste land for the gathering of sods and for grazing (Vervloet, 1984). The waste land locally suffered from wind erosion. The threat of drifting

sand in the region has been reported by many authors for several periods (see Van Mourik 1985, 1988).

During this period improved agricultural techniques were introduced, parallel with a slight orientation to the growing town-markets of Brabant and Limburg. For some labor-intensive crops, agricultural production in small well-kept parcels reached levels in the High Middle Ages that are not equalled even today (Slicher van Bath 1963). While much land use was based on subsistence farming, contributions to a wider market occurred (Hoppenbrouwers *et al.* 1986), as is shown, for example, by

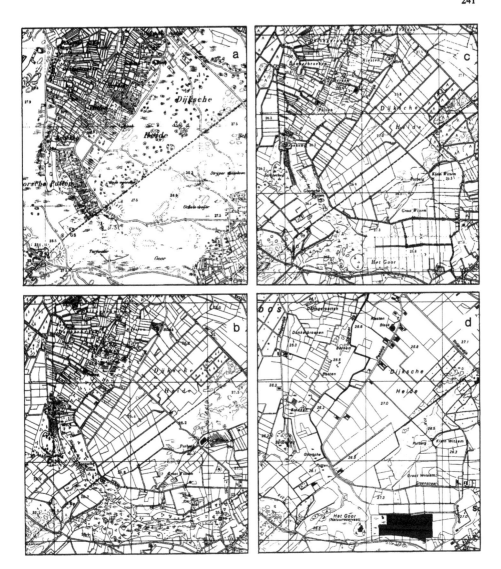

Fig. 3. Topographic maps. One gridsquare equals 1 km² (original scale 1:25,000, Dutch Topographic Service).
a. Edition 1928, surveyed 1898–1922. b. Edition 1952, surveyed 1950. c. Edition 1963, surveyed 1961. d. Edition 1985, surveyed 1980/1981.

242

the existence of centers for falconry in the region (Swaen 1937; Goris 1966).

Modern times

The land use system can be recognized on the first detailed maps of the area. At the end of the 18th century, the map of Hendrik Verhees (1794) was the first to show the location of the cultivated area, clearly associated with the course of the brooks (Fig. 2a). The settlements had an open character, grouped along SSW-NNE axes. The remaining area was waste land, which seems to have consisted mainly of wet heathland, except for some drier hilly parts with drift sands. All forests were cleared. On later maps (*e.g.*, Fig. 2b) it can be observed, that the largest part of the cultivated area drawn by Verhees (1794), consisted of meadows and brookland coppice.

Essentially this picture did not change in the course of the last century. Only in the last quarter of the 19th century and the first quarter of the present century, the beginning decay of the agricultural system becomes visible on the topographic maps. In some areas, forest was planted on heathland soils, and some agricultural reclamation was made, *e.g.*, to the east of the brooklands in Fig. 3a. This was possible because by the introduction of the French legislation at the beginning of the 19th century, the heath was no longer common property but belonged to newly established municipalities (Hendrikx 1989).

Recent

On the topographic maps after World War II, the landscape appears completely changed (Fig. 3bcd). Between 1920 and 1950 approximately 80% of the heathland had been reclaimed, either as forest plantations or as agricultural land. Several moorland pools have disappeared and many paths have been paved and straightened. The Strijper Aa brook now drains an area in Belgium as well.

In the 1970's a comprehensive land reallotment took place. The whole landscape was reconstructed to allow for large-scale agricultural production (Fig. 3d), including industrial pig-breeding and as-

sociated corn-growing. As a compromise, the polluted water from beyond the Belgian border was directed through a derivation canal, protecting the original source area of the Strijper Aa in the central valley from further contamination.

Thus, the development of the Strijper Aa landscape is characterized by the gradual conversion of the land use system from subsistence to commercial farming, with a rapid acceleration of change in the present century. All through time, there has been an interaction between the potential of the landscape and the change brought about by man. Water management, which was associated with the use of the wetlands, had an especially crucial role in this landscape. To illustrate this role, the history of the wetland area in the Strijper Aa valley is discussed below.

The history of human interactions with the hydrology of the Strijper Aa area

Historical interactions

Since the clearing of the original forests, man has probably always tried to come to terms with the water in his landscape. Initially low dikes were made to cross wetland areas. To judge from the abundance of *dijk* toponomics in present day dry heathland, the landscape in the Middle Ages must have been much wetter than during the present century (*dijken* mentioned within wet land in Arch. Leende, inv. no. 7, 1527).

On old maps (Fig. 2), in the Strijper Aa valley a wetland complex is indicated with the names *Het Ghoor*, *Turfwater* and *Witsel*, near the village of Soerendonk. *Het Goor* had already existed a long time as a wetland area, since it was mentioned in 1440 as *'t Goere* in connection with delineation of a boundary (Iven and Van Gerwen 1974, p. 6). In the same period, an indication of nearby peat-digging is found in the toponymic *Cattenput*, where *put* stands for a peat digging pond (Iven and Van Gerwen 1974, p. 45). Also in Maarheeze, a village in the vicinity (Fig. 1), in 1483 a dispute was settled about peat-digging in a fen (Arch. Maarheeze, Soerendonk en Gastel, inv. no. 14).

Het Ghoor must have been dug out for peat prior

to its first appearance on a map as an open water-surface, *i.e.*, on Verhees' map in 1794 (Fig. 2a). In this part of the landscape, the discharge of relatively nutrient-rich groundwater (Pedroli 1990a) enhances the natural conversion of open water by marshland within about 100 years (Moore and Bellamy 1974; Overbeck 1975; Gore 1983). This is confirmed by observations of Schimmel (1952) and the 1953 topographic map (Fig. 3b), where *Het Goor* appears as marshland.

Moorland pools on the higher ground may also have been covered with bog. The palynological records of a moorland pool in the area (Van Leeuwaarden 1982), indicate that omʰ trophic peat closed a cycle of peat forı   ɔn since ʈʰe last glaciation. This pool is not shown on Verhees' map, possibly implying that it was still covered with bog in late 18th century.

On the 1:25,000 minute map surveyed 1836, the *Goor* has new extensions to the northwest, named *Goorsche Putten*, and to the west, named *Turfwater* (Fig. 2b), indicating peat-digging ponds dating from the first half of the 19th century or earlier. Peat-digging took place until WW II (Iven and Van Gerwen 1974), when the *Strijper Heg*, to the west of the *Goorsche Putten* was dug (Fig. 3b). Also several *putten* are indicated further to the north.

Since the Middle Ages, people kept ponds for fish (Clason 1980). In the area, several toponymics of *Weijer* occur. *Weijer* often means an artificially dammed pond for a water mill or for fishing (Fig. 2b). Fishing was practiced by local people in the area at least around 1800 (Arch. Maarheeze, Soerendonk en Gastel, inv. no. 11, 1807), but was often reserved for clerical possessors (Clason, 1980). There is no evidence of a water mill in the upper Strijper Aa, although one may have been associated with the *Molenschut* (millpond), still existing as a toponymic to the east of Leende (mentioned in 1527, Arch. Leende, inv. no. 7).

Recent interferences

The most far reaching changes were initiated only in the second half of the last century. The first regulation works on the course of the Strijper Aa date

from 1884 (Hazendonk and Veen 1988). A radical change took place in the 30s of this century, when the Belgian drainage from the small city of Hamont was connected with the southernmost ponds of the wetland area (Iven and Van Gerwen 1974, p. 81).

The reclamation of *Het Goor* and the drainage of *Witsel* in the 50s seemed attractive for several reasons. First, a general lowering of the groundwater table had occurred in the area because of increased artificial drainage, increased evapotranspiration due to coniferous forest plantations and the droughtiness of the preceding decade. In 1955, local foresters reported a lowering of the water surface of moorland pools of about 50 cm 20 years after afforestation (Local Forest Statistics). Secondly, installation of a pumping-engine had become relatively simple, and further there existed an increased demand for agricultural land. The 60 ha reclamation of *Het Goor* was accomplished in 1957 (Fig. 3c), but it had no success because of serious continuing wetness due to groundwater. Since 1964, the area of reclamation has been part of a state nature reserve. With the land reallotment of 1974, the area was inundated again to allow for its restoration as wetland area (Heyink 1975).

Interpretation: four stages in landscape development

Prehistory: ecologically dynamic starting point

In *prehistoric times*, the higher grounds in the area were covered with deciduous forest, while the lower lands were covered with fens and local patches of peat-bog (Fig. 4a). In between, meandering peatbrooks had their courses through wet brookforests. Man did not play a significant role in landscape development. The rainwater that infiltrated through the rootzone of the forests on the higher grounds reached a relatively low groundwater table, due to the high evapotranspiration of the vegetation. It had a composition that was largely determined by mineralization and decomposition processes in the root zone and may therefore be called *rhizocline*. Mineral-rich *lithocline* groundwater, which had acquired chemical characteristics of the lithologic en-

244

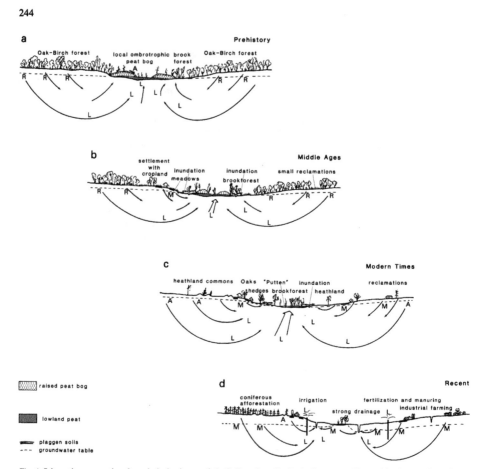

Fig. 4. Schematic cross-section through the landscape of the Strijper Aa-valley in the four stages discussed in the text. Groundwater quality: A = atmocline, R = rhizocline, L = lithocline, M = metacline.

vironment, was discharged in the lower grounds and brook valleys. It was diluted by nutrient-poor rainwater-like *atmocline* waters draining rapidly from the raised peat-bogs, in which dystrophic water accumulated to levels higher than in the surrounding area due to the low permeability of the peat. For reference to the terms atmocline and lithocline see Van Wirdum (1979).

From the Iron Age to the Middle Ages, this picture essentially did not change, although ecosystem development never was stationary (Jansen 1972; Van Leeuwaarden 1982). Since Roman times and the Early Middle Ages the region was permanently inhabited (Slofstra 1982; Theuws 1988), and man gradually introduced changes by settlement, agricultural development and grazing. Possibly, also some peat bogs were reclaimed for fuel. Due to decreased evapotranspiration, these changes led to a slight rise in groundwater levels at the higher elevations and an associated increase in the discharge of lithocline groundwater in the valleys (Fig. 4b).

Middle Ages: reclamation

During the *High Middle Ages* radical changes took place in the landscape (Hendrikx, 1989). Agricultural land was extended, and the deciduous forest was cleared and converted into open rangeland. The small peat-bogs were reclaimed as well. Where formerly rhizocline water had recharged the groundwater, now atmocline water took its place. The groundwater table at the higher elevations was higher due to lower evaporation rates. Further, at the highest parts of the landscape, drifting sands emerged, at places obstructing drainage courses. The groundwater draining from the agricultural lands must have been richer in nutrients than the water from uncultivated grounds. Such water has been called *metacline* (Pedroli 1990b). The higher rate of discharge of lithocline groundwater led to a higher biological production in the wetlands in the form of hay, fire-wood and peat (Fig. 4c).

Modern Times: water retention

From the *Middle Ages to the 20th century*, the landscape pattern did not change drastically. Social organization and population pressure continuously varied, resulting in changing proportions of cultivated and abandoned land. This period may be characterized by the retention of water within the area, as is illustrated by the presence of many fish-ponds. Croplands benefited from high summer groundwater, although inundation was frequent.

Up to recent times, the wetland complex must have been the source of the Strijper Aa (Schimmel 1952), although most 17th century maps draw a watercourse comparable to the Strijper Aa, which originates either further east, or further south in present Belgium. Still, from the more detailed 19th century maps it can be deduced that the Strijper Aa did not drain lands east of Soerendonk (Fig. 2b). Only the westernmost cultivated lands of Gastel and Soerendonk drained to the Strijper Aa, partly east and north around the village of Soerendonk. Also, the contribution of water from the *Elsbroek*, just to the south of the Belgian border, to the Strijper Aa probably was not important in the 19th cen-

tury, since near the Belgian border the Strijper Aa was clearly not a main drainage course at that time, and most Belgian water of the adjacent area probably was artificially drained to the west (Fig. 2a).

So, the *Goor* wetland area was mainly a discharge area of deeper groundwater. As such it represented one of the few nearly natural source areas for lowland brooks still preserved in the Netherlands around 1950 (Schimmel, 1952). A schematic hydrologic representation of this situation is given in Fig. 4c.

Recent: artificial drainage

In *recent* times, a landscape-ecological revolution completely changed the landscape. The remaining wetlands were drained and reclaimed, which led to a drastic lowering of water tables. Especially in summer, even the traditional croplands suffer from dryness, which is compensated by irrigation. The heathlands were reclaimed as well and largely converted to coniferous plantations or agricultural lands. This had also an effect of lowering the groundwater table.

The recent interferences led to an increase in stream discharge and the input of allochtonous water into the wetland area: in summer, water draining through the Strijper Aa was partly derived from irrigation of poplar-stands by water from a navigation canal 10 km to the south. However, still in 1953, it was observed that the groundwater isohypses in the area of *Het Goor* showed a gradient of water table depth from south to north, *i.e.* the water table was not influenced by the course of the Strijper Aa (Visser 1953). This means that discharge of deeper groundwater dominated over the effect of drainage. Also, the residence time of the water in the marshland was sufficient to purify it (Iven and Van Gerwen 1974).

The measures described above resulted in a further lowering of the water table around *Het Goor*, and also in the diversion of agricultural drainage water by a drainage canal. Now polluted metacline groundwater dominates throughout the area and alien lithocline groundwater is pumped from deep layers for irrigation. Together, these changes led to

246

a strongly decreased discharge of pure mineral-rich lithocline groundwater in the centre of the valley (Fig. 4d).

Discussion: man and landscape

The Strijper Aa wetland area for centuries was a large, very poorly drained area, dominated both in water quantity and in water quality by discharging deep groundwater. This deep groundwater originated from the higher wastelands, which were increasingly deprived of nutrients. From about 1935 to 1975 the groundwater was seriously affected by pollution, both from household sewage and from agriculture, and incidentally by industrial waste (Pedroli *et al.* 1990). During this period, the water table in the surrounding area probably was lowered by about 1 m, which decreased the discharge of deep groundwater. The Strijper Aa wetland area was largely converted into the Strijper Aa reclamation area.

It is clear from this study that, although the landscape is continuously in transformation, some of its characteristics remain stable throughout the ages. In this case, the general topography represents a polarity between the higher and the lower parts, associated with recharge and discharge of groundwater. How this polarity worked was determined by dynamic processes in landscape development, in which human activities played a significant role (Pedroli 1989).

As soon as regular human settlement occurred in the area, there were interferences within the range of human perception. The poor village communities restricted themselves to their own lands, using them as commons. With the introduction of the French legislation at the beginning of the 19th century, the commons became municipal property. At this time the task of the municipality changed from care of the environment of the owners of the land, to care for the well-being of all inhabitants. This made later redistribution and sale of the former commons to individuals possible. This was the beginning of a trend reducing the role of the village community as the basis for the organisation of land use, and replacing it with an anonymous body

without direct personal ties to the landscape. For example, in 1863 an official regional body for surface water management was installed, which acted mainly in favor of increased drainage (Hazendonk and Veen 1988).

In the present century, the negative side-effects of the decreasing personal connection of man with landscape has led to an increased awareness of naturalistic landscape values (Bockemühl 1985). It is mainly the interference of particular individuals that permitted promotion of economically independent development in the landscape. For example, the acquisition of part of *Het Goor* wetland, and therefore its conservation as wetland, can be attributed to the perseverence of the official responsible for nature values in the land reallotment discussions in the 1970's (pers. comm. J.A. Hendrikx 1989).

During the last 15 years, a restricted part of the area was hydrologically isolated from the surroundings by diversion of drainage water, which decreased the influence of contamination. To compensate for the lowered groundwater table, at some places fens were dug out. As a result of these conservation measures, recovery had taken place in some locations with mesotrophic vegetation at the border of *Het Goor* by 1981.

On the other hand, the view of the landscape as reflected in the public works accompanying the land reallotment, lacked any consistency or ecological vision (Hazendonk and Veen 1988). This study shows that landscape planners must assist in development of the landscape to assure that present or potential diversity in landscape pattern and structure, and a vision of its nature is included in the program. Historical information can be of value in the evolution of that vision. In the particular landscape studied here, the ecological and perceptional polarity between the higher elevation with nutrient-poor groundwater and the lower elevation with relatively nutrient-rich groundwater should be part of such a vision.

Acknowledgements

The valuable suggestions of Theo Bakker, Hans Joosten, Pim Jungerius and Annejet Rümke are gratefully acknowledged. We are indebted to the Editor for improving the manuscript.

References

Archives consulted:
- Streekarchief Zuid-Oost Brabant, Eindhoven: Municipal Archives of Leende, Inventory by Goossens (1985).
- Streekarchief Peelland, Deurne: Archief van het voormalige corpus Maarheeze, Soerendonk en Gastel. Inventory in typescript.

Arts, N. 1987. De kwaliteit en de toekomst van archeologische en cultuurhistorische objecten in de boswachterij 'Leende' e.o., Provincie Noord-Brabant. Fysisch Geografisch en Bodemkundig Laboratorium Amsterdam/Staatsbosbeheer Tilburg Report Nr. 20-11-87. 68 pp.

Billinge, M. 1977. In search of negativism: Phenomenology and historical geography. J. Histor. Geogr. 3: 55–67.

Bockemühl, J. 1985. Erkennen und Handeln im Gewebe der Naturreiche oder Woran krankt der Wald. Elemente d. Naturw. 43: 49–66, 44: 3–21.

Borman, F.H. and Likens, G.E. 1979. Pattern and process in a forest ecosystem: disturbance, development, and the steady state based on the Hubbard Brook Ecosystem Study. Springer, New York, NY. 253 pp.

Clason, A.T. 1980. Jager, visser, veehouder, vogellijmer. In N. Chamalaun and H.T. Waterbolk: Voltooid verleden tijd? Een hedendaagse kijk op de prehistorie. Intermediair, Amsterdam. pp. 131–146.

Coopmans, J.P.A. 1987. De betekenis van de gemeynten voor de dorpswetgeving in de Meijerij van 's Hertogenbosch (1250–1650). In Streefkerk, C. and Faber, S. (Eds): Ter recognitie. Bundel H. van der Linden. Hilversum, pp. 139–154.

Ellenberg, H., Mayer, R. and Schauermann, J. 1986: Ökosystemforschung – Ergebnisse des Sollingprojekts: 1966–1986. Ulmer, Stuttgart. 507 pp.

Gardner, R.H., Milne, B.T., Turner, M.G. and O'Neill, R.V. 1987. Neutral models for the analysis of broad-scale landscape pattern. Landsc. Ecol. 1: 19–28.

Golley, F.B. 1987. Introducing landscape ecology. Landsc. Ecol. 1: 1–3.

Goossens, R.J.I. 1985. Inventaris van de archieven van de gemeente Leende, 1563–1945. Streekarchief Zuid-Oost Brabant, Eindhoven.

Gore, A.J.P., (Ed.) 1983. Mires: swamp, bog, fen and moor. General studies. Ecosystems of the World 4A. Elsevier, Amsterdam. 516 pp.

Goris, J. 1966. Aloude valkerij in de Kempen en aan de vorstelijke hoven. Arendonk. 171 pp.

Gregory, D. 1978. The discourse of the past: phenomenology, structuralism and historical geography. J. Histor. Geogr. 4: 161–173.

Hazendonk, N.F.C. and Veen, P.J. 1988. Het landschap van de zandgebieden. Probleemverkenning in de ruilverkaveling Strijper Aa-Budel. Bos- en Landschapsbouw, Utrecht. 158 pp. + app.

Hendrikx, J.A. 1989. De ontginning van Nederland. Beschrijving van het ontstaan van de agrarische cultuurlandschappen in Nederland. Reeks BAELL Nr. 11, Directie Bos- en Landschapsbouw, Utrecht. 201 pp.

Heyink, J. 1975. Een ontwerp-beheersplan van het Soerendonckse Goor en het Turfwater. Rapport 328, Vakgroep Natuurbeheer, Landbouwhogeschool, Wageningen/Rijksinstituut voor Natuurbeheer, Leersum. 41 pp.

Hoppenbrouwers, P.C.M., Lesger, Cl., Joor, J., van Zanden, J.L. and Peys, R. 1986. Agrarische geschiedenis van Nederland van prehistorie tot heden. Staatsuitgeverij, The Hague. 168 pp.

Iven, W. and van Gerwen, T. 1974. Lind dè is de sgonste plats. Natuur en landschap van Leende, een Oost-Brabants dorp. Leende. 183 pp.

Ivanov, K.E. 1981. Water movement in mirelands. Academic Press, London. 276 pp.

Iverson, L.R. 1988. Land-use changes in Illinois, USA: The influence of landscape attributes on current and historic land use. Landsc. Ecol. 2: 45–61.

Janssen, C.R. 1972. The palaeoecology plant communities in the Dommelvalley, North Brabant, The Netherlands. J. Ecol. 60: 411–437.

Joosten, J.H.J. and Bakker, T.W.M. 1987. De Groote Peel in verleden, heden en toekomst. SBB rapport 88-4, Utrecht. 291 pp. + app.

Kopp, D., Jäger, K.-D. and Succow M. 1982. Naturräumliche Grundlagen der Landnutzung. Akademie-Verlag, Berlin. 339 pp.

Kragt, F. 1985. Hydrologie van het Strijper Aa gebied. Unpublished manuscript, Research Institute for Nature Management, Leersum.

Lambert, A.M. 1985 (2nd Edition). The making of the Dutch landscape. An historical geography of the Netherlands. Academic, London. 372 pp.

Leenders, K.A.H.W. 1986. Die Brabanter Torfkanäle. Siedlungsforschung Arch. Gesch. Geogr. 4: 103–125.

Londo, G. 1968. Rapport over het te stichten Natuurreservaat 'Kranenveld' (bovenloop van de Strijper Aa met omgeving). RIVON, Zeist. 15 pp., 5 maps.

Martens, M. 1943. Actes relatifs à l'administration des revenues domaniaux du Duc de Brabant (1271–1408). Commission Royale d'Histoire, Bruxelles. 355 pp.

Mertens, J. 1982. De landbouwers in het Zuiden 1100–1300. In Algemene Geschiedenis der Nederlanden, Vol. II, pp. 105–122. Edited by D.P. Blok, W. Prevenier and D.J. Roorda. Fibula-Van Dishoek, Haarlem.

Moore, P.D. and Bellamy, D.J. 1974. Peatlands. Elck Science, London. 221 pp.

Neef, E. 1982a. Geographie – einmal anders gesehen. Geogr. Zeitschr. 70: 241–260.

Neef, E. 1982b. Stages in the development of landscape ecology. Proc. Int. Congr. Neth. Soc. Landscape Ecol., Pudoc, Wage-

248

ningen, pp. 19–27.

O'Neill, R.V., Milne, B.T., Turner, M.G. and Gardner, R.H. 1988. Resource utilization scales and landscape pattern. Landsc. Ecol. 2: 63–69.

Opdam, P., Rijsdijk, G. and Hustings, F. 1985. Bird communities in small woods in an agricultural landscape: Effects of area and isolation. Biol. Conserv. 34: 333–352.

Overbeck, F. 1975. Botanisch-geologische Moorkunde. Wachholtz, Neumünster. 719 pp.

Pape, J.C. 1970. Plaggen soils in the Netherlands. Geoderma 4: 229–255.

Pedroli, B. 1989. Die Sprache der Landschaft. Elemente d. Naturw. 51: 25–49.

Pedroli, G.B.M. 1990a. Classification of shallow groundwater types in a Dutch coversand landscape. J. Hydrol., in press.

Pedroli, G.B.M. 1990b. Ecohydrologic parameters indicating different types of shallow groundwater. J. Hydrol., in press.

Pedroli, G.B.M., Maasdam, W.A.C. and Verstraten, J.M. 1990. Zinc in poor sandy soils and associated groundwater. A case study. Sci. Tot. Envir. 91: 59–77.

Renes, J. 1987. Urban influences on rural areas: Peat digging in the western part of the Dutch Province of North Brabant from the thirteenth to the eighteenth century. In The Medieval and Early-Modern rural landscape under the impact of the commercial economy, pp. 49–73. Edited by H.-J. Nitz. Dept. Geography, Univ. Göttingen.

Schaller, J. and Haber, W. 1988. Ecological balancing of network structures and land use patterns for land-consolidation by using GIS-technology. In Connectivity in landscape ecology, pp. 181–191. Edited by K.-F. Schreiber. Münstersche Geogr. Arb. 29, Schöningh, Paderborn.

Schimmel, H.J.W. 1952. Rapport betreffende het 'Goor', het 'Turfwater' en de 'Grote Heide' onder Maarheeze-Leende. Afd. Natuurbescherming en Landschap, Staatsbosbeheer, Utrecht. 17 pp., 3 app.

Slicher van Bath, B.H. 1963. The agrarian history of Western Europe A.D. 500–1850. Trans. Olive Ordish. Edward Arnold, London. 364 pp.

Slofstra, J. 1982. De archeologische landesaufnahme. In Het Kempenprojekt: een regionaal-archeologisch onderzoeksprogramma, pp. 84–92. Edited by J. Slofstra, H.H. van Regteren Altena, N. Roymans and F. Theuws. IPP Publ. 306, Amsterdam.

Slofstra, J., Van Regteren Altena, H.H. and Theuws, F. 1985. Het Kempenprojekt 2. Een regionaal-archeologisch onderzoek in uitvoering. Stichting Brabants Heem, Waalre. 93 pp.

Swaen, A.E.H. 1937. De valkerij in de Nederlanden. Thieme, Zutphen. 132 pp.

Theuws, F. 1988. De archeologie van de periferie. Thesis, University Amsterdam. 178 pp.

Van Dorp, D. and Opdam, P.P.M. 1987. Effects of size, isolation and regional abundance on forest bird communities. Landsc. Ecol. 1: 59–73.

Van Leeuwaarden, W. 1982. Palynological and macropalaeobotanical studies in the development of the vegetation mosaic in eastern Noord-Brabant (the Netherlands) during lateglacial and early holocene times. Thesis, University of Utrecht, Utrecht. 134 pp.

Van Leeuwen, Chr.G. 1966. A relation-theoretical approach to pattern and process in vegetation. Wentia 15: 25–46.

Van Mourik, J.M. 1985. Een eeuwen jong podzolprofiel op de Schaijksche Heide palynologisch bekeken (with English summary). K.N.A.G. Geogr. Tijdschr. XIX: 105–112.

Van Mourik, J.M. (Ed.) 1988. Landschap in beweging: ontwikkeling en bewoning van een stuifzandgebied in de Kempen. Netherlands Geographical Studies 74, KNAG/FGBL, Amsterdam. 191 pp.

Van Uytven, R. 1958. Kloosterstichtingen en stedelijke politiek van Godfried I van Leuven (1095–1139). Bijdragen voor de Geschiedenis der Nederlanden XIII: 177–185.

Van Winden, A. and Opdam, P. 1987. Chorologische relaties via pendelende vogels: een onderzoekmethode (with English summary: Landscape ecology relations created by commuting birds: a method). Landschap 4(4): 307–319.

Van Wirdum, G. 1979. Dynamic aspects of trophic gradients in a mire complex. In Proceedings and Information 25, TNO Committee on Hydrological Research, The Hague, pp. 66–82.

Verhees, H. 1794. Kaart van Brabant [appr. scale 1:180,000]. Mortier, Covens & Zn, Amsterdam.

Vervloet, J.A.J. 1984. Inleiding tot de historische geografie van de Nederlandse cultuurlandschappen. Pudoc, Wageningen. 136 pp.

Visser, W.C. 1953. Rapport betreffende een oriënterend onderzoek naar de mogelijkheden van ontginning van het moerasgebied 'Het Goor' te Soerendonk. Cultuurtechnische Dienst, Afdeling Onderzoek, Utrecht. 12 pp., 8 app.

Williams, M. 1989. Historical geography and the concept of landscape. J. Histor. Geogr. 15: 92–104.

Landscape Ecology vol. 2 no. 1 pp 23-44 (1988)
SPB Academic Publishing, The Hague

Quaternary landscape ecology: Relevant scales in space and time

Hazel R. Delcourt[1] and Paul A. Delcourt[2]

[1]Department of Botany, Graduate Program in Ecology and Center for Quaternary Studies of the Southeastern United States, University of Tennessee, Knoxville, TN 37996; [2]Department of Geological Sciences, Graduate Program in Ecology and Center for Quaternary Studies of the Southeastern United States, University of Tennessee, Knoxville, TN 37996

Keywords: archaeology, hierarchy, long-term data sets, paleoecology, Southeastern United States

Abstract

Two primary goals of landscape ecologists are to (1) evaluate changes in ecological pattern and process on natural landscapes through time and (2) determine the ecological consequences of transforming natural landscapes to cultural ones. Paleoecological techniques can be used to reconstruct past landscapes and their changes through time; use of paleoecological methods of investigation in combination with geomorphic and paleoethnobiological data, historical records, and shorter-term ecological data sets makes it possible to integrate long-term ecological pattern and process on a nested series of temporal and spatial scales. 'Natural experiments' of the past can be used to test alternative hypotheses about the relative influences of environmental change, biological interactions, and human activities in structuring biotic communities within landscape mosaics.

On the absolute time scale of the Quaternary Period, spanning the past 1.8 million years, current distributional ranges of the biota have taken shape and modern biotic communities have assembled. Quaternary environmental changes have influenced the development of natural landscapes over time scales of centuries to hundreds of thousands of years; human cultural evolution has resulted in the transformation of much of the biosphere from natural to cultural landscapes over the past 5,000 years. The Quaternary extends to and includes the present and the immediate future. Knowledge of landscape changes on a Quaternary time scale is essential to landscape ecologists who wish to have a context for predicting future trends on local, regional, and global scales.

Introduction

The Quaternary Period encompasses approximately the past 1.8 million years of geologic history (Bowen 1985). The Quaternary includes two subdivisions of time: (1) the Pleistocene Epoch, which lasted from about 1.8 million years ago until 10,000 years ago; and (2) the Holocene Epoch, from 10,000 years ago to the present (Fairbridge 1983). The Quaternary is characterized by alternating episodes of glacial and interglacial climates. Each

glacial-interglacial cycle lasts about 100,000 years (CLIMAP 1976), of which 90,000 years are relatively cold, and during which continental glaciers develop and expand at middle and high latitudes. Approximately 10,000 years of each climatic cycle are characterized by warm climate, for example the current, or Holocene, interglacial interval. The Quaternary Period is also defined on the basis of minimal evolution or extinction of both terrestrial and marine flora and fauna (Lyell 1834; Kurtén and Anderson 1980). Two major exceptions to this are

24

the evolution of *Homo sapiens* about 400,000 years ago (Bowen 1985) and the late-Quaternary extinctions of numerous species of megafauna, for example, occurring in North America at the end of the Pleistocene, 10,000 years ago (Martin and Klein 1984). For the late-Quaternary time interval, corresponding to the most recent glacial-interglacial cycle, an absolute chronology of events is available through radiocarbon dating, an isotopic-dating technique useful to about 80,000 years Before Present (yr B.P.). The late Quaternary is characterized by major changes in species distributions and composition of biotic communities; only during the Holocene interval have modern distributional ranges of the biota taken shape and contemporary biotic communities assembled (Kurtén and Anderson 1980; Davis 1981, 1983; Huntley and Birks 1983; Nilsson 1983; Graham 1986; Graham *et al.* 1987; Delcourt and Delcourt 1987a; Ritchie 1987).

The Quaternary Period represents not only a portion of the history of life on earth, but it also includes the present day and will extend into both the immediate and the distant future. Paleoecological studies that reconstruct late-Quaternary landscape changes (Berglund 1986) thus can be designed to dovetail with other methods of investigation, including historical records and shorter-term data sets, to integrate landscape patterns and processes on a nested series of temporal and spatial scales in ecological time. An appreciation of past changes in both environment and biota is imperative for evaluating the current trajectory of change in landscapes (Delcourt *et al.* 1983a; Jacobson and Grimm 1986; Clark 1986). Histories of landscape development that can be reconstructed using paleoecological techniques provide long-term 'experiments of the past' ('natural experiments' *sensu* Diamond 1986) that give a basis for testing alternative hypotheses about the relative influences of environmental change, biological interactions, and human activities in structuring biotic communities within landscape mosaics (Delcourt and Delcourt 1987a, b).

Changes in landscapes on a Quaternary time scale have been two-fold: (1) changes in climate and in geomorphic processes have affected vegetational patterns and processes on natural landscapes on the scale of hundreds of years to hundreds of thousands of years (Wright 1984; Kutzbach and Wright 1985); and (2) human activities have greatly modified the biosphere, particularly over the past 5,000 years, resulting in the widespread conversion of natural landscapes to cultural ones (Barker 1985; Behre 1986). Long-term changes in natural landscapes, which are defined here as those landscapes without substantial human intervention, include shifts in species ranges and ecotones between communities, changes in the mosaic of vegetation patches and community composition, and dynamic interactions of geomorphic processes with vegetational processes. Human modification of landscapes has involved changes in community composition, extensions or truncations in the ranges of plant and animal species, changes in the proportion of forest to nonforest land, and changes in disturbance regimes that have favored perpetuation of invasive, weedy (ruderal) species (Delcourt 1987).

In Europe, the ecological effects of prehistoric as well as historic human activities are widely appreciated by landscape ecologists (Naveh and Lieberman 1984). In North America, primary emphasis has been placed on understanding of energy transfer, nutrient flow, and changes in heterogeneity of natural landscapes on a short time frame of years to decades (Risser *et al.* 1984). Over a longer time scale of the past several hundred years, the effects of land conversion and forest fragmentation since Euro-American colonization also have been considered integral to developing concepts in North American landscape ecology (Burgess and Sharpe 1981; Turner 1987).

In this paper, we emphasize the potential for the Quaternary paleoecological record to contribute a wealth of information relevant to the developing subfield in Quaternary Landscape Ecology. First, we review the specific environmental forcing functions and biotic responses that result in landscape-scale ecological patterns and processes. Then we consider several aspects of development of an integrated, interdisciplinary research design that are necessary for successful integration of data within a nested hierarchy of space-time domains. We illustrate the relevance of this approach to understanding the dynamics of both natural and culturally influenced landscapes using examples from our

SCALE PARADIGM

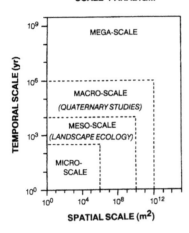

Fig. 1. Spatial-temporal domains for a hierarchical characterization of environmental forcing functions, biological responses, and vegetational patterns.

research in the southern Appalachian Mountains within the southeastern United States. Finally, we identify several fundamental issues in ecology that can be resolved by an integrated, interdisciplinary approach to Quaternary Landscape Ecology.

Spatial and temporal scale as an organizational paradigm in landscape ecology

Environmental forcing functions, biotic responses, and patterns of organization of communities on terrestrial landscapes vary on all scales in space and time (Delcourt *et al.* 1983a). A successful research design first defines the scale at which the phenomenon of interest can be observed, and then selects methods of analysis appropriate to resolving ecological pattern and process at that spatial-temporal scale. An operational scale paradigm into which landscape ecology can be incorporated includes micro-scale, meso-scale, macro-scale, and mega-scale spatial-temporal domains (Figs 1 and 2). These domains, as defined here, are modified from the hierarchical scheme first developed in Delcourt

et al. (1983a). It should be noted that the bounds placed on the dimensions of these domains represent a generalized overview for the purpose of illustrating relationships. In any particular case study, the dimensions chosen for study may be within one of these arbitrary domains, or they may cross the boundaries in order to arrive at an appropriate scaling relative to the generation times of the organisms studied or to the recurrence intervals for disturbances or of environmental changes relative to the ecosystems investigated.

Micro-scale domain

The micro-scale domain (Figs 1 and 2) has a duration of from 1 yr to 500 yr, and a spatial dimension of 1 m^2 to 10^6 m^2 (100 ha). The micro-scale is the domain of interest for the landscape manager, the process geomorphologist, and population and plant-succession ecologists. On this spatial-temporal scale, seasonal patterns of temperature and precipitation as well as longer-term weather trends and climatic fluctuations of decades to centuries are important stimuli to both plant and animal populations. Local to widespread disturbances of relatively short duration, such as wildfire, windthrow, and clearcutting have immediate effects on community composition (Pickett and White 1985). Geomorphic processes operative on this scale include soil creep, movement of sand dunes, debris avalanches, slumps, fluvial transport and deposition, and cryoturbation (Swanson *et al.* 1988). Biological responses to weather and climate changes, as well as to geomorphic and other kinds of disturbance events, include cyclic changes in animal populations, gap-phase replacement of forest trees, and plant succession on abandoned agricultural fields. These events thus affect vegetation at levels from individual plants to large forest stands, and occur on areas extending from the size of sample plots up to first-order stream watersheds, for example, the experimental watersheds within the Coweeta Basin (Fig. 3) in western North Carolina (Swank and Crossley 1988), or small second-order stream watersheds. Examples of changes in the landscape mosaic on this time scale include forest

26

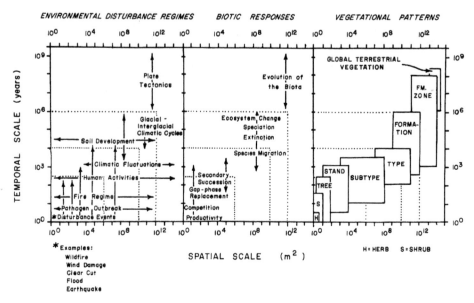

Fig. 2. Environmental disturbance regimes, biotic responses, and vegetational patterns viewed in the context of four space-time domains (modified from Delcourt et al. 1983a).

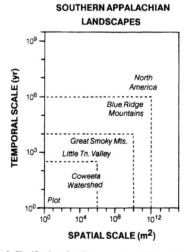

Fig. 3. Classification of study areas nested within the hierarchical framework of spatial-temporal domains.

fragmentation, with changes in relative size of forest and nonforest patches, increases in forest edge, and changes in available corridors due to land clearance (Forman and Godron 1981; Burgess and Sharpe 1981).

Meso-scale domain

The meso-scale domain (Figs 1 and 2) extends in time from 500 yr to 10,000 yr and in space from 10^6 m² (a physical feature averaging 1 km in width) to 10^{10} m² (100 km width). This domain encompasses events occurring over the last interglacial interval, the Holocene, and on watersheds of most second-order streams such as (Fig. 3) the Coweeta Basin (Swank and Crossley 1988), the Little Tennessee River of East Tennessee (Delcourt et al. 1986), mountain ranges such as within the Great Smoky Mountains National Park (White and Wofford 1984), and extending up to 1° latitude × 1° longitude (1/2 of a U.S.G.S. 1:250,000 topographic or geologic quadrangle; Olson *et al.* 1976). On the meso-scale, changes in geomorphic and climatic regimes effect changes in dynamics of patches in a landscape mosaic. Species migrations and ecotone displacements occur on this scale in response to changes in environmental gradients and predominant disturbance regimes. Long-term changes in the landscape mosaic occur on second-order and larger stream watersheds as well as on other large landforms. Within this domain, human cultural evolution has resulted in the transformation of natural landscapes to cultural ones.

Macro-scale domain

The macro-scale domain (Figs 1 and 2) lies largely within the research sphere of the Quaternary scientist. Within this domain, natural phenomena operate at temporal scales from 10,000 yr to 1,000,000 yr and at spatial scales ranging from 10^{10} m^2 (physical features averaging 100 km width) up to 10^{12} m^2 (1,000 km width). This domain spans in time from one to many glacial-interglacial cycles and in space an area of a physiographic province, such as the Blue Ridge Mountains of eastern North America (Fig. 3), to that of a subcontinent (Delcourt and Delcourt 1987a). On this scale, speciation and extinction become important biotic responses along with subcontinental-scale migrations and displacements of biomes. Changes in landscape heterogeneity occur on a macro-scale across entire physiographic regions, with consequent changes in the make-up of ecoregions *sensu* Bailey (1976).

Mega-scale domain

The mega-scale domain (Figs 1 and 2) encompasses 1 million yr to 4.6 billion yr (the age of the Earth) and includes areas > 10^{12} m^2 (land features > 1,000 km in average width). This scale ranges from continental, for example, North America (Fig. 3), to hemispheric and global and includes the majority of geologic time during which plate tectonics have changed the configurations of continents and ocean

basins, the biota has undergone major episodes of evolution and extinction, and the linkages between the lithosphere, cryosphere, and biosphere have developed (Frakes 1979).

Hierarchical relationships

This scale paradigm inherently is built upon an implied set of hierarchical relationships (Allen and Starr 1982; Delcourt *et al.* 1983a; O'Neill *et al.* 1986; Urban *et al.* 1986; Delcourt and Delcourt 1987a). Micro-scale, meso-scale, macro-scale, and megascale domains, as we define them in this paper, are a nested series of spatial-temporal configurations, each bounded by the next larger scale and each integrating all the patterns and processes ongoing at lower levels within the hierarchy.

Many ecological patterns and processes of direct relevance to landscape ecologists are resolvable at the interface between the micro-scale and the meso-scale domains. Whereas it seems of less immediate interest for some landscape ecologists to understand the macro-scale biogeographic changes that occurred in the Pleistocene (Urban *et al.* 1986), meso-scale events occurring during the Holocene are clearly relevant to all landscape ecologists, as these events of the present interglacial period have shaped the landscapes that were observed at the time of Euro-American settlement of North America (Williams 1982; Clark 1986). Landscape ecologists include not only the subset of ecologists concerned with integration of short-term ecological patterns and processes over relatively broad areas (Turner 1987); landscape ecologists must also concern themselves with understanding the long-term changes in ecological patterns and processes on landscapes (Risser 1987).

Developing an integrated research design

If Quaternary Landscape Ecology is to become a productive interface between neo-ecology and Quaternary paleoecology, it will be important for landscape ecologists to be aware of and to use effectively the full spectrum of investigative tools available for evaluating landscape change.

28

The history of forest stands and disturbance regimes

A number of methods are in common use for determining the modern composition and history of individual forest stands and the disturbance regimes that have influenced their development (Mueller-Dombois and Ellenberg 1974; Pickett and White 1985). Inferences about stand history can be made from age-class data and present stand composition (Harcombe and Marks 1978; Peet and Christensen 1980). Systematic excavation of the forest floor reveals not only changes in stand composition but the history of windthrow disturbance (Henry and Swan 1974; Oliver and Stephens 1978). Historical documents give insights about the influences of human disturbance on stand composition (Pyle and Schafale 1988). Tree ring counts and fire scars are used to establish recurrence intervals of fire that affect forest stand composition (Heinselman 1973; Romme 1982; Romme and Knight 1982). Analysis of macroscopic and microscopic plant remains including pollen grains and charcoal particles preserved in woodland hollows (Bradshaw 1981a, 1981b; Heide 1984) can extend the known history of individual forest stands over time intervals as long as nine thousand years. A series of individual site-intensive studies can be summarized to formulate patterns of forest history on a regional scale (Pyle and Schafale 1988).

The presettlement vegetation mosaic

Mapping landscape patterns on the meso-scale to interpret the presettlement vegetation mosaic as a point of departure for land-use history studies (Marschner 1959) can be accomplished with several complementary kinds of data sets. The most widely used are the records of the General Land Office Surveys (Bourdo 1956, 1983; Delcourt and Delcourt 1977; Grimm 1984). The GLOS records are limited in geographic coverage to the region west of the Appalachian Mountains and beyond the original 13 colonies (Pattison 1970). Other kinds of historical documents including ethnohistoric accounts and diaries can provide useful in recon-

structing landscape changes since the time of Euro-American settlement (Cronon 1983). Pollen assemblages from presettlement horizons in lake sediments constitute another complementary form of information concerning the landscape mosaic (McAndrews 1966; Davis *et al*. 1986). All of these kinds of data require careful interpretation within the limitations of uncertainty inherent in the methods. Special training may be required in the correct use of historical documents (Cronon 1983) or in the proper interpretation of the fossil pollen record (Faegri and Iversen 1975; Birks and Gordon 1985; Berglund 1986) in order to avoid pitfalls.

Landscape changes prior to Euro-American colonization of North America

Geomorphic and other geologic evidence of change in configuration of hillslopes, stream courses, terraces, and in recurrence intervals of geomorphic disturbance events can be obtained through radiocarbon dating and stratigraphic interpretation of sedimentary deposits (Mills and Delcourt 1989). Changes in the vegetation mosaic through the Holocene interval are interpretable from pollen and plant-macrofossil evidence preserved in lake sediments (see Bryant and Holloway 1985 for complete bibliographies to all North American literature in Quaternary pollen analysis). Paleolimnologic evidence of changes in water quality and coupled watershed-lake ecosystems is available in the form of fossil diatoms, cladocera, biochemical pigments, and in stratigraphic profiles of elements such as nitrogen and phosphorus (Likens 1985; Binford *et al*. 1987; Davis 1987). Evidence of prehistoric land use can be gleaned from the archaeological and paleoethnobiological (both floral and faunal) record (Butzer 1982; Dincauze 1987; McAndrews 1988).

An interdisciplinary approach

Lists of the kinds of tools and observations that can be made, especially when explicitly tied to the appropriate scale of resolution, can serve as helpful guides to developing a comprehensive research design for gathering and tabulating data for inclusion in a time-series of Geographic Information System (GIS) data bases for subsequent mapping (Meentemeyer and Box 1987). Integration and understanding of Quaternary landscape history requires development of a well-designed research strategy, followed by synthesis and effective interpretation. Dincauze (1987) has emphasized the need for effective interdisciplinary research designs in environmental archaeology ('prehistoric human ecology' *sensu* Butzer 1982). The need for anthropologists to have a thorough understanding of the underlying assumptions in other paleoenvironmental disciplines and to be cautious about borrowing data sets from other fields (Dincauze 1987) is also applicable to the field of landscape ecology. Landscape ecologists must become broadly based if our field is to mature from a descriptive/correlative phase of scientific endeavor to an experimental/hypothesis-testing phase (Risser *et al.* 1984).

An integrated, interdisciplinary approach to Quaternary Landscape Ecology includes the following elements: (1) identification of the problem to be addressed (specific hypothesis to be tested); (2) bounding the problem in a spatial-temporal domain appropriate to its resolution; (3) selection of methods appropriate to investigation of the problem, employing specialists to develop data sets or provide advice when necessary (a comprehensive summary of paleoecologic techniques is found in Berglund 1986); (4) collection of data in a statistically valid way; (5) interpretation of results of each independent line of evidence within the limitations of the techniques available; (6) integration of the data from an ecological perspective in terms of inferred dynamics of landscape patterns and processes, using map series, descriptive or quantitative models, and scenarios; and (7) use of resulting scenarios as tests of hypotheses originally proposed.

The examples that follow are from our recent work in the central and southern Appalachian Mountain region. These case studies illustrate the potential of the interdisciplinary approach of the Quaternary scientist for reconstructing long-term changes in natural landscape mosaics, as well as for evaluating the ecological consequences of prehistoric human activities.

Late-Quaternary landscape history in the central and southern Appalachian Mountains

Methods of investigation and site selection

We have studied the late-Quaternary history of the central and southern Appalachian Mountains from a transect of sites distributed from 500 m to over 1,500 m elevation from eastern West Virginia south to western North Carolina. Methods of investigation included (1) field mapping of local and regional vegetation in each study area; (2) field mapping of geomorphic landform features and geologic deposits in each study area; (3) compilation of historic records of logging and other disturbances at each site; (4) paleoecological analyses of peat or lake sediment cores collected from bogs and fens in West Virginia and North Carolina, from the former lake at Saltville, Virginia, and from natural lakes in East Tennessee; (5) establishment of absolute chronologies for late-Quaternary vegetation and geomorphic events in each study area. These paleoecological records extend back in time from 1,500 yr B.P. to 17,000 yr B.P.

Our series of coordinated studies was designed to gather available evidence to test the long-held hypothesis that the modern species richness and composition of biotic communities in the southern Appalachians can be explained by the two major factors most frequently cited in scientific literature: (1) the great length of time (260 million yr) since the Appalachian Mountain chain was affected by significant tectonic uplift or by marine submergence (thereby providing habitat continuity for persistence of vegetation types of presumed great antiquity, such as the proposed Tertiary relict cove hardwoods forest of Cain [1943]); and (2) the present-day configuration of elevation and slope aspect of the landforms, the modern disturbance regime, and the resulting diversity of habitats distributed along an environmental gradient of more than 2,000 m elevational range (Whittaker 1956). Alternative hypotheses that we proposed included

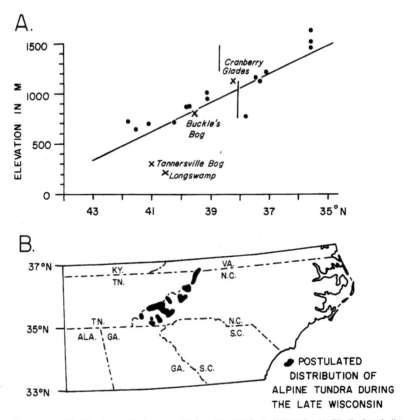

Fig. 4. Late-Quaternary periglacial environments of the central and southern Appalachian Mountains: (A) Distribution of relict, perigla-cial sorted, patterned-ground features (designated by solid dots) and selected montane bog sites (designated by X's), plotted along an elevational and latitudinal transect; (B) Postulated distribution of alpine tundra during the full-glacial interval (from Delcourt and Del-court 1985).

(1) climatic changes during the past 10,000 to 100,000 years that would have induced both north-ward and southward migrations of plant and animal species as well as speciation events; and (2) changes in geomorphic disturbance regimes during the late-Quaternary interval (linked to changes in temperature and precipitation) that would have provided opportunities for rare species to persist within the prevailing vegetation mosaic.

In the course of this research project, we, our students, and other colleagues have documented the quantitative relationship of both the local and regional composition of modern forests to pollen rain (Delcourt *et al.* 1983b; Delcourt and Pittillo 1986), evaluated the geomorphic and vegetational history at individual sites (Delcourt and Delcourt 1986; Larabee 1986; Shafer 1984, 1985, 1988), and prepared summaries of the major biogeographic changes that have occurred during the late-Quater-nary interval for the central and southern Ap-palachians (Delcourt 1985; Delcourt and Delcourt 1984, 1985, 1986, 1987a).

Relationship of geomorphic changes to landscape history

At mid- and high elevations in the central and southern Appalachian Mountains, conditions during the last glacial maximum of the Pleistocene (20,000 yr B.P. to 16,500 yr B.P.) were characterized by intense periglacial environments in which the ground was perennially frozen. Evidence for discontinuous permafrost during the Pleistocene is in the form of both paleovegetation localities for alpine tundra (Maxwell and Davis 1972; Watts 1979; Larabee 1986) and relict geomorphic features that were produced under a former climatic regime with mean annual temperatures as low as −8°C (Péwé 1985; Mills and Delcourt 1989). Periglacial geomorphic features including stone polygons, stone stripes, block fields, and boulder streams have been mapped throughout the Appalachian region (Michalek 1968; Clark 1968; Shafer 1984, 1988). The elevational and latitudinal distribution of these relict, periglacial features corresponds well with the distribution of paleoecological localities with plant-fossil assemblages representing tundra vegetation that are radiocarbon-dated from the last glacial maximum (Fig. 4a). By extrapolation of this elevation-latitude relationship for the upper tree-line limit during full-glacial time, we have inferred (Delcourt 1985; Delcourt and Delcourt 1985, 1987a) that the potential area of alpine tundra in the southern Appalachians extended to the summits of the highest mountain peaks above approximately 1,500 m elevation (Fig. 4b).

With climatic warming in the late-glacial interval, which began as early as 16,500 yr B.P. at 35°N latitude (Delcourt 1979), mean annual temperatures at high elevations increased from −8°C up to about 0°C (Shafer 1984, 1988; Delcourt and Delcourt 1985; Fig. 5), crossing a climatic threshold governing geomorphic processes (Mills and Delcourt 1989). With an increase in mean annual temperature, both the frequency and intensity of freeze-thaw cycles would have been augmented during the late-glacial interval from 16,500 yr B.P. to about 10,000 yr B.P. (Péwé 1985). Sediment par-

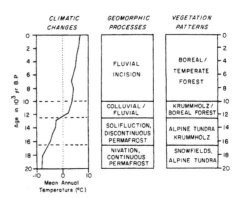

Fig. 5. Late-Quaternary landscape model for interactions of climate, geomorphology, and vegetation at elevations above 1,500 m in the southern Blue Ridge Physiographic Province (from Delcourt and Delcourt 1985).

ticles in all size ranges from boulders to silt and clay were produced during late-Pleistocene periglacial climates, with intense freeze-thaw cycles creating sorted patterned-ground features such as stone polygons and stone stripes (Clark 1968; Péwé 1985). Most of the periglacial sediments remained perennially frozen in place until the late-glacial climate warmed sufficiently for processes such as solifluction to move the materials downslope (Mills and Delcourt 1989).

Solifluction and active colluvial movement of boulder streams were the predominant geomorphic processes through the late-glacial interval (Fig. 5). At Flat Laurel Gap, in western North Carolina, reactivation of solifluction lobes and boulder streams persisted into the early Holocene to as recently as 7,800 yr B.P. (Shafer 1984, 1988; Mills and Delcourt 1989). This geomorphic activity would have been a major form of physical disturbance that may have affected the rate of reestablishment of both boreal and temperate trees on hillslopes throughout the region. Boulder streams have been mapped to as low an elevation as 600 m in the Great Smoky Mountains of East Tennessee and western North Carolina (Michalek 1968).

With the onset of Holocene interglacial climatic

32

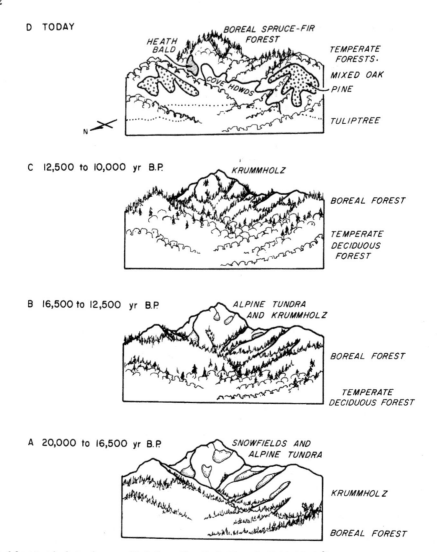

Fig. 6. Long-term landscape changes on Mt. LeConte, Great Smoky Mountains National Park (from Delcourt and Delcourt 1985).

conditions, a major geomorphic threshold was passed, resulting in the replacement of periglacial colluvial processes by temperate fluvial processes (Fig. 5). A decrease in the frequency and intensity of freeze-thaw cycles resulted in stabilization of slopes and inactivation of boulder streams. Fluvial processes became predominant, influencing the distributions of plants on the montane terrain (Hack and Goodlett 1960), but only limited stream incision has occurred during the Holocene (Mills and Delcourt 1989).

Consequences of late-Quaternary history for landscape ecology: tests of hypotheses

Through the late-Quaternary interval, climate change has been a major forcing function for landscape change in the southern Appalachian Mountains. During times of periglacial climate, cryoturbation was the primary geomorphic disturbance regime. The combination of cold temperatures and freeze-thaw churning of the soil resulted in a landscape mosaic that consisted of permanent snowfields and alpine tundra at mid- to high elevations in the central and southern Appalachian Mountains (Delcourt and Delcourt 1985; Figs 5 and 6). Paleoecological sites at relatively low elevations, such as Saltville, Virginia (Delcourt and Delcourt 1986), document species-rich boreal forest below 500 m elevation during full-glacial times.

With late-glacial climatic warming, thresholds were crossed for both fundamental geomorphic processes and cold-hardiness tolerances of plant species. Krummholz and boreal forest began to establish populations farther upslope, interfingering with alpine tundra (Delcourt and Delcourt 1985; Figs 5 and 6). Solifluction moved large quantities of sediment downslope, and active boulder streams funneled mineral debris through mountain ravines and coves. Both faunal and paleovegetational records indicate that the late-glacial transition was characterized by increased patchiness of the landscape mosaic, resulting in coexistence of species of small mammals no longer sympatric (Graham 1986) and intermingling of boreal and temperate trees in communities unlike those of today (Delcourt and Delcourt 1986).

With postglacial climatic warming, hillslopes stabilized and fluvial activity became predominant in the southern Appalachian Mountains. Herbaceous plant species formerly characteristic of alpine tundra communities either became locally extinct or else were restricted to high-elevation sites kept open because of disturbances (White 1984). In the Holo-

cene, natural disturbances maintaining grassy balds and populations of relict tundra species included fire, rock fall, and debris avalanches (Grant 1988) that resulted from storms associated with passage of hurricanes emanating from the Gulf of Mexico. During the Holocene, boreal forest spread upslope to the summits of the highest mountain peaks, and deciduous forest replaced boreal forest at mid- and low elevations (Figs 5 and 6). Modern cove hardwoods communities (Cain 1943; mixed mesophytic forest *sensu* Braun 1950) assembled as recently as 10,000 yr B.P. to 8,000 yr B.P.

It is now clear that climatic changes of the Quaternary have had profound effects on the landscapes of the southern Appalachians, not only in redistributing species across landforms, but in changing the rates and kinds of surficial geomorphic processes. Glacial-interglacial climatic fluctuations have enhanced the species richness of the Great Smoky Mountains and other mountain ranges of the southern Blue Ridge Province. The 'great length of time' (260 million years) potentially available for gradual accumulation and evolution of species can no longer be used as the major explanation for biotic diversity in the southern Appalachians. Cove hardwoods communities can no longer be viewed as relicts of an Arcto-Tertiary Geoflora (as proposed by Cain (1943) and Braun (1950)). Many of the sites now occupied by cove hardwoods forest were locations continually disturbed by active boulder streams until 10,000 yr B.P. to 8,000 yr B.P. During the last glacial-interglacial cycle, severe geomorphic disturbance and cold periglacial climate would not have favored establishment of temperate trees on Appalachian Mountain slopes until the present Holocene interglacial. The forest communities as we know them have assembled only recently, even relative to the maximum lifetimes of individual trees. The variation in community composition from site to site described as characteristic of mixed mesophytic forest and cove hardwoods forest (Cain 1943; Braun 1950) is now better explained by chance dispersal as deciduous forest species spread northward from glacial refuges in the Gulf Coastal Plain and throughout the southern Appalachian region in postglacial times (Delcourt and Delcourt 1987a).

34

Natural and cultural landscapes in the Little Tennessee River Valley

Dynamic changes in environment in the southern Blue Ridge Mountains on a late-Quaternary time scale have had important consequences for changing landscapes in adjacent physiographic provinces such as the Ridge and Valley Province of East Tennessee. Rock debris produced at high elevations during full-glacial times was eroded from hillslopes and transported by streams to lower elevations during the late-glacial and Holocene (Delcourt 1980; Delcourt *et al.* 1986). Consequently, in the Valley and Ridge Province of East Tennessee, streams such as the Little Tennessee River aggraded their floodplains during this time interval from 15,000 yr B.P. to 4,000 yr B.P. (Delcourt *et al.* 1986). Changing fluvial geomorphology during the late-glacial transition and the Holocene not only changed the nature of vegetation patches and corridors for movement of animals, but was central to the developing cultures of prehistoric Native Americans. Paleo-Indians first immigrated to the southern Appalachian region 12,000 yr B.P.; Native Americans subsequently lived primarily on aggrading floodplains and on adjacent, topographically higher stream terraces (Chapman 1985). Our investigations of landscape history in the Little Tennessee River Valley have concentrated upon testing whether changes in climate and geomorphology have been the major agents of vegetational change, or, alternatively, whether human activities beginning in late Paleo-Indian and early Archaic times about 10,000 yr B.P. and continuing to the present have had a significant role in altering the landscape mosaic through the Holocene interval.

Interdisciplinary methods of investigation

Geologic framework
Reconstruction of dynamic lanscapes during the time of prehistoric human occupation of the Little Tennessee River Valley was accomplished through geomorphic mapping of stream terraces and radiocarbon dating of Quaternary alluvial deposits underlying the terrace surfaces (Delcourt 1980). Nine

sets of alluvial terraces occur within the study area; highest, oldest terraces have undulating land surfaces modified topographically by karst solution and collapse in underlying carbonate bedrock. Fluvial sediments associated with each stream terrace are derived from local sedimentary rocks in the Valley and Ridge Province, as well as metasedimentary rocks transported from the Blue Ridge Province, located to the east. The two youngest stream terraces contained wood and wood charcoal datable within the limits of the radiocarbon method. Wood from Terrace 2 (T2) was dated between about 27,000 and 32,000 yr B.P., indicating that sediment deposition in the T2 floodplain occurred at the transition from a relatively cold (Altonian stadial) time to a relatively warm (Farmdalian interstadial) interval. Wood and wood charcoal contained in sediments underlying the land surface of the youngest terrace (T1) date from 15,000 yr B.P. to 4,000 yr B.P.; the modern floodplain was formed when the T1 was incised 4,000 yr B.P. The timing of deposition of the T1 corresponds with the transition from maximum cold (Late Wisconsin glacial) conditions to maximum warm (mid-Holocene interglacial) conditions (Delcourt 1980). The beginning of deposition of T1 sediments corresponds with the late-glacial pulse of downslope solifluction and the early-Holocene change to predominance of fluvial processes in the adjacent Blue Ridge Province (Shafer 1984, 1988).

Archaeological record
The late-Pleistocene and Holocene chronology of geomorphic events in the Little Tennessee River Valley provides an environmental context that is critical for interpreting the archaeological record. The deeply stratified alluvial deposits of Terrace 1 contain a continuous sequence of human occupation of the Valley that extends back 10,000 years (Chapman 1985 and references cited therein). Using early-historic maps showing the distribution of major Indian towns (Timberlake 1762, *in* Williams 1927), systematic testing was conducted on over 60 archaeological sites, of which some 25 sites were completely excavated (extensive summary and bibliography of archaeological studies in Chapman 1985). Extensive use of radiocarbon dating of or-

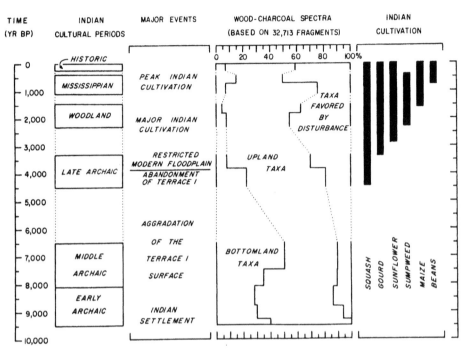

Fig. 7. Major Holocene cultural, environmental, and vegetational events within the Little Tennessee River Valley. The vertical bar for each cultigen corresponds to its temporal range, as preserved within stratified archaeological deposits (from Delcourt and Delcourt 1985).

ganic material (wood charcoal and carbonized fruits and seeds) from the stratigraphic horizons of human occupation provided an absolute chronology for the long-term cultural transformations reflected in changing types of projectile points and ceramic pottery (Chapman 1985).

Locations of hearths, post-holes representing habitation structures, and other archaeological features were systematically mapped at each buried, stratigraphic level (each representing a temporal horizon for a different culture) for subsequent spatial analysis. Ethnobotanical remains (13.7 kg of macroscopic wood charcoal and 7.7 kg of charred fruits and seeds from 956 excavated features) and zooarchaeological remains (animal bones and teeth) recovered from hearths and other archaeological contexts were identified, quantified, and in-

terpreted in the context of Native American utilization of natural resources gleaned from the nearby landscape (Chapman and Shea 1981; Bogan 1982).

The ethnobotanical remains represent plant materials gathered for use as food, fuel, or for construction. Therefore, these plant remains represent a culturally biased record of vegetation composition. However, if it is assumed that people were gathering firewood from an area as close to the hearth site as possible, then the changes in relative representation of woody species in major ecological groupings can be used as a general indicator of changing landscape conditions in the vicinity of human habitations (Chapman et al. 1982). We used 32,713 fragments of wood charcoal that were identified to species to summarize the 10,000-year record of changing dominance of local bottomland

36

(riparian) forest, upland forest, and early-successional (disturbed) forest within the valley (Fig. 7;
Chapman *et al.* 1982).

Paleoecological record

As an independent record of changes in the vegetation mosaic through time, the fossil-pollen, plant-
macrofossil, and charcoal-particle records from
sediments of two ponds were documented for comparison with the archaeological record (Cridlebaugh 1984). The two sites chosen for paleoecological analysis (Figs 8 and 9) were (1) Tuskegee Pond,
located near the Historic Cherokee town of Tuskegee on a mid-level Quaternary stream terrace 1.5
km from a major archaeological excavation of the
T1 (Icehouse Bottom); and (2) Black Pond, located
in the karst uplands about 4 km northeast of Tuskegee Pond. Small pond sites were chosen to reflect
primarily the local (within 20 m radius of the site)
and extralocal (20 m to 2 km radius of the site;
Prentice 1985; Jackson 1988) contribution of fossil
plant remains, with only a minor regional component to the pollen assemblages (Jacobson and
Bradshaw 1981). Cross-comparison of the paleoecological records from the two sites, along with
comparison of corresponding prehistoric changes
in the ethnobotanical record, allows estimates of
the areal extent of land clearance and vegetational
change during the Holocene along a gradient of human modification of the landscape, extending from
the floodplain sites of continuous human settlement across higher alluvial terraces and into the uplands. The charcoal-particle record in the pond
sediments provides an indication of changes in fire
frequency through time, and changes in sedimentation rate indicate changes in erosion of hillslopes
surrounding the ponds (Delcourt *et al.* 1986).

Synthesis of Holocene landscape history

More than 50 scientific publications, theses, and

dissertations provide the evidence for particular
aspects of the Holocene environmental, biotic, and
prehistoric cultural data from the Little Tennessee
River Valley (see references in Chapman 1985 and
Delcourt *et al.* 1986). Here we characterize the
major environmental and human forcing functions, landscape responses, and resulting changes in
landscape patterns through time (Fig. 8) that can be
inferred from both published and unpublished
summaries of the geomorphic, archaeological, and
paleoecological data.

In the early Holocene, in adjustment to the influx
of sediment, the floodplain of the Little Tennessee
River was aggrading in response to climatic warming that triggered erosion of sediments from nearby
hillslopes in the Blue Ridge Province. Where the
Little Tennessee River emerged from its relatively
restricted, meandering channel carved through the
Blue Ridge Province and flowed westward into the
Ridge and Valley Province, its floodplain broadened and it became a braided stream, creating an
anastomosing series of wide, riparian corridors
(Fig. 8a). At the Icehouse Bottom archaeological
site, pollen evidence from gley soils buried beneath
alluvium indicates that by 9,500 yr B.P., the landscape was predominantly forested, with deciduous
forest of oak (*Quercus*), chestnut (*Castanea denta-
ta*), and hickory (*Carya*) probably widespread
across the uplands of the Ridge and Valley Province. Species of ash (*Fraxinus*), elm (*Ulmus*), and
willow (*Salix*) occupied bottomlands along stream
courses and point bars of the aggrading floodplain.
The pollen record from the early Holocene contains
very little evidence of herbs such as grasses, sedges,
and composites, indicating that openings in the
forest were infrequent and probably confined to
small, frequently disturbed areas along point bars,
river-eroded outcrops of steep bluffs, and temporary hunting camps of PaleoIndian and early Archaic people. The incidence of wildfire was probably low in the mesic forested environment of the

→

Fig. 8. Landscape reconstructions for the Little Tennessee River Valley, East Tennessee, showing the distribution of stream corridors
and the mosaic of forest and nonforest vegetation. (A.) PaleoIndian cultural period, 12,000 yr B.P. to 10,000 yr B.P. (B.) Archaic cultural period, 10,000 yr B.P. to 2,800 yr B.P. (C.) Woodland and Mississippian cultural periods, 2,800 yr B.P. to 500 yr B.P. (A.D.
1,500). (D.) Historic cultural period, 500 yr B.P. (A.D. 1,500) to present.

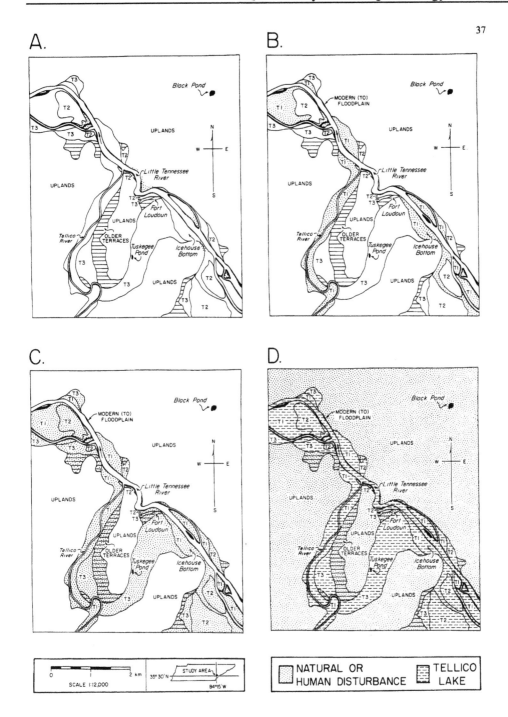

A.

B.

C.

D.

SCALE 1:12,000

STUDY AREA
35° 30' N
84° 15' W

NATURAL OR HUMAN DISTURBANCE

TELLICO LAKE

38

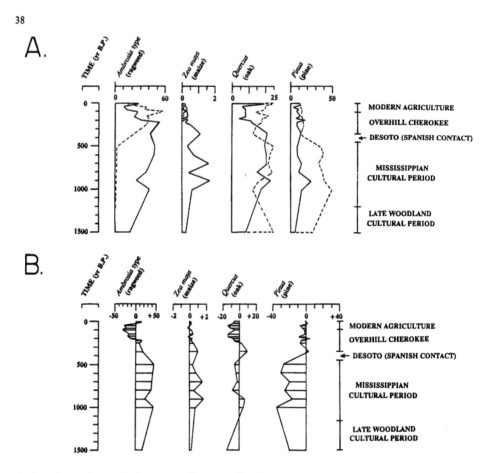

Fig. 9. (A.) Contrast diagram of Tuskegee Pond (solid curve) and Black Pond (dashed curve), Little Tennessee River Valley, for 1,500 yr B.P. to the present. Percentages are expressed as percent of arboreal pollen grains plus nonarboreal pollen grains and spores, excluding grains of obligate aquatic plants. (B.) Difference diagram of Tuskegee Pond minus Black Pond, Little Tennessee River Valley, for 1,500 yr B.P. to the present. Percentages are expressed as percent of arboreal pollen grains plus nonarboreal pollen grains and spores, excluding grains of obligate aquatic plants.

aggrading floodplain, although fire may have been used deliberately by Native Americans for hunting in game drives or to open the understory of upland oak-chestnut-history forests.

Through the Archaic cultural period (10,000 yr B.P. to 2,800 yr B.P.; Fig. 8b), the wood-charcoal record reflects primarily the utilization of bottomland trees that would have been readily available in

the local environment of the aggrading floodplain (T1) surface. The ethnobotanical record of charred fruits and seeds was comprised mainly of mast trees, including husks of hickory nuts (*Carya*) and walnuts (*Juglans nigra*) as well as acorns of oak (*Quercus*). Hazelnuts (*Corylus*) and chestnuts were used to a lesser degree as plant foods by the aboriginal populations of Native Americans. After 4,500 yr B.P., the first cultigens were introduced, including squash (*Cucurbita pepo*) and gourd (*Lagenaria siceraria*). After 4,000 yr B.P., the T1 alluvial surface was incised and abandoned, forming the restricted modern floodplain; representatives of bottomland forest taxa diminished in the wood-charcoal record at that time, corresponding with the reduction in area of the bottomland habitat. However, as the T1 terrace became more well-drained, populations of early-successional plant taxa increased, including red cedar (*Juniperus virginiana*), pine (*Pinus virginiana*), cane (*Arundinaria gigantea*), and tuliptree (*Liriodendron tulipifera*).

The incision and abandonment of Terrace 1 by the Little Tennessee River may have been a hydrologic response to climatic change at the transition from middle to late-Holocene times. Alternatively, it may have resulted from human modification of the landscape, including both deforestation and cultivation of garden plots. Based on the spatial distribution of habitations and other archaeological features, by 2,800 yr B.P., the end of the Archaic cultural period, we infer that all of the T1 surface was disturbed by the activities of Native Americans, no longer in original forest, but converted to clearings and in early stages of secondary succession (Fig. 8b). The landscape may have been kept open through the deliberate use of fire, and hearth fires may have escaped accidentally from control as well.

During the Woodland cultural period (2,800 yr B.P. to about A.D. 900; Fig. 8c), lifeways became more sedentary. Pottery was developed, and garden plots were cultivated. Crops included a diverse group of species often referred to as the 'eastern agricultural complex' (Ford 1985), such as sunflower (*Helianthus annuus*), marsh elder (*Iva annua*), goosefoot (*Chenopodium*), and canary grass (*Phalaris caroliniana*). Fire would have been used by Woodland people to clear garden plots, fire pottery, and to cook food. Maize (*Zea mays*) was introduced by 1,775 yr B.P., and in the Mississippian cultural period, after A.D. 900, maize became the primary plant food in the diet. Increasing dependence on maize required more intensive cultivation of the land in order to sustain increasing human populations. By A.D. 1,500, the proportion of early-successional plant species recorded in the wood-charcoal record increased to as much as 50% of all wood utilized by Mississippian people.

The pollen record from Tuskegee Pond (Delcourt *et al.* 1986) shows that by 1,500 yr B.P. maize cultivation extended to mid-level terraces at distances up to at least 1.5 km from the modern floodplain. Pollen of herbs today associated with agricultural fields (*i.e.*, ragweed (*Ambrosia* type)) was as much as 30% of the upland pollen assemblage from Tuskegee Pond as early as 1,500 yr B.P., indicating a very open landscape by late Woodland times. The record of charcoal particles from Tuskegee Pond sediments shows order-of-magnitude increases in the use of fire corresponding with the changes in major cultural periods of prehistoric and historic Native Americans. The paleoecological record from Black Pond, located in the uplands about 4 km away from the river, contained less than 5% ragweed pollen until about 400 years ago, after which ragweed percentages increased dramatically. With the records from the two paleoecological sites, it is possible to realistically constrain the area of human impact on the forest/nonforest mosaic through time (Figs 8c and 9a, b).

The paleoecological records from Black and Tuskegee ponds can be compared directly by means of a 'contrast diagram' (Janssen *et al.* 1985; Fig. 9a) and a 'difference diagram' (Jacobson 1979; Fig. 9b). These diagrams illustrate graphically that the two paleoecological sites have recorded distinctively different histories of vegetational change through the past 1,500 years that can be attributed to differences in land use by prehistoric Native Americans. Prior to Spanish contact in A.D. 1,540, deforestation and agriculture was restricted to lowlands along the Little Tennessee River, whereas after Euro-American settlement, the uplands were deforested of pine and converted to agricultural fields. Pollen percentages of ragweed (*Ambrosia* type) subsequently diminished in the lowlands, tracking the historic decline of Overhill Cherokee populations that formerly occupied valley bottoms and river terraces (Chapman, 1985). The dramatic differences in the pollen records from these two small pond sites indicates that their pollen source

40

areas do not overlap to a substantial extent. Thus, small sites such as Black Pond and Tuskegee Pond, situated along an elevational transect and spaced 4.5 km apart in settings with different land use histories, can be used in combination to place spatial constraints on the area impacted by long-term human activities on a watershed of the scale of the Little Tennessee River (Figs 3 and 8).

Both archaeological and paleoecological evidence indicates intensified land use along the riparian corridor and on the lower and mid-level stream terraces of the Little Tennessee through the Woodland and Mississippian cultural periods. With progressive Indian clearance of forests on the floodplain and lower terraces, and with intensification of crop cultivation, the late-Holocene landscape was gradually converted into a mosaic consisting of permanent Indian settlements and cultivated fields, early-successional forests invading abandoned Indian old-fields, and remnants of original deciduous forest in the uplands. As the forests continued to be fragmented by the ever-enlarging agricultural fields, forest-edge habitat would have increased. This in turn would have provided additional browse that together with the greater extent of forest edge would have led to increases in the populations of deer and wild turkey, both abundantly represented in the zooarchaeological record (Bogan 1982).

In Historic times, with introduction of new technology such as the iron hoe, cultivation of the land became more efficient. This is reflected in the sediment record of Tuskegee Pond by an order-of-magnitude increase in sedimentation rate. Beginning with the Spanish, Euro-Americans introduced a series of domesticated plants, animals, and weeds, once again transforming the lifeways of the Historic Cherokee. Today's agricultural landscape (Fig. 8d) includes old-fields that are invaded by a mixture of native and introduced species. All of the landscape within the study area mapped in Fig. 8d is either in secondary succession or flooded beneath the Tellico lake impoundment.

Over the past 10,000 years of the late-Quaternary interval, the vegetation mosaic of the Little Tennessee watershed has undergone long-term and progressive modification, gradually being transformed from a natural landscape to a culturally maintained one. We conclude that, in the Little Tennessee River Valley, Native Americans were not living in balance with their environment, although at any given moment they may have perceived it as such, but rather they were part of and active agents within a dynamically changing landscape.

Priorities for future research

The case studies reviewed in this paper illustrate the potential for collaborative research among Quaternary paleoecologists, geomorphologists, archaeologists, and landscape ecologists to provide a productive approach to examining the long-term changes in both natural and cultural landscapes. Priorities for future research in Quaternary Landscape Ecology include:

1. Additional refinements in research design to make it possible to develop more quantitative reconstructions of the changes in area, shape, configuration, heterogeneity, and connectivity of landscape patches and corridors on a Quaternary time scale. This basic documentation will be useful as boundary conditions for studies of more recent landscape changes at, for example, Long-Term Ecological Research sites such as Coweeta in western North Carolina (Swank and Crossley 1988).

2. Development of definitive tests of landscape hypotheses such as Godron and Forman's (1983) predictions of changes in patch characteristics, corridors, and other landscape features along a gradient of human modification. Temporal tests of this landscape hypothesis would not only complement the spatial tests that can be accomplished by comparing the patterning of modern natural, managed, agricultural, suburban, and urban landscapes, but would also yield new insights such as timing of introduction of invasive weedy plant species. Such events of biological invasion may have major ramifications for subsequent changes in ecological relationships within communities (Mooney and Drake 1986; Crosby 1986), but the process by which these changes have occurred in prehistoric times may be knowable only by using paleoecological or archaeological methods of investigation.

3. Using available tools to help solve fundamental ecological problems at the interface between meso-scale and micro-scale domains. One example is investigating the nature of presettlement vegetation and its prior development. The notion of a 'virgin' and 'stable' presettlement vegetation (*circa* A.D. 1,500 to A.D. 1,800) is a baseline assumption for most contemporary research in North American plant ecology and for management of natural resources. Was the presettlement vegetation mosaic changing at the time of Euro-American settlement? If so, was it responding to climatic change, prehistoric human impacts, or some combination of forcing functions? What were the spatial and temporal patterns of these landscape dynamics on a meso-scale, and what were their relationships to sensitive ecotones such as the prairie-forest border?

4. Challenge to develop new testable hypotheses in landscape ecology that have central significance to all of ecology. These might center around scale-dependent interactions of perturbations, processes, and patterns such as: (a) the effects of changes in landscape heterogeneity on community assembly and population dynamics; (b) the relative influence of environmental change, biological interactions, and human activities in structuring communities; and (c) the effects of changes in the patchwork of the landscape mosaic for microevolutionary processes including selection for demographic characteristics in plant populations (investment in seed production and mechanisms for seed dispersal), the importance of changes in connectivity of corridors for animal migrations and/or extinctions, or the effects of changing size, shape, and configuration of vegetation patches for long-term sympatry of small mammals within a region.

Acknowledgements

Our Quaternary geological and paleoecological research in the central and southern Appalachian Mountain region was supported the Ecology Program of the National Science Foundation (grants BSR-83-00345 and BSR-84-15652). Additional funding for research in the Little Tennessee River Valley, East Tennessee, was provided by the Tennessee Valley Authority and the National Geographic Society through grants to Jefferson Chapman. We thank Frank Golley, Michael Binford, and Steve Jackson for productive discussions and two anonymous reviewers for their constructive comments on this manuscript. Contribution #49, Center for Quaternary Studies of the Southeastern United States, University of Tennessee, Knoxville, TN, U.S.A. This manuscript was based upon the invited Plenary talk for the 3rd Annual Landscape Ecology Symposium 'Observations Across Scales: The Structure, Function, and Management of Landscapes' March 16–19, 1988, The University of New Mexico, Albuquerque.

References

Allen, T.F.H. and Starr, T.B. 1982. Hierarchy, perspectives for ecological complexity. University of Chicago Press, Chicago.

Bailey, R.G. 1978. Description of the ecoregions of the United States. United States Department of Agriculture, Forest Service, Odgen, Utah.

Barker, G. 1985. Prehistoric farming in Europe. Cambridge University Press, Cambridge.

Behre, K.-E. (Ed.) 1986. Anthropogenic indicators in pollen diagrams. A.A. Balkema, Rotterdam.

Berglund, B.E. (Ed.) 1986. Handbook of Holocene paleoecology and palaeohydrology. Wiley, New York.

Binford, M.W., Brenner, M., Whitmore, T.J., Higuera-Gundy, A., Deevey, E.S. and Leyden, B. 1987. Ecosystems, paleoecology and human disturbance in subtropical and tropical America. Quat. Sci. Rev. 6: 115–128.

Birks, H.J.B. and Gordon, A.D. 1985. Numerical methods in Quaternary pollen analysis. Academic Press, New York.

Bogan, A.E. 1982. Archaeological evidence for subsistence patterns in the Little Tennessee River Valley. Tennessee Anthropol. 7: 38–50.

Bourdo, E.A., Jr. 1956. A review of the General Land Office Survey and of its use in quantitative studies of former forests. Ecology 37: 754–768.

Bourdo, E.A., Jr. 1983. The forest the settlers saw. *In* The Great Lakes forest: an environmental and social history, pp.3–16, Edited by S.L. Flader. University of Minnesota Press, Minneapolis.

Bowen, D.Q. 1985. Quaternary geology, a stratigraphic framework for multidisciplinary work. Pergamon, Oxford.

Bradshaw, R.H.W. 1981a. Quantitative reconstruction of local woodland vegetation using pollen analysis from two small basins in Norfolk, England. J. Ecol. 69: 941–955.

Bradshaw, R.H.W. 1981b. Modern pollen-representation factors for woods in south-east England. J. Ecol. 69: 45–70.

Bryant, V.M., Jr. and Holloway, R. (Eds) 1985. Pollen records

42

of late-Quaternary North American sediments. American Association of Stratigraphic Palynologists Foundation, Dallas.

Burgess, R.L. and Sharpe, D.M. 1981. Forest island dynamics in man-dominated landscapes. Springer-Verlag, New York.

Braun, E.L. 1950. (Reprinted in 1974). Deciduous forests of eastern North America. Hafner Press, Macmillan, New York.

Butzer, K.W. 1982. Archaeology as human ecology. Cambridge University Press, Cambridge.

Cain, S. 1943. The Tertiary character of the cove hardwood forests of the Great Smoky Mountains National Park. Bull Torrey Bot. Club 70: 213–235.

Chapman, J. 1985. Tellico archaeology: 12,000 years of Native American history. University of Tennessee Press, Knoxville.

Chapman, J. and Shea, A.B. 1981. The archaeobotanical record: Early Archaic period to Contact in the Lower Little Tennessee River Valley. Tennessee Anthropol. 6: 61–84.

Chapman, J., Delcourt, P.A., Cridlebaugh, P.A., Shea, A.B. and Delcourt, H.R. 1982. Man-land interaction: 10,000 years of American Indian impact on native ecosystems in the Lower Little Tennessee River Valley, East Tennessee. Southeastern Archaeol. 1: 115–121.

Clark, G.M. 1968. Sorted patterned ground: new Appalachian localities south of the glacial border. Science 161: 355–356.

Clark, J.S. 1986. Dynamism in the barrier-beach vegetation of Great South Beach, New York. Ecol. Monogr. 56: 97–126.

CLIMAP, 1976. The surface of the Ice-Age Earth. Science 191: 1131–1137.

Cridlebaugh, P.A. 1984. American Indian and Euro-American impact upon Holocene vegetation in the Lower Little Tennessee River Valley, East Tennessee. Ph.D Dissertation, University of Tennessee, Knoxville.

Cronon, W. 1983. Changes in the land: Indians, colonists, and the ecology of New England. Hill and Wang, New York.

Crosby, A.W. 1986. Ecological imperialism: the biological expansion of Europe, 900–1900. Cambridge University Press, Cambridge.

Davis, M.B. 1981. Quaternary history and the stability of forest communities. In Forest succession, concepts and application, pp. 132–153, Edited by D.C. West, H.H. Shugart and D.B. Botkin. Springer-Verlag, New York.

Davis, M.B. 1983. Quaternary history of deciduous forests of eastern North America and Europe. Ann. Mo. Bot. Gard. 70: 550–563.

Davis, M.B., Woods, K.D. and Futyma, R.P. 1986. Dispersal versus climate: expansion of Fagus and Tsuga into the Upper Great Lakes region. Vegetatio 67: 93–103.

Davis, R.B. 1987. Paleolimnological diatom studies of acidification of lakes by acid rain: an application of Quaternary science. Quat. Sci. Rev. 6: 147–163.

Delcourt, H.R. 1979. Late-Quaternary vegetation history of the eastern Highland Rim and adjacent Cumberland Plateau of Tennessee. Ecol. Monogr. 49: 255–280.

Delcourt, H.R. 1985. Holocene vegetational changes in the southern Appalachian Mountains, U.S.A. Ecol. Mediterranea 11: 9–16.

Delcourt, H.R. 1987. The impact of prehistoric agriculture and land occupation on natural vegetation. Trends Ecol. Evol. 2: 39–44.

Delcourt, H.R. and Delcourt, P.A. 1977. Presettlement Magnolia-Beech Climax of the Gulf Coastal Plain: quantitative evidence from the Apalachicola River Bluffs, north-central Florida. Ecology 58: 1085–1093.

Delcourt, H.R. and Delcourt, P.A. 1986. Late-Quaternary vegetational history of the central Atlantic states. In The Quaternary of Virginia, pp. 23–35, Edited by J. McDonald and S.O. Bird. Virginia Commonwealth Division of Mineral Resources, Charlottesville.

Delcourt, H.R. and Pittillo, J.D. 1986. Comparison of contemporary vegetation and pollen assemblages: an altitudinal transect in the Balsam Mountains, Blue Ridge Province, western North Carolina, USA. Grana 25: 131–141.

Delcourt, H.R., Delcourt, P.A. and Webb, T. III. 1983a. Dynamic plant ecology: the spectrum of vegetational change in space and time. Quat. Sci. Rev. 1: 153–175.

Delcourt, P.A. 1980. Quaternary alluvial terraces of the Little Tennessee River Valley, East Tennessee. University of Tennessee, Knoxville, Department of Anthropology, Report of Investigations, No. 29: 110–121, 175–212.

Delcourt, P.A. and Delcourt, H.R. 1985. Dynamic landscapes of East Tennessee: an integration of paleoecology, geomorphology, and archaeology. University of Tennessee, Knoxville. Department of Geological Sciences, Studies in Geology 9: 191–220.

Delcourt, P.A. and Delcourt, H.R. 1987a. Long-term forest dynamics of the Temperate Zone, Ecological Studies 63. Springer-Verlag, New York.

Delcourt, P.A. and Delcourt, H.R. 1987b. Late-Quaternary dynamics of temperate forests: applications of paleoecology to issues of global environmental change. Quat. Sci. Rev. 6: 129–146.

Delcourt, P.A., Delcourt, H.R. and Davidson, J.L. 1983b. Mapping and calibration of modern pollen-vegetation relationships in the southeastern United States. Rev. Palaeobot. Palynol. 39: 1–45.

Delcourt, P.A., Delcourt, H.R., Cridlebaugh, P.A. and Chapman, J. 1986. Holocene ethnobotanical and paleoecological record of human impact on vegetation in the Little Tennessee River Valley, Tennessee. Quat. Res. 25: 330–349.

Diamond, J. 1986. Overview: laboratory experiments, field experiments, and natural experiments. In Community ecology, pp. 3–22. Edited by J. Diamond and T.J. Case. Harper and Row, New York.

Dincauze, D.F. 1987. Strategies for paleoenvironmental reconstruction in archaeology. Adv. Archaeol. Method Theory 11: 255–336.

Faegri, K. and Iversen, J. 1975. Textbook of pollen analysis (3rd rev. ed.). Hafner Press, Macmillan, New York.

Fairbridge, R.W. 1983. The Pleistocene-Holocene boundary. Quat. Sci. Rev. 1: 215–244.

Ford, R.I. 1985. The processes of plant food production in prehistoric North America. In Prehistoric food production in North America, pp. 1–18. Edited by R.I. Ford. Museum of Anthropology, University of Michigan, Anthropological Papers No. 75.

Forman, R.T.T. and Godron, M. 1981. Patches and structural components for a landscape ecology. Bioscience 31: 733–740.

Frakes, L.A. 1979. Climates throughout geologic time. Elsevier, New York.

Godron, M. and Forman, R.T.T. 1983. Landscape modifications and changing ecological characteristics. In Disturbance and ecosystems: components of response, pp. 12–28. Edited by H.A. Mooney and M. Godron. Springer-Verlag, New York.

Graham, R.W. 1986. Response of mammalian communities to environmental changes during the late Quaternary. In Community ecology, pp. 300–313. Edited by J. Diamond and T.J. Case. Harper and Row, New York.

Graham, R.W., Semken, H.A., Jr. and Graham, M.A. (Eds). 1987. Late Quaternary mammalian biogeography and environments of the Great Plains and prairies. Illinois State Museum, Springfield, Scientific Papers 22.

Grant, W.H. 1988. Debris avalanches and the origin of first-order streams. In Forest hydrology and ecology at Coweeta, Ecological Studies 66, pp. 103–110. Edited by W.T. Swank and D.A. Crossley, Jr. Springer-Verlag, New York.

Grimm, E.C. 1984. Fire and other factors controlling the vegetation of the Big Woods region of Minnesota. Ecol. Monogr. 54: 291–311.

Hack, J.T. and Goodlett, J.C. 1960. Geomorphology and forest ecology of a mountain region in the central Appalachians. United States Geol. Surv. Prof. Paper 347: 1–64.

Harcombe, P.A. and Marks, P.L. 1978. Tree diameter distributions and replacement processes in southeast Texas forests. For. Sci. 24: 153–166.

Heide, K. 1984. Holocene pollen stratigraphy from a lake and small hollow in north-central Wisconsin, USA. Palynology 8: 3–20.

Heinselman, M. 1973. Fire in the virgin forests of the Boundary Waters Canoe Area, Minnesota. Quat. Res. 3: 329–382.

Henry, J.D. and Swan, J.M.A. 1974. Reconstructing forest history from live and dead plant material – an approach to the study of forest succession in southwest New Hampshire. Ecology 55: 772–783.

Huntley, B. and Birks, H.J.B. 1983. An atlas of past and present pollen maps for Europe: 0–13,000 years ago. Cambridge University Press, Cambridge.

Jackson, S.T. 1988. Pollen-vegetation relationships in small lake basins: evidence for varying pollen source areas within and among taxa. American Quaternary Association Program and Abstracts of the 10th Biennial Meeting, University of Massachusetts, Amherst: 76.

Jacobson, G.L. 1979. The palaeoecology of white pine (Pinus strobus) in Minnesota. J. Ecol. 67: 697–726.

Jacobson, G.L.and Bradshaw, R.H.W. 1981. The selection of sites for paleovegetational studies. Quat. Res. 16: 80–96.

Jacobson, G.L. and Grimm, E.C. 1986. A numerical analysis of Holocene forest and prairie vegetation in central Minnesota. Ecology 67: 958–966.

Janssen, C.R., Braber, F.I., Bunnik, F.P.N., Delibrias, G., Kalis, A.J. and Mook, W.G. 1985. The significance of chronology in the ecological interpretation of pollen assemblages of contrasting sites in the Vosges. Ecol. Mediterranea 11: 39–43.

Kurtén, B. and Anderson, E. 1980. Pleistocene mammals of North America. Columbia University Press, New York.

Kutzbach, J.E. and Wright, H.E., Jr. 1985. Simulation of the climate of 18,000 years BP: results for the North American/North Atlantic/European Sector and comparison with the geologic record of North America. Quat. Sci. Rev. 4: 147–187.

Larabee, P.A. 1986. Late-Quaternary vegetational and geomorphic history of the Allegheny Plateau at Big Run Bog, Tucker County, West Virginia. MS Thesis, University of Tennessee, Knoxville.

Likens, G.E. (Ed.) 1985. An ecosystem approach to aquatic ecology: Mirror Lake and its environment. Springer-Verlag, New York.

Lyell, G. 1834. Principles of Geology, 3rd Edition. London.

McAndrews, J.H. 1966. Postglacial history of prairie, savanna, and forest in northwestern Minnesota. Mem. Torrey Bot. Club 22: 1–72.

McAndrews, J.H. 1988. Human disturbance of North American forests and grasslands: the fossil pollen record. In Handbook of vegetation science, vol. 7 – vegetation history (in press). Edited by B. Huntley and T. Webb III. Dr W. Junk Publishers, Dordrecht.

Marschner, F.J. 1959. Land use and its patterns in the United States. US Dept. Agric., Agric. Handbook 153.

Martin, P.S. and Klein, R.G. 1984. Quaternary extinctions, a prehistoric revolution. University of Arizona Press, Tucson.

Maxwell, J.A. and Davis, M.B. 1972. Pollen evidence of Pleistocene and Holocene vegetation on the Allegheny Plateau, Maryland. Quat.Res. 2: 506–530.

Meentemeyer, V. and Box, E.O. 1987. Scale effects in landscape studies. In Landscape heterogeneity and disturbance, Ecological Studies 64, pp. 15–34. Edited by M.G. Turner. Springer-Verlag, New York.

Michalek, D.D. 1968. Fanlike features and related periglacial phenomena of the southern Blue Ridge. PhD. Dissertation, University of North Carolina, Chapel Hill.

Mills, H.H.and Delcourt, P.A. 1989. Appalachian Highlands and Interior Low Plateaus. In Quaternary non-glacial geology of the conterminous United States, Volume K2, Decade of North American Geology. Edited by R.B. Morrison. Geological Society of America, Boulder, Colorado (in press).

Mooney, H.A. and Drake, J.A. (Eds) 1986. Ecology of biological invasions of North America and Hawaii, Ecological Studies 58. Springer-Verlag, New York.

Mueller-Dombois, D. and Ellenberg, H. 1974. Aims and methods of vegetation ecology. Wiley, New York.

Naveh, Z. and Lieberman, A.S. 1984. Landscape ecology, theory and application. Springer-Verlag, New York.

Nilsson, T. 1983. The Pleistocene, Geology and Life in the Quaternary Ice Age (English Edition). D. Reidel Publishing Company, Dordrecht, The Netherlands.

Oliver, C.D. and Stephens, E.P. 1977. Reconstruction of a mixed species forest in central New England. Ecology 58: 562–572.

Olson, R.J., Goff, F.G. and Olson, J.S. 1976. Development and

44

applications of spatial data resources in energy related assessment and planning. *In* Advancements in retrieval technology as related to information systems. Advisory Group for Aerospace Research and Development, Neuilly sur Seine, France, AGARD Conference Proceedings No. 207, North Atlantic Treaty Organization 12–1 to 12–7.

O'Neill, R.V., DeAngelis, D.L., Waide, J.B. and Allen, T.F.H. 1986. A Hierarchical Concept of Ecosystems. Princeton University Press, Princeton.

Pattison, W.D. 1970. Beginnings of the American rectangular land survey system, 1784–1800. Ohio Historical Society, Columbus.

Peet, R.K. and Christensen, N.L. 1980. Succession: a population process. Vegetatio 43: 131–140.

Péwé, T.L. 1983. The periglacial environment in North America during Wisconsin time. *In* Late-Quaternary environments of the United States, Volume 1, The Late Pleistocene, pp. 157–189. Edited by S.C. Porter. University of Minnesota Press, Minneapolis.

Pickett, S.T.A. and White, P.S. (Eds.) 1985. The ecology of natural disturbance and patch dynamics. Academic Press, New York.

Prentice, I.C. 1985. Pollen representation, source area, and basin size: toward a unified theory of pollen analysis. Quat. Res. 23: 76–86.

Pyle, C. and M.P. Schafale. 1988. Land use history of three spruce-fir forest sites in Southern Appalachia. J. For. Hist. 32: 4–21.

Risser, P.G. 1987. Landscape ecology: state of the art. *In* Landscape heterogeneity and disturbance, Ecological Studies 64, pp. 3–14. Edited by M.G. Turner. Springer-Verlag, New York.

Risser, P.G., Karr, J.R. and Forman, R.T.T. 1984. Landscape ecology: directions and approaches. Illinois Natural History Survey Special Publication No. 2.

Ritchie, J.C. 1987. Postglacial vegetation of Canada. Cambridge University Press, Cambridge.

Romme, W.H. 1982. Fire and landscape diversity in subalpine forests of Yellowstone National Park. Ecol. Monogr. 52: 199–221.

Romme, W.H. and Knight, D.H. 1982. Landscape diversity: the concept applied to Yellowstone Park. Bioscience 32: 644–670.

Shafer, D.S. 1984. Late-Quaternary paleoecologic, geomorphic, and paleoclimatic history of Flat Laurel Gap, Blue Ridge Mountains, North Carolina. MS Thesis, University of Tennessee, Knoxville.

Shafer, D.S. 1985. Flat Laurel Gap Bog, Pisgah Ridge, North Carolina: late-Holocene development of a high-elevation heath bald. Castanea 51: 1–10.

Shafer, D.S. 1988. Late Quaternary landscape evolution at Flat Laurel Gap, Blue Ridge Mountains, North Carolina. Quat. Res. 30: 7–11.

Swank, W.T. and Crossley, D.A., Jr. 1988. Introduction and site description. *In* Forest hydrology and ecology at Coweeta, Ecological Studies 66, pp. 3–16. Edited by W.T. Swank and D.A. Crossley, Jr. Springer-Verlag, New York.

Swanson, F.J., Fratz, T.K., Caine, N. and Woodmansee, R.G. 1988. Landform effects on ecosystem patterns and processes. Bioscience 38: 92–98.

Turner, M.G. 1987. Preface. *In* Landscape heterogeneity and disturbance, Ecological Studies 64, pp. 3–14. Edited by M.G. Turner. Springer-Verlag, New York.

Urban, D.L., O'Neill, R.V. and Shugart, H.H., Jr. 1987. Landscape ecology: a hierarchical perspective can help scientists understand spatial patterns. Bioscience 37: 119–127.

Watts, W.A. 1979. Late Quaternary vegetation of central Appalachia and the New Jersey coastal plain. Ecol. Monogr. 49: 427–469.

White, P.S. 1984. The southern Appalachian spruce-fir ecosystem, an introduction. *In* The southern Appalachian spruce-fir ecosystem: its biology and threats, pp. 1–21. Edited by P.S. White. United States National Park Service Research/Resources Management Rept. SER-71.

White, P.S. and Wofford, B.E. 1984. Rare native Tennessee vascular plants in the flora of Great Smoky Mountain National Park. J. Tennessee Acad. Sci. 59: 61–64.

Whittaker, R.H. 1956. Vegetation of the Great Smoky Mountains. Ecol. Monogr. 26: 1–80.

Williams, M. 1982. Clearing the United States forests: pivotal years 1810–1860. J. Hist. Geogr. 8: 12–28.

Williams, S.C. 1927. Lieut. Henry Timberlake's Memoirs. Watauga Press, Johnson City, Tennessee.

Wright, H.E., Jr. 1984. Sensitivity and response time of natural systems to climatic change in the Late Quaternary. Quat. Sci. Rev. 3: 91–131.

5

The Analysis
of Landscape Patterns

Introduction and Review

The essential feature of landscapes, the place from which to begin deeper investigations of causes and consequences, is their spatial pattern. Considerations of the effects of changing spatial scale, or of the temporal dynamics of landscapes over various scales, are based on comparisons and analyses of landscape patterns. This emphasis has only gained force with the development of sophisticated technologies of remote sensing (e.g., increased spectral resolution of satellite imagery), computerized image analysis (e.g., geographic information systems, GIS; Longley et al. 2001), and advanced and easily accessible spatial statistics (e.g., Haining 2003; Fortin and Dale 2005).

At the foundation of these technological and analytical advances, however, is the basic need to quantify landscape patterns. Landscape ecologists recognized this need early in the development of the discipline, although they were slow to replace qualitative descriptions of landscape patterns with more quantitative descriptions (Wiens 1992; Hobbs 1997). Ecologists became interested in quantifying landscape patterns when they began to take spatial heterogeneity seriously. Both realized that pattern quantification is essential to relate patterns to processes; to detect changes in spatial heterogeneity through time; to determine whether different landscapes are, in fact, different in their spatial structure; and to explore the spatial implications of alternative management strategies. A variety of landscape metrics has been developed and applied to different landscapes (e.g., O'Neill et al. 1988; Turner 1990; Baker and Cai 1992; for more details, see McGarigal and Marks 1995; Haines-Young and Chopping 1996; Gustafson 1998; Hargis et al. 1997; Turner et al. 2001; Gergel and Turner 2002).

Most of these approaches used gridded representations of space, often because they could be linked smoothly with raster-based imagery and GIS. Beginning in the 1950s and gaining momentum especially during the 1970s and 1980s, statisticians in a variety of disciplines began to develop an increasing array of statistical tools for analyzing spatial patterns, usually using data derived from spatially referenced sampling points. By the late 1980s, a powerful yet somewhat bewildering array of statistics had become available, yet most ecologists and landscape ecologists had either not heard of or had not considered spatial statistics. **Pierre Legendre and Marie-Josée Fortin** addressed this deficiency in a comprehensive yet comprehendible review of the array of statistical approaches that could be taken to analyze and quantify spatial patterns. They began from Tobler's (1970) "First Law of Geography," that near things are more related than are distant things, what statisticians recognize as spatial autocorrelation. Although Meentemeyer (part IV) called this "one of the more esoteric problems in spatial analyses" (1989:167), Legendre and Fortin showed not only that there could be dire consequences if one ignored it but also that, properly used, it could be a powerful tool in assessing not just the existence but the form of spatial structuring of data. The clear conclusion was that spatial structuring is the rule rather than the exception; one should therefore assume from the outset that spatial data are autocorrelated in some way and conduct the appropriate tests. Legendre and Fortin went on to describe several ways of testing for more complex spatial patterns using locational statistics. Although the arsenal of spatial statistics available has expanded dramatically, Legendre and Fortin's review (primer, really) remains a useful introduction.

One other quantitative approach was more explicitly related to the challenge of translating spatial patterns across spatial scales. Mandelbrot had introduced the concept and the term of fractals in 1977, and the idea (or hope) that fractals might be used to assess the scale dependence of landscape patterns, or even that certain fractal structures might be a general characteristic of landscapes, received considerable attention during the 1980s. The paper by **Peter Burrough** laid out many of these ideas. Readers should be alert to the parallels with the semivariogram approach described in more detail by Legendre and Fortin and for the potential relationships to the scaling thresholds or transmutations discussed by Wiens and O'Neill (part IV). Burrough's paper stimulated considerable research on fractals, and fractals were subsequently used as a metric of spatial structure (e.g., Krummel et al. 1987; Milne 1988; O'Neill et al. 1988), a means for characterizing the trajectories of movement across landscapes (e.g., Crist et al. 1992; Johnson, Milne, et al. 1992; Johnson, Wiens, et al. 1992), and eventually as part of the generators used in neu-

tral landscape models (e.g., With and King 1997; Gardner 1999). Although the promise that fractals might be a solution to scaling challenges was greater than what was delivered, fractals have been useful as a metric of spatial complexity.

Several papers included in other sections of this volume also contributed to the quantification of landscape pattern. For example, the Romme and Knight paper included in part IV illustrated the first application of diversity indexes derived from information theory to landscape pattern. The Franklin and Forman paper included in part VI used quantitative measures of patches and edges to evaluate the effects of alternative forest cutting patterns on landscape patterns and habitat connectivity. Opdam et al. (part VI) used quantitative estimates of spatial pattern to understand the variation in bird communities in woodland fragments in The Netherlands. These examples underscored the incorporation of new methods for spatial analysis in many of the early papers, and the review by Turner (part VII) summarized the state of these methods in the late 1980s. Most synthetic papers on quantifying landscape patterns, however, were published after the 1990 cutoff for inclusion of papers in this volume. Readers interested in knowing more about the initial concepts and methods developed to quantify landscape patterns should peruse Turner and Gardner (1991), an edited volume that offered the first compilation of the quantitative approaches being developed and applied during the 1980s; Baker and Cai (1992), which described one of the several analysis programs written by individual researchers to analyze spatial data and return values to describe both composition and configuration; and the excellent documentation for FRAGSTATS provided by McGarigal and Marks (1995), developers of the analysis program that is now most widely used in landscape ecology for the analysis of categorical data.

References

Baker, W. L., and Y. Cai. 1992. The r.le programs for multiscale analysis of landscape structure using the GRASS geographic information system. *Landscape Ecology* 7:291–302.

Crist, T. O., D. S. Guertin, J. A. Wiens, and B. T. Milne. 1992. Animal movement in heterogeneous landscapes: An experiment with *Eleodes* beetles in short-grass prairie. *Functional Ecology* 6:536–544.

Fortin, M. -J., and M. Dale. 2005. *Spatial analysis: A guide for ecologists*. Cambridge: Cambridge University Press.

Gardner, R. H. 1999. RULE: Map generation and a spatial analysis program. In *Landscape Ecological Analysis: Issues and Applications*, eds. J. M. Klopatek and R. H. Gardner, 280–303. New York: Springer-Verlag.

Gergel, S. E., and M. G. Turner. 2002. *Learning landscape ecology*. New York: Springer-Verlag.

Gustafson, E. J. 1998. Quantifying landscape spatial pattern: What is the state of the art? *Ecosystems* 1:143–156.

Haines-Young, R., and M. Chopping. 1996. Quantifying landscape structure: A review of landscape indices and their application to forested landscapes. *Progress in Physical Geography* 20:418–445.

Haining, R. 2003. Spatial data analysis: Theory and practice. Cambridge: Cambridge University Press.

Hargis, C. D., J. A. Bissonette, and J. L. David. 1997. Understanding measures of landscape pattern. In *Wildlife and Landscape Ecology: Effects of Pattern and Scale*, ed. J. A. Bissonette, 231–261. New York: Springer-Verlag.

Hobbs, R. J. 1997. Future landscapes and the future of landscape ecology. *Landscape and Urban Planning* 37:1–9.

Johnson, A. R., B. T. Milne, and J. A. Wiens. 1992. Diffusion in fractal landscapes—simulations and experimental studies of tenebrionid beetle movements. *Ecology* 73:1968–1983.

Johnson, A. R., J. A. Wiens, B. T. Milne, and T. O. Crist. 1992. Animal movements and population dynamics in heterogeneous landscapes. *Landscape Ecology* 7:63–75.

Krummel, J. R., R. H. Gardner, G. Sugihara, R. V. O'Neill, and P. R. Coleman. 1987. Landscape patterns in a disturbed environment. *Oikos* 48:321–324.

Longley, P. A., M. F. Goodchild, D. J. Maguire, and D. W. Rhind. 2001. *Geographic information systems and science*. New York: Wiley.

McGarigal, K., and B. J. Marks. 1995. *FRAGSTATS: Spatial analysis program for quantifying landscape structure*. Portland, OR: U.S. Dept. of Agriculture, Forest Service, Pacific Northwest Research Station.

Meentemeyer, V. 1989. Geographical perspectives of space, time, and scale. *Landscape Ecology* 3 (4/5): 163–173.

Milne, B. T. 1988. Measuring the fractal geometry of landscapes. *Applications in Mathematics and Computation* 27:67–79.

O'Neill, R. V., J. R. Krummel, R. H. Gardner, G. Sugihara, B. Jackson, D. L. DeAngelis, B. T. Milne, et al. 1988. Indices of landscape pattern. *Landscape Ecology* 1:153–162.

Tobler, W. R. 1970. A computer movie: Simulation of population change in the Detroit region. *Economic Geography* 46:234–240.

Turner, M. G. 1990. Spatial and temporal analysis of landscape patterns. *Landscape Ecology* 4:21–30.

Turner, M. G., and R. H. Gardner. 1991. *Quantitative methods in landscape ecology*. New York: Springer-Verlag.

Turner, M. G., R. H. Gardner, and R. V. O'Neill. 2001. *Landscape ecology in theory and practice*. New York: Springer-Verlag.

Wiens, J. A. 1992. What is landscape ecology, really? *Landscape Ecology* 7:149–150.

With, K. A., and A. W. King. 1997. The use and misuse of neutral landscape models in ecology. *Oikos* 97:219–229.

Vegetatio 80: 107–138, 1989.
© 1989 *Kluwer Academic Publishers. Printed in Belgium.*

Spatial pattern and ecological analysis

Pierre Legendre[1] & Marie-Josée Fortin[2]
[1] *Département de sciences biologiques, Université de Montréal, C.P. 6128, Succursale A, Montréal, Québec, Canada H3C 3J7;* [2] *Department of Ecology and Evolution, State University of New York, Stony Brook, NY 11794-5245, USA*

Accepted 17.1.1989

Keywords: Ecological theory, Mantel test, Mapping, Model, Spatial analysis, Spatial autocorrelation, Vegetation map

Abstract

The spatial heterogeneity of populations and communities plays a central role in many ecological theories, for instance the theories of succession, adaptation, maintenance of species diversity, community stability, competition, predator-prey interactions, parasitism, epidemics and other natural catastrophes, ergoclines, and so on. This paper will review how the spatial structure of biological populations and communities can be studied. We first demonstrate that many of the basic statistical methods used in ecological studies are impaired by autocorrelated data. Most if not all environmental data fall in this category. We will look briefly at ways of performing valid statistical tests in the presence of spatial autocorrelation. Methods now available for analysing the spatial structure of biological populations are described, and illustrated by vegetation data. These include various methods to test for the presence of spatial autocorrelation in the data: univariate methods (all-directional and two-dimensional spatial correlograms, and two-dimensional spectral analysis), and the multivariate Mantel test and Mantel correlogram; other descriptive methods of spatial structure: the univariate variogram, and the multivariate methods of clustering with spatial contiguity constraint; the partial Mantel test, presented here as a way of studying causal models that include space as an explanatory variable; and finally, various methods for mapping ecological variables and producing either univariate maps (interpolation, trend surface analysis, kriging) or maps of truly multivariate data (produced by constrained clustering). A table shows the methods classified in terms of the ecological questions they allow to resolve. Reference is made to available computer programs.

Introduction

In nature, living beings are distributed neither uniformly nor at random. Rather, they are aggregated in patches, or they form gradients or other kinds of spatial structures.

The importance of spatial heterogeneity comes from its central role in ecological theories and its practical role in population sampling theory.

Actually, several ecological theories and models assume that elements of an ecosystem that are close to one another in space or in time are more likely to be influenced by the same generating process. Such is the case, for instance, for models of epidemics or other catastrophes, for the theories of competition, succession, evolution and adaptations, maintenance of species diversity, parasitism, population genetics, population growth, predator-prey interactions, and social behaviour. Other theories assume that discontinuities between homogeneous zones are important for the structure of ecosystems (succession, species-environment relationships: Allen et al. 1977; Allen & Starr 1982; Legendre et al. 1985) or for ecosystem dynamics (ergoclines: Legendre & Demers 1985). Moreover, the important contribution of spatial heterogeneity to ecological stability seems well established (Huffaker 1958; May 1974; Hassell & May 1974; Levin 1984). This shows clearly that the spatial or temporal structure of ecosystems is an important element of most ecological theories.

Not much has been learned up to now about the spatial variability of communities. Most 19th century quantitative ecological studies were assuming a uniform distribution of living organisms in their geographic distribution area (Darwin 1881; Hensen 1884), and several ecological models still assume, for simplicity, that biological organisms and their controlling variables are distributed in nature in a random or a uniform way (e.g., simple models of population dynamics, some models of forest or fisheries exploitation, or of ecosystem productivity). This assumption is actually quite remote from reality since the environment is spatially structured by various energy inputs, resulting in patchy structures or gradients. In fluid environments for instance (water, inhabited by aquatic macrophytes and phytoplankton, and air, inhabited by terrestrial plants), energy inputs of thermal, mechanical, gravitational, chemical and even radioactive origins are found, besides light energy which lies at the basis of most trophic chains; the spatio-temporal heterogeneity of energy inputs induces convection and advection movements in the fluid, leading to the formation of spatial or temporal discontinuities (interfaces) between relatively homogeneous zones. In soils, heterogeneity and discontinuities are the result of geomorphologic processes. From there, then, the spatio-temporal structuring of the physical environment induces a similar organization of living beings and of biological processes, spatially as well as temporally. Strong biological activity takes place particularly in interface zones

(Legendre & Demers 1985). Within homogeneous zones, biotic processes often produce an aggregation of organisms, following various spatio-temporal scales, and these can be measured (Legendre et al. 1985). The spatial heterogeneity of the physical environment thus generates a diversity in communities of living beings, as well as in the biological and ecological processes that can be observed at various points in space.

This paper includes methodological aspects.

Table 1. Methods for spatial surface pattern analysis, classified by ecological questions and objectives.

1) Objective: Testing for the presence of spatial autocorrelation.
    1.1) Establish that there is no significant spatial autocorrelation in the data, in order to use parametric statistical tests.
    1.2) Establish that there is significant spatial autocorrelation and determine the kind of pattern, or shape.

Method 1: Correlograms for a single variable, using Moran's I or Geary's c; two-dimensional spectral analysis.
Method 2: Mantel test between a variable (or multidimensional matrix) and space (geographical distance matrix); Mantel test between a variable and a model.
Method 3: Mantel correlogram, for multivariate data.

2) Objective: Description of the spatial structure.
Method 1: Correlograms (see above), variograms.
Method 2: Clustering and ordination with spatial or temporal constraint.

3) Objective: Test causal models that include space as a predictor.
Method: Partial Mantel test, using three dissimilarity matrices, A, B et C.

4) Objectives: Estimation (interpolation) and mapping.
Method 1: Interpolated map for a single variable: trend surface analysis, that provides also the regression residuals; other interpolation methods.
Method 2: Interpolation taking into account a spatial autocorrelation structure function (variogram): kriging map, for a single variable; programs give also the standard deviations of the estimations, that may help decide where to add sampling locations.
Method 3: Multidimensional mapping: clustering and ordination with spatial constraint (see above).

We shall define first what spatial autocorrelation is, and discuss its influence on classical statistical methods. Then we shall describe the univariate and multivariate methods that we have had experience with for the analysis of the spatial structure of ecological communities (list not necessarily exhaustive), and illustrate this description with actual plant community data. Finally, recent developments in spatial analysis will be presented, that make it possible to test simple interrelation models that include space as an explanatory variable. The methods described in this paper are also applicable to geology, pedology, geography, the earth sciences, or to the study of spatial aspects of the genetic heterogeneity of populations. These sciences have in common the study of observations positioned in geographic space; such observations are related to one another by their geographic distances, which are the basic relations in that space. This paper is organized around a series of questions, of increasing refinement, that ecologists can ask when they suspect their data to be structured by some underlying spatial phenomenon (Table 1).

Classical statistics and spatial structure

We will first try to show that the methods of classical statistics are not always adequate to study space-structured ecological phenomena. This will justify the use of other methods (below) when the very nature of the spatial structure (autocorrelation) is of interest.

In classical inferential statistical analysis, one of the most fundamental assumptions in hypothesis testing is the independence of the observations (objects, plots, cases, elements). The very existence of a spatial structure in the sample space implies that this fundamental assumption is not satisfied, because any ecological phenomenon located at a given sampling point may have an influence on other points located close by, or even some distance away. The spatial structures we find in nature are, most of the time, gradients or patches. In such cases, when one draws a first sample (A), and then another sample (B) located

anywhere near the first, this cannot be seen as a random draw of elements; the reason is that the value of the variable observed in (A) is now known, so that if the existence and the shape of the spatial structure are also known, one can foresee approximately the value of the variable in (B), even before the observation is made. This shows that observations at neighbouring points are not independent from one another. Random or systematic sampling designs have been advocated as a way of preventing this possibility of dependence among observations (Cochran 1977; Green 1979; Scherrer 1982). This was then believed to be a necessary and sufficient safeguard against violations of the assumption of independence of errors. It is adequate, of course, when one is trying for instance to estimate the parameters of a local population. In such a case, a random or systematic sample of points is suitable to achieve unbiased estimation of the parameters, since each point a priori has the same probability of being included in the sample; we know of course that the variance, and consequently also the standard error of the mean, will be larger if the distribution is patchy, but their estimation remains unbiased. On the other hand, we know now that despite the random or systematic allocation of samples through space, observations may retain some degree of spatial dependence if the average distance between samples is smaller than the zone of spatial influence of the underlying ecological phenomenon; in the case of large-scale spatial gradients, no sampling point is far enough to lie outside this zone of spatial influence.

A variable is said to be *autocorrelated* (or *regionalized*) when it is possible to predict the values of this variable at some points of space [or time], from the known values at other sampling points, whose spatial [or temporal] positions are also known. Spatial [or temporal] autocorrelation can be described by a mathematical function, called *structure function*; a spatial autocorrelogram and a semi-variogram (below) are examples of such functions.

Autocorrelation is not the same for all distance classes between sampling points (Table 2). It can be positive or negative. Most often in ecology,

110

Table 2. Examples of spatial autocorrelation in ecology (non-exhaustive list). Modified from Sokal (1979).

Sign of spatial autocorrelation	Significant autocorrelation for	
	short distances	large distances
+	Very often: any phenomenon that is contagious at short distance (if the sampling step is small enough).	Aggregates or other structures (e.g., furrows) repeating themselves trough space.
–	Avoidance (e.g., regularly spaced plants); sampling step too wide.	Spatial gradient (if also significantly positive at short distance).

autocorrelation is positive (which means that the variable takes similar values) for short distances among points. In gradients, this positive auto-correlation at short distances is coupled with negative autocorrelation for long distances, as points located far apart take very different values. Similarly, an aggregated structure recurring at intervals will show positive autocorrelation for distances corresponding to the gap between patch centers. Negative autocorrelation for short distances can reflect either an avoidance phenomenon (such as found among regularly spaced plants and solitary animals), or the fact that the sampling step (interval) is too large compared to patch size, so that any given patch does not contain more than one sample, the next sample falling in the interval between patches. Notice finally that if no spatial autocorrelation is found at a given scale of perception (i.e., a given intensity of sampling), it does not mean that autocorrelation may not be found at some other scale.

In classical tests of hypotheses, statisticians count one degree of freedom for each independent observation, which allows them to choose the statistical distribution appropriate for testing. This is why it is important to take the lack of independence into account (that results from the presence of autocorrelation) when performing a test of statistical hypothesis. Since the value of the observed variable is at least partially known in advance, each new observation contributes but a fraction of a degree of freedom. The size of this fraction cannot be determined, however, so that statisticians do not know the proper reference distribution for the test. All we know for certain is that positive autocorrelation at short distance distorts statistical tests such as correlation, regression, or analysis of variance, and that this distortion is on the 'liberal' side (Bivand 1980; Cliff & Ord 1981); this means that when positive spatial autocorrelation is present in the small distance classes, classical statistical tests determine too often that correlations, regression coefficients, or differences among groups are significant, when in fact they are not. Solutions to these problems include randomization tests, the corrected *t*-test proposed by Cliff & Ord (1981), the analysis of variance in the presence of spatial autocorrelation developed by Legendre *et al.* (submitted), etc. See Edgington (1987) for a general presentation of randomization tests; see also Upton & Fingleton (1985) as well as the other references in the present paper, for applications to spatial analysis. Another way out, when the spatial structure is simple (e.g., a linear gradient), is to extract the spatial component first and conduct the analysis on the residuals (see: trend surface analysis, below), after verifying that no spatial autocorrelation remains in the data.

The situation described above also applies to classical multivariate data analysis, which has been used extensively by ecologists for more than two decades (Orlóci 1978; Gauch 1982; Legendre & Legendre 1983a, 1984a; Pielou 1984). The spatial and temporal coordinates of the data points are usually neglected during the search for ecological structures, which aims at bringing out processes and relations among observations. Given the importance of the space and/or time component in ecological theory, as argued in the Introduction, ecologists are now beginning to study these important relationships. Ordination and clustering methods in particular are often used to detect and analyse spatial structures in vegetation analysis (e.g., Andersson 1988), even though these techniques were not designed specifically for this purpose. Methods are also being developed that take spatial or temporal relationships into account during multivariate data analysis. These include the methods of constrained clustering presented below, as well as the methods of constrained ordination developed by Lee (1981), Wartenberg (1985a,b) and ter Braak (1986, 1987) where one may use the geographical coordinates of the data points as constraints.

Spatial analysis is divided by geographers into *point pattern analysis*, which concerns the distribution of physical points (discontinuous phenomena) in space – for instance, individual plants and animals; *line pattern analysis*, a topological approach to the study of networks of connections among points; and *surface pattern analysis* for the study of spatially continuous phenomena, where one or several variables are attached to the observation points, and each point is considered to represent its surrounding portion of space. Point pattern analysis is intended to establish whether the geographic distribution of data points is random or not, and to describe the type of pattern; this can then be used for inferring processes that might have led to the observed structure. Graphs of interconnections among points, that have been introduced by point pattern analysis, are now widely used also in surface pattern analysis (below), where they serve for instance as basic networks of relationships for constrained clustering, spatial autocorrelation analysis, etc. The methods of point pattern analysis, and in particular the quadrat-density and the nearest-neighbour methods, have been widely used in vegetation science (e.g., Galiano 1982; Carpenter & Chaney 1983) and need not be expounded any further here. These methods have been summarized by a number of authors, including Pielou (1977), Getis & Boots (1978), Cicéri et al. (1977) and Ripley (1981, 1987). The exposé that follows will then concentrate on the methods for surface pattern analysis, that ecologists are presently experimenting with.

Testing for the presence of a spatial structure

Let us first study one variable at a time. If the map of a variable (see Estimation and mapping, below) suggests that a spatial structure is present, ecologists will want to test statistically whether there is any significant spatial autocorrelation, and to establish its type unambiguously (gradient, patches, etc.). This can be done for two diametrically opposed purposes: either (1) one wishes to show that there is no spatial autocorrelation, because one wants to perform parametric statistical hypothesis tests; or (2) on the contrary one hopes to show that there is a spatial structure in order to study it more thoroughly. In either case, a spatial autocorrelation study is conducted. Besides *testing* for the presence of a spatial structure, the various types of correlograms, as well as periodograms, provide *a description* of the spatial structure, as will be seen.

Spatial autocorrelation coefficients

In the case of quantitative variables, spatial autocorrelation can be measured by either Moran's I (1950) or Geary's c (1954) spatial autocorrelation coefficients. Formulas are presented in App. 1. Moran's I formula behaves mainly like Pearson's correlation coefficient since its numerator consists of a sum of cross-products of centered values (which is a covariance term), comparing in turn the values found at all pairs of points in the given distance class. This coefficient is sensitive to extreme values, just like a covariance or a Pearson's correlation coefficient. On the contrary, Geary's c coefficient is a distance-type function, since the numerator sums the squared differences between values found at the various pairs of points being compared.

The statistical significance of these coefficients can be tested, and confidence intervals can be computed, that highlight the distance classes showing significant positive or negative autocorrelation, as we shall see in the following examples. More detailed descriptions of the ways of computing and testing these coefficients can be found

112

in Sokal & Oden (1978), Cliff & Ord (1981) or Legendre & Legendre (1984a). Autocorrelation coefficients also exist for qualitative (nominal) variables (Cliff & Ord 1981); they have been used to analyse for instance spatial patterns of sexes in plants (Sakai & Oden 1983; Sokal & Thomson 1987). Special types of spatial autocorrelation coefficients have been developed to answer specific problems (e.g., Galiano 1983; Estabrook & Gates 1984).

A *correlogram* is a graph where autocorrelation values are plotted in ordinate, against distances (d) among localities (in abscissa). When computing a spatial correlogram, one must be able to assume that a single 'dominant' spatial structure exists over the whole area under study, or in other words, that the main large-scale structure is the same everywhere. This assumption must actually be made for any structure function one wishes to compute; other well-known functions, also used to characterize spatial patterns, include the variogram (below), Goodall's (1974) paired-quadrat variance function, the two-dimensional correlogram and periodogram (below), the multivariate Mantel correlogram (below), and Ibanez' (1981) auto-D^2 function.

In correlograms, the result of a test of significance is associated with each autocorrelation coefficient; the null hypothesis of this test is that the coefficient is not significantly different from zero. Before examining each significant value in the correlogram, however, we must first perform a global test, taking into account the fact that several tests (v) are done at the same time, for a given overall significance level α. The global test is made by checking whether the correlogram contains at least one value which is significant at the $\alpha' = \alpha/v$ significance level, according to the Bonferroni method of correcting for multiple tests (Cooper 1968; Miller 1977; Oden 1984). The analogy in time series analysis is the Portmanteau Q-test (Box & Jenkins 1970). Simulations in Oden's 1984 paper show that the power of Oden's Q-test, which is an extension for spatial series of the Portmanteau test, is not appreciably greater than the power of the Bonferroni procedure, which is computationally a lot simpler.

Readers already familiar with the use of correlograms in time series analysis will be reassured to know that whenever the problem is reduced to one physical dimension only (time, or a physical transect) instead of a bi- or polydimensional space, calculating the coefficients for different distance classes turns out to be equivalent to computing the autocorrelation coefficients of time series analysis.

All-directional correlogram

When a single correlogram is computed over all directions of the area under investigation, one must make the further assumption that the phenomenon is isotropic, which means that the autocorrelation function is the same whatever the direction considered. In anisotropic situations, structure functions can be computed in one direction at a time; this is the case for instance with two-dimensional correlograms, two-dimensional spectral analysis, and variograms, all of which are presented below.

Example 1 – Correlograms are analysed mostly by looking at their shape, since characteristic shapes are associated with types of spatial structures; determining the spatial structure can provide information about the underlying generating process. Sokal (1979) has generated a number of spatial patterns, and published the pictures of the resulting correlograms. We have also done so here, for a variety of artificial-data structures similar to those commonly encountered in ecology (Fig. 1). Fig. 1a illustrates a surface made of 9 bi-normal bumps. 100 points were sampled following a regular grid of 10 × 10 points. The variable 'height' was noted at each point and a correlogram of these values was computed, taking into account the geographic position of the sampled points. The correlogram (Fig. 1b) is globally significant at the $\alpha = 5\%$ level since several individual values are significant at the Bonferroni-corrected level $\alpha' = 0.05/12 = 0.00417$. Examining the individual significant values, can we find the structure's main elements from the correlo-

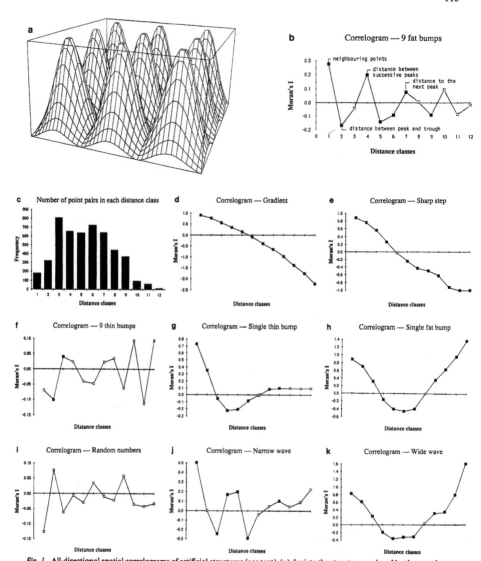

Fig. 1. All-directional spatial correlograms of artificial structures (see text). (a) depicts the structure analysed by the correlogram in (b). (c) displays the number of distances (between pairs of points) in each distance class, for all the correlograms in this figure. In the correlograms (b, d–k), black squares represent significant values at the $\alpha = 5\%$ level, before applying the Bonferroni correction to test the overall significance of the correlograms; white squares are non-significant values.

114

gram? Indeed, since the alternation of positive and negative values is precisely an indication of patchiness (Table 2). The first value of spatial autocorrelation (distance class 1), corresponding to pairs of neighbouring points on the sampling grid, is positive and significant; this means that the patch size is larger than the distance between 2 neighbouring points. The next significant positive value is found at distance class 4: this one gives the approximate distance between successive peaks. (Since the values are grouped into 12 distance classes, class 4 includes distances between 3.18 and 4.24, the unit being the distance between 2 neighbouring points of the grid; the actual distance between neighbours is 3.4 units). Negative significant values give the distance between peaks and troughs; the first of these values, found at distance class 2, corresponds here to the radius of the basis of the bumps. Notice that if the bumps were unevenly spaced, they could produce a correlogram with the same significant structure in the small distance classes, but with no other significant values afterwards. Since this correlogram was constructed with equal distance classes, the last autocorrelation coefficients cannot be interpreted, because they are based upon too few pairs of localities (see histogram, Fig. 1c).

The other artificial structures analysed in Fig. 1 were also sampled using a 10×10 regular grid of points. They are:

- Linear gradient (Fig. 1d). The correlogram has an overall 5% level significance (Bonferroni correction).
- Sharp step between 2 flat surfaces (Fig. 1e). The correlogram has an overall 5% level significance. Comparing with Fig. 1d shows that correlogram analysis cannot distinguish between real data presenting a sharp step and a gradient respectively.
- 9 thin bumps (Fig. 1f); each is narrower than in Fig. 1a. Even though 2 of the autocorrelation coefficients are significant at the $\alpha = 5\%$ level, the correlogram is not, since none of the coefficients is significant at the Bonferroni-corrected level $\alpha' = 0.00417$. In other words, 2 autocorrelation coefficients as extreme as those

encountered here could have been found among 12 tests of a random structure, for an overall significance level $\alpha = 5\%$. 100 sampling points are probably not sufficient to bring out unambiguously a geometric structure of 9 thin bumps, since most of the data points do fall in the flat area in-between the bumps.

- Single thin bumps (Fig. 1g), about the same size as one of the bumps in Fig. 1a. The correlogram has an overall 5% level significance. Notice that the 'zone of influence' of this single bump spreads into more distance classes than in (b) because the phenomenon here is not limited by the rise of adjacent bumps.
- Single fat bump (Fig. 1h): a single bi-normal curve occupying the whole sampling surface. The correlogram has an overall 5% level significance. The 'zone of influence' of this very large bump is not much larger on the correlogram than for the single thin bump (g).
- 100 random numbers, drawn from a normal distribution, were generated and used as the variable to be analysed on the same regular geographic grid of 100 points (Fig. 1i). None of the individual values are significant at the 5% level of significance.
- Narrow wave (Fig. 1j): there are 4 steps between crests, so that there are 2.5 waves across the sampling surface. The correlogram has overall 5% level significance. The distance between successive crests of the wave show up in the significant value at $d = 4$, just as in (b).
- Wide wave (Fig. 1k): a single wave across the sampling surface. The correlogram has overall 5% level significance. The correlogram is the same as for the single fat bump (h). This shows that bumps, holes and waves cannot be distinguished using correlograms; maps are necessary. •

Ecologists are often capable of formulating hypotheses as to the underlying mechanisms or processes that may determine the spatial phenomenon under study; they can then deduct the shape the spatial structure will display if these hypotheses are true. It is a simple matter then to construct an artificial model-surface corresponding to these hypotheses, as we have done in Fig. 1, and to analyse that surface with a correlogram. Although a test of significance of the difference between 2 correlograms is not easy to construct, because of the non-independence of the values in each correlogram, simply looking at the 2 correlograms – the one obtained from the real data, and that from the model data – suffices in many cases to find support for, or to reject the correspondence of the model-data to the real data.

Material: Vegetation data – These data were gathered during a multidisciplinary ecological study of the terrestrial ecosystem of the Municipalité Régionale de Comté du Haut-Saint-Laurent (Bouchard *et al.* 1985). An area of approximately 0.5 km^2 was sampled, in a sector a few km north of the Canada-USA border, in southwestern

Québec. A systematic sampling design was used to survey 200 vegetation quadrats (Fig. 2) each 10 by 20 m in size. The quadrats were placed at 50-m intervals along staggered rows separated also by 50 m. Trees with more than 5 cm diameter at breast height were noted and identified at species level. The data to be analysed here consist of the abundance of the 28 tree species present in this territory, plus geomorphological data about the 200 sampling sites, and of course the geographical locations of the quadrats. This data set will be used as the basis for all the remaining examples presented in this paper.

Example 2 – The correlogram in Fig. 3 describes the spatial autocorrelation (Moran's I) of the hemlock, *Tsuga canadensis*. It is globally significant (Bonferroni-corrected test, $\alpha = 5\%$). We can then proceed to examining significant individual values: can we find the structure's main elements from this correlogram? The first value of spatial autocorrelation (distance class 1, including distances from 0 to 57 m), corresponding to pairs of neighbouring points on the sampling grid, is positive and significant; this means that the patch size is larger than the distance between two neighbouring sampling points. The second peak of this correlogram (distance class 9, whose center is the 485 m distance) can be readily interpreted as the distance among peak centers, in the spatial distribution of the hemlock; see Fig. 10, where groups 3, 7 and 11 have high densities of hemlocks and have their centers located at about that

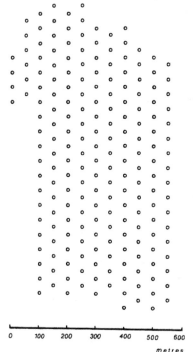

Fig. 2. Position of the 200 vegetation quadrats, systematically sampled in Herdman (Québec), during the summer of 1983. From Fortin (1985).

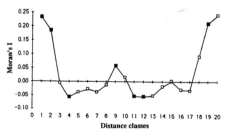

Fig. 3. All-directional spatial correlogram of the hemlock densities (*Tsuga canadensis*). Abscissa: distance classes; the width of each distance class is 57 m. Ordinate: Moran's I statistics. Symbols as in Fig. 1.

116

distance. The last few distance classes cannot be interpreted, because they each contain < 1% of all pairs of localities. ●

Two-dimensional correlogram

All-directional correlograms assume the phenomenon to be isotropic, as mentioned above. Spatial autocorrelation coefficients, computed as described in App. 1 for all pairs of data points, irrespective of the direction, produce a mean value of autocorrelation, smoothed over all directions. Indeed, a spatial autocorrelation coefficient gives a single value for each distance class, which is fine when studying a transect, but may not be appropriate for phenomena occupying several geographic dimensions (typically 2). Anisotropy is however often encountered in ecological field data, because spatial patterns are often generated by directional geophysical phenomena. Oden &

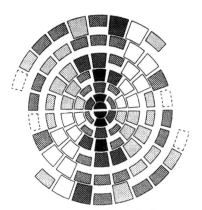

Fig. 4. Two-dimensional correlogram for the sugar-maple *Acer saccharum.* The directions are geographic and are the same as in Fig. 2. The lower half of the correlogram is symmetric to the upper half. Each ring represents a 100-m distance class. Symbols are as follows: full boxes are significant Moran's *I* coefficients, half-boxes are non-significant values; dashed boxes are based on too few pairs and are not considered. Shades of gray represent the values taken by Moran's *I*: from black (+ 0.5 to + 0.2) through hachured (+ 0.2 to + 0.1), heavy dots (+ 0.1 to − 0.1), light dots (− 0.1 to − 0.2), to white (− 0.2 to − 0.5).

Sokal (1986) have proposed to compute correlograms only for object pairs oriented in pre-specified directions, and to represent either a single, or several of these correlograms together, as seems fit for the problem at hand. Computing structure functions in pre-specified directions is not new, and has traditionally been done in variogram analysis (below). Fig. 4 displays a two-dimensional spatial correlogram, computed for the sugar-maple *Acer saccharum* from our test vegetation data. Calculations were made with the very program used by Oden & Sokal (1986); the same information could also have been represented by a set of standard correlograms, each one corresponding to one of the aiming directions. In any case, Fig. 4 clearly shows the presence of anisotropy in the structure, which could not possibly have been detected in an all-directional correlogram: the north-south range of *A. saccharum* is much larger (ca 500 m) than the east-west range (200 m).

Two-dimensional spectral analysis

This method, described by Priestly (1964), Rayner (1971), Ford (1976), Ripley (1981) and Renshaw & Ford (1984), differs from spatial autocorrelation analysis in the structure function it uses. As in regular time-series spectral analysis, the method assumes the data to be stationary (no spatial gradient), and made of a combination of sine patterns. An autocorrelation function r_{gh}, as well as a periodogram with intensity $I(p, q)$, are computed.

Just as with Moran's I, the autocorrelation values are a sum of cross products of lagged data; in the present case, one computes the values of the function r_{gh} for all possible combinations of lags (g, h) along the 2 geographic sampling directions (App. 1); in Moran's I on the contrary, the lag d is the same in all geographic directions. Besides the autocorrelation function, one computes a Schuster two-dimensional periodogram, for all combinations of spatial frequencies (p, q) (App. 1), as well as graphs (first proposed by Renshaw & Ford. 1983) called the *R*-spectrum and the Θ-spectrum that summarize respectively the frequencies and directions of the dominant waves that form the spatial pattern. See App. 1 for computational details.

Two-dimensional spectral analysis has recently been used to analyse spatial patterns in crop plants (McBratney & Webster 1981), in forest canopies (Ford 1976; Renshaw & Ford 1983; Newbery *et al.* 1986) and in other plants (Ford & Renshaw 1984). The advantage of this technique is that it allows analysis of anisotropic data, which are frequent in ecology. Its main disadvantage is that, like spectral analysis for time series, it requires a large data base; this has prevented the technique from being applied to a wider array of problems. Finally, one should notice that although the autocorrelogram can be interpreted essentially in the same way as a Moran's correlogram, the periodogram assumes on the contrary the spatial pattern to result from a combination of repeatable patterns; the periodogram and its R and Θ spectra are very sensitive to repeatabilities in the data, but they do not detect other types of spatial patterns which do not involve repeatabilities.

Example 3 – Fig. 5a shows the two-dimensional periodogram of our vegetation data for *Acer saccharum*. For the sake of this example, and since this method requires the data to form a regular, rectangular grid, we interpolated sugar-maple abundance data by kriging (see below) to obtain a rectangular data grid of 20 rows and 12 columns. The periodogram (Fig. 5a) has an overall 5% significance, since 4 values exceed the critical Bonferroni-corrected value of 6.78; these 4 values explain together 72% of the spatial variance of our variable, which is an appreciable amount.

The most prominent values are the tall blocks located at $(p, q) = (0, 1)$ and $(0, -1)$; together, they represent 62% of the spatial variance and they indicate that the dominant phenomenon is an east-west wave with a frequency of 1 (which means that the phenomenon occurs once in the east-west direction across the map). This structure has an angle of $\Theta = \tan^{-1}$ (0/[1 or

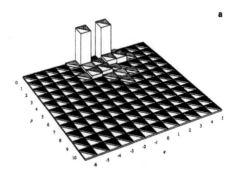

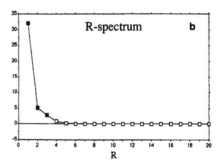

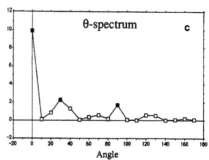

Fig. 5. (a) Two-dimensional periodogram. The ordinate represents the intensity of the periodogram. (b) R-spectrum. (c) Θ-spectrum. Bonferroni-corrected significant values in the spectra are represented by dark squares, for an overall significance level of 5%.

$-1]) = 0°$ and is the dominant feature of the Θ-spectrum; with its frequency $R = \sqrt{(0^2 + 1^2)} = 1$, it also dominates the R-spectrum. This east-west wave, with its crest

118

elongated in the north-south direction, is clearly visible on the map of Fig. 13a.

The next 2 values, that ought to be considered together, are the blocks (1, 2) and (1, 1) in the periodogram. The corresponding angles are $\Theta = 26.6°$ and $45°$ (they form the 4th and 5th values in the Θ-spectrum), for an average angle of about $35°$; the R frequencies of the structure they represent are $\sqrt{(p^2 + q^2)} = 2.24$ and 1.41, for an average of 1.8. Notice that the values of p and q have been standardized as if the 2 geographic axes (the vertical and horizontal directions in Fig. 13) were of equal lengths, as explained in App. 1; these periodogram values indicate very likely the direction of the axis that crosses the centers of the 2 patches of sugar-maple in the middle and bottom of Fig. 13a.

Two other periodogram values are relatively high (5.91 and 5.54) but do not pass the Bonferroni-corrected test of significance, probably because the number of blocks of data in our regular grid is on the low side for this method. In any case, the angle they correspond to is $90°$, which is a significant value in the Θ-spectrum. These periodogram values indicate obviously the north-south direction crossing the centers of the 2 large patches in the upper and middle parts of Fig. 13a ($R = 2$).

These results are consistent with the two-dimensional correlogram (Fig. 4) and with the variograms (Fig. 9), and confirm the presence of anisotropy in the *A. saccharum* data. They were computed using the program of Renshaw & Ford (1984). Ford (1976) presents examples of vegetation data with clearer periodic components. ●

The Mantel test

Since one of the scopes of community ecology is the study of relationships between a number of biological variables – the species – on the one hand, and many abiotic variables describing the environment on the other, it is often necessary to deal with these problems in multivariate terms, to study for instance the simultaneous abundance fluctuations of several species. A method of carry-

ing out such analyses is the Mantel test (1967). This method deals with 2 distance matrices, or 2 similarity matrices, obtained independently, and describing the relationships among the same sampling stations (or, more generally, among the same objects). This type of analysis has two chief domains of application in community ecology.

Let us consider a set of n sampling stations. In the first kind of application, we want to compare a matrix of ecological distances among stations (**X**) with a matrix of geographic distances (**Y**) among the same stations. The ecological distances in matrix **X** can be obtained for instance by comparing all pairs of stations, with respect to their faunistic or floristic composition, using one of the numerous association coefficients available in the literature; notice that qualitative (nominal) data can be handled as easily as quantitative data, since a number of coefficients of association exist for this type of data, and even for mixtures of quantitative, semi-quantitative and qualitative data. These coefficients have been reviewed for instance by Orlóci (1978), by Legendre & Legendre (1983a and 1984a), and by several others; see also Gower & Legendre (1986) for a comparison of coefficients. Matrix **Y** contains only geographic distances among pairs of stations, that is, their distances in m, km, or other units of measurement. The scope of the study is to determine whether the ecological distance increases as the samples get to be geographically farther apart, i.e., if there is a spatial gradient in the multivariate ecological data. In order to do this, the Mantel statistic is computed and tested as described in App. 2. Examples of Mantel tests in the context of spatial analysis are found in Ex. 8 in this paper, as well as in Upton & Fingleton's book (1985).

The Mantel test can be used not only in spatial analysis, but also to check the goodness-of-fit of data to a model. Of course, this test is valid only if the model in matrix **Y** is obtained independently from the similarity measures in matrix **X** – either by ecological hypothesis, or else if it derives from an analysis of a different data set than the one used in elaborating matrix **X**. The Mantel test cannot be used to check the conformity to a matrix **X** of a model derived from the **X** data. Goodness-of-fit Mantel tests have been used recently in vegetation studies to investigate very precise hypotheses related to questions of importance, like the concept of climax (McCune & Allen 1985) and the environmental control model (Burgman 1987). Another application can be found in Hudon & Lamarche (in press) who studied competition between lobsters and crabs.

Example 4 – In the vegetation area under study, 2 tree species are dominant, the sugar-maple *Acer saccharum* and the red-maple *A. rubrum*. One of these species, or both, are present in almost all of the 200 vegetation quadrats. In such a case, the hypothesis of niche segregation comes to mind. It can be tested by stating the null hypothesis that the habitat of the 2 species is the same, and the alternative hypothesis that there is a difference. We are going to test this hypothesis by comparing the environmental data to a model corresponding to the alternative hypothesis (Fig. 6), using a Mantel test. The environmental data were chosen to represent factors likely to influence the growth of these species. The 6 descriptors are: quality of drainage (7 semi-quantitative classes), stoniness of the soil (7 semi-quantitative classes), topography (11 unordered qualitative classes), directional exposure (the 8 sectors of the compass card, plus class 9 = flat land), texture of horizon 1 of the soil (8 unordered qualitative classes), and geomorphology (6 unordered qualitative classes, described in Example 8 below). These data were

used to compute an Estabrook-Rogers similarity coefficient among quadrats (Estabrook & Rogers 1966; Legendre & Legendre 1983a, 1984a). The Estabrook & Rogers similarity coefficient makes it possible to assemble mixtures of quantitative, semi-quantitative and qualitative data into an overall measure of similarity; for the descriptors of directional exposure and soil texture, the partial similarities contributing to the overall coefficient were drawn from a set of partial similarity values that we established, as ecologists, to represent how similar are the various pairs of semi-ordered or unordered classes, considered from the point of view of tree growth. The environmental similarity matrix is represented as **X** in Fig. 6.

The ecological hypothesis of niche segregation between *A. saccharum* and *A. rubrum* can be translated into a model-matrix of the alternative hypothesis as follows: each of the 200 quadrats was coded as having either *A. saccharum* or *A. rubrum* dominant. Then, a model similarity matrix among quadrats was constructed, containing 1's for pairs of quadrats that were dominant for the same species (maximum similarity), and 0's for pairs of quadrats differing as to the dominant species (null similarity). This model matrix is represented as **Y** in Fig. 6, where it is shown as if all the *A. saccharum*-dominated quadrats came first, and all the *A. rubrum*-dominated quadrats came last; in practice, the order of the quadrats does not make any difference, insofar as it is the same in matrices **X** and **Y**.

One can obtain the sampling distribution of the Mantel statistic by repeatedly simulating realizations of the null hypothesis, through permutations of the quadrats (corresponding to the lines and columns) in the **Y** matrix, and recomputing the Mantel statistic between **X** and **Y** (App. 2). If indeed there is no relationship between matrices **X** and **Y**, we can expect the Mantel statistic to have a value located near the centre of this sampling distribution, while if such a relation does exist, we expect the Mantel statistic to be more extreme than most of the values obtained after random permutation of the model matrix. The Mantel statistic was computed and found to be significant at $p < 0.00001$, using in the present

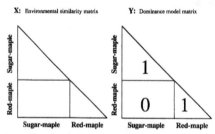

Fig. 6. Comparison of environmental data (matrix **X**) to the model (matrix **Y**), to test the hypothesis of niche segregation between the sugar-maple and the red-maple.

120

case Mantel's t test, mentioned in the remarks of App. 2, instead of the permutation test. So, we must reject the null hypothesis and accept the idea that there is some measurable niche differentiation between *A. saccharum* and *A. rubrum*. Notice that the objective of this analysis is the same as in classical discriminant analysis. With a Mantel test, however, one does not have to comply with the restrictive assumptions of discriminant analysis, assumptions that are rarely met by ecological data; furthermore, one can model at will the relationships among plants (or animals) by computing matrix **X** with a similarity measure appropriate to the ecological data, as well as to the nature of the problem, instead of being imposed the use of an Euclidean, a Mahalanobis or a chi-square distance, as it is the case in most of the classical multivariate methods. In the present case, the Mantel test made it possible to use a mixture of semi-quantitative and qualitative variables, in a rather elegant analysis.

To what environmental variable(s) do these tree species react? This was tested by a series of a posteriori tests, where each of the 6 environmental variables was tested in turn against the model-matrix **Y**, after computing an Estabrook & Rogers similarity matrix for that environmental variable only. Notice that these a posteriori tests could have been conducted by contingency table analysis, since they involve a single semi-quantitative or qualitative variable at a time; they were done by Mantel testing here to illustrate the domain of application of the method. In any case, these a posteriori tests show that 3 of the environmental variables are significantly related to the model-matrix: stoniness ($p < 0.00001$), topography ($p = 0.00028$) and geomorphology ($p < 0.00001$); the other 3 variables were not significantly related to **Y**. So the three first variables are likely candidates, either for studies of the physiological or other adaptive differences between these 2 maple species, or for further spatial analyses. One such analysis is presented as Ex. 8 below, for the geomorphology descriptor. •

The Mantel correlogram

Relying on a Mantel test between data and a model, Sokal (1986) and Oden & Sokal (1986) found an ingenious way of computing a correlogram for multivariate data; such data are often encountered in ecology and in population genetics. The principle is to express ecological relationships among sampling stations by means of an **X** matrix of multivariate distances, and then to compare **X** to a **Y** model matrix, different for each distance class; for distance class 1, for instance, neighbouring station pairs (that belong to the first class of geographic distances) are linked by 1's, while the remainder of the matrix contains zeros only. A first normalized Mantel statistic (r) is calculated for this distance class. The process is repeated for each distance class, building each time a new model-matrix **Y**, and recomputing the normalized Mantel statistic. The graph of the values of the normalized Mantel statistic against distance classes gives a multivariate correlogram; each value is tested for significance in the usual way, either by permutation, or using Mantel's normal approximation (remark in App. 2). [Notice that if the values in the **X** matrix are similarities instead of distances, or else if the 1's and the 0's are interchanged in matrix **Y**, then the sign of each Mantel statistic is changed.] Just as with a univariate correlogram (above), one is advised to carry out a global test of significance of the Mantel correlogram using the Bonferroni method, before trying to interpret the response of the Mantel statistic for specific distance classes.

Example 5 – A similarity matrix among sampling stations was computed from the 28 tree species abundance data, using the Steinhaus coefficient of similarity (also called the Odum, or the Bray and Curtis coefficient: Legendre & Legendre 1983a, 1984a), and the Mantel correlogram was computed (Fig. 7). There is overall significance in this correlogram, since many of the individual values exceed the Bonferroni-corrected level $\alpha' = 0.05/20 = 0.0025$. Since there is significant positive autocorrelation in the small distance classes and significant negative autocorrelation in

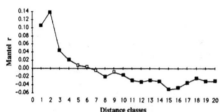

Fig. 7. Mantel correlogram for the 28-species tree community structure. See text. Abscissa: distance classes (one unit of distance is 57 m); ordinate: standardized Mantel statistic. Dark squares represent significant values of the Mantel statistic ($p \leq 0.05$).

the large distances, the overall shape of this correlogram could be attributed either to a vegetation gradient (Fig. 1d) or to a structure with steps (Fig. 1e). In any case, the zone of positive autocorrelation lasts up to distance class 4, so that the average size of the 'zone of influence' of multivariate autocorrelation (the mean size of associations) is about 4 distance classes, or (4 classes × 57 m) ≈ 230 m. This estimation is confirmed by the maps in Fig. 10, where many of the associations delimited by clustering have about that size. ●

Detection and description of spatial structures

As mentioned above, the different types of correlograms, outlined in the section entitled 'Testing for the presence of a spatial structure', do provide a description of spatial structures. Other methods, that are more exclusively descriptive, can also be used for this purpose. They are presented in this section.

The variogram

The semi-variogram (Matheron 1962), often called variogram for simplicity, is related to spatial correlograms. It is another structure function, allowing to study the autocorrelation phenomenon as a function of distance; however this method, on which the kriging contouring method

is based (below), does not lend itself to any statistical test of hypothesis. The variogram is a univariate method, limited to quantitative variables, allowing to analyse phenomena that occur in one, 2 or 3 geographic dimensions. Burrough (1987) gives an introduction to variogram analysis for ecologists.

Before using the variogram, one must make sure that the data are stationary, which means that the statistical properties (mean and variance) of the data are the same in the various parts of the area under study, or at least that they follow the 'intrinsic hypothesis', which means that the increments between all pairs of points located a given distance d apart have a man zero and a finite variance that remains the same in the various parts of the area under study; this value of variance, for distance class d, is twice the value of the semi-variance function $\gamma(d)$. This relaxed form of the stationarity assumption makes it possible to use the variogram, or for that matter any other structure function (for instance spatial autocorrelograms), with ecological data. Of course, a large-scale spatial structure, if present, will necessarily be picked up by the structure function and may mask finer spatial structures; large-scale trends, in particular, should be removed by regression (trend surface analysis) or some other form of modelling before the presence of other, finer structures can be investigated.

There are two types of variograms: the experimental and the theoretical. The experimental variogram (semi-variogram) is computed from the data using the formula in App. 1. It is presented as a plot of $\gamma(d)$ (ordinate) as a function of distance classes (d), just like a correlogram. As noticed in App. 1, $\gamma(d)$ is a distance-type function, so that it is related to Geary's c coefficient. The experimental variogram can be used as a description of the structure function of the spatial phenomenon and in this way it is of help in understanding the spatial structure.

The variogram was originally designed by mining engineers, as a basis for the contouring method known as kriging (below). This is how it became known to ecologists, among whom its use is spreading (Burrough 1987). To be useful for kriging, a theoretical variogram has to be fitted to the experimental one; the adjustment of a theoretical variogram to the experimental function provides the parameters used by the kriging method. The most important of these parameters are (1) the *range* of influence of the spatial structure, which is the distance where the variogram stops increasing; (2) the *sill*, which is the ordinate value of the flat portion of the variogram, where the semi-variance is no longer a function of direction and distance, and corresponds to the variance of the samples; and eventually (3) the *nugget effect* (see below). As in any type of nonlinear curve fitting, the user must decide what type of nonlinear function is wanted to adjust to his experimental variogram; this step requires both experience, and insight into the ecological process under study. Several types of theoretic functions are often used for this adjustment. 4 of them, the most common ones, are described in App. 1 and illustrated in Fig. 8. Differences between these theoretic functions lie mostly in the shape of the left-hand part of the curves, near the origin. A *linear variogram* indicates a linear spatial gradient; this model has no sill. *Gaussian, exponential* and *spherical variograms* give a measure of the size of the spatial influence of the process (patch size, if the phenomenon is patchy), as well as the shape of the drop of this influence as one gets farther away from the center of the phenomenon; the exponential model does not necessarily have a sill. A flat variogram, also called 'pure nugget effect', indicates the absence of a spatial structure in the data, at least at the scale the observations were made. The so-called *nugget effect* refers to variograms that do not go

through the origin of the graph, but display some amount of variance even at distance zero; this effect may be caused by some intrinsic random variability in the data (sampling variance), or it may suggest that the sampling has not been performed at the right spatial scale. Variograms have recently been used to measure the fractal dimension of environmental gradients (Phillips 1985).

Mining engineers compute separate variograms for different spatial directions, to determine if the spatial structure is isotropic or not. We have seen above that this procedure has now been extended to correlograms as well. The spatial structure is said to be *isotropic* when the variograms are the same regardless of the direction of measurement. 2 different kinds of anisotropy can be detected: geometric anisotropy and stratified anisotropy. Geometric anisotropy (same sill, different ranges) is measured by the *anisotropy ratio*, which is equal to the range of the variogram in the direction producing the longest range, divided by the range in the direction with the smallest range. Stratified (or zonal) anisotropy (different sills, same range) refers to the fact that the sills of the variograms may not be the same in different directions. In the presence of one or the other type of anisotropy, or both, there are three solutions to obtain acceptable interpolated maps by kriging: one can compute compromise variogram parameters, using the formulas in David (1977) or in Journel & Huijbregts (1978); secondly, one can use a kriging program that makes use of the parameters of variograms computed separately in different directions of the physical space (2 or 3, depending on the problem); or finally, one can use 'generalized intrinsic random functions of order k' (Matheron 1973) that allow for linear or quadratic trends in the data.

Example 6 – Experimental variograms were computed by Fortin (1985), for *A. saccharum*, in the 45° and 90° directions (window: 22°), and in all directions (Fig. 9). Comparing the 45° and 90° variograms shows the presence of anisotropy, as was observed in Fig. 4. The range in the 45° variogram (dashed line) is about 445 m, while the range in the 90° variogram is about 685 m, so that

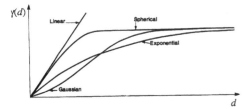

Fig. 8. Four of the most common theoretic variogram models.

The Analysis of Landscape Patterns

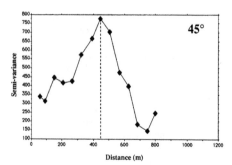

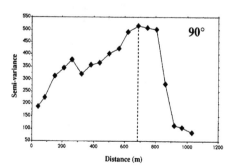

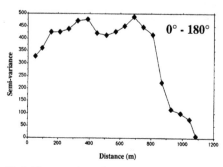

Fig. 9. Three experimental variograms computed for the *Acer saccharum* data. See text. Abscissa: distance classes. Ordinate: values of the semi-variance function $\gamma(d)$. Dashed lines: ranges. Modified from Fortin (1985).

Clustering methods with spatial contiguity constraint

Describing multivariate structures can be done by the methods of clustering, which are classical methods of multivariate data analysis, and in particular by clustering with spatial contiguity constraint. If the clustering results are represented on a map, the multivariate structure of the data – plant associations for instance – will be clearly described by the map.

Clustering with spatial contiguity constraint has been suggested by many authors since 1966 (e.g., Ray & Berry 1966; Webster & Burrough 1972; Lefkovitch 1978, 1980; and others), in such different fields as pedology, political science, economy, psychometry and ecology. Starting from multivariate data, the common need of these authors was to establish geographical regions made of adjacent sites (i.e., a choropleth map: see 'Estimation and mapping' below) which would be homogeneous with respect to certain variables. In order to do this, it is necessary (1) to compute a matrix of similarity among sites from the variables on which these homogeneous regions have to be based (of course, this step applies only to cluster-ing methods that are similarity-based), then (2) proceed with any of the usual clustering methods, with the difference that one constrains the algorithm to cluster only these sites or site groups that are geographically contiguous. The constraint is provided to the program in the form of a list of connections, or spatial links, among neighbouring localities. Connections may be established in a variety of ways: see App. 1. Adding such constraints to existing programs raises algorithmic problems which we will not discuss here. Clustering with constraint has inter-esting properties. On the one hand, it reduces the set of mathematically possible solutions to those that are geographically meaningful; this avoids the well-known problem of clustering methods, where different solutions may be obtained after applying different clustering algorithms to the same data set; constraining all these algorithms to produce results that are geographically consistent forces them to converge towards very similar solutions. On the other hand, the partitions

the anisotropy ratio can be computed as $685/445 \approx 1.5$. The all-directions variogram does not clearly render this information. •

124

obtained in this way reproduce a larger fraction of the structure's spatial information than equivalent partitions obtained without constraint (Legendre 1987). Finally, constrained agglomerative clustering is faster with large data sets than the unconstrained equivalent, because the search for 'the next pair to join' is limited to adjacent groups only (Openshaw 1974; Lebart 1978).

Example 7 – A vegetation map was constructed from our test data, as follows. (1) The same Steinhaus similarity matrix among stations was used as in Ex. 5; it is based upon the 28 tree species abundance data. (2) The spatial relationships among sampling quadrats were represented by a list of connections among close neighbours; the list was established in the present case by the Delaunay triangulation method (App. 1). The presence of a connection between 2 quadrats tells the clustering programs that these 2 localities are located close to one another and thus may eventually be included in the same cluster, if their ecological similarity allows. (3) Agglomerative clustering with spatial contiguity constraint was conducted on the similarity matrix. The spatial contiguity constraint was read by the program from the list of connections, or 'link edges', described above. We used a proportional-link linkage agglomerative algorithm (with 50% connectedness: Sneath 1966), that produced a series of maps, one for each clustering level (Legendre & Legendre 1984b). The map with 13 groups was retained as being ecologically the most meaningful (Fig. 10a); 5 quadrats remain unclustered at that level. Recognizing 13 groups implies that the mean area per association is $740\,000\ m^2/13 = 56\,923\ m^2$/association, corresponding to an average area diameter of $(56\,923)^{1/2} = 238.6$ m; this compares very well with the average size of the zone of influence of our species associations found in the Mantel correlogram, 230 m (Ex. 5).

Agglomerative clustering may have produced small distortions of the resulting map, because of the hierarchical nature of the classification that results from such sequential algorithms. So, we tried to render our 13 groups as homogeneous as possible in terms of vegetation composition, using a k-means algorithm (MacQueen 1967) with spatial contiguity constraint. A k-means algorithm uses an iterative procedure of object reallocation to minimize the sum of within-group dispersions. This type of algorithm tends to produce compact clusters in the variable space (here, the vegetation data), which is exactly what we are looking for; there is no reason however to expect this phenomenon to affect the shape of the clusters in *geographic* space. We provided our program with the list of constraining connections computed in step 2 above, with the 13-group classification obtained in step 3 to be used as the starting configuration (temporarily allocating the 5 unclustered quadrats to the group that enclosed them geographically), and with a set of principal coordinates computed from the Steinhaus similarity matrix (since our k-means program computes within-group variances from raw variables, and not from a similarity or distance matrix). The

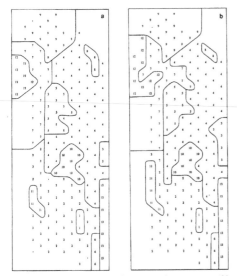

Fig. 10. Map of the multivariate vegetation structure (28 species), obtained by constrained clustering. (a) Space-constrained agglomerative proportional-link linkage, at the level where 13 groups were obtained; the five unclustered quadrats are materialized by dots. (b) Optimization of the previous map by space-constrained k-means clustering.

map of the optimized groups is shown in Fig. 10b. The number of groups remained the same, of course, but 19 objects out of 200 changed group (10%). 4 groups remained unmodified: groups number 1, 6, 10 and 13 in Fig. 10.

The 2 13-group classifications were compared to the raw species abundance data in a series of contingency tables. This work was facilitated by dividing first each species' abundance range into a few classes, following the method described by Legendre & Legendre (1983b). Comparing the interpretations of the 2 classifications, the groups produced by the *k*-means classification were slightly easier to characterize than those produced by the agglomerative classification. Their main biotic characteristics are the following:

– Open area, with rare *A. saccharum*: Group 1.
– *A. rubrum* stands, Group 2.
– Oldfield-birch stands, *Betula populifolia*, located between the *A. rubrum* and *A. saccharum* areas: Group 10.
– *A. saccharum* stands: Groups 4 and 12.
– Stands dominated by white pine *Pinus strobus* and aspen *Populus tremuloides*: Group 6.
– Hemlock stands, *Tsuga canadensis*: Groups 3, 7 and 11.
– Species diversity is highest in the three following groups of stands, dominated by black ash *Fraxinus nigra* and yellow birch *Betula alleghaniensis*:
  – In the bottom of a kettle, with aspen *Populus tremuloides*, white cedar *Thuja occidentalis* and American elm *Ulmus americana*: Group 5.
  – With red ash *Fraxinus pennsylvanica* and basswood *Tilia americana*: Groups 8 and 9.
– Fence-shaped region (formerly cleared land) characterized by white cedar *Thuja occidentalis* and American elm *Ulmus americana* but, contrary to group 5, with few *F. nigra* and *B. alleghaniensis*: Group 13. ●

Univariate or multivariate data that form a transect in space, instead of covering a surface, often need to be summarized by identifying breaking points along the series. Several authors have proposed to use clustering methods with contiguity constraint in a single dimension (space or

time). One such program was developed in P.L.'s lab to analyse ecological successions, with the explicit purpose of locating the abrupt changes that may occur along successional series of community structure; before each group fusion, a statistical permutation test is performed, that translates into statistical terms the ecological model of the development of communities by abrupt structure jumps (Legendre *et al.* 1985). Since then, this method has been used to segment spatial transects of ecological data (Galzin & Legendre 1988), as well as paleontological series (Bell & Legendre 1987). Other applications are in progress, including the reconstruction of climatic fluctuations by studying tree rings, and the segmenting of pollen stratigraphic data. Other methods for segmenting such series, taking into account the spatial or temporal contiguity of samples, have been proposed by Fisher (1958) for univariate economic data, by Webster (1973) for soil data, by Hawkins & Merriam for univariate (1973) and for multivariate (1974) geologic data, by Gordon & Birks (1972, 1974) and by Gordon (1973) for pollen stratigraphic data. This work has been summarized by Legendre (1987).

Causal modelling

Although empirical models are used by ecologists and have their usefulness, modelers often prefer to include only the specific (ecological) hypotheses they may have about the factors and mechanisms determining the process under study. The purpose of modelling is then to verify that experimental or field data do support these hypotheses ('causes'), and to confirm the relational way in which they are assembled into the model. Given the importance of space in our ecological theories, this review of spatial analysis methods would not be complete without mentioning how space can be included in the calculation of relationships among variables. 2 variables may appear related if both of them are linked to a common third one; space is a good candidate for creating such false correlations, since 2 variables may actually seem to be linked because they are driven by a

126

common spatial gradient. Even if correlation does not mean causality, the absence of correlation, monotonic or linear, is sufficient to abandon the hypothesis of a causal link between 2 variables. It is thus important for ecologists interested in causal relationships to check whether the spatial gradient of A could be explained, at least in part, by a spatially structured variable B, or if an apparent correlation between 2 variables is not to be ascribed to a common spatial structure (an unmeasured or untested space-structured variable causing A and B independently). There is still some way to go before space can be included as a variable in complex ecological models, but we will show how it can at least be included in simple models.

Partial Mantel test

How can a partial correlation between two variables be calculated, controlling for a space effect? Smouse *et al.* (1986) dealt with this problem and suggested expressing the variations of each of the two variables by matrices (**A** and **B**) that contain the differences in values between all sampling station pairs. On the other hand, as in the Mantel test, the 'space' variable is expressed by a matrix of geographic distances among stations (matrix **C**). Actually, matrices **A** and **B** could as well be multivariate distance matrices. A partial Mantel statistic is calculated between **A** and **B**, controlling for the effect of matrix **C**. The Smouse *et al.* partial Mantel statistic has the same formula as a partial product-moment correlation coefficient, computed from standardized Mantel statistics. Actually, the computations are done as follows in order to test the partial Mantel statistic between **A** and **B**, controlling for the effect of matrix **C**: (1) compute matrix **A'** that contains the residuals of the linear regression of the values of **A** over the values of **C**; (2) likewise, compute matrix **B'** of the residuals of the linear regression of the values of **B** over the values of **C**; (3) compute the Mantel statistic between **A'** and **B'** (which is just another way of obtaining the partial Mantel statistic between **A** and **B** controlling for

C, as in Pearson partial correlations). (4) Test as usual, either by permuting **A'** or **B'**, or by Mantel's normal approximation. This is equivalent to what would be obtained by permuting all 3 matrices. Partial Mantel tests are not easy to interpret; Legendre & Troussellier (1988) have shown the consequences, in terms of significant Mantel and partial Mantel statistics, of all the possible three-matrices models implying space. As in the case of the Mantel test (App. 2), the restrictive influence of the linearity assumption has not been fully investigated yet for partial Mantel tests.

This type of analysis has numerous applications for studying variables distributed in space. Actually, 3 other forms of test of partial association involving 3 distance matrices have been proposed. 2 of these are based upon the Mantel test, one by anthropologists (Dow & Cheverud 1985), the second one in the field of psychometry (Hubert 1985); the third one involves multiple regressions on distance matrices (Manly 1986; Krackhardt 1988).

Example 8 – We will use our vegetation data to study the much debated question of the environmental control of vegetation structures. We will study in particular the relationship between vegetation structure and the geomorphology of the sampling sites. Of course, vegetation structures are most often autocorrelated, and this can be due either to the fact that biological reproduction is a contagious process, or to some linkage between vegetation and substrate conditions, since soil composition, geomorphology, and so on, are autocorrelated. So, if we find a relationship between vegetation and geomorphology, we will ask the following additional question: do the data support the hypothesis of a causal link between vegetation structure and geomorphology, or is the observed correlation spurious, resulting from the fact that both vegetation and geomorphology follow a common spatial structure, through some unstudied factor that could affect both?

Since our vegetation data are multivariate (28 tree species), they will be represented in the computations by a matrix of multivariate Steinhaus ecological similarities, as in Ex. 5. Space is repre-

Table 3. Above the diagonal: simple standardized Mantel statistics and associated probabilities. Below the diagonal: partial Mantel statistics and associated probabilities. Tests of significance are one-tailed.

Mantel tests / Partial Mantel tests	Vegetation structure	Geomorphology	Space
Vegetation structure	–	0.15054 $p = 0.000$	0.17053 $p = 0.000$
Geomorphology	0.09397 $p = 0.000$	–	0.38073 $p = 0.000$
Space	0.12384 $p = 0.000$	0.36449 $p = 0.000$	–

sented by a matrix of geographic distances among quadrats. The geomorphology variable (6 unordered qualitative classes: moraine ridge, stratified till ridge, reworked till, kettle, relict channel, Champlain sea deposits) was used to compute a simple matching similarity coefficient. Similarities were transformed into distances ($D = 1 - S$) before computing the Mantel tests.

The results of the simple and partial Mantel tests are presented in Table 3. The 3 simple Mantel tests (above the diagonal) show that both the vegetation structure and the geomorphology are autocorrelated, as expected, and also that there exists a significant relation between vegetation and geomorphology. Notice that the Mantel statistic values do not behave like product-moment correlation coefficients, and do not have to be large in absolute value to be significant. All 3 partial Mantel tests (Smouse *et al.* 1986) are significant at the Bonferroni-corrected level $\alpha' = 0.05/3 = 0.01667$. Of special interest to us is the unique influence of geomorphology on the vegetation structure, compared to the influence of space. To decide among the various possible models of interrelations among these 3 groups of variables, we have to consider in turn all 3 possible competing models, and proceed by elimination, as follows. (1) The first model states that the vegetation spatial structure is caused by the spatial structure of geomorphology [Space →

Geomorphology → Vegetation structure]. If this model were supported by the data, then we would expect the partial Mantel statistic (Space · Vegetation), controlling for the effect of Geomorphology, not to be significantly different from zero; this condition is not met in Table 3. (2) The second model states that there is a spatial component in the vegetation data, which is independent from the spatial structure in geomorphology [Geomorphology ← Space → Vegetation structure]. If this model were supported by the data, we would expect the partial Mantel statistic (Geomorphology · Vegetation), controlling for the effect of Space, not to differ significantly from zero, a condition that is not met in Table 3. (3) The third possible model (Fig. 11) claims that the spatial structure in the vegetation data is partly determined by the spatial gradient in the geomorphology, and partly by other factors not explicitly identified in the model. According to this model, all 3 simple and all 3 partial Mantel tests should be significantly different from zero. This is indeed what we find in Table 3.

Although this decomposition of the correlation would best be accomplished by computing standard partial regression-type coefficients (as in path analysis), we can draw some conclusions by looking at the partial Mantel statistics. They show that the Mantel statistic describing the influence of geomorphology on vegetation structure is reduced from 0.15 to 0.09 when controlling for the effect of space. The proper influence of geomorphology on vegetation is then 0.09, while the difference (0.06) is the part of the influence of geomorphology on vegetation that corresponds to the spatial component of geomorphology ($0.15 \times 0.38 = 0.06$). On the other hand, the partial Mantel statistic describing the spatial determi-

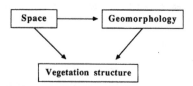

Fig. 11. Diagram of interrelationships between vegetation structure, geomorphology and space.

nation of the vegetation structure not accounted for by geomorphology is still large (0.12) and significant; this shows that other space-related factors do influence the vegetation structure, which is then not entirely spatially determined by geomorphology. Work is in progress on other hypotheses to fill the gap. ●

Estimation and mapping

Any quantitative study of spatially structured phenomena usually starts with mapping the variables. Ecologists, like geographers, usually satisfy themselves with rather unsophisticated kinds of map representations. The 2 most common kinds are (1) divisions of the study area into non-overlapping regions, since 'many areal phenomena studied by geographers [and ecologists] can be represented in 2 dimensions as a set of contiguous, nonoverlapping, space-exhaustive polygons' (Boots 1977), and (2) isoline maps, or contoured maps, used for instance by geographers to represent altitudes on topographic maps, where the nested isolines represent different intensities of some continuous variable. Both types can be produced by computer software. Before attempting to produce a map, especially by computer, ecologists must make sure that they satisfy the following assumption: all parts of the 'active' study area must have a non-null probability of being found in one of the states of the variable to be mapped. For instance, in a study of terrestrial plants, the 'active' area of the map must be defined in such a way as to exclude water masses, roads, large rocky outcrops, and the like.

Since the map derives in most cases from samples obtained from a surface, intermediate values have to be *estimated* by interpolation; or, in the case of a regular sampling grid, one can map the surface as a juxtaposition of regular tiles whose values are given by the points in the center of the tiles. One should notice that interpolated maps can only represent one variable at a time; thus these methods are not multivariate, although it is possible in some cases to superpose two or three maps. When it does not seem desirable or

practicable to map each variable or each species separately, it remains then possible to map, instead, synthetic environmental variables such as species diversity, or else the first few principal axes from a principal components or a correspondence analysis, for instance.

Several methods exist for interpolated mapping. These include trend surface analysis, local weighted averaging, Fourier series modelling, spline, moving average, kriging, kernel estimators, and interpolation by drawing boundaries (in which case the resulting maps may be called 'choropleth maps' or 'tessellations'). They have been reviewed by several authors, including Tapia & Thompson (1978), Ripley (1981, ch. 4), Lam (1983), Bennett *et al.* (1984), Burrough (1986, ch. 8), Davis (1986) and Silverman (1986). Computer programs can provide an estimate of the variable at all points of the surface considered; the density of reconstructed points is either selected by the user or set by the program. Contouring algorithms are used to draw maps from the fine grid of interpolated points.

Besides simple linear interpolation between closest neighbours, trend surface analysis is perhaps the oldest form of spatial interpolation used by ecologists (Gittins 1968; Curtis & Bignal 1985). It consists in fitting to the data, by regression, a polynomial equation of the x and y coordinates of the sampling localities. The order of the polynomial is determined by the user; increasing the order increases the number of parameters to be fitted and so it produces a better-fitting map, with the inconvenient that these parameters become more and more difficult to interpret ecologically. For instance, the commonly used equation of degree one is written:

$$\hat{z} = b_0 + b_1 x + b_2 y \tag{1}$$

where $\hat{z}$ is the estimated value of the response variable z (the one that was measured and is to be mapped), while the b's are the three regression parameters. A second-degree polynomial model is:

$$\hat{z} = b_0 + b_1 x + b_2 y + b_3 x^2 + b_4 xy + b_5 y^2 \tag{2}$$

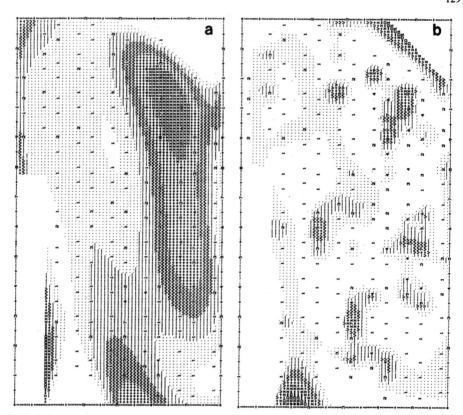

Fig. 12. (a) Trend surface map of *Acer saccharum*, sixth order polynomial. The observation points are identified by numbers. Shades of gray and numbers form a scale that represents the estimated frequencies of sugar-maples. (b) Map of the regression residuals. From Fortin (1985).

Besides the map of the fitted values ($\hat{z}$), trend surface analysis programs usually provide also a map of residuals ($z - \hat{z}$), representing the variation left undescribed by the interpolated map. Fig. 12a illustrates the map of the 6th order polynomial adjusted to the *A. saccharum* data. Compared to Fig. 13 (kriged map) the contouring obtained is still crude, although 28 parameters have been adjusted. Fig. 12b is the map of regression residuals, showing the variations in *A. saccharum* frequencies not expressed by the trend surface map. Burrough (1987) presents an example of trend surface analysis of soil data. Since trend surface analysis computes a single poly-

nomial regression equation for the whole surface, the resulting map cannot have the precision that, more local criteria can provide. For that reason, it is used in ecology mostly to compute and remove large-scale trends, using the first degree equation in most cases, prior to further spatial analyses that can be conducted on the residual values. Trends can also be detected and modelled by autoregressive methods (e.g., Edwards & Coull 1987). Another valid use of trend surface analysis is the predictive modelling of spatial distributions of organisms, using geographic coordinates alone as predictors; or, one can use other predictive variables to build such a model, alone

130

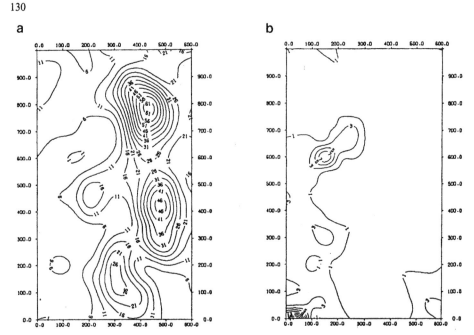

Fig. 13. (a) Map of *Acer saccharum* obtained by kriging, and (b) map of the standard deviations of the estimations. From Fortin (1985).

or in conjunction with geographic coordinates, using multiple regression or some other form of modelling.

Kriging, developed by mining engineers and named after Krige (1966) to estimate mineral resources, usually produces a more detailed map than ordinary interpolation. Contrary to trend surface analysis, kriging uses a local estimator that takes into account only data points located in the vicinity of the point to be estimated, as well as the autocorrelation structure of the phenomenon; this information can be provided either by the variogram (see above), or by generalized intrinsic random functions of order k (Matheron 1973) that allow to make valid interpolation in the case of non-stationary variables (Journel & Huijbregts 1978). The variogram is used as follows during kriging: the kriging interpolation method estimates a point by considering all the other data points located in the observation cone of the variogram (given by the direction and

window aperture angles), and weighs them using the values read on the adjusted theoretic variogram at the appropriate distances; furthermore, kriging splits this weight among neighbouring points, so that the result does not depend upon the local density of points. Kriging programs produce not only a map of resource estimates but also one of the standard deviations of these estimations (David 1977; Journel & Huijbregts 1978); this map may help identify the regions where sampling should be intensified, the map being often obtained from a much smaller number of samples than in Fig. 13.

The problem of mapping multivariate phenomena is all the more acute because cartography seems essential to reach an understanding of the structures brought to light for instance by correlogram analysis. What could be done in the multivariate case? How could one combine the variability of a large number of variables into a single, simple and understandable map? Since

Table 4. The following programs are available to compute the various methods of spatial analysis described in this paper. This list of programs is not exhaustive.

Package	Methods of spatial analysis
BLUEPACK	Variogram, kriging.
CANOCO	Constained ordinations: canonical correspondence analysis, redundancy analysis.
CORR2D	Two-dimensional correlogram.
GEOSTAT	Variogram, kriging.
Kellogg's	Variogram, kriging.
NTSYS-PC	Simple Mantel test.
'R'	Spatial autocorrelation (quantitative and nominal data), simple Mantel test, partial Mantel tests, Mantel correlogram, clustering with spatial contiguity constraint, clustering with time constraint. A variety of connecting networks.
SAAP	Spatial autocorrelograms (Moran's I and Geary's c).
SASP	Two-dimensional spectal analysis.
SYMAP	Trend surface analysis; other interpolation methods.
UNIMAP	Variogram, kriging; other interpolation methods.

- The BLUEPACK package is available from: Centre de géostatistique et de morphologie mathématique, 35 rue Saint-Honoré, F-77305 Fontainebleau Cedex, France.
- The CANOCO program is available from Cajo J.F. ter Braak, Agricultural Mathematics Group, TNO Institute for Applied Computer Science, Box 100, NL-6700 AC Wageningen, The Netherlands.
- The CORR2D program written by Geoffrey M. Jacquez is available from Applied Biostatistics Inc., 100 North Country Road, Bldg. B, Setauket, New York 11733, USA.
- The GEOSTAT package is available from: Geostat Systems International Inc., 4385 rue Saint-Hubert, Suite 1, Montréal, Québec, Canada H2J 2X1.
- The Kellogg's programs are available from the Computer Laboratory, W.K. Kellogg Biological Station, Michigan State University, Hickory Corners, Michigan 49060, USA.
- The NTSYS package, developed by F. James Rohlf, is available in PC version from Applied Biostatistics Inc., 100 North Country Road, Bldg. B, Setauket, New York 11733, USA.
- 'The R package for multivariate data analysis', developed by Alain Vaudor (P. Legendre's lab.: see title page), is

available for Macintosh microcomputers, VAX, and IBM mainframes. English and French speaking versions.
- The SAAP package is a set of FORTRAN programs available from Daniel Wartenberg, Department of Environmental and Community Medicine, Robert Wood Johnson Medical School, 675 Hoes Lane, Piscataway, New Jersey 08854, USA.
- The SASP program is available from E. Renshaw, Department of Statistics, University of Edinburgh, King's Buildings, Mayfield Road, Edinburgh EH9 3JZ, United Kingdom.
- SYMAP is not distributed any longer by Laboratory for Computer Graphics and Spatial Analysis, Harvard University, USA. It is however still available at many computing centers.
- UNIMAP is available from: European Software Contractors A/S, Nørregade, DK-2800 Lyngby, Denmark.

constrained clustering, explained in some detail above, produces groups that can be mapped – and indeed constrained clustering programs can be made to draw these maps directly (Fig. 10) – we have here a way of producing heuristic maps out of multivariate data. The methods of constrained ordination developed by Lee (1981), by Wartenberg (1985a, b) and by ter Braak (1986, 1987) are other ways of accomplishing this. They differ from the simple mapping of principal components or correspondence analysis scores, mentioned at the beginning of this section, in that they take into account the spatial relationships among samples; they resemble them in that it may be necessary to draw several maps in order to represent the variability extracted by all the important but orthogonal axes. MacDonald & Waters (1988) give examples of palynological maps obtained using Lee's *Most Predictable Surface Analysis* (MPS); other examples are found in Wartenberg (1985a,b). These methods should find ample use among community ecologists, who study essentially multivariate (multi-species) phenomena.

Conclusion

Where should ecologists stand? As we have seen, the physical environment is not homogeneous, and most ecological theories are based on precise

132

assumptions about the spatial structure of populations and communities. If we rely upon models that assume, as many still do for simplicity, that biological populations are distributed uniformly or at random in space, chances of obtaining valid predictions are small since the ecological reality is quite different. So, in the descriptive or hypothesis-generating phase of a research, ecologists who sample spatial distributions of organisms should consider *a priori* that their data are structured in space (i.e., are autocorrelated); they should then test for the presence of spatial autocorrelation, and describe the spatial structure using maps and spatial structure functions. In some cases, it may be adequate to remove large-scale spatial structures by regression or model-fitting in order to carry out classical statistical analyses on residuals, but in doing so, one must be careful not to remove one of the important determinants of the processes under study, since heterogeneity is *functional* in ecosystems. In the hypothesis-testing (model-testing) phase of a research, when two variables or groups of variables linked by a causal hypothesis are both autocorrelated, one should test whether their correlation, if significant, could be spurious and due to a similar spatial structure present in both. This in turn could give clues as to the identity of some other spatially autocorrelated causal variable that may have given them their common autocorrelated structure. In a world where spatial structuring is the rule rather than the exception, this precaution can prevent one from reaching unwarranted conclusions.

Statistical methods of spatial analysis (descriptive or inferential) are currently under development, and already they offer a way of answering many relevant questions about populations and communities (Table 1): demonstration of the existence of spatial or temporal structures, description of these structures, univariate or multivariate mapping, comparison with models, analysis of the influence of spatial structures on assumed causal links between variables, statistical analyses which do not assume the independence of the observations. Programs available for spatial analysis are becoming widely available. Some are listed in Table 4; this list is not exhaustive.

We can expect the spatial approach to ecological problems to bring about a quantic jump for ecologists and population geneticists who had learned a type of statistics where one had to hide space or time structures. It is now possible to use these structures and to integrate them into our analyses as fully-fledged controlled variables.

Acknowledgements

This is publication No. 339 from the Groupe d'Écologie des Eaux douces, Université de Montréal, and contribution No. 689 in Ecology and Evolution from the State University of New York at Stony Brook. We are indebted to Pierre Drapeau, Ph.D. student at Université de Montréal, who directed the sampling program that produced the data used for the various examples that illustrate this paper. Dr Michel David, École Polytechnique de Montréal, gave us instructions for and access to his GEOSTAT computer package, that we used for kriging. Geoffrey M. Jacquez, State University of New York at Stony Brook, revised the sections on two-dimensional spectral analysis. Alain Vaudor, computer analyst, developed some of the programs of 'The R Package for Multivariate Data Analysis' during and for the present study. We are grateful to Dr Robert R. Sokal, State University of New York at Stony Brook, who gave us access to Dr Neal L. Oden's two-dimensional correlogram program and provided computing funds to produce Fig. 4 of this paper. We are also indebted to Dr E. David Ford and Dr Cajo J.F. ter Braak for very helpful comments. This study was supported by NSERC grant No. A7738 to P. Legendre, and by a NSERC scholarship to M.-J. Fortin.

Appendix 1
Formulas and technical points

Spatial autocorrelation analysis

H_0: there is no spatial autocorrelation. The values of the variable are spatially independent. Each value of the I coefficient is equal to $E(I) = -(n-1)^{-1} \approx 0$, where $E(I)$ is the expectation of I and n is the number of data points; each value of the c coefficient equals $E(c) = 1$.

H_1: there is significant spatial autocorrelation. The values of the variable are spatially dependent. The value of the I coefficient is significantly different from $E(I) = -(n-1)^{-1} \approx 0$; the value of c is significantly different from $E(c) = 1$.

$$I(d) = [n \sum \sum w_{ij}(y_i - \bar{y})(y_j - \bar{y})]/[W \sum (y_i - \bar{y})^2] \quad (1)$$

$$c(d) = [(n-1) \sum \sum w_{ij}(y_i - y_j)^2]/[2W \sum (y_i - \bar{y})^2] \quad (2)$$

These coefficients are computed for each distance class d. The values of the variable are the y's. All summations are for i and j varying from 1 to n, the number of data points, but exclude the cases where $i = j$. The w_{ij}'s take the value 1 when the pair (i, j) pertains to distance class d (the one for which the coefficient is computed), and 0 otherwise. W is the sum of the w_{ij}'s, or in other words the number of pairs (in the whole *square* matrix of distances among points) taken into account when computing the coefficients for the given distance class. Moran's coefficient varies generally from -1 to 1, but sometimes it can exceed -1 or $+1$ (Fig. 1d, h, k); positive values of Moran's I correspond to positive autocorrelation. Geary's coefficient varies from 0 to some indeterminate positive value which rarely exceeds 3 in real cases; values of c smaller than 1 correspond to positive autocorrelation. These coefficients can be tested for significance; formulas for computing the standard error of the estimated statistics can be found in Cliff & Ord (1981), Sokal & Oden (1978) and Legendre & Legendre (1984a). A special form of spatial autocorrelation coefficient for nominal (qualitative) data is described by Cliff & Ord (1981) and by Sokal & Oden (1978).

Technical points:

- Spatial autocorrelation analysis should not be performed with fewer than ca. 30 localities, because the number of pairs of localities in each distance class then becomes too small to produce significant results.
- There are two ways of dividing distances into classes: either by forming equal distance classes, or classes with equal frequencies. This last solution makes it possible to compute valid coefficients even in the right-hand part of the correlogram (Sokal 1983); with equal distance classes on the contrary, the number of pairs of points becomes too small for valid testing in the large distance classes (Fig. 1c).
- Spatial autocorrelation analysis cannot be performed with a data set that contains a lot of double zeros, because the degree of autocorrelation would then be overestimated and would reflect the fact that the localities share their absence for that variable, which is not what is intended in most applications.
- Euclidean distances between pairs of localities may not be the best way of expressing geographic relationships when

analysing ecological data. Instead, one could use $1/d$ or $1/d^2$ (Mantel 1967; Jumars *et al.* 1977), or some other appropriate transformation (Estabrook & Gates 1984).
- In cases where the Euclidean distance is felt to be meaningless, one can use instead some topological network of connections between localities (see: Connecting networks, below) and compute distances in terms of number of edges along this network.

Two-dimensional spectral analysis

The spatial autocorrelation matrix contains all pairs of sample autocorrelation values r_{gh}, corresponding to all possible lags (g, h) where g is the lag along the x geographic axis of sampling and h is the lag along the y axis. Each value r_{gh} is the ratio of the sample autocovariance at lag (g, h) to the sample variance of the y_{ij}'s. The sample autocovariance s_{gh} is computed as

$$s_{gh} = (1/mn) \sum_{i=1}^{m-g} \sum_{j} (y_{(i,j)} - \bar{y})(y_{(i+g, j+h)} - \bar{y}) \quad (3)$$

where $0 \leq g < m$ and $-n < h < n$; m and n are respectively the number of rows and columns of the geographic sampling grid. The second summation is taken over $j = 1, \ldots, n - h$ if $h \geq 0$ and over $j = |h| + 1, \ldots, n$ if $h < 0$. There is no need to compute the whole autocorrelation surface $(-m < g < m)$ since the surface is a reverse image of itself round either of the zero lag axes.

The Schuster periodogram can also be computed, again for all possible combinations of lags (g, h). The periodogram is a more compact description of the spatial pattern than the full two-dimensional correlogram. Periodograms and power spectra are often expressed as functions of frequencies instead of periods (frequency = 1/period). For convenience, frequencies are here multiplied by the size of the series (m or n) so that a wave that occupies the whole length (m or n) of a side of the sampling area has a frequency (p or q) of 1. The range of frequencies considered is then $p = 0, \ldots, (m/2)$ and $q = (-n/2), \ldots, ((n/2) - 1)$ where $(m/2)$ and $(n/2)$ are respectively the Nyquist frequencies (highest frequency in the observation window) in directions x and y of the sampling surface. The sign of q gives the direction of travel of the sine wave under study. As in time series analysis, the intensity of the periodogram $I(p, q)$, for each frequency combination, measures the amount of variance of variable y that is explained by the given combination of frequencies (p, q), after fitting to the data, by least squares, a Fourier series (sum of sines and cosines) with the given combination of frequencies. See formulas in Renshaw & Ford (1984), for instance. The periodogram is presented as a three-dimensional plot, with frequencies (p, q) along the axes of the controlling plane, and the intensity of the periodogram $I(p, q)$ as the response variable.

The polar spectrum of the data aims at measuring the *frequencies* and *angular directions* of the dominant wave pat-

134

terns present in the data. 2 graphs, first proposed by Renshaw & Ford (1983), are produced. The first one, called the R-spectrum, measures the frequencies of the waves forming the spatial pattern. The R-spectrum is a graph of the average response $I(p, q)$ of all the elements in the periodogram that have approximately the same frequency magnitude $R = \sqrt{(p^2 + q^2)}$. The second one, called the Θ-spectrum, measures the directions (angles) of the waves. It is presented as a graph of the average response $I(p, q)$ of all the elements in the periodogram that have approximately the same angle $\Theta = \tan^{-1}(p/q)$ $(0° \leq \Theta \leq 180°)$. In these graphs, the values along the abscissa (that is, the various R and Θ values) are first divided into a manageable number of classes before the graphs are drawn.

The $I(p, q)$ values have been scaled to have an average value of 1, so that a data set with no spatial structure should produce an R-spectrum and a Θ-spectrum with values close to 1. Since the individual values of $I(p, q)$ in the periodogram are approximately distributed as $(100/mn)\chi^2_{(2)}$, then they can be tested for significance against a critical value of $(100/mn)\chi^2_{(\alpha, 2)}$. In the same way, particular values in the graphs of the R- and Θ-spectra that correspond to intervals containing, say, k individual values of I, can be tested for significance against a critical value of $I = [1/(2k)]\chi^2_{(\alpha, 2k)}$. As in all cases of multiple testing, one should apply the usual Bonferroni correction and use the corrected significance level $\alpha' = \alpha/v$ where v is the number of tests performed simultaneously; this point had not been emphasized by the above-mentioned authors. Actual use of two-dimensional spectral analysis shows that the spectra are the most useful instruments for interpreting the spatial structure; the periodogram has more of a descriptive value.

Variogram

The experimental semi-variogram (often called the variogram) is a plot of the values of semi-variance as a function of distance. The estimator of the semi-variance function is

$$\gamma(d) = [1/(2n_d)] \sum [y_{(i+d)} - y_{(i)})^2] \qquad (4)$$

where n_d is the number of pairs of points located at distance d from one another. The summation is for i varying from 1 to n_d. Just like Geary's c autocorrelation coefficient (above), this structure function is a distance-type function; the difference lies mainly in the denominator of the function.

Some of the most often used theoretic variogram models are the following (Fig. 8). Other models are proposed by Journel & Huijbregts (1978).

Linear model: $\gamma(d) = C_0 + bd$ where b is the slope and C_0 is the intercept (nugget effect).

Exponential model: $\gamma(d) = C_0 + C [1 - \exp(-|d|/a)]$ where C_0 is the nugget effect and $C =$ sill $- C_0$; a is the range.

Spherical model: $\gamma(d) = C_0 + C [(3d/2a) - (d^3/2a^3)]$ if $d \leq a$, while $\gamma(d) = C_0 + C$ if $d > a$.

Gaussian model: $\gamma(d) = C_0 + C [1 - \exp(-d^2/a^2)]$.

Technical points:

– As in correlograms, variograms are computed for distance classes, which implies that the number of pairs of points used in the computation decreases as distance increases (Fig. 1c). Thus, only about the first two-thirds of a variogram should be taken into account when describing the spatial structure.
– With ecological data, the stationarity property is rare and the data often contain some overall trend, called 'drift' in the kriging jargon; drift can affect significantly the accuracy of kriging. Thus in the presence of non-stationarity and drift, the use of 'generalized intrinsic random functions of order k' is recommended, instead of a variogram, to estimate the autocorrelation structure.

Connecting networks

Graphs of interconnections among points are used to describe spatial interrelations for such data analysis methods as constrained clustering, spatial autocorrelation analysis, and other methods that require information about neighbouring localities. In the case of a regular square grid of sampling locations, the solution is simple, since one can connect each point to its neighbours in all 4 directions ('rook's move'), or else in all 8 directions ('queen's move') if he so chooses. If the regular sampling design has the form of staggered rows, as in Fig. 2 for instance, connections (also called 'link edges') may be established with neighbours in all 6 directions. If the sampling localities are irregular tiles that touch one another and cover the whole surface under study, a natural choice is to connect localities that have a border in common.

It often happens, however, that the sampling localities do not form a regular pattern. In such cases, one should wonder first if the ecological problem under study would not provide a natural way of deciding what the close neighbours are. If no such ecological criterion can be found, then one can rely on the more arbitrary geometric criteria. The most commonly used graph-theoretic criteria are the minimum spanning tree (Gower & Ross 1969), the Gabriel graph (Gabriel & Sokal 1969), or the Delaunay triangulation which is simply an algorithmic method of dividing a plane into triangles that obey some precise set of rules (Miles 1970; Ripley 1981; Watson 1981). It is interesting to note that the minimum spanning tree is a subset of the Gabriel graph, which in turn is a subset of the Delaunay triangulation.

Appendix 2
Theory of the Mantel test

Hypotheses

H_0: Distances among points in matrix **X** are not linearly related to the corresponding distances in matrix **Y**. When **Y** represents geographic distances, H_0 reads as follows: the variable (or the multivariate data) in **X** is not autocorrelated as a gradient.

H_1: Distances among points in matrix **X** are correlated to the corresponding distances in matrix **Y**.

Statistics

— Mantel (1967) statistic:

$$z = \sum_i \sum_j x_{ij} y_{ij} \qquad (5)$$

for $i \neq j$, where i and j are row and column indices.

— Normalized Mantel statistic:

$$r = [1/(n-1)] \sum_i \sum_j [(x_{ij} - \bar{x})/s_x] [(y_{ij} - \bar{y})/s_y] \qquad (6)$$

for $i \neq j$, where i and j are row and column indices, and n is the number of distances in one of the matrices (diagonal excluded).

Distribution of the auxiliary variable

— According to H_0, the values observed at any one point could have been obtained at any other point.
— H_0 is thus realized by permuting the points, holding with them their vectors of values for the observed variables.
— An equivalent result is obtained by permuting at random the rows of matrix **X** as well as the corresponding columns.
— Either **X** or **Y** can be permuted at random, with the same net effect.
— Repeating this operation, the different permutations produce a set of values of the Mantel statistic, z or r, obtained under H_0. These values represent the sampling distribution of z or r under H_0.

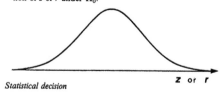

Statistical decision

As in any other statistical test, the decision is made by comparing the actual value of the auxiliary variable (z or r)

to the reference distribution obtained under H_0. If the actual value of the Mantel statistic is one likely to have been obtained under the null hypothesis (no relation between **X** and **Y**), then H_0 is accepted; if it is too extreme to be considered a likely result under H_0, then H_0 is rejected.

Remarks

The z or the r statistics can be transformed into another statistic, called t by Mantel (1967), which can be tested by referring to a table of the standard normal distribution. This test gives a good approximation of the probability when the number of objects is large.

Like Pearson's correlation coefficient, the Mantel statistic formula is a linear model, that brings out the linear component of the relationship between the values in the two distance matrices. Strong non-linearity can probably prevent relationships from expressing themselves through the Mantel test; this led Dietz (1983) to suggest the use of a non-parametric correlation formula. The influence of lack of linearity, and of transformations in one or both of the distance matrices, has not yet been fully investigated.

References

Allen, T.F.H., Bartell, S.M. & Koonce, J.F. 1977. Multiple stable configurations in ordination of phytoplankton community change rates. Ecology 58: 1076–1084.

Allen, T.F.H. & Starr, T.B. 1982. Hierarchy – Perspectives for ecological complexity. Univ. of Chicago Press, Chicago.

Andersson, P.-A. 1988. Ordination and classification of operational geographic units in Southwest Sweden. Vegetatio 74: 95–106.

Bell, M.A. & Legendre, P. 1987. Multicharacter chronological clustering in a sequence of fossil sticklebacks. Syst. Zool. 36: 52–61.

Bennett, R.J., Haining, R.P. & Griffith, D.A. 1984. The problem of missing data on spatial surfaces. Ann. Assoc. Am. Geogr. 74: 138–156.

Bivand, R. 1980. A Monte Carlo study of correlation coefficient estimation with spatially autocorrelated observations. Quaest. Geogr. 6: 5–10.

Boots, B.N. 1977. Contact number properties in the study of cellular networks. Geogr. Anal. 9: 379–387.

Bouchard, A., Bergeron, Y., Camiré, C., Gangloff, P. & Gariépy, M. 1985. Proposition d'une méthodologie d'inventaire et de cartographie écologique: le cas de la MRC du Haut-Saint-Laurent. Cah. Géogr. Qué. 29: 79–95.

Box, G.E.P. & Jenkins, G.M. 1970. Time series analysis, forecasting and control. Holden-Day, San Francisco.

Burgman, M.A. 1987. An analysis of the distribution of plants on granite outcrops in southern Western Australia using Mantel tests. Vegetatio 71: 79–86.

136

Burrough, P.A. 1986. Principles of geographical information systems for land resources assessment. Clarendon Press, Oxford.

Burrough, P.A. 1987. Spatial aspects of ecological data. In: Jongman, R.H.G., ter Braak, C.J.F. & van Tongeren, O.F.R. (eds), Data analysis in community and landscape ecology, pp 213–251. Pudoc, Wageningen.

Carpenter, S.R. & Chaney, J.E. 1983. Scale of spatial pattern: four methods compared. Vegetatio 53: 153–160.

Cicéri, M.-F., Marchand, B. & Rimbert, S. 1977. Introduction à l'analyse de l'espace. Collection de Géographie applicable. Masson, Paris.

Cliff, A.D. & Ord, J.K. 1981. Spatial processes: models and applications. Pion Limited, London.

Cochran, W.G. 1977. Sampling techniques, 3rd ed. John Wiley & Sons, New York.

Cooper, D.W. 1968. The significance level in multiple tests made simultaneously. Heredity 23: 614–617.

Curtis, D.J. & Bignal, E.M. 1985. Quantitative description of vegetation physiognomy using vertical quadrats. Vegetatio 63: 97–104.

Darwin, C. 1881. The formation of vegetable mould through the action of worms. John Murray, London.

David, M. 1977. Geostatistical ore reserve estimation. Developments in geomathematics, 2. Elsevier, Amsterdam.

Davis, J.C. 1986. Statistics and data analysis in geology, 2nd ed. John Wiley & Sons, New York.

Dietz, E.J. 1983. Permutation tests for association between two distance matrices. Syst. Zool. 32: 21–26.

Dow, M.M. & Cheverud, J.M. 1985. Comparison of distance matrices in studies of population structure and genetic microdifferentiation: quadratic assignment. Am. J. Phys. Anthropol. 68: 367–373.

Edgington, E.S. 1987. Randomization tests, 2nd ed. Marcel Dekker Inc., New York.

Edwards, D. & Coull, B.C. 1987. Autoregressive trend analysis: an example using long-term ecological data. Oikos 50: 95–102.

Estabrook, G.F. & Gates, B. 1984. Character analysis of the Banisteriopsis campestris complex (Malpighiaceae), using spatial autocorrelation. Taxon 33: 13–25.

Estabrook, G.F. & Rogers, D.J. 1966. A general method of taxonomic description for a computed similarity measure. BioScience 16: 789–793.

Fisher, W.D. 1958. On grouping for maximum homogeneity. J. Amer. Stat. Ass. 53: 789–798.

Ford, E.D. 1976. The canopy of a Scots pine forest: description of a surface of complex roughness. Agric. Meteorol. 17: 9–32.

Ford, E.D. & Renshaw, E. 1984. The interpretation of process from pattern using two-dimensional spectral analysis: modelling single species patterns in vegetation. Vegetatio 56: 113–123.

Fortin, M.-J. 1985. Analyse spatiale de la répartition des phénomènes écologiques: méthodes d'analyse spatiale, théorie de l'échantillonnage. Mémoire de Maîtrise, Université de Montréal.

Gabriel, K.R. & Sokal, R.R. 1969. A new statistical approach to geographic variation analysis. Syst. Zool. 18: 259–278.

Galiano, E.F. 1982. Pattern detection in plant populations through the analysis of plant-to-all-plants distances. Vegetatio 49: 39–43.

Galiano, E.F. 1983. Detection of multi-species patterns in plant populations. Vegetatio 53: 129–138.

Galzin, R. & Legendre, P. 1988. The fish communities of a coral reef transect. Pac. Sci. 41: 158–165.

Gauch Jr., H.G. 1982. Multivariate analysis in community ecology. Cambridge Univ. Press, Cambridge.

Geary, R.C. 1954. The contiguity ratio and statistical mapping. Incorp. Statist. 5: 115–145.

Getis, A. & Boots, B. 1978. Models of spatial processes: an approach to the study of point, line and area patterns. Cambridge Univ. Press, Cambridge.

Gittins, R. 1968. Trend-surface analysis of ecological data. J. Ecol. 56: 845–869.

Goodall, D.W. 1974. A new method for the analysis of spatial pattern by random pairing of quadrats. Vegetatio 29: 135–146.

Gordon, A.D. 1973. Classification in the presence of constraints. Biometrics 29: 821–827.

Gordon, A.D. & Birks, H.J.B. 1972. Numerical methods in Quaternary palaeoecology. I. Zonation of pollen diagrams. New Phytol. 71: 961–979.

Gordon, A.D. & Birks, H.J.B. 1974. Numerical methods in Quaternary palaeoecology. II. Comparison of pollen diagrams. New Phytol. 73: 221–249.

Gower, J.C. & Legendre, P. 1986. Metric and Euclidean properties of dissimilarity coefficients. J. Classif. 3: 5–48.

Gower, J.C. & Ross, G.J.S. 1969. Minimum spanning trees and single linkage cluster analysis. Appl. Statist. 18: 54–64.

Green, R.H. 1979. Sampling design and statistical methods for environmental biologists. John Wiley & Sons, New York.

Hassell, M.P. & May, R.M. 1974. Aggregation in predators and insect parasites and its effect on stability. J. Anim. Ecol. 43: 567–594.

Hawkins, D.M. & Merriam, D.F. 1973. Optimal zonation of digitized sequential data. J. Int. Assoc. Math. Geology 5: 389–395.

Hawkins, D.M. & Merriam, D.F. 1974. Zonation of multivariate sequences of digitized geologic data. J. Int. Assoc. Math. Geology 6: 263–269.

Hensen, V. 1884. Ueber die Bestimmung des Planktons oder des im Meer triebenden Materials an Pflanzen und Tieren. Ber. Comm. Wiss. Unters. Deutsch. Meere 5.

Hubert, L.J. 1985. Combinatorial data analysis: association and partial association. Psychometrika 50: 449–467.

Hudon, C. & Lamarche, G. in press. Niche segregation between the American lobster (Homarus americanus) and the rock crab (Cancer irroratus). Mar. Ecol. Prog. Ser.

Huffaker, C.B. 1958. Experimental studies on predation: dispersion factors and predator-prey oscillations. Hilgardia 27: 343–383.

Ibanez, F. 1981. Immediate detection of heterogeneities in

continuous multivariate, oceanographic recordings. Application to time series analysis of changes in the bay of Villefranche sur Mer. Limnol. Oceanogr. 26: 336–349.

Journel, A.G. & Huijbregts, C. 1978. Mining geostatistics. Academic Press, London.

Jumars, P.A., Thistle, D. & Jones, M.L. 1977. Detecting two-dimensional spatial structure in biological data. Oecologia (Berl.) 28: 109–123.

Krackhardt, D. 1988. Predicting with networks: nonparametric multiple regression analysis of dyadic data. Social Networks 10: 359–381.

Krige, D.G. 1966. Two dimensional weighted moving average trend surfaces for ore evaluation. Jour. S. Afr. Inst. Min. Metall. 66: 13–38.

Lam, N.S. 1983. Spatial interpolation methods: a review. Am. Cartogr. 10: 129–149.

Lebart, L. 1978. Programme d'agrégation avec contraintes (C.A.H. contiguïté). Cah. Anal. Données 3: 275–287.

Lee, P.J. 1981. The most predictable surface (MPS) mapping method in petroleum exploration. Bull. Can. Petrol. Geol. 29: 224–240.

Lefkovitch, L.P. 1978. Cluster generation and grouping using mathematical programming. Math. Biosci. 41: 91–110.

Lefkovitch, L.P. 1980. Conditional clustering. Biometrics 36: 43–58.

Legendre, L. & Demers, S. 1985. Auxiliary energy, ergoclines and aquatic biological production. Naturaliste can. (Qué.) 112: 5–14.

Legendre, L. & Legendre, P. 1983a. Numerical ecology. Developments in environmental modelling, 3. Elsevier, Amsterdam.

Legendre, L. & Legendre, P. 1983b. Partitioning ordered variables into discrete states for discriminant analysis of ecological classifications. Can. J. Zool. 61: 1002–1010.

Legendre, L. & Legendre, P. 1984a. Écologie numérique, 2ième éd. Tome 2: La structure des données écologiques. Masson, Paris et les Presses de l'Université du Québec.

Legendre, P. 1987. Constrained clustering. In: Legendre, P. & Legendre, L. (eds), Developments in numerical ecology. NATO ASI Series, Vol. G 14, pp. 289–307. Springer, Berlin.

Legendre, P., Dallot, S. & Legendre, L. 1985. Succession of species within a community: chronological clustering, with applications to marine and freshwater zooplankton. Am. Nat. 125: 257–288.

Legendre, P. & Legendre, V. 1984b. Postglacial dispersal of freshwater fishes in the Québec peninsula. Can J. Fish. Aquat. Sci. 41: 1781–1802.

Legendre, P. & Troussellier, M. 1988. Aquatic heterotrophic bacteria: modeling in the presence of spatial autocorrelation. Limnol. Oceanogr. 33: 1055–1067.

Levin, L.A. 1984. Life history and dispersal patterns in a dense infaunal polychaete assemblage: community structure and response to disturbance. Ecology 65: 1185–1200.

MacDonald, G.M. & Waters, N.M. 1988. The use of most predictable surfaces for the classification and mapping of taxon assemblages. Vegetatio 74: 125–135.

MacQueen, J.B. 1967. Some methods for classification and analysis of multivariate observations. In: Le Cam, L.M. & Neyman, J. (eds), Proc. Fifth Berkeley Symp. Math. Stat. Probab., Vol. 1, pp. 281–297. University of California Press, Berkeley.

Manly, B.F.J. 1986. Randomization and regression methods for testing for associations with geographical, environmental and biological distances between populations. Res. Popul. Ecol. 28: 201–218.

Mantel, N. 1967. The detection of disease clustering and a generalized regression approach. Cancer Res. 27: 209–220.

Matheron, G. 1962. Traité de géostatistique appliquée. Tome 1. Éditions Technip, Paris.

Matheron, G. 1973. The intrinsic random functions and their applications. Adv. Appl. Prob. 5: 439–468.

May, R.M. 1974. Stability and complexity in model ecosystems, 2nd ed. Princeton Univ. Press, Princeton, New Jersey.

McBratney, A.B. & Webster, R. 1981. Detection of ridge and furrow pattern by spectral analysis of crop yield. Int. Stat. Rev. 49: 45–52.

McCune, B. & Allen, T.F.H. 1985. Will similar forest develop on similar sites? Can. J. Bot. 63: 367–376.

Miles, R.E. 1970. On the homogeneous planar Poisson point process. Math. Biosci. 6: 85–127.

Miller Jr, R.G. 1977. Developments in multiple comparisons. J. Amer. Stat. Ass. 72: 779–788.

Moran, P.A.P. 1950. Notes on continuous stochastic phenomena. Biometrika 37: 17–23.

Newbery, D.McC., Renshaw, E. & Brünig, E.F. 1986. Spatial pattern of trees in kerangas forest, Sarawak. Vegetatio 65: 77–89.

Oden, N.L. 1984. Assessing the significance of a spatial correlogram. Geogr. Anal. 16: 1–16.

Oden, N.L. & Sokal, R.R. 1986. Directional autocorrelation: an extension of spatial correlograms to two dimensions. Syst. Zool. 35: 608–617.

Openshaw, S. 1974. A regionalisation program for large data sets. Computer Appl. 3-4: 136–160.

Orlóci, L. 1978. Multivariate analysis in vegetation research, 2nd ed. Junk, The Hague.

Phillips, J.D. 1985. Measuring complexity of environmental gradients. Vegetatio 64: 95–102.

Pielou, E.C. 1977. Mathematical ecology, 2nd ed. J. Wiley & Sons, New York.

Pielou, E.C. 1984. The interpretation of ecological data. A primer on classification and ordination. John Wiley & Sons, New York.

Priestly, M.B. 1964. The analysis of two dimensional stationary processes with discontinuous spectra. Biometrika 51: 195–217.

Ray, D.M. & Berry, B.J.L. 1966. Multivariate socioeconomic regionalization: a pilot study in central Canada. In: Ostry, S. & Rymes, T. (eds), Papers on regional statistical studies, pp. 75–130. Univ. of Toronto Press, Toronto.

Rayner, J.N. 1971. An introduction to spectral analysis. Pion

138

Ltd., London.

Renshaw, E. & Ford, E.D. 1983. The interpretation of process from pattern using two-dimensional spectral analysis: methods and problems of interpretation. Appl. Statist. 32: 51–63.

Renshaw, E. & Ford, E.D. 1984. The description of spatial pattern using two-dimensional spectral analysis. Vegetatio 56: 75–85.

Ripley, B.D. 1981. Spatial statistics. John Wiley & Sons, New York.

Ripley, B.D. 1987. Spatial point pattern analysis in ecology. In: Legendre, P. & Legendre, L. (eds), Developments in numerical ecology. NATO ASI Series, Vol. G 14, pp. 407–429. Springer-Verlag, Berlin.

Sakai, A.K. & Oden, N.L. 1983. Spatial pattern of sex expression in silver maple (*Acer saccharinum* L.): Morisita's index and spatial autocorrelation. Am. Nat. 122: 489–508.

Scherrer, B. 1982. Techniques de sondage en écologie. In: Frontier, S. (ed.), Stratégies d'échantillonnage en écologie, pp. 63–162. Collection d'Écologie, 17. Masson, Paris et les Presses de l'Université Laval, Québec.

Silverman, B.W. 1986. Density estimation for statistics and data analysis. Chapman and Hall, London.

Smouse, P.E., Long, J.C. & Sokal, R.R. 1986. Multiple regression and correlation extensions of the Mantel test of matrix correspondence. Syst. Zool. 35: 627–632.

Sneath, P.H.A. 1966. A comparison of different clustering methods as applied to randomly-spaced points. Classification Soc. Bull. 1: 2–18.

Sokal, R.R. 1979. Ecological parameters inferred from spatial correlograms. In: Patil, G.P. & Rosenzweig, M.L. (eds), Contemporary quantitative ecology and related ecometrics. Statistical Ecology Series, Vol. 12, pp. 167–196. International Co-operative Publ. House, Fairland, Maryland.

Sokal, R.R. 1983. Analyzing character variation in geographic space. In: Felsenstein, J. (ed.), Numerical taxonomy. NATO ASI Series, Vol. G1, pp. 384–403. Springer-Verlag, Berlin.

Sokal, R.R. 1986. Spatial data analysis and historical processes. In: Diday, E. *et al.* (eds), Data analysis and informatics, IV. Proc. Fourth Int. Symp. Data Anal. Informatics, Versailles, France, 1985, pp. 29–43. North-Holland, Amsterdam.

Sokal, R.R. & Oden, N.L. 1978. Spatial autocorrelation in biology. 1. Methodology. Biol. J. Linnean Soc. 10: 199–228.

Sokal, R.R. & Thomson, J.D. 1987. Applications of spatial autocorrelation in ecology. In: Legendre, P. & Legendre, L. (eds), Developments in numerical ecology. NATO ASI Series, Vol. G 14, pp. 431–466. Springer-Verlag, Berlin.

Tapia, R.A. & Thompson, J.A. 1978. Nonparametric probability density estimation. Johns Hopkins Univ. Press, Baltimore.

ter Braak, C.J.F. 1986. Canonical correspondence analysis: a new eigenvector technique for multivariate direct gradient analysis. Ecology 67: 1167–1179.

ter Braak, C.J.F. 1987. The analysis of vegetation-environment relationships by canonical correspondence analysis. Vegetatio 69: 69–77.

Upton, G.J.G. & Fingleton, B. 1985. Spatial data analysis by example. Vol. 1: Point pattern and quantitative data. John Wiley & Sons, Chichester.

Wartenberg, D. 1985a. Canonical trend surface analysis: a method for describing geographic patterns. Syst. Zool. 34: 259–279.

Wartenberg, D. 1985b. Multivariate spatial correlation: a method for exploratory geographical analysis. Geogr. Anal. 17: 263–283.

Watson, D.F. 1981. Computing the n-dimensional Delaunay tesselation with application to Voronoi polygones. Computer J. 24: 167–172.

Webster, R. 1973. Automatic soil-boundary location from transect data. J. Int. Assoc. Math. Geol. 5: 27–37.

Webster, R. & Burrough, P.A. 1972. Computer-based soil mapping of small areas from sample data. I. Multivariate classification and ordination. II. Classification smoothing. J. Soil Sci. 23: 210–221, 222–234.

PAPER 27

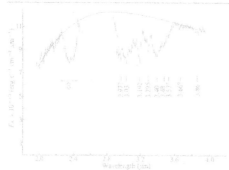

Fig. 2 The 2.9–4.0 µm spectrum of IRS7. The solid curve is the estimated position of the continuum (see text). The dashed curve is the July 1980 spectrum where it deviates from the May 1981 spectrum.

bands stretching from 2.9 to 3.6 µm. Two dominant bands lie at 3.03 and 3.40 µm, and have optical depths of 0.41 and 0.32 respectively. The increased optical depth of the 3.4 µm feature relative to our previous result reflects the improved definition of the continuum. This optical depth would be further increased if the redness between 3.6 and 3.95 µm were due to absorption rather than to the addition of cool dust.

Figure 2 shows all of the absorption features we believe to be real, and a few (marked ?) which are less certain. The absorption clearly breaks up into many discrete features and two of these (at 3.102 and 3.295 µm) are quite narrow. Whilst the broad features centred at 3.03 µm and 3.40 µm presumably arise in solid grains, the narrow features may have a gaseous origin. We have found no convincing identification for either, although many hydrocarbons have stretching absorptions in this spectral range. However, laboratory data are not available for even quite simple, incompletely bonded molecules (for example CH₃) such as might be found in the interstellar environment. Quite narrow absorption bands are found in the absorption spectra of some solids, such as methane and methyl alcohol.

Suitable data on the absorption properties of solid materials are not available, and laboratory work is needed to match the present spectra. We see no evidence for water ice in the available data. The absorption band at 3.03 µm is displaced from that of water (3.06 µm) and is narrower. If water ice is present, it contributes little to the absorption. Similarly, solid ammonia (2.91 µm) is not present. Molecules involving carbon and hydrogen, on the other hand, can produce absorption near both 3.0 and 3.4 µm, and are thus particularly attractive identifications. Organic molecules containing OH and NH bonds also cause absorption near 3.0 (refs 6, 7). The wavelengths of CH, NH and OH vibration depend critically on the composition of the solid and only a few cases of astronomical interest have been discussed in the literature. The H bonded solid complex of H₂O, CH₃OH and NH₃ (ref. 5) has a feature due to NH at 2.97 µm but there is no correspondence with the rest of the spectrum.

It has been suggested recently, that surface functional groups attached to reactive sites on small carbon grains may be responsible for the IR features seen in IRS7. Although there are some interesting wavelength coincidences (aromatic – CH (3.3 µm), –CH₂ (3.4, 3.5 µm), –CHO (3.5, 3.65 µm)) no information has been given on band strengths and shapes to enable a detailed comparison.

We have looked at published spectra of specific grain models involving organic polymers which could be possibilities for the organic component of interstellar grains. The polymer-like material in carbonaceous chondrites show absorption due to CH near 3.3 µm and weak absorption at 3.0 µm. Clearly this material will not produce a fit to the astronomical data even over a restricted wavelength region. However, hydrated silicates of

meteoritic origin show a broad absorption feature centred near 3 µm. This may be relevant in view of the possible identifications of the 9.7 µm and 18 µm features in the galactic centre with silicates. We consider it significant that the yellow component of UV tholin looks qualitatively similar to that of IRS7 in the 3.4 µm region. In particular the shoulder at 3.36 µm, the central component at 3.4 µm and the component at 3.48 µm are present in the laboratory data with wavelengths agreeing to within ± 0.02 µm. This could indicate the existence of complex organic molecules in grains. However, this material does not produce significant absorption near 3 µm as required by the IRS7 data. It is interesting that the satellite features at 3.36, 3.48 µm are also seen in the spectrum of polyformaldehyde suggesting that H₂CO may be present as a structural unit.

We cannot rule out the possibility that a mix of simple organics might match the observed spectrum. This requires further investigation using laboratory data. However, as our data are of relevance mainly to the properties of dust in the diffuse interstellar medium it will be more appropriate to look at refractory materials related to the UV component of tholins such as the non-volatile residue produced in the laboratory experiments designed to simulate conditions in the interstellar medium described by Greenberg and with predictions from the Hoyle–Wickramasinghe model. The latter comparison has shown a remarkable similarity between the spectrum of IRS7 and that of dried bacteria (*Escherichia coli*) which will be reported elsewhere.

Finally emission features are found in this waveband in some H II regions and carbon-rich planetary nebulae (for example, NGC 7027). These normally comprise a strong, narrow peak at 3.29 µm and a weaker, broader feature peaking at 3.40 µm. In this respect they mimic features in the relevant portion of the absorption spectrum of IRS7. However, the 3.03 and 3.192 µm features are not seen in emission in any objects. An interpretation of the absorption spectrum of IRS7 which simultaneously predicted the more restricted range of the emission bands would be very satisfying.

Received 25 June; accepted 2 October 1981.

1. Wickramasinghe, D. T. & Allen, D. A. *Nature* 287, 518 (1980).
2. Sagan, C. & Khare, B. N. *Nature* 277, 102 (1979).
3. Jones, T. J. (Hyland, A. R. *Mon. Not. R. astr. Soc.* 192, 359 (1980).
4. Allamandola, L. J. & Norman, C. A. *Astr. Astrophys.* 63, L23 (1978).
5. Hagen, W., Allamandola, L. J. & Greenberg, J. M. *Astr. Astrophys.* 86, L3 (1980).
6. Herzberg, G. *Infrared and Raman Spectra of Polyatomic Molecules* (van Nostrand, Amsterdam, 1951).
7. Wexler, A. S. *Appl. Spectrosc. Rev.* 1, 29 (1967).
8. Duley, W. W. & Williams, D. A. *Mon. Not. R. astr. Soc.* 196, 269 (1981).
9. Knacke, R. F. *Nature* 269, 132 (1977).
10. Knacke, R. F. & Kratschmer, C. *Astr. Astrophys.* 92, 281 (1981).
11. Hoyle, F. & Wickramasinghe, N. C. *Astrophys. Space Sci.* 66, 77 (1979).
12. Greenberg, J. M. in *Molecules in the Galactic Environment* (eds Gordon, M. & Snyder, L.) (Wiley, New York, 1973).
13. Wickramasinghe, N. C. *Nature* 252, 462 (1974).
14. Hoyle, F. H., Wickramasinghe, N. C., Al-Mufti, S., Olavesen, A. H. & Wickramasinghe, D. T. *Astrophys. Space Sci.* (in the press).
15. Wdowiak, A. T., Puetter, R. C., Russell, R. W. & Sadler, B. T. *Astrophys. Space Sci.* 66, 75 (1979).

Fractal dimensions of landscapes and other environmental data

P. A. Burrough

Department of Regional Soil Science, Agricultural University, Wageningen, The Netherlands

Mandelbrot[1] has introduced the term 'fractal' specifically for temporal or spatial phenomena that are continuous but not differentiable, and that exhibit partial correlations over many scales. The term fractal strictly defined refers to a series in which the Hausdorf–Besicovitch dimension exceeds the topological dimension. A continuous series, such as a polynomial, is differentiable because it can be split up into an infinite number of absolutely smooth straight lines. A non-differentiable continuous series cannot be so resolved. Every attempt to split it up into smaller parts results in the resolution of still more structure or

Nature Vol. 294   19 November 1981

241

Table 1 Estimated D values for various environmental series

Location	Property	Lag	D as lag→0	D at max. slope	Ref.
Wales	Soil—sodium content	15.2 m	1.7–1.9*	—	7
	—stone content	15.2 m	1.1–1.8*	—	7
	(both over four directions)				
England	Soil—thickness of cover loam	20 m	1.6*	—	7
England	Soil—electrical resistivity (4 directions)	1 m	1.4–1.6*	—	7
England	Surface of airport runway	30 cm	1.5†	—	8
Deserts in Africa	Soil – mean cone index	~1 km	1.9‡	—	9
and America	—silt + clay in 0–15 cm layer	~1 km	1.8‡	—	9
	—mean diameter of surface stones	~1 km	1.8‡	—	9
	—coarse sand fraction in 0–15 cm layer	~1 km	1.8‡	—	9
	Vegetation cover	~1 km	1.6‡	—	9
South Africa	Gold	Various	1.9*	—	2
Australia	Soil—phosphorus level	5 m	2.0‡	—	10
	—pH	5 m	1.5‡	—	10
	—potassium level	5 m	1.6‡	1.1‡	10
	—bulk density	5 m	1.5‡	—	10
	—0.1 bar water	5 m	1.5‡	—	10
France	Iron ore in rocks				
	—chlorite	15 µm	1.6*	—	11
	—quartz	15 µm	1.9*	—	11
	—quartz	5 cm	1.6*	—	11
	—iron	5 cm	1.5*	—	11
	—iron (E–W)	100 m	1.7*	—	11
	—iron (N–S)	100 m	1.8*	—	11
	—iron (E–W)	500 m	1.6*	—	11
	—iron (N–S)	500 m	1.9*	—	11
France	Sea anemones	10 cm	1.6§	—	12
Chad	Rainfall	1 km	1.7*	—	13
Mauritania	Iron ore	3 m	1.4*	—	2
Ivory Coast	Groundwater levels				
	Piezometer 1	1 day	1.6*	—	2
	2	1 day	1.7*	—	2
	3	1 day	1.8*	1.3*	2
	4	1 day	1.3*	1.1*	2
Canada	Oil grades	60 cm	1.7*	—	2
Chile	Copper grades	2 m	1.7*	—	2
France	Topographic heights	10 m	1.5*	1.1*	2
USA	Soil—sand content	10 m	1.6–1.8*	—	14
	—pH	10 m	2.0*	—	14
Worldwide	Crop yields	1–1,000 m	1.6–1.8‡	—	15
India	Water table depth	250 m	1.6*	—	16

* Estimated from variogram. ‡ Estimated from block variance. † Estimated from power spectrum. § Estimated from covariance.

roughness. For a linear fractal function, the Hausdorf–Besicovitch dimension D may vary between 1 (completely differentiable) and 2 (so rough and irregular that it effectively takes up the whole of a two-dimensional topological space). For surfaces, the corresponding range for D lies between 2 (absolutely smooth) and 3 (infinitely crumpled). Because the degree of roughness of spatial data is important when trying to make interpolations from point data such as by least-squares fitting or kriging[2], it is worth examining them beforehand to see if the data contain evidence of variation over different scales, and how important these scales might be. Mandelbrot's work[1] suggests that the fractal dimensions of coastlines and other linear natural phenomena are of the order of $D = 1.2$–1.3, implying that long range effects dominate. I show here that published data on many environmental variables suggest that not only are they fractals, but that they may have a wide range of fractal dimensions, including values that imply that interpolation mapping may not be appropriate in certain cases.

Berry and Lewis[3] have shown that the Weierstrass–Mandelbrot fractal function (WMF)

$$W(t) \equiv \sum_{n=-\infty}^{\infty} \frac{[(1-e^{i\gamma^n t})e^{i\phi_n}]}{\gamma^{(2-D)n}}$$

$(1 < D < 2, \gamma > 1, \phi_n = \text{arbitrary phases})$

has a power spectrum $P(\omega)$ that varies approximately as $\omega^{-(5-2D)}$, and a variance of increments $V(t) = \langle[W(t_0) - W(t_0 + t)]^2\rangle$ that varies as t^{4-2D} at the origin. If $D > 1.5$, $V(t)$ is itself a fractal function. These results allow us to estimate the fractional dimension D of a real series, either from the slope of the log–log plot of the power spectrum as $t \to 0$, or from

$$\frac{d \log V(t)}{d \log t} = 4 - 2D \qquad (t \to 0)$$

The variance of increments (or the half thereof, the semivariance) is much used in geostatistical studies where, computed over distances, it is referred to as the variogram[2]. Computing the variogram is the first step in the interpolation procedure known as kriging which is used to assist estimation of mineral reserves, contouring groundwater surfaces, and so on. Thus many published data are available in this form and it is then a simple matter to calculate their dimensions D relative to the sampling interval used.

For a second-order stationary series, the variance of increments at a given lag is equal to twice the difference between the variance of the series and the covariance. Thus, D values may also be computed from covariances. In agriculture and soil science, many data have been published in the form of block variances; that is the variance within blocks of equal size plotted against block size. Because Yates[4] showed that the variance for a block of a given size k is

$$s_k^2 = \frac{2}{k(k+1)} \sum_{h=1}^{k} (k-h+1)Vh$$

where s_k^2 is the block variance of block length k, and Vh the variance increment for lag h, the slope of log Vh versus log $h (h \to 0)$ can also be estimated from these data.

242

Nature Vol. 294 19 November 1981

Table 1 presents a selection of data so analysed, giving the study location, the type of environmental variable, the lag interval used for sampling and the estimated D value assuming that the real data are but a series of regularly spaced samples of a realization of the Weierstrass–Mandelbrot function over one-dimensional space or time.

The data support Mandelbrot's[1] assertion that D values of landscape and other data may range over many values. It is evident that most of the values reported here exceed 1.5, and many are greater than 1.8. Note that one of the smoothest surfaces imaginable in a landscape, a new airport runway, has a relatively high D, presumably because variations over long distance are low in amplitude. These data do not conform to a single roughness model as proposed by Sayles and Thomas,[5] indeed, as Berry and Hannay[6] have commented, much wider ranges of roughness or randomness are to be expected.

These results should not be accepted uncritically, however. First, it is important to realize that the Weierstrass–Mandelbrot function is just one of a class of 'model' fractals. Its peculiarity is that it has a discrete and geometric spectrum, and this might make some of its properties non-universal among other fractals with the same D (M. V. Berry, personal communication).

Second, although some environmental data do appear to display the fractal property of statistical self-similarity at all scales, there are also many that show self-similarity over a limited range of scales, or over a few widely separated scales. For example, a variable with a highly regular spatial variability has a variogram that exhibits parabolic behaviour near the origin[2]. This is a good example where $d \log V(h)/d \log h$ becomes less steep as $h \to 0$, and it indicates that the ultra-short, short and middle range variations are trivial compared with the main variations seen at larger scales. In fact, computing D as $h \to 0$ will in this case estimate D for the ultra-short range scales of variation.

Third, the form of the variogram is often highly dependent on sampling direction and sampling interval. It is well known that when a particular sample spacing tends to match the scale of a spatial pattern, the variance of increments can fall rapidly. Accordingly, when a sampling interval matches a particular scale of a phenomenon in the landscape it perceives an apparently lower D. If this is so, it would seem more appropriate to estimate D values from the parts of the variogram showing maximum slope. Table 1 also contains these data for those phenomena showing a maximum slope at positions other than $h \to 0$.

The results presented here suggest that Mandelbrot's D can be used as a useful indicator of the complexity of autocorrelations over many scales for natural phenomena. However, although many natural phenomena do display certain degrees of statistical self-similarity over many spatial scales, there are others that seem to be structured and have their levels of variability clustered at particular scales. This behaviour does not exclude them from the fractal concept. Mandelbrot[1] considers that it is quite acceptable to have a series of zones of distinct dimensions connected by transition zones. If this is reasonable, it means that the examination of D values would be useful for trying to separate scales of variation that might be the result of particular natural processes. Moreover, identifying such scales could be of enormous practical value because one could then tailor sampling to a particular scale range of the phenomenon in question, thereby improving the efficiency of expensive field investigations and the resulting interpolations. The high level of the D values for some soil and geological data reported here would seem to question the wisdom of interpolation mapping in certain instances, however, and it would seem worthwhile to use D values as a guide to how further mapping and interpolation should proceed.

Received 15 June; accepted 30 September 1981.

1. Mandelbrot, B. *Fractals, Form, Chance and Dimension* (Freeman, San Francisco, 1977).
2. Journal, A. G. & Huijbregts, Ch. J. *Mining Geostatistics* (Academic, London, 1978).
3. Berry, M. V. & Lewis, Z. V. *Proc. R. Soc.* A370, 459–484 (1980).
4. Yates, F. *Phil. Trans. R. Soc.* A241, 345–77 (1948).
5. Sayles, R. A. & Thomas, T. R. *Nature* 271, 431–434 (1978).
6. Berry, M. V. & Hannay, J. H. *Nature* 273, 573 (1978).
7. Burgess, T. & Webster, R. *J. Soil Sci.* 31, 315–42 (1980).
8. Jenkins, G. M. & Watts, D. G. *Spectral Analysis* (Holden Day, London, 1968).
9. Mitchell, C. W., Webster, R., Beckett, P. H. T. & Clifford, C. *Geogr. J.* 145, 72–85 (1979).
10. Webster, R. & Butler, B. E. *Aust. J. Soil Res.* 14, 1–24 (1976).
11. Serra, J. *Miner. Deposita* 3, 135–154 (1968).
12. Kooijman, S. A. L. M. *Spatial and Temporal Analysis in Ecology* Vol. 8 (eds Cormack, R. M. & Ord, J. K.) 305–332 (International Cooperative Publishing House, Maryland, 1979).
13. Delfiner, P. & Delhomme, J. P. *Display and Analysis of Spatial Data* (eds Davis, J. C. & McCullagh, M. J.) (Wiley, London, 1975).
14. Campbell, J. *Soil Sci. Soc. Am. J.* 42, 460–4 (1978).
15. Fairfield Smith, A. *J. Agric. Sci.* 28, 1–23 (1938).
16. Hillel, D. *Applications of Soil Physics* (Academic, New York, 1980).

Reactor-released radionuclides in Susquehanna River sediments

C. R. Olsen*, I. L. Larsen*, N. H. Cutshall*,
J. F. Donoghue, O. P. Bricker & H. J. Simpson§

* Environmental Sciences Division, Oak Ridge National Laboratory, Oak Ridge, Tennessee 37830, USA
† Smithsonian Institution, Washington DC 20560, USA
‡ Maryland Geological Survey, Baltimore, Maryland, USA
§ Lamont-Doherty Geological Observatory and the Department of Geological Sciences of Columbia University, Palisades, New York 10964, USA

Three Mile Island (TMI) and Peach Bottom (PB) reactors have introduced ^{137}Cs, ^{134}Cs, ^{60}Co, ^{58}Co and several other anthropogenic radionuclides into the lower Susquehanna River. Here we present the release history for these nuclides (Table 1) and radionuclide concentration data (Table 2) for sediment samples collected in the river and upper portions of the Chesapeake Bay (Fig. 1) within a few months after the 28 March 1979 loss-of-coolant problem at TMI. Although we found no evidence for nuclides characteristic of a ruptured fuel element, we did find nuclides characteristic of routine operations. Despite the TMI incident, more than 95% of the total ^{137}Cs input to the Susquehanna has been a result of controlled low-level releases from the PB site. ^{134}Cs activity released into the river is effectively trapped by sediments with the major zones of reactor-nuclide accumulation behind Conowingo Dam and in the upper portions of Chesapeake Bay. The reported distributions document the fate of reactor-released radionuclides and their extent of environmental contamination in the Susquehanna–Upper Chesapeake Bay system.

During the past 14 yr, five nuclear power reactors have operated for varying lengths of time at two sites on the Susquehanna River (Fig. 1). The first reactor, PB 1, was a 40-MW, gas-cooled reactor and operated from March 1966 to October 1974. Three others, PB units 2 and 3 (boiling-water reactors), and TMI unit 1 (a pressurized-water reactor), began power production in 1974 and each produces ~1,000 MW. TMI unit 2, a twin of TMI 1, was activated on 30 December 1978 but shut down on 28 March 1979 after the loss-of-coolant incident. As with the other nuclear power stations, minor amounts of ^{137}Cs, ^{134}Cs, ^{60}Co, ^{58}Co and other radionuclides are released with the coolant-water effluent. The radionuclide release history of these reactors has been compiled using Nuclear Regulatory Commission documents[1] and is summarized in Table 1. The Peach Bottom plant has contributed most of the reactor-produced radionuclides introduced into the Susquehanna River and from 1975 to 1979, the PB reactors have released >95% of the total ^{137}Cs input to the river.

In addition to reactor releases ^{137}Cs (half-life of ~30 yr) has been introduced into the Susquehanna–Chesapeake Bay system as global fallout from atmospheric nuclear weapons testing. The major influx of fallout ^{137}Cs occurred between 1962 and 1964. Rell[2] used the distribution of fallout ^{137}Cs in Chesapeake Bay to study the hydrography of the estuary. Although other fission products and neutron activation nuclides, such as ^{134}Cs, ^{60}Co and ^{58}Co are also produced during weapons tests, their short half-lives (~2, 5 and 0.2 yr, respectively) and low yields cause the

6 Linking Models with Empiricism: Landscape Boundaries and Connectivity

Introduction and Review

As landscape ecology coalesced around some central themes during the 1980s, both models and empirical field studies explicitly designed to address these themes began to appear. In ecology, both analytical and simulation modeling had developed rapidly in the 1960s and 1970s, the former primarily associated with theoretical population and community ecology, the latter with ecosystem studies. Nearly all of these models were nonspatial; all places in the model domain were considered as equivalent, or if they were differentiated, locational relationships were not addressed. The emergence of landscape ecology, with its focus on spatial heterogeneity, demanded that new modeling approaches be developed, approaches that represented space either explicitly or abstractly (Baker 1989; Sklar and Costanza 1991; Mladenoff 2005). Concurrently (but often independently), scientists working in the field were beginning to overcome their reluctance to tackle the complexities posed by spatial heterogeneity and designed studies that attempted to document, quantify, and test the effects of one or another aspect of landscape pattern. The marriage of ecology with landscape ecology was being consummated.

The study of **L. P. Lefkovitch and Lenore Fahrig** illustrates how simulation models can be used in a systematic approach to help understand how populations respond to spatial heterogeneity. Lefkovitch and Fahrig modeled population sizes and population persistence in a variety of arrays of habitat patches with different levels and forms of connectivity, using population parameters appropriate to the white-footed mouse *(Peromyscus leucopus)* in an agricultural landscape. The modeling approach was simple, based on transition matrices between patches, and the results were not surprising: Survival rates for populations in isolated patches were lower than those in connected patches, and survival rates of populations in connected patches increased with the number of interpatch connections. Despite their simplicity, models such as this led the way for later, more sophisticated modeling of such things as metapopulation dynamics (Hanski 1999) and generated testable predictions for field studies.

The patches in Lefkovitch and Fahrig's model were all the same, differing only in configuration and connections. Drawing his inspiration from both theoretical and empirical population studies, **Ronald Pulliam** developed an analytical model to address the situation in which local populations in a patchy landscape have different demographic responses to spatial variation in habitat characteristics between patches. Pulliam distinguished between habitat patches in which birthrates exceeded death rates, thereby leading to the production of excess individuals; that is, emigration rates exceed immigration rates (source patches). Sink habitat patches were those with the opposite characteristics. Over the long run, the persistence of populations in sink patches could only be maintained by immigration from source patches, so patch connectivity as well as variations in patch quality were clearly important to the dynamics of the metapopulation over the landscape as a whole. Conceptually, Pulliam's model expressed ideas put forth some years before by Fretwell and Lucas (1969) and Lidicker (1975), but it gave these ideas a firm theoretical footing. It also reinforced Van Horne's (1983) argument that, under some conditions, population density might actually be greater in poor-quality habitats (= sinks) than in high-quality habitats (= sources). This conclusion has an important implication: A species may occur and even reproduce in sink patches, but because populations in these patches are not self-sustaining without immigration, management of these habitat patches that does not also consider source patches in the landscape could lead to localized extinction, even though habitat was apparently being conserved. Moreover, because what is a source for one species may be a sink for another, the persistence of an assemblage of species in an area (what community ecologists have long studied) may be as much a product of the types and proximity of neighboring patches in a landscape as of the resources and conditions at that site. Interestingly, Pulliam's paper had a more immediate impact on population ecology than on landscape ecology, perhaps because of the preoccupation of landscape ecologists with spatial patterns on the ground rather than the more functional way of looking at spatially linked population processes.

Perhaps the epitome of a patchy landscape in the real world is the sort of mosaic created by clear-cut timber harvesting in forested landscapes. **Jerry Franklin and**

Richard Forman developed a cell-based simulation model to address the patchwork disturbance of forested landscapes created by timber harvesting in the Pacific Northwest of North America. They showed that characteristics of the harvest regime, such as the size of a harvest unit, the cutting rotation, or the spatial arrangement of harvests, are conspicuous generators of spatial heterogeneity and can be considered as analogous to attributes of natural disturbance regimes. Franklin and Forman realized that the implications of varying these components of a harvest regime could be explored using simple simulation models, such as a checkerboard distribution of clearcut patches in differing configurations. One clear message of the modeling results was that there were thresholds in landscape structure (patch density and interpatch distances) with increasing cutover area, a finding reinforced by neutral landscape models based on percolation theory (e.g., Gardner et al. 1987). The study was conducted when concerns about fragmentation of old-growth forests that resulted from dispersed patch clear-cutting in the Pacific Northwest were gaining momentum. By contrasting the effects of dispersed versus aggregated cutting patterns, Franklin and Forman showed how landscape approaches and modeling could be applied to resource management issues. This linkage has now become central to both landscape ecology and resource management (e.g., Lindenmayer and Franklin 2002; Liu and Taylor 2002).

Cell-based simulation models have evolved to include more complex, spatially dynamic processes and have become a staple item in the tool kit of spatial modelers and geographic information systems (GIS) analysts (Mladenoff and Baker 1999; Mladenoff 2005). Such models usually operate by, in effect, embedding a dynamic simulation model in each cell of a large matrix, with spatial structuring incorporated by varying the parameter values of the model between different types of cells (= patch types in a landscape). Cells may also be linked so that the dynamics of any given cell are contingent on the properties and dynamics of neighboring cells, potentially over a range of scales. **Robert Costanza, Fred Sklar, and Mary White** used this approach to model the spatial dynamics of water and nutrient fluxes in the Atchafalaya Basin in Louisiana. An ecosystem simulation model operated in each grid cell, which was connected by lateral fluxes to the four nearest neighboring cells. Costanza et al. used this model to predict changes in land-cover patterns across a large geographic region over long time scales that might result from various scenarios of site-specific management and natural changes. This paper demonstrated the power of this sort of spatial modeling, particularly for assessing the effects of landscape structure on ecosystem functions. Of course, the power of this approach has increased dramatically with advances in computational capacity (Costanza and Voinov 2004).

Constructing spatial models is a fine game, one that has clearly produced some important insights about possible landscape dynamics and the effects of spatial patterns on ecological processes and management alternatives. But it is also a rather academic game if it does not relate to real-world situations, either through empirical demonstration of underlying mechanisms or empirical validation of the model predictions. One of the premises addressed in the models of Lefkovitch and Fahrig and (indirectly) of Pulliam, for example, is that connectivity between patches in a landscape does indeed influence the movements of individuals. **John Wegner and Gray Merriam's** study of movements of birds and mammals in an agricultural landscape in Ontario exemplifies how field studies began to quantify the differential use of habitats and the effects of spatial connections between habitats. Using trapline sampling and direct observations, they recorded how individuals moved within woodland patches and

between woods, connecting wooded fencerows, and the surrounding agricultural matrix. Their results clearly demonstrated how the fencerows concentrated the movements of small mammals and birds in ways that could lessen the isolation of woodlots surrounded by agriculture.

As landscape ecologists grappled with understanding how organisms and populations responded to spatial heterogeneity, the ongoing loss and fragmentation of natural habitats became a growing focus of research. One consequence of fragmenting large, continuous areas of habitat into smaller, separate patches is an increase in the amount of edge habitat or patch boundaries. Forman and Godron (part III) called attention to the importance of edge configuration, and Wegner and Merriam noted that woodland birds moved more frequently from fencerows into adjacent agricultural fields than across an equal length of woodland boundary. **Lennart Hansson's** study, conducted in a coniferous forest–clear-cut mosaic in central Sweden, was designed specifically to determine how birds from both patch types responded to the sharp boundaries between them. By recording bird abundances along transects extending from the forest interior across the boundary well into the clear-cut, Hansson was able to show that the distributions of most species were sharply truncated at the habitat boundaries. Interestingly, several forest species exhibited greatest abundances close to the boundary. Hansson attributed this response to a weakening of the trees by increased exposure, which in turn led to increased insect attacks and greater resource availability for the birds. In a similar study of bird distributions across the edge between riparian woodland and adjacent desert habitats in Arizona, Szaro and Jakle (1985) found a similar elevation of densities at the woodland edge; in this case, abundance dropped sharply in the adjacent desert upland but fell off gradually with increasing distance from the boundary in adjacent desert wash habitats. In both studies, patch edges had influences all their own, but the patterns differed, depending on what was on the other side of the boundary.

Habitat fragmentation was also the focus of the work of **Paul Opdam, G. Rijsdijk, and F. Hustings,** although their study was conducted squarely in the conceptual tradition of island biogeography theory (MacArthur and Wilson 1967). Opdam and his colleagues surveyed bird communities in sixty-eight small woodlots in two agricultural landscapes in The Netherlands, considering the effects of woodland area, habitat variation, interpatch distance, and distance from more extensive woodlands. In contrast to the predictions of island biogeography theory, however, the occurrence of species in the woodland fragments was determined largely by their abundance in the surrounding forests, population size in the woodlot, and the minimum patch size of a species' habitat. Although later work, such as that of Addicott et al. (part IV) and Pulliam (part VI), might lead one to ask additional questions about the distributional dynamics of birds among forest fragments, the fundamental finding of Opdam et al., that the structure of the landscape was every bit as important as the characteristics of the woodland fragments considered by themselves, helped bolster the empirical foundation of landscape ecology.

These field studies considered population and community responses to landscape boundaries, connectivity, and structure, but landscape effects on ecosystem properties were not being ignored. Ecosystem ecologists were becoming aware of how land use might influence process rates and, especially, the lateral fluxes of water and nutrients across terrestrial landscapes into surface waters. The runoff of nutrients from fertilized agricultural landscapes was attracting particular attention. **William Peterjohn and David Correll** developed budgets for the pools and fluxes of nitrogen and phosphorus from upland agriculture through a bordering riparian forest. The paper was among the first to highlight the potential role of riparian buffers in filtering nutrients and reducing the load to surface waters, and it represented the beginnings of spatially explicit approaches to quantifying nutrient dynamics within the context of the well-established field of environmental or landscape biogeochemistry (Fortescue 1980). An extensive literature has developed in the subsequent decades (see, e.g., Naiman and Décamps 1997 for an extensive review; Carpenter et al. 1998), and basic and applied questions about nutrient fluxes to streams, lakes, and coastal waters at a range of scales remain prominent in contemporary landscape studies.

The last paper in this section, by **Robert Naiman, Henri Décamps, John Pastor, and Carol Johnston,** was one of the first to apply the emerging concepts of landscape ecology and spatial patterns of nutrient and material flux to aquatic systems (what some have called *aquascapes*). Most ecological work on fluvial ecosystems at this time was being conducted in the framework of the river continuum concept (Vannote et al. 1980), which treated the stream-river profile as an unbroken continuum of physical gradients and organismic responses. Naiman and his colleagues proposed instead that insights might emerge by considering stream-river systems as a series of resource patches separated by boundaries. They suggested that boundaries have attributes that may buffer a landscape against the spread of disturbance between patches and that disturbance events that are frequent or large may act to sharpen the contrast between the boundary and adjacent patches. Boundaries are also often characterized by greater variability and uncertainty than are the adjacent patches, and Naiman et al. suggested that chaos theory might be helpful in understanding this variability. The paper set the stage for the integration of aquatic systems into the mainstream of landscape ecology, although it would be some years before many players appeared on the stage (Malanson 1993; Poff 1997; Wiens 2002; and the papers in *Freshwater Biology* volume 47, number 4 [2002]).

References

Baker, W. L. 1989. A review of models of landscape change. *Landscape Ecology* 2:111–133.

Carpenter, S. R., N. F. Caraco, D. L. Cornell, R. W. Howarth, A. N. Sharpley, and V. H. Smith. 1998. Nonpoint pollution of surface waters with phosphorus and nitrogen. *Ecological Applications* 8:559–568.

Costanza, R., and A. Voinov. 2004. *Landscape simulation modeling.* New York: Springer-Verlag.

Fortescue, J. A. C. 1980. *Environmental geochemistry: A holistic approach.* New York: Springer-Verlag.

Fretwell, S. D., and H. L. Lucas Jr. 1969. On territorial behaviour and other factors influencing habitat distribution in birds: I. Theoretical development. *Acta Biotheoretica* 19:16–36.

Gardner, R. H., B. T. Milne, M. G. Turner, and R. V. O'Neill. 1987. Neutral models for the analysis of broad-scale landscape patterns. *Landscape Ecology* 1:19–28.

Hanski, I. 1999. *Metapopulation ecology.* Oxford, U.K.: Oxford University Press.

Lidicker, W. Z., Jr. 1975. The role of dispersal in the demography of small mammals. In *Small Mammals: Their Productivity and Population Dynamics,* eds. F. B. Golley, K. Petrusewicz, and L Ryszkowski, 103–128. Cambridge: Cambridge University Press.

Lindenmayer, D. B., and J. F. Franklin. 2002. *Conserving forest biodiversity: A comprehensive multiscaled approach.* Washington, DC: Island Press.

Liu, J., and W. W. Taylor. 2002. *Integrating landscape ecology into natural resource management.* Cambridge: Cambridge University Press.

MacArthur, R. H., and E. O. Wilson. 1967. *The theory of island biogeography.* Princeton, NJ: Princeton University Press.

Malanson, G. P. 1993. *Riparian landscapes.* Cambridge: Cambridge University Press.

Mladenoff, D. J. 2005. The promise of landscape modeling: Successes, failures, and evolution. In *Issues and Perspectives in Landscape Ecology,* eds. J. A. Wiens and M. R. Moss, 90–100. Cambridge: Cambridge University Press.

Mladenoff, D. J., and W. L. Baker. 1999. *Spatial modeling of forest landscape change.* Cambridge: Cambridge University Press.

Naiman, R. J., and H. Décamps. 1997. The ecology of interfaces: Riparian zones. *Annual Review of Ecology and Systematics* 28:621–658.

Poff, N. L. 1997. Landscape filters and species traits: Towards mechanistic understanding and prediction in stream ecology. *Journal of the North American Benthological Society* 16:391–409.

Sklar, F. H., and R. Costanza. 1991. The development of dynamic spatial models for landscape ecology: A review and prognosis. In *Quantitative Methods in Landscape Ecology,* eds. M. G. Turner and R. H. Gardner, 239–288. New York: Springer-Verlag.

Szaro, R. C., and M. D. Jakle. 1985. Avian use of a desert riparian island and its adjacent scrub habitat. *Condor* 87:511–519.

Van Horne, B. 1983. Density as a misleading indicator of habitat quality. *Journal of Wildlife Management* 47:893–901.

Vannote, R. L., G. W. Minshall, K. W. Cummins, J. R. Sedell, and C. E. Cushing. 1980. The river continuum concept. *Canadian Journal of Fisheries and Aquatic Sciences* 37:130–137.

Wiens, J. A. 2002. Riverine landscapes: Taking landscape ecology into the water. *Freshwater Biology* 47:501–515.

Ecological Modelling, 30 (1985) 297–308
Elsevier Science Publishers B.V., Amsterdam – Printed in The Netherlands

SPATIAL CHARACTERISTICS OF HABITAT PATCHES AND POPULATION SURVIVAL

L.P. LEFKOVITCH [1] and LENORE FAHRIG [2]*

[1] *Engineering and Statistical Research Institute, Research Branch, Agriculture Canada, Central Experimental Farm, Ottawa, Ont. K1A 0C6 (Canada)*
[2] *Department of Biology, Carleton University, Ottawa, Ont. (Canada)*

(Accepted 28 May 1985)

ABSTRACT

Lefkovitch, L.P. and Fahrig, L., 1985. Spatial characteristics of habitat patches and population survival. *Ecol. Modelling*, 30: 297–308.

The effect of the spatial arrangement of habitat patches on the survival of resident populations were considered in a stochastic model using population parameters appropriate to *Peromyscus leucopus*. The 34 possible arrangements of connections among five otherwise identical patches were simulated in order to determine the survival probabilities and population sizes.

The main findings are that populations in completely isolated patches have lower survival probabilities than those in patches that are connected to other patches, and that the survival probabilities of populations in connected patches increases with the size of the largest geometric figure of which the patch is a part. The results are discussed in the context of resource management.

INTRODUCTION

Although most species live in heterogeneous environments composed of 'patches' of suitable and unsuitable habitats (Emmel, 1976; Cowie and Krebs, 1979; McNamara, 1982), most ecological models assume that the environment is homogeneous (e.g., see Lotka, 1925; Leslie, 1945; Wiegert, 1974; Beddington et al., 1975; Anderson, 1979; Hansen and Tuckwell, 1981). It has been shown, both theoretically and empirically, that this assumption produces inaccurate predictions about population survival (Levin, 1976;

* Present address: Department of Zoology, University of Toronto, Toronto, Ont., Canada. ESRI Contribution No. I-653.

298

Goldstein, 1977; Southwood, 1977; Fahrig and Merriam, in press). However, apart from a preliminary study by Fahrig et al. (1983), it appears that there is no description of the way in which habitat heterogeneity changes model predictions. The problem is complex since every heterogeneous environment in nature is a unique spatial arrangement of habitat patches.

Based on simulations of all eleven arrangements of four woodlots, Fahrig et al. (1983) suggested the importance of isolation, but explanations for differences in population survival among the connected patches were not offered. The present study of the 34 arrangements of five patches, which clearly include those of four as a subset, was undertaken to remedy this. It does not seem likely that a further study of the arrangements of six, of which there are 156, will yield any further insights.

The purpose of this study was to understand which aspects of the spatial characteristics of groups of habitat patches are important predictors of the survival of resident populations.

METHODS

To study the problem, a number of arrangements of ha⅛·ʼɪt patches were chosen, and the population sizes in the patches were simuɪated over time. The simulation results were then analysed to determine which features of the spatial arrangement of patches appear to be important for population survival.

Choice of arrangements

The spatial arrangements of habitat patches (see Fig. 1) were chosen to satisfy the following conditions: (a) there are five habitat patches; (b) each patch may or may not be 'connected' to one or more of the other four, where the existence of a connection allows individuals to move through the inter-patch region, e.g. the patches may be sufficiently close together to allow this movement, or a physical 'corridor' of suitable habitat may link them; (c) all connections are of equal 'strength', i.e. the probability that individuals move between two connected patches is the same for all connections; (d) all patches are otherwise equivalent.

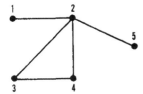

Fig. 1. A suggested arrangement of five habitat patches; circles are patches, lines are connections.

Under these restrictions, all 34 distinct spatial arrangements (Riordan, 1958, p. 146) of the five patches were examined.

Simulation experiments

To make the simulations biologically meaningful, published information was used on the white-footed mouse *Peromyscus leucopus* for populations in southeastern Ontario, which is the northern edge of its range. *P. leucopus* appears to prefer habitats with vegetational cover in the form of shrubs and trees and to avoid open areas such as fields (Bendell, 1961; Whitaker, 1967; M'Closkey and Lajoie, 1975; Hansen and Warnock, 1978). The hypothetical habitat patches can be thought of as woodlots separated from each other by fields, but sometimes linked by 'fencerows', which are fencelines along which trees and shrubs are growing; the latter have been suggested as corridors for movement of *P. leucopus* (Wegner and Merriam, 1979; Fahrig and Merriam, in press).

The model (presented in detail in Fahrig et al., 1983) is stochastic and is structured to follow the population sizes in four age classes in a series of interconnected patches through time. Within a patch, additions to the youngest age class occur through births, and to the other classes through aging, while losses occur through deaths. Between-patch dynamics consist of additions due to immigration from connected patches, and losses due to emigration. Since the age classes are of unequal duration, the underlying deterministic model is intermediate between the equal age class model of Leslie (1945) and the stage class model of Lefkovitch (1965).

The basic operator of the model is a transition matrix, for each time period. Each matrix has a block form. An element of the off-diagonal block, $p(i, j, k, t)$, is the proportion of organisms in class k moving from patch j to patch i during the time t to $t + 1$.

The probabilities of remaining are elements in the diagonal blocks (i.e. $i = j$), which also include birth and death rates, so that a typical diagonal element is

$$p(i, j, k, t) = 1 - \sum_{\substack{j=1 \\ j \neq 1}}^{m} p(j, i, k, t) + b(i, k, t) - d(i, k, t)$$

where $b(i, k, t)$ is the birth rate in patch i for class k during t to $t + 1$, $d(i, k, t)$ is the death rate, and m is the total number of patches. Parameters estimated from the literature were used as means. In the simulations for each year a new random estimate was chosen, assuming a normal distribution, with standard deviation equal to 10% of the mean. The vector of population sizes of the various age classes in the different patches at time $t + 1$ was then

300

obtained as the product of the (random) transition matrix for time t with the vector of population sizes at time t.

Detailed discussion of the parameter values is found in Fahrig et al. (1983), but the following points should be noted. First, there is no evidence that any of the probabilities of giving birth, dying, and dispersing from a woodlot are density-dependent for *P. leucopus*. Second, as no evidence to the contrary was found, it was assumed that *P. leucopus* disperse irrespective of whether or not a dispersal corridor exists. Individuals dispersing in the absence of a corridor are assumed to be lost due to predation. Third, the adult population in each woodlot in the spring is assumed to be 15% of the population in the previous fall (Taylor, 1978; Middleton, 1979). Fourth, all connections between woodlots are assumed to be biologically equi-distant (same length, width, plant species composition, etc.) and all woodlots are biologically equivalent. Fifth, the mice leaving a woodlot are divided equally among the connections.

The population sizes were calculated for each week from the 3rd week of March to the 2nd week of October for each woodlot in each spatial arrangement, and are summarized by the population sizes at the end of the year.

The average yearly final population sizes (in the 2nd week of October), and their variances were calculated for each woodlot. The simulations were terminated at 100 years; if a population did not survive for 100 years, the year in which it died out was noted.

Analysis of results

Those aspects of the spatial arrangement important for population survival were determined by analysis of variance of three dependent variables, which, in appropriately transformed scales, are:

(1) YEAR = log[population survival in years/(100 − survival in years)], i.e. the logit transformation of the probability of surviving 100 years.

(2) PRES = log(average yearly final population size), i.e. the mean number of mice at the end of the year in a logarithmic scale.

(3) PREV = log(variance of PRES).

The following features of spatial arrangement were considered to be possible explanatory factors. For each patch:

(1) DEGRE: the degree of a patch is defined as the number of other patches to which it is connected.

(2) NEIGH: the neighbouring degree of a patch is the average degree of all patches to which it is connected.

(3) ISO: a patch is either isolated (DEGRE = 0) or is connected (DEGRE > 0).

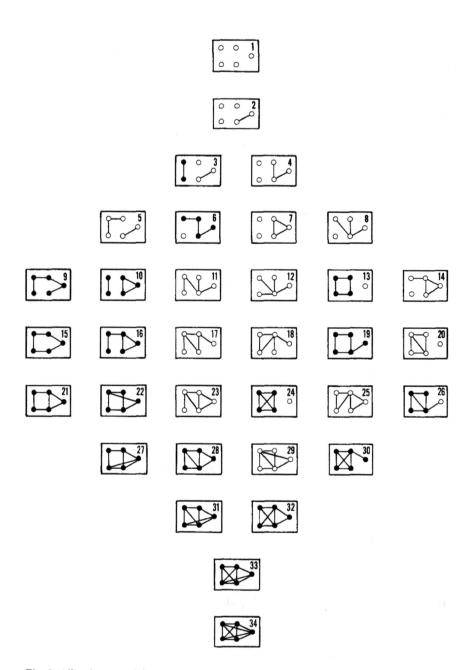

Fig. 2. All unique spatial arrangements of five unlabelled habitat patches; circles are patches, lines are connections. Populations in patches shown as open circles died out within the 100-year simulation.

302

(4) CRIT: a critical patch is one which, if removed from the spatial arrangement, separates the arrangement into two or more components. For example, in Fig. 1, only patch 2 is critical.

(5) GG: connected patches may be distinguished according to the largest geometrical figure of which they are a part. The only possibilities for five patches are a line, triangle, square, and pentagon. For example, in Fig. 1, patches 1 and 5 belong to a line, and patches 2, 3 and 4 belong to a triangle.

These explanatory factors are not completely independent, and, for example, CRIT and GG may be regarded as one factor with five levels. Two other predictors, namely, the orbits of the graphs and their structural information content (Mowshowitz, 1968) were also considered but found not to play a detectable role, and are not considered further in this paper.

Analyses of variance were performed to determine which of the factors, if any, were important sources of variation for the dependent variables described above. Because of the lack of balance, the analyses used instrumental variables.

RESULTS

All possible arrangements of five patches, under the constraints listed in Section Choice of arrangements, are illustrated in Fig. 2. A summary of the

TABLE 1

Summary analysis of variance giving mean squares for the (transformed) dependent variables

Transformation used	DF	YEAR (Survival) logit	PRES (Abundance) logarithm	PREV (Variability) logarithm
Source of variation				
ISO	1	237.664 *	90.923 *	69.881 *
GG within ISO	3	94.158 *	4.197 *	0.726
CRIT within GG within ISO	3	5.405	1.171	0.594
Covariates (NEIGH, DEGRE)	2	5.600	0.907	3.852
Residual	160	4.331	0.666	1.077
Means (back-transformed)		Probability of surviving 100 years	Mean numbers at end of year	Estimated variance at end of year
ISO 1 (isolated)		0.0085	0.1554	1.1219
ISO 2 (connected)				
GG2 (line)		0.6660	10.0744	61.992
GG3 (triangle)		0.6770	10.0442	47.323
GG4 (square)		0.9773	19.9654	67.357
GG5		0.9942	25.0031	68.855

* $p < 0.05$.

simulation results after transformations, together with the possible explanatory factors used in the analysis are given in Appendix 1.

The results of the analyses of variance are given in Table 1. Most of the variation in YEAR, PRES and PREV is explained by the contrast between isolated and nonisolated patches (i.e., factor ISO). Populations in isolated patches died out much earlier, had lower average population sizes and variances, but higher coefficients of variability than those in connected patches.

For YEAR and PRES, a significant portion of the variance not explained by ISO was attributed to GG (the largest geometrical figure). The means (Table 1) show that woodlots forming part of a square (GG = 4) or a pentagon (GG = 5) had higher population sizes and probabilities of survival for 100 years than those which formed part of a line (GG = 2) or a triangle (GG = 3). None of the other factors nor the covariates could be shown to play an additional significant role for any of the dependent variables.

DISCUSSION

The factors identified in the analyses as being the most important for population survival are whether the patch is isolated or not and, if not, then the size of the largest geometrical figure to which it belongs. The set of predictor variables used in the analyses of variance, however, are not independent. For example, degree zero corresponds to isolation, and so an isolated patch can never be critical, and the mean degree of its neighbours is necessarily zero (or undefined). Furthermore, since isolated patches represent the smallest possible geometrical figure, ISO and GG although treated separately here, are better regarded as a single factor of five levels. Finally, if an arrangement is not closed, there is a positive association between DEGRE and CRIT. Thus the explanatory factors imply both more and less than their original definitions.

Data for a decisive test of the results are difficult to obtain. This would require studies of *Peromyscus leucopus* population in a very large number of woodlots, which are interconnected in a variety of ways. However, Fahrig and Merriam (in press) studied *P. leucopus* populations in six woodlots, of which two were apparently completely isolated, while the other four were inter-connected by fencerows to form a square. The populations were studied over a summer using smoked paper tracking (Justice, 1961). Since the number of tracks (per 100 tracking stations per week) was shown to be correlated with the number of animals trapped in live traps (Fahrig and Merriam, in press), it is a measure of relative population size. The number of tracks per week for 12 weeks in the six woodlots, given in Table 2, are consistent with the results of the present study, since all four inter-connected

304

TABLE 2

(a) Number of tracks per 100 tracking stations in each of 12 weeks, June–September, 1982, in six woodlots

	Week												Mean
	1	2	3	4	5	6	7	8	9	10	11	12	
Isolated													
1	0	3	6	2	0	4	10	12	2	2	9	8	4.8 ⎫ 5.5
2	0	1	0	2	5	8	14	12	13	13	6	5	6.3 ⎭
Interconnected													
1	0	1	1	1	1	4	5	19	16	5	15	11	6.6 ⎫
2	0	1	2	0	9	19	19	18	14	3	3	20	9.0 ⎪ 8.4
3	0	2	2	1	5	5	9	10	27	6	8	7	6.8 ⎬
4	1	18	1	2	26	26	8	19	13	2	9	14	11.2 ⎭

(b) Analysis of deviance for the above data using a log–linear model and Poisson errors

Model	Model		Change in	
	DF	Deviance	DF	Deviance (X^2)
Mean	71	477.517		
+Weeks (W)	60	206.794	11	270.723 *
+Connections (C)	59	188.374	1	18.420 *
+Interaction (W.C)	48	163.508	11	24.866 *

* $p < 0.01$.

woodlots had higher mean population sizes than either of the two isolated ones. A log-linear analysis, rating the number of tracks as a Poisson variate (see Table 2), not surprisingly shows a significant time effect, but also an isolation and an interaction effect at the 0.1% level.

The sensitivity of any simulation model to the input values is important for its refining and interpretation. While insensitive parameters are perhaps superfluous, those which are sensitive may require further study. Although an extended analysis has not been performed, the only sensitive parameter appears to be the overwintering survival rate; small increases result in a population explosion, while small decreases result in rapid extinction. Since mortality in the model is not density dependent, the sensitivity to over-wintering survival is not surprising.

In interpreting the results of this study, the assumptions in the model and the special characteristics of the population dynamics of *P. leucopus* must be remembered:

(1) There is an assumption that a proportion of each population dispersed, irrespective of the number of connections. In the absence of connec-

tions, mice are supposed to have dispersed into the open fields where they had no chance of survival. This explains the importance of ISO (isolation) in the results of the simulation. For a species whose populations do not disperse when isolated, ISO may not be as important.

(2) The birth, mortality and dispersal rates of *P. leucopus* are all assumed to be density-independent. If, for another species, at low population densities, birth rate is high, mortality and dispersal rates low, with the reverse for high population densities, then populations will tend to persist for longer. Populations of such species, in arrangements such as no. 25 in Fig. 2, may not then die out. As the relative importance of stabilizing density-dependent factors increases, it must be expected that the probability of survival of populations in isolated habitats will increase.

(3) The assumptions that all patches are equivalent and all connections equally strong, allow other aspects of spatial arrangement to be considered. In nature, neither of these assumptions need be true, so that further work will be necessary to determine their importance.

Although interpretation of the results must recognize these limitations, three inferences have important implications for applied ecology:

(1) If the species is a pest (e.g. apple tree leaf mites), so far as is possible, its habitat patches (e.g. apple trees) should be arranged in patterns that minimize not only the possible connections between trees but also the sizes of figures formed by connections between neighbouring patches (e.g., two trees sufficiently close to allow mite dispersal constitute a connection). Isolation is best, but if it is not possible to arrange this, separated rows or small figures are to be preferred.

(2) If the species is harvestable or produces a harvestable product (e.g. honey bees), the opposite is true. The patches (e.g. bee hives) should be arranged in well-connected, large closed figures. This also follows from the studies of linear arrangements in which drift to an adjacent position occurs with non-zero probability in a finite number of positions (Jay, 1965).

(3) For rare species needing conservation, the provision of a number of suitable habitats each of sufficient size to maintain a population is obvious, but this study also suggests that establishing links able to be traversed by individuals can be a major factor in their long-term survival.

If the model described in this paper is a reasonable picture of reality for a species, it follows that the recognition of the connection possibilities among patchily distributed populations is essential to an understanding of its dynamics, survival, and gene flow.

306

REFERENCES

Anderson, R.M., 1979. The influence of parasitiç infestation on the dynamics of host population growth. In: R.M. Anderson, B.D. Turner and L.R. Taylor (Editors), Population Dynamics. Blackwell, London, pp. 245–281.

Beddington, J.R., Free, C.A. and Lawton, J.H., 1975. Dynamic complexity in predator–prey models framed in difference equations. Nature, 255: 58–60.

Bendell, J.F., 1961. Some factors affecting the habitat selection of the white-footed mouse, Can. Field-Nat., 75: 244–255.

Cowie, R.J. and Krebs, J.R., 1979. Optimal foraging in patchy environments. In: R.M. Anderson, B.D. Turner and L.R. Taylor (Editors), Population Dynamics. Blackwell, London, pp. 183–205.

Emmel, T.C., 1976. Population biology. Harper and Row, New York, 371 pp.

Fahrig, L. and Merriam, H.G., in press. Habitat patch connectivity and population survival. Ecology.

Fahrig, L., Lefkovitch, L.P. and Merriam, H.G., 1983. Population stability in a patchy environment. In: W.K. Lauenroth, G.V. Skogerboe and M. Flug (Editors), Analysis of Ecological Systems: State-of-the-Art in Ecological Modelling. Proc. Symp., 24–28 May 1982, Fort Collins, CO. Developments in Environmental Modelling, 5. Elsevier, Amsterdam/Oxford/New York, pp. 61–67.

Goldstein, R.A., 1977. Reality and models: difficulties associated with applying general ecological models to specific situations. In: S.A. Levin (Editor), Mathematical Models in Biological Discovery, Lecture Notes in Biomathematics, 13. Springer, New York, NY, pp. 206–215.

Hansen, F.B. and Tuckwell, H.C., 1981. Logistic growth with random density independent disasters. Theor. Popul. Biol., 19: 1–8.

Hansen, L.P. and Warnock, J.E., 1978. Response of two species of Peromyscus to vegetational succession on land strip-mined for coal. Am. Midl. Nat., 100: 416–423.

Jay, S.C., 1965. Drifting of honeybees in commercial apiaries. 1. Effects on various environmental factors. J. Apic. Res., 4: 167–175.

Justice, K.E., 1961. A new method for measuring home ranges of small mammals. J. Mammal., 42: 462–470.

Lefkovitch, L.P., 1965. The study of population growth in organisms grouped by stages. Biometrics, 21: 1–18.

Leslie, P.H., 1945. On the use of matrices in certain population mathematics. Biometrika, 33: 182–212.

Levin, S.A., 1976. Population dynamic models in heterogeneous environments. Annu. Rev. Ecol. Syst., 7: 287–310.

Lotka, A.J., 1925. Elements of Physical Biology. Williams and Wilkins, Baltimore, MD, 465 pp.

M'Closkey, R.T. and Lajoie, D.T., 1975. Determinants of local distribution and abundance in white-footed mice. Ecology, 56: 467–472.

McNamara, J., 1982. Optimal patch use in a stochastic environment. Theor. Popul. Biol., 21: 269–288.

Middleton, J.D., 1979. Insular biogeography in a rural mosaic: the evidence of Peromyscus leucopus. M. Sc. Thesis, Carleton University, Ottawa, Ont., 86 pp.

Mowshowitz, A., 1968. Entropy and the complexity of graphs. Bull. Math. Biophys., 30: 175–204, 225–240, 387–414, 533–546.

Myton, B., 1974. Utilization of space by Peromyscus leucopus and other small mammals. Ecology, 55: 277–290.

Riordan, J., 1958. An Introduction to Combinatorial Analysis. Wiley, New York, NY, 244 pp.

Southwood, T.R.E., 1977. Habitat, the templet for ecological strategies? J. Anim. Ecol., 46: 337–366.

Taylor, D.G., 1978. The population biology of white-footed mice in an isolated and a non-isolated woodlot in southeastern Ontario. M.Sc. Thesis, Carleton University, Ottawa, Ont., 134 pp.

Wegner, J.F. and Merriam, H.G., 1979. Movements by birds and small mammals between a wood and adjoining farmland habitats. J. Appl. Ecol., 16: 349–358.

Wiegert, R.G., 1974. Competition: a theory based on realistic, general equations of population growth. Science, 185: 539–542.

Whitaker, J.O., Jr., 1967. Habitat relationships of four species of mice in Vigo Country, Indiana, Ecology, 48: 867–872.

APPENDIX 1

Summary of simulation results after transformations YEAR, PRES, PREV, CFFV), and possible explanatory factors used in the analyses of variance (DEGRE, NEIGH, ORBIT, SI, CRIT, ISO, GG):

YEAR	PRES	PREV	CFFV	DEGRE	NEIGH	ORBIT	STRUC	CRIT	ISO	GEOG
-3.4864	-1.4692	0.9981	7.1584	0	0.000	1	0.0000	0	1	1
-3.4864	-2.3036	-1.0495	5.9229	0	0.000	1	0.0000	0	1	1
-3.4864	-2.9937	-2.4357	5.9055	0	0.000	1	0.0000	0	1	1
-3.4864	-2.9937	-2.4357	5.9055	0	0.000	1	0.0000	0	1	1
-3.4864	-2.9937	-2.4357	5.9055	0	0.000	1	0.0000	0	1	1
-3.4864	-1.6089	0.6761	7.0077	0	0.000	1	0.0000	0	1	1
-3.4864	-2.5245	-1.2396	6.7172	0	0.000	1	0.9710	0	1	1
-3.4864	-2.3036	-1.0495	5.9229	0	0.000	1	0.9710	0	1	1
-3.4864	-0.3285	0.2384	1.2328	0	0.000	2	0.9710	0	2	2
-3.4864	-0.1744	-0.5698	0.8954	1	1.000	2	0.9710	0	2	2
4.6052	3.4944	5.0285	0.3906	1	1.000	2	0.6748	0	2	2
-3.4864	-2.6593	-0.9970	8.6779	0	0.000	1	0.6748	0	2	2
4.6052	3.4720	4.9917	0.3693	1	1.000	2	0.6748	0	2	2
-1.7463	1.1537	4.4504	2.9196	1	1.000	2	0.6748	0	2	2
-1.7463	1.2147	4.5295	2.8514	1	1.000	2	0.6748	0	2	2
-3.4864	-2.8134	-1.9561	6.2675	0	2.000	1	1.5219	0	1	1
-3.4864	-0.3711	0.6099	1.9650	1	1.000	1	1.5219	0	2	2
-3.4864	-0.7540	0.0129	3.5298	0	2.000	1	1.5219	0	1	1
-3.4864	0.0954	-0.1103	0.8602	2	1.000	18	1.5219	0	1	1
-3.4864	-0.2358	0.7103	0.8976	1	2.000	18	1.5219	1	2	2
2.2083	3.5924	5.8477	0.5124	1	2.000	18	1.5219	0	2	2
2.3246	2.7443	4.4220	0.4803	2	2.000	3	1.5219	0	2	2
2.1019	2.9098	4.3346	0.4757	1	1.000	3	1.5219	0	2	2
-0.8615	2.2507	5.6148	1.7449	1	1.000	2	1.5219	0	2	2
-0.8615	2.2642	5.7423	1.8349	1	1.000	2	1.5219	0	2	2
4.6052	3.0625	4.3450	0.4107	1	1.000	2	1.5219	0	2	2
4.6052	3.6344	5.1376	0.3445	2	1.000	19	1.5219	1	2	2
-3.4864	-2.5297	-0.6960	8.8265	0	1.000	19	1.5219	0	1	1
4.6052	3.6859	5.3063	0.3560	2	1.000	19	1.5219	0	2	2
4.6052	3.1497	4.2897	0.3659	1	2.000	4	1.5219	0	2	2
-3.4864	-2.0410	-0.3660	6.4110	2	0.000	1	0.9710	0	1	1
-3.4864	-0.0620	2.3132	3.3824	2	0.000	20	0.9710	0	2	2
-3.4864	-1.0217	-0.2384	2.4656	2	0.000	1	0.9710	0	2	2
-3.4864	0.1223	2.6380	3.3092	2	0.000	20	0.9710	0	2	2
-3.4864	-0.2779	2.4616	3.8913	2	0.000	20	0.9710	0	2	2
-2.1019	0.7883	2.4176	1.5228	1	3.000	5	1.3356	0	2	2
-2.0037	0.9837	1.8687	1.0518	1	3.000	5	1.3356	0	2	2
-3.4864	-2.0405	-0.3917	6.3260	0	0.000	5	1.3356	0	1	1
-2.0037	1.7699	3.3818	0.9240	3	0.000	44	1.3356	0	2	2
-2.0037	0.8918	2.3994	1.3405	1	3.000	21	1.3356	0	2	2
4.6052	3.4654	5.5101	0.4915	2	2.000	22	1.5219	0	2	2
4.6052	3.4423	4.6464	0.3266	1	2.000	6	1.5219	0	2	2
4.6052	2.7688	4.6955	0.6563	2	1.000	6	1.5219	1	2	2
4.6052	2.9194	4.4443	0.4980	2	2.000	21	0.9710	0	2	2
4.6052	3.5659	5.8823	0.5354	2	1.000	2	0.9710	0	2	2
4.6052	3.4865	4.6150	0.3076	1	2.000	20	0.9710	0	2	2
4.6052	3.4187	4.7164	0.3463	1	2.000	2	0.9710	0	2	2
4.6052	3.4992	4.7012	0.3171	1	2.000	20	0.9710	0	2	2
4.6052	3.4727	5.0236	0.3825	2	2.000	20	0.9710	0	2	2
4.6052	3.4282	4.6263	0.3289	2	2.000	23	0.9710	0	2	2
-0.1787	2.4835	3.9652	0.6060	2	3.000	23	1.9219	0	2	2
-0.3399	1.1962	2.2909	1.5671	1	2.000	8	1.9219	0	2	2
-2.2187	2.0398	3.3503	0.6951	3	2.000	45	1.9219	0	2	2
-0.3399	2.5567	4.5718	0.7350	3	1.333	45	1.9219	0	2	2
-0.3399	2.4371	4.3561	1.2599	1	4.000	8	1.9219	0	2	2
-1.0082	2.6837	4.6156	0.6867	1	4.000	9	0.6748	0	2	2
-1.1119	2.4549	4.5875	0.8493	1	4.000	9	0.6748	0	2	2
-1.1119	3.3528	4.7109	0.8209	1	1.000	9	0.6748	0	2	2
-0.9583	3.8686	6.8334	0.6364	4	1.000	66	0.6748	0	2	2
-1.0082	2.6257	4.7092	0.7526	1	4.000	66	0.6748	1	2	2
4.6052	2.0175	3.4486	0.7459	1	4.000	24	0.6748	0	2	4
4.6052	0.0127	2.6499	0.8289	2	2.000	24	0.6748	0	2	4
4.6052	2.0072	2.9516	0.5878	2	2.000	24	0.6748	0	2	4

308

```
 4.6052   1.9834   3.2374   0.6944    0  2.000   24  0.6748    0   2   4
-3.4864  -2.7364  -0.0540  15.0209    0  0.000    1  0.6748    0   1   1
-0.4220   1.9302   3.6197   0.8866    1  3.000   10  1.9219    0   2   2
-0.3808  -2.9069   4.2563   0.4590    3  0.000   46  1.9219    1   2   3
-3.9020  -2.1875  -0.2744   7.6932    0  2.500    1  1.9219    0   1   3
-0.3808   2.4484   3.3737   0.4504    2  2.500   25  1.9219    0   2   3
-0.3808   3.5085   3.8068   0.5460    2  2.500   25  1.9219    0   2   3
 4.6052   3.4952   4.6653   0.3158    2  2.000   26  0.0000    0   2   5
 4.6052   3.4513   4.5636   0.3105    2  2.000   26  0.0000    0   2   5
 4.6052   3.5290   5.3706   0.4301    2  2.000   26  0.0000    0   2   5
 4.6052   3.4797   4.8790   0.3534    2  2.000   26  0.0000    0   2   5
 4.6052   3.4015   4.4130   0.3027    2  2.000   26  0.0000    0   2   5
 4.6052   3.5369   5.3660   0.4257    2  2.000   27  1.9219    0   2   5
 4.6052   3.6886   5.0006   0.3047    3  2.000   47  1.9219    1   2   5
 4.6052   2.9591   4.2870   0.4424    1  2.000   11  1.9219    0   2   5
 4.6052   3.3383   4.8552   0.4022    2  2.500   28  1.9219    0   2   5
 4.6052   3.3164   4.5030   0.3448    3  2.500   28  1.9219    1   2   3
-0.8615   2.1736   3.9584   0.8233    3  2.000   48  1.9219    1   2   3
-0.8615   1.7959   3.9472   0.9779    1  3.000   48  1.9219    0   2   3
-0.9583   1.3137   2.5971   0.9847    2  3.000   12  1.9219    0   2   3
-0.8615   1.9228   3.4337   0.8139    1  3.000   29  1.9219    0   2   3
-1.0082   0.4575   2.4419   2.1457    2  3.000   12  1.9219    0   2   3
-1.1658   1.7699   3.4875   0.9742    4  1.500   30  1.9219    1   2   3
-1.1658   2.2607   4.9743   1.2541    2  3.000   67  1.9219    0   2   3
-1.1119   1.7210   3.2397   0.9038    1  4.000   30  1.9219    0   2   4
-1.7463   0.5710   3.2030   2.8025    2  4.000   13  1.9219    0   2   4
-1.2784   0.9400   3.3861   2.1235    2  3.000   13  1.9219    0   2   4
 4.6052   3.3992   4.7501   0.3591    2  4.000   31  1.9219    0   2   4
 4.6052   3.3290   4.2769   0.3041    2  2.500   32  1.9219    0   2   4
 4.6052   3.3481   4.6793   0.3648    3  2.667   49  1.9219    1   2   4
 4.6052   3.6447   5.0511   0.3266    1  3.000   14  1.9219    0   2   4
 4.6052   2.5855   5.9442   0.5470    3  3.000   50  1.9219    0   2   4
 1.0082   1.1908   5.0650   0.5177    2  3.000   33  1.9219    0   2   4
 0.8145   2.8305   4.5067   0.5615    2  3.000   33  1.9219    0   2   4
 0.5583   2.8555   4.2611   0.4944    3  2.333   50  1.9219    0   2   4
 1.0082   2.1693   4.6871   0.4379    0  1.500    1  1.9219    0   1   1
-3.9020  -2.6593  -1.0022   8.6555    2  1.500   34  1.9219    0   2   5
 4.6052   3.2515   4.1764   0.3124    1  2.333   51  1.9219    0   2   5
 4.6052   3.4687   4.6230   0.3143    2  1.500   34  1.9219    0   2   5
 4.6052   3.2492   4.2283   0.3214    3  2.333   51  1.9219    1   2   5
 4.6052   3.5053   4.5821   0.2969    2  2.000   35  1.9219    0   2   5
 4.6052   3.1705   4.2954   0.3596    3  2.000   52  0.9710    0   2   4
 4.6052   3.6275   4.9781   0.3203    2  3.000   36  0.9710    0   2   4
 4.6052   3.2438   3.9765   0.2849    2  3.000   36  0.9710    0   2   4
 4.6052   3.2669   4.1071   0.2972    2  2.000   52  0.9710    0   2   4
 4.6052   3.5074   4.6849   0.2880    2  3.000   36  0.9710    0   2   4
 4.6052   3.2367   4.4540   0.3643    2  3.000   36  0.9710    0   2   4
 2.5974   3.3262   4.7357   0.3835    3  2.333   53  1.9219    1   2   4
 2.3246   3.3621   5.3954   0.5146    2  3.000   54  1.9219    0   2   4
 2.3246   2.9935   4.1434   0.3978    1  2.667   15  1.9219    0   2   4
 2.5974   2.2978   4.5967   0.3662    3  3.000   54  1.9219    0   2   4
 1.9124   2.2437   4.3322   0.9253    2  3.000   37  1.9219    0   2   4
 4.6052   3.4272   4.6376   0.3301    3  3.000   55  0.6748    0   2   4
 4.6052   3.4262   4.5262   0.3125    3  3.000   55  0.6748    0   2   4
 4.6052   3.3989   4.3498   0.2941    3  3.000   55  0.6748    0   2   4
 4.6052   3.4217   4.7414   0.3496    3  3.000   55  0.6748    0   2   4
-3.4864  -1.3863   1.0637   6.8082    0  0.000    1  0.6748    0   1   1
 0.7230   2.7252   4.2237   0.5415    2  3.000   38  0.6748    0   2   3
 0.7230   3.3057   4.9481   0.4353    4  3.000   60  0.6748    1   2   3
 0.7230   2.7036   4.1221   0.5260    2  3.000   38  0.6748    0   2   3
 0.7230   3.8032   3.9851   0.4446    2  3.000   38  0.6748    0   2   3
 0.7230   2.8304   2.4424   0.4920    2  3.000   38  0.6748    0   2   3
 4.6052   3.6109   4.9992   0.3291    3  3.333   57  1.9219    0   2   4
 4.6052   2.2316   4.2033   0.3231    3  3.500   39  1.9219    0   2   4
 4.6052   3.2684   3.3585   0.3365    2  3.500   39  1.9219    0   2   4
 4.6052   2.7977   5.2449   0.3087    3  2.000   69  1.9219    1   2   4
 3.9020   3.5825   3.9897   0.5556    2  4.000   16  1.9219    0   2   2
 4.6052   3.1851   4.3577   0.3656    3  3.000   40  1.9219    0   2   4
 4.6052   3.4651   4.7908   0.3431    3  2.667   57  1.9219    0   2   4
 4.6052   3.4223   4.7256   0.3466    3  2.667   57  1.9219    0   2   4
 4.6052   3.4148   4.3881   0.2950    3  3.000   58  1.9219    0   2   4
 4.6052   3.3861   4.3972   0.3050    3  3.000   58  1.9219    0   2   4
 4.6052   3.4032   4.7068   0.3500    3  3.000   59  1.9219    0   2   4
 4.6052   3.4382   4.2765   0.2726    3  3.000   59  1.9219    0   2   4
 4.6052   3.0459   3.9416   0.3412    4  3.500   41  1.9219    0   2   4
 4.6052   3.6413   5.0226   0.3230    4  2.500   70  1.9219    0   2   4
 0.4636   3.0828   4.0752   0.3516    4  3.000   41  1.9219    0   2   4
 0.4636   2.2103   5.3817   0.5948    4  2.500   71  0.9710    0   2   4
 0.4636   2.6553   4.1135   0.5496    2  4.000   42  0.9710    0   2   4
 0.4636   2.6341   3.6673   0.4491    4  4.000   42  0.9710    0   2   4
 0.3808   3.1302   4.8901   0.5040    2  2.500   71  0.9710    0   2   4
 0.3399   2.5321   4.2598   0.6689    4  4.000   42  0.9710    0   2   4
 4.6052   3.3160   4.6728   0.3755    3  3.333   61  1.3356    0   2   4
 4.6052   3.4751   4.9530   0.3684    3  3.500   72  1.3356    0   2   4
 4.6052   3.3164   4.0685   0.2775    3  3.500   60  1.3356    0   2   4
 4.6052   2.2923   4.5898   0.3688    3  3.000   17  1.3356    0   2   4
 4.6052   1.9707   4.4941   1.1728    3  3.333   62  0.6748    0   2   4
 4.6052   3.2445   4.5212   0.3316    3  3.333   62  0.6748    0   2   4
 4.6052   3.2474   4.4532   0.3260    3  3.333   63  0.6748    0   2   4
 4.6052   3.3670   4.3538   0.3042    4  3.000   73  0.6748    0   2   4
 4.6052   3.5747   4.9815   0.3382    4  3.333   63  0.6748    0   2   4
 4.6052   3.3386   4.1895   0.2883    3  3.333   32  1.5219    0   2   4
 4.6052   3.3932   4.7647   0.3639    3  3.667   74  1.5219    0   2   4
 4.6052   3.6112   5.1573   0.3561    3  3.000   32  1.5219    0   2   4
 4.6052   3.3975   4.3328   0.2920    3  3.667   74  1.5219    0   2   4
 4.6052   3.5946   5.1035   0.3525    3  3.000   43  1.5219    0   2   4
 4.6052   2.9760   3.9659   0.3704    2  4.000   65  0.9710    0   2   4
 4.6052   3.2966   5.1823   0.4939    4  3.000   75  0.9710    0   2   4
 4.6052   3.5522   5.3107   0.4079    4  3.500   75  0.9710    0   2   4
 4.6052   3.4974   5.2062   0.4089    4  3.500   75  0.9710    0   2   4
 4.6052   3.5278   5.0670   0.3700    4  3.500   75  0.9710    0   2   4
 4.6052   3.2592   4.3807   0.3434    4  4.000   65  0.9710    0   2   4
 4.6052   3.4941   4.7622   0.3286    4  4.000   76  0.0000    0   2   5
 4.6052   3.6548   5.0569   0.3242    4  4.000   76  0.0000    0   2   5
 4.6052   3.4720   4.6793   0.3223    4  4.000   76  0.0000    0   2   5
 4.6052   3.5184   4.9156   0.3463    4  4.000   76  0.0000    0   2   5
 4.6052   3.5109   4.7816   0.3262    4  4.000   76  0.0000    0   2   5
```

Landscape Ecology vol. 1 no. 1 pp 5-18 (1987)
SPB Academic Publishing, The Hague

Creating landscape patterns by forest cutting: Ecological consequences and principles

Jerry F. Franklin* and Richard T.T. Forman**
*USDA Forest Service, Pacific Northwest Research Station, 3200 Jefferson Way, Corvallis, OR 97331, USA, and University of Washington, College of Forest Resources, Seattle, WA 98195, USA
**Harvard University, Graduate School of Design, Cambridge, MA 02138, USA

Keywords: landscape pattern, patch size, forestry, forest cutting, forest management, game populations

Abstract

Landscape structural characteristics, such as patch size, edge length, and configuration, are altered markedly when management regimes are imposed on primeval landscapes. The ecological consequences of clearcutting patterns were explored by using a model of the dispersed patch or checkerboard system currently practiced on federal forest lands in the western United States. Thresholds in landscape structure were observed on a gradient of percentages of landscape cutover. Probability of disturbance, *e.g.*, wildfire and windthrow, and biotic components, *e.g.*, species diversity and game populations, are highly sensitive to these structural changes. Altering the spatial configuration and size of clearcuts provides an opportunity to create alternative landscapes that differ significantly in their ecological characteristics. Both ecosystem and heterogeneous landscape perspectives are critical in resource management.

Introduction

How many red spots make a white cow red?
How many clearings make a forest, prairie?
A score? More? A coalescing core?
A threshold reached?
(*Landscape Ecology*, Forman and Godron 1986)

Geomorphic processes, natural disturbances, and human activities combine to create the richness in spatial pattern visible from the air. In some landscapes, this pattern has been stable for centuries; in others − *e.g.*, primeval forests of the tropics, boreal regions, and the western United States − an onslaught of changes has resulted from forest cutting (Harris 1984; Lovejoy *et al.* 1984). Many specific variables, such as location of roads and patch size, are issues in the cutting operations for both economic and ecological reasons. But emerging ideas in landscape ecology (Risser *et al.* 1984; Forman and

Godron 1986), suggest the pattern created on the landscape by a sequence of cutting operations may be more critical than these specific variables.

Assume that a primeval landscape, say of coniferous or tropical forest, is to be progressively cut. Should we cut small patches and disperse them as much as possible? Should we start at one side and clearcut progressively to the other? Or should we use some other cutting strategy? The answer must reflect the economics of harvesting wood products, the probabilities of major disturbance in the landscape (*e.g.*, windthrow, wildfire, and pests), and desired levels of biotic components (*e.g.*, game, rare species, and threatened habitats).

In this paper we explore the ecological consequences of cutting patterns along a gradient of forest conditions from primeval to completely clearcut landscape. We begin with a simple geometric model to see how spatial patterns (size and density

6

Fig. 1. Nearly a half-billion board feet of timber blew down on 900 ha of forest in the Bull Run drainage of the Mount Hood National Forest, Oregon, during a storm in December 1983; most of the blown down patches were associated with existing clearcuts and roads.

of the patches and length of border) change when small clearcuts are dispersed over the landscape. We examine how susceptibility to disturbance and biotic components, *e.g.*, species diversity, might change along the cutting gradient. Finally, we analyze how these patterns might be controlled by size of cutting unit or different spatial combinations of cutting.

The dispersed patch or checkerboard model is particularly important because variations on this model are widely used in forestry. We thus include some illustrative data from the primeval Douglas-fir (*Pseudotsuga menziesii* (Mirb.) Franco) forests on federal lands in the Pacific Northwest (USA). For 40 years these landscapes have been cut extensively with a staggered-setting system of clearcutting; 10- to 20-ha patches are interspersed with uncut forest areas of at least equal size (cover) (Smith 1985). Ease of forest regeneration, slash disposal, and access road development were the reasons this approach was originally adopted; the ecological and

economic appropriateness of this system should be reexamined, however, as objectives and techniques change and the fragmentation of the forest landscape continues (Harris 1984). Evidence is accumulating that this system also increases the risk of some types of catastrophic disturbance (Fig. 1).

Ecosystem theory (Odum 1959; Woodwell and Whittaker 1968; Bormann and Likens 1979), island biogeographic theory (MacArthur and Wilson 1967; Simberloff and Wilson 1970), and geographic spatial theory (Chorley and Haggett 1970; Haggett *et al.* 1977; Ripley 1981) are important foundations for our analysis. Landscape ecology, in contrast, focuses on the ecological structure, functioning, and change of a landscape (Neef 1967; Forman and Godron 1981; Brandt and Agger 1984; Naveh and Liebermann 1984; Risser *et al.* 1984; Forman and Godron 1986) from the small homogenous patch to the large, heterogenous area. In this paper we focus specifically on landscape change, and our

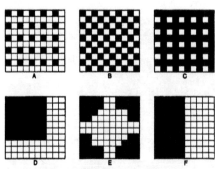

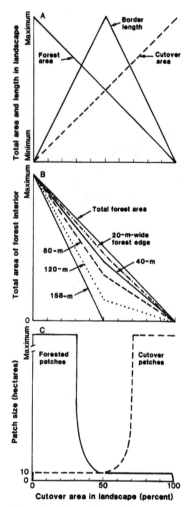

Fig. 2. Patterns of clearcutting developed under various models. (A–C) Progression of clearcutting using the dispersed patch model in which areas are selected for cutting so as to be distributed regularly across the landscape; shown are 25, 50, and 75% cutover points. (D–E) Pattern of cutting at 50% point using single-nucleus, four-nucleus, and progressive-parallel cutting systems.

analysis shows patterns of ecological response to landscape structure, including evidence for thresholds along the continuum of forest cutting intensities. Principles from such analyses can contribute to wise management of our finite remaining forest landscape.

Spatial model

The checkerboard model is designed to disperse the cuttings as evenly as possible across the landscape (Fig. 2). To do that, we made some initial assumptions for our model. The sample area is a 1000-ha² representative of a larger landscape. It is divided into 10-ha cells, one of which is removed by each logging operation. Cells are selected for cutting so that each new cut minimizes the variance in interpatch distance between it and all other cut patches. The model begins with a completely forested landscape; passes through the 25%, 50% (checkerboard), and 75% cutover points (Fig. 2); and ends with a completely cutover landscape. Only two types of cells are possible — forest and cutover — and once a cell is cut, it remains in that state. The road system is assumed to be complete at the 50% cutover point.

In real landscapes, new forests establish and grow on cutover areas. The pattern in an actual forest is

Fig. 3. Changes in landscape characteristics along a landscape cutting gradient based on the regularly distributed dispersed patch cutting or checkerboard model. (A) Total forest area and border length. (B) Total area of forest interior environment for different edge widths. (C) Forest and cutover patch sizes.

also complicated by areas reserved to provide wildlife habitat and recreation and to protect streams. Cutting of all forests within a National Forest landscape will probably never occur.

In the checkerboard model, the total length of

8

border between forest and cutover areas rises linearly to a peak when the landscape is 50% cut, and drops linearly thereafter (Fig. 3A). This curve approximates the total length of forest edge (Leopold 1933; Dierschke 1974; Thomas 1979; Ranney *et al.* 1981; Morgan and Gates 1982; Forman and Godron 1986) in the landscape if the width of the forest edge is narrow. The forest edge differs significantly in environmental conditions from the forest interior and is a concentrated area for edge species, including many game species.

The total area of forest interior in the landscape, where interior species are concentrated, is related to both the total length of border and the width of forest edge. The forest interior area disappears much more rapidly in cutting landscapes with wide edges (*e.g.*, 120 m) than in those with narrow edges (20 m) (Fig. 3B). With an edge width of over 160 m, no forest interior environment remains after the landscape is 50% cut over; *i.e.*, the 10-ha patches are entirely edge habitat. At boundaries between open areas and forest, the edge width is generally related to the height and structure of the forest. If a measure equivalent to two tree heights is used to estimate the width of recently exposed edges (a conservative rule-of-thumb), the landscape of 80-m-high Douglas-fir trees has no unmodified forest interior conditions left by the time it is 50% cut.

Distinctive patterns for patch size (area) in the landscape also emerge from the model. In the early period of cutting, the average forested-patch size remains equal to the total size of the landscape because the 10-ha cuts are scattered regularly as gaps or clearings within continuous, although increasingly porous, forest. At about the 30% cutover point, the average forest-patch size begins to drop sharply (Fig. 3C) because cuts coalesce into continuous lines of patches dividing the previously continuous forest into sections. Average forested-patch size thus drops rapidly from the 30% threshold to the 50% cutover point; after 50%, the forested-patch size remains constant at 10 ha until the last patch is cut. Variation in forest-patch size − *i.e.*, having different-sized patches in the landscape − exists only in the 30 to 50% cutover interval. Patch size is of central importance in many fields including considerations of biological diversity, nature reserve design, and logging operations

(MacArthur and Wilson 1967; Wilson and Willis 1975; Forman *et al.* 1976; Pickett and Thompson 1978; Gilbert 1980; Cubbage 1983).

The size pattern for cutover patches is the mirror image of the forest pattern; *i.e.*, cutover patches are 10 ha until the checkerboard point when cutover patches begin to coalesce (Fig. 3C). The process continues until a threshold is reached at about 70%; after 70%, the landscape is a continuous cutover area containing remnant 10-ha forested patches. Similarly, different-sized cutover patches are present only in the 50 to 70% cutover interval.

Thresholds for patch density and interpatch distance also appear in the model. At about the 30% cutover point (Fig. 3C), a sharp rise occurs in forest-patch density peaking at the 50% cutover or checkerboard point. Thereafter, patch density decreases linearly to the completely cutover state. The curve for the density of clearcut patches is a mirror image of that for forested patches.

Interpatch distance − the distance that fire, dust, seeds, or mammals must cross to spread to a neighboring forest patch − is zero until the 50% checkerboard point; the corners of all forested patches then touch their neighbors. The first cut thereafter begins the process of isolation or increased interpatch distance. Again, the pattern for the distance between clearings is a mirror image of that for forests.

Several patterns and thresholds of potential major ecological significance appear in the theoretical model when a cutting pattern of regularly distributed squares is assumed. Thresholds are present at 30% (average patch size and patch density), 50% (virtually all characteristics), and 70% (average patch size and patch density).

Development of the patchwork on National Forest lands in the Pacific Northwest shows increasing deviation from the theoretical model as cutting progresses; the results are lower patch densities and border lengths and larger clearcut patch sizes in the real landscape than the model predicts (Table 1). The departures from the dispersed patch cutting pattern in the real landscape result from clearcutting of adjacent forest patches early in the cutting cycle; this is because of road accessibility of these stands and of programs to salvage timber killed by wind or escaped slash fires.

Table 1. Number and mean size of clearcut patches and amount of edge in three National Forest landscapes in relation to percentage of landscape cutover; multiple, adjacent clearcuts are counted as a single patch.

Area and landscape characteristic	Percent landscape cutover				
	10	20	30	40	50
H.J. Andrews, Oregon					
Cutover patches (no.)	51	87	93	–	–
Cutover patch zone (ha)	11.9	13.8	16.9	–	–
Edge length (km)	70.6	128.2	167.5	–	–
Bull Run, Oregon*					
Cutover patches (no.)	47	60	–	–	–
Cutover patch size (ha)	17.0	22.9	–	–	–
Edge length (km)	76.9	118.4	–	–	–
South Fork McKenzie, Oregon					
Cutover patches (no.)	26	55	55	53	45
Cutover patch size (ha)	25.4	22.4	36.0	49.8	73.3
Edge length (km)	50.4	124.0	157.2	186.9	204.2

*Mainstem subbasin.

Disturbance susceptibility

Patterns of patches across the landscape can significantly affect the potential for major forest disturbances, although the nature of the response varies with type of disturbance (Pickett and Thompson 1978; Pickett and White 1985). Here we relate disturbance and ecological response to the geometry of developing clearcut patchwork. Some general patterns of response are apparent, and others are specific to the example from the Pacific Northwest, thus illustrating some of the variability resulting from environment and biota.

The potential for catastrophic windthrow in residual stands of primeval forest relates strongly to the developing landscape patchwork (Fig. 4A) (Ruth and Yoder 1953; Gratkowski 1956; Holtam 1971; DeWalle 1983; Savill 1983). Windthrow susceptibility increases in the model with the amount of edge between cutover area and primeval forest, the isolation of primeval forest in small patches, and increasing wind fetches (Flemming 1968). The initial increase in windthrow potential results from more edge. As cutting progresses and the primeval forest remnants are isolated as 10-ha patches, windthrow potential increases because forest patches are exposed on all sides. Windthrow occurs on corners and along the sides of exposed forest blocks. Wind

fetches progressively lengthen, especially after half of the landscape is cut over. After about 80% of the landscape is cut over, windthrow risk to all remaining patches of primeval forest should be at a maximum, assuming that no significant increase in the wind-firmness of the exposed trees or stands occurs.

The effect of the staggered-setting pattern on fire potential is more complex because it involves both ignition and spread of fire (Fig. 4B). Ignition sources include lightning, human accidents, campfires, cigarettes, and sparks from equipment, and planned ignitions (escaped slash burns). In the primeval landscape, ignition potential is based on natural sources, which can be defined to either include or exclude aboriginal people. In the Douglas-fir region, the potential of natural ignition is relatively low and is primarily from lightning. Sources of wildfire increase rapidly with development of access roads and creation of logging slash from both accidental and planned ignitions; these human-caused ignitions level off once the road system is essentially completed (Fig. 4B).

Once a fire is ignited, the probability of its spreading in residual primeval forest refects the extent of edge influence in the patch (Fig. 3B). The hypothesized relation assumes that the microclimate of the primeval forest is, because of cooler

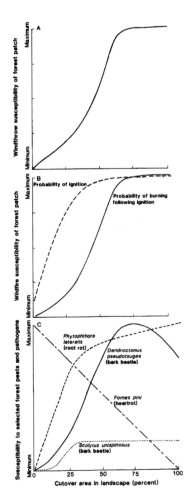

Fig. 4. Susceptibility of forests in the Douglas-fir region to various damaging agents along a landscape cutting gradient as shown by the checkerboard model. (A) Potential for windthrow in residual forest patches. (B) Potential for wildfire ignition and spread. (C) Susceptibility to insect and fungus pests.

and moister conditions, less favorable for fire spread than the cutover area. Drier and warmer conditions induced by adjacent cutover areas would, therefore, increase the probability of a fire spreading to a residual forest patch. The potential for a fire across the landscape is also affected by the cutting pattern; *e.g.*, wildfire control should be easier in a landscape with greater heterogeneity (including firebreaks, such as roads) or less acreage of primeval forest with its large fuel accumulations.

The responses of pests and pathogens to a developing patchwork are highly species specific, which reflects the interactions of their life histories (including dispersal mechanisms) and environment (including host distribution) (Fig. 4C). In the Douglas-fir landscape of the Pacific Northwest, some bark beetles, such as *Dendroctonus pseudotsugae*, require mature or old-growth trees as host and often reach outbreak populations after broadscale blowdown events. Populations breed in freshly felled trees and attack standing trees. The probability of outbreaks of such a pest should parallel the risk of blowdown (Fig. 3C) until late in the cutting cycle when host trees of the necessary size become increasingly scarce (Fig. 4C). Other beetles — such as *Scolytus unispinosus*, which typically uses smaller, competitively stressed trees — might show an increase proportional to the area in regenerated stands of susceptible age (Fig. 4B). Heart rot fungi, such as *Fomes pini*, are most common in old forests and should decrease in importance with decreasing area of primeval forests (Figs. 3A and 4C). But a root rot fungus dispersed by human activities, such as *Phytopthora lateralis*, might expand its role as the road system is developed and the landscape is cut over (Zobel *et al.* 1985) (Fig. 4C). The point is that no general pattern of pest or pathogen response to the developing landscape pattern emerges. Each pest has its own response to changing landscape geometry.

Cutting patterns and intensities affect other types of disturbances, such as landslides. Potential for landslides is affected by both road networks and clearcuts; in the Pacific Northwest, the relative contributions may be nearly equal in the long run (Swanson and Dyrness 1975; Ziemer 1981). Major impacts of checkerboard cutting are that it requires essentially the entire road system to be operational throughout the cutting cycle and maintains a constant proportion of the landscape in clearcuts of an age vulnerable to landslides; the period between loss of the old root network and establishment of a new root network typically has the highest risk of landslides.

Disturbances observed in landscapes of dispersed clearcuts in the Pacific Northwest appear to substantiate at least some of the hypothetical patterns. Windthrow has long been recognized as a chronic problem in such landscapes (Ruth and Yoder 1953; Gratkowski 1956). The possibility that dispersed patchcutting might lead to extensive catastrophic blowdown (Fig. 1) has not been generally understood. On the H.J. Andrews Experimental Forest in western Oregon, about 4% (250 ha) of the residual forest area has suffered serious windthrow (> 10% of the stand volume) during the last 35 years. All this windthrow was adjacent to existing clearcuts or road clearings. The only significant wildfire (4 ha) in this landscape during the last 35 years occurred when a slash fire escaped from a clearcut.

The Bull Run River drainage in the Mount Hood National Forest in western Oregon provides strong evidence of the potential for catastrophic disturbance created by the checkerboard pattern (Fig. 1). We analyzed windthrow patterns on a 37,000-ha area including the Bull Run and adjacent tracts. Major windstorms in this area in December 1973 and 1983 blew down forests of 482 ha and 899 ha, respectively. Nearly 1/2-billion board feet of timber fell in the 1983 blowdown. About 48% of the 1973 and 81% of the 1983 blowdowns were adjacent to existing clearcuts and roads; both are statistically significant relations. About one-fourth of the clearcut-related 1983 blowdown was associated with clearcuts created in previously unlogged landscapes to salvage the 1973 blowdown; such areas might be considered natural rather than management-related blowdown. All significant wildfire damage to forests within Bull Run has resulted from escaped slash burns, further indicating the contribution of management activities to disturbance potential. Similar patterns of catastrophic windthrow and wildfire have occurred in the patchcut landscapes of the southern Gifford Pinchot National Forest in southern Washington; the major wildfire in this region during the last 40 years was the West Point fire, which burned both regenerated forest and residual old-growth patches within a highly fragmented area.

Large, protected reserves within such fragmented landscapes also suffer disturbance (Pickett and Thompson 1978: Forman 1979). In 1980. forty-five

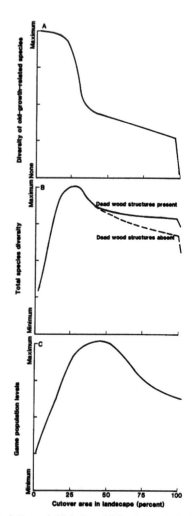

Fig. 5. Changes in biotic characteristics of a Douglas-fir forest landscape along a cutting gradient as shown by the checkerboard model. (A) Diversity of species dependent on a forest interior environment. (B) Total species diversity in the landscape (including forest and cutover area) when standing dead trees and fallen logs are present and when they are not within cutover areas of the landscape. (C) Total game populations in the landscape.

216-ha 'spotted owl management areas' were set aside for protection of *Strix occidentalis* (Xantus de Vesev) in the Gifford Pinchot National Forest.

12

A windstorm in 1983 caused '... 215 acres [86 ha] of blowdown in six spotted owl management areas, including 75 acres [30 ha] in one core ... [resulting in] having to make adjustments in five areas, combine two other areas into one, and relocate one management area. Blowdown appeared greatest adjacent to harvest units, particularly in locations where habitat had already been fragmented by timber harvesting' (Ruediger 1985).

Biotic components

The checkerboard system of clearcutting affects many ecological characteristics of the landscape and creates potential for major disturbances. It may, for example, affect diversity of species, occurrence of species requiring interior forest conditions, and populations of game animals. Biological diversity and maintaining interior species dependent on old-growth forest are increasingly important issues for National Forests (*e.g.*, Franklin *et al.* 1981; Harris 1984; Meehan *et al.* 1984; Brown 1985).

The diversity of forest-interior species declines with progressive cutting of the primeval forest (Fig. 5A). An initial, gradual decline is likely as the result of a loss of solitary predators during the early cuttings. We expect that the first large loss of species considered obligate interior-forest dwellers will coincide with the major reduction in the size of primeval forest patches that occurs between 30 and 50% cutover (Fig. 3C). This theory assumes that the 10-ha patch size is under the minimum area required for some interior species (Forman *et al.* 1976; Robbins 1980; Lovejoy *et al.* 1984; Gutiérrez and Carey 1985). Species loss may be the result of an unsuitable forest microenvironment, competitive interactions with edge or opening species, or an insufficient total area of suitable foraging habitat. The lichen *Lobaria oregana*, for example, depends on a strictly interior microclimate (Franklin *et al.* 1981), and the spotted owl requires a minimum total old-growth area for foraging (Gutiérrez and Carey 1985). The second large loss of species will occur when the last primeval forest patch is cut. The assumption is that some forest interior species, such as certain salamanders and forest herbs, can survive in a 10-ha patch. The gradual decline in species diversi-

ty between the two major episodes of species loss reflects the combined effects of chance extinction of some species from patches and the declining number of forest patches.

Some variation in the shape of the curve for loss of interior species (Fig. 5A) is possible, depending on minimum patch sizes (total or interior) for the species present, the patch size selected for cutting, and the width of critical edge conditions. If each interior species could persist in a 10-ha patch of primeval forest, a continuous curve of interior species loss would be expected as random events and declining patch numbers take their toll. The magnitude of interior species loss would be lessened if some species could colonize regenerating forests on cutover land.

Total species diversity should increase during the initial phases of cutting an old-growth forest landscape (Fig. 5B). Clearcutting results in the addition of many pioneer organisms that find the disturbed, open environment suitable for colonization. Species diversity for vascular plants in the Douglas-fir region is, for example, highest in the early successional stages preceding canopy closure (Franklin and Dyrness 1973). An initial episode of total species loss occurs with fragmentation of the primeval forest, and a second episode occurs when the last primeval forest patch is eliminated, as described above. The presence of dead-wood structures (*i.e.*, standing dead trees and fallen logs) is a key factor in survival of many animal species (*e.g.*, Brown 1985; Harmon *et al.* 1986). Thus, the magnitude of the loss of total species diversity depends on the presence or absence of dead-wood structures in the cutover landscape (Fig. 5B).

Most game species favor the open, early stages of forest succession and, therefore, probably will increase rapidly with the early cuttings (Fig. 5C) (Leopold 1933; Thomas 1979). Many of these species also make heavy use of edges or use two ecosystem types, forest and open. The maximum game population in the forest and cutover patchwork is expected at about the 50% cutover point. Population declines follow as the amount of edge decreases and the high-quality protective cover and winter habitat provided by the primeval forest are lost (see Alaback 1984 and Schoen *et al.* 1984; Brown 1985).

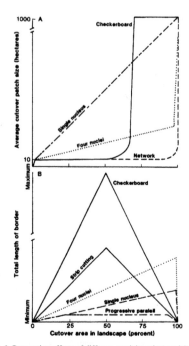

Fig. 6. Contrasting effects of different spatial cutting models for a landscape. The single-nucleus model indicates progressive, contiguous cutting from an initial central point, and the four-nucleus model is similar with four regularly distributed initial points. The network model makes 10-ha cuts regularly distributed in an initially established network of narrow forest corridors. The progressive-parallel model has progressive contiguous cuts of strips parallel to one side of the landscape. (A) Average cutover patch size. (B) Total length of forest border between forest and cutover.

Patch configuration and size

The analyses to this point are based on the checkerboard model – equal-sized cutting units distributed regularly across the landscape. We now generate alternative major cutting models to compare with the checkerboard. We consider five alternative spatial configurations of the basic 10-ha cutting unit: (a) 'single-nucleus model', where an initial central patch is progressively enlarged by contiguous cutting; (b) 'four-nucleus model', where four evenly spaced initial-cut patches are similarly enlarged; (c)

'network model', where cuts are regularly distributed within an initially defined network of narrow, forested line corridors; (d) 'strip-cutting model', where regularly distributed parallel strips, 316 m wide, are cut; and (e) 'progressive parallel model', where 316-m-wide contiguous strips are progressively cut parallel to one edge of the landscape (Fig. 2). We also explore the effects of different clearcut sizes.

The cutting configuration from the checkerboard model produces a constant average cutover-patch size of 10 ha until the landscape is 50% cut; this is followed by a steep rise and leveling off of clearcut patch size near the 70% point, where a cutover landscape contains remnant 10-ha forest patches (Fig. 6A). In contrast, the single-nucleus model produces a steady increase in size of the single cutover patch, and the four-nucleus model produces a steady but smaller increase in size of the four cutover patches until they coalesce near the end (Fig. 6A). Average size of cutover patches in the network model remains at 10 ha until the forested corridors of the network itself are cut. Thus, if minimizing the average size of cutover patches (*e.g.*, to minimize wind fetch or desiccation of soil) is an important planning and management criterion, then clear differences among the models emerge. In addition, individual models differ in response between a slightly cutover and an extensively cutover landscape.

The average patch size during a cutting cycle also differs markedly for the different cutting approaches. In the checkerboard model, the average patch size remains high up to the 30% cutover point, drops steeply, and levels off at 10 ha at the 50%-cutover point (Fig. 3C). The single-nucleus and the progressive-parallel models produce a steady decrease in forest patch size from beginning to end. The four-nucleus model also follows this curve initially because the four cuts are simply gaps within a single, huge forest patch. At some point beyond the 50%-cutover point, the cut patches in this model coalesce, and the average size of the remaining forest patches drops sharply. If maintaining large forest patches is critical in management to protect interior species or headwaters of stream systems, then clear differences in cutting configuration are evident.

14

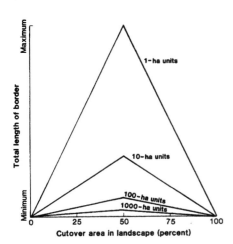

Fig. 7. Total length of border between forest and cutover area with different cutting-unit sizes. Based on the checkerboard model and an unlimited-size grid.

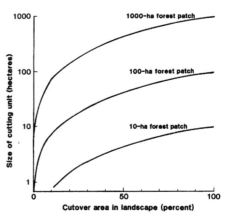

Fig. 8. Disappearance of large patches as a result of using different cutting-unit sizes in the checkerboard model. Cuts are distributed regularly by using an unlimited-sized grid. For example, the last square, 100-ha forest patch (middle curve) disappears when the landscape is only 1% cutover, if 1-ha cutting units are used. Using 10-ha cutting units removes the last large patch (same curve) at 15% cutover, also early in a cutting cycle. Using 100-ha cuts retains a 100-ha forest patch until the last cut in a landscape grid.

The cutting configurations have major effects on the total length of border between forest and cutover area, which is often critical in managing for game populations or windthrow susceptibility. Strip cutting results in half of the border length of the checkerboard model (Fig. 6B). The single- and four-nucleus models produce less total border length except just before coalescence at the end, and the progressive-parallel model produces the minimal average total length of border.

We can also vary the size of the cutting units from 1 to 1000 ha in a checkerboard model and gauge possible effects. The average forest-patch size remains unchanged up to 30%-cutover point because the cuts are gaps in an extensive landscape (Fig. 3C). Forest-patch size drops from the 30 to 50% points, and then levels off as remnant woods equal to the cutting unit size are scattered through the cutover landscape. The total length of border between forest and cutover areas peaks at the 50% point for all cutting unit sizes (Fig. 7). Much less border results, however, from the larger cutting-unit sizes than from using 1- or 10-ha cuts for a given area of cutover. These patterns of patch size and border length resulting from different cutting-unit sizes have major implications for critical ecological processes,

including disturbances and biotic components.

Because many interior species need large patches (Forman *et al.* 1976; Robbins 1980; see Alaback 1984 and Schoen *et al.* 1984), knowing when the last large patch of a certain size will disappear along a landscape cutover gradient is critical. The checkerboard model told us that the last square 100-ha forest patch without cut gaps will disappear when the landscape is only 15% cut (Fig. 8). Using a 100-ha cutting unit, the last 1000-ha forest patch will disappear at this same 15%-cutover threshold. The initial steep slope of the curves up to the 15% threshold (Fig. 8) reflects the rapid loss of large, intact forest patches. The single-nucleus, four-nucleus, and progressive-parallel models retain large patches much farther into the cutting cycle than does the checkerboard model.

A critical but unknown number of large patches appear necessary, however, to maintain maximal species diversity, retain genetic diversity, and provide stability in the face of disturbance to a single patch (Forman *et al.* 1976; Pickett and Thompson 1978; Game and Peterken 1984). As noted above, the last large patch disappears at about the 15%-cutover point. In addition, the last two, three, or several large square patches disappear sequentially just before the 15% cutover point. Thus, in a dispersed-patch cutting pattern, major spatial landscape characteristics with attendant ecological effects change suddenly and rapidly early in the cutting cycle.

Yet another cutting strategy is to use more than one size of cutting unit in a landscape. When 1-ha and 10-ha cuts are mixed such that equal areas result (*e.g.*, forty 1-ha and four 10-ha cuts each equal 40 ha), the total length of border between forest and cutover areas is midway between the border length for just 1-ha cuts and for just 10-ha cuts. Thus, many planning and management options for minimizing border length (to control windthrow) or maximizing border length (to enhance game) are available by using a mixture of clearcut sizes.

Differences in cutting configuration and basic cutting-unit sizes (or mixtures of different sizes) are critical in understanding the ecology of a landscape and in forest planning and management. Changes in these factors drastically alter the spatial patterns of a landscape along a cutover gradient.

Landscape ecology and forest management implications

These models are simplified for identifying underlying patterns and principles. In the real world, many constraints limit the creation of particular land-use patterns. Topography (*e.g.*, drainage pattern and the length, evenness, and steepness of slopes) limits the possibilities. Broad geographic patterns (*e.g.*, differences in frequency of high winds between coastal and inland regions) and localized site conditions (*e.g.*, poorly drained soils) strongly affect the absolute risk of catastrophic events like windthrow and fire (Ruth and Yoder 1953; Forman 1979). The existing vegetation mosaic also influences the size and distribution of patches imposed on the landscape, particularly where much of the landscape is composed of forests currently unsuited to cutting. Nevertheless, the theoretical analyses we have presented clearly show

that ecological consequences can differ drastically depending on the pattern imposed on a landscape by land-use activities. Data from at least one forest region support us.

The number, size, and arrangement of the patches in a mosaic created by forest managers strongly influence the degree to which management objectives are fulfilled. Our checkerboard approach maximizes the high-contrast edge between primeval forest and cutover areas and demonstrates thresholds in patch size and interior environment when the original forest matrix is fragmented between 30 and 50% cutover. Disturbances and biota may be altered dramatically with these structural changes, certainly at the scale of forest patch (10 ha) and tree heights (dominant heights of 80 m) typical of the Douglas-fir region. Alternative approaches will result in different ecological responses.

We suggest a two-point guide for managers of natural landscapes (*e.g.*, Federal timberlands in the Douglas-fir region) based on this landscape analysis and on other ecological considerations. First, reduce the emphasis on dispersing small clearcut patches through the forest landscape. The fragmentation that results does not enhance many resource values. Approaches featuring progressive or clustered cuts from scattered nuclei should be considered. In this way, the risks of disturbance associated with edges and the amount of maintained road systems can be reduced. The size of a cluster of cuts would depend on management objectives and landscape characteristics. Networks of corridors and small forest patches should be retained within the clustered cutover areas to provide additional cover and edge for game species, reduce wind fetches and erosion, and enhance interpatch movement of species.

Second, identify and reserve large patches of primeval forest in the landscape for maintenance of interior species and amenity values. It is urgent because many current cutting programs are rapidly reducing the size of available patches. Clearcutting generally must be avoided within the reserved patches because of the substantial vulnerability that results from placing even small cuts within a reserved tract (Ruediger 1985). Finally, other evidence suggests that connections between reserved

16

tracts, such as forest corridors, may be critical to long-term protection of the landscape (Neef 1967; Forman and Godron 1981; Brandt and Agger 1984; Naveh and Liebermann 1984; Forman and Godron 1986).

We have demonstrated the importance of evaluating the spatial and long-term consequences of human-imposed landscape patterns. As land-management prescriptions are developed, consideration must be given to how ecological conditions in a patch are affected by the position of the patch within the landscape: How do the surroundings affect a patch and vice versa? This contrasts with an exclusive focus on the individual stand or homogeneous ecosystem. Management programs must similarly balance long- and short-term considerations. The checkerboard approach worked reasonably well during the first decades of cutting in the Douglas-fir region but appears increasingly unsatisfactory thereafter. The circumstances originally leading to its adoption have also changed. Basic road systems are now largely present, and tree planting has replaced natural regeneration. The consequences of the patch-cutting pattern will also have a strong and persistent effect on management of future cutting cycles.

Summary

Economic, sociological, and ecological concerns must be considered in any analysis of cutting systems. Significantly higher management costs, for example, are associated with the checkerboard system. A basic road network must be created for the entire landscape early in the cutting cycle and be continuously maintained because silvicultural activities are dispersed rather than concentrated. Unit-area costs for management activities in the Douglas-fir region, such as slash burning, tree planting, and stand thinning, can be reduced by geographical concentration and increased patch size. On the other hand, extensive clearcutting of a drainage basin over a short time could create severe problems, such as floods after rain-on-snow storms (Harr and Berris 1983).

Across major portions of the globe, human activities are imposing new patterns on natural or seminatural landscapes. The effects of agricultural clearing and forest cutting in the tropical forests of South America and southeastern Asia are critical examples, as are arid lands being placed under irrigation. Major collaborations are needed among scientists, planners, managers, and the public to utilize landscape perspectives in development of land-use policies and imposition of management regimes.

Acknowledgements

F.J. Swanson, T. Spies, and B.T. Milne contributed to the development of the concepts presented in this paper during numerous discussions and reviews of earlier manuscripts. We also thank W.A. Reiners, S.A. Levin, E.O. Wilson, M. Godron, F. Golley, and F. Sampson for their valuable manuscript reviews. A Harvard University Bullard Fellowship in Forest Research permitted this collaboration. Support for this work was also provided by National Science Foundation Grants DEB 80-12162, BSR-8514325, and BSR-8315174; U.S. Department of Agriculture, Forest Service, Pacific Northwest Research Station; and Harvard University Graduate School of Design. This is a contribution to the H.J. Andrews Ecosystem Study and has received policy approval by USDA Forest Service.

References

Alaback, P.B. 1984. A comparison of old-growth and second-growth forest structure in the western hemlock-Sitka spruce forests of southeast Alaska. *In* Fish and Wildlife Relationships in Old-Growth Forests. pp. 219–226. Edited by W.R. Meehan, T.R. Merrell Jr. and T.A. Hanley. American Institute of Fisheries Research Biologists, Morehead City, North Carolina.

Bormann, F.H. and Likens, G.E. 1979. Pattern and Process in a Forested Ecosystem. Springer-Verlag, New York.

Brandt, J. and Agger, P. (eds.) 1984. Proceedings of the First International Seminar on Methodology in Landscape Ecological Research and Planning. Roskilde Universitetsforlag Geo-Ruc., Roskilde, Denmark.

Brown, E.R. (ed.) 1985. Management of Wildlife and Fish Habitats in the Forests of Western Oregon and Washington. Publ. R6-F&WL-192-1985. U.S. Department of Agriculture,

Forest Service, Pacific Northwest Region, Portland, Oregon.

Chorley, G.A. and Haggett, R.J. 1970. Social and Economic Models in Geography. Methuen, London.

Cubbage, F.W. 1983. Tract size and logging costs in southern pine. J. For. 81: 430–433.

DeWalle, D.R. 1983. Wind damage around clearcuts in the ridge and valley province of Pennsylvania. J. For. 81: 158–159.

Dierschke, H. 1974. Die naturraumliche Gliederung der Verdener Geest. Forsch. Deutsch. Landeskunde 177: 1–113.

Flemming, G. 1968. Die windgeschwindigkeit auf waldumgebenen freiflachen. Arch. Forstwes. 17: 5–16.

Forman, R.T.T. (ed.) 1979. Pine Barrens: Ecosystem and Landscape. Academic Press, New York.

Forman, R.T.T., Galli, A.E. and Leck, C.F. 1976. Forest size and avian diversity in New Jersey woodlots with some land use implications. Oecologia 26: 1–8.

Forman, R.T.T. and Godron, M. 1981. Patches and structural components for a landscape ecology. BioScience 31: 733–740.

Forman, R.T.T. and Godron, M. 1986. Landscape Ecology. John Wiley & Sons, New York.

Franklin, J.F. and Dyrness, C.T. 1973. Natural Vegetation of Oregon and Washington. Gen. Tech. Rep. PNW-8. U.S. Department of Agriculture, Forest Service, Pacific Northwest Forest and Range Experiment Station, Portland, Oregon.

Franklin, J.F., Cromack, K., Jr., Denison, W., McKee, A., Maser, C., Sedell, J., Swanson, F. and Juday, G. 1981. Ecological Characteristics of Old-Growth Douglas-Fir Forests. Gen. Tech. Rep. PNW-118. U.S. Department of Agriculture, Forest Service, Pacific Northwest Forest and Range Experiment Station, Portland, Oregon.

Game, M. and Peterken, G.F. 1984. Nature reserve selection strategies in the woodlands of central Lincolnshire, England. Biol. Conserv. 29: 157–181.

Gilbert, L.E. 1980. The equilibrium theory of island biogeography: fact or fiction? J. Biogeogr. 7: 209–235.

Gratkowski, H.J. 1956. Windthrow around staggered settings in old-growth Douglas-fir. For. Sci. 2: 60–74.

Gutiérrez, R.J. and Carey, A.B. (eds.) 1985. Ecology and Management of the Spotted Owl in the Pacific Northwest. Gen. Tech. Rep. PNW-185. U.S. Department of Agriculture, Forest Service, Pacific Northwest Forest and Range Experiment Station, Portland, Oregon.

Haggett, P., Cliff, A.D. and Frey, A. 1977. Locational Analysis in Human Geography. Wiley, New York.

Harmon, M.E., Franklin, J.F., Swanson, F.J., Sollins, P., Gregory, S.V., Lattin, J.D., Anderson, N.H., Cline, S.P., Aumen, N.G., Sedell, J.R., Lienkaemper, G.W., Cromack Jr., K. and Cummins, K.W. 1986. Ecology of coarse woody debris in temperate ecosystems. In Advances in Ecological Research, Vol. 15, pp. 133–302. Academic Press, New York.

Harr, R.D. and Berris, S.N. 1983. Snow accumulation and subsequent melt during rainfall in forested and clearcut plots in western Oregon. In Proceedings of Western Snow Conference, 51st Annual Meeting, Vancouver, Washington, 19–21 April 1983, pp. 38–45. Colorado State University, Ft. Collins, Colorado.

Harris, L.D. 1984. The Fragmented Forest. University of Chicago Press, Chicago.

Holtam, B.W. (ed.) 1971. Windblow of Scottish Forests in January 1968. Bull. 45. [Brit.] Forestry Commission.

Leopold, A. 1933. Game Management. Scribners, New York.

Lovejoy, T.E., Rankin, J.M., Bierregaard, R.D., Jr., Brown, K.S., Jr., Emmons, L.H. and Van der Voort, M.E. 1984. Ecosystem decay of Amazon forest remnants. In Extinctions. pp. 295–325. Edited by M.H. Nitecki. University of Chicago Press, Chicago.

MacArthur, R.H. and Wilson, E.O. 1967. The Theory of Island Biogeography. Princeton University Press, Princeton, New Jersey.

Meehan, W.R., Merrell, T.R., Jr. and Hanley, T.A. (eds.) 1984. Fish and Wildlife Relationships in Old-Growth Forests. American Institute of Fisheries Research Biologists, Morehead City, North Carolina.

Morgan, K.A. and Gates, J.E. 1982. Bird population patterns in forest edge and strip vegetation at Remington Farms, Maryland. J. Wildl. Manage. 46: 933–944.

Naveh, Z. and Liebermann, A.S. 1984. Landscape Ecology, Theory and Application. Springer-Verlag, New York.

Neef, E. 1967. Die Theoretischen Grundlagen der Landschaftslehre. Geographisch- Kartographische Anstalt Gotha. Hermann Haack, Leipzig.

Odum, E.P. 1959. Fundamentals of Ecology. Saunders, Philadelphia, Pennsylvania.

Pickett, S.T.A. and Thompson, J.N. 1978. Patch dynamics and the design of nature reserves. Biol. Conserv. 13: 27–37.

Pickett, S.T.A. and White, P.S. (eds.) 1985. The Ecology of Natural Disturbance and Patch Dynamics. Academic Press, New York.

Ranney, J.W., Bruner, M.C. and Levenson, J.B. 1981. The importance of edge in the structure and dynamics of forest islands. In Forest Island Dynamics in Man-Dominated Landscapes. pp. 67–96. Edited by R.L. Burgess and D.M. Sharpe. Springer-Verlag, New York.

Ripley, B.D. 1981. Spatial Statistics. Wiley, New York.

Risser, P.G., Karr, J.R. and Forman, R.T.T. 1984. Landscape Ecology: Directions and Approaches. Special Publ. 2. Illinois Natural History Survey, Urbana, Illinois.

Robbins, C.S. 1980. Effect of forest fragmentation on bird populations. In Management of North Central and Northeastern Forests for Nongame Birds. pp. 198–212. Edited by R.M. DeGraaf and K.E. Evans. Gen. Tech. Rep. NC-51. U.S. Department of Agriculture, Forest Service, North Central Experiment Station, St. Paul, Minnesota.

Ruediger, W.C. 1985. Implementing a spotted owl management plan: the Gifford Pinchot National Forest experience. In Ecology and Management of the Spotted Owl in the Pacific Northwest. pp. 10–13. Edited by R.J. Gutiérrez and A.B. Carey. Gen. Tech. Rep. PNW-185. U.S. Department of Agriculture, Forest Service, Pacific Northwest Forest and Range Experiment Station, Portland, Oregon.

Ruth, R.H. and Yoder, R.A. 1953. Reducing Wind Damage in the Forests of the Oregon Coast Range. Res. Pap. 7. U.S. Department of Agriculture, Forest Service, Pacific Northwest

18

Forest and Range Experiment Station, Portland, Oregon.

Savill, P.S. 1983. Silviculture in windy climates. For. Abstr. 44: 473–488.

Schoen, J.W., Kirchhof, M.D. and Wallmo, O.C. 1984. Sitka black-tailed deer/old-growth forest relationships in southeast Alaska: implications for management. *In* Fish and Wildlife Relationships in Old-Growth Forests. pp. 315–319. Edited by W.R. Meehan, T.R. Merrell, Jr. and T.A. Hanley. American Institute of Fisheries Research Biologists, Morehead City, North Carolina.

Simberloff, D.S. and Wilson, E.O. 1970. Experimental zoogeography of islands: a two-year record of colonization. Ecology 51: 934–937.

Smith, D.E. 1985. Principles of Silviculture. 8th ed. John Wiley & Sons, New York.

Swanson, F.J. and Dyrness, C.T. 1975. Impact of clear-cutting and road construction on soil erosion by landslides in the western Cascade Range. Geology 3: 393–396.

Thomas, J.W. (ed.) 1979. Wildlife Habitats in Managed Forests:

The Blue Mountains of Oregon and Washington. Handbook 553. U.S. Department of Agriculture, Washington, D.C.

Wilson, E.O. and Willis, E.O. 1975. Applied biogeography. *In* Ecology and Evolution of Communities. pp. 522–534. Edited by M.L. Cody and J.M. Diamond. Belknap Press, Cambridge, Massachusetts.

Woodwell, G.M. and Whittaker, R.H. 1968. Primary production in terrestrial ecosystems. Amer. Zool. 8: 19–30.

Ziemer, R.R. 1981. Roots and the stability of forested slopes. *In* Proceedings, Erosion and Sediment Transport in Pacific Rim Steeplands Symposium, Christchurch, New Zealand, 1981 January, Publ. 132, pp. 343. International Association for the Advancement of Science, Institute of Hydrology, Washington, D.C.

Zobel, D.B., Roth, L.F. and Hawk, G.M. 1985. Ecology, Pathology, and Management of Port-Orford-Cedar (*Chamaecyparis lawsoniana*). Gen. Tech. Rep. PWN-184. U.S. Department of Agriculture, Forest Service, Pacific Northwest Forest and Range Experiment Station, Portland, Oregon.

Vol. 132, No. 5   The American Naturalist   November 1988

SOURCES, SINKS, AND POPULATION REGULATION

H. Ronald Pulliam

Institute of Ecology and Department of Zoology, University of Georgia, Athens, Georgia 30602

Submitted December 9, 1986; Revised August 3, 1987; Accepted February 11, 1988

Many animal and plant species can regularly be found in a variety of habitats within a local geographical region. Even so, ecologists often study population growth and regulation with little or no attention paid to the differences in birth and death rates that occur in different habitats. This paper is concerned with the impact of habitat-specific demographic rates on population growth and regulation. I argue that, for many populations, a large fraction of the individuals may regularly occur in "sink" habitats, where within-habitat reproduction is insufficient to balance local mortality; nevertheless, populations may persist in such habitats, being locally maintained by continued immigration from more-productive "source" areas nearby. If this is commonly the case for natural populations, I maintain that some basic ecological notions concerning niche size, population regulation, and community structure must be reconsidered.

Several authors (Lidicker 1975; Van Horne 1983) have discussed the need to distinguish between source and sink habitats in field studies of population regulation; however, most theoretical treatments (Gadgil 1971; Levin 1976; McMurtie 1978; Vance 1984) of the dynamics of single-species populations in spatially subdivided habitats have not explicitly addressed the maintenance of populations in habitats where reproduction fails to keep pace with local mortality. Holt (1985) considered the dynamics of a food-limited predator that occupied both a source habitat containing prey and a sink habitat with no prey. He demonstrated that passive dispersal from the source can maintain a population in the sink and that the joint sink and source populations can exceed what could be maintained in the source alone. Furthermore, he showed that "time-lagged" dispersal back into the source from the sink can stabilize an otherwise unstable predator-prey interaction. Holt argued, however, that passive dispersal between source and sink habitats in a temporally constant environment is usually selectively disadvantageous, implying that sink populations will be transient in evolutionary time.

In this paper, I consider the consequences of active dispersal (i.e., habitat selection based on differences in habitat quality) on the dynamics of single-species populations in spatially heterogeneous environments. I argue that active dispersal from source habitats can maintain large sink populations and that such dispersal may be evolutionarily stable.

Am. Nat. 1988. Vol. 132, pp. 652–661.

SOURCES, SINKS, AND POPULATION REGULATION 653

BIDE MODELS

One approach to modeling spatially heterogeneous populations is to employ *BIDE* models (Cohen 1969, 1971), which simultaneously consider birth (*B*), immigration (*I*), death (*D*), and emigration (*E*). Normally, in *BIDE* models, the parameters are considered random variables but not spatially heterogeneous. In this paper, I make the opposite assumptions, namely, that rates of birth, death, immigration, and emigration are deterministic but may differ between habitats.

First, consider a spatially distributed population with *m* subpopulations, each occupying a discrete habitat or "compartment." If b_j and d_j are, respectively, the number of births and the number of deaths occurring over the course of a year in compartment *j*, then the total number of births and deaths during that year in all compartments is given, respectively, by

$$B = \sum_{j=1}^{m} b_j \quad \text{and} \quad D = \sum_{j=1}^{m} d_j, \tag{1}$$

since every birth and every death takes place in some compartment.

Now, let i_{jk} be the number of individuals immigrating from compartment *k into* compartment *j*. Each immigrant into *j* must come from one of the other *m* − 1 compartments or come into *j* from outside the *m* compartments that constitute the ensemble of interest. That is, immigration into compartment *j* is given by

$$i_j = \sum_{k=1}^{m} i_{jk} + i_{j0} = \sum_{k=0}^{m} i_{jk},$$

where i_{j0} represents immigration from outside the ensemble into compartment *j* and i_{jj} is zero.

Similarly, if e_{jk} represents the number of emigrants from *j* into *k*, then

$$e_j = \sum_{k=1}^{m} e_{jk} + e_{j0} = \sum_{k=0}^{m} e_{jk}.$$

Note that $e_{kj} = i_{jk}$ for all *j*, $k \neq 0$. Finally, to complete the definitions of the *BIDE* parameters, let

$$I = \sum_{j=1}^{m} i_{j0} \quad \text{and} \quad E = \sum_{k=1}^{m} e_{k0}.$$

The ensemble of all compartments is said to be in dynamic equilibrium in ecological time when the number of individuals (n_j) in each and every compartment does not change from year to year. This occurs only if the number of births plus the number of immigrants exactly equals the number of deaths plus the number of emigrants for every compartment. That is,

$$b_j + i_j - d_j - e_j = (bide)_j = 0, \tag{2}$$

for every *j*, and *BIDE* = 0. Source and sink compartments can now be defined in terms of the *BIDE* parameters. A source compartment (or habitat) is one for which

$$b_j > d_j \quad \text{and} \quad e_j > i_j \tag{3}$$

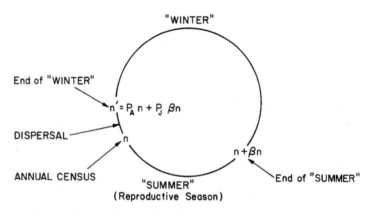

Fig. 1.—An annual census is taken in each habitat or "compartment" in the spring at the initiation of the breeding season (summer). Each individual breeding in the habitat produces β juveniles that are alive at the end of the breeding season. There is no adult mortality during the breeding season; adults survive the nonbreeding (winter) season with probability P_A and juveniles survive with probability P_J.

when $(bide)_j = 0$. A sink compartment (or habitat) is one for which

$$b_j < d_j \quad \text{and} \quad e_j < i_j \tag{4}$$

when $(bide)_j = 0$.

The above definitions apply strictly for equilibrium populations only. A more general definition of a source is a compartment that, over a large period of time (e.g., several generations), shows no net change in population size but, nonetheless, is a net exporter of individuals. Similarly, a sink is a net importer of individuals.

HABITAT-SPECIFIC DEMOGRAPHICS

To see how the *BIDE* parameters relate to habitat-specific survival probabilities and per capita birthrates, consider a simple annual cycle for a population in a seasonal environment (see fig. 1). An annual census is taken in the spring at the initiation of the breeding season. Each individual breeding in habitat 1 produces (on the average) β_1 juveniles that are alive at the end of the breeding season. There is no adult mortality during the breeding season; adults survive the nonbreeding (winter) season with probability P_A and juveniles survive with probability P_J. Thus, the expected number of individuals alive at the end of the winter and just before spring dispersal is given by

$$n_1(t + 1) = P_A n_1(t) + P_J \beta_1 n_1(t) = \lambda_1 n_1.$$

If there were only one compartment (habitat), λ_1 for a small population would be the finite rate of increase for the population. In a multi-compartment model, the λ_j's indicate which compartments are sources and which are sinks. For a simple example, consider two habitats that do not differ in either adult or juvenile survival probabilities but that do differ in per capita reproductive success. If habitat 1 is a source and habitat 2 a sink, then by definition,

$$\lambda_1 = P_A + P_J\beta_1 > 1 \tag{5a}$$

and

$$\lambda_2 = P_A + P_J\beta_2 < 1. \tag{5b}$$

The finite rate of increase for a multi-compartment model depends on the fraction of the population in each habitat. This, in turn, depends on how individuals distribute themselves among available habitats at the time of dispersal, before the onset of breeding (fig. 1).

Before considering how habitat dispersal between source and sink habitats influences population regulation, I briefly discuss population regulation in a source habitat in the absence of a sink habitat. To do this, I must specify the nature of density dependence in the source habitat. For the model discussed below, the critical feature of density dependence is that some individuals in the source habitat do predictably better than others in terms of fitness. Simple assumptions reflecting this feature are that the number of breeding sites is limited and that some individuals obtain breeding sites and others do not. A more general, and more realistic, model of the distribution of habitat quality is discussed briefly in the next section.

I assume that there are only $\hat{n}$ breeding sites available in the source habitat. If the total population size is N and no other breeding sites are available, $N - \hat{n}$ individuals either stay in the source habitat as nonbreeding "floaters" or migrate to nearby sink habitats. In either case, they fail to reproduce but survive with the same probability (P_A) as do breeding individuals. (The qualitative features of the model are unchanged by the assumption that nonbreeding individuals survive with higher or lower probability than breeders.) Since the average reproductive success of an individual securing a breeding site is β_1, the average reproductive success for the entire population is given by

$$\beta(N) = \begin{cases} \beta_1 & \text{if } N \le \hat{n}, \\ (\hat{n}/N)\beta_1 & \text{if } N > \hat{n}. \end{cases} \tag{6}$$

Thus, according to the definition of a source habitat (eq. 5a), the population increases when rare and continues to grow at the rate $\lambda_1 = P_A + P_J\beta_1$ until all breeding sites are occupied. The population will be regulated when

$$\lambda(N) = P_A + (\hat{n}/N)P_J\beta_1 = 1$$

or

$$N^* = \hat{n}P_J\beta_1/(1 - P_A). \tag{7}$$

Again, from the definition of a source, $P_J\beta_1/(1 - P_A)$ is greater than one; thus, the equilibrium population density (N^*) exceeds the number of breeding sites ($\hat{n}$), implying the existence of a nonbreeding surplus.

Assume that, adjacent to the source habitat, is a large sink habitat, where breeding sites are abundant but of poor quality. According to the definition of a sink (eq. 5b), $\beta_2 < (1 - P_A)/P_J$; thus, the sink population declines and eventually disappears altogether in the absence of immigration from the source. Individuals unable to find a breeding site in the source emigrate to the sink because a poor-quality breeding site is better than none at all. If the source is saturated and there are sufficient breeding sites in the sink, the entire nonbreeding surplus from the source emigrates, yielding an increase in the growth rate of the total population:

$$\begin{aligned}\lambda(N) &= (n_1/N)\lambda_1 + (n_2/N)\lambda_2 \\ &= (\hat{n}/N)\lambda_1 + \lambda_2(N - \hat{n})/N \\ &= \lambda_2 + (\hat{n}/N)P_J(\beta_1 - \beta_2).\end{aligned} \qquad (8)$$

The total population equilibrates when $\lambda(N) = 1$; and, according to equation (8),

$$N^* = P_J\hat{n}(\beta_1 - \beta_2)/(1 - P_A - P_J\beta_2). \qquad (9)$$

A relatively simple way to determine the equilibrium populations that will inhabit the source and sink habitats under this model is to note that, since the annual census is taken after the emigration of the reproductive surplus, $\hat{n}$ individuals remain in the source and $\hat{n}(\lambda_1 - 1)$ immigrate. Therefore, in terms of the BIDE model, $i_{21} = \hat{n}(P_A + P_J\beta_1 - 1)$. The local reproduction and survival in the sink is supplemented by this immigration, so that

$$n_2(t + 1) = (P_A + P_J\beta_2)n_2(t) + i_{21} = \lambda_2 n_2(t) + \hat{n}(\lambda_1 - 1).$$

At equilibrium, $n_2^* = i_{21}/(1 - \lambda_2)$, or

$$n_2^* = \hat{n}(\lambda_1 - 1)/(1 - \lambda_2). \qquad (10)$$

Notice that $\lambda_1 - 1$ is the per capita reproductive surplus in the source and $1 - \lambda_2$ is the per capita reproductive deficit in the sink.

If there are many habitats, the total population reaches an equilibrium when the total surplus in all source habitats equals the total deficit in all sink habitats. That is,

$$\sum_{j=1}^{m_1} e_j = \sum_{j=1}^{m_1} n_j^*(\lambda_j - 1) = \sum_{k=1}^{m_2} n_k^*(1 - \lambda_k) = \sum_{k=1}^{m_2} i_k,$$

where there are m_1 source habitats and m_2 sink habitats.

ECOLOGICAL AND EVOLUTIONARY STABILITY

In the preceding analysis, I calculated the equilibrium population sizes in source and sink habitats without addressing the stability of this equilibrium. A local-stability analysis involves finding the slope (b) of $\lambda(N)$ evaluated at the equilibrium population size N^*. If $-bN$ is less than one, the equilibrium is locally stable and approached monotonically (Maynard Smith 1968). The rate of increase for the combined source-sink population is given by equation (8). Differentiating, one obtains

$$d\lambda(N)/dN = -\hat{n}P_J(\beta_1 - \beta_2)/N^2.$$

SOURCES, SINKS, AND POPULATION REGULATION 657

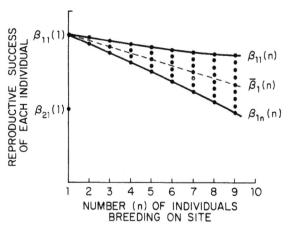

Fig. 2.—The reproductive success of each individual breeding in a particular habitat depends on the number of other individuals in that habitat. $\beta_{1j}(n)$, the expected reproductive success of the individual occupying the jth-best breeding site in habitat 1 when there are a total of n individuals breeding there; $\bar{\beta}_1(n)$, the average reproductive success in the habitat.

Noting the equilibrium population size given in equation (9), the value of $-bN$ is easily seen to be $1 - P_A - P_J\beta_2$ or $1 - \lambda_2$. Since habitat 2 is a sink, λ_2 is less than 1, and, therefore, $-bN$ is also less than one. As for the case of a source habitat with no sink, β_2 equals zero and $-bN$ equals $1 - P_A$. Thus, with or without a sink habitat, the equilibrium population size is locally stable. For the simple cases analyzed above, there is only one nonzero equilibrium, and this equilibrium is approached monotonically from any positive initial population size.

A different question of stability concerns the evolutionary stability of the dispersal rule that determines the proportion of individuals in each habitat. Holt (1985) argued that passive dispersal between a source and a sink is evolutionarily unstable. Two essential differences between the current model and that of Holt are passive versus active dispersal and unequal versus equal fitnesses within a habitat. In my model, individuals choose to leave the source whenever their expected reproductive success is higher in the sink. This never happens in Holt's model because all individuals in the source have equal fitness and the mean fitness in the source never drops to less than one. Since the mean fitness in the sink is always less than one, it always pays for individuals not to immigrate and the evolutionarily stable strategy is no dispersal.

In my model, when the local population in the source exceeds the number of breeding sites available, it pays all surplus individuals to emigrate because they can achieve a higher fitness by doing so. The habitat-selection rule built into the model is a special case of the more general rule "never occupy a poorer breeding site when a better one is still available." Assuming no habitat-specific differences in survival probability, this is the evolutionarily stable habitat-selection rule because no individual can do better by changing habitats (see Pulliam and Caraco 1984; Pulliam 1989).

A more general application of an evolutionarily stable habitat rule that also results in stable occupancy of sink habitats is illustrated in figure 2. In this figure,

$\beta_{11}(n)$ is the expected breeding success of the individual using the best breeding site in habitat 1, and $\beta_{1n}(n)$ is the expected success of the individual using the poorest site occupied when there are n individuals in habitat 1 (the source). Assuming that individuals never occupy a poorer site when a better one is still available, habitat 2 (the sink) will not be occupied as long as $\beta_{1N}(N) > \beta_{21}(1)$. That is to say, the sink habitat will not be occupied as long as all N members of the population can enjoy greater reproductive success in the source. However, if $\beta_{21}(1)$ exceeds $\beta_{1N}(N)$ before the average reproductive success in the source reaches one, the sink will be occupied and the habitat distribution will be evolutionarily stable. Of course, the relative numbers of individuals in the source and sink habitats depend on details of how reproductive success changes with crowding in each habitat. If good breeding sites in the source are rare and poor sites in the sink are relatively common, a large population may occur in the sink.

IMPLICATIONS

Sink habitats may support very large populations despite the obvious fact that the sink population would eventually disappear without continued immigration. Consider the simple situation in which each year i individuals are released into a habitat where local reproduction is incapable of keeping up with local mortality. The equilibrium population maintained in this sink habitat would be $i/(1 - \lambda)$. Thus, if no individuals survived the winter ($\lambda = 0$), only the i recently released individuals would be censused each year. If adults survived winter with probability $\frac{1}{2}$, $2i$ individuals would be censused each year. If, in addition, each adult produced an average of 0.4 juveniles that survived to the following spring, the equilibrium population would be 10 times i, even though the population could not be maintained without an annual subsidy.

In some circumstances, only a small fraction of the population may be breeding in a source habitat. Figure 3 shows the fraction of the equilibrium population in source habitat based on the assumptions of the model developed above and calculated according to equation (10). Clearly, if the reproductive surplus of the source is large and the reproductive deficit of the sink is small, a great majority of the population may occur in the sink habitat. For example, with a per capita source surplus of 1.0 and a sink deficit of 0.1, less than 10% of the population occurs in habitats where reproductive success is sufficient to balance annual mortality.

The concept of niche.—Joseph Grinnell is often credited with introducing the niche concept into ecology. James et al. defined the Grinnellian niche as "the range of values of environmental factors that are necessary and sufficient to allow a species to carry out its life history"; under normal conditions, "the species is expected to occupy a geographic region that is directly congruent with the distribution of its niche" (1984, p. 18). Though James et al. suggested that a species with limited dispersal may not occur in some areas where its niche is found, they clearly implied that the species will not occur where its niche is absent. A sink habitat is by definition an area where factors are not sufficient for a species to carry out its life history, but as discussed above, some species may be more

SOURCES, SINKS, AND POPULATION REGULATION   659

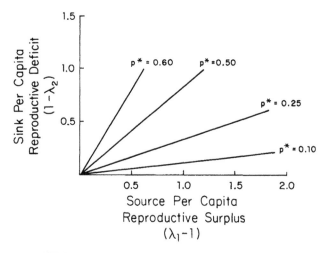

FIG. 3.—The equilibrium proportion ($p*$) of the population in the source habitat depends on both the per capita surplus in the source and the per capita deficit in the sink. A large proportion of the population may occur in the sink habitat if the source surplus is large and the sink deficit is small.

common in sink habitats than in the source habitats on which sink populations depend.

Hutchinson's (1958) particularly influential formulation of the niche concept differentiated between the fundamental niche and the realized niche. Hutchinson argued that the realized niche of most species would be smaller than the fundamental niche as a result of interspecific competition. I have argued in this paper that reproductive surpluses from productive sources may immigrate into and maintain populations in population sinks. If this is commonly true in nature, many species occur where conditions are not sufficient to maintain a population without continued immigration. Thus, in such cases, it can be said that the fundamental niche is smaller than the realized niche.

Species conservation.—Given that a species may commonly occur and successfully breed in sink habitats, an investigator could easily be misled about the habitat requirements of a species. Furthermore, autecological studies of populations in sink habitats may yield little information on the factors regulating population size if population size in the sink is determined largely by the size and proximity of sources.

Population-management decisions based on studies in sink habitats could lead to undesirable results. For example, 90% of a population might occur in one habitat. On the basis of the relative abundance and breeding status of individuals in this habitat, one might conclude that destruction of a nearby alternative habitat would have relatively little impact on the population. However, if the former habitat were a sink and the alternative a source, destruction of a relatively small habitat could lead to local population extinction.

Community structure.—What is a sink habitat for one species may be a source for other species. Thus, a "community" may be a mixture of populations, some of which are self-maintaining and some of which are not. Attempts to understand phenomena such as the local coexistence of species should, therefore, begin with a determination of the extent to which the persistence of populations depends on continued immigration.

Many attempts to understand community structure have focused on resource partitioning and the local diversity of food types. The diversity and relative abundance of the organisms in any particular habitat may depend as much on the regional diversity of habitats as on the diversity of resources locally available. In extreme cases, the local assemblage of species may be an artifact of the type and proximity of neighboring habitats and have little to do with the resources and conditions at a study site. This is not to imply that local studies of the mechanisms of population regulation and species coexistence are unnecessary, but rather that they need to be done in concert with "landscape" studies of the availability of habitat types on a regional basis.

My goal is to draw attention to some of the implications of habitat-specific demographic rates. In many ways, they may be ecologically more important than the age-specific demographic rates that have received so much attention in the ecological and evolutionary literature.

SUMMARY

Animal and plant populations often occupy a variety of local areas and may experience different local birth and death rates in different areas. When this occurs, reproductive surpluses from productive source habitats may maintain populations in sink habitats, where local reproductive success fails to keep pace with local mortality. For animals with active habitat selection, an equilibrium with both source and sink habitats occupied can be both ecologically and evolutionarily stable. If the surplus population of the source is large and the per capita deficit in the sink is small, only a small fraction of the total population will occur in areas where local reproduction is sufficient to compensate for local mortality. In this sense, the realized niche may be larger than the fundamental niche. Consequently, the particular species assemblage occupying any local study site may consist of a mixture of source and sink populations and may be as much or more influenced by the type and proximity of other habitats as by the resources and other conditions at the site.

ACKNOWLEDGMENTS

I wish to acknowledge the assistance of G. Reynolds and J. Nelms in the preparation of the manuscript and the financial support of the National Science Foundation (BSR-8415770).

LITERATURE CITED

Cohen, J. 1969. Natural primate troops and a stochastic population model. Am. Nat. 103:455–477.
———. 1971. Casual groups of monkeys and men: stochastic models of elemental social systems. Oxford University Press, London.

SOURCES, SINKS, AND POPULATION REGULATION   661

Gadgil, M. 1971. Dispersal: population consequences and evolution. Ecology 52:253–261.

Holt, R. D. 1985. Population dynamics in two-patch environments: some anomalous consequences of an optimal habitat distribution. Theor. Popul. Biol. 28:181–208.

Hutchinson, G. E. 1958. Concluding remarks. Cold Spring Harbor Symp. Quant. Biol. 22:415–427.

James, F. C., R. F. Johnston, G. J. Niemi, and W. J. Boecklen. 1984. The Grinnellian niche of the wood thrush. Am. Nat. 124:17–47.

Levin, S. A. 1976. Population dynamic models in heterogeneous environments. Annu. Rev. Ecol. Syst. 7:287–310.

Lidicker, W. Z., Jr. 1975. The role of dispersal in the demography of small mammals. Pages 103–128 *in* F. B. Golley, K. Petrusewicz, and L. Ryszkowski, eds. Small mammals: their productivity and population dynamics. Cambridge University Press, New York.

Maynard Smith, J. 1968. Mathematical ideas in biology. Cambridge University Press, Cambridge.

McMurtie, R. 1978. Persistence and stability of single-species and predator-prey systems in spatially heterogeneous environments. Math. Biosci. 39:11–51.

Pulliam, H. R. 1989. Individual behavior and the procurement of essential resources. Pages 25–38 *in* J. Roughgarden, R. M. May, and S. Levin, eds. Perspectives in ecological theory. Princeton University Press, Princeton, N.J.

Pulliam, H. R., and T. Caraco. 1984. Living in groups: is there an optimal group size? Pages 122–147 *in* J. R. Krebs and N. B. Davies, eds. Behavioural ecology: an evolutionary approach, 2d ed. Sinauer, Sunderland, Mass.

Vance, R. R. 1984. The effect of dispersal on population stability in one-species, discrete-space population growth models. Am. Nat. 123:230–254.

van Horne, B. 1983. Density as a misleading indicator of habitat quality. J. Wildl. Manage. 47:893–901.

Modeling Coastal Landscape Dynamics

Process-based dynamic spatial ecosystem simulation can examine long-term natural changes and human impacts

Robert Costanza, Fred H. Sklar, and Mary L. White

Predicting the way ecological systems respond to human modifications has been a primary goal of ecosystem ecology (Hall and DeAngelis 1985). Ecosystems represent an economic resource whose value to society is only now becoming recognized. Coastal ecosystems in particular provide valuable marketed and nonmarketed services, including fish and wildlife resources, storm protection, and recreation. The average value to society of coastal wetlands has been estimated as $2000–$10,000/acre, even though their market price is only $200–$400/acre (Costanza et al. in press, Farber and Costanza 1987, Turner et al. 1988).

Coastal ecosystems are being threatened by a host of human activities, including oil and gas exploration, urban development, and sediment diversion. The potential for sea level rise due to global greenhouse-effect warming is also of concern. Protecting and preserving these ecosystems requires the ability to predict the direct and indirect, temporal, and

Robert Costanza is an associate professor in the Coastal and Environmental Policy Program, Chesapeake Biological Laboratory, Center for Environmental and Estuarine Studies, University of Maryland, Solomons, MD 20688-0038. Fred H. Sklar is an assistant professor at the Belle Baruch Institute, University of South Carolina, Columbia, SC 29211. Mary L. White is a research associate at the Coastal Ecology Institute, Center for Wetland Resources, Louisiana State University, Baton Rouge, LA 70803. © 1990 American Institute of Biological Sciences.

> **Management options can be effective at mitigating impacts of proposed alterations**

spatial effects of proposed human activities, the ability to separate these effects from natural changes, and the ability to appropriately modify the short-term incentive structures that guide local decision-making to better reflect these impacts (Costanza 1987a). Adequately predicting ecosystem impacts requires sophisticated computer simulation models that represent a synthesis of the best available understanding of the way these complex coastal ecosystems function.

The more general objectives of landscape modeling are to predict changes in land cover patterns across large geographic regions (tens to hundreds of kilometers) over long time scales (tens to hundreds of years) as a result of various site-specific management alternatives and natural changes. Development of this capability is needed for regional ecosystem management and also for modeling regional and global ecosystem response to regional and global climate change, sea level rise resulting from atmospheric CO_2 enrichment, acid precipitation, toxic waste dumping, and a host of other potential impacts.

Two recent developments make this type of modeling feasible. First, the ready accessibility of extensive

spatial and temporal databases from such sources as remote sensing and historical aerial photography make it possible to measure the behavior of real landscapes over large spatial and long temporal scales. Second, advances in computer power and convenience make it possible to build and run predictive landscape models at the necessary levels of spatial and temporal resolution.

Coastal marshes

The Atchafalaya delta and adjacent Terrebonne Parish marshes represent one of the most rapidly changing landscapes in the world. Figure 1 shows the study area. Figure 2 shows the historical sequence of Mississippi River main distributaries that have deposited sediments to form the current Mississippi deltaic plain marshes. This delta switching cycle (on average lasting 1500 years) sets the historical context of this landscape. At present, the river is in the process of changing from the current channel to the much shorter Atchafalaya River.

The US Army Corps of Engineers maintains a control structure at Old River (see Figure 1) to control the percentage of Mississippi River flow going down the Atchafalaya. For the last 40 years, this percentage has been set at approximately 30%. Sediment borne by the Atchafalaya River first filled in open water areas in the upper Atchafalaya basin, and more recently it has begun to build a delta in Atchafalaya Bay (Roberts et al. 1980, Van Heerden and Roberts 1980a,b). Atchafalaya-borne sediments are dispersed to Fourleague Bay (Figure 3) and are contributing to marsh building in the eastern part of the study area as well (Baumann and Adams 1981, Baumann et al. 1984). During the next few decades, new delta is projected to form at the mouth of the river, and plant community succession will occur on the recently formed delta and in the existing marshes.

In contrast, the leveeing of the Mississippi and Atchafalaya rivers, along with the damming of distributaries, has virtually eliminated riverine sediment input to most Louisiana coastal marshes. This change has broken the deltaic cycle and greatly accelerated land loss. The overall Louisiana coastal zone has been projected to lose a net of approximately 100 km^2/yr due to sediment starvation and salt water intrusion (Gagliano et al. 1981). Only in the area of the Atchafalaya delta is sediment-laden water flowing into wetland areas and land gain occurring (Roberts et al. 1980, Van Heerden and Roberts 1980a,b).

Primary human activities that potentially contribute to wetland loss are flood control, canals, spoil banks, land reclamation, fluids withdrawal, and highway construction. There is evidence that canals and levees are an important factor in wetland loss in coastal Louisiana, but there is much disagreement about the magnitude of the indirect loss caused by them (Cleveland et al. 1981, Craig et al. 1979, Deegan et al. 1984, Leibowitz 1989, Scaife et al. 1983). Natural channels are generally not deep enough for the needs of oil recovery, navigation, pipelines, and drainage, so a vast network of canals has been built. In the Deltaic Plain of Louisiana, canals and their associated spoil banks of dredged material currently comprise 8% of the total marsh area compared to 2% in 1955. The construction of canals leads to direct loss of marsh by dredging and spoil deposition and indirect loss by changing hydrology, sedimentation, and productivity. Canals are thought to lead to more rapid salinity intrusion, causing the death of freshwater vegetation. Canal spoil banks also limit water exchange with wetlands, thereby decreasing deposition of suspended sediments.

Proposed human activities can have a dramatic impact on the distribution

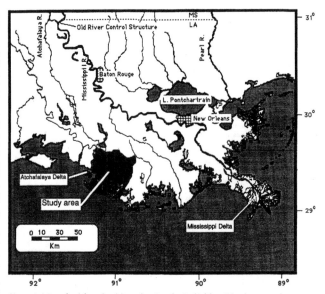

Figure 1. Map of southern Louisiana showing the Atchafalaya/Terrebonne study area.

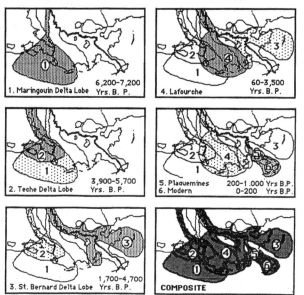

Figure 2. Historical sequence of major Mississippi River distributary changes. (After Baumann and Adams 1981.)

of water and sediments from the Atchafalaya River, and consequently on the development of the Atchafalaya landscape. For example, the Corps of Engineers is considering extending a levee along the east bank of the Atchafalaya that would restrict water and sediment flow into the Terrebonne marshes (Figure 3). This situation represents a unique opportunity to study landscape dynamics. The Atchafalaya landscape is changing rapidly enough to provide time-series observations that can be used to test basic hypotheses about how coastal landscapes develop. In addition to short-term observations, there is a uniquely long and detailed history of field and remotely sensed data available on the study area (Bahr et al. 1983, Costanza et al. 1983).

Solutions to the land loss problem in Louisiana all have far-reaching implications. Outside forces (such as rates of sea level rise) also influence the effectiveness of any proposed solution. In the past, suggested solutions have been evaluated independently of each other and in an ad hoc manner. To more objectively evaluate the many interdependent implications of the various natural changes, management strategies, and specific projects that have been suggested to remedy the coastal erosion problem, an integrated spatial simulation modeling approach was developed (Costanza et al. 1988, Sklar et al. 1985, in press). Using this approach, we first demonstrated the ability to simulate the past behavior of the system, and then we projected future conditions as a function of various management alternatives and natural changes, both individually and in various combinations. Our approach simulates both the dynamic and spatial behavior of the system, and it keeps track of several of the important landscape level variables in the system, such as ecosystem type, water level and flow, sediment levels and sedimentation, subsidence, salinity, primary production, nutrient levels, and elevation.

Process-based dynamic spatial ecosystem simulation models

Ecosystem models can be differentiated from population models in that the former include biotic and abiotic components whereas the latter in-

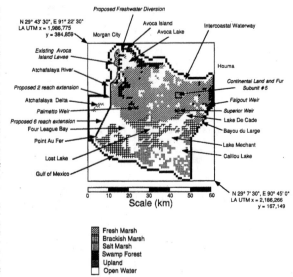

Figure 3. Atchafalaya/Terrebonne study area showing major geographic features, ecosystem types in 1983, and the location of the management options analyzed.

clude only populations of organisms. Ecosystem models tend to be more complex and realistic, and population models tend to be more simple and general. The majority of ecosystem models in the literature are designed to predict dynamic behavior while treating the system as spatially homogeneous (Costanza and Sklar 1985). Many existing ecosystem models are process based, in the sense of attempting to mimic (at least in an aggregated way) the underlying physical and ecological processes in the system, as opposed to statistical or probabilistic models, which are based directly on observed correlations in the data, generally without specifying mechanisms.

One way to extend the process-based approach to model spatial dynamics is to arrange a spatial array of point ecosystem models and connect them with fluxes of, for example, water and nutrients, employing rules to govern successional, evolutionary, or other changes in the structure of the system. This approach is some-

what analogous to that employed in general atmospheric circulation models used in long-term climate modeling (Potter et al. 1979, Schlesinger and Zhao 1989, Washington and Williamson 1977, Williams et al. 1974), but it also incorporates elements of cellular automata and expert systems modeling incorporated in the successional rules.

Richard Levins (1966) first described the fundamental trade-offs in modeling between realism, precision, and generality. No single model can maximize all three goals, and the choice of which objectives to pursue depends on the fundamental purposes of the modeling study. Our approach favors higher realism and precision (the ability to accurately and realistically depict specific processes and ecosystems) at the expense of generality (direct and easy applicability to a broad range of systems). It is a relatively expensive and time-consuming approach to modeling that requires much site-specific data and a significant amount of effort to calibrate the model to local conditions. But the payoffs are significant in light of the model's ability to simulate realistically a specific system. This approach was essential to achieve the analytical and management goals of our study.

The spatial process-based approach had been attempted in only a few previous cases for ecosystem modeling (Botkin et al. 1972, Costanza and Sklar 1985, Phipps 1979). In general, past applications were relatively successful, and their rarity is probably due to the size and complexity of the resultant models and the difficulty of assembling the necessary databases for calibrating and verifying the models. These limitations are decreasing with the increasing availability of remote sensing data and supercomputers, and we expect the relative expense of the spatial process–based approach to continue to decrease in the future.

The CELSS model

We developed a process-based spatial simulation model for the Atchafalaya/Terrebonne marsh/estuarine complex in south Louisiana (Figure 3) called the coastal ecological landscape spa-

tial simulation (CELSS) model (Costanza et al. 1988, Sklar et al. 1989). The model consists of 2479 interconnected square cells, each representing 1 square kilometer. Each cell contains a dynamic ecosystem simulation model (Figure 4), and each cell is connected to its four nearest neighbors by the exchange of water and suspended materials (salts, nitrogen, and suspended organic and inorganic sediments). The buildup of land or the development of open water in a cell depends on the balance between net inputs of sediments and local organic peat deposition on the one hand and outputs due to erosion and subsidence on the other hand. The balance of sediment inputs and outputs is critical for predicting how marsh succession and productivity is affected by natural and human activities.

Inputs are specified in the form of time series during the simulation period. Weekly values of Atchafalaya and Mississippi river discharges, Gulf of Mexico salinity, river sediments and nutrients, rainfall, sea level, runoff, temperature, and winds are supplied to the simulation with each iteration. The location, time of con-

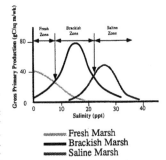

Figure 5. General relationship between gross primary production and salinity for three habitat types in the CELSS model. Data used to calibrate these relationships were taken from Conner and Day (1987).

struction, and characteristics of the major waterways, canals, and levees are also supplied as inputs to the simulation.

The change in water level in each cell is determined in the model by water exchanges in both directions across all four boundaries plus surplus rainfall (precipitation minus evapotranspiration). The hydrologic component of the model is, in essence, a two-dimensional, finite-difference, mass-balance model. Although this mass-balance approach does not accurately simulate short-term hydrodynamics (because it leaves out momentum transfers, vertical dynamics, and Coriolis forces), it does approximate the major longer-term effects in our flat, shallow, and well-mixed study area.

It was necessary to give up some hydrologic precision because we needed to do continuous simulations for 58 years, which would have been prohibitively expensive using a complete hydrodynamic model, even on the fastest supercomputer. In addition, our primary goal in this model was to simulate plant community succession and productivity, conditions that respond more strongly to longer-term water flow patterns than to short-term flooding. Our results for water flow were similar to those produced for a few selected points in time by a more elaborate two-dimensional hydrodynamic model of the same area.

In our model, water can exchange

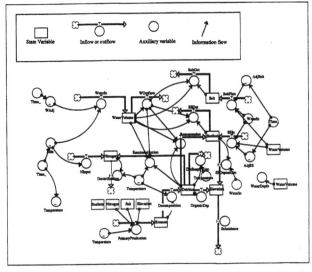

Figure 4. Diagram of a typical unit cell model in the CELSS landscape model in the STELLA modeling language (Costanza 1987b). All ecosystem types used a unit model with the same general structure, but with unique parameter settings.

94

with adjacent cells via canals, natural bayous, and overland flow, or it may be prevented from exchanging with adjacent cells by the presence of levees. An overall water flow connectivity parameter (K) is adjusted during the model run to reflect the presence and size of waterways or levees at the cell boundaries. Connectivity is a function of habitat type, drainage density, waterway orientation, and levee height. If a waterway is present at a cell boundary, a large K value is used, increasing with the size of the waterway. If a levee is present, a K value of 0 is used until water level exceeds the height of the levee. The model's canal and levee network is updated each year during a simulation run (i.e., dredged canals and levees are added to the model's hydrologic structure at the beginning of the year they were built).

Water crossing from one cell to another carries both organic and inorganic sediments. This sediment is partitioned between being deposited, resuspended, lost due to subsidence, and carried to the next cell. The relative rates of each of these sediment exchanges in each location is a function of ecosystem type. Plants and nutrients within each cell also influence these exchanges and flows. Changes in other abiotic material concentrations (i.e., salts and nitrogen) are also a function of water flow between cells and concentration of materials in the cells, along with internal deposition and resuspension (Figure 4).

The primary production in a cell is related to flooding regimes, turbidity, temperature, and elevation, according to functions like the one illustrated in Figure 5, which shows the model's relationship between weekly gross primary productivity and salinity for three of the habitat types. Data to calibrate the productivity functions was taken from Conner and Day (1987). In this type of function, maximal output occurs at optimal levels of inputs, and output is reduced with increased deviation from the optimum. The response of primary production to different nitrogen concentrations was stimulated with a Michaelis-Menten type rate equation, in which production continues to increase with increasing nitrogen, but at a decreasing rate. Data to calibrate

this function were taken from Hopkinson (1978).

Ecosystem succession occurs in the model (after a time lag) when the physical conditions in a cell become appropriate to a different ecosystem type. The program monitors the state variables in each cell and checks the physical environment (e.g., salinity, elevation, and water level). If the values of the state variables change to the extent that the environment in the cells is outside the range for its currently designated ecosystem type, then the cell's ecosystem type and all the associated parameter settings are switched to a new, better-adapted set. For example, if salinity in a cell that is fresh marsh rises beyond a certain threshold value and remains at this high level long enough, then the model converts the cell to brackish marsh and changes all the associated parameters.

The square cells have exchanges across their four sides. We did not use a hexagonal or triangular grid or a square one that allowed exchanges across the diagonals mainly because

we wanted the simplest arrangement that would work reasonably well for our purposes and was easy to program. We used a simple Euler numerical integration technique with a variable time step. The maximal time step is one week. The model checks the rates of change of each variable, and if any of these is above a predetermined maximum it reduces the time step in increments until the rates of change are below the threshold, down to a minimum of one day.

The model was written in standard FORTRAN (3021 lines of code), and was run on a variety of computers, including a VAX 11/780, IBM 3034, and CRAY X/MP. A typical 22-year run at a weekly time step (1144 total time steps) for all 2479 cells with eight state variables each (19,832 total simultaneous difference equations) takes approximately 24 hours of CPU time on the VAX, 2 hours on the IBM, and 15 minutes on the CRAY. Most of the full-scale runs of the model were done on the CRAY, with the other computers used mainly in the early model development stages.

Table 1. Analysis of variance table for selected individual parameters and interactions of parameters on total fit of the model with 1978 conditions.

Parameter	Mean square	F value	Pr > F
Individual parameters			
Water flow (fresh marsh)	2.30404601	5.31	0.0266
Sedimentation	0.12118223	0.28	0.6002
Primary production	8.31807987	19.16	0.0001
Organic matter	6.31000576	14.54	0.0005
Water flow parameters	0.00848915	0.02	0.8895
Turbulent resuspension	1.59570305	3.68	0.0625
Water flow (open water)	0.04134081	0.10	0.7592
Subsidence	3.59999056	8.29	0.0064
Two-way interactions			
Water flow (fresh marsh) and sedimentation	1.05885098	2.44	0.1264
Water flow (fresh marsh) and primary production	0.71286012	1.64	0.2076
Water flow (fresh marsh) and organic matter	0.19934273	0.46	0.5020
Water flow (fresh marsh) and turbulent resuspension	0.24456926	0.56	0.4574
Water flow (fresh marsh) and subsidence	0.51497506	1.19	0.2827
Sedimentation and primary production	2.00028590	4.61	0.0381
Sedimentation and organic matter	1.66397862	3.83	0.0574
Sedimentation and turbulent resuspension	2.91185842	6.71	0.0134
Sedimentation and subsidence	0.34385175	0.79	0.3789
Primary production and organic matter	4.65211481	10.72	0.0022
Primary production and turbulent resuspension	5.94659088	13.70	0.0007
Primary production and subsidence	2.23346004	5.15	0.0289
Organic matter and turbulent resuspension	0.54097224	1.25	0.2711
Turbulent resuspension and subsidence	0.09995870	0.23	0.6340
Water flow (open water) and subsidence	0.00524133	0.01	0.9131
Three-way interactions			
Sedimentation and primary production and turbulent resuspension	5.05771666	11.65	0.0015
Error (degrees of freedom = 39)	0.43402680		

Apple Macintoshs were used extensively for unit model design and debugging using STELLA (Costanza 1987b), for telecommunicating with the CRAY, and for mapping, analysis, and animation of results.

Input data. Primary input data for the model were: detailed, digitized ecosystem type maps prepared by the US Fish and Wildlife Service for 1956, 1978, and 1983; the history of canal and levee construction in the area, obtained from analysis of state permit records for the 1956–1978 period; a weekly record of climate variables, including rainfall, temperature, wind, river flow, and sediment and nutrient concentrations; and water level and salinity in the Gulf of Mexico. In addition, field measurements of productivity, biomass, and nutrient uptake were available for setting parameters for the productivity component of the model.

The model was initialized using the 1956 ecosystem type map. Because no direct measurements for the initial values for the model's other state variables in each cell were available, they were set at the middle of the range for the corresponding ecosystem type.

Calibration based on ecosystem type maps. Our first objective was to replicate the behavior of the system during the period from 1956 to 1978 and quantify the degree of fit between various alternative models and the data. The CELSS model was calibrated by starting it with the 1956 ecosystem type conditions and associated values for the other variables and simulating the evolution of the area during the intervening 22 years to 1978 with a maximum of weekly time steps. A measured ecosystem type map was available for 1978 for comparison with the model's predictions. The simulated and real maps were then compared for degree of fit. The model's parameters were iteratively adjusted within predetermined ranges of uncertainty to maximize the fit. The model was then continued to 1983 (for which a third map was available), and the fit was again calculated to estimate the model's predictive accuracy.

The CELSS model contains more than 130 parameters. In calibrating the model, we first choose initial values and ranges of uncertainty for these parameters, based on available data and literature values. We then ran calibration runs of the model from 1956 to 1978 and compared the goodness of fit of the model's ecosystem type predictions with the 1978 data using visual inspection and the procedures described below. We also looked at the other variables in the model to make sure they were approximately replicating what we knew to have happened in the system.

But we concentrated on the ecosystem type predictions because ecosystem type is the result (in the model as well as in the real world) of the integration of the behavior of the other variables. Ecosystem type is also the most easily and accurately observed variable in the real system at the landscape scale.

We performed a preliminary parameter optimization by manually adjusting parameters within the predetermined ranges of uncertainty to improve the ecosystem type fit. When

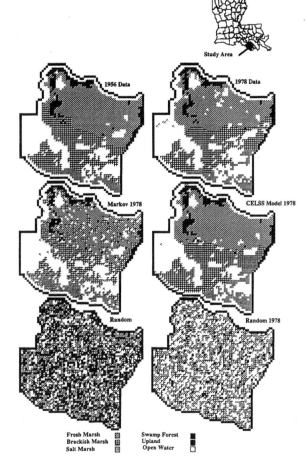

Figure 6. CELSS model and various alternative null model predictions for 1978 ecosystem type using the 1956 conditions as the starting point.

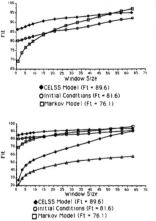

Figure 7. Example plots of fit versus window size for the various models in Figure 6 compared with the 1978 data. Upper graph shows the CELSS model, the Marov model, and the initial (1956) conditions on an expanded y-axis.

this procedure began to yield negligible improvements, we began a more formal parameter optimization procedure.

Because the model is nonlinear and has many discontinuities (due in part to the ecosystem succession algorithm), it is not amenable to traditional linear parameter optimization procedures, such as linear regression. We therefore used a procedure that is applicable to any model structure. This procedure treats the model as if it were an experiment (with the parameters as the controllable factors in the experiment) and uses traditional analysis-of-variance procedures to analyze the response of the model's fit to various combinations of changes in the parameters (factors). This method required a large number of simulation runs (one for each combination of parameter changes chosen) and was only possible for a model of this size because the speed of the supercomputer made generating the requisite number of runs feasible. A by-product of this approach is a fairly elaborate sensitivity analysis of the model's response to changes in the parameters, which gave us significant

insight into the model's behavior.

Table 1 shows the results of an analysis of the variance in fit due to the manipulated parameters and combinations of parameters. These results indicate that, although the physical and hydrologic parameters (such as water flow and subsidence) are important, the biological parameters (such as primary productivity), alone and in combination with the physical parameters, contribute significantly to controlling the fit of the model. It also appeared that our decision to use a process-based, whole-ecosystem approach was justified, because no single parameter explained most of the variation in fit and the higher order interactions between physical and biological parameters were significant.

We limited the parameters manipulated in this procedure to those of which we were most unsure, and, based on preliminary sensitivity analysis, those with the largest impact on the fit. Based on an analysis of the parameters, we could derive a set of changes in the parameter settings that would maximally improve the fit. We then implemented these changes and repeated the factorial experimental procedure. We iterated this process until significant improvements in fit were no longer being made. The base case CELSS model results reflect these optimized parameter settings.

Degree of fit of predicted ecosystem type with data. For relatively simple purposes, there are well-established statistical measures of goodness of fit. For example, a standard chi-square test can determine the fit between two categorical variables, but it ignores the spatial or other patterns of the variables. It could be used to determine if the total number of cells in each land use category were significantly different between two maps, but not whether there was a significant difference in the pattern of two maps that both had the same total number of cells in each land use category.

Landscape analysis requires quantifying the degree of matching of complex spatial and temporal patterns. Spatial pattern matching is not a straightforward statistical procedure. Even quantifying the degree to which nonspatial ecosystem modeling time-series results match real patterns of

ecosystem behavior is difficult and there is no agreed-on procedure (Gardner et al. 1982, Jørgensen 1982).

We used a multiple resolution approach to test the goodness-of-fit between the model and data. This approach allows a more complete analysis of the way the spatial patterns match (Costanza 1989, Turner et al. 1989). The algorithm gradually degrades the resolution with which the fit is measured by gradually increasing the size of a sampling window in which the fit is calculated. A plot of it versus window size (resolution) yields information on the way the patterns match. The total fit is estimated as a weighted average of the fit at all the window sizes, with the smaller window sizes given the most weight.

Figure 6 shows various model predictions for 1978 (including several null models) based on simulations using the 1956 conditions as the starting point. Figure 7 shows the variable resolution procedure applied to the spatial modeling results summarized in Figure 6. Fit versus window size is plotted for the 1978 data compared with the 1978 CELSS simulation model prediction (labeled "CELSS model") and with several alternative and null models.

The most simple-minded null model is a completely random distribution of habitats with all the categories being equally probable (labeled "random" in Figure 6). A slightly more reasonable null model is a random distribution but with the same

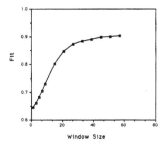

Figure 8. Comparison of CELSS model predictions for salinity with the WES two-dimensional hydrodynamic model (Donnell and Letter 1989, Letter 1982) for low-flow conditions over a range of window sizes.

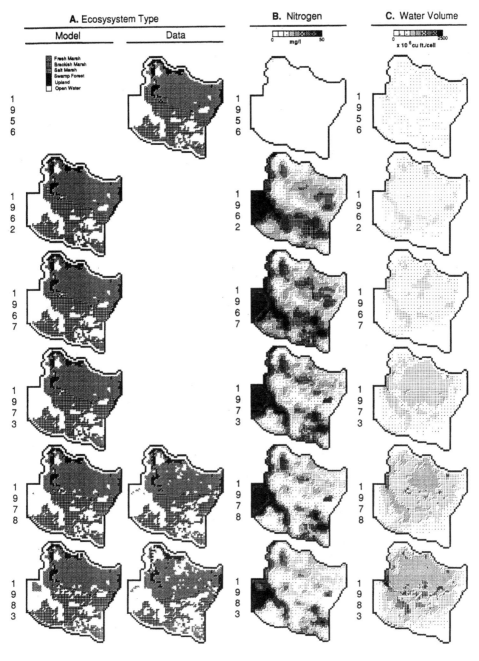

A. Ecosystem Type

Model   Data

Fresh Marsh
Brackish Marsh
Salt Marsh
Swamp Forest
Upland
Open Water

B. Nitrogen

0   mg/l   50

C. Water Volume

0   2500
x 10⁶ cu ft./cell

1956

1962

1967

1973

1978

1983

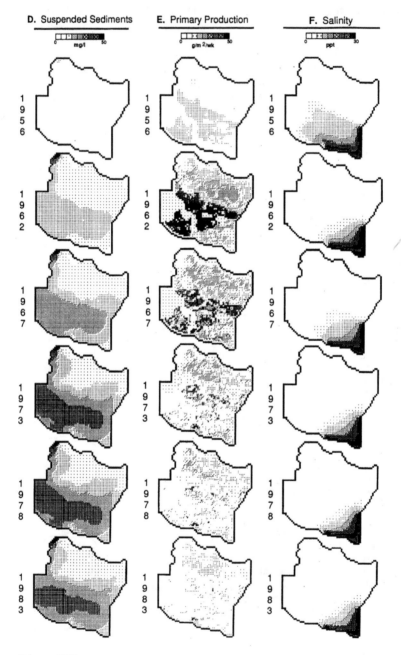

D. Suspended Sediments   **E.** Primary Production   **F.** Salinity

Figure 9. Sample model prediction output from the CELSS model, starting with the 1956 initial conditions. a. Ecosystem type distribution along with real ecosystem type data for 1978 and 1983. b. Nitrogen. c. Water volume. d. Suspended sediments. e. Primary production. f. Salinity for selected years.

overall land cover probabilities (frequencies) as the 1978 data (labeled "random 1978"). A simple Markov chain model is a slightly more sophisticated null model that incorporates the statistical trend in habitat changes in the area in the form of transition probabilities (or frequencies) of each habitat type into each other type during the 1956–1978 interval. The fit for various window sizes for the Markov model is labeled "Markov model." Finally, the 1956 initial conditions represent a null model that predicts no change (labeled "initial conditions").

The CELSS simulation model performs better than any of the alternative or null models, but its performance can best be judged in relation to the null models. The simulation model fits the 1978 data better than "random" (F_t = 89.6% versus 32.5%) and "random 1978" (F_t = 89.6% versus 50%). The fit for "random 1978" approaches 100% as window size increases, indicating that the overall percentages of habitat types are the same, but the pattern is not well matched.

If only the fit at window size 1 is considered, the 1956 initial conditions null model does not appear to be much worse than the CELSS model (86% versus 81%). The initial conditions null model appears to fit fairly well because only approximately 20% of the cells actually changed type between 1956 and 1978. But the pattern of fit is revealed by looking at the plots of fit versus window size in Figure 7. The CELSS model plot increases rapidly as window size increases indicating that the patterns of the model and data are similar. The initial conditions null model exhibits a flatter plot, indicating the patterns are not as similar. Total weighted fit (F_t) reflects this added pattern matching (89.6% versus 81.6%).

The Markov null model has a lower fit (69%) at window size 1 than the initial conditions null model (81%), but the fit increases more rapidly as window size increases. At the maximal window size, it fits better than the initial conditions model because the total number of cells of each type is closer to the 1978 data. Total weighted fit for the Markov model is less than the initial conditions model (81.6% versus 76.1%) because, by

randomly placing the transition cells, the detailed spatial pattern is disrupted even while the fit is increased at larger window sizes.

The CELSS model is the only alternative that also predicts changes in the underlying physical and ecological processes and that links these changes in a causal way to ecosystem type changes. Thus, even if it were only as good as the alternative models at predicting ecosystem type changes, the CELSS model would be preferred for management applications because it can address the underlying reasons for habitat changes and the relative impacts of human changes on these processes.

This is not to say that the CELSS model approach to spatial ecosystem modeling is in all aspects superior to other possible approaches. Like anything else, it has advantages and disadvantages. Its advantages are the ability to deal with a large, complex ecological system in a relatively realistic way. If the goal of the modeling effort is to make specific predictions about specific management alternatives in a specific region, then these advantages are crucial.

In achieving realism, however, this approach must sacrifice some generality. The model's parameters must be calibrated to historical conditions for each application of the model. Thus, while the CELSS modeling approach can be applied to any coastal area, its specific results in one area are not directly generalizable to other areas.

We compared the CELSS model simulations for some of the physical

variables with a more elaborate hydrodynamic model of the area constructed by the Corps of Engineers' Waterways Experiment Station (WES; Donnell and Letter in press, Letter 1982). The WES model deals with fewer variables (water, salt, and sediments) over much shorter time scales (maximum of one week continuous runs with time steps of minutes or seconds). It gives a more precise picture of short-term water, salt, and sediment levels and fluxes, whereas the CELSS model is more concerned with long-term ecosystem succession as driven by such factors as long-term water, salt, and sediment fluxes.

In general, the two models agree fairly well for their areas of overlap. Figure 8 shows the fit between the CELSS model and the WES model for salinity over several window sizes. The weighted average fit for this comparison was approximately 80%.

No one model can provide a complete picture of a complex phenomenon such as Louisiana's coastal marshes. Like the blind men all feeling different parts of an elephant and coming to different conclusions about what it is, we must take the results of a suite of modeling and experimental studies and integrate them to provide a reasonable overall picture.

Base case output. The CELSS model can produce a huge amount of output. Some of the most useful pieces of output are contour maps for each of the eight state variables and for internal flows (i.e., sedimentation) and ex-

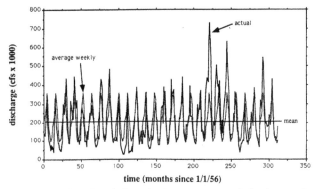

Figure 10. Atchafalaya River discharge data for 1956–1983 period. Also shown are the weekly average data and the mean for the period.

changes between cells (i.e., water flow) for each week of the simulation. The results of the model are best comprehended by viewing animations of the model's time series mapped output for each of the variables. Here, we can only present a few example snapshots of the model's output and discuss some of the findings.

Figure 9 shows some sample output from the base-case run of the CELSS model, covering the period 1956–1983. Starting with the 1956 initial conditions, the model's predictions of average annual ecosystem distribution (the ecosystem that was present in the cell for the majority of the year), salinity, suspended sediments, water volume, nitrogen, and primary production for several years are shown, along with the real ecosystem type data for 1978 and 1983 for comparison with the model's output.

The CELSS model accurately reproduced the gradual intrusion of salt into the system from the southeastern part of the study area with the concurrent freshening in the southwestern sector due to increased Atchafalaya River input (Figure 9f). It also shows a loss of elevation in the north and an increase in elevation in the south. Both of these trends are indicative of river water and sediments moving further south in recent times, as well as a lack of connectivity with the more northern fresh marsh areas due to levee construction. Predicted water volume (Figure 9c) and suspended sediments (Figure 9d) behaved in a similar way and are generally consistent with what is known about the historical behavior of these variables in the area.

Scenario analysis

Once the model was calibrated, we analyzed a series of future and historical scenarios. The scenarios can be divided into four categories: *climate scenarios,* which address the impacts of climate variations on the model's behavior; *management scenarios,* which address the impacts of various internal site-specific manipulations to the system; *historical scenarios,* which address what the system would have looked like if the environment had not been altered by human intervention or if climatic conditions had been different, and *boundary scenar-*

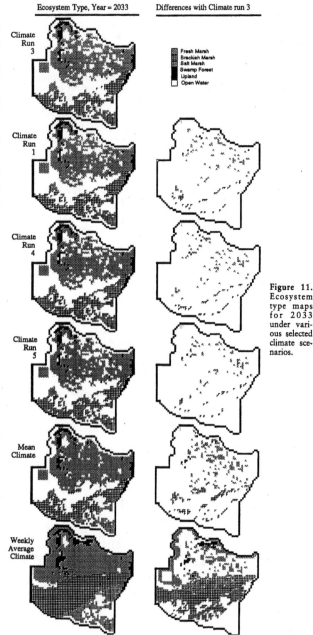

Figure 11. Ecosystem type maps for 2033 under various selected climate scenarios.

Table 2. Number of square kilometers of each ecosystem type for the three years for which data is available and for 2033 for various scenarios. Changes from the base case (in sqare kilometer) are indicated in parentheses.

	Swamp		Fresh marsh		Brackish marsh		Saline marsh		Upland		Total land		Open water	
1956	130		864		632		98		13		1737		742	
1978	113		766		554		150		18		1601		878	
1983	116		845		347		155		18		1481		998	
2033 Scenarios*														
Climate scenarios[†]														
Climate run 3 (climate base case)	84		871		338		120		10		1423		1056	
Climate run 1	79	(−5)	874	(+3)	337	(−1)	127	(+7)	10	(0)	1427	(+4)	1052	(−4)
Climate run 4	85	(+1)	900	(+29)	355	(+17)	130	(+10)	10	(0)	1480	(+57)	999	(−57)
Climate run 5	83	(−1)	891	(+20)	332	(+6)	126	(+6)	10	(0)	1442	(+19)	1037	(−19)
Mean climate	94	(+10)	974	(+103)	402	(+64)	136	(+16)	11	(+1)	1617	(+194)	862	(−194)
Weekly average climate	128	(+44)	961	(+90)	813	(+475)	300	(+180)	11	(+1)	2213	(+790)	266	(−790)
Management scenarios														
No levee extension (base case)	100		796		410		123		15		1444		1035	
Two-reach levee extension	98	(−2)	804	(+8)	399	(−11)	123	(0)	15	(0)	1439	(−5)	1040	(+5)
CLF marsh management	102	(+2)	798	(+2)	409	(−1)	123	(0)	15	(0)	1447	(+3)	1032	(−3)
Falgout Weir	104	(+4)	799	(+3)	403	(−7)	122	(−1)	16	(+1)	1444	(0)	1035	(0)
Full six-reach levee extension	103	(+3)	790	(−6)	362	(−48)	122	(−1)	15	(0)	1392	(−52)	1087	(+52)
Fresh-water diversion (FWD)	103	(+3)	803	(+7)	404	(−6)	123	(0)	15	(0)	1448	(+4)	1031	(−4)
FWD and Palmetto Weir	102	(+2)	802	(+6)	407	(−3)	123	(0)	15	(0)	1449	(+5)	1030	(−5)
FWD and Superior Weir	104	(+4)	799	(+3)	404	(−6)	123	(0)	15	(0)	1445	(+1)	1034	(−1)
FWD and Superior and Palmetto weirs	104	(+4)	792	(−4)	407	(−3)	123	(0)	15	(0)	1441	(−3)	1038	(+3)
FWD and Superior and Falgout weirs	104	(+4)	803	(+7)	407	(−3)	122	(−1)	15	(0)	1451	(+7)	1028	(−7)
Boundary scenarios[‡]														
EPA low sea level rise	104	(+4)	800	(+4)	411	(+1)	124	(+1)	15	(0)	1454	(+10)	1025	(−10)
EPA high sea level rise	89	(−11)	794	(−2)	396	(−14)	131	(+8)	15	(0)	1425	(−19)	1054	(+19)
Historical scenarios[§]														
No original Avoca Levee	84		951		350		126		13		1524		955	
No effects	130		863		401		144		12		1550		929	

*The summary maps and this table indicate the dominant habitat type for each cell (i.e., the ecosystem type present in the cell for the largest amount of time during the year). Alternatively, we could have added the total number of cells of each ecosystem type for each week of the simulated year and divided the totals by 52. Although this gives a somewhat more accurate picture of the habitat distribution, it is inconsistent with the totals from the maps.
[†]The climate analysis scenarios used a slightly different set of parameters for the model than the other scenarios. See text for details.
[‡]EPA low scenario is a 50-centimeter rise by 2100. We used 0.46 cm/yr, which is double the historical rate of eustatic sea level rise in the study area of 0.23 cm/yr. Subsidence in the study area varies horizontally from 0.57 to 1.17 cm/yr, giving historical rates of apparent sea level rise (eustatic rise plus subsidence) of 0.8 to 1.4 cm/yr. EPA high scenario is a 200-centimeter rise by 2100 (1.67 cm/yr eustatic or 2.24 to 2.84 cm/yr apparent). Base case for comparison was no levee extension.
[§]No comparisons with a base case are given for the historical scenarios because these runs started in 1956 rather than 1983.

ios, which address the impacts of natural and human-induced variations in the systems' boundary conditions (i.e., sea level rise).

Management and boundary scenarios were run by restarting the model with the real 1983 ecosystem type map (rather than the model's predicted 1983 map) to ensure maximal realism. Climate and historical scenarios were run starting with 1956, so that the full impacts of climate and historical variations could be analyzed. After 1978, canal and levee construction for oil and gas exploration slowed significantly in the area compared with the 1956–1978 period. Therefore, all scenarios assumed that no additional canals or levees were constructed after 1978, except those specifically mentioned in the scenarios.

We present summary maps of the ecosystem type in the year 2033 for comparison of alternatives (Figures 11, 13, and 14). We also further summarize the results in Table 2, which lists for each alternative the total area of each ecosystem type in 2033, along with differences between each scenario and the base case.

Climate scenarios. Predicting future marsh succession required that we provide the model with predictions of future climate and boundary conditions. By *climate* we mean all the model's external forcing functions, including rainfall, Atchafalaya River flow, wind, and sea level.

For example, Figure 10 is a plot of Atchafalaya River discharge during the 1956–1983 period showing the weekly data used in calibrating the model. To project the base case climate into the future, we selected at random one-, two-, or three-year blocks from this record, modified them by randomly increasing or decreasing each point a maximum of 10%, and added this block to the end of the projected time series. This exercise provided a series of future years that were statistically identical to the past record and included the seasonal timing of the historical signals, but allowed for some novel random variations. We produced several future climate runs with this algorithm and tested the model's response to each. Some examples are shown in Figure 11.

We were also interested in what

part of the behavior of the system was due to random, aperiodic variations in climatic variables; what part was due to the annual periodic signal; and what part was due to the mean or average signal. We therefore analyzed two additional climate scenarios. These were the weekly average climate scenario, in which each week's value for each climate variable for the entire run from 1956 to 2033 was set at the average value for that week from the 1956 to 1983 data, and the mean-climate scenario, in which the long-term average for each variable was used for all weeks. For example, Figure 10 shows both the weekly average and mean signals for Atchafalaya River discharge.

Figures 11 and 12 show the projected ecosystem type maps for the year 2033 and a summary of the land/water balance in the system during the 1956–2033 period for the various climate scenarios. These figures clearly show the major effect of differing climate assumptions on the model. The base case runs show a moderate amount of difference due to the different random components in each run, but the general behavior was approximately the same. In all other future scenarios, we used the climate run labeled 3, because this case produced intermediate results.

The degree to which the other climate scenarios affected the results was surprising. The weekly average climate scenario produced a drastic reversal in land loss in the area. The mean climate scenario produced only a moderate decrease in land loss. This result indicates that the annual flood cycle and other annual periodicities in the climate are important to the land building process, but that random, aperiodic events (such as major storms and floods) tend to have a net erosional effect on marshland. A periodic but predictable climate seems to maximize marsh building. If global climate becomes less predictable in the future due to global warming, it may bode ill for the stability of coastal marshes.

Management alternatives. A base case run and several management alternatives were analyzed using the model. The results for 2033 for selected alternatives are summarized in Figure 13a and Table 2. The figure shows

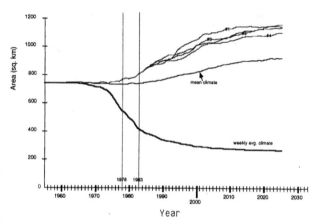

Figure 12. Summary time series of total open water area predictions under various climate scenarios.

both the ecosystem type map resulting from each alternative in 2033 and the cells that are different from those in the base case run, regardless of the direction of change. All the alternatives (see Figure 3) except the base case assume the two-reach Avoca Island levee extension (a 15-kilometer extension to the current levee system on the Atchafalaya River) has been built. The results for the other management options therefore indicate the amount of remaining (unmitigated) impact from the two-reach levee extension. All alternatives used climate run 3.

BASE CASE. The base case of the model assumed the canal network fixed as it was in 1978. As shown in Table 2, the base case run indicates a net land loss to open water of 37 km² from 1983 to 2033. Most of this loss occurred in fresh (−49 km²) or saline marsh (−32 km²) that was either converted to open water or to brackish marsh (+63 km²) due to continued fresh water inputs via the Atchafalaya River.

The overall rate of land loss predicted by the model to occur after 1983 (0.75 km²/yr) was much lower than the rates between 1956 and 1978 (6.2 km²/yr) and between 1978 and 1983 (24 km²/yr). There was therefore some concern that the model may be seriously underestimating the future rate of land loss.

To test the model's predictions, we performed a careful visual inspection of infrared aerial photographs taken in December 1988, which indicated that the rates of land loss have indeed slowed dramatically in the study area. Areas which appeared stressed in 1983 and which might have been expected to succeed to open water habitat have remained as marsh, as the model predicted. There were no apparent areas of significant marsh loss in the study area between 1983 and 1988, as predicted by the base case run of the model. The 1988 photographs are currently being interpreted and digitized to allow a quantitative assessment of the model's predictions, but it is clear that the model provided a better prediction of the status of the area in 1988 than predictions based on simple extrapolation of past trends. This finding underscores the effectiveness of our spatial, process-based approach to landscape modeling.

TWO-REACH LEVEE EXTENSION. The model was then modified in structure to allow for the construction of the first two reaches of the proposed Avoca Island levee extension. All other model parameters were held constant and identical climate inputs were used. The predicted effects are the indirect effects of the construction and do not include the acreage directly affected by construction.

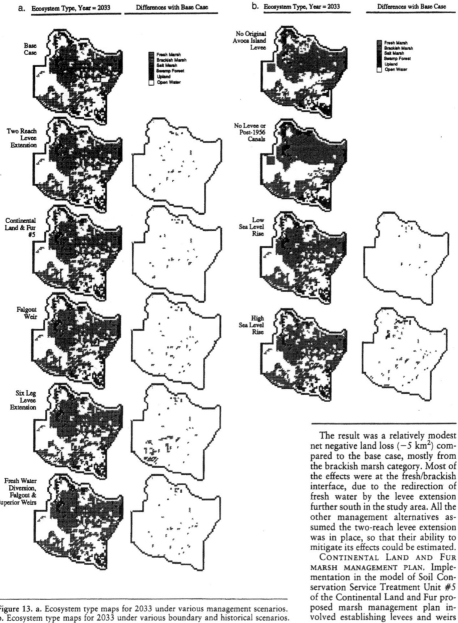

Figure 13. a. Ecosystem type maps for 2033 under various management scenarios. b. Ecosystem type maps for 2033 under various boundary and historical scenarios.

The result was a relatively modest net negative land loss (-5 km^2) compared to the base case, mostly from the brackish marsh category. Most of the effects were at the fresh/brackish interface, due to the redirection of fresh water by the levee extension further south in the study area. All the other management alternatives assumed the two-reach levee extension was in place, so that their ability to mitigate its effects could be estimated.

CONTINENTAL LAND AND FUR MARSH MANAGEMENT PLAN. Implementation in the model of Soil Conservation Service Treatment Unit #5 of the Continental Land and Fur proposed marsh management plan involved establishing levees and weirs to control water levels in an approximately 4-square-kilometer area (see Figure 3). The plan was installed in 1988, three years after the proposed levee extension was implemented in the model. This alternative resulted in a slight net gain of land (+3 km²) relative to the base case. It caused modest gains in fresh marsh and swamp (+2 km² each) balanced by smaller losses in brackish marsh (−1 km²) due to indirect effects on water flow patterns.

FALGOUT WEIR. Implementation of this alternative involved a run of the model with a weir at the entrance of the Falgout Canal, beginning in model year 1985. This alternative resulted in zero net gain of land relative to the base case. It caused increases in fresh marsh (probably due to reduction of salt water intrusion) and swamp, but also caused losses in brackish marsh that canceled out these gains.

FULL SIX-REACH LEVEE EXTENSION. We also analyzed the full six-reach (25 km) levee extension that had been under consideration at one time (see Figure 3). This scenario involved introducing the levee extension beginning in 1983, with completion of the final reach in 1988. This alternative results in a relatively large net loss of land (−52 km²) relative to the base case. The majority of the loss was from the brackish marsh (−48 km²) and, to a lesser extent, from the fresh marsh (−6 km²). The full levee extension prevented sediment-laden water from reaching the brackish marshes bordering Four League Bay, where most of the loss occurs.

FRESH-WATER DIVERSION STRUCTURE. This run of the model included a fresh-water diversion structure located in the northwest corner of Avoca Lake (see Figure 3), installed in 1985. The fresh-water diversion structure would allow 2% of the Atchafalaya River flow to be diverted into the Terrebonne Marsh, up to a river flow of 370,000 cubic feet per second. At and above this flow, the diversion flow is turned off to prevent backwater flooding.

This alternative resulted in a net gain of land (+4 km²) relative to the base case. The filling of Avoca Lake with sediments, which would be expected to occur with a water diversion structure, did not take place in

the model. The flow of water did not allow for significant sedimentation because the turbulence is too high near the division structure. The CELSS model does not, however, currently distinguish sediment grain sizes, and it may not adequately account for large-grain sediments that fall out quickly near a diversion site. If coarse-grained sediments make up a significant fraction of the total, the net land gain under this scenario may therefore be higher than the model predicts.

PALMETTO WEIR AND FRESH-WATER DIVERSION. Implementation of this alternative involved a run of the model with a weir on Palmetto Bayou (see Figure 3). The weir was installed in model year 1985 and allowed flow one way to the south. The fresh-water diversion structure was installed as described above and run with the same limits on the river flow conditions.

This alternative resulted in a 5-square-kilometer net land gain relative to the base case and only a one-cell land gain relative to the simulation with fresh-water diversion as the only mitigation. Because the weir is preventing the newly diverted water from flowing out Palmetto Bayou, redirected water has beneficial effects in the brackish marsh. It does, however, flood some of the fresh marsh just north of the weir.

SUPERIOR WEIR AND FRESH-WATER DIVERSION. Implementation of this alternative involved a run of the model with a weir on Superior Canal at the location of an existing plug (see Figure 3). The weir was installed in model year 1985 and allowed flow one way to the south. The fresh-water diversion structure was installed as described above and run with the same limits on the river flow conditions.

This alternative resulted in a 1-square-kilometer net land gain relative to the base case. The swamp habitat in the north is prevented from succeeding to open water, and land loss occurs just to the north. The retardation of the diverted water, which is beneficial to the swamp, waterlogs the marsh in the south and causes land loss there.

FRESH-WATER DIVERSION STRUCTURE AND BOTH PALMETTO AND SUPERIOR WEIRS. This run included the fresh-water diversion structure and

both the Palmetto Bayou Weir and the Superior Canal Weir, as described above. This model run illustrates the importance of cumulative impacts. In this scenario there is a net loss of land (−3 km²) relative to the base case. This net loss occurs even though the structures each produced a predicted net gain when used individually. The effects are not additive, however. The positive effects of the sediments in the fresh water are overcome by the negative effects of waterlogging when both weirs are used, but not when either one is used by itself. There is some optimal balance of water-diversion rate and location and weir location that can best be determined by iterative simulations of a model of this type.

FRESH-WATER DIVERSION STRUCTURE AND BOTH FALGOUT AND SUPERIOR WEIRS. This scenario included the Falgout Canal Weir, the Palmetto Bayou Weir, and the fresh-water diversion structure. This alternative resulted in a net gain of land (+7 km²) relative to the base case. This net gain occurred mainly in the swamp and fresh marsh ecosystem types, due mainly to increased sediment flows to these areas as a result of the diversion. This combination of management alternatives yielded the largest positive net land gain of those we have investigated thus far.

Boundary scenarios. One important boundary scenario was analyzed—the effect of increasing rates of future sea level rise due to global warming. Projected rates of future sea level rise are in the range of 50 to 200 cm during the next 120 years (or 0.42 to 1.67 cm/yr; Titus 1988) Coastal Louisiana has historically experienced high rates of apparent sea level rise due to regional subsidence. In the study area, this rate has ranged from 0.57 to 1.17 cm/yr (0.23 cm/yr from eustatic or general oceanic sea level rise and the remainder from subsidence; Penland et al. 1987). Thus coastal Louisiana represents an appropriate laboratory for studying the response of coastal wetlands to high rates of sea level rise (Day and Templet 1989).

We experimented with the CELSS model to study the effects on the study area of the projected higher rates of sea level rise. Figure 13b and Table 2 show the results for rates of eustatic sea level rise of 0.46 and 1.67 cm/yr (1.03 to 1.63 and 2.24 to 2.84 cm/yr apparent rates including subsidence) starting in model year 1983.

The results were somewhat surprising. Doubling the rate of eustatic sea level rise (from 0.23 to 0.46 cm/yr) actually caused a slight net gain in land in the study area (+10 km², or 0.7% of the total land area) relative to the base case. We believe this increase occurs because, in the presence of high sediment loads, healthy coastal marshes can keep up with moderate rates of sea level rise. In fact, the entire existing coastal marsh system in Louisiana has developed in the presence of high apparent rates of sea level rise (Day and Templet 1989). If sea level rise rates become high enough, however, the marshes can no longer keep up. The Environmental Protection Agency's higher rate of sea level rise (1.67 cm/yr eustatic, 2.24 cm/yr apparent) caused a net loss of land of 19 km² (or 1.3% of the total land area).

These results indicate that healthy coastal marshes with adequate sediment inputs may act as a buffer against moderate rates of sea level rise. The marshes build coastal marshland as fast as the sea can rise to inundate it.

Historical scenarios. Two historical scenarios were analyzed. We looked at how the system would have evolved if the original Avoca Island levee had not been built or if neither the levee nor the the post-1956 canals had been built. Results are summarized in Figure 13b and Table 2. These results indicate that both the original levee and the canals had a major effect on the evolution of the system. Results of these runs are not directly comparable with the base case 2033 scenario, because the base case was restarted with actual 1983 data, whereas the historical scenarios had to use model predictions for 1983. Nevertheless, it is clear that the post-1956 canals and the original levee caused increased land loss in the area compared to what would have happened without them.

Conclusions

The CELSS model simulation of long-term ecosystem changes demonstrates

that spatially linked ecological and physical processes can be realistically modeled on modern supercomputers. The results of the CELSS model indicate complex and often counterintuitive behavior that, like the real system, is difficult to summarize. This observation is not surprising, given that the CELSS model trades off generality for realism. The model is most enlightening in its elaboration of the effects of specific cases. Nevertheless, a few general conclusions can be drawn from this modeling exercise.

● Past and future climate variations are important, especially severe deviations from average conditions (i.e., severe droughts, floods, and hurricanes). Random climate variations with no change in the historical trends can cause changes in the region of approximately equal magnitude to that of the human modifications analyzed. A comprehensive, process-based model such as the CELSS model is essential to separate these effects. Projected future climate changes (particularly changes in the predictability of climates) due to global warming can cause major changes in the region.

● Ecological variables, such as primary production, have significant feedback to physical and chemical processes and significantly affect the model's ability to replicate historical changes in ecosystem type distribution.

● Coastal marshes can adapt to future projected higher rates of sea level rise and serve as a buffer against future global sea level rise if the rates are moderate, sediment sources are available, and the marshes remain healthy.

● Past activities in the area that modified the hydrologic and sediment flow patterns, such as canal and levee construction, at critical stages had significant influence on the evolution of the region. Proposed future modifications seem to have smaller effects, partly because they will occur at less critical times in this landscape's evolution.

● Management options in the area can be effective at mitigating the

effects of proposed human alterations, but cumulative effects make analysis of individual options in isolation risky. Process-based spatial models such as CELSS may be necessary to incorporate these cumulative effects, to effectively weigh the costs and benefits of major projects, and to design optimal coastal management strategies.

At present, ecosystem models of the size and complexity of the CELSS model are fairly new and expensive to build and run. But as we gain experience, and as supercomputers and parallel processors become more readily available, models of this type will become more practical tools for understanding and managing ecosystems. Future studies might employ parallel computers (e.g., the Connection Machine[1] or microcomputer-based transputer systems such as TransLink[2] boards) to radically improve the speed, complexity, and ease of use of spatial ecosystem models.

We are currently applying the CELSS approach at multiple scales to other coastal areas in Louisiana, Maryland, and South Carolina, and we plan to continue to modify, improve, and experiment with the existing model toward the goal of better environmental management. We hope to learn enough from these applications and experiments to allow both scaling up of this approach for global ecosystem modeling to assess the impacts of global climate change and scaling down to assess local effects in more detail.

Acknowledgments

This research was supported by the US Fish and Wildlife Service in part with funds from the US Army Corps of Engineers under cooperative agreement #14-16-0009-84-921 titled "Spatial Simulation Modeling of Coastal Wetland Systems for Evaluating Management Alternatives" (E. Pendleton, project officer) and by from the National Science Foundation under project BSR-8906269 titled "Landscape Modeling: The Synthesis

---

[1] Thinking Machine Corp., 215 First Street, Cambridge, MA 02112.
[2] Levco Corp., 6160 Lusk Boulevard, San Diego, CA 92121.

of Ecological Processes Over Large Geographic Regions and Long Time Scales." Computer time on the CRAY supercomputer at the University of Illinois was provided by the National Center for Supercomputer Applications (NCSA) and by an NSF grant on modeling landscape dynamics (D. Jameson, principal investigator). The NCSA staff was extremely helpful in the performance of this research. We thank J. W. Day Jr. for his critical role in the birth and success of this project and for his input on model formulation and evaluation of results, W. Kitchens for getting this project off the ground, T. Ozman, K. Sampey, S. Leibowitz, P. Nepalli, S. Nellore, M. Ahmed, and P. Templet Jr. for their contributions to computer programming and data assembly, and three anonymous reviewers for their helpful comments on earlier drafts of the manuscript.

References cited

Bahr, L. M., R. Costanza, J. W. Day Jr., S. E. Bayley, C. Neill, S. G. Leibowitz, and J. Fruci. 1983. Ecological characterization of the Mississippi Deltaic Plain Region: a narrative with management recommendations. FWS/OBS-82/69, US Fish and Wildlife Service Division of Biological Service, Washington, DC.

Baumann, R. H., and R. Adams. 1981. The creation and restoration of wetlands by natural processes in the lower Atchafalaya River system: possible conflicts with navigation and flood control management. Pages 1–24 in R. H. Stovall, ed. Proceedings of the Eighth Annual Conference on Wetlands Restoration and Creation. Hillsborough, FL.

Baumann, R. H., J. W. Day Jr., and C. A. Miller. 1984. Mississippi deltaic wetland survival: sedimentation versus coastal submergence. Science 224: 1093–1095.

Botkin, D. B., J. F. Janak, and J. R. Wallis. 1972. Some ecological consequences of a computer model of forest growth. J. Ecol. 60: 849–872.

Cleveland, C. J., C. Neill, and J. W. Day Jr. 1981. The impact of artificial canals on land loss in the Barataria Basin, Louisiana. Pages 425–434 in W. J. Mitsch, R. W. Bosserman, and J. M. Klopatek, eds. Energy and Ecological Modeling. Elsevier, Amsterdam.

Conner, W. H., and J. W. Day Jr., eds. 1987. The ecology of the Barataria Basin, Louisiana: an estuarine profile. US Fish and Wildlife Service biology report 85(7.13).

Costanza, R. 1987a. Social traps and environmental policy. BioScience. 37: 407–412.

———. 1987b. Simulation modeling on the Macintosh using STELLA. BioScience. 37: 129–132.

———. 1989. Model goodness of fit: a multiple resolution procedure. Ecol. Model. 47: 199–215.

Costanza, R., S. C. Farber, and J. Maxwell. In press. The valuation and management of wetland ecosystems. Ecological Economics.

Costanza, R., C. Neill, S. G. Leibowitz, J. R. Fruci, L. M. Bahr Jr., and J. W. Day Jr. 1983. Ecological models of the Mississippi deltaic plain region: data collection and presentation. FWS/OBS-82/68, US Fish and Wildlife Service, Division of Biological Service, Washington, DC.

Costanza, R., and F. H. Sklar. 1985. Articulation, accuracy, and effectiveness of mathematical models: a review of freshwater wetland applications. Ecol. Model. 27: 45–68.

Costanza, R., F. H. Sklar, M. L. White, and J. W. Day Jr. 1988. A dynamic spatial simulation model of land loss and marsh succession in coastal Louisiana. Page 99–114 in W. J. Mitsch, M. Straskraba, and S. E. Jørgensen, eds. Wetland Modelling. Elsevier, Amsterdam.

Craig, N. J., R. E. Turner, and J. W. Day Jr. 1979. Land loss in coastal Louisiana (U.S.A.) Environ. Manage. 3: 133–144.

Day, J. W. Jr., and P. H. Templet. 1989. Consequences of sea level rise: implications from the Mississippi Delta. Coastal Management 17: 241–257.

Deegan, L. A., H. M. Kennedy, and C. Neill. 1984. Natural factors and human modifications contributing to marshloss in Louisiana's Mississippi River deltaic plain. Environ. Manage. 8: 519–528.

Donnell, B., and J. V. Letter. In press. The Atchafalaya River delta. Report 11. Two dimensional model verification. Waterways Experiment Station, Vicksburg, MS.

Farber, S., and R. Costanza. 1987. The economic value of wetlands system. J. Environ. Manage. 24: 41–51.

Gagliano, S. M., K. J. Meyer-Arendt, and K. M. Wicker. 1981. Land loss in the Mississippi River deltaic plain. Trans. Gulf Coast Assoc. Geol. Soc. 31: 285–300.

Gardner, R. H., W. G. Cale, and R. V. O'Neill. 1982. Robust analysis of aggregation error. Ecology 63: 1771–1779.

Hall, C. A. S., and D. L. DeAngelis. 1985. Models in ecology: paradigms found or paradigms lost? Bulletin of the Ecological Society of America 66: 339–346.

Hopkinson, C. S. 1978. The relation of man and nature in Barataria Basin, Louisiana. Ph.D. dissertation, Louisiana State University, Baton Rouge.

Jørgensen, S. E. 1982. Modelling the eutrophication of shallow lakes. Pages 125–155 in D. O. Logofet and N. K. Luckyanov, eds. Ecosystem Dynamics in Freshwater Wetlands and Shallow Water Bodies. vol 2. UNEP/SCOPE, USSSR Academy of Sciences, Moscow.

Leibowitz, S. 1989. The pattern and process of land loss in coastal Louisiana: a landscape ecological analysis. Ph.D. dissertation, Louisiana State University, Baton Rouge.

Letter, J. V. Jr. 1982. The Atchafalaya River deta: extrapolation of delta growth. Waterways Experiment Station, technical report HL-82-15. Vicksburg, MS.

Levins, R. 1966. The strategy of model building in population biology. Am. Sci. 54: 421–431.

Penland, S., K. E. Ramsey, R. A. McBride, T. F. Moslow, and K. A. Westphal. 1987. Relative sea level rise and subsidence in Louisiana and the Gulf of Mexico. Louisiana Geological Survey, Coastal Geology Section, Baton Rouge.

Phipps, R. L. 1979. Simulation of wetland forest vegetation dynamics. Ecol. Model. 7: 257–288.

Potter, G. L., H. W. Ellsaesser, M. C. MacCracken, and F. M. Luther. 1979. Performance of the Lawrence Livermore Laboratory zonal atmospheric model. Pages 852–871 in W. L. Gates, ed. Report of the JOC study conference on climate models: performance, intercomparison and sensitivity studies. Global Atmospheric Research Programme Series no. 22, Washington, DC.

Roberts, H. H., R. D. Adams, and R. H. Cunningham. 1980. Evolution of the sand-dominated subaerial phase, Atchafalaya Delta, Louisiana. Am Assoc. Pet. Geol. Bull. 64: 264–279.

Scaife, W. W., R. E. Turner, and R. Costanza. 1983. Coastal Louisiana recent land loss and canal impacts. Environ. Manage. 7: 433–442.

Schlesinger, M. E., and Z. C. Zhao, 1989. Seasonal climatic changes induced by doubled CO_2 as simulated by the OSU atmospheric GCM/mixed-layer ocean model. Journal of Climate 2: 463–499.

Sklar, F. H., R. Costanza, and J. W. Day Jr. 1985. Dynamic spatial simulation modeling of coastal wetland habitat succession. Ecol. Modell. 29: 261–281.

Sklar, F. H., M. L. White, and R. Costanza. In press. The Coastal Ecological Landscape Spatial Simulation (CELSS) model: structure and results for the Atchafalaya/Terrebonne study area. NWRC open file report. US Fish and Wildlife Service, Washington, DC.

Titus, J. G., ed. 1988. Greenhouse effect, sea level rise and coastal wetlands. EPA-230-05-86-013, Office of Policy, Planning, and Evaluation, Washington, DC.

Turner, M. G., R. Costanza, and F. H. Sklar. 1989. Methods to evaluate the performance of spatial simulation models. Ecol. Modell. 48: 1–18.

Turner, M. G., R. Costanza, T. M. Springer, and E. P. Odum. 1988. Market and nonmarket values of the Georgia landscape. Environ. Manage. 12: 209–217.

Washington, W. M., and D. L. Williamson. 1977. A description of the NCAR global circulation models. Pages 111–172 in J. Chang, ed. Methods of Comparative Physics. vol. II. General Circulation Models of the Atmosphere. Academic Press, New York.

Williams, J., R. G. Barry, and W. M. Washington. 1974. Simulation of the atmospheric circulation using the NCAR global circulation model with ice age boundary conditions. Journal of Applied Meteorology 11: 305–317.

Van Heerden, I. L., and H. H. Roberts. 1980a. The Atchafalaya Delta—Louisiana's new prograding coast. Trans. Gulf Coast Assoc. Geol. Soc. 30: 497–506.

———. 1980b. The Atchafalaya Delta—rapid progradation along a traditionally retreating coast (South Central Louisiana). Zeitschrift Geomorphologie N.F. 34: 186–201.

MOVEMENTS BY BIRDS AND SMALL MAMMALS BETWEEN A WOOD AND ADJOINING FARMLAND HABITATS

By JOHN F. WEGNER AND GRAY MERRIAM

Department of Biology, Carleton University, Ottawa, Ontario K1S 5B6 Canada

SUMMARY

(1) White-footed mice (*Peromyscus leucopus*) and chipmunks (*Tamias striatus*) moved between a beech-maple wood and connecting fencerows four times as often as they moved between traplines within the wood. They seldom moved between the wood and adjacent perennial grass fields or across the fields.

(2) Birds seldom flew directly across open fields between woods. More species of birds moved more frequently between the wood and fencerows than between any other habitats. Wood-nesting birds moved more frequently from well-vegetated fencerows into fields to forage than from an equal length of wood border. Poorly developed fencerow vegetation restricted foraging by wood-nesters into fields. None of tree species diversity, line intercept measures, dendrograms, or foliage height diversity satisfactorily distinguished among the vegetation structure of different fencerows.

(3) The results indicate that fencerows connect the wood to the surrounding agricultural mosaic and concentrate the activity of small mammals and birds into a habitat corridor that may relieve the isolating effect of farmland surrounding the wood.

INTRODUCTION

In some agricultural regions scattered woods may provide useful and possibly vital refuges for woodland communities. Studies that clarify the importance of such woods in North America have appeared only recently. Forman, Galli & Leck (1976) and Galli, Leck & Forman (1976) have reported relationships between the number of bird species and area of the woodlot inhabited. R. Whitcomb and co-workers (R. Whitcomb 1977; Whitcomb, Whitcomb & Bystrak 1977; MacClintock, Whitcomb & Whitcomb 1977) have investigated some effects of fragmentation of Maryland piedmont forest on bird species diversity.

In intensively farmed regions, woods are often habitat islands (Usher 1973; Moore & Hooper 1975; Levenson 1976) isolated by other habitat patches of the agricultural mosaic. Gromadzki (1970) showed that the bird communities in mid-field afforestations could not be characterized as either forest or field communities but that the afforestations were linked to open field communities by their bird species. Few other studies have reported how animals living in such isolated woods use the adjacent habitats and thus couple the wood to the surrounding mosaic.

We investigated the connections between a beech-maple wood and the adjacent fields and fencerows as indicated by movements of birds and small mammals.

0021–8901/79/0800–0349$02·00 © 1979 Blackwell Scientific Publications

350   *Woodland isolation*

METHODS

Study site vegetation

The study wood (Fig. 1) is on 12·2 ha of a glacial ridge of Grenville sandy loam, 16·5 km south of Ottawa, Canada. Tree cover is primarily beech (*Fagus grandifolia*) and sugar maple (*Acer saccharum*) with some basswood (*Tilia americana*) and ironwood (*Ostrya virginiana*). The wood was surrounded by pasture fields (Fig. 1). The most common vegetation on the close-cropped pastures to the N and W was short grasses, thistles (*Cirsium arvense*) and common mullein (*Verbascum thapsus*). The old hay fields on the S and E side of the wood were mainly timothy (*Phleum pratense*) grazed seasonally. The whole area, including the wood, was grazed by cattle during the study.

'Fencerow' as used here refers to a line originally followed by a fence but which subsequently often develops a distinguishable line of vegetation that may outlast the original fence. Fencerow vegetation is distinct from vegetation of British hedgerows which originally were cultured and therefore have lower species diversity, are structurally more dense and homogeneous and generally are much wider.

Beginning at the junction of each fencerow and the wood, a line was strung through each fencerow along its axis. Along fencerows 1 and 4, 50 m was sampled; an additional 50 m was sampled in fencerows 2 and 3 (Fig. 1). At 2 m intervals along each axial line a cross line perpendicular to the fencerow was extended to the margins of the fencerow and all plants whose stems intersected the cross line were measured. Data recorded were: species, length of the axial line intercepted by the canopy of each individual, and maximum width of the plant canopy perpendicular to the axial line. Succulents were classed as grasses or forbs. Plant names follow Gleason & Cronquist (1963).

Small mammals

Live-trapping was done from 6 May to 24 June, from 8 July to 25 August and from 4 September to 3 November 1975. Each of these periods including four trapping sessions

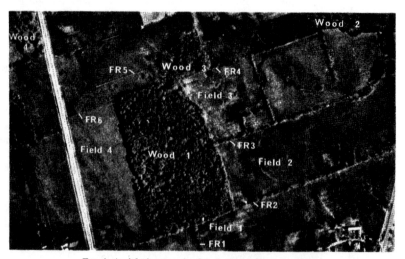

FIG. 1. Aerial photograph of study area and surroundings.

J. F. Wegner and G. Merriam   351

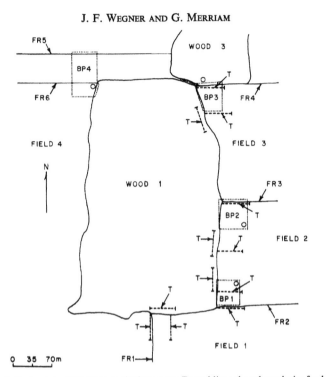

Fig. 2. Study plots. FR1–FR6 marks fencerows. Dotted lines show boundaries for bird plots and circles denote associated observations posts. T indicates mammal trapline.

of seven nights each. During each trapping period two habitat combinations were trapped. Each habitat combination consisted of three traplines one in each of three habitats: wood, fencerow and field (Fig. 2). Each trapline consisted of twenty Sherman-type live traps (8 × 8 × 31 cm). Traps were placed in fencerows in pairs separated by half the width of the fencerow and 5 m apart along the fencerow beginning at the wood margin. The field trapline was parallel to the associated fencerow trapline. Its junction with the wood margin was placed randomly within 75 m of the fencerow. Traps were placed in the field in the same pattern as in the fencerow. The wood traps were similarly paired in a 5 m wide transect parallel to the field edge and 10 m inside the wood.

Traps were baited with peanut butter and rolled oats and were checked each morning. Captured animals were tagged with numbered monel fingerling tags and released after recording species, sex, date and location. One movement was counted for each sequential pair of different capture locations of an individual. Mammal names follow Hall & Kelson (1959).

Bird observations

Forty-two h of preliminary observation during the previous summer revealed the greatest distance at which bird movements could be observed, maximum plot size and workable plot locations. Four plots, each with one fixed observation post, were used to standardize the amount of each habitat (wood, field, fencerow) observed in each combination (see Fig. 2). Preliminary observations also showed that a blind was not needed if there was a 5 min waiting period after the initial disturbance and if the observer remained seated throughout the one hour observation period. Morning observations began 15 min before sunrise and took about 3 h and 25 min. Three plots were observed two mornings per week from 6 May to 31 July 1975 and five mornings per week from 4 August to 31 October 1975. Every other week from 4 August to 31 October five plots were sampled each morning which extended observations another 2 h and 15 min into the morning. Evening observations began 3 h before sunset and plots were observed in the evening once per week in every month except October. The order in which plots were sampled was rotated regularly.

For each individual bird movement, origin and destination, species, observation post, time and date were recorded. Each plot was observed for 47 to 54 h from May to October and totals were standardized to 42 h (7 h/plot/month). Bird names follow Godfrey (1966).

RESULTS

Fencerow structure

Fencerow 1 (Fig. 1) was the most open of the fencerows based on samples along the first 50 m from the wood. It had alternating patches of grasses and hawthorn (*Crataegus* sp.), one sugar maple and a wire fence along its length. The first 50 m of fencerow 2 had the most vegetation with more forbs than grass and more forb area than any other fencerow; the shrub layer comprised hawthorn and choke cherry (*Prunus pensylvanica*) and the trees were mainly basswood, with one sugar maple, one white oak (*Quercus alba*), one white ash (*Fraxinus americana*) and some dead elms (*Ulmus americana*). The ground flora of fencerow 3 had less forbs than grass which was continuous with the adjoining pastures. Hawthorn dominated the shrub layer, with sugar maples and a few dead elms in the extensive tree cover. Fencerow 4 contained no grass but had several narrow strips of short forbs and mosses along cowpaths paralleling the central wire fence. There were few shrubs and some sugar maple, basswood and elm. This fencerow was narrower than the other fencerows and without its tree cover would have average less than a metre wide. Fencerow 4 had the least vegetation except fencerows 5 and 6 which were wire fences with vegetation like that of the surrounding fields.

Vegetation was extremely variable along each well vegetated fencerow (fencerows 2 and 3). Species occurred in clumps and none occurred in every 2 m interval. For example, *A. saccharum* occurred commonly along more than two-thirds of fencerow 3 nearest to the wood but was absent in the last third. In fencerow 2, *Tilia* occurred in clumps near the wood and was replaced by *Crataegus* further out.

Tree species diversity, percentage of the line intercepted by herbs, shrubs or trees, Dansereau's (1957) dendrograms, and foliar height diversity (MacArthur 1964) were calculated in an attempt to distinguish the fencerows from one another. Even for fencerows with the greatest apparent differences, none of these methods was adequate for quantifying differences in the vegetation of the fencerows. This was due to the tremendous amount of variation within each fencerow. For distinguishing among the fencerows, a technique is needed that will incorporate this spatial heterogeneity in it.

Small mammals

Capture records (Table 1) reflect habitat use by the small mammals. Eastern chipmunks (*Tamias striatus*) used woods and fencerows but were never caught in fields. Meadow

J. F. WEGNER AND G. MERRIAM   353

TABLE 1. Number of captures and number of individuals ( ) in each habitat standardized to 3360 trap-nights per habitat. (Data from traplines in Fig. 2)

	Wood	Fencerows	Fields*
Peromyscus leucopus	206	231	27
	(59)	(68)	(19)
Tamias striatus	153	93	0
	(36)	(29)	
Zapus hudsonius	1	22	37
	(1)	(15)	(20)

* Exludes traps within 10 m of wood on one trapline where site manipulation invalidated results.

jumping mice (*Zapus hudsonius*) used fencerows and fields but rarely used woods. White-footed mice (*Peromyscus leucopus*) were caught predominantly, and nearly equally, in woods and fencerows. The infrequent use of fields by *P. leucopus* also is shown by movements between traps in the field which were 2% of those between traps in the wood and occurred only in autumn (Table 2). Movements between fields and wood were also recorded primarily in autumn and half were restricted to the 5 m of field next to the wood. Movements between fields and fencerows showed the same seasonal pattern but were even less frequent. No differences in movement patterns were noted between sexes or age classes.

Peromyscus leucopus moved between traps in fencerows as commonly as they moved between traps in the wood (Table 2). They also moved between traplines in the wood and traplines in the fencerows much more frequently than they moved between separate traplines in the wood (wood to wood in Table 2). The five movements between traplines in separate fencerows illustrates that these mice can move between distant habitats. The route between fencerows was not determined but the infrequent use of fields by these mice suggests that the longer fencerow to wood to fencerow route was more likely than a direct route across an open field.

TABLE 2. Absolute number of movements between traps within and between habitats recorded in 1120 trap-nights/habitat/period

	Movement within:			Movement between:					
	Wood	Fencerow	Field†	Wood and fencerow	Wood and field†	Wood and wood*	Fencerow and field†	Fencerow and fencerow	Total
P. leucopus									
May–June	16	12	0	8	1	5	1	0	43
July–Aug.	35	25	0	9	0	1	0	0	70
Sept.–Oct.‡	61	88	2	24	14	6	3	5	203
Total	112	125	2	41	15	12	4	5	316
T. striatus									
May–June	20	10	0	10	0	5	0	3	48
July–Aug.	55	33	0	27	0	1	0	0	116
Sept.–Oct.‡	16	0	0	4	0	4	0	0	24
Total	91	43	0	41	0	10	0	3	188

* Between traplines in the wood.
† Excludes traps within 10 m of wood on one trapline where site manipulation invalidated results.
‡ Includes 420 extra trap-nights, a fencerow and a field.

354   *Woodland isolation*

The movements by *T. striatus* between fencerow traplines (Table 2) are also unlikely to have been directly across an open field. *Tamias striatus* were never recorded in fields or moving between fields and any other habitat. They moved between traps in the fencerows about half as often as they moved between individual traps in the wood. This differs from the pattern for *P. leucopus*. However, *T. striatus* moved between traplines in the fencerows and traplines in the wood much more often than they moved between separate traplines within the wood, much as the *P. leucopus* did. Again, there were no sex- or age-related differences.

The increased number of movements of all kinds by *P. leucopus* in September and October (Table 2) is related to increasing numbers of juveniles which resulted in increasing numbers of tagged individuals. The peak in movements of *T. striatus* in July and August is not explained by changes in population size or in numbers tagged. The decrease in movements of *T. striatus* in September and October is explained by hibernation.

Of individuals captured more than once, 44% of *P. leucopus* and 44% of *T. striatus* had moved between habitats over a 6 month period. This is a minimum estimate because some animals may have moved between habitats without being captured in both.

Birds

Movements were recorded for all species moving in the observation plots. The area, types and amounts of habitat that could be observed from each observation post within each plot varied considerably (Fig. 2). BP1 and BP2 included 50 m each of wood and of fencerow bordering the field. The movements that could be observed were: between wood and fencerow; between wood and field; and between fencerow and field; also movements between two woods were observed in BP3 and BP4. BP3 contained an equal amount of wood-field, fencerow-field, and wood-wood boundaries.

Where fencerows had well-developed vegetation (fencerow 2 in BP1, fencerow 3 in BP2) movements between wood and fencerow were double the total of all other movements in the plot (Table 3). Movements between fencerow and wood in BP1 were half those in BP2. This difference may be due to differences in composition of fencerow 2 and fencerow 3 or to the connection which fencerow 3 provides between two woods (Fig. 1).

The number of movements (and species) from 50 m of fencerow to the field in each of BP1 and BP2 also was much greater than the number moving directly to the field from 50 m of wood edge. Most of this activity was due to red-winged blackbirds (*Agelaius phoeniceus*), which nested in the field and used the fencerow for song posts or for roosting,

TABLE 3. Number of movements by birds between habitat units (number of species in parentheses)*

Plot**	Wood ↔ Wood	Wood ↔ Fencerow	Wood ↔ Field	Fencerow ↔ Field
BP1	No data	261 (30)	28 (4)	92 (8)
BP2	No data	511 (41)	41 (10)	104 (15)
BP3	631 (36)	94 (21)	106 (13)	24 (6)
BP4	132 (17)	52 (14)	227 (14)	188 (11)
Total		918	402	408

* Number of movements per 42 h observation per plot and flocks of more than twenty-five of one species are counted as one individual.
** See Fig. 2.

J. F. Wegner and G. Merriam   355

but several wood-nesting species also moved between the fencerow and field. Species such as eastern wood pewee (*Contopus virens*), red-eyed vireo (*Vireo olivaceus*), and wood warblers (Parulidae) were observed moving from wood to fencerows and thence to fields.

This pattern is not clear in BP3 (Table 3) because wood edge in this plot was effectively double because there were two borders and because fencerow 4 had poorly developed vegetation. In BP4 fencerows 5 and 6 were wire fences with no fencerow vegetation and were rarely used by birds enroute from wood to field although they were used for perching during foraging in the field. The lack of woody vegetation along these fencerows restricted the number of wood species that foraged in the field; only American robins (*Turdus migratorius*), starlings (*Sturnus vulgaris*), and yellow-shafted flickers (*Colaptes auratus*) commonly did so.

It was possible to observe wood to wood movements only in BP3 and BP4 (Figs 1 and 2). The 10 m gap between wood 1 and wood 3 did not affect movements and the number of movements between woods in BP3 (Table 3) actually measures movements within a wood. The movements between woods in BP4 involved flying at least 300 m across open fields (wood 1 to wood 4, Fig. 1). These movements show that long overflights between woods occur but were uncommon and were mainly by *Turdus migratorius*, *Sturnus vulgaris* and blue jays (*Cyanocitta cristata*).

The proportion of movements between wood and fencerows compared to other movements changed seasonally but, in all months, movements between woods and fencerows exceeded all others (Table 4). The proportion of movements between wood and fencerows was lower in May and June when migrants were just returning than during the rest of the season; also in May and June there were many movements between fencerows and fields especially by *Agelaius phoeniceus* and other field-nesting species that used the fencerows as singing posts and perches.

The number of movements between wood and fencerows was high in May and June but dropped sharply in July. Causes were complex and included fewer species moving between wood and fencerows especially and a general reduction in activity by birds in July. The decrease in movements between fencerow and field in July was due to absence of *A. phoeniceus* that had moved away to migrate. In August the number of movements between wood and fencerow increased again and reached a maximum in September. This was caused by wood-nesting species that had completed raising broods and were

TABLE 4. Seasonal pattern of movements by birds between habitats. Number of monthly movements in plots 1 and 2* with percentage of movements in parentheses**

	Wood ↔ Fencerow	Wood ↔ Field	Fencerow ↔ Field
May	121 (58·2)	16 (7·7)	71 (34·1)
June	100 (52·4)	18 (9·4)	73 (38·2)
July	64 (80·0)	2 (2·5)	14 (17·5)
August	130 (88·4)	6 (4·1)	11 (7·5)
September	291 (89·0)	24 (7·3)	12 (3·7)
October	66 (78·6)	3 (3·5)	15 (17·9)
Total	772	69	196

* See Fig. 2.
** Number of movements per 42 h observation per plot (see Methods) and flocks of more than twenty-five of one species are counted as one individual.

no longer territorial. Small flocks of transient migrants from further north also began to move commonly between wood and fencerows in August. Local species that nest in the wood and transient migrants both used the fencerows intensively from late summer until the end of October. The movements between woods and fields in September and between fencerows and fields in October were by foraging migrating warblers.

DISCUSSION

The movement between habitats by *Peromyscus leucopus* and *Tamias striatus* suggest that fencerows provide important corridors into and out of the wood. Oxley, Fenton & Carmody (1974) reported that both species avoid travelling across areas with no vegetation cover and that a roadway 90 m wide stops them. The limited activity of these species in pasture fields supports previous studies (Bider 1968; Forsyth & Smith 1973, D. A. Smith & D. G. Taylor, personal communication; Baker 1968). Fencerows apparently relieve the barrier created by open fields.

More birds of more species moved between wood and fencerows than between any other habitats (Table 3). Birds moved along well-vegetated fencerows and then foraged from them into fields. Poorly-vegetated fencerows restricted foraging by wood-nesting species.

Although foraging from fencerows into fields declined in August and September (Table 4), use of well-vegetated fencerows for movements from the wood increased as wood-species such as *Vireo olivaceous*, *Contopus virens*, and white-breasted nuthatch (*Sitta carolinensis*) became less territorial and fledged their broods. In late summer, northern migrants such as black-capped chickade (*Parus atricapillus*), *Cyanocitta cristata*, and warblers joined the wood-species moving along the fencerows and there was renewal of foraging into fields.

Birds moving along a fencerow usually flew short distances along it from perch to perch. It was therefore difficult to observe whether a bird moving along the fencerow from wood to wood, except for some large colourful species such as baltimore oriole (*Icterus galbula*), scarlet tanger (*Piranga olivacea*) and *Cyanocitta cristata*. Imbalance between the movements into and out of the study wood along a fencerow implied movement along the fencerows into different woods.

That the bird movements were so restricted and that most were channelled along fencerows confirms other evidence of habitat disconformities acting as barriers. For example, Willis (1974) and Diamond (1973) reported that many bird species will not fly across a small water gap and Terborgh (1975) found that distances of 1 to 2 km of unsuitable habitat kept out half of the tropical bird species that would colonize openings in the rain forest. Even the width of a major river can be a barrier to birds (Willis 1974).

We have also shown that for the small mammals *Peromyscus leucopus* and *Temias striatus* that they commonly move between wood and fencerows and that the fencerows have resident populations.

If the species richness of a terrestrial habitat island is a function of size and isolation as it is for oceanic islands (MacArthur & Wilson 1967), the nature of the isolation of terrestrial islands should be examined in more detail. Isolation of habitat patches in a terrestrial mosaic will be a function of the degree to which each habitat acts as a barrier to a particular species. Simple measures of distance are unlikely to suffice. Our results show that corridors of a particular habitat type may cross barriers in the isolating mosaic. Although the idea of corridors has been applied to terrestrial islands (Wilson & Willis

J. F. WEGNER AND G. MERRIAM   357

1975; Diamond 1975) there has been very little evidence of corridor effect in terrestrial habitats (cf. Prestt 1971) since Preston (1962) suggested it on theoretical grounds. Further studies of connections between habitat patches are required to understand the operation of terrestrial environmental mosaics and to assess the utility of island biogeographic theory in understanding these common terrestrial systems.

REFERENCES

Baker, R. H. (1968). Habitats and distribution. *Biology of Peromyscus.* (Ed. by J. A. King), pp. 98–126. Special Publication No. 2 American Society of Mammalogists, Stillwater, Okla.

Bider, J. R. (1968). Animal activity in uncontrolled terrestrial communities as determined by a sand transect technique. *Ecological Monographs,* **38,** 269–308.

Dansereau, P. M. (1957). *Biogeography: An Ecological Perspective.* Ronald Press Co., New York.

Diamond, J. M. (1973). Distributional ecology of New Guinea birds. *Science, New York,* **179,** 759–769.

Diamond, J. M. (1975). The island dilemma: Lessons of modern biogeographic studies for the design of natural reserves. *Biological Conservation,* **7,** 129–146.

Forman, R. T. T., Galli, A. E. & Leck, C. F. (1976). Forest size and avian diversity in New Jersey woodlots with some land use implications. *Oceologia,* **26,** 1–8.

Forsyth, D. J. & Smith, D. A. (1973). Temporal variability in home ranges of eastern chipmunks (*Tamius striatus*) in a southeastern Ontario woodlot. *American Midland Naturalist,* **90,** 107–117.

Galli, A. E., Leck, C. F. & Forman, R. T. T. (1976). Avian distribution patterns in forest islands of different sizes in central New Jersey. *The Auk,* **93,** 356–364.

Gleason, H. A. & Cronquist, A. (1963). *Manual of Vascular Plants.* Van Nostrand Reinhold Co., Toronto.

Godfrey, W. E. (1966). *The Birds of Canada.* National Museum of Canada Bulletin No. 203 Biological Series No. 73. Queen's Printer, Ottawa.

Gromadzi, M. (1970). Breeding communities of birds in mid-field afforested areas. *Ekologia Polska,* **18,** 307–350.

Hall, E. R. & Kelson, K. R. (1959). *The Mammals of North America.* Ronald Press Co., New York.

Levenson, J. B. (1976). *Forested woodlots as biogeographic islands in an urban-agricultural matrix.* D.Phil. thesis, University of Wisconsin-Milwaukee.

MacArthur, R. (1964). Environmental factors affecting bird species diversity. *American Naturalist,* **96,** 387–397.

MacArthur, R. & Wilson, E. O. (1967). *The Theory of Island Biogeography.* Princeton University Press, Princeton, New Jersey.

MacClintock, L., Whitcomb, R. F. & Whitcomb, B. L. (1977). II. Evidence for the value of corridors and minimization of isolation in preservation of biotic diversity. *American Birds,* **31,** 6–16.

Madison, D. M. (1977). Movements and habitat use among interacting *Peromyscus leucopus* as revealed by radiotelemetry. *Canadian Field-Naturalist,* **91,** 273–281.

Moore, N. W. & Hooper, M. D. (1975). On the number of bird species in British woods. *Biological Conservation,* **8,** 239–250.

Oxley, D. J., Fenton, M. B. & Carmody, G. R. (1974). The effects of roads on populations of small mammals. *Journal of applied Ecology,* **11,** 51–59.

Preston, F. W. (1962). The canonical distribution of commones and rarity: Part II. *Ecology,* **43,** 410–431.

Prestt, I. (1971). An ecological study of the viper (*Vipera berus*) in southern Britain. *Journal of Zoology, London,* **164,** 373–418.

Terborgh, J. (1975). Faunal equilibria and the design of wildlife preserves. *Tropical Ecological Systems: Trends in Terrestrial and Aquatic Research.* (Ed. by F. B. Golley and E. Medina), pp. 369–380. Ecological Studies II. Springer-Verlag, New York.

Whitcomb, B. L., Whitcomb, R. F. & Bystrak, D. (1977). III Long-term turnover and effects of selective logging on the avifauna of forest fragments. *American Birds,* **31,** 17–23.

Whitcomb, R. F. (1977). I. Island biogeography and "habitat islands" of eastern forest. *American Birds,* **31,** 3–5.

Usher, M. D. (1973). *Biological Management and Conservation.* Chapman and Hill, London.

Willis, E. O. (1974). Populations and local extinctions of birds on Barro Colorado Island, Panama. *Ecological Monographs,* **44,** 153–169.

Wilson, E. O. & Willis, E. O. (1975). Applied biogeography. *Ecology and the Evolution of Communities.* (Ed. by M. L. Cody and J. M. Diamond), pp. 522–536. Belknap Press, Cambridge, Mass.

(*Received* 18 *July* 1978)

ORNIS SCANDINAVICA 14: 97–103. Copenhagen 1983

Bird numbers across edges between mature conifer forest and clearcuts in Central Sweden

Lennart Hansson

Hansson, L. 1983. Bird numbers across edges between mature conifer forest and clearcuts in Central Sweden. – Ornis Scand. 14: 97–103.

Bird distributions were examined along transects extending from forest edges 250 m out on clearcuts and the same distance inside forests. Inventories were performed in spring, summer and autumn 1980, winter 1980/81 and spring 1981.
67% of all observations were made in forest parts. However, there was a seasonal change with relatively more observations on clearcuts in summer and autumn. There were considerably more observations and higher densities in an outer forest belt of c. 50 m than deeper inside the forest. This was apparent for most species but particularly in the common species *Erithacus rubecula*, *Fringilla coelebs* and certain *Parus* spp. There were few pronounced edge species (*Anthus trivialis* and *Parus major* dominating in numbers). Fewer individuals of open land species were observed in the first 100 m of clearcuts than in their centres. There the bird fauna differed completely from that in the forest, consisting of just a few species of open habitats, such as *Emberiza citrinella* and *Lanius collurio*.
The high occurrence of tree-gleaning species in the outer forest belt is suggested to depend on a rich supply of insects. The low density of open land species on clearcuts close to the forest may be due to several factors generally important on habitat islands as varying intra- and interspecific competition and risks for ambush predation.

L. Hansson, Dept of Wildlife Ecology, Swedish University of Agricultural Sciences, S-750 07 Uppsala, Sweden.

1. Introduction

Most bird species have increased in numbers during the last 50 years in South Finland (Järvinen and Väisänen 1978), and presumably also in Central Scandinavia. This trend has mainly been attributed to changes in habitats, especially in the forests (Järvinen and Väisänen 1979, Haila et al. 1980).

Changes in forest habitat may be due to changed areal extension of a certain forest type, alteration in habitat quality (e.g. due to grazing livestock) and effects of the composition in the total landscape of various types or stages of forest (cf. Hansson 1977). One aspect of forest composition (the third effect) will be treated here.

In the taiga landscape of Central Fennoscandia the clear-cutting of forest is the main contributor to the habitat mosaic. Ecotones between two habitats are usually assumed (e.g. Odum 1971) to contain both more species and individuals than pure communities on either side due to greater structural complexity. However, the edges between forests and clearcuts and following reforestations are very sharp. Thus, my null hypothesis is that such forest edges do not show any "ectotone effects" and that the respective bird communities of forests and clearcuts stop abruptly at the forest edge.

2. Study area and methods

The studies were performed in an extensive area with comparatively old forest (c. 100 yr) and large clearcuts, c. 30 km W of Uppsala (59°55′N, 17°20′E) in Central Sweden. The fertility of the moraine soil was inter-

Accepted 20 December 1982

© ORNIS SCANDINAVICA

mediate-poor with partly exposed bedrock. The forest consisted of spruce *Picea abies* and pine *Pinus silvestris* in roughly equal numbers. On the clearcuts, forest was regenerated from a few remaining seed pines. However, small birches *Betula* spp. were more common than pine seedlings on the clearcuts. Some tall aspens *Populus tremula* were the only other trees on the clearcuts besides the seed pines but they had been killed by debarking for silvicultural sanitation.

Bird censuses were performed at the largest clearcuts available, being five-eight years old. The latter had radii of about 300 m, and birds were censused along lines extending 250 m into the clearcut and 250 m into the forest from the forest edge (Fig. 1). Six such transects were marked out with intersections every 25 m and border flags 25 m to each side of the line. The inventories were technically performed according to Järvinen and Väisänen (1975), with notes on pairs and singing or warning birds in the springs (May-June) of 1980 and 1981. However, similar inventories were also performed in summer (July) and autumn (September) 1980 and in winter (February) 1980/81, when all bird observations were taken into account. The only exceptions were birds flying above the tree tops, as mainly happened in autumn. The observations were marked out on maps and either entered for the various sections of the transect (inside the "main belt" according to Järvinen and Väisänen 1975) or as outside the transect (in the "supplementary belt"). In the latter case only observations in forest and on clearcuts were distinguished. In spring, censuses were performed before 09.00 hr and at all seasons before noon. About 15 in-

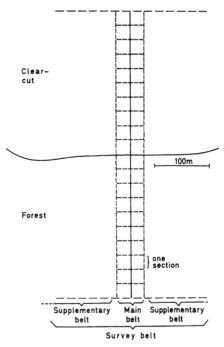

Fig. 1. Design of transects and definition of terms used for describing the inventories.

Tab. 1. Distributions of all observations with regard to habitat and season.

Species	Spring 1980 14 inventories		Summer 1980 15 inventories		Autumn 1980 14 inventories		Winter 1980 15 inventories		Spring 1981 13 inventories	
	Forest	Clearcut	Forest	Clearcut	Forest	Clearcut	Forest	Clearcut	Forest	Clearcut
Columba palumbus L.	33	4	27	3	4	0	0	0	27	0
Dendrocopos major (L.)	2	11	32	30	21	19	24	6	6	6
Anthus trivialis (L.)	12	39	19	37	3	28	0	0	30	36
Erithacus rubecula (L.)	143	11	50	5	22	8	0	0	81	3
Turdus merula L.	19	5	22	3	1	1	0	0	7	0
Turdus philomelos C. L. Brehm	34	9	31	12	44	21	0	0	25	5
Phylloscopus trochilus (L.)	54	17	60	20	4	0	0	0	59	7
Regulus regulus (L.)	70	0	86	0	132	6	71	3	42	0
Parus montanus Conrad	15	2	48	14	27	9	20	5	2	1
Parus cristatus L.	30	0	44	1	33	2	18	0	18	0
Parus ater L.	18	1	13	0	7	4	6	0	15	1
Parus major L.	7	13	14	21	7	12	1	2	13	22
Lanius collurio L.	0	6	0	70	0	2	0	0	0	5
Fringilla coelebs L.	245	21	80	13	45	32	0	0	180	3
Carduelis spinus (L.)	76	30	13	11	5	9	0	3	53	20
Loxia spp.	6	12	3	18	8	22	11	23	2	8
Emberiza citrinella L.	1	72	0	47	0	37	3	4	0	70
Remaining 39 species	64	58	40	44	53	68	8	5	39	69
All bird species	829	311	582	349	416	280	162	51	599	256

98

ventories of each of the six transects were made in each season.

There is one serious source of error when examining animal distributions on the basis of several inventories along the same transects: Observations of resident animals with home range radii greater than the length of each section may accumulate in the end sections at repeated censuses. Such apparent surpluses were obvious for some species (e.g. *Erithacus rubecula* in Fig. 2A) but not for others. However, they constitute a theoretical reason not to expect random distributions inside homogeneous habitats, are called "edge effect" in trapping studies (cf. Stenseth and Hansson 1979) and can be corrected for either by tedious computations or by excluding the outer (end) belts or sections. The latter method will be used here.

3. Results

3.1. Bird distributions with regard to habitat and season

67% of all bird observations (3835 from the "survey belt") derive from the forest part of the transects (Tab. 1). Also the number of common species was larger in the forest. However, the distribution between habitats

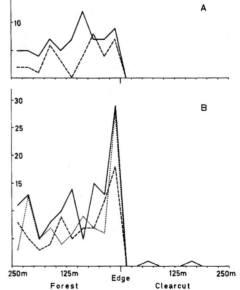

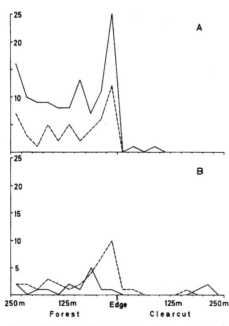

Fig. 2. Distribution of all observations in the main belt of *Erithacus rubecula*. A. Spring 1980 (———) and spring 1981 (— — —). B. Summer (— — —) and autumn (———) 1980.

Fig. 3. Distribution of all observations in the main belt of *Regulus regulus*. A. Spring 1980 (———) and spring 1981 (— — —). B. Summer 1980 (— — —), autumn 1980 (———) and winter 1980/81 (· · · · ·).

changed between seasons for certain species, as established by χ^2-tests.

Dendrocopus major was more restricted to the forest in winter than in the other seasons (P < 0.01). *Anthus trivialis* (P < 0.01) and *E. rubecula* (P < 0.01) showed a higher proportion of observations on clearcuts in autumn and *Fringilla coelebs* (P < 0.001) and *Carduelis spinus* (P < 0.01) were comparatively more abundant on clearcuts in both summer and autumn. In the whole material (P < 0.001) there was also a significant change in the proportions of observations in forest and on clearcuts, with an increase of observations on clearcuts in summer and autumn.

3.2. Distributions in relation to the forest edge

The observations in the main belt were used to examine the detailed distribution of the birds. In order to correct for possible accumulations of observations in the transect ends, two sections (cf. Figs 2–8) were excluded at each end in the χ^2-tests of the homogeneity of the observations with regard to various transect sections.

Figs 2–5 show clearly for a series of forest species a much higher number of observations close to the forest edge than further inside the forest. The numbers in the two first sections from the edge were compared with

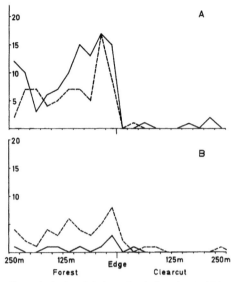

Fig. 4. Distribution of all observations in the main belt of *Fringilla coelebs*. A. Spring 1980 (———) and spring 1981 (– – – –). B. Summer 1980 (– – – –) and autumn 1980 (———).

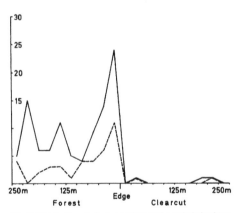

Fig. 5. Distribution of all observations in the main belt of two forest titmice: *Parus cristatus* (———) and *Parus ater* (– – – –).

no significant differences in the springs of either 1980 or 1981 but with clear differences in summer ($P < 0.001$), autumn ($P < 0.001$) and also in winter ($P < 0.001$). *F. coelebs* (Fig. 4) showed a surplus at the forest edge both in spring 1980 ($P < 0.01$) and 1981 ($P < 0.001$) as well as in summer and autumn combined ($P < 0.01$). Two species of titmice showed a similar distribution (Fig. 5), i.e. in all observations of *Parus cristatus* ($P < 0.001$) and *P. ater* ($P < 0.001$).

However, two other species of titmice showed another distribution with more occurrences on the clearcut (Fig. 6A). In *P. montanus* the total data showed the same edge pattern as in the previous forest species ($P < 0.001$) but there were also a substantial number of occurrences far out on the clearcut. *P. major* showed a peak in observations on both sides of the edge and it may be a species especially favoured by forest edges. The homogeneity of observations in sections 1–2 vs 3–8 was examined for both forest and clearcut and both deviated significantly ($P < 0.01$ and $P < 0.001$ respectively). Also thrushes appeared to a fairly great extent on the clearcuts (Fig. 6B) but the most obvious pattern was an excess of observations just inside the forest edge in *Turdus philomelos* ($P < 0.01$). Both *D. major* and *A. trivialis* appeared both in forest and on clearcuts (Fig. 7) but their respective peak abundances were in forest and clearcut parts of the edge. They were more common in

numbers on sections 3–8 inside the forest. In *E. rubecula* (Fig. 2) significant differences appeared in spring 1980 ($P < 0.001$) and spring 1981 ($P < 0.001$) and in summer ($P < 0.001$) but not in autumn. In *Regulus regulus* (Fig. 3) the pattern was the opposite, with

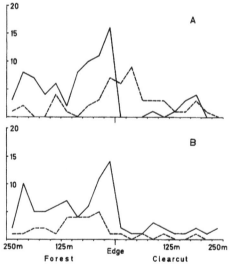

Fig. 6. Distribution of all observations in the main belt of bird species appearing both in forest and on clearcuts. A. Two titmice species: *Parus montanus* (———) and *Parus major* (– – – –). B. Two thrush species: *Turdus philomelos* (———) and *Turdus merula* (– – – –).

100

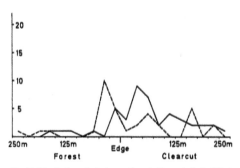

Fig. 7. Distribution of all observations in the main belt of bird species particularly common at the forest edge: *Anthus trivilalis* (———) and *Dendrocopos major* (– – – –).

the centres of the clearcuts than far inside the forest. However, the latter two species showed also seasonal heterogeneity in abundance in the two habitats (Tab. 1).

Only few species were restricted to the clearcuts. The pattern in this habitat deviated considerably from that in the forest. Few birds from these species were observed in the four first sections close to the forest edge while the number was higher at the centres of the clearcuts (Fig. 8). Therefore, the homogeneity in distributions between sections 1–4 and 5–8 was evaluated. *Emberiza citrinella* and *Lanius collurio* were most common with clearly significant clumping, with regard to all observations, near the clearcut centres (P < 0.001 in both cases). When observations of some sparse clearcut species, as *Lullula arborea* and *Saxicola rubetra*,

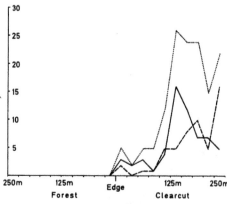

Fig. 8. Distribution of all observations in the main belt of bird species restricted to the clearcut: *Emberiza citrinella* (———), *Lanius collurio* (– – – –) and these two species plus *Lullula arborea* and *Saxicola rubetra* (· · · · · ·).

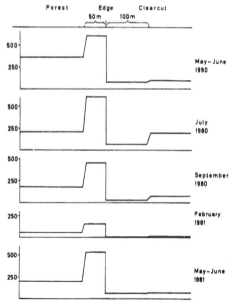

Fig. 9. Distribution of all observations in the main belt in relation to distances from the forest edge.

were added to the previous two species this pattern was further enforced (P < 0.001).

The total distribution of observations showed a pronounced peak just inside the forest edge (Fig. 9) due to the numerical dominance of forest species. The distribution on the clearcuts was fairly even which depended on two compensatory factors, i.e. forest species appearing in small numbers close to the forest and species of open habitats appearing in abundance only about a hundred meters from the forest edge. However, a higher abundance on clearcut centres was especially obvious in summer and at the same time the forest side of the edge showed much higher abundance than the "deep forest". Also in the winter the abundance of birds was especially high a short distance inside the forest edge.

3.3. Bird densities at various distances from the forest edge

The total number of birds observed within a belt of 25 m on each side of an observer has been used for density estimation (Alatalo 1978, Nilsson 1979). However, then a decrease emerges at least in the deep forest until summer and autumn densities. This decline is hardly real as the production of juveniles ought to cause an increase. Therefore, spring densities were estimated from the means of each transect according to Järvinen

Tab. 2. Spring densities (pairs km⁻²) and total weight (g km⁻²) (according to Järvinen and Väisänen 1977) of birds in the first 50 m of forest from the edge ("forest edge"), further inside the forest ("deep forest"), on the first 100 m of clearcut from the edge ("clearcut edge") and further out on the clearcuts ("clearcut centres").

Species	Deep forest 1980	Deep forest 1981	Forest edge 1980	Forest edge 1981	Clearcut edge 1980	Clearcut edge 1981	Clearcut centre 1980	Clearcut centre 1981
Columba palumbus L.	3	2	3	2	0	0	0	0
Dendrocopos major (L.)	0	0	0	7	2	2	4	0
Anthus trivialis (L.)	4	5	0	14	12	13	5	4
Erithacus rubecula (L.)	36	18	70	51	8	0	0	0
Turdus merula L.	8	2	3	3	0	0	3	0
Turdus philomelos C. L. Brehm . .	6	4	10	6	3	0	0	2
Phylloscopus trochilus (L.)	4	6	17	16	4	0	0	0
Regulus regulus (L.)	30	19	34	27	0	0	0	0
Parus montanus Conrad	8	1	4	0	1	0	0	0
Parus cristatus L.	11	3	13	23	0	0	0	0
Parus ater L.	8	4	14	20	1	0	0	0
Parus major L.	0	0	11	14	5	12	6	5
Lanius collurio L.	0	0	0	0	0	3	8	0
Fringilla coelebs L.	43	36	78	80	3	2	3	0
Emberiza citrinella L.	0	0	0	0	9	0	20	43
Carduelis spinus (L.)	12	9	12	9	3	3	3	3
All bird species	182	114	278	276	65	51	66	73
Weight of bird community	10100	6100	12800	12800	2800	1900	3800	3500

and Väisänen (1975), using their linear model. Although not giving absolute densities this method enables comparisons to be made with other investigations.

The most common birds in the forest were, according to these estimates, *E. rubecula*, *R. regulus* and *F. coelebs* and on the clearcut *A. trivialis*, *P. major* and *E. citrinella* (Tab. 2). *C. spinus* was also fairly common but as most individuals of this species were observed flying at or below tree-top level it was impossible to establish the exact location of breeding pairs. The only other species with similar behaviour was *Loxia* sp. (*curvirostris?*). A higher density just inside the forest edge than in the deep forest was apparent for most species except *T. merula* and *P. montanus*. However, the difference was not so pronounced in *R. regulus* as in *E. rubecula* and *F. coelebs*. Several common forest species showed a low density on the clearcut close to the forest while *E. citrinella* had its highest density in clearcut centres.

The total number of bird pairs was 3–4 times as large in the forest as on the clearcuts. A similar or higher relation applies to their weights, computed after body weights given by Järvinen and Väisänen (1977). However, these estimates were heavily influenced by the abundance of *Columba palumbus*. Small densities of rare, heavy species such as *Tetrao tetrix* L. on the clearcuts and *Tetrao urogallus* L. in the forest were excluded in this computation.

The bird densities were generally lower in spring 1981 than in spring 1980. However, the densities in the forest close to the edge and on the clearcuts centres were similar while large differences appeared deep inside the forest and on the clearcut close to the edge. This may indicate a higher stability in optimal habitats.

4. Discussion

The observations give a fairly good picture of the main bird composition. Only night-active species (owls, nightjar) were definitely underrepresented. In spring almost only singing and warning birds were detected. Thus, spring observations should be comparable to those from the other seasons. However, present correction factors for density estimation seem only applicable to spring observations as the mean number of birds observed declined towards summer. Other and higher correction factors should be developed for the remaining seasons. Present spring density estimates agree in a general way with estimates from northern coniferous forests obtained with other methods (e.g. Hogstad 1975).

The clearcuts do not contribute much to the observed increase in species number and density of the Fennoscandian bird fauna. In fact, they will cause a decrease, except in certain species adapted to open areas, such as *E. citrinella* and *L. collurio*. The latter are favoured by large clearcut areas and were usually not found on very small clearcuts. Thus, the increase in species number with area found in theoretical island biogeography (MacArthur and Wilson 1967) seems to apply. However, the mechanisms may be different. This relation depends on a presumed balance between immigration and extinction but there might be several simpler reasons for the present area-dependence. It can be explained by ambush predation from forest (possibly by Jays *Garrulus glandarius* L.), by interspecific competition (possibly by forest species exploiting clear-cuts close to the forest edge), with lessened intraspecific competition at edges and little territorial defense there or with the appearance of the open habitat species mainly at activity centres of necessarily large territories. The latter three explanations are applicable only if the species of open habitats show stronger habitat selection than the forest species. All four factors will affect estimates of both species number and density on areas of varying sizes. They will also be generally important for explaining area-dependence on habitat islands as an important topic in the theory of nature reserve designs (cf. Simberloff and Abele 1982).

However, clearcutting may have contributed indirectly to the great increase in bird numbers. The density of many bird species becomes greater in the forest close to the clearcut than further inside the forest. The causal relationships may be a weakening of the trees by sudden exposure to sun and winds causing drying and severe insect attacks. It is well-known to entomologists (e.g. B. Ehnström, pers. comm.) that edge trees harbour a richer insect fauna than the deep forest but no hard data seem available. In addition, higher primary productivity has been measured in edge belts than further inside forests (Ranney et al. 1981).

Forest islands and edge belts contain fewer bird species than extensive forest tracts of deciduous forest in the eastern USA (Galli et al. 1976, Whithcomb et al. 1981). This was not any pervading feature of this study but some rare species such as *T. urogallus*, *Phylloscopus collybita* (Vieill.) and *Fringilla montifringilla* L. were only observed during the breeding season deep inside the forest. Another deviation from the American study was that "real edge species" were few in Central Sweden. However, the American forests were surrounded by bushy edge zones and thus showed an ectone effect. The latter may well consist of both increased number of species due to structural complexity and increased density of e.g. forest species due to an increased resource base.

The ground-feeding and mainly forest-living thrushes did not show any pronounced change in numbers in relation to the forest edge. The edge effect was largest in the tree-gleaning species (*Parus* spp., *Ph. trochilus*, *F. coelebs*). A similar change could be expected for *R. regulus* but appeared only outside the breeding seasons. Ulfstrand (1976) similarly found that the Goldcrest was strongly bound to spruce at breeding but foraged in other trees in winter and that there might be a limitation of suitable nest or forage trees for this species during the breeding season or possibly competition from other species.

Several species used clearcuts to a greater extent in summer-autumn than in spring. This agrees with findings by Alatalo (1981), who found that habitat breadths were generally narrowest during the breeding time. Still, the clearcuts will be important to these species if survival was improved as a result of clearcut use and if territoriality is not limiting breeding numbers. Breeding density varied strongly so the second condition does not

seem to be preventing. Clearcuts may be especially accessible during migration but their importance in this respect has to be examined further.

Acknowledgements – I thank S. Ulfstrand and S. Svensson for comments.

References

Alatalo, R. V. 1978. Bird community energetics in a boreal coniferous forest. – Holarct. Ecol. 1: 367–376.
– 1981. Habitat selection of forest birds in the seasonal environment of Finland. – Ann. Zool. Fennici 18: 103–114.
Galli, A. E., Leck, C. F. and Forman, R. T. T. 1976. Avian distribution patterns in forest islands of different sizes in central New Yersey. – Auk 93: 356–364.
Haila, Y., Järvinen, O., and Väisänen, R. V. 1980. Effects of changing forest structure on long-term trends in bird populations in SW Finland. – Ornis Scand. 11: 12–22.
Hansson, L. 1977. Landscape ecology and stability of populations. – Landscape Plann. 4: 85–93.
Hogstad, O. 1975. Interspecific relations between willow warbler (*Phylloscopus trochilus*) and brambling *Fringilla montifringilla*) in subalpine forests. – Norw. J. Zool. 23: 223–234.
Järvinen, O. and Väisänen, R. A. 1975. Estimating relative densities of breeding birds by the line transect method. – Oikos 26: 316–322.
– and Väisänen, R. A. 1977. Constants and formulae for analysing line transect data. Helsinki, 10 pp.
– and Väisänen, R. A. 1978. Long-term population changes of the most abundant south Finnish forest birds during the past 50 years. – J. Orn. 119: 441–449.
– and Väisänen, R. A. 1979. Climatic changes, habitat changes, and competition: dynamics of geographical overlap in two pairs of congeneric bird species in Finland. – Oikos 33: 261–271.
MacArthur, R. H. and Wilson, E. O. 1967. The theory of island biogeography. – Princeton Univ. Press, Princeton, N.J.
Nilsson, S. G. 1979. Seed density, cover, predation and the distribution of birds in a beech wood in southern Sweden. – Ibis 121: 177–185.
Odum, E. P. 1971. Fundamentals of ecology, 3rd ed.– Philadelphia.
Ranney, J. W., Bruner, M. C. and Levenson, J. B. 1981. The importance of edge in the structure and dynamics of forest islands. – In: Burgess, R. L. and Sharpe, D. M. (eds.). Forest islands dynamics in man-dominated landscapes. New York, pp. 67–96.
Simberloff, D. and Abele, L. G. 1982. Refuge design and island biogeographic theory: effects of fragmentation. – Am. Nat. 120: 41–50.
Stenseth, N. C. and Hansson, L. 1979. Correcting for the edge effect in density estimation: explorations around a new method. – Oikos 32: 337–348.
Ulfstrand, S. 1976. Feeding niches of some passerine birds in a South Swedish coniferous plantation in winter and summer. – Ornis Scand. 7: 21–27.
Whitcomb, R. F., Robbins, C. S., Lynch, J. F., Whitcomb, B. L., Klimkiewicz, M. K. and Bystrak, D. 1981. Effects of forest fragmentation on avifauna of the eastern deciduous forest. – In: Burgess, R. L. and Sharpe, D. M. (eds.). Forest island dynamics in man-dominated landscapes. New York, pp. 125–206.

Biological Conservation **34** (1985) 333–352

Bird Communities in Small Woods in an Agricultural Landscape: Effects of Area and Isolation

P. Opdam, G. Rijsdijk & F. Hustings

Research Institute for Nature Management, PO Box 46, 3956 ZR Leersum,
The Netherlands

ABSTRACT

The distribution of breeding bird species in 68 small woodlots in two areas of agricultural landscape was investigated. Effects of area, habitat variation, interpatch distance, and distance from extensive woods were analysed with the help of multivariate techniques. No correlation was found between number of breeding bird species and isolation variables, but the degree of isolation was shown to affect the number of bird species restricted to mature woods. Species showed different responses to changes of area or isolation variables. It is concluded that, for several species of woodland birds, patches of mature woodland can be regarded as habitat islands. Some indications are given to apply the regression models found to landscape planning.

INTRODUCTION

The distribution of birds in a patchy environment is determined primarily by the presence of suitable habitat. The size of the habitat patches may, however, affect the presence of species, since patch size is related to population size and thereby to extinction rate (Jones & Diamond, 1976; Wright & Hubbell, 1983). At the same time, the distance between the patches may be inversely related to the frequency of interpatch dispersal (Whitcomb *et al.*, 1981; Opdam, 1983; Lynch & Whigham, 1984).

The equilibrium theory of MacArthur & Wilson (1967) combined extinction and immigration rates in a dynamic model for the number of species on oceanic islands, and various authors later extended the

333

Biol. Conserv. 0006-3207/85/$03·30 © Elsevier Applied Science Publishers Ltd, England, 1985. Printed in Great Britain predictions of the model to habitat patches on the mainland. The empirical support for this theory and the implications for reserve design have been discussed exhaustively (Diamond, 1975; Connor & McCoy, 1979; Gilbert, 1980; Margules *et al.*, 1982). However, evidence supporting an area-dependent extinction rate and a distance-dependent immigration rate is scanty (Gilbert, 1980; Margules *et al.*, 1982). For mobile species, such as most birds, Margules *et al.* (1982) and Ambuel & Temple (1983) even postulate that isolation of reserves will not affect local immigration.

In this paper the distribution of breeding birds is discussed in relation to the size and degree of isolation of wooded patches. It is clear that in the modern cultivated landscape, patches of habitat tend to decrease in size and become more and more widely separated from each other. Quantification of the relation between landscape structure and the distribution and survival of fauna groups is an important task for nature conservationists. We chose birds for the present analysis because they can be censused quickly, and woods because great numbers of birds breed in this type of habitat.

Effects of patch size on the number of breeding-bird species have been studied often, but in most of these studies, the effects of size *per se* were not separated from those of habitat diversity increasing with size (Connor & McCoy, 1979, but see Ambuel & Temple, 1983). Furthermore, in most studies the variable 'number of species' encompassed all the species of a taxonomic order. However, the study done by Humphreys & Kitchener (1982) indicated, as expected, that species restricted to habitats in isolated patches may show species–area relations differing strongly from those of species that use adjacent habitats as well, and Galli *et al.* (1976), Whitcomb *et al.* (1981) and Opdam & Retel Helmrich (1984) have also pointed to species-specific relations with patch size.

The impact of isolation on birds in habitat patches is not quite clear. An effect has been claimed by Fritz (1979), Whitcomb *et al.* (1981) and Opdam *et al.* (1984). In the first-mentioned study the spruce grouse *Canachites canadensis* was shown to occupy habitat remnants more often the closer these patches were to another patch. In the other studies, large sets of data were used to analyse the island effect of small woods on the number of breeding birds. However, the results of Whitcomb *et al.* (1981) were not unambiguous because in their data wood area and degree of isolation were strongly interdependant, whereas Opdam *et al.* (1984) could not account for possible habitat variation affecting the distribution of bird species. In a study on bird-species numbers in British woods, Helliwell (1976) found no effect of isolation, probably because he had included in the analysis too many species for which the woods did not represent isolated habitat patches. In Vuilleumier's (1970) analysis of birds on mountain tops the number of species was even positively correlated with distance, but after re-examination of the data Mauriello & Roskoski (1974) claimed that the coefficient of distance should have been negative.

The results of the studies done so far are not very convincing, the more so because differences between species were not quantified. In a recent paper, however, Lynch & Whigham (1984) presented an excellent analysis of the avian communities of 185 forest patches ranging in size from 5 to more than 1000 ha and divided over two regions. They could demonstrate the impact of isolation on the distribution of most of the species they studied. In our study we followed more or less the same strategy and selected in an agricultural landscape 68 small woodlots, ranging in size from less than 1 ha to 20 ha. The woodlots differed substantially as to degree of isolation, but otherwise we tried to keep habitat characteristics as uniform as possible. The remaining habitat variation was measured. The presence of all breeding bird species was determined and the following predictions were tested:

(a) Woodlot size and degree of isolation affect the number of bird species.
(b) Species that are restricted to woods occur less often in small woodlots than in large ones, even after corrections for differences in habitat structure.
(c) These species are absent most often in the most isolated woodlots.
(d) Species differ as to the degree to which they are affected by area and isolation.

METHODS

Two study areas were chosen (Fig. 1). One of them, the Gelderse Vallei (Fig. 1a), an agricultural area of about 400 km^2 in the central part of The Netherlands, is surrounded on three sides by extensive mixed forests. Many woods, wooded banks, hedgerows and small groups of trees are scattered over the area, but most of the woods are coppice or half-grown. Within this area 40 mature deciduous woods were selected. The second

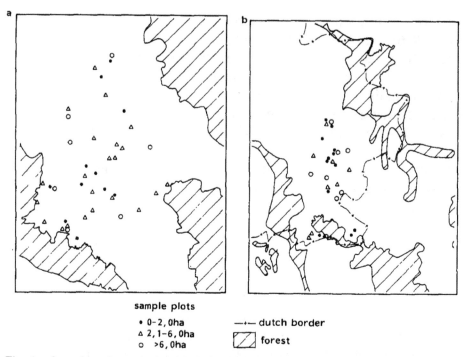

Fig. 1. Location of sample sites in (a) the Gelderse Vallei area and (b) the southern part of the province of Limburg (Zuid Limburg), showing the outline of surrounding forests. (Scale: 1 cm equals 22 km).

area is situated in the southernmost part of the province of Limburg and extends over 200 km². On the north and south, extensive mixed forests border the area. In this agriculturally-used landscape deciduous woods are scattered; among these 28 mature woods were suitable for our study. The Limburg area (Fig. 1b) is more open than the Gelderse Vallei, with fewer wooded sites and groups of trees between the selected woods.

Although the sampling plots were selected for uniformity of the tree layer, some divergence between the sites had to be accepted. Moreover, the coverage, structural heterogeneity, and species composition of the shrub layer varied considerably. Therefore, habitat diversity variables were recorded in the field (Table 1). Most edges of the woodlots were sharp changes from trees to cultivated fields.

From maps 1:25 000, isolation variables were calculated, and a map 1:10 000 was used to estimate the area of the woodlots (Table 1). Because oblong woods have a higher edge-to-interior ratio than compact woods and the forest-edge may represent a different habitat type, the woods were

TABLE 1

Variables for Habitat and Landscape Attributes Determined for Each Sample Site

Habitat variables	
Trunk	Trunk diameter, a measure of age differences between trees (2 classes)
CovH	Coverage of the herb layer (3 classes)
CovS	Coverage of the shrub layer (3 classes)
StrDiv	Structural diversity of tree and shrub layers (3 classes)
	Dominant tree species (9 types)
Variables of size and landscape	
SW	Shape of the woodlot (compact or oblong)
A	Area of mature deciduous wood (in ha)
TA	Total area of wood containing the selected site
DLF	Distance from large forest areas, measured as the shortest distance between the edge of the woodlot and the edge of an extensive forest area
DW	Distance from nearest mature deciduous wood larger than 20 ha (in km)
DAW	Distance from nearest wood of any type larger than 20 ha (in km)
NDW	Number of mature deciduous woods within a 3 km radius from the centre of the woodlot
AAW	Area of wood of any type within a 3 km radius from the centre of the woodlot (in ha)
WB	Proximity of wooded banks or rows of mature trees

classified as oblong or compact. Some of the woodlot attributes appeared to be correlated (Table 2).

Each woodlot was visited five times, between March and the middle of June of 1983, for periods of 15 min each. During each visit all parts of the wood were scanned for birds showing behaviour indicating the presence of a territory. In the largest woodlots more time was needed to cross the whole area, at most between 30 and 45 min. For all species, the first relevant observation was recorded on each visit and the presence of a territory was derived from the combined visits on the basis of the criteria in Hustings *et al.* (1985). For most sedentary species at least two observations made on separate visits were required, but for some species for which observation efficiency was low we accepted a territory on the basis of only one record in the post-migration period. In this way a list of breeding birds was obtained for each woodlot.

Three sets of analyses were carried out: (a) with the number of species restricted to woods; (b) with the number of species restricted to mature deciduous woods; and (c) with the presence/absence of particular species. The last of these variables assumes either 0 or 1, the other two range from

338   *P. Opdam, G. Rijsdijk, F. Hustings*

TABLE 2
Significant Correlations between Woodlot Attributes pertaining to Area and Isolation

Gelderse Vallei

	A	DAW	WB	TA	DLF	DW	NDW	AAW
A[a]								
DAW	—							
WB	—	5%						
TA	0·1%	—	—					
DLF	—	0·1%	—	—				
DW	—	—	—	—	0·1%			
NDW	—	5%	—	—	—	—		
AAW	—	1%	—	—	0·1%	—	—	

Limburg

	A	TA	DLF	DW	DAW	NDW	AAW	WB
A								
TA	0·1%							
DLF	—	—						
DW	—	—	1%					
DAW	—	—	1%	0·1%				
NDW	—	—	0·1%	—	—			
AAW	—	—	0·1%	0·1%	1%	1%		
WB	5%	5%	—	—	—	—	—	

[a] See Table 1.

7 to 39 and from 1 to 11, respectively. The variation of these variables in dependence of the other woodlot attributes was analysed by stepwise multiple linear regression using GENSTAT (Alvey *et al.*, 1982). For several variables (e.g. size and most isolation variables) we assumed non-linearity in their relation to species number and therefore applied logarithmically transformed values.

RESULTS

All woodland species combined

As a rule, in studies on the island effect island attributes are related to the total number of species on the (habitat) island, excluding species that also use the habitats surrounding the island. Likewise, we excluded those birds feeding mainly on surrounding agricultural land, and studied the island effect on the remaining species, which we called woodlot species. Since these birds can inhabit different layers, i.e., the ground layer, the shrub layer, the canopy, or a combination of layers, some of them are not restricted to the type of wood under study, but could also occur in nearby or even adjacent plantations, coppices, hedgerows, or gardens.

A significant amount of variation in the number of woodlot species is explained by woodlot area and a variable of tree-layer diversity that accounts for the combination of oaks and beeches (Table 3). This habitat variable was strongly correlated with oakwood and the presence of large trees. At the 5% level of significance, none of the isolation parameters were selected.

Hence, for all species combined, an effect of isolation could not be demonstrated. Probably the isolation effect was too small to be measurable, because, for most of the species in the set, the woods did not represent true habitat islands. Woodlot size might be partially related to habitat diversity, especially as to the coverage, structural heterogeneity, and species composition in the shrub layer.

Although species–area relations are generally described in a double

TABLE 3

Summary of Stepwise Multiple Regression Analysis, Showing the Best Over All Fits for the Number of Bird Species in the Woodlots (Standard error between brackets)

$$S_{WG} = 16 \cdot 56 + 2 \cdot 42 \ln A + 2 \cdot 53 \, BO \qquad V^2 = 39 \cdot 3 \quad R^2 = 42 \cdot 3$$
$$\phantom{S_{WG} =} (0 \cdot 95) \quad (0 \cdot 75) \quad (1 \cdot 22)$$
$$S_{WL} = 19 \cdot 96 + 5 \cdot 16 \ln A - 9 \cdot 27 \, POP \qquad V^2 = 78 \cdot 1 \quad R^2 = 79 \cdot 8$$
$$S_{MG} = 5 \cdot 24 - 0 \cdot 55 \ln DLF + 0 \cdot 87 \ln A + 1 \cdot 30 \, BO \qquad V^2 = 47 \cdot 2 \quad R^2 = 51 \cdot 2$$
$$\phantom{S_{MG} =} (0 \cdot 48) \, (0 \cdot 20) \quad (0 \cdot 31) \quad (0 \cdot 51)$$
$$S_{ML} = 6 \cdot 00 + 1 \cdot 61 \ln A - 0 \cdot 85 \, DLF - 2 \cdot 69 \, POP - 1 \cdot 37 \, BIO \qquad V^2 = 86 \cdot 5 \quad R^2 = 88 \cdot 5$$
$$\phantom{S_{ML} =} (0 \cdot 33) \quad (0 \cdot 16) \quad (0 \cdot 22) \quad (1 \cdot 06)' \quad (0 \cdot 63)$$
$$S_{MG} = 4 \cdot 11 + 0 \cdot 96 \, BO + 0 \cdot 94 \ln A - 0 \cdot 73 \ln DNF + 1 \cdot 85 \, WB \qquad V^2 = 55 \cdot 3 \quad R^2 = 61 \cdot 0$$

S_W, All woodlot species
S_M, Species of mature deciduous woods
G, Gelderse Vallei
L, Limburg
A, Area
BIO, birch-oakwood
BO, beech-oakwood
POP, poplar-alder
DLF, Distance from large forest
DNF, Distance from the nearest forest larger than 20 ha

logarithmic form, we preferred a semi-logarithmic form because the regression on S gave a slightly better fit ($R^2 = 29.2$ against 27.5).

Species restricted to mature woods

Since we selected mature deciduous woods as habitat islands, isolation might be particularly relevant with respect to species restricted to woodlots of this type. Therefore a selection was made of 15 species that are found as breeding birds particularly in mature woods, at least in the areas under study (Table 4). Most of these species are numerous in the extensive forests surrounding the study areas. The effects of woodlot size and isolation on the number of these species (S_M) were assessed in the two areas separately. The results of the regression analysis (Table 3) show that in both areas woodlot size and distance to a large forest are significantly correlated with species number. Additionally, the regressions included variables for habitat differences. In the Gelderse Vallei beech–oakwood contained more species than pure oak stands, presumably because

TABLE 4

Occurrence of Bird Species of Mature Deciduous Woods in the 68 Plots. (Values are numbers of plots in which each species was recorded)

Species	Values
Black woodpecker *Dryocopus martius*	2[a]
Woodwarbler *Phylloscopus sibilatrix*	4[b]
Redstart *Phoenicurus phoenicurus*	4
Golden oriole *Oriolus oriolus*	5
Green woodpecker *Picus viridus*	7
Pied flycatcher *Ficedula hypoleuca*	8[a]
Hawfinch *Coccothraustes coccothraustes*	12[b]
Lesser spotted woodpecker *Dendrocopos minor*	15
Marsh tit *Parus palustris*	29
Nuthatch *Sitta europaea*	37
Great spotted woodpecker *Dendrocopos major*	40
Spotted flycatcher *Muscicapa striata*	54
Tree creeper *Certhia brachydactyla*	59
Blue tit *Parus caeruleus*	66
Chaffinch *Fringilla coelebs*	66

[a] In the Gelderse Vallei only
[b] In Limburg only

beeches offer more nesting holes. In the Limburg area two types of wood (poplar–alder and oak–birch) had fewer species than oakwood and beech–oakwood. These types of wood had mainly half-grown trees and presumably offered fewer feeding sites for bark-feeding birds and fewer nesting holes. Again, because of the better fit, we used a linear-dependent rather than a logarithmic-dependent variable.

The variable distance to large forest in the Gelderse Vallei regression model could be replaced by a combination of three variables. The total amount of variation accounted for by this more complicated model did not differ much from that with the simple one. Two of these variables are regarded as indicators of isolation (presence of tree rows and wooded banks around the woodlot and amount of woods within a 3 km radius). Shape of wood, the third variable, is considered as a habitat variable. Oblong-shaped woods, which contained fewer species, are affected most by the surrounding environment (e.g. microclimate) and may therefore be less suitable for typical woodland birds.

The coefficients of woodlot size did not differ significantly (t-test, $t = 1.17$) between the two areas. The greater amount of variation explained by this variable in the Limburg area is attributed to its greater range compared to the Gelderse Vallei (Fig. 2).

Since the sample of wooded sites was selected on the basis of tree-layer homogeneity, habitat diversity is rather small. None of the habitat variables was correlated with woodlot size. Hence we suggest that most, if not all, of the effect of woodlot size is due to the mere effect of area as such.

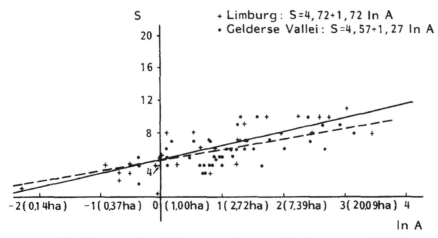

Fig. 2. Species-area plot of the two study areas. (Scale: 1 cm equals 22 km).

Distance to large forest has a somewhat greater impact in the Limburg area as compared with the Gelderse Vallei. This isolation variable ranged between 0·1 and 6·5 km in Limburg and between 0·1 and 9·8 km in the Gelderse Vallei. Calculated over the maximal range, the number of bird species diminished from 8 to 4·4 and from 6·5 to 4, respectively, all other variables being constant. However, the difference between the regression coefficients is not statistically significant (t-test, $t = 1·01$). A difference between the two areas might be due to the greater amount of woodland and rows or groups of trees in the Gelderse Vallei, which might increase the dispersal rate of woodland birds.

We conclude that distance from large forest area is a significant factor determining the distribution of birds inhabiting mature deciduous woods.

Analysis of separate species

Thus far, the number of species has been used to analyse the island effect in isolated woodlots. The underlying assumption was that all species are equal with respect to the relationship between woodlot area, population size, and extinction rate, as well as the extent to which their dispersal is hampered by interpatch distance. However, differences between species are plausible and may give more insight into the mechanism leading to the observed species–area relations. Therefore, we undertook an analysis at the species level. For this purpose, we used the presence/absence of a species in the woodlots as a dependent variable in logistic regression analysis. This meant, however, that the amount of variation became much smaller, especially for species occurring either in a very few or in almost all woodlots. The best possible solution to this problem was to combine all woodlots, which was acceptable because of the small differences between the two regions as to the effects of area and isolation.

The occurrence of the species in the woodlots is shown in Table 4. Species found in less that 10 % or more than 90 % of the samples were omitted from further analysis. The regression analysis was restricted to simple models, each containing one variable for size of the woodlot, one isolation variable and two habitat variables (oak-beech type and other woodland types). We also added a region variable (a dummy variable) to account for possible differences between the two areas under study. Three sets of regressions were run, each with a different isolation variable, namely distance from large forest, amount of woodland within 3 km radius, and proximity of wooded banks or rows of trees.

For most of the species none of the habitat variables explained a significant part of the variation, with the exception of the nuthatch and the great spotted woodpecker, which occurred more often in beech–oak woodland than in any of the other types. The coefficients for woodlot size are shown in Table 5. The ranking of the species according to these coefficients in the three sets of regressions remains roughly the same, especially if only significant coefficients are considered. This concordance can be seen as support for the reliability of these results.

TABLE 5
Area Effect in Woodland Birds
(Coefficients of correlation between frequency of occurrence and woodlot size, allowing for isolation and habitat differences. Only significant coefficients ($P < 0.10$) are given. The coefficients were calculated for three sets, using as an isolation measure area of woodland within 3 km radius (1). distance from large wood (2), and proximity of wooded banks and rows of trees (3).)

Species	*(1)*	*(2)*	*(3)*
Great spotted woodpecker	2·28*	2·42*	2·21*
Tree creeper	1·57*	1·62*	1·84*
Redstart	2·11	1·14	—
Nuthatch	1·10*	1·09*	0·99
Hawfinch	0·97*	0·93*	1·92
Marsh tit	0·87*	0·90*	1·02*
Lesser spotted woodpecker	0·70	0·69	0·62

* P smaller than 0·05

The regression coefficients for the isolation variables are presented in Table 6, except for the proximity of wooded banks, which showed very low T values. The probability of occurrence of nuthatch and marsh tit is significantly affected by the two isolation variables used and that of the redstart by the amount of nearby forest only, whereas for the lesser spotted woodpecker the coefficient of this isolation variable was almost significant. The coefficients of the two variables show a positive correlation.

A more detailed consideration of the relation between presence/absence of species and woodlot size may shed more light on the causal

344 *P. Opdam, G. Rijsdijk, F. Hustings*

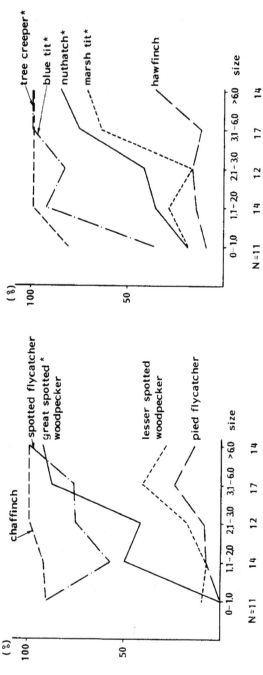

Fig. 3. Incidence curves showing the frequency of occurrence (in % of maximal frequency) according to five classes of wood size (in ha) of ten species restricted to mature deciduous woodland (Gelderse Vallei + Zuid Limburg). *Significantly differing from random distribution (X^2 test, $P < 0.05$).

mechanism underlying the distribution patterns found. In Fig. 3 the woodlots have been classified according to five size classes, which were chosen such that the woods were more or less evenly distributed over the size classes. The blue tit and chaffinch are absent in only a few of the smallest woods. The incidence curve of the spotted flycatcher suggests an effect of habitat rather than of woodlot size. In the Gelderse Vallei this species was recorded more often in beech–oak woods than in oak woods. The great spotted woodpecker and tree creeper occur in almost all woods exceeding 3 ha and 1 ha, respectively, but the woodpecker was not recorded in woodlots smaller than 1 ha. The stagnation in the upward trends in class 3 may be due to lack of suitable habitat. The nuthatch and marsh tit show no clear minimal habitat size: apparently they can live in even the smallest woodlots. Their absence in some of the woods must be due to lack of suitable habitat (cf. distribution of the great spotted woodpecker and tree creeper), to isolation or extinction. The distribution of the lesser spotted woodpecker is difficult to explain, except by referring to its very clumped distribution pattern in the Gelderse Vallei. The hawfinch was not recorded in the Gelderse Vallei woods, though it occurs in considerable numbers in the surrounding extensive forests of the Veluwe and the Utrechtse Heuvelrug. In the remaining woods the distribution is partially explained by an effect of woodlot size. The pied flycatcher is too scarce to draw reliable conclusions from its incidence curve.

We conclude that at the species level there is a good evidence for effects of woodlot size, especially in the great spotted woodpecker (minimal area) and in the tree creeper, nuthatch, marsh tit, and hawfinch. Isolation of woodlots is a significant factor in the distribution of the nuthatch, marsh tit and probably the redstart and lesser spotted woodpecker. Because the species are affected to different degrees, they show species-specific distribution patterns.

DISCUSSION

Evaluation of the bird survey method

Some researchers have been concerned about the influence of sampling methods on species-area relationships. Woolhouse (1983) argued that sampling time per unit area must be constant for all sites surveyed. For a qualitative survey of breeding-bird territories, this would mean for each wood and for each bird species, equal time spent in one territory. Because the position of territories cannot be known in advance, this condition is hardly attainable in practice, if at all. Further problems are associated with the differences in territory size and observing chance between the species and from the fact that isolated territory owners in a small woodlot may be less active in territorial display than birds in territories surrounded by neighbours. Also, one territory out of a cluster of several is more likely to be recorded solely by pure chance than an isolated one is, even if the sampling effort is the same.

We attempted to deal with these problems in several ways. In the first place, observation time was held constant in the smallest woods (up to 4–5 ha, depending on the shape). These woods were so small that it was highly probable that any bird singing would be noticed. In the larger woodlots, which were less easy to survey at a glance, observation time was increased, up to 45 min for a 30 ha wood, depending on the time needed to take a quick look at every section of the wood. Thus, sampling effort (time per unit area) was highest in the smallest woodlots, in agreement with the decreased territorial activities of isolated birds and the increasing chance of recording a single specimen with increasing population size. With this method, however, single small territories in large woods were under-estimated, and therefore differences between small and large woods may be somewhat underrated. Because in large woods most of the species in question were represented by several breeding pairs, this bias is assumed to be very small. Differences in recording chance between species were accounted for by the species-specific criteria for the number of records (in combination with recording date) required for the acceptance of a territory (Hustings *et al.*, 1985). These criteria were chosen such that all species were surveyed at an accuracy level of at least 90% (Kwak & Meijer, 1985).

The role of patch size

Lynch & Whigham (1984) concluded that the occurrence of many species can be predicted better from structural and floristic characteristics of woods than from patch size and isolation. However, in the uniform series of woods we have intentionally chosen, habitat differences play a minor role and the distribution of most bird species, as well as the composition of avian communities, are predicted best from patch size. The incidence curves (Fig. 3) indicate that with increasing woodlot size species are added in a fairly predictable sequence. Several explanations can be offered for this pattern:

(1) The bird community in a wood is a random sample of that in the nearest larger woods, and with increasing patch size species are added in order of their overall abundance. Species do not differ as to immigration or extinction rate. This 'random sampling hypothesis' does not imply any biological process and can be used as a null hypothesis to test alternative models (Connor & McCoy, 1979).

(2) Forest-interior species cannot persist in the smallest woods because of competition with edge species for nest sites or for food, or become extinct as a result of predation by predators from the surrounding landscape (Ambuel & Temple, 1983).

(3) In the smallest woods, populations have the highest extinction rates due to stochastic perturbations of the population number (MacArthur & Wilson, 1967).

(4) Species differ as to the minimal critical area they need to satisfy their food requirements during the breeding season.

The random-sampling hypothesis predicts that the slopes of the incidence curves are equal for all species. However, Fig. 3 suggests that this is not the case. Some of the species patterns seem to follow predictions from this hypothesis. Chaffinches and blue tits, the first species to be expected in the smallest woods, are more numerous than other species in the two regions, whereas the black woodpecker, green woodpecker and golden oriole are among the least abundant.

The hypothesis of Ambuel & Temple (1983) can only be tested experimentally by manipulation of competitors and predators. However, because the selected woodlots are so small that even their centre is reached by edge species or by predators coming from outside, hypothesis (2) does not offer a plausible explanation of the area effect we found.

According to Wright & Hubbell (1983), the extinction rate hypothesis (3) predicts sigmoid incidence curves. Because territories left open by the death of individuals may be filled by offspring as well as by immigrants, differences in fecundity and vagility of species should result in a different likelihood of their presence in equally sized woods not differing in isolation. The sigmoid incidence curves (Fig. 3) suggest that these predictions hold for the woodland birds in the study area. The immigration rate may affect the duration of the extinction interval, both in small woods, irrespective of isolation, and in remote woods, irrespective of their size. Arrangement of species according to the regression coefficient of area and distance variables (Tables 5 and 6) gives a remarkably similar sequence, which suggests that different immigration rates of species account for the effects of area and distance in a similar way.

TABLE 6
Isolation Effect in Woodland Birds
(Partial correlation coefficients for the logistic regression of presence of species on distance from large forest (DLF) and area of forest within a 3 km radius (AF3), allowing for variation in woodlot size and habitat differences)

Species	DLF	AF3
Redstart	—	0·0077*
Nuthatch	−0·94*	0·0031**
Hawfinch	−0·80	0·0037
Marsh tit	−0·61*	0·0023*
Green woodpecker	−0·49	0·0027
Great spotted woodpecker	−0·42	0·0005
Spotted flycatcher	−0·39	0·0014
Lesser spotted woodpecker	−0·23	0·0018°
Pied flycatcher	−0·14	0·0020
Tree creeper	−0·25	−0·0003

** $p < 0.02$ * $p < 0.05$ ° $p < 0.10$

That territorial breeding pairs require a minimum area for existence (hypothesis (4)) is obvious. One would expect to find this type of area effect in species with the largest territories, irrespective of the degree of isolation. This is confirmed by the data for the great spotted woodpecker and the tree creeper, showing a significant area effect but no effect of isolation.

In sum, our data suggest that the occurrence of species in wooded isolates is governed by the abundance of the species in surrounding forests, the population size in the woodlot, and the minimum patch size of a species habitat.

The role of isolation

If death of individuals in small isolated woods is rapidly compensated for by immigration, these individuals are likely to be replaced before the next breeding season, and extinction will not be recorded by breeding bird censuses. With decreasing dispersal rates the interval between extinction and immigration increases and extinction is more likely to be recorded during the breeding season. Since remote islands might be reached less often by trespassers than less isolated woods, low-dispersive species will be absent more often in the remote isolates, whereas for highly dispersive species such differences will not be found. Thus, the sequence of species in Table 6 can be explained by specific dispersal rates. Both distance from large forest area and area of woodland within 3 km radius are good predictors for the isolation effect, but this can be attributed to the fact that they are correlated.

An isolation effect on the number of species was only found in the restricted set of 15 species of mature woodland. This would mean that the variation in the total species number attributable to isolation is small, and that the species not included in the final data set are not influenced by distance from other woods. Four of these species (long-tailed tit, blackcap, willow tit, and robin) which were not very widely spread in the woodlots were tested for area and isolation effect. Significant area variables were not found for any of them, and only the occurrence of the long-tailed tit was significantly correlated with area of woodland within a 3 km radius.

Are mature woodlots habitat islands for birds?

The present results suggest that in small isolated woods the immigration rates of several species are not high enough to compensate for extinctions within a short time. Immigration rates seem to be affected by distance. Hence, for at least some breeding bird species, isolated woodlots can be regarded as habitat islands.

Because most bird species in a woodlot seem to be unaffected by the island effect, MacArthur & Wilson's (1967) dynamic equilibrium theory does not seem to offer an appropriate description of avian community dynamics. Even the species which are affected seem to have different extinction and immigration rates and would therefore have a different likelihood of being present in a particular isolated site. We conclude that our results do not support MacArthur & Wilson's dynamic equilibrium theory, but are consistent with their assumption that in small isolated populations local extinction and immigration are important processes that may influence species distribution.

Applications in landscape planning

The presence of a species in a habitat isolate is primarily correlated with habitat attributes such as vegetation structure and plant species composition. However, our results indicate that patch size and isolation significantly influence the distribution of species in a patchy environment. In the modern agricultural landscape, many species find their habitat as scattered isolates in a matrix of unsuitable patches. In fact, the areas studied are fairly representative of this landscape type. Therefore, the results of this study can be extrapolated to other areas in so far as to select wood size and distance to large forest in relation to a preferred composition of the bird fauna. For a given number of bird species, Fig. 4 shows the relation between wood size and distance from nearest large forest area. These size-isolation lines are useful for conservation and

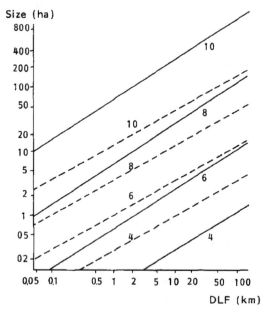

Fig. 4. Size-isolation lines showing, for a given number of bird species, the relation between woodlot size and distance from large forest area (————, Gelderse Vallei; – – – –, Limburg).

planning, e.g. to determine the desirable woodlot size given a fixed degree of isolation and a chosen species number. The predictable sequence of species extinctions with decreasing woodlot size makes it possible roughly to associate species number with species composition. Generalisation of our results to other habitat types is not possible, however. Mature woodland can be regarded as one of the least dynamic habitats in the man-dominated landscape. Birds inhabiting stable habitats are thought to be less dispersive than species in dynamic environments (Wiens, 1976; Andersson, 1980; Mikkonen, 1983).

REFERENCES

Alvey, N., Galway, N. & Lane, P. (1982). *An introduction to GENSTAT.* London, New York, Paris, Academic Press.

Ambuel, B. & Temple, S. A. (1983). Area dependent changes in the bird communities and vegetation of southern Wisconsin forests. *Ecology,* **64**, 1057–68.

Andersson, H. (1980). Nomadism and site tenacity as alternative reproductive tactics in birds. *J. Anim. Ecol.,* **49**, 175–84.

Connor, E. F. & McCoy, E. D. (1979). The statistics and biology of the species–area relationship. *Am. Nat.,* **113**, 791–833.

Diamond, J. M. (1975). The island dilemma: lessons of modern biogeographic studies for the design of natural reserves. *Biol. Conserv.,* **7**, 129–46.

Fritz, R. S. (1979). Consequences of insular population structure: Distribution and extinction of spruce grouse populations. *Oecologia,* **42**, 57–65.

Galli, A. E., Leck, C. F. & Forman, R. T. T. (1976). Avian distribution patterns in forest islands of different sizes in central New Jersey. *Auk,* **93**, 356–64.

Gilbert, F. S. (1980). The equilibrium theory of island biogeography: Fact or fiction? *J. Biogeogr.,* **7**, 209–35.

Helliwell, D. R. (1976). The effect of size and isolation on the conservation value of wooded sites in Britain. *J. Biogeogr.,* **3**, 407–16.

Humphreys, W. F. & Kitchener, D. J. (1982). The effect of habitat utilization on species–area curves: Implications for optimal reserve area. *J. Biogeogr.,* **9**, 391–6.

Hustings, M. F. H., Kwak, R. G. M., Opdam, P. F. M. & Reijnen, M.J.S.M. (1985). *Natuurbeheer in Nederland,* **3**. *Vogelinventarisatie.* Wageningen, Pudoc.

Jones, H. L. & Diamond, J. H. (1976). Short-time-base studies on turnover in breeding bird populations on the California Channel islands. *Condor,* **78**, 526–49.

Kwak, R. & Meijer, R. (1985). Interpretatiecriteria voor broedvogelinventarisaties met de territoriumkartering. *Limosa,* **58**, 97–108.

352   *P. Opdam, G. Rijsdijk, F. Hustings*

Lynch, J. F. & Whigham, D. F. (1984). Effects of forest fragmentation on breeding bird communities in Maryland, USA. *Biol. Conserv.*, **28**, 287–324.

MacArthur, R. H. & Wilson, E. O. (1967). *The theory of island biogeography*. Princeton, University Press.

Margules, C., Higgs, A. J. & Rafe, R. W. (1982). Modern biogeographic theory: Are there any lessons for nature reserve design? *Biol. Conserv.*, **24**, 115–28.

Mauriello, D. & Roskoski, J. P. (1974). A re-analysis of Vuilleumier's data. *Am. Nat.*, **108**, 711–14.

Mikkonen, A. V. (1983). Breeding site tenacity of the chaffinch *Fringilla coelebs* and the brambling *F. montifringilla* in northern Finland. *Orn. Scand.*, **14**, 36–47.

Opdam, P. (1983). Verspreiding van broedvogels in het cultuurlandschap: de betekenis van oppervlakte en isolatie van ecotopen. *Mededelingen WLO*, **10**, 179–89.

Opdam, P. & Retel Helmrich, V. (1984). Vogelgemeenschappen van heide en hoogveen: een typologische beschrijving. *Limosa*, **57**, 47–63.

Opdam, P., Van Dorp, D. & Ter Braak, C. J. F. (1984). The effect of isolation on the number of woodland birds of small woods in the Netherlands. *J. Biogeogr.*, **11**, 473–8.

Vuilleumier, F. (1970). Insular biogeography in continental regions, I. The northern Andes of South America. *Am. Nat.*, **104**, 373–88.

Whitcomb, R. F., Robbins, L. S., Lynch, J. F., Whitcomb, B. L., Klimkiewicz, M. K. & Bystrak, D. (1981). Effects of forest fragmentation on avifauna of the eastern deciduous forest. In *Forest island dynamics in man dominated landscapes*, ed. by R. L. Burgess and D. M. Sharpe, 125–205. Springer, New York.

Wiens, J. A. (1976). Population responses to patchy environments. *Annu. Rev. Ecol. Syst.*, **7**, 81–120.

Woolhouse, M. E. J. (1983). The theory and practice of the species-area effect, applied to the breeding birds of the British woods. *Biol. Conserv.*, **27**, 315–32.

Wright, S. J. & Hubbell, S. P. (1983). Stochastic extinction and reserve size: A focal species approach. *Oikos*, **41**, 466–76.

Ecology, 65(5), 1984, pp. 1466–1475
© 1984 by the Ecological Society of America

NUTRIENT DYNAMICS IN AN AGRICULTURAL WATERSHED: OBSERVATIONS ON THE ROLE OF A RIPARIAN FOREST[1]

WILLIAM T. PETERJOHN AND DAVID L. CORRELL
Smithsonian Environmental Research Center, P.O. Box 28,
Edgewater, Maryland 21037 USA

Abstract. Nutrient (C, N, and P) concentration changes were measured in surface runoff and shallow groundwater as they moved through a small agricultural (cropland) watershed located in Maryland. During the study period (March 1981 to March 1982), dramatic changes in water-borne nutrient loads occurred in the riparian forest of the watershed. From surface runoff waters that had transited ≈50 m of riparian forest, an estimated 4.1 Mg of particulates, 11 kg of particulate organic-N, 0.83 kg of ammonium-N, 2.7 kg of nitrate-N and 3.0 kg of total particulate-P per ha of riparian forest were removed during the study year. In addition, an estimated removal of 45 kg·ha^{-1}·yr^{-1} of nitrate-N occurred in subsurface flow as it moved through the riparian zone.

Nutrient uptake rates for the cropland and riparian forest were estimated. These systems were then compared with respect to their pathways of nutrient flow and ability to retain nutrients. The cropland appeared to retain fewer nutrients than the riparian forest and is thought to incur the majority of its nutrient losses in harvested crop. The dominant pathway of total-N loss from the riparian forest seemed to be subsurface flux. Total phosphorus loss from the riparian forest appeared almost evenly divided between surface and subsurface losses.

Nutrient removals in the riparian forest are thought to be of ecological significance to receiving waters and indicate that coupling natural systems and managed habitats within a watershed may reduce diffuse-source pollution.

Key words: carbon; cropland; diffuse source pollution; groundwater; mass balance; nitrogen; nutrient cycling; phosphorus; riparian forest; surface runoff; watershed.

INTRODUCTION

Excessive nutrient loading can have significant ecological effects on the receiving waters of lakes (Edmondson 1972, Powers et al. 1972, Schelske and Stoermer 1972), streams (Hynes 1969), and estuaries (Fraser and Wilcox 1981, Myers and Iverson 1981). Nutrient loadings from managed watersheds have contributed an increasing amount of nutrients to receiving waters as agricultural practices have intensified and residential development has expanded to accommodate a growing human population. Furthermore, nutrient losses from agricultural lands represent both a monetary and energy loss to society.

In an effort to understand how nutrient loss is affected by internal watershed structure, Correll (1984) attempted to relate surface soil and subsoil composition to nutrient discharge for three watersheds of differing land use. Little or no correlation was found, and this was thought to indicate that significant nutrient transformations were occurring as water moved through the watersheds as surface runoff and groundwater. Thus, one objective of this study was to investigate nutrient (N, P, and C) transformations occurring as water moved through an agricultural (cropland and riparian forest) watershed as surface runoff and shallow groundwater. Another objective was to synthesize past data about this watershed to gain a better perspective of the overall nutrient dynamics of an agricultural ecosystem.

SITE DESCRIPTION

The study site is a small subwatershed of the 3,332-ha Rhode River drainage basin. The Rhode River lies within the mid-Atlantic Coastal Plain along the western shore of Chesapeake Bay ≈20 km south of Annapolis, Maryland (38°53′N, 76°33′W). The subwatershed studied (Fig. 1) is a 16.3-ha basin of which 10.4 ha was planted with corn. Cropland was tilled prior to planting and preemergent herbicides were used to control subsequent weed growth. Riparian forest and hedgerows were composed of broadleaved, deciduous trees and accounted for the remaining 5.9 ha. The soils are a fine sandy loam and are extremely deep (>600 m). An underlying clay layer (the Marlboro Clay) is thought to create an effective aquiclude near sea level for the entire basin (Chirlin and Schaffner 1977). At the weir the clay layer is at a depth of ≈2.0 m. Therefore, the site contains a perched, shallow aquifer. Soils above the aquiclude are noncalcareous and contain no phosphate minerals (Pierce 1982). Surface soil organic matter content and pH in the cultivated fields were 1.9% and 5.6, respectively, when measured in 1977. The basin slope is 5.44% and the channel slope is 2.65%.

MATERIAL AND METHODS

Sampling

Bulk precipitation for chemical analysis was continuously sampled at an elevation of 13 m at the central

[1] Manuscript received 31 January 1983; revised 14 September 1983; accepted 16 September 1983.

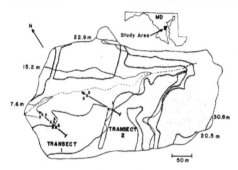

FIG. 1. General location of the study area (inset), and map of the watershed showing the positions of sampling transects (‑) and goundwater wells (×). Shading indicates cultivated fields.

weather station for the Rhode River watershed (2.3 km from the study site). Collection was made in a clean polyethylene bottle fitted to a 28 cm diameter polyethylene funnel. Fiberglass window screen covered the funnel to keep insects out. Samples were collected following each rain event, and the sampler was cleaned and returned for use. The amount of rainfall was measured with a Belfort mass-recording rain gauge and a standard weather bureau manual gauge at the central weather station, and with a Steven's tipping-bucket gauge at the study site itself.

Stream discharge was monitored and water samples taken at a 120° sharp-crested V-notch weir, the foundation of which rested upon the Marlboro Clay layer. The associated instrument shed was equipped with a stilling well, depth monitor, and flow meter. The weir was built and instrumented in the spring of 1976 and both water and nutrient discharges have been measured since that time. Depth measurements were taken every 5 min and recorded as digital data on a punch tape. The flow meter was modified to close a sampling switch every 38 m^3 of discharge. Switch closure activated a sampling cycle in which a fixed volume of stream water was pumped from the base of the V-notch into sample containers. One sample bottle contained sulfuric acid preservative for biologically labile chemical species, the other contained no preservative. Samples were composited over weekly intervals for the 1-yr study period beginning in March 1981 and ending in March 1982. During a typical storm week over 2000 m^3 are discharged and over 50 samples taken. During a single, intense, summer storm that occurred during the study period 46 samples were taken in <12 h. During periods of very low discharge (<10 m^3/wk) weekly grab samples were taken. Quickflow and slowflow rates were calculated graphically from the hydrograph (Barnes 1940). In this paper we equate surface runoff with quickflow and subsurface or groundwater flows with slowflow. Subsurface (groundwater) flows in

the system represent soil water percolation above the aquiclude. Seasons are defined in this paper as 3-mo periods (winter = December, January, and February).

Two transects of groundwater wells, ≈180 m apart, were established in the watershed (Fig. 1). Each transect consisted of several clusters of samplers at different elevations along the expected direction of flow. Each cluster consisted of a central groundwater well (piezometer) and two lateral wells located ≈10 m to either side and normal to the transect axis. Surface-water collectors located to collect only overland flow were associated with each groundwater well along transect 1. A total of 18 wells and nine surface water collectors were used. Surface-water collectors consisted of 4-L polyethylene bottles placed into holes in the soil in an inverted position. A rectangular slot was cut in the uphill side of the bottle at ground level and a short plastic apron was glued to the bottom of the slot and spread uphill to funnel surface flow into the bottle. Plastic tubing sealed into the cap of the inverted bottle allowed samples to be withdrawn without disturbing the sampler. Air was briefly bubbled through the sample to suspend particulates before the sample was withdrawn. Samplers were emptied prior to expected storm events and periodically cleaned. Groundwater wells consisted of 3.8 cm (inside diameter) polyvinyl chloride pipe perforated with 2.5-mm holes for ≈8 cm on the lower end. The bottom of the pipe was capped. Well holes were bored with a bucket auger either to the top of the Marlboro Clay or to a 4 m depth, whichever came first. Most wells were between 1.5 and 3.0 m deep. The pipes were inserted into the augered holes and clay packed around the pipe at the soil surface. The pipes extended ≈0.4 m above the ground and a loose cap was placed over the top. In all but one case the cluster of wells also had a fourth, shallower, groundwater well. On the day following a storm, water table heights were measured and wells pumped free of water. Samples were taken of the fresh groundwater the following day. When groundwater levels were low, equal volumes from each well within a cluster were composited to obtain enough sample for chemical analysis. Samples were returned to the laboratory and stored in the dark at 1.1°C. Aliquots for dissolved-nutrient determinations were filtered through prewashed Millipore HA membrane filters (0.45 µm nominal pore size) the day of collection and then returned to cold storage.

Nutrient analysis

Unfiltered rainfall samples were analyzed for nitrate, total Kjeldahl-N, ammonium-N, total-P, orthophosphate-P and organic matter concentrations. Surface runoff and groundwater samples were filtered and the filtrate analyzed for "dissolved" nitrite, nitrate, total Kjeldahl-N, ammonium-N, total-P, orthophosphate-P and organic matter concentrations. Unfiltered (whole) surface runoff samples were also analyzed for exchangeable ammonium-N, total Kjeldahl-N, ex-

1468   W. T. PETERJOHN AND D. L. CORRELL   Ecology, Vol. 65, No. 5

changeable orthophosphate-P, total-P, total suspended particulate, and organic matter concentrations. Unfiltered, acidified, stream samples were analyzed for total Kjeldahl-N, ammonium-N, nitrate, total-P, orthophosphate-P, and organic matter.

Dissolved nitrite and nitrate were determined in nonacidified samples with a Dionex model 16 ion chromatograph. Nitrite and nitrate were determined in acidified samples by reduction on cadmium amalgam and colorimetry (APHA 1976). Since nitrite was present only in trace amounts, nitrite and nitrate are routinely summed and referred to as nitrate. Dissolved ammonium-N was determined by the oxidation of ammonia and labile amino compounds to nitrite and the subsequent colorimetric determination of nitrite (Richards and Kletsch 1964). Total Kjeldahl nitrogen, which includes ammonium nitrogen and organic amines, was determined by digestion with sulfuric acid and hydrogen peroxide (Martin 1972), distillation, and Nesslerization (APHA 1976). Organic nitrogen was calculated as the difference between total Kjeldahl nitrogen and ammonium nitrogen.

Dissolved total phosphorus was determined by a perchloric acid digestion (King 1932) and reaction with ammonium molybdate and stannous chloride (APHA 1976). Total phosphorus in unfiltered samples was determined by perchloric acid digestion, the development of a phosphomolybdic acid-blue complex and extraction of the blue complex with isobutyl alcohol (Correll and Miklas 1976). Determination of dissolved orthophosphate followed the same procedure used for dissolved total phosphorus except that samples were not digested.

Organic matter was measured as chemical oxygen demand (in milligrams per litre) (Maciolek 1962). An empirical conversion factor of 0.417 was used to obtain organic carbon concentration, in milligrams per litre. To determine extractable ammonium-N and orthophosphate-P, measured volumes of whole samples were filtered through polycarbonate nucleopore filters (0.4-μm pore size) and the particulate residue was washed with 1 mol/L KCl for ammonium-N or 0.5 mol/L HCl for orthophosphate-P. Following extraction and neutralization the same procedures used for the determination of dissolved ammonia and orthophosphate were followed. Suspended particulate concentration was calculated from the gain in mass of a prewashed, dried, and weighed millipore filter after filtration of a measured volume of sample. Filters both before and after filtration were dried for at least a week in a desiccator. All particulate parameters were calculated as the difference between concentrations found in unfiltered and filtered aliquots.

Vegetative sampling

In early April, 1982, prior to leafing out, $\approx$0.4 ha of the riparian forest near transect 1 was intensively sampled. The point-centered quarter technique (Cottam

and Curtis 1956) was used to determine the relative dominance, density, frequency, size distribution, and importance value of the tree species present. Two increment cores were also taken from 130 of the 134 trees sampled. The average ring width was recorded for each tree and the dbh of the previous year calculated. The mean dbh for each of 13 size classes was then determined for the beginning and end of the study year. Weighted averages were calculated for both time periods using the number of trees in each size class as the weighting factor. These values and the allometric relations developed by Harris et al. (1973) were used to develop a rough estimate of the standing biomass of various tree components (branches, boles, large roots, and leaves). The difference in branch, bole, and root biomass prior and subsequent to the study year was considered the production of each component. Leaf production was considered to be the leaf biomass calculated by using the dbh at the end of the study period. Estimates of component biomass production were multiplied by the total density of trees and expressed in units of kilograms per hectare per year.

Cropland management, production, and nutrient uptake

Estimates of fertilizer applications and crop yields were obtained by direct inquiry from the farmer. These estimates were used in conjunction with data available from an intensive study in 1976–1978 that directly measured management inputs, crop nutrient uptake, crop production, and agricultural exports. A comparison of data obtained from the farmer with actual measurements of application rates showed an average difference of $\approx$2% for nitrogen and phosphorus.

RESULTS

Mass balance

During the study period 10 040 m³/ha of rainfall were measured. Of this 2540, 2970, 1890, and 2650 m³/ha fell in the spring, summer, fall, and winter seasons, respectively (Table 1). The long-term (160-yr) average for seasonal precipitation in the Rhode River watershed is 2800, 3140, 2450, and 2460 m³/ha in the spring, summer, fall, and winter, respectively (Higman and Correll 1982). Thus, seasonal precipitation was slightly above average in the winter, but below average for all other seasons. The net result was that annual rainfall was 800 m³/ha below the long-term annual average.

Of the 10 040 m³/ha of water that fell on the watershed during the study year, 23% was discharged in stream flow. Seasonally, 27, 13, 3.5, and 44% of precipitation inputs were discharged during the spring, summer, fall, and winter, respectively. Quickflow comprised 7.1% of the annual discharge and was greatest during the summer. Annual slow flow and quick flow measured during the study period were both below

October 1984   DYNAMICS OF CROPLAND AND RIPARIAN FOREST   1469

TABLE 1. Watershed mass balances for the whole watershed in 1981–1982.

Season	Precipitation (100 m³/ha)*	Nutrient flux for watershed					
		Nitrate-N (kg/ha)	Ammonium-N (kg/ha)	Organic-N (kg/ha)	Total-P (kg/ha)	Orthophosphate-P (kg/ha)	Organic-C (kg/ha)
Bulk precipitation							
Spring	25.4	1.20	1.12	3.39	0.284	0.0584	13.9
Summer	29.7	1.26	0.785	1.43	0.0675	0.0252	11.2
Fall	18.9	0.95	0.540	0.111	0.0549	0.0349	5.67
Winter	26.5	1.36	0.385	1.67	0.0290	0.0190	5.58
Year	100.4	4.77	2.83	6.60	0.435	0.138	36.4
Long-term annual mean†	108.0 ± 21.8	4.79 ± 1.18	2.59 ± 0.56	6.01 ± 1.15	0.810 ± 0.278	···‡	43.1 ± 6.81
Management inputs							
Year	···	10.1	14.9	42.2	12.5	12.5	···
Watershed discharge							
	Slow flow Quick flow (100 m³/ha)*						
Spring	6.26 0.687	0.532	0.197	0.577	0.251	0.116	153
Summer	3.20 0.804	0.428	0.0890	1.04	0.926	0.111	9.26
Fall	0.620 0.037	0.00675	0.00662	0.0264	0.0102	0.00656	0.424
Winter	11.5 0.119	1.53	0.0605	0.387	0.104	0.0548	3.56
Year	21.6 1.64	2.50	0.353	2.03	1.29	0.288	166
Long-term annual mean†	29.6 ± 4.97 ± 13.9 4.51	4.15 ± 3.06	0.434 ± 0.124	9.17 ± 10.2	1.72 ± 1.47	0.622 ± 0.407	58.2 ± 61.3

* Precipitation and stream flow expressed as 100 m³/ha is equivalent to the more traditional units of cm.

† 1973–1980 yearly means ± 1 SD for nitrogen bulk precipitation parameters (Correll and Ford 1982).
    1973–1978 yearly means ± 1 SD for phosphorus and organic-C bulk precipitation parameters (D. L. Correll, *personal communication*).
    1977–1981 yearly means ± 1 SD for nutrient and water discharge (D. L. Correll, *personal communication*).
    Long-term (160-yr) yearly precipitation mean ± 1 SD for the vicinity of the Rhode River watershed (Higman and Correll 1982).

‡ ··· indicates no data are available.

their respective long-term means, but did fall within one standard deviation of their expected values (Table 1).

During the study year 14.2 kg/ha of total nitrogen (nitrate-N + ammonium-N + organic-N) was delivered to the watershed by bulk precipitation. Of this input of total nitrogen, 34% was nitrate-N, 20% ammonium-N, and 46% organic N. Input of total phosphorus by bulk precipitation was 0.435 kg/ha, of which 32% was orthophosphate-P. Organic-C dominated the measured inputs from bulk precipitation with an influx of 36.4 kg/ha. Nutrient inputs by bulk precipitation were all similar (±1 SD) to their respective long-term means, except for total phosphorus, which was almost half its expected value. When compared to the nutrient influx from bulk precipitation, management inputs were the most important source of nutrients to the system except for organic-C (Table 1). Management inputs occurred in three pulses of fertilizer application during the spring and summer of the study year.

The watershed as a whole discharged 17, 2, 4, and 10% of the total inputs of nitrate-N, ammonium-N, organic-N and total-P, respectively (Table 1). Output of organic-C was over four times the input in bulk precipitation. Seasonal area yield discharges ranged over

an order of magnitude for all nutrients. Peak output occurred in the winter for nitrate-N; the spring for ammonium-N, orthophosphate-P, and organic-C; and the summer for organic-N and total-P.

Intrawatershed changes

Annual and seasonal average nutrient concentrations in surface runoff and groundwater are presented in Tables 2 and 3. In surface runoff flowing between the first and last clusters of samplers, reductions of 94, 78, 86, 84, 74 and 64% were observed in the mean annual total particulate, exchangeable ammonium-N, particulate organic-N, total particulate-P, exchangeable orthophosphate-P, and particulate organic carbon concentrations, respectively. Most of the total changes in concentration occurred within the first 19 m of riparian habitat. Dissolved nitrogen compounds in surface runoff also declined in concentration after traversing the riparian forest, with the greatest change occurring in the first 19 m. Total reductions of 79% for nitrate, 73% for ammonium-N, and 62% for organic-N were observed. The mean annual concentration of dissolved total phosphorus changed little in surface runoff. Dissolved organic carbon concentration increased 2.8-fold, but only 41% of this change occurred within the initial

1470   W. T. PETERJOHN AND D. L. CORRELL   Ecology, Vol. 65, No. 5

TABLE 2. Seasonal and yearly mean nutrient concentrations in mg/L for surface runoff in watershed 109 (transect 1).

Position	Season	Total Sus. Part.	Nitrate-N	Ammonium-N Exch. Part.	Ammonium-N Diss.	Organic-N Part.	Organic-N Diss.	Total-P Part.	Total-P Diss.	Exch. Part. Ortho-phosphate-P	Organic-C Part.	Organic-C Diss.
Entering riparian forest	Spring	8 840	3.73	0.734	3.63	27.7	1.47	3.22	0.256	0.354	67.2	12.1
	Summer	11 500	10.5	0.524	1.17	32.1	2.72	11.9	0.127	0.740	148.1	10.0
	Fall	3 830	1.57	0.301	0.896	16.8	0.779	3.29	0.128	0.863	101.1	6.75
	Winter*	1 760	1.99	0.048	0.250	1.32	2.04	0.860	0.320	0.675	63.2	19.1
	Year	6 480	4.45	0.402	1.49	19.5	1.75	4.82	0.208	0.658	94.9	12.0
19 m into riparian forest	Spring	1 380	2.60	0.218	1.23	6.47	1.18	2.31	0.081	0.456	35.9	12.0
	Summer	966	1.93	0.120	0.409	5.06	1.44	2.09	0.093	0.406	72.4	9.90
	Fall	122	0.343	0.038	0.069	2.61	0.529	0.604	0.393	0.134	5.97	4.09
	Winter*	176	2.18	0.042	0.158	0.37	1.33	0.065	0.375	0.108	···†	56.6
	Year	661	1.76	0.104	0.466	3.63	1.12	1.27	0.236	0.276	38.1	20.6
Leaving riparian forest	Spring	372	0.742	0.076	0.404	2.54	1.18	0.449	0.251	0.163	27.9	23.8
	Summer	524	1.03	0.108	0.175	3.46	0.713	1.04	0.183	0.244	45.6	16.0
	Fall	···	···	···	···	···	···	···	···	···	···	···
	Winter*	360	1.05	0.078	0.651	2.02	0.081	···	···	0.109	29.9	59.2
	Year	419	0.941	0.087	0.410	2.67	0.658	0.744	0.217	0.172	34.5	33.0

* Data from winter 1981. No samples were taken in winter 1982.
† ··· indicates no data are available.

portion of riparian forest. Although mean annual particulate concentrations of phosphorus, carbon, and organic-N in surface runoff decreased after moving through the riparian zone, the concentrations of these nutrients per unit of sediment increased. The effect of the riparian forest on particulate composition in surface runoff was to increase the proportion that was organic-C from 1.5 to 8.2%, organic-P (particulate total-P − exchangeable orthophosphate-P) from 0.064 to 0.14%, and organic-N from 0.30 to 0.64%. Furthermore, mean annual concentrations of exchangeable ammonium-N and orthophosphate-P per unit of sediment increased approximately two-fold after transit through the riparian zone. These results indicate that the particulates leaving the forest were more organic in composition and had a greater exchange capacity.

A t test was used to compare annual mean concentrations between the clusters of surface runoff collectors along transect 1. Noncomposited samples were used in the statistical analyses so that both spatial and temporal variability were included. Mean annual concentrations were significantly different ($P < .05$) between the first cluster (on the edge of the cropland) and the second cluster (≈ 19 m into the riparian forest) for total suspended particulates, exchangeable ammonium-N, dissolved organic-N, total particulate-P, exchangeable orthophosphate-P, and particulate organic-C. Significant differences in mean annual concentrations between the second and third clusters were found for only nitrate-N and total particulate-P.

Another way of assessing the role of the riparian forest is to estimate the amount of nutrients or total particulates trapped or released per hectare. This was done by multiplying the change in mean concentration (concentration entering − concentration leaving) by the volume of surface runoff (quickflow) or groundwater flow (slow flow) measured at the weir for each season (see Tables 1–3). For surface runoff the result was that during the study year an estimated 4.13 Mg of particulates were trapped along with 0.219, 11.2, 2.98, 0.172 and 32.2 kg/ha of exchangeable ammonium-N, particulate organic-N, total particulate-P, exchangeable orthophosphate-P, and particulate organic-C, respectively. Calculated removals of dissolved nutrients in surface waters were 2.71, 0.827, and 0.568 kg/ha, respectively, for nitrate-N, ammonium-N, and organic-N, whereas 0.0160 kg/ha of dissolved total-P and 4.84 kg/ha of dissolved organic-C were released. These values are probably underestimates of the actual change in surface runoff loads because of reductions in surface flow due to evapotranspiration and/or infiltration in the riparian forest.

Surface runoff leaving the cropland had peak concentrations of total particulates, total particulate-P, and all nitrogen parameters in either the spring or summer. These concentration peaks thus corresponded with the time of fertilizer application and intense storm activity. Peak spring or summer concentrations were also apparent after runoff had traversed the 1st 19 m of riparian forest, but were less pronounced.

Changes in all mean annual groundwater concentrations while traversing the riparian zone were qualitatively similar for both transects, although differences in magnitude were evident (Table 3). In general, nitrate concentrations declined dramatically while ammonium-N increased in concentration. The remaining nutrient concentrations were relatively constant except for a four-fold increase in total-P along transect 1 and

TABLE 3. Seasonal and yearly mean nutrient concentrations in mg/L for groundwater in watershed 109.

Position	Season	Nitrate-N Transect 1	Transect 2	Ammonium-N Transect 1	Transect 2	Organic-N Transect 1	Transect 2	Total-P Transect 1	Transect 2	Organic-C Transect 1	Transect 2
Entering riparian forest	Spring	5.43	5.78	0.097	0.058	0.312	0.226	0.024	0.120	1.41	1.36
	Summer	6.96	6.21	0.119	0.076	0.145	0.122	0.006	0.106	2.38	1.76
	Fall	6.89	7.22	0.045	0.132	0.169	0.121	0.016	0.152	1.97	1.90
	Winter	10.3	7.81	0.038	0.028	0.202	0.115	0.013	0.144	2.46	1.50
	Year	7.40	6.76	0.075	0.074	0.207	0.146	0.015	0.130	2.06	1.63
19 m into riparian forest	Spring	0.000	···*	0.098	···	0.213	···	0.012	···	2.43	···
	Summer	0.355	···	0.165	···	0.173	···	0.005	···	2.72	···
	Fall	0.282	···	0.114	···	0.214	···	0.018	···	1.82	···
	Winter	1.44	···	0.035	···	0.125	···	0.007	···	2.14	···
	Year	0.519	···	0.103	···	0.181	···	0.010	···	2.28	···
Leaving riparian forest	Spring	0.153	0.248	0.196	0.269	0.487	0.291	0.074	0.082	1.72	2.86
	Summer	0.453	0.000	0.498	0.478	0.223	0.419	0.065	0.378	2.81	4.58
	Fall	1.47	0.052	0.248	0.621	0.213	0.136	0.076	0.345	3.08	3.26
	Winter	0.978	0.105	0.156	0.396	0.144	0.125	0.032	0.183	2.77	3.08
	Year	0.764	0.101	0.274	0.441	0.267	0.243	0.062	0.247	2.60	3.44

* ··· indicates no data are available.

a two-fold increase in organic-C along transect 2. A 90% total decrease in the mean annual nitrate concentration was observed along transect 1 and a 98% decrease was observed along transect 2. Essentially all nitrate loss occurred within the first portion of riparian forest. Mean annual concentrations of ammonium-N increased in transect 1, with the overall increase being greater than three-fold. A six-fold increase occurred in transect 2. Overall, the annual average nitrate concentration declined 6.6 mg/L while traversing the forest, while the average concentration of combined forms of reduced nitrogen only increased by 0.36 mg/L. Thus, nitrate loss was apparently not due to its conversion into reduced forms of nitrogen that were measured.

Using a *t* test, comparisons were made to detect significant differences in mean annual groundwater concentrations between clusters of wells along each transect. Only noncomposited samples were used in the statistical analyses. Along transect 1 significant differences ($P < .05$) in mean annual concentrations between the first and second clusters were found only for nitrate-N, and between the second and third clusters for ammonium-N and total P. Significant differences between the two clusters along transect 2 were found for nitrate-N, ammonium-N, total-P, and organic-C.

The changes in concentration along transect 1, when related to seasonal groundwater discharge (slow flow; Table 1) give an estimated annual removal of 45.5 kg·ha⁻¹·yr⁻¹ of nitrate-N and an estimated release of 0.917, 0.194, 0.209, and 2.09 kg·ha⁻¹·yr⁻¹ of ammonium-N, organic-N, total-P, and organic-C, respectively, from the riparian forest. These values are probably underestimates because of reductions in subsurface flow due to evapotranspiration in the riparian forest.

Riparian vegetative production and nutrient uptake

Seventy-two percent of the sampled trees in the riparian forest were <22 cm in dbh, and the forest was clearly dominated by sweetgum (43.2% Importance Value) and red maple (16.0% I.V.) (Peterjohn 1982). Net primary production (NPP) was calculated as 18 600 kg·ha⁻¹·yr⁻¹ and was the sum of 5650 leaf, 3770 branch, 8100 bole, and 1030 large root production per hectare per year.

Estimates for the net production of respective tree components were multiplied by the mean midsummer (June, July, and August) nutrient concentrations found in a nearby forest (Table 4) and summed to estimate total nitrogen and phosphorus requirements. Sapwood concentrations were used for both branch and bole components. Uptake is defined as the annual elemental increment associated with the bole, branches, and large roots, plus the annual loss in leaf litterfall and throughfall. Nitrogen uptake by forest trees was calculated as ≈77 kg·ha⁻¹·yr⁻¹. Nitrogen return in leaf litter was calculated as 62 kg·ha⁻¹·yr⁻¹. Throughfall was not measured at the study site, so a literature value of 0.04 for the ratio of annual net nitrogen removal to foliage nitrogen content was used as an estimator (Henderson et al. 1977). Using this ratio, net nitrogen flux from leaves as throughfall was found to be 3.7 kg·ha⁻¹·yr⁻¹.

Using the same procedure employed for nitrogen, total phosphorus uptake by forest trees was calculated to be ≈9.9 kg·ha⁻¹·yr⁻¹. Phosphorus return in leaf litter was calculated as 7.8 kg·ha⁻¹·yr⁻¹. The flux of phosphorus from leaves as throughfall was estimated as 0.53 kg·ha⁻¹·yr⁻¹, using 0.06 as the ratio of annual net phosphorus removal to foliage phosphorus content

1472   W. T. PETERJOHN AND D. L. CORRELL   Ecology, Vol. 65, No. 5

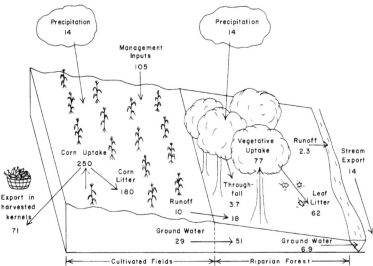

Fig. 2. Diagram of total-N flux and cycling in the study watershed from March 1981 to March 1982. All values are kilograms per hectare of the respective habitats (cropland and riparian forest/streams).

(Henderson et al. 1977). It should be emphasized that the uptake and production values presented are rough estimates. Nutrient uptake and turnover in the understory shrubs, herbaceous vegetation, and fine roots; nutrient return in branch, fruit, and flower litter; and whole-tree mortality were all factors not considered. Although rough, our production and uptake estimates are surprisingly reasonable when compared to literature values. Whittaker and Likens (1975) report 6000–25 000 kg·ha⁻¹·yr⁻¹ as the normal range of NPP in temperate deciduous forests, and the mean uptake for 14 temperate deciduous International Biological Program sites has been reported as 75.4 kg·ha⁻¹·yr⁻¹ of N

and 5.6 kg·ha⁻¹·yr⁻¹ of P by Cole and Rapp (1981). A measurement of 6350 kg·ha⁻¹·yr⁻¹ for litter production in a nearby forest also compared well with our estimate of 5650 kg·ha⁻¹·yr⁻¹ (D. F. Whigham, *personal communication*).

Cropland management, production, and nutrient uptake

Limestone (1680 kg/ha) was applied to the agricultural fields in early April, granular 2-10-13 fertilizer in mid-April, liquid N and herbicides in mid-May and urea in mid-July. Of the 105 kg/ha of total-N applied; 63.0% was organic-N, 22.3% was ammonium-N, and 15.0% was nitrate-N. Total-P loading was 19.6 kg/ha. In 1976 and 1977, the respective loadings were 142 and 189 kg/ha of total-N, and 39.7 and 33.7 kg/ha of total-P. Thus, nutrient loadings for this study year were low when compared to previous years.

Crop nitrogen and phosphorus uptake, losses of nitrogen and phosphorus as annual harvest, and crop NPP were estimated using data from a previous study of the same watershed. The same farmer has cultivated this basin since 1970 and little change in farming practices has been observed. Nutrient uptake by corn plants (above and belowground) was measured directly in 1976 and 1977; N uptake was found to be ≈250 kg and P uptake 45.0 kg·ha⁻¹·yr⁻¹ (Peterjohn 1982). The NPP of corn plants was also directly measured and found to be ≈24,000 kg·ha⁻¹·yr⁻¹. Nitrogen loss in harvested crop was estimated as 71.3 kg·ha⁻¹·yr⁻¹ by multiplying the yield of 7.8 m³ (222 bushels)/ha in 1981 by a 1976

TABLE 4. Mean Kjeldahl-N and total-P concentrations (μg/g) in trees in a mature deciduous forest in the Rhode River watershed, Maryland.*

Tissue	Kjeldahl-N	Total-P
Midsummer (June, July and August)		
sapwood	814	99.9
large roots	1 567	441
leaves	16 300	1 551
Winter (November and January)		
sapwood	1 080	159
large roots	2 070	435
leaf litter (November only)	10 996	1 386

* Samples are from 1977. Species represented are *Liriodendron tulipifera, Fagus grandifolia, Acer rubrum, Liquidambar styraciflua, Nyssa sylvaticus*, and *Quercus falcata.*

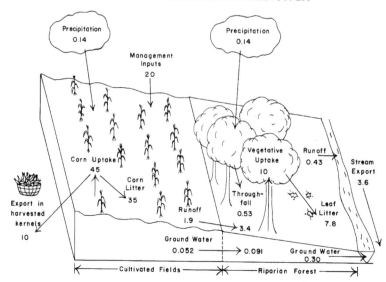

FIG. 3. Diagram of total-P flux and cycling in the study watershed from March 1981 to March 1982. All values are kilograms per hectare of the respective habitats (cropland and riparian forest/stream).

measurement of 33.3 kg of kernels per bushel, then multiplying the resultant product by a 1977 nitrogen measurement of 9.65 g/kg of kernels. Phosphorus loss in the harvested crop was estimated as 10.3 kg·ha^{-1}·yr^{-1} using the same procedure. The value used for P content of kernels was 1.40 g/kg. Thus, only 28% of the total nitrogen and 22.9% of the total phosphorus uptake by corn plants was exported from the system in harvested kernels.

DISCUSSION

To synthesize information about the N and P dynamics of this watershed, two schematic diagrams were drawn (Figs. 2 and 3). Although the values presented are estimates with unknown error terms, these diagrams are useful to compare pathways of nutrient flux between and within habitats, and to develop falsifiable hypotheses for future research. It is important to note that flux estimates are given per unit area of each component habitat (cultivated fields or riparian forest). Thus, because there is 1.76× more cultivated land than riparian forest, N export from the agricultural habitat in groundwater was 29 kg/cultivated ha, whereas 51 kg/ha were gained by the riparian forest (Fig. 2).

The nitrogen dynamics for both the cultivated and riparian habitats are illustrated in Fig. 2. For the cultivated fields, input of total-N from bulk precipitation was small (12% of estimated total inputs) when compared to management applications. Of the estimated total nitrogen exports from the cropland, 64% (28% of

corn uptake) was in harvested crop, 9.2% in surface runoff, and 26% in groundwater flow. Thus, the harvest seems to be the major pathway of total-nitrogen loss from the cropland. Groundwater appears to be the dominant pathway of total nitrogen flux between the cropland and riparian forest. Nitrogen retention (1 − outputs/inputs) for the cropland was calculated from our flux estimates and found to be low (8%), which is consistent with general ideas about disturbed ecosystems.

For the riparian forest, 17% of the estimated total-N inputs came in bulk precipitation, 61% in groundwater, and 22% in surface runoff. Of the estimated total-N losses from the riparian forest, 75% was lost in groundwater flow. Thus, it appears that the major pathway of nitrogen loss from the riparian forest was in subsurface flow. The calculated nitrogen retention by the riparian forest was 89%, and much higher than the retention of the cropland (8%).

The phosphorus dynamics for both the cultivated and riparian habitats are depicted in Fig. 3. For the cultivated fields, input of total-P from bulk precipitation was only 0.7% of the estimated total inputs. Of the estimated total phosphorus exports from the cropland, 84% (22% of corn uptake) was apparently lost in harvested crop, 16% in surface runoff and <1% in groundwater flow. Thus, the harvest seems to be the dominant pathway of phosphorus loss from the cropland, and surface runoff the dominant pathway of phosphorus flux between cropland and riparian forest. The

1474 W. T. PETERJOHN AND D. L. CORRELL Ecology, Vol. 65, No. 5

calculated phosphorus retention by the cropland was 41% and higher than the same value for total nitrogen.

For the riparian forest, 3.8% of the estimated total phosphorus inputs came in bulk precipitation, 94% in surface runoff, and 2.5% in groundwater flow. Unlike the loss of phosphorus from the cropland, phosphorus export from the riparian forest was nearly evenly divided between the estimated losses in surface runoff (59%) and groundwater flow (41%). The calculated phosphorus retention by the riparian forest was 80%. This value is twice as high as the retention by the cropland and slightly lower than the calculated total-nitrogen retention for the riparian zone.

From the observations made in this study two important questions arise: (1) What caused the decrease in groundwater nitrate concentration in the riparian forest? (2) How widely applicable are the effects we observed? Data from this study are insufficient to determine conclusively the reason for nitrate loss from groundwater in the riparian forest. Only two theoretical possibilities, however, appear likely: uptake by the vegetation and denitrification. To assess the approximate effect of vegetative uptake we estimated the annual incremental storage of N in the boles, branches, and large roots of the riparian trees. This was done by multiplying our production estimates by the mean winter tissue concentrations presented in Table 4. Winter tissue concentrations were used to correct for the effects of any retranslocation of nutrients from the leaves. From this calculation we estimate that N at the rate of 15 kg·ha^{-1}·yr^{-1} was removed due to incremental growth, which, if true, accounts for only 33% of the nitrate-N removed from the groundwater. Considering that groundwater was sampled below the fine root zone at a depth of at least 1 m, it would seem that the occurrence of significant denitrification is a viable hypothesis.

An insufficient number of studies have been conducted to assess adequately how common the effects we observed are. Reductions in sediment loads and their associated nutrients in surface runoff should be a fairly universal effect of riparian forests because of the physical nature of the processes involved; a few studies present evidence that riparian zones reduce sediment and phosphorus loads in adjacent streams (McColl 1978, Schlosser and Karr 1981a, b). Nitrate removal from shallow groundwater is probably a biological process requiring an appropriate environment. Therefore, the extent to which riparian forests serve as sites of subsurface nutrient change is difficult to predict. It is encouraging to note, however, that results similar to those presented here (particularly subsurface nitrate-N removal) were obtained by a closely analogous study conducted in Georgia (Lowrance et al. 1984).

Nutrient loss from diffuse sources (such as cornfields) can have significant ecological effects which are generally understood as a threat to most bodies of water. Therefore, the estimated removal in surface runoff of

4.1 Mg of particulates, 11 kg of particulate organic-N, 0.83 kg of dissolved ammonium-N, 2.7 kg of nitrate-N and 3.0 kg of total particulate-P per hectare of riparian forest is potentially an extremely important ecological function. In addition, the removal of nitrate-N at an estimated rate of 45 kg/ha in subsurface flow is especially important. Release of this amount would have doubled the amount of nitrate-N discharged during the study year.

ACKNOWLEDGMENTS

We thank the staff at the Smithsonian Environmental Research Center for their assistance and technical support throughout this study. For financial support we are grateful to Miami University, the Smithsonian Institution's Environmental Sciences Program, and National Science Foundation grant DEB-79-11563.

LITERATURE CITED

APHA (American Public Health Association). 1976. Standard methods for the examination of water and waste water. 14th edition. American Public Health Association, New York, New York, USA.

Barnes, B. S. 1940. Discussion on analysis of runoff characteristics by O. H. Meyers. Transactions of the American Society of Civil Engineers 105:104–106.

Chirlin, G. R., and R. W. Schaffner. 1977. Observations on the water balance for seven sub-basins of Rhode River, Maryland. Pages 277–306 in D. L. Correll, editor. Watershed research in eastern North America. Volume 1. Smithsonian Press, Washington, D.C., USA.

Cole, D. W., and M. Rapp. 1981. Elemental cycling in forest ecosystems. Pages 341–409 in D. E. Reichle, editor. Dynamic properties of forest ecosystems. Cambridge University Press, Cambridge, England.

Correll, D. L. 1984, in press. N and P in soils and runoff of three coastal plain land uses. In R. L. Todd, R. Leonard, and L. Assmussen, editors. Nutrient cycling in agroecosystems. University of Georgia Press, Athens, Georgia, USA.

Correll, D. L., and D. Ford. 1982. Comparison of precipitation and land runoff as sources of estuarine nitrogen. Estuarine, Coastal and Shelf Science 15:45–56.

Correll, D. L., and J. J. Miklas. 1976. Phosphorus cycling in a Maryland deciduous forest subjected to various levels of mineral nutrient loading. Pages 642–657 in F. G. Howell, J. B. Gentry, and M. H. Smith, editors. Mineral cycling in southeastern ecosystems. Symposium Series (CONF-740513), Energy Research and Development Administration, Washington, D.C., USA.

Cottam, G., and J. T. Curtis. 1956. The use of distance measures in phytosociological sampling. Ecology 37:451–460.

Edmondson, W. T. 1972. Nutrients and phytoplankton in Lake Washington. American Society of Limnology and Oceanography Special Symposium 1:172–188.

Fraser, T. H., and W. H. Wilcox. 1981. Enrichment of a subtropical estuary with nitrogen, phosphorus and silica. Pages 481–498 in B. J. Neilson and L. E. Cronin, editors. Estuaries and nutrients. Humana, Clifton, New Jersey, USA.

Harris, W. F., R. A. Goldstein, and G. S. Henderson. 1973. Analysis of forest biomass pools, annual primary production and turnover of biomass for a mixed deciduous forest watershed. Pages 41–46 in H. Young, editor. Proceedings of the working party on forest biomass of IUFRO. University of Maine Press, Orono, Maine, USA.

Henderson, G. S., W. F. Harris, D. E. Todd, and T. Grizzard. 1977. Quantity and chemistry of throughfall as influenced by forest-type and season. Journal of Ecology 65:365–374.

Higman, D., and D. L. Correll. 1982. Seasonal and yearly variation in meterological parameters at the Chesapeake Bay Center for Environmental Studies. Pages 1–159 *in* D. L. Correll, editor. Environmental data summary for the Rhode River ecosystem. Volume A, Part I. Chesapeake Bay Center for Environmental Studies, Edgewater, Maryland, USA.

Hynes, H. B. N. 1969. The enrichment of streams. Pages 188–196 *in* Eutrophication: causes, consequences, correctives. National Academy of Sciences, Washington, D.C., USA.

King, E. J. 1932. The colorimetric determination of phosphorus. Biochemical Journal **26**:292–297.

Lowrance, R. R., R. L. Todd, and L. E. Asmussen. 1984. Nutrient cycling in an agricultural watershed: I. Phreatic movement. Journal of Environmental Quality **13**:22–27.

Maciolek, J. A. 1962. Limnological organic analyses by quantitative dichromate oxidation. Fish and Wildlife Service Publication, Washington, D.C., USA.

Martin, D. F. 1972. Marine chemistry. Volume 1. Marcel Dekker, New York, New York, USA.

McColl, R. H. S. 1978. Chemical runoff from pastures: the influence of fertilizer and riparian zones. New Zealand Journal of Marine and Freshwater Research **12**:371–380.

Myers, V. B., and R. I. Iverson. 1981. Phosphorus and nitrogen limited phytoplankton productivity in northeastern Gulf of Mexico coastal estuaries. Pages 569–582 *in* B. J. Neilson and L. E. Cronin, editors. Estuaries and nutrients. Humana, Clifton, New Jersey, USA.

Peterjohn, W. T. 1982. Nutrient transformations in three single-land-use watersheds. Thesis. Miami University of Ohio, Oxford, Ohio, USA.

Pierce, J. W. 1982. Geology and soils of the Rhode River Watershed. Pages 181–216 *in* D. L. Correll, editor. Environmental data summary for the Rhode River ecosystem. Volume A, Part I. Chesapeake Bay Center for Environmental Studies, Edgewater, Maryland, USA.

Powers, C. F., D. W. Schults, K. W. Malueg, R. M. Brice, and M. D. Schuldt. 1972. Algal responses to nutrient additions in natural waters. American Society of Limnology and Oceanography Special Symposium **1**:141–154.

Richards, F. A., and R. A. Kletsch. 1964. The spectrophotometric determination of ammonia and labile amino compounds in fresh- and seawater by oxidation to nitrite. Pages 65–81 *in* Sugawara festival volume. Maruza, Tokyo, Japan.

Schelske, C. L., and E. F. Stoermer. 1972. Phosphorus, silica, and eutrophication in Lake Michigan. American Society of Limnology and Oceanography Special Symposium **1**:157–170.

Schlosser, I. J., and J. R. Karr. 1981*a*. Water quality in agricultural watersheds: impact of riparian vegetation during base flow. Water Resources Bulletin **17**:233–240.

Schlosser, I. J., and J. R. Karr. 1981*b*. Riparian vegetation and channel morphology impact on spatial patterns of water quality in agricultural watersheds. Environmental Management **5**:233–243.

Whittaker, R. H., and G. E. Likens. 1975. The biosphere and man. Page 306 *in* H. Lieth and R. H. Whittaker, editors. Primary productivity of the biosphere. Springer-Verlag, New York, New York, USA.

J. N. Am. Benthol. Soc., 1988, 7(4):289-306
© 1988 by The North American Benthological Society

The potential importance of boundaries to fluvial ecosystems*

ROBERT J. NAIMAN[1,3], HENRI DÉCAMPS[2], JOHN PASTOR[1],
AND CAROL A. JOHNSTON[1]

[1] Center for Water and The Environment, Natural Resources Research Institute,
University of Minnesota, Duluth, Minnesota 55811 USA
[2] Centre Nationale de la Recherche Scientifique, Centre d'Ecologie des Ressources Renouvelables,
29, rue Jeanne Marvig, 31055 Toulouse, France
[3] Present address: Center for Streamside Studies, AR-10, University of Washington,
Seattle, Washington 98195 USA

Abstract.   Boundaries separating adjacent resource patches are dynamic components of the aquatic landscape. This article addresses some fundamental questions about boundary structure and function in lotic ecosystems. We give examples of longitudinal and lateral boundaries associated with stream systems, demonstrate the application of chaos theory to understanding the inherent variability of boundary properties, and compare characteristics of boundaries in an arctic–tropical transect. We conclude that studies of resource patches, their boundaries, and the nature of exchange with adjacent patches will improve our perspective of drainage basin dynamics over a range of temporal and spatial scales.

Key words: boundary, ecotone, lotic, patch, stream, river.

The river continuum concept (Vannote et al. 1980) and its corollaries (Elwood et al. 1983, Ward and Stanford 1983) treat the stream–river profile as a continuum of physical gradients and associated biotic adjustments. Streams are envisioned as longitudinally linked systems where ecosystem processes in downstream areas are linked to those in upstream areas by the unidirectional flow of water and materials. The original concept and subsequent modifications (Minshall et al. 1985, Statzner and Higler 1985) provide a strong theoretical base for developing a holistic perspective for lotic ecosystems.

Nevertheless, there are also compelling reasons for examining the stream–river profile as a series of discrete patches or communities with reasonably distinct boundaries, rather than a gradual gradient or a continuum (Frissell et al. 1986, Hawkins 1985, Huet 1949, 1954, Illies 1961, Illies and Botosaneanu 1963). Most populations and processes in nature are arranged in discrete patches (Pickett and White 1985); boundaries between these patches are readily detected on various spatial and temporal scales (Forman and Godron 1986, Frissell et al. 1986, Wiens et al. 1985), and the landscape is being increasingly divided into patches with clearly defined boundaries. It is difficult to apply the river continuum concept to river systems where sharply defined zones exist naturally or to systems experiencing sustained anthropogenic alterations.

The river continuum or the river mosaic?

We suggest it may be informative to view lotic systems as a collection of resource patches separated by boundaries. This perspective is not proposed as a replacement for the river continuum concept but rather as complementary to it, providing a perspective that operates over various spatial and temporal scales. This boundary perspective has been implicitly expressed by others concerned with biotic zonation (Hawkins 1985, Huet 1949, 1954, Illies 1961, Illies and Botosaneanu 1963) and by those conversant with the diversity of communities and interactions encountered on some of the world's large rivers (Rzoska 1978).

Our boundary perspective also addresses lateral linkages (e.g., channel–riparian forest exchanges) within the fluvial corridor rather than only longitudinal ones (e.g., upstream–downstream linkages; Ward and Stanford 1988). This perspective also overcomes some of the immediate difficulties raised by the river continuum concept of quantifying upstream–down-

* Paper presented at a Symposium on "Community structure and function in temperate and tropical streams" held 24–28 April 1987 at Flathead Lake Biological Station, University of Montana, Polson, USA.

stream linkages over long distances. Upstream–downstream influences do occur in lotic ecosystems as evidenced by the "self-purification" of stream segments receiving organic pollution (Hynes 1960, Wuhrmann 1974). Yet, aside from extreme cases of material loading (N, P, sediments), little is known about how ecosystem processes (e.g., primary production, microbial decomposition, or nutrient cycling) in upstream reaches influence downstream communities or processes. Further, little is known about how far such influences may extend, although limited data suggest that the distance is short for essential nutrients (Mulholland et al. 1985, Naiman 1983a, Naiman et al. 1987). Viewing stream systems as a collection of resource patches separated by boundaries allows the relative importance of upstream–downstream versus lateral linkages to be examined, and provides a better understanding of factors regulating the exchange of energy and materials between identifiable resource patches.

The boundary perspective

A boundary has been defined as a zone of transition between adjacent ecological systems, having a set of characteristics uniquely defined by space and time scales and by the strength of the interactions between adjacent ecological systems—i.e., an ecotone (Holland 1988). The term "ecological system" is meant to include such commonly described biological assemblages as communities, ecosystems, and biomes. Thus, a boundary can be described as a type of transitional zone which, like ecosystems and biomes, possesses specific physical and chemical attributes, biotic properties, and energy and material flow processes but is unique in its interactions with adjacent ecological systems. The strength of these interactions, which may vary over wide temporal and spatial scales, appears to be controlled by the contrast between adjacent resource patches or ecological units. In general, a boundary may be thought of as analogous to a semipermeable membrane regulating the flow of energy and material between adjacent resource patches.

Despite a relative paucity of data on boundaries themselves, some putative characteristics of boundaries thought to be important in understanding ecological systems include: elevat-

ed abundance of some resources (Blake and Hoppes 1986); important control points for energy and material flux (Chauvet and Fustec 1988, Jordan et al. 1986, Peterjohn and Correll 1984, Whigham et al. 1986), potentially sensitive sites for studying interactions between biological populations and their controlling variables (Clements 1905, Gates and Gysel 1978, Johnston and Naiman 1987); support of relatively high biological diversity (Lynch and Whigham 1984); maintenance of critical habitat for a number of threatened species (Salo et al. 1986); and refuge and source regions for agricultural pests and predators (McDonnell 1984, Slezak 1976). Other characteristics that may be specific to a particular type of boundary include: sites for longitudinal migration (e.g., along windbreaks; Décamps et al. 1987), influence on the climatic regime of a surrounding area or on the development of soil conditions (e.g., forest/grasslands ecotones), and as genetic pools or sites for active microevolution (e.g., forest/agricultural field ecotone; Forman and Godron 1986).

Boundaries are potentially important in maintaining the stability of the biosphere because of these characteristics. For example, riparian forests are effective in controlling the movement of nitrogen and phosphorus within riverine landscapes, significantly influencing the accumulation and transformation of these materials (Peterjohn and Correll 1984, Pinay and Décamps 1988, Swank and Caskey 1982). Similarly, boundaries harboring a particularly large number of species or diverse genomes are important in maintaining landscape biodiversity (Salo et al. 1986).

As cultivation and civilization expand and contract there are concomitant changes in ratios of boundary to horizontal surface area and ratios of boundary to patch volume (Johnston and Naiman 1987). This relationship can be seen in the changing proportions of riparian forests to agricultural fields and river channels over the last several centuries (Fortuné 1988, Sedell and Froggatt 1984). An important but yet unanswered question is whether human-created boundaries have the same properties as natural boundaries. Alarmingly, changes in the boundary/surface and boundary/volume ratios of the landscape are proceeding without sufficient information on the environmental and biotic implications of these anthropogenic alterations.

It is known, however, that communities at the boundary between terrestrial and freshwater ecosystems appear to be particularly sensitive to landscape change. Examples include riparian forests, marginal wetlands, littoral lake zones, floodplain lakes and forests, and areas with significant groundwater–surface water exchange (Correll 1986, Pinay 1986). It is postulated that these boundaries may show early responses to human influences on the environment and, therefore, may be important areas in which to examine processes involved in global change (Holland 1988, Naiman et al. 1988a). Nevertheless, boundaries, with their inherent heterogeneity, are not generally appreciated as significant ecological components of the riverine landscape.

A hierarchical perspective on boundary characteristics

An immediate question raised by implying the importance of boundaries is whether they can be classified to facilitate comparative studies. Events or factors that create or maintain boundaries may be internal to the boundary or external to it. For instance, the boundary between spruce forest and treeless muskegs is partly determined by the depth of soil permafrost (Bryson et al. 1965). This is a natural process within the boundary. In contrast, forest gaps in some areas of the moist tropics are predominantly created and maintained by frequent hurricanes (Weaver 1986), an external control. In both cases a gradient in specific biotic variables between two resource patches is being maintained by either internal or external mechanisms. Differences between boundaries created or maintained by internal or external mechanisms are partially due to differences in scale between these phenomena. Events that are relatively frequent, large in area, or unusually intense may maintain greater contrast between the boundary and adjacent ecological systems than will events that are of lesser scale. Thus, boundaries may be classified by quantifying the scale and intensity of events creating or maintaining them, where the appropriate scale is chosen by the nature of the question being asked.

A relationship may exist between spatial and temporal scales for boundaries occurring in riv-er systems. Boundaries covering a larger area may persist longer as distinct features of the landscape than those on smaller spatial scales, and they probably have different functional properties. Yet in specific cases, such as in bedrock-controlled streams, small boundaries are persistent because of essentially fixed topography. Therefore, a hierarchical classification scheme for boundaries should consider both descriptive attributes, such as size, shape, biotic composition, and structural contrast, as well as functional attributes. Size refers to the area or volume of the boundary relative to the adjacent resource patches and the spatial scales of energy and material flows between the two resource patches. Patch shape has been shown to influence the rate of flow across and along patches (Forman and Godron 1981) and, as such, seems particularly applicable to boundaries. Biotic composition addresses the biomass and density of dominant organisms. Contrast quantifies differences between the boundary and the adjacent resource patches in biotic structure and available resources.

Functional variables to be considered should include stability, resiliency, energetics, and contrast in rates or processes. These variables address, sequentially, the tendency of the boundary to resist change under stress, rates at which boundaries return to an initial condition following stress, nutrient cycling or productivity rates associated with the dominant organisms, and quantification of functional variables occurring between boundaries and adjacent resource patches. Frissel et al. (1986) successfully developed a framework for a hierarchical classification of streams using a similarly organized view of spatial and temporal variation within a regional biogeoclimatic perspective. They classified streams as hierarchically organized systems incorporating, on successively lower levels, stream segment, reach, pool/riffle, and microhabitat subsystems. Each level develops and persists predominantly at a specified spatiotemporal scale. They were able to delineate boundaries between subsystems and describe how subsystems are similar or dissimilar to the total population. The same approach is applicable to classifying lotic boundaries, but the exact mechanisms for doing this may require more specific information on the structural and functional attributes of contrasting boundaries (i.e.,

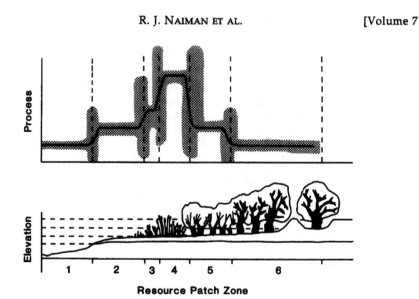

Fig. 1. Process rates in resource patch zones change abruptly at the boundary. Shown are changes in relative rates of nitrogen fixation occurring over several resource patches between the river channel and the uplands.

riparian zones along main channels and those occurring on secondary channels).

Predicting the unpredictability of boundaries

Boundaries often mark abrupt changes in many ecological processes such as litter fall, primary production, nutrient cycling, and decomposition rates. To illustrate this we show in Figure 1 relative changes in expected nitrogen fixation along a gradient from submerged habitat (zone 1) through an upland forest (zone 6). Nitrogen fixation is highest in the alder dominated zone 4 and is lowest in submerged sediments and in the mature forest. However, many ecosystem parameters measured at the boundary often demonstrate considerable variability because this region is a transition zone between adjacent, more homogenous resource patches. Boundaries may be mapped according to distribution of relatively homogenous resource patches, designating the transition zone of high variability between them as the boundary. However, most ecosystem theory is based on observations within the homogenous units; few detailed observations are made in the transition zone because of the complex sampling problems. A paradox of modeling boundaries is predicting when and where ecosystem parameters become unpredictable relative to adjacent homogenous patches. One approach to overcoming this problem is to use models dealing with uncertainty.

Chaos theory is a mathematical concept of predicting unpredictability. The premise of chaos theory is that chaotic, or seemingly unpredictable, behavior is in reality a special manifestation of an underlying structure. One of the best examples comes from medicine: cardiac arrhythmia is the chaotic contraction of the heart muscle which normally contracts regularly. The system, in this case the heart, is not replaced by another system but simply enters another phase. May (1974, 1975, 1976) discusses chaotic and stable behaviors of populations as special cases of a continuum of behaviors.

Models of chaos are usually in the form of difference equations which exhibit regular behavior at some values of one or more variables and irregular behavior at others. An example is:

$$Z(x) = r \cdot Z(x - 1) \cdot [1 - Z(x - 1)]$$

where Z is, say, the value of an ecosystem variable at some point x along a transect. Z at point x is predicted from its value at some preceding point, x − 1, and parameter, r. Small values of r yield stable values of Z along a transect (Fig. 2A). This outcome would occur in a homogenous ecosystem, such as an oak forest or a tall grass prairie, where there is little or no variation in ecosystem properties from point to point. Indeed, this is simply the measurement protocol defining homogeneity.

As the value of r increases, this equation becomes increasingly chaotic (Fig. 2B). Now, ecosystem parameters take on very different values from point to point along the transect. Such would be the case along a transect from prairie to forest: aboveground biomass would vary greatly depending on whether the measurement was made in a patch of prairie or forest. The high variability and apparent chaotic behavior is due to the "packing" of patches with two possible states (grass or tree) into a small zone. This bifurcation of possible states is the structural reason for apparent chaotic behavior (Holden 1986).

If r is allowed to vary along a transect, the value of an ecosystem variable can be stable in some portions of the transect and unstable in other portions (Fig. 2C). In this case, r was set equal to 1.75 for the first 10 m, increased to 4.00 for the second 10 m, then decreased to 1.25 for the final 10 m. Thus, the y-axis could represent average aboveground biomass in a transect from forest (first 10 m) through ecotonal savanna (second 10 m), to prairie (final 10 m). This equation represents a generic model which could predict changes in resource patch properties from one homogenous patch through a boundary to another homogenous patch solely as a function of r. It is therefore possible to model the unifying behavior of two resource patches and the intervening boundary. For the example cited, r could be a climatic parameter expressing "unfavorableness" for forest and prairie. In the boundary the climate is unfavorable for both trees and grasses, but at the beginning and end of the transect it is favorable for either trees or grasses.

Because the value of Z at any point depends not only on the value of r at that point but also on the value of Z at adjacent points, equations of this sort incorporate spatial relationships into their predictions. These spatial relationships can be either stabilizing (first and last 10 m), or destabilizing (middle 10 m). Note that Z in the boundary is not restricted to values intermediate between those of the adjacent ecosystems.

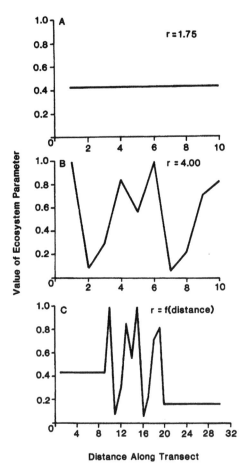

FIG. 2. Predictions of the chaos model to changing levels of r. In (A), r is at a value imparting stability to the process being measured while in (B), r imparts considerable variability. In (C) the chaos model illustrates the variability one would expect to encounter moving from one resource patch through the boundary into an adjacent resource patch.

This equation therefore predicts that boundaries can behave in ways that are not simple averages of adjacent resource patches, implying that the spatial patterning of resources (i.e., r) and interactions between patches within the boundary are as important determinants of boundary behavior as the adjacent patches. Often the variability will be much greater than the additive properties of the adjacent resource

TABLE 1. Some fundamental research questions pertaining to the ecological role of lotic ecotones (adapted from Holland 1988 and Naiman et al. 1988a).

1. Do ecotones provide stability for the resource patches they separate and, if so, at what spatial and temporal hierarchical scales do they operate?

2. What are the key attributes (processes and components) of ecotones which impart resistance and resilience to disturbance in the adjacent resource patches?

3. Is there a predictable pattern to change in ecotone characteristics under naturally dynamic conditions?

4. How are the characteristics and processes of ecotones sensitive to changes in the global environment?

5. What is the importance of ecotones in maintaining local, regional, and global biodiversity?

patches. Yet, the extreme variability may not in fact be chaotic. It is possible that underlying patterns could be identified with more extensive studies as has been demonstrated for the distribution of aquatic insects in streams.

Application of the boundary perspective to lotic systems

When viewed at an appropriate scale, lotic ecosystems become a complex biophysical mosaic containing interdigitating patches of different types (sensu Wiens et al. 1985). This view raises several fundamental questions (Table 1) on the role of boundaries in providing system-level stability, the attributes that impart stability, our ability to predict successional changes in structure and function, sensitivity of boundaries to long-term climatic changes, and their role in maintaining biodiversity along the fluvial corridor. In lotic systems, however, these questions are complicated by biophysical adjustments to increasing discharge from the headwaters to the sea. Consider, for example, boundaries associated with changes in channel gradient, channel depth, floodplain edges, and the sediment–water interface. Certain channel characteristics (gradient, depth) usually develop over long distances and are relatively persistent. In contrast, other channel features (such as the floodplain edge and the sediment–water interface) are often dynamic with respect to discharge and channel width.

Boundaries along the longitudinal profile of the river are not always as distinct as those occurring laterally to the main channel, and may extend over several stream orders. For example, in landscapes with high topographic relief, there are often substantial changes in channel gradient and depth associated with landscape features (Fig. 3A). These changes, even though they may extend over long distances, can be treated as boundaries, depending upon the question being asked. Similarly, discontinuities in longitudinal temperature gradients (Ward and Stanford 1983) and the edges of eddies in large rivers (Adams 1986) may be considered as boundaries between discrete resource patches. Similar changes also exist for functional processes. For example, as small streams gradually coalesce into rivers, concomitant adjustments are seen in benthic and pelagic primary production (Bott et al. 1985, Naiman 1983b). Stream reaches where rates of benthic and pelagic (water column) primary production change gradually to stable levels may also be viewed as boundaries even though they may range over relatively long distances (Fig. 3B).

Lateral boundaries undergo profound changes in a downstream direction in response to increasing magnitude and variability in discharge volume and channel width. In headwater streams the boundary between the stream edge is relatively sharp with only a short distance separating the channel from upland vegetation (Fig. 3C). The riparian boundary can be distinct from the stream channel and from the uplands but with considerable overlap in vegetation. Boundaries between riparian resource patches are better defined in terms of human perception in middle order streams experiencing strong seasonal shifts in water levels (represented by broken lines in Fig. 3D). Seasonal water level fluctuations result in plant communities ranging from submerged aquatic vascular plants (zone 1), to emergent aquatic plants and sedges (zones 2 and 3), to early successional woody plants such as willow (*Salix*) and alder (*Alnus*) (zone 4), and to various upland species (zones 5 and 6). In larger rivers with several channels (Fig. 3E), boundaries exist between communities in a manner similar to those in middle order streams. However, the riparian resource patches either overlap or become expanded in area, depending upon local erosion, sedimentation, and water level.

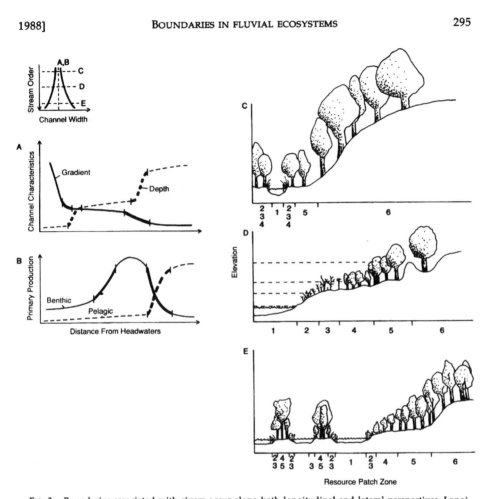

FIG. 3. Boundaries associated with rivers occur along both longitudinal and lateral perspectives. Longitudinal boundaries, such as transitional zones (bold lines) in (A) channel characteristics and (B) primary production, may occur over relatively long distances. Lateral boundaries occur over relatively shorter distances but are complicated by longitudinal shifts in channel dimensions (C–E). This may be seen as shifts in the distinctiveness of resource patch zones which change in a downstream direction. Dashed lines in D refer to ecologically important flood levels. Numbered resource patch zones refer to distinct biotic assemblages. In this example they are: (1) submerged aquatic macrophytes, (2) emergent aquatic plants, (3) sedges and grasses, (4) alder and willow, (5) pioneering trees (e.g., ash), and established trees (e.g., oak).

Stability

Implicit within a boundary perspective is that boundaries provide something for the resource patches they separate and from which they were formed (Table 1). Do they improve system-level stability for the fluvial corridor? Probably not in all cases; but there are situations where ecotonal communities do provide either resistance or resilience to some external factor. For example, riparian plant communities slow flood waters, trap sediments and nutrients, retard erosion, and provide seasonal habitat for many aquatic organisms (Décamps and Naiman 1989). This boundary community adjusts over long time periods to patterns of deposition and erosion as well as variations in water volume (Décamps et al. 1987, Salo et al. 1986). This is well

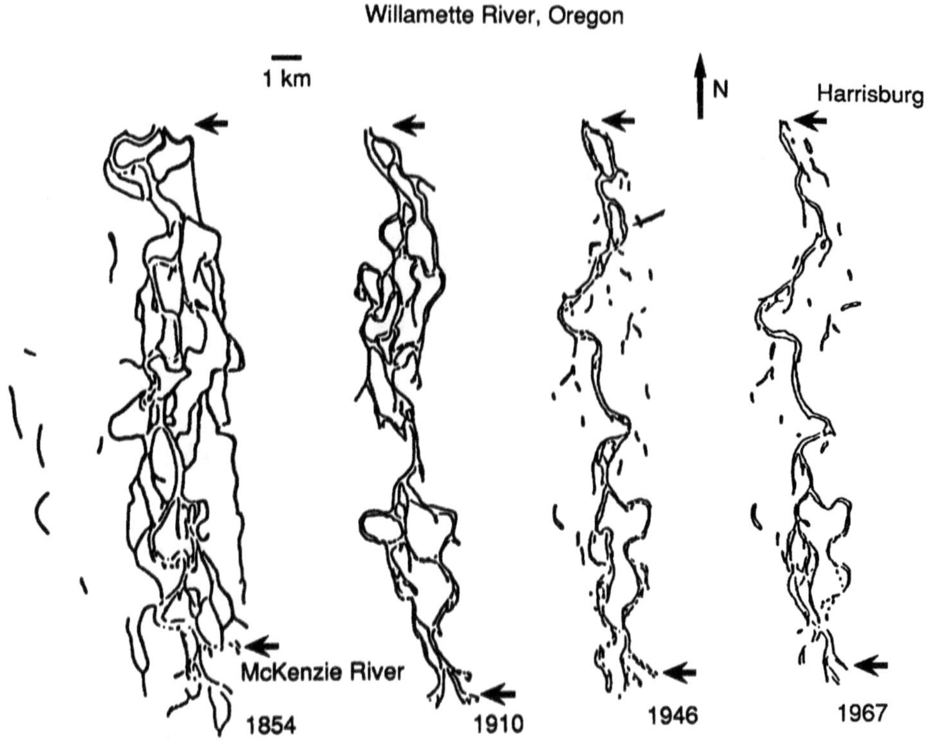

Willamette River, Oregon

FIG. 4.   Changes in channel dimensions occurring between 1854 and 1967 for the Willamette River, Oregon (Sedell and Froggatt 1984).

documented for the Rhine River in Germany where the size, shape, position, and migration rates of oxbow lakes along the fluvial corridor respond to changes in sediment load and discharge (Schäfer 1973). Over the past several centuries the hydrologic interaction between the Rhine River and the upland has been negated by channelization of the river.

The disappearance of similar riparian boundaries along major rivers should be cause for serious concern among scientists and managers. Sedell and Froggatt (1984) dramatically documented such a change for 25 km of the Willamette River, Oregon (Fig. 4). Between 1854 and 1967 the Willamette River became increasingly isolated from its floodplain as a result of channelization and agricultural modification of the riparian forests. In 1854, >250 km of river edge was in contact with the riparian forest. By 1967 the length of river edge was systematically de-creased to 64 km, a reduction of 74% (Table 2). Much of the change in the ratio of edge length to river length was completed by 1910 owing to snag removal and river navigation improvement programs. The implications of these changes is that biotic diversity and the ability of the riparian forest ecotone to retain sediment and nutrients have been severely lessened, resulting in deteriorating water quality and an increased probability of economic losses from floods.

Yet, there are situations where the ratio of boundary to surface area is increasing in lotic ecosystems. In North America, beaver (*Castor canadensis*) have multiplied substantially since their near extinction 90 yr ago (Naiman et al. 1986, 1988b). They have, in many areas, significantly increased water surface area and associated boundaries with the upland by building dams and selectively cutting vegetation. On the

294-km² Kabetogama Peninsula in northern Minnesota (USA) beaver have increased water surface area 14.5-fold since 1940. At the same time, length of water edge has increased 13.6-fold and the area of the riparian forest boundary 12.5-fold (Naiman et al. 1988b; C. A. Johnston and R. J. Naiman, unpublished data). This beaver-induced modification provides long-term stability for aquatic communities by creating an ecological situation relatively resistant to most disturbances. Further, the implications are that management programs directed at certain species, such as beaver, may have repercussions on the stability of the system which go far beyond their intended target.

Boundary attributes

It should follow, therefore, that boundaries possess key attributes (processes and components) which impart resistance and resilience to disturbance of adjacent resource patches. The full implications of this statement are that each structural/functional type of boundary is related to a unique pattern and frequency of disturbance. The statement also implies that boundaries perform a function for the resource patches they separate. There is increasing evidence that some boundaries (e.g., riparian vegetation, floodplain lakes) may act as semipermeable membranes between ecological systems modifying the direction, character, and magnitude of materials and information exchanged by the adjacent ecological systems. Peterjohn and Correll (1984) and Jordan et al. (1986) have shown strong nutrient interactions between agricultural land, natural land, and the Rhode River, Maryland (Fig. 5). Agriculture provides the largest input of N and P to the watershed but harvests remove <50% of those inputs. Large amounts of nutrients leave the cropland and are either accumulated elsewhere in the landscape or are transformed to gases. Only 1% of the N and 7% of the P entering the watershed is discharged to the Rhode River; most nutrients are intercepted by adjacent riparian forests. These forests reduce total particulate concentrations in overland flow by 94% and reduce nitrate in groundwater by 85%. The nature and magnitude of the efficiency of riparian use is determined by the structural and functional properties of the boundary. The boundary, as a selective filter, acts to modify disturbances as

TABLE 2. Changes in total edge length and the ratio of river length (25 km) to edge length for the Willamette River, Oregon, between 1854 and 1967 (see Sedell and Froggatt 1984).

Year	Edge Length (km)	Ratio Edge to River Length
1854	>250	>10.0
1910	120	4.8
1946	82	3.3
1967	64	2.6

well as the response of adjacent resource patches to that disturbance. Nevertheless, components and processes that differentially act to impart resistance and resilience to a given disturbance are seldom specifically identified.

Predicting long-term changes

Boundaries, by definition, are dynamic components of the landscape. But are biotic adjustments associated with dynamic change predictable? We suggest that changes in the composition and functioning of boundaries can be predicted for some spatial and temporal scales provided the nature of the interaction between adjacent ecological systems is known. The predicted biotic changes, however, may follow multisuccessional pathways depending upon the timing and magnitude of forcing factors.

Riparian vegetation along the Garonne River near Toulouse, France, is evolving from a pioneer forest to an oak hardwood forest in response to changing flood regimes which affect erosional and depositional processes (Fig. 6). This is a multisuccessional pathway operating over a temporal scale ranging up to 1000 yr and over nearly 200 km of the river's length (Décamps et al. 1988). Within the fluvial corridor, deposition and erosion interact at varying intensities and frequencies with the riparian vegetation over 100-yr cycles. These interactions keep the boundary riparian communities in early stages of succession. Normally, the communities are dominated by sedge (Carex spp.), willow (Salix alba), and alder (Alnus glutinosa) with occasionally oak (Quercus spp.) on the terrestrial edge of the boundary. A gradual decrease in seasonal flooding of the riparian community results in changes via several possible successional pathways. Over 100–1000 yr

Nitrogen:

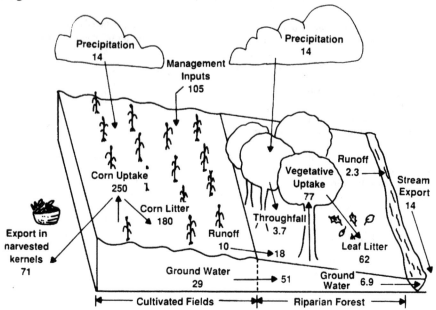

Phosphorus:

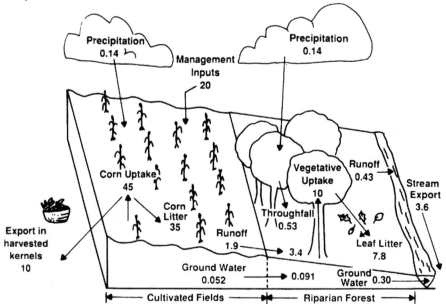

Fig. 5. Nitrogen and phosphorus budgets for agricultural fields and riparian zones associated with the Rhode River, Maryland (Peterjohn and Correll 1984). Units are kg/ha/yr.

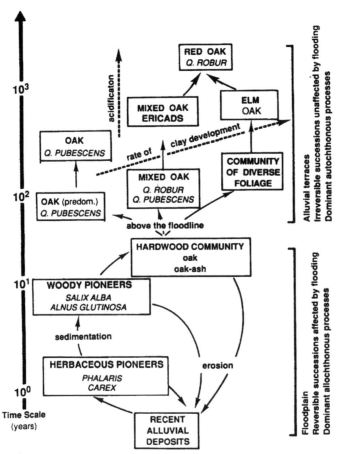

FIG. 6. Multisuccessional pathway development of riparian vegetation along the Garonne River, France, in response to changing hydrologic conditions occurring over several centuries (Décamps et al. 1988).

the terrestrial community becomes increasingly dominated by *Quercus robur* with associated changes in soil characteristics. In this sequence of events the boundary riparian community changes from a system influenced by seasonal flooding to a predominantly terrestrial system no longer heavily influenced by flooding. Thus, seemingly insignificant hydrologic changes have strong implications for the effective management of biodiversity, and for energy and material fluxes over both short and long time periods. Salo et al. (1986) recently described a similar situation for riparian forests in the Upper Amazon River, Peru.

Sensitivity to global change

Boundary characteristics become especially important when viewed in the context of global change (National Research Council 1986). Predicted changes in climate will have substantial effects on boundaries associated with streams, lakes, wetlands, and their riparian communities. Unfortunately, there have been no comprehensive attempts to understand the sensitivity of boundaries to major changes in the hydrologic cycle. Even general responses are difficult to predict. For example, one may postulate that because boundaries contain biotic

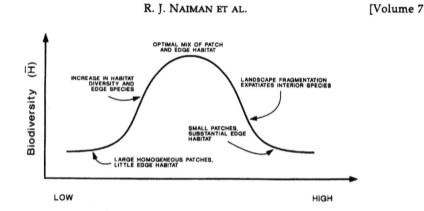

Frequency of Ecotones

FIG. 7. We predict that the frequency of boundaries can affect biodiversity in a predictable manner by affecting the ratio between patch size and boundary dimensions.

elements at the margins of their distributions and, therefore, are under stress, biodiversity would be especially sensitive to changes in the global environment. Alternatively, an equally plausible argument would be that boundary conditions are inherently variable and containing biotic elements relatively immune to a changing global environment. The net result would be that boundaries would only adjust in area or volume in response to a global change in the hydrologic cycle. The immediate challenge is to model boundary sensitivity to changes in the global environment using an approach similar to that of Solomon (1986). He used a specific atmospheric CO_2 climate-change scenario developed by Mitchell (1983) to predict terrestrial vegetation shifts in eastern North America. At finer scales, models are available to accommodate specific characteristics and processes associated with boundaries, such as ecosystem processes associated with forest gaps (Botkin et al. 1972, Shugart 1984) and biotic dispersal (DeAngelis and Waterhouse 1987).

Maintaining biodiversity

One commonly noted phenomenon in studies of faunal communities is the "edge-effect"—the tendency for communities to be more dense and often more diverse in ecotonal situations (Leopold 1933). The edge-effect is a generality and probably does not apply to all taxa or to all boundaries. Nevertheless, in cases where the edge-effect does occur, biodiversity can be af-

fected by the extent and quality of boundaries (Holland 1988, Salo et al. 1986). Within boundary communities some species are characteristic of those areas while other species perform activities there essential for their survival. The abundance and survival of these species are related to the amount and quality of boundary space. The potential importance of boundaries for certain species, and the need for large unfragmented reserves for other species, creates a difficult management problem. Fragmented landscapes have an abundance of ecotonal space but may not provide suitable habitats for non-boundary requiring species (Harris 1984). Consequently, local diversity may not be maximized if the ratio of edge to patch area is reduced below some threshold value (Fig. 7). This would lead to a reduction in regional diversity due to a decrease in edge-avoiding species (Lovejoy et al. 1986). These factors have yet to be considered by those investigating biodiversity relationships in lotic systems. Yet, many species require more than one ecological system in which to complete their life cycle. Amphibians breed and lay eggs in water but live as adults on land; waterfowl and fish often feed in one ecological system but rest, nest, or hide from predators in another. Many species either pass through boundaries or require boundaries during critical periods of their life cycle. For example, salmonids use rivers for spawning whereas the newly emerged fry and young-of-the-year quickly move into flooded riparian zones or to stream edges for feeding and predator avoidance (Bustard and Narver 1975, Walsh et al. 1988). During this critical period salmonids use the stream edge boundary in a manner that significantly affects their abundance, growth, and survival.

Other considerations

Together, the questions in Table 1 represent an array of problems fundamental to understanding lotic systems from both individualistic and holistic perspectives. Viewing stream systems as a collection of resource patches separated by boundaries, offers a perspective different from that of the river continuum concept. The boundary perspective, however, is still in an early stage of development, especially as it relates to lotic ecology (but see Huet 1949, 1954, Illies 1961, Illies and Botosaneanu 1963). Impediments to further progress are the lack of a central body of theory and the failure to place boundary dynamics in an evolutionary perspective. Nevertheless, it is recognized that the spatial and temporal arrangements of resource patches and their boundaries are important in the functioning of the landscape. Patch resource characteristics and their associated boundaries exhibit considerable contrast across biomes, providing a potentially rewarding perspective for comparative studies. To illustrate this we offer the following discussion of fundamental ecosystem changes in patch dynamics and boundaries in a latitudinal gradient.

Latitudinal gradients in boundary characteristics

Characteristics of resource patches

Comparison of selected river–riparian boundaries along a latitudinal gradient, from the tundra to moist tropical forests, offers some interesting contrasts. Fundamental shifts in plant diversity, vegetative patch size, plant biomass, soils, ion exchange mechanisms, nutrient cycling rates, and nutrient limitation occur along this gradient (Table 3).

There is an increase in species diversity and a corresponding decrease in the patch size of the forest mosaic from the tundra to the tropics (Bormann and Likens 1979, Heinselmann 1981, Romme 1982, Whittaker 1975). In northern forests, species diversity is low, and large mono-

typic stands of ha or even km² in area are common. Fire may be a major control over the size and distribution of these stands, and single fire patches can be as large as 5×10^5 ha, although most are on the order of 10^3–10^4 ha (Heinselmann 1981, Romme 1982). Thus, streams flowing through northern forests are influenced by large, relatively stable stands. Changes to the surrounding forest are large and catastrophic with a recurrence interval of 100–200 yr. In contrast, Hubbell and Foster (1986) argue for a different spatial scale in tropical forests, where patches caused by individual elimination are the size of a single tree and result from competition, with continuous replacement from the surrounding forest. In contrast to northern ecosystems, where species such as black spruce (*Picea mariana*) may dominate square kilometers, in tropical forests individuals of any given species may be hundreds of meters or more apart (Hubbell and Foster 1986). A stream flowing through such a forest would encounter a variety of small patches and associated boundaries that are changing on time scales of years rather than centuries.

Soil diversity also increases along this latitudinal gradient (Buol et al. 1973). The dominant soil process in the tundra is the accumulation of organic matter due to slow decomposition rates. This leads to large expanses of organic soils, or histosols. In the boreal zone, podzolization, or the complexing of iron and aluminum with humic acids, becomes dominant and spodosols become interspersed with histosols. In the temperate zone, the dominant soil process is the weathering of particles to clays, with clay accumulation in the subsoils. Such soils, known as alfisols and ultisols, dominate the landscape, but are interspersed with histosols in wetlands and spodosols beneath conifers. All these soils are also present in the tropics, with the addition of the highly weathered, sequioxide-rich oxisols. Thus, streams in northern latitudes are influenced by a few types of vegetation and only one or two soil types. Streams in the tropics may flow through many different soil types, each with different weathering regimes and nutrient retention capacities.

Nitrogen and phosphorus cycling in tropical forests differ qualitatively as well as quantitatively from those of northern forests. Quantitatively, the cycles of N and P are much more rapid in tropical forests (Anderson and Swift

TABLE 3. Comparison of fundamental terrestrial ecosystem-level characteristics in a latitudinal gradient from tundra to tropical regions.

Parameter	Biome				References
	Tundra	Boreal	Temperate	Moist Tropical	
Plant diversity	Low	Low	Moderate	High	Whittaker 1975
Average distance between individuals of same species (mosaic patch size)	<0.5 m	1-2 m	2-10 m	>10 m	Hubbell and Foster 1986
Patch size of same species (homogeneity)	Large	Large	Medium	Small	Hubbell and Foster 1986
Biomass	Low	Low-Mod	Moderate	High	Art and Marks 1971
Soils	Histosols, Entisols	Spodosols, Histosols, Entisols	Alfisols, Ultisols, Spodosols, Histosols, Entisols	Oxisols, Ultisols, Alfisols, Entisols, Histosols, Spodosols	Buol et al. 1973
Predominant soil ion exchange	Cation	Cation	Cation	Anion	Brady 1974, Uehara and Gilman 1981
Nutrient cycling rate	Low	Low	Moderate	High	Flanagan and Van Cleve 1983, Pastor et al. 1984, Vitousek 1984
Limiting nutrient	N	N	N, P	P	Flanagan and Van Cleve 1983, Pastor et al. 1984, Vitousek 1984

1983, Bernhard-Reversat 1977, Matson et al. 1987, Robertson 1984, Vitousek 1984, Vitousek and Denslow 1986, Vitousek and Sanford 1986) compared with northern forests (Gosz 1981, Pastor et al. 1984, Van Cleve and Alexander 1981, Vitousek et al. 1982).

The qualitative differences in nutrient cycles between northern and tropical forests have perhaps the strongest impact on ecosystem functioning and terrestrial–aquatic couplings. Tropical forest soils differ qualitatively from northern forest soils in two ways: first, variable charge clays in tropical soils can have anion-exchange capacity rather than solely cation-exchange capacity as in northern forests; second, phosphorus is strongly absorbed on iron and aluminum oxides which predominate in the tropics (Kinjo and Pratt 1971, Uehara and Gillman 1981).

These different behaviors of tropical soils strongly affect other ecosystem properties. First, nitrate is much less mobile in tropical forests than in northern forests. Whereas nitrate leaches from the soils of many northern forests after disturbance (Vitousek et al. 1982), it is retained in the soils of disturbed tropical forests (Matson et al. 1987). Ammonium, being a cation, may be the prevalent form of inorganic nitrogen loss from tropical forests to streams. Second, the strong phosphate absorption on anion exchange sites and sesquioxides may cause tropical forests to be P limited (Jordan 1985, Vitousek 1984). In contrast, competition with free-living heterotrophs for nitrogen and more recalcitrant carbon compounds in litter causes nitrogen limitation in northern forests (Gosz 1981, Pastor et al. 1984, Van Cleve and Alexander 1981). Manipulation of tropical forests, particularly fertilization, may result in high ammonium and low nitrate and phosphorus inputs to aquatic ecosystems, rather than the opposite which is the norm for northern forests.

These fundamental changes in ecosystem-level properties occurring along the tundra-tropic transect offer a robust test for many of the postulated properties of aquatic and terrestrial/aquatic boundaries. Models predicting boundary behavior in relation to adjacent resource patches can be tested and comparisons made between similar classes of boundaries as to properties influencing ecotonal integrity and longevity. Together, these factors in combination with the river continuum and river mosaic concepts may be able to provide an unparalleled understanding of lotic ecosystems.

Acknowledgements

We thank M. Dyer, M. M. Holland, P. G. Risser, and H. H. Shugart for stimulating discussions on boundary dynamics, and V. H. Resh and J. G. Blake for constructive suggestions on the content of the manuscript. This article was presented at a National Science Foundation sponsored workshop on future directions in riverine research at the University of Montana, 23–28 April 1987. Research was supported by the National Science Foundation (BSR 85-16284), UNESCO's Man and the Biosphere Programme, and the Centre Nationale de la Recherche Scientifique (CNRS), France.

Literature Cited

ADAMS, J. R. 1986. Mechanics of a large eddy in the Mississippi River. Pages 645–652 in R. E. Arndt, H. G. Stefan, C. Farell, and S. M. Peterson (editors). Advancements in aerodynamics, fluid mechanics, and hydraulics. American Society of Civil Engineers, New York.

ANDERSON, J. M., AND M. J. SWIFT. 1983. Decomposition in tropical forests. Pages 287–339 in S. L. Sutton, F. C. Whitmore, and A. C. Chadwick (editors). Tropical rain forest: ecology and management. Blackwell Scientific Publications, Oxford.

ART, H. W., AND P. L. MARKS. 1971. A summary table of biomass and net annual primary production in forest ecosystems of the world. Pages 3–34 in H. E. Young (editor). Forest biomass studies. University of Maine, Orono.

BERNHARD-REVERSAT, F. 1977. Recherches sur les variations stationelles des cycles biogéochimiques en forêt ombrophile de Côte d'Ivoire. Cahiers ORSTROM (Office de la Recherche Scientifique et Technique Outre-Mer), Série Pédologie 15:175–189.

BLAKE, J. G., AND W. G. HOPPES. 1986. Influence of resource abundance on use of tree-fall gaps by birds in an isolated woodlot. Auk 103:328–340.

BORMANN, F. H., AND G. E. LIKENS. 1979. Pattern and process in a forested ecosystem. Springer-Verlag, New York.

BOTKIN, D. B., J. F. JANAK, AND J. R. WALLIS. 1972. Some ecological consequences of a computer model of forest growth. Journal of Ecology 60: 849–872.

BOTT, T. L., J. T. BROCK, C. S. DUNN, R. J. NAIMAN, R. W. OVINK, AND R. C. PETERSEN. 1985. Benthic community metabolism in four temperate stream systems: an inter-biome comparison and evaluation of the river continuum concept. Hydrobiologia 123:3–45.

BRADY, N. C. 1974. The nature and properties of soils. MacMillan Publishing Co., New York.

BRYSON, R. A., M. W. IRVING, AND J. A. LARSEN. 1965. Radiocarbon and soil evidence of former forest in the southern Canadian tundra. Science 147: 46–48.

BUOL, S., F. D. HOLE, AND R. J. McCRAKEN. 1973. Soil genesis and classification. Iowa State University Press, Ames.

BUSTARD, D. R., AND D. W. NARVER. 1975. Aspects of the winter ecology of juvenile coho salmon (*Oncorhynchus kisutch*) and steelhead trout (*Salmo gairdneri*). Journal of the Fisheries Research Board of Canada 32:667–680.

CHAUVET, E., AND E. FUSTEC. 1988. Importance of the transfer and the fate of terrestrial organic matter in large river systems: the case of the Garonne Valley. Pages 119–126 in L. Lauga, H. Décamps, and M. M. Holland (editors). Land use impacts on aquatic ecosystems: the use of scientific information. MAB/UNESCO, Toulouse.

CLEMENTS, F. E. 1905. Research methods in ecology. University Publishing Company, Lincoln, Nebraska.

CORRELL, D. L. (editor). 1986. Watershed research perspectives. Smithsonian Institution Press, Washington, D.C.

DeANGELIS, D. L., AND J. C. WATERHOUSE. 1987. Equilibrium and nonequilibrium concepts in ecological models. Ecological Monographs 57:1–21.

DÉCAMPS, H., M. FORTUNÉ, F. GAZELLE, AND G. PAUTOU. 1988. Historical influence of man on the riparian dynamics in a fluvial landscape. Landscape Ecology 1:163–173.

DÉCAMPS, H., L. JOACHIM, AND L. LAUGA. 1987. The importance for birds of the riparian woodlands within the alluvial corridor of the River Garonne, SW France. Regulated Rivers 1:301–316.

DÉCAMPS, H., AND R. J. NAIMAN. 1989. Les fleuves de l'avenir. La Recherche (in press).

ELWOOD, J. W., J. D. NEWBOLD, R. V. O'NEILL, AND W. VAN WINKLE. 1983. Resource spiralling: an operational paradigm for analyzing lotic ecosystems. Pages 3–28 in T. D. Fontaine and S. M. Bartell (editors). Dynamics of lotic ecosystems. Ann Arbor Science Publishers, Ann Arbor, Michigan.

FLANAGAN, P. W., AND K. VAN CLEVE. 1983. Nutrient cycling in relation to decomposition and organic matter quality in taiga ecosystems. Canadian Journal of Forest Research 13:795–817.

FORMAN, R. T. T., AND M. GODRON. 1981. Patches and structural components for a landscape ecology. BioScience 31:733–740.

FORMAN, R. T. T., AND M. GODRON. 1986. Landscape ecology. John Wiley and Sons, New York.

FORTUNÉ, M. 1988. Historical changes in a large river in an urban area: the Garonne River, Toulouse, France. Regulated Rivers 2:179–186.

FRISSELL, C. A., W. J. LISS, C. E. WARREN, AND M. D. HURLEY. 1986. A hierarchical framework for stream habitat classification: viewing streams in a watershed context. Environmental Management 10:199–214.

GATES, J. E., AND L. W. GYSEL. 1978. Avian nest dispersion and fledging success in field-forest ecotones. Ecology 59:871–883.

GOSZ, J. M. 1981. Nitrogen cycling in coniferous ecosystems. Pages 405–426 in T. Rosswall and F. E. Clark (editors). The terrestrial nitrogen cycle. Ecological Bulletin 33. Stockholm.

HARRIS, L. D. 1984. The fragmented forest. University of Chicago Press, Chicago.

HAWKINS, C. P. 1985. Substrate associations and longitudinal distributions in species of Ephemerellidae (Ephemeroptera:Insecta) from western Oregon. Freshwater Invertebrate Biology 4:181–188.

HEINSELMAN, M. L. 1981. Fire and succession in the conifer forests of North America. Pages 374–405 in D. C. West, H. H. Shugart, and D. B. Botkin (editors). Forest succession. Academic Press, New York.

HOLDEN, A. V. (editor). 1986. Chaos. Princeton University Press, Princeton.

HOLLAND, M. M. 1988. SCOPE/MAB technical consultations on landscape boundaries. Pages 47–104 in F. di Castri et al. (editors). A new look at ecotones: merging international projects on landscape boundaries. Biology International Special Issue 17.

HUBBELL, S. P., AND R. B. FOSTER. 1986. Biology, chance, and history and the structure of tropical rain forests. Pages 314–330 in J. Diamond and T. Case (editors). Community ecology. Harper and Row, New York.

HUET, M. 1949. Aperçu des relations entre la pente et les populations piscicoles des eaux courantes. Schweizerische Zeitschrift für Hydrologie 11:333–351.

HUET, M. 1954. Biologie, profils en long et en travers des eaux courantes. Bulletin Français de Pisciculture 175:41–53.

HYNES, H. B. N. 1960. The biology of polluted waters. Liverpool University Press, Liverpool.

ILLIES, J. 1961. Versuch einer allgemeinen biozonotischen Gliederung der Fliessgewässer. Internationale Revue der gesamten Hydrobiologie 46: 205–213.

ILLIES, J., AND L. BOTOSANEANU. 1963. Problèmes et méthodes de la classification et de la zonation écologique des eaux courantes, considérées surtout du point de vue faunistique. Mitteilungen der Internationalen Vereinigung für Theoretische und Angewandte Limnologie 12:1–57.

JOHNSTON, C. A., AND R. J. NAIMAN. 1987. Boundary dynamics at the aquatic–terrestrial interface: the influence of beaver and geomorphology. Landscape Ecology 1:45–57.

JORDAN, C. F. 1985. Nutrient cycling in tropical forest ecosystems: principles and their application in management and conservation. John Wiley and Sons, New York.

JORDAN, T. E., D. L. CORRELL, W. T. PETERJOHN, AND D. E. WELLER. 1986. Nutrient flux in a landscape: the Rhode River and receiving waters. Pages 57–76 in D. L. Correll (editor). Watershed research perspectives. Smithsonian Press, Washington, D.C.

KINJO, T., AND P. F. PRATT. 1971. Nitrate absorption. II. In competition with chloride, sulfate, and phosphate. Soil Science Society of America Proceedings 35:725–728.

LEOPOLD, A. 1933. Game management. Scribner, New York.

LOVEJOY, T. E., R. O. BIERRAGAARD, A. B. RYLANDS, J. R. MALCOLM, C. E. QUINTELA, L. H. HARPER, K. S. BROWN, A. H. POWELL, G. V. N. POWELL, H. O. R. SCHUBERT, AND M. H. HAYS. 1986. Edge and other effects of isolation on Amazon forest fragments. Pages 257–285 in M. E. Soulé (editor). Conservation biology: the science of scarcity and extinction. Sinauer Associates, Sunderland, Massachusetts.

LYNCH, J. F., AND D. F. WHIGHAM. 1984. Effects of forest fragmentation on breeding bird communities in Maryland, USA. Biological Conservation 28:287–324.

MATSON, P. A., P. M. VITOUSEK, J. J. EWEL, M. J. MAZZARINO, AND G. P. ROBERTSON. 1987. Nitrogen transformations following tropical forest felling and burning on a volcanic soil. Ecology 68:491–502.

MAY, R. M. 1974. Biological populations with non-overlapping generations: stable points, stable cycles, and chaos. Science 186:645–647.

MAY, R. M. 1975. Biological populations obeying difference equations: stable points, stable cycles, and chaos. Journal of Theoretical Biology 49:511–524.

MAY, R. M. 1976. Models for single populations. Pages 4–25 in R. M. May (editor). Theoretical ecology, principles and applications. W. B. Saunders, Philadelphia.

McDONNELL, M. J. 1984. Interactions between landscape elements: dispersal of bird-disseminated plants in post-agricultural landscapes. Pages 47–58 in J. Brandt and P. Agger (editors). Methodology in landscape ecological research and plan-

ning. Proceedings, 1st International Seminar, International Association of Landscape Ecology, Roskilde, Denmark.

MINSHALL, G. W., K. W. CUMMINS, R. C. PETERSEN, C. E. CUSHING, D. A. BRUNS, J. R. SEDELL, AND R. L. VANNOTE. 1985. Developments in stream ecosystem theory. Canadian Journal of Fisheries and Aquatic Sciences 42:1045–1055.

MITCHELL, J. F. B. 1983. The seasonal response of a general circulation model to changes in CO_2 and sea temperatures. QJR Meteorological Society 109: 113–152.

MULHOLLAND, P. J., J. D. NEWBOLD, J. W. ELWOOD, L. A. FERREN, AND J. R. WEBSTER. 1985. Phosphorus spiralling in a woodland stream: seasonal variations. Ecology 66:1012–1023.

NAIMAN, R. J. 1983a. The influence of stream size on the food quality of seston. Canadian Journal of Zoology 61:1995–2010.

NAIMAN, R. J. 1983b. The annual pattern and spatial distribution of aquatic oxygen metabolism in boreal forest watersheds. Ecological Monographs 53:73–94.

NAIMAN, R. J., H. DÉCAMPS, J. PASTOR, AND C. A. JOHNSTON. 1988a. A new UNESCO programme: research and management of land/inland water ecotones. Pages 107–136 in F. di Castri et al. (editors). A new look at ecotones: merging international projects on landscape boundaries. Biology International Special Issue 17.

NAIMAN, R. J., C. A. JOHNSTON, AND J. C. KELLEY. 1988b. Alteration of North American streams by beaver. BioScience (in press).

NAIMAN, R. J., J. M. MELILLO, AND J. E. HOBBIE. 1986. Ecosystem alteration of boreal forest streams by beaver (Castor canadensis). Ecology 67:1254–1269.

NAIMAN, R. J., J. M. MELILLO, M. A. LOCK, T. E. FORD, AND S. R. REICE. 1987. Longitudinal patterns of ecosystem processes and community structure in a subarctic river continuum. Ecology 68:1138–1156.

NATIONAL RESEARCH COUNCIL. 1986. Global change in the geosphere-biosphere. National Academy Press, Washington, D.C.

PASTOR, J., J. D. ABER, C. A. McCLAUGHERTY, AND J. M. MELILLO. 1984. Aboveground production and N and P cycling along a nitrogen mineralization gradient on Blackhawk Island, Wisconsin. Ecology 65:256–268.

PETERJOHN, W. T., AND D. L. CORRELL. 1984. Nutrient dynamics in an agricultural watershed: observation of a riparian forest. Ecology 65:1466–1475.

PICKETT, S. T. A., AND P. S. WHITE (editors). 1985. The ecology of natural disturbance and patch dynamics. Academic Press, Orlando, Florida.

PINAY, G. 1986. Relations sol-nappe dans les bois riverains de la Garonne. Etude de la dénitrifica-

306   R. J. Naiman et al.   [Volume 7

tion. Thèse de doctorat, Université Claude Bernard, Lyon, France.

Pinay, G., and H. Décamps. 1988. The role of riparian woods in regulating nutrient fluxes between the alluvial aquifer and surface water: a conceptual model. Regulated Rivers (in press).

Robertson, G. P. 1984. Nitrification and nitrogen mineralization in a lowland rainforest succession in Costa Rica, Central America. Oecologia 61:91–104.

Romme, W. H. 1982. Fire and landscape diversity in subalpine forests of Yellowstone National Park. Ecological Monographs 52:199–211.

Rzoska, J. 1978. On the nature of rivers with case stories of Nile, Zaire and Amazon. Dr. W. Junk Publishers, The Hague.

Salo, J., R. Kalliola, I. Hakkinen, Y. Makinen, P. Niemela, M. Puhakka, and P. D. Coley. 1986. River dynamics and the diversity of Amazon lowland forest. Nature 322:254–258.

Schäfer, W. 1973. Alkrhein verbund am nordlichen oberrhein. Courier Forschunginstitut Senckenberg 7:1–63.

Sedell, J. R., and J. L. Froggatt. 1984. Importance of streamside vegetation to large rivers: the isolation of the Willamette River, Oregon, USA from its floodplain. Verhandlungen der Internationalen Vereinigung für Theoretische und Angewandte Limnologie 22:1828–1834.

Shugart, H. H. 1984. A theory of forest dynamics. Springer-Verlag, New York.

Slezak, W. F. 1976. *Lonicera japonica* Thunb., an aggressive introduced species in a mature forest ecosystem. M.S. Thesis, Rutgers University, New Brunswick, New Jersey.

Solomon, A. M. 1986. Transient response of forests to CO_2-induced climate change: simulation modeling experiments in eastern North America. Oecologia 68:567–579.

Statzner, B., and B. Higler. 1985. Questions and comments on the river continuum concept. Canadian Journal of Fisheries and Aquatic Sciences 42:1038–1044.

Swank, W. T., and W. H. Caskey. 1982. Nitrate depletion in a second-order mountain stream. Journal of Environmental Quality 11:581–584.

Uehara, G., and G. Gillman. 1981. The mineralogy, chemistry, and physics of tropical soils with variable charge clays. Westview Press, Boulder, Colorado.

Van Cleve, K., and V. Alexander. 1981. Nitrogen cycling in tundra boreal ecosystems. Pages 375–404 *in* T. Rosswall and F. E. Clark (editors). The terrestrial nitrogen cycle. Ecological Bulletin 33. Stockholm.

Vannote, R. L., G. W. Minshall, K. W. Cummins, J. R. Sedell, and C. E. Cushing. 1980. The river continuum concept. Canadian Journal of Fisheries and Aquatic Sciences 37:130–137.

Vitousek, P. M. 1984. Litterfall, nutrient cycling, and nutrient limitation in tropical forests. Ecology 65:285–289.

Vitousek, P. M., and J. S. Denslow. 1986. Nitrogen and phosphorus availability in treefall gaps in a lowland tropical rainforest. Journal of Ecology 74:1167–1178.

Vitousek, P. M., J. R. Gosz, C. C. Grier, J. M. Melillo, and W. A. Reiners. 1982. A comparative analysis of potential nitrification and nitrate mobility in forest ecosystems. Ecological Monographs 52:155–177.

Vitousek, P. M., and R. L. Sanford. 1986. Nutrient cycling in moist tropical forest. Annual Review of Ecology and Systematics 17:137–167.

Walsh, G., R. Morin, and R. J. Naiman. 1988. Daily rations, diel feeding activity and distribution of age-0 brook charr (*Salvelinus fontinalis*) in two subarctic streams. Environmental Biology of Fishes 21:195–205.

Ward, J. V., and J. A. Stanford. 1983. The serial discontinuity concept of lotic ecosystems. Pages 29–42 *in* T. D. Fontaine and S. M. Bartell (editors). Dynamics of lotic ecosystems. Ann Arbor Science Publishers, Ann Arbor, Michigan.

Ward, J. V., and J. A. Stanford. 1988. Riverine ecosystems: the influence of man on catchment dynamics and fish ecology. Canadian Journal of Fisheries and Aquatic Sciences (in press).

Weaver, P. L. 1986. Structure and dynamics in the Colorado forest of the Luquillo Mountains of Puerto Rico. Ph.D. Dissertation, Michigan State University, East Lansing.

Whigham, D. F., C. Chitterling, B. Palmer, and J. O'Neill. 1986. Modification of runoff from upland watersheds: the influence of a diverse riparian ecosystem. Pages 305–332 *in* D. L. Correll (editor). Watershed research perspectives. Smithsonian Press, Washington, D.C.

Whittaker, R. H. 1975. Communities and ecosystems. MacMillan Publishing Co., New York.

Wiens, J. A., C. S. Crawford, and J. R. Gosz. 1985. Boundary dynamics: a conceptual framework for studying landscape ecosystems. Oikos 45:421–427.

Wuhrmann, K. 1974. Some problems and perspectives in applied limnology. Mitteilungen der Internationalen Vereinigung für Theoretische und Angewandte Limnologie 20:324–402.

7 Synthesis

Introduction and Review

The 1980s were a period of flowering and growth of landscape ecology. Textbooks began to appear that summarized both European and North American perspectives of the discipline (e.g., Forman and Godron 1986; Naveh and Lieberman 1984), and the first cohort of students trained explicitly as landscape ecologists rather than something else began to emerge. The field was ripe for a clear and compelling synthesis of what landscape ecology was really all about and why ecologists in particular could no longer ignore the influences of spatial patterns on ecological processes. **Monica Turner** provided just this synthesis in her 1989 paper, which was published in *Annual Review of Ecology and Systematics* rather than a journal aimed at landscape ecologists (who, presumably, already knew the importance of spatial patterns).

Although written from a distinctly North American perspective, Turner's paper hit on many of the themes that had emerged in landscape ecology during the 1980s: scale, quantifying patterns, modeling, disturbance spread, boundaries, movement of organisms, population dynamics, redistribution of matter and nutrients, and ecosystem processes.

She gave less emphasis to the role of humans and human cultures in shaping and being shaped by landscapes, the sorts of things that were central to European landscape ecology. Turner's objective was primarily to bring landscape ecology into the awareness of basic and applied ecologists, however, and in that she succeeded admirably.

Landscape ecology has exploded dramatically from its disparate beginnings in separate disciplines half a century ago. A number of recent books provide detailed guides through the discipline; we list several of these in the introduction to this collection. Collectively, these treatments demonstrate the power and the promise of landscape ecology that were clear in Turner's synthesis. Ultimately, they trace back to the foundations established by the papers assembled in this volume.

References

Forman, R. T. T., and M. Godron. 1986. *Landscape ecology.* New York: Wiley.

Naveh, Z., and A. Lieberman. 1984. *Landscape ecology: Theory and application.* New York: Springer-Verlag.

Annu. Rev. Ecol. Syst. 1989. 20:171–97

LANDSCAPE ECOLOGY: The Effect of Pattern on Process[1]

Monica Goigel Turner

Environmental Sciences Division, Oak Ridge National Laboratory, Oak Ridge, TN 37831

INTRODUCTION

A Historical Perspective

Ecology and natural history have a long tradition of interest in the spatial patterning and geographic distribution of organisms. The latitudinal and altitudinal distribution of vegetative zones was described by Von Humboldt (154), whose work provided a major impetus to studies of the geographic distribution of plants and animals (74). Throughout the nineteenth century, botanists and zoologists described the spatial distributions of various taxa, particularly as they related to macroclimatic factors such as temperature and precipitation (e.g. 21, 82, 83, 156). The emerging view was that strong interdependencies among climate, biota, and soil lead to long-term stability of the landscape in the absence of climatic changes (95). The early biogeographical studies also influenced Clements' theory of successional dynamics, in which a stable endpoint, the climax vegetation, was determined by macroclimate over a broad region (14, 15).

Clements stressed temporal dynamics but did not emphasize spatial patterning. Gleason (36–38) argued that spatially heterogeneous patterns were important and should be interpreted as individualistic responses to spatial gradients in the environment. The development of gradient analysis (e.g. 17, 164) allowed description of the continuous distribution of species along environmental gradients. Abrupt discontinuities in vegetation patterns were believed to be associated with abrupt discontinuities in the physical environment (165), and the spatial patterns of climax vegetation were thought to reflect localized intersections of species responding to complex environmental gradients.

172   TURNER

A revised concept of vegetation patterns in space and time was presented by Watt (157). The distribution of the entire temporal progression of successional stages was described as a pattern of patches across a landscape. The orderly sequence of phases at each point in space accounted for the persistence of the overall pattern. The complex spatial pattern across the landscape was constant, but this constancy in the pattern was maintained by the temporal changes at each point. Thus, space and time were linked by Watt (157) for the first time at the broader scale that is now termed the landscape. The concept of the shifting steady-state mosaic (3), which incorporates natural disturbance processes, is related to Watt's conceptualization.

Consideration of spatial dynamics in many areas of ecology has received increased attention during the past decade (e.g. 1, 89, 99, 103, 135, 161). For example, the role of disturbance in creating and maintaining a spatial mosaic in the rocky intertidal zone was studied by Paine & Levin (99). Patch size could be predicted very well by using a model based on past patterns of disturbance and on measured patterns of mussel movement and recruitment. The dynamics of many natural disturbances and their effects on the spatial mosaic have received considerable study in a variety of terrestrial and aquatic systems (e.g. 103).

This brief overview demonstrates that a long history of ecological studies provides a basis for the study of spatial patterns and landscape-level processes. However, the emphasis previously was on describing the processes that created the patterns observed in the biota. The explicit effects of spatial patterns *on* ecological processes have not been well studied; the emphasis on pattern and process is what differentiates landscape ecology from other ecological disciplines. Therefore, this review focuses on the characterization of landscape patterns and their effects on ecological processes.

Landscape Ecology

Landscape ecology emphasizes broad spatial scales and the ecological effects of the spatial patterning of ecosystems. Specifically, it considers (a) the development and dynamics of spatial heterogeneity, (b) interactions and exchanges across heterogenous landscapes, (c) the influences of spatial heterogeneity on biotic and abiotic processes, and (d) the management of spatial heterogeneity (107).

The term "landscape ecology" was first used by Troll (138); it arose from European traditions of regional geography and vegetation science (the historical development is reviewed in 90, 91). Many disciplines have contributed to the recent development of landscape ecology. For example, economists and geographers have developed many of the techniques to link pattern and process at broad scales (e.g. 53, 172), such as the development of spatial models to address questions of human geography (reviewed in 42). Landscape ecology is well integrated into land-use planning and decision-making in Europe (e.g. 7, 111, 112, 121, 151, 153, 169). In Czechoslovakia, for example, landscape-level studies serve as a basis for determining the optimal uses of land across whole regions (113). Landscape ecology is also developing along more theoretical avenues of research with an emphasis on ecological processes (e.g. 29, 61, 107, 140, 150), and a variety of practical applications are being developed concurrently (e.g. 2, 26, 48, 56, 93).

Landscapes can be observed from many points of view, and ecological processes in landscapes can be studied at different spatial and temporal scales (106). "Landscape" commonly refers to the landforms of a region in the aggregate (Webster's New Collegiate Dictionary 1980) or to the land surface and its associated habitats at scales of hectares to many square kilometers. Most simply, a landscape can be considered a spatially heterogeneous area. Three landscape characteristics useful to consider are structure, function, and change (29). "Structure" refers to the spatial relationships between distinctive ecosystems, that is, the distribution of energy, materials, and species in relation to the sizes, shapes, numbers, kinds and configurations of components. "Function" refers to the interactions between the spatial elements, that is, the flow of energy, materials, and organisms among the component ecosystems. "Change" refers to alteration in the structure and function of the ecological mosaic through time.

Consideration of Scale

The effects of spatial and temporal scale must be considered in landscape ecology (e.g. 81, 86, 145, 150). Because landscapes are spatially heterogeneous areas (i.e. environmental mosaics), the structure, function, and change of landscapes are themselves scale-dependent. The measurement of spatial pattern and heterogeneity is dependent upon the scale at which the measurements are made. For example, Gardner et al (34) demonstrated that the number, sizes, and shapes of patches in a landscape were dependent upon the linear dimension of the map. Observations of landscape function, such as the flow of organisms, also depend on scale. The scale at which humans perceive boundaries and patches in the landscape may have little relevance for numberous flows or fluxes. For example, if we are interested in a particular organism, we are unlikely to discern the important elements of patch structure or dynamics unless we adopt an organism-centered view of the environment (165). Similarly, abiotic processes such as gas fluxes may be controlled by spatial heterogeneity that is not intuitively obvious nor visually apparent to a human observer. Finally, changes in landscape structure or function are scale-dependent. For example, a dynamic landscape may exhibit a stable mosaic at one spatial scale but not at another.

The scale at which studies are conducted may profoundly influence the conclusions: Processes and parameters important at one scale may not be as important or predictive at another scale. For example, most of the variance in litter decomposition rates at local scales is explained by properties of the litter and the decomposer community, whereas climatic variables explain most of the variance at regional scales (79, 80). The distribution of oak seedlings is also explained differently at different scales (92). Seedling mortality at local scales decreases with increasing precipitation, whereas mortality at regional scales is lowest in the drier latitudes. Thus, conclusions or inferences regarding landscape patterns and processes must be drawn with an acute awareness of scale.

CHARACTERIZING LANDSCAPE STRUCTURE

Landscape structure must be identified and quantified in meaningful ways before the interactions between landscape patterns and ecological processes can be understood. The spatial patterns observed in landscapes result from complex interactions between physical, biological, and social forces. Most landscapes have been influenced by human land use, and the resulting landscape mosaic is a mixture of natural and human-managed patches that vary in size, shape, and arrangement (e.g. 5, 8, 28, 29, 61, 148). This spatial patterning is a unique phenomenon that emerges at the landscape level (59). In this section, current approaches to the analysis of landscape structure are reviewed.

Quantifying Landscape Patterns

Quantitative methods are required to compare different landscapes, identify significant changes through time, and relate landscape patterns to ecological function. Considerable progress in analyzing and interpreting changes in landscape structure has already been made (for detailed methods and applications, see 146; statistical approaches are reviewed in 149). Table 1 reviews several methods that have been applied successfully in recent studies.

Landscape indexes derived from information theory (Table 1) have been applied in several landscape studies. Indexes of landscape richness, evenness, and patchiness were calculated for a subalpine portion of Yellowstone National Park and related to the fire history of the site since 1600 (109, 110). The trends observed in the landscape pattern and the disturbance regime suggested that Yellowstone Park is a non-steady-state system characterized by long-term cyclic changes in landscape composition and diversity. Changes in landscape diversity were also hypothesized to have effects on species diversity, habitat use by wildlife, and the nutrient content and productivity of aquatic systems (110).

The indexes developed by Romme (109) were adapted by Hoover (51) and applied to six study areas in Georgia. Landscape patterns in sites with relatively little human influence were compared along a gradient from the mountains to the coastal plain. Results showed that landscape diversity increased southward from the mountains to the coastal plain, whereas the diversity of plant species decreased. However, a study that included human land-use patterns revealed a general trend of decreasing landscape diversity from the mountains to the coastal plain of Georgia (148). This apparent contradiction illustrates the sensitivity of these indexes to the scheme that is used to classify the different components of the landscape.

Shapes and boundaries in the landscape have been quantified by using fractals, which provide a measure of the complexity of the spatial patterns. Fractal geometry (71, 72) was introduced as a method to study shapes that are partially correlated over many scales. Fractals have been used to compare simulated and actual landscapes (34, 141), to compare the geometry of different landscapes (61, 85, 96, 148), and to judge the relative benefits to be gained by changing scales in a model or data set (10). It has been suggested that human-influenced landscapes exhibit simpler patterns than natural landscapes, as measured by the fractal dimension (61, 96, 148). Landscapes influenced by natural rather than anthropogenic disturbances may respond differently, with natural disturbances increasing landscape complexity. The fractal dimension has also been hypothesized to reflect the scale of the factors causing the pattern (61, 85). Landscape complexity has not been shown to be constant across a wide range of spatial scales (i.e. self-similarity). This lack of constancy probably reflects the effects of processes that operate at different scales; however, it remains a focus of current research. Applying predictions made at one scale to other scales may be difficult if landscape structure varies with scale (84).

The use of three complementary landscape indexes (dominance, contagion, and fractal dimension) in the eastern United States discriminated between major landscape types, such as urban coastal, mountain forest, and agricultural areas (96). The three indexes also appeared to provide information at different scales, with the fractal dimension and dominance indexes reflecting broad-scale pattern and the contagion index reflecting the fine-scale attributes that incorporate the adjacency of different habitats. This type of scale sensitivity could prove useful in selecting measures of pattern that can be easily monitored through time (e.g. by means of remote sensing) and that can be related to different processes.

The size and distribution of patches in the landscape is another measure of landscape structure. These characteristics may be of particular importance for species that require habitat patches of a minimum size or specific arrangement [e.g. the spotted owl *(Strix occidentalis)* in the Pacific northwest (41)]. The potential effects that the changes in patch structure created by forest clear-

Table 1 Some measures of spatial pattern that have been applied in the analysis of landscape structure

Measure	Equation	Conditions	Source
Relative richness (R)	$R = \dfrac{s}{s_{max}} \times 100$	s = number of different habitat types present s_{max} = maximum number of habitat types possible	(109)
Relative evenness (E)	$E = \dfrac{H2(j)}{H2(\text{max})} \times 100$ $H2 = -\ln \sum\limits_{i=1}^{s} pk^2$	$H2(j)$ = modified Simpson's dominance index (104) for landscape j $H2(\text{max})$ = maximum possible $H2$ for s community types p_k = proportion of total landscape area covered by habitat k s = number of habitats present	(109)
Relative patchiness (P)	$P = \dfrac{\sum\limits_{i=1}^{N} D_i}{N} \times 100$	N = number of boundaries between adjacent cells D_i = dissimilarity value for the ith boundary between adjacent cells	(109)
Diversity (H)	$H = -\sum\limits_{k=1}^{s} (Pk)\ \ln(Pk)$	P_k = the proportion of the landscape in habitat k s = the number of habitats observed	(96)

177

Measure	Equation	Definitions	References
Dominance (D_o)	$D_o = H_{max} + \sum_{k}^{s} (P_k) \ln(P_k)$	s = number of habitats observed P_k = the proportion of the landscape in habitat k H_{max} = ln (s), the maximum diversity when habitats occur in equal proportions	(96)
Fractal dimension (d)	$\log(A) \sim d \log(P)$	A = the area of a two-dimensional patch P = the perimeter of the patch at a particular length-scale d = the fractal dimension	(10, 34, 61, 72, 85, 96, 134, 148)
Nearest neighbor probabilities (q)	$q_{i,j} = n_{i,j}/n_i$	$n_{i,j}$ = the number of cells of type i adjacent to type j n_i = the number of cells of type i.	(63, 143)
Contagion (C)	$C = 2 s \log s + \sum^{m} \sum^{n} q_{i,j} \log q_{i,j}$	$q_{i,j}$ = the probability of habitat i being adjacent to habitat j s = number of habitats observed	(96)
Edges (E)	$E_{i,j} = \sum e_{i,j} \times l$	$e_{i,j}$ = the number of horizontal and vertical interfaces between cells of types i and j l = the length of the edge of a cell	(34, 141, 143, 148)

cutting patterns have on the persistence of interior and edge species were analyzed by Franklin & Forman (32). Patch size and arrangement may also reflect environmental factors, such as topography or soil type. The size and isolation of forest patches in southern Wisconsin were correlated with groups of environmental variables—for example, soil type, drainage, slope, and disturbance regime (126). The pattern of presettlement forests was closely related to topography and the pattern of natural disturbances, especially fire; the subsequent deforestation that accompanied human settlement was selective (126). Small patches of forest (i.e. woodlots) have also been studied as biogeographic islands for both flora and fauna (e.g. 5, 8, 27, 47, 163).

A variety of other techniques are available for quantifying landscape structure. The amount of edge between different landscape elements may be important for the movement of organisms or materials across boundaries (e.g. 44, 73, 144, 168), and the importance of edge habitat for various species is well known (e.g. 62). Thus, it may be important to monitor changes in edges when one quantifies spatial patterns and integrates pattern with function. Fine-scale measures of adjacency patterns and the directionality of individual cover types can be quantified by using nearest neighbor probabilities. Nearest neighbor probabilities reflect the degree of fragmentation in the landscape and, indirectly, the complexity of patch boundaries. Directionality in the landscape pattern, which may reflect topographic or other physical constraints, can be measured by calculating nearest neighbor probabilities both vertically and horizontally (or even diagonally).

The quantitative measures reviewed here could be easily applied to remotely sensed data, which would permit broad-scale monitoring of landscape changes, and to data in a geographic information system (GIS). However, it is important to note that the value of any measurement is a function of how the landscape units were classified (e.g. land use categories vs successional stages) and the spatial scale of the analysis (e.g. grain and extent). "Grain" refers to the level of spatial or temporal resolution within a data set, and "extent" refers to the size or area of the study. For example, an analysis might be conducted for a 10,000-ha study site (extent) by using data with a resolution of 1 ha (grain). Measurements of landscape pattern do not respond in the same way to changes in grain and extent. Therefore, both classification and scale must be carefully considered in analyses of landscape structure.

Important questions remain about landscape patterns and their changes. For example, what constitutes a significant change in landscape structure? Which measures best relate to ecological processes? How do the measurements of pattern relate to the scale of the underlying processes? Which measures of structure give the best indications of landscape change; that is, can any serve as "early warning" signals? Answers to these and other questions are necessary for the development of broad-scale experiments and for the design of strategies to monitor landscape responses to global change.

Predicting Changes in Landscape Structure

Models are necessary for landscape studies because experiments frequently cannot be performed at the ideal spatial or temporal scale. Because most ecological modeling has focused on temporal changes, spatial stimulation modeling is not yet well developed (16). Yet, the linking of models with geographic information systems and remote sensing technologies has begun (e.g. 9, 43, 57), and functional models are being constructed. A review of simulation modeling as applied to landscape ecology is beyond the scope of this article (see 133), but some recent developments are highlighted.

Three general classes of ecological model are presently being applied in the prediction of changes in landscape structure: (a) individual-based models; (b) transition probability models; and (c) process models. Individual-based models incorporate the properties of individual organisms and the mechanisms by which they interact with their environment (52). The JABOWA-FORET models used to predict forest succession are examples (4, 127). Multiple simulations can be done with these models to represent a variety of environmental conditions in the landscape (9, 129–131, 159). Individual-based models can be linked together spatially in a transect or grid-cell format to represent a heterogeneous landscape (e.g. 128), and methods are available to assess the error associated with the broad-scale applications (20). In a somewhat different application, Pastor & Post (101) combined an individual-based model with a nutrient cycling model and demonstrated that the patterns of soil heterogeneity in the landscape had a strong influence on forest responses to global climatic change.

Transition probability models have been used in a spatial framework to predict changing landscape patterns in natural (e.g. 43) and human-dominated landscapes (e.g. 50, 55, 141, 143). Transition models may be particularly useful when factors causing landscape change (e.g. socioeconomics) are difficult to represent mechanistically. Process-based simulation models are also being developed. For example, a model that combines hydrology, nutrient dynamics, and biotic responses into a grid-cell based spatial model has been used successfully to predict changes in a coastal landscape (132).

Simulation modeling will continue to play an important role in predicting landscape changes and in developing our understanding of basic landscape dynamics. The development of new computer architectures should facilitate the simulation of landscape dynamics (e.g. 12). In addition, many opportunities now exist for linking ecosystem models to geographic information systems to study landscape processes. For example, Burke et al (9) used a GIS to develop a regional application of an ecosystem model. The variability of soil organic carbon across the US central grasslands was studied through the use of a GIS model of macroclimate, soil texture, and management status. Soil organic carbon increased with precipitation, decreased with temperature, and was lowest in sandy soils. From a regional soils data base, regression analysis was used to examine predictive variables at different spatial scales. Net primary production was driven primarily by precipitation and exhibited a linear relationship. Predictions of soil organic matter, however, were driven by soil texture, and responses were nonlinear. The need to understand the spatial relationships between driving variables and output variables was demonstrated.

RELATING LANDSCAPE PATTERNS AND ECOLOGICAL PROCESSES

Elucidating the relationship between landscape pattern and ecological processes is a primary goal of ecological research on landscapes. This goal is difficult to accomplish, however, because the broad spatial-temporal scales involved make experimentation and hypothesis testing more challenging. Thus, achieving this goal may require the extrapolation of results obtained from small-scale experiments to broad scales (e.g. 140). This section first reviews the use of neutral models to predict the effects of pattern or process and then examines current research addressing ecological processes for which landscape pattern is important.

Neutral Models of Pattern and Process

An expected pattern in the absence of a specific process has been termed a "neutral model" (13). The use of neutral models in landscape ecology is a promising approach for testing the relationship between landscape patterns and ecological processes (34).

Percolation theory (98, 134) was used by Gardner et al (34) to develop neutral models of landscape patterns. Methods developed from percolation theory provide a means of generating and analyzing patterns of two-dimensional arrays, which are similar to maps of landscape patterns. A two-dimensional percolating network within an m by m array is formed by randomly choosing the occupation of the m^2 sites with probability p. This is analogous to generating a spatial pattern of sites occupied by a particular habitat, such as forest or grassland, at random. A "cluster" (i.e. patch) is defined as a group of sites of similar type that have at least one edge in common. The number, size distribution, and fractal dimension of clusters on these random maps vary as a function of the size of the map and the fraction of the landscape occupied by the habitat. Cluster characteristics change most rapidly near the critical probability, p_c, which is the probability at which the largest cluster will "percolate" or connect the map continuously from one side to the other ($p_c = 0.5928$ for very large arrays). Thus, for example, a hypothetical animal restricted to a single habitat type might be expected to disperse successfully across a random landscape if the probability of occurrence of habitat exceeded 0.5928.

Neutral models can be used as a baseline from which to measure the improvement in predicting landscape patterns that can be achieved when topographic, climatic, or disturbance effects are included. Neutral models need not be restricted to purely random maps. For example, maps with known connectivity, hierarchical structure, or patterns of environmental characteristics might be used. It is also possible to generate the expected patterns of other ecological phenomena, such as the spatial distribution of wildlife (e.g. 88), by using a neutral model approach.

Landscape Heterogeneity and Disturbance

The spread of disturbance across a landscape is an important ecological process that is influenced by spatial heterogeneity (e.g. 107, 109, 140). Disturbance can be defined as "any relatively discrete event in time that disrupts ecosystem, community, or population structure and changes resources, substrate availability, or the physical environment" (103). Ecological disturbance regimes can be described by a variety of characteristics, including spatial distribution, frequency, return interval, rotation period, predictability, area, intensity, severity, and synergism (e.g. 114, 162).

Disturbances operate in a heterogeneous manner in the landscape—gradients of frequency, severity, and type are often controlled by physical and vegetational features. The differential exposure to disturbance, in concert with previous history and edaphic conditions, leads to the vegetation mosaic observed in the landscape. For example, a study of the disturbance history of old-growth forests in New England between 1905 and 1985 found that site susceptibility to frequent natural disturbances (e.g. windstorms, lightning, pathogens, and fire) was controlled by slope position and aspect (30). No evidence was found that the last 350 years have provided the stability, species dominance, or growth patterns expected in a steady-state forest (30). This result demonstrates the need for a better understanding of the geographic role of disturbance, not only in New England but elsewhere. It should be possible to determine susceptibility to disturbance across the landscape. For example, Foster has also shown that wind damage in forest stands produces predictable patterns based on the age of the trees (31). Similarly, mature coniferous forest stands in Yellowstone National Park are generally most susceptible to fire, whereas younger forests are least susceptible (109, 110, 123).

Landscapes respond to multiple disturbances, and the interactive effects of disturbances are important but difficult to predict (e.g. 60, 144). In forested landscapes of the southeastern United States, a low-level disturbance of individual pine trees (by lightning), may be propagated to the landscape level by bark beetles (115). With this propagation, disturbance effects change from physiological damage of an individual tree to the creation of forest patches (bark beetle spots) in which gap phase succession is initiated. Under conditions favorable for the beetle (stressful conditions for the trees), the beetle populations can expand to become an epidemic with quite different effects on the landscape. Rykiel et al (115) suggests that the bark beetles are amplifying the original disturbance of lightning strikes.

Estimation of the cumulative impacts of disturbances in a landscape is important for protecting sensitive habitats or environmental quality. A comparison of the arctic landscape in 1949 and 1983 demonstrated that indirect impacts of anthropogenic disturbances may have substantial time lags; furthermore, the total area influenced by both direct and indirect effects can greatly exceed the area of planned development (155). This suggests a strong need for comprehensive landscape planning through the use of current technologies (e.g. geographic information systems) to address such cumulative or synergistic disturbance effects.

The spatial spread of disturbance may be enhanced or retarded by landscape heterogeneity. In forests of the Pacific Northwest, increased landscape heterogeneity due to "checkerboard" clear-cutting patterns enhances the susceptibility of old growth forest to catastrophic windthrow (32). On a barrier island, the unusually close proximity of different habitats in the landscape appeared to enhance the disturbance effects that resulted from introduced ungulate grazers in mature maritime forest (144). Landscape heterogeneity may also retard the spread of disturbance. In some coniferous forests, heterogeneity in the spatial patterns of forest by age class tends to retard the spread of fires (e.g. 35). Other examples of landscape heterogeneity impeding the spread of disturbance include pest outbreaks and erosional problems in agricultural landscapes, in which disturbance is generally enhanced by homogeneity.

Can the relationship between landscape heterogeneity and disturbance be generalized? Disturbances can be further characterized by their mode of propagation: (a) those that spread within the same habitat type (e.g. the spread of a species-specific parasite through a forest); and (b) those that cross boundaries and spread between different habitat types (e.g. fire spreading from a field to a forest). Whether landscape heterogeneity enhances or retards the spread of disturbance may depend on which of these two modes of propagation is dominant. If the disturbance is likely to propagate within a community, high landscape heterogeneity should retard the spread of the disturbance. If the disturbance is likely to move between communities, increased landscape heterogeneity should enhance the spread of disturbance. Furthermore, the rate of disturbance propagation should be directly proportional to landscape heterogeneity for disturbances that spread between communities, but inversely proportional for disturbances that spread within the same community.

Another approach to generalizing the spread of disturbance across a heterogeneous landscape is to characterize the landscape in terms of habitat that is susceptible to the disturbance (e.g. pine forests susceptible to bark beetle infestations) and habitat that is not susceptible to the disturbance (e.g. pine forest that is too young to be infested, hardwood forest, grassland, etc). A neutral model approach can then be used to provide predictions of the spread of disturbance that can be tested against observations, as by Turner et al (147). Disturbance was simulated as a function of (a) the proportion of the landscape occupied by habitat susceptible to the disturbance; (b) disturbance frequency, the probability of disturbance initiation; and (c) disturbance intensity, the probability that a disturbance, once initiated, would spread to an adjacent site. The propagation of disturbance and the associated effects on landscape pattern were qualitatively different when the proportion of the landscape occupied by disturbance-susceptible habitat was above or beyond the percolation threshold (p_c). Habitats occupying less than p_c tended to be fragmented, with numerous, small patches, and low connectivity (34). The spread of a disturbance was constrained by this fragmented spatial pattern, and the sizes and numbers of clusters were not substantially affected by the intensity of disturbance. Habitats occupying more the p_c tended to be highly connected, forming continuous clusters (34), and disturbances spread through the landscape even when frequency was relatively low.

The relationship between landscape pattern and disturbance regimes must be studied further, particularly in light of potential global climatic change. Disturbances operate at many scales simultaneously, and their interactions contribute to the observed landscape mosaic. The interactive effects of disturbances are not well known, partly because we often tend to study single disturbances in small areas rather than multiple disturbances in whole landscapes. Natural disturbances are likely to vary with a changing global environment, and altered disturbance frequency or intensity may be the proximal cause of substantial changes in the landscape. A better understanding of how disturbance regimes vary through time and space is needed.

Movement and Persistence of Organisms

The spatial patterns of biological diversity have long been of concern in ecology (e.g. 67, 68, 166), and biogeographical studies have examined the regional abundance and distribution patterns of many species (e.g. 92). Landscape ecological studies focus on the effects that spatial patterning and changes in landscape structure (e.g. habitat fragmentation) have on the distribution, movement, and persistance of species.

Landscape connectivity may be quite important for species persistence. The landscape can be considered as a mosaic of habitat patches and interconnections. For example, birds and small mammals in an agricultural landscape use fencerows between woodlots more than they travel across open fields, suggesting that well-vegetated fencerows may provide interconnections between patches of suitable habitat (158). It has been sug-

184   TURNER

gested that, because the survival of populations in a landscape depends on both the rate of local extinctions (in patches) and the rate of organism movement among patches (22), species in isolated patches should have a lower probability of persistence. Several studies support this idea. Local extinctions of small mammals from individual forest patches were readily recolonized by animals from other patches when fencerows were present (49). In simulations and field studies, Fahrig & Merriam (24) demonstrated that the survival of populations in individual woody patches was enhanced when patches had more corridors connecting to other patches. Simulation of numerous possible network configurations further showed that one linkage with another patch accounted for most of the variance in survival and that more than two linkages had no significant effects, regardless of network configuration (24). Another study reported that small forest patches connected by a corridor to a nearby forest system were characterized by typical forest interior avifauna, whereas similar but isolated forests were not (69).

Within a neutral model framework, the effects of patch isolation were studied by Milne et al (88), who examined the effects of landscape fragmentation on the wintering areas of white-tailed deer *(Odocoileus virginianus)*. A model was developed by using Bayesian probabilities conditional on 12 landscape variables, including soil type, canopy closure, and woody species composition. Deer habitat was predicted independently at each of 22,750 contiguous 0.4-ha locations. Comparison of the predictions of the neutral model with observed habitat-use data demonstrated that sites containing suitable habitat but isolated from other suitable patches were not used by the deer (88).

Modifications of habitat connectivity or patch sizes can have strong influences on species abundance and movement patterns. The effects of road development on grizzly bear movements within a 274-km^2 area of the Rocky Mountains were studied for seven years (75). Bears used habitat within 100 m of roads significantly less than expected. Furthermore, avoidance of roads was independent of traffic volume. Because roads often followed valley bottoms, passing through riparian areas frequently used by grizzlies, the road development represented approximately an 8.7% loss of habitat.

Theoretical approaches are being developed to identify scale-dependent patterns of resource utilization by organisms on a landscape. This approach may allow the connectivity of a landscape to be described for a variety of species. Minimal scales for resource utilization were predicted by O'Neill et al (97) by considering the spatial distribution of resources. The minimal requirement is that organisms be able to move across a landscape in a path of length *n* with a high probability of locating a resource. Every point need not contain a critical resource, but the resource must occur with high probability along the path. The path length will vary for different organisms (e.g. ants and antelope would have different scales of resource utilization). Linear corridors stretching across the landscape would permit percolation (i.e. resources spanning the landscape) at lower values of p. If resources are clumped, organisms must adjust their scale of resource utilization and operate at larger scales in order to move from one resource patch to another.

The size, shape, and diversity of patches also influence patterns of species abundance. In a study of forest fragments in an agricultural landscape, larger and more heterogeneous forests had more species and bird pairs, suggesting that regional conservation strategies should maximize both patch size and forest heterogeneity (33). Nonrandom use of patches by shrubsteppe birds was reported by Wiens (167). Studies of patch characteristics and use by two sparrows (*Amphispiza belli* and *Spizella breweri*) suggested that the birds may select relatively large patches for foraging. In areas containing large patches, use was indiscriminate with respect to size, but where smaller patches predominated, overall patch use was shifted toward the larger patches (167). Woodlot size was also found to be the best single predictor of bird species richness in the Netherlands (152).

The shape of patch may also influence patterns of species diversity within the patch. For example, more of the variance in the richness of woody plant species on peninsulas in Maine was explained by sample position in relation to the base of the peninsula than by distance from mainland (87). Another study demonstrated that revegetation patterns on reclaimed strip mines in Maryland and West Virginia differed, depending on whether the adjacent forest boundary was convex, concave, or straight. Mines near concave forest boundaries had 2.5 times more colonizing stems and greater evidence of browsing than mines adjacent to convex forest boundaries (46).

The interaction between dispersal processes and landscape pattern influences the temporal dynamics of populations. From their studies in the Netherlands, Van Dorp & Opdam (152) concluded that the distribution of forest birds in a landscape results from a combination of dispersal flow, governed by local and regional patch density, resistance of the landscape (i.e. barrier effects), and population characteristics, such as birth rate and death rate. Wolff (170) suggested that southerly populations of the snowshoe hare *(Lepus americanus)* may not be cyclical because of habitat discontinuities resulting from the wide spacing of suitable habitat patches, which prevents interpatch dispersal. In contrast, in the cyclic northerly populations, patches of suitable habitat may provide refuges from predators during population crashes, protecting the local populations from extinction. A similar effect of landscape heterogeneity on cyclic populations of *Microtus* was also suggested by Hansson (45).

Local populations of organisms with large dispersal distances may not be as strongly affected by the spatial arrangement of habitat patches. The effect of

186   TURNER

spatial arrangement of host-plant patches on the local abundance of cabbage butterfly *(Pieris rapae)* was studied by Fahrig & Paloheimo (25) through the use of models and field experiments. Results suggested that if an organism disperses along corridors, then the spatial relationships between habitat patches are important. If, however, the organism disperses large distances in random directions from patches and does not detect patches from a distance, then the spatial arrangement of habitat will have less effect on population dynamics. In a study of revegetation of debris avalanches on Mount St. Helens, Dale (18) reported that absolute distance to a seed source (i.e. dispersal distance) did not correlate with either seed abundance or plant density in revegetated study sites.

Regional-scale studies of the dominance patterns of six native grass species in the central United States suggested that the spatial patterns of these grasses were limited primarily by dispersal processes or resistance barriers caused by competition from other grasses (6). Graphic and geographic migration models were used to examine the relationship between present dominance patterns and presumed source areas for the six species. The spatial patterns supported a migrating-wave hypothesis of grass species dominance and did not support the idea that grass species distributions were controlled primarily by climatic factors. Results also suggested that the Plains grasses are probably not yet in equilibrium with their environment.

The effect that the spatial structure of habitats has on populations is also a focus of conservation biology. For example, in an experimentally fragmented California winter grassland, species richness increased with habitat subdivision, whereas extinction, immigration, and turnover rates were relatively independent of habitat subdivision (108). In an urban habitat, Dickman (23) found that two small patches retained more species than one large patch of equal area. These results contrast with predictions that habitat subdivision necessarily results in greater rates of extinction. Experimental approaches (e.g. 108) would be extremely valuable in studies of landscape heterogeneity and species persistence. Furthermore, a blending of concepts developed in conservation biology and landscape ecology could yield much insight into these issues (e.g. 105). It remains a challenge to predict quantitatively the dynamic distribution of a species from the spatial arrangement of habitat patches and the landscape structure of the surrounding region.

Redistribution of Matter and Nutrients

The redistribution of matter and nutrients across heterogeneous landscape is not well known, although input-output studies of whole ecosystems and watersheds have been extensive. For example, it is well known that increased nutrient loadings in water bodies can result from agricultural practices, forestry, or urban development (e.g. 3, 160). However, few studies have addressed the influence that spatial pattern may have upon the flow of matter and nutrients, although there is increasing recognition that such influence is important (e.g. 39).

The horizontal flow of nutrients or sediment in surface waters of human-modified landscapes may be affected by spatial patterning. Research has shown that riparian forests reduce sediment and nutrient loads in surface runoff (64, 118, 119). For example, Peterjohn & Correll (102) studied concentrations of nutrients (carbon, nitrogen and phosphorus) in surface runoff and shallow groundwater in an agricultural watershed that contained both cropland and riparian forest. Their study demonstrated that nutrient removal had occurred in the riparian forest. Nutrient removal is significant to receiving waters; the coupling of natural and managed systems within a watershed may reduce non-point-source pollution (102). Kesner (57) used a grid-cell model to study the spatial variability in the loss, gain, and storage of total nitrogen across an agricultural landscape. Total nitrogen output (kg/ha) was subtracted from total nitrogen input for each cell in a geographic information system (GIS). Results indicated that upland agricultural areas were exporting nitrogen to the surface flow, whereas the riparian habitats were removing nitrogen from the surface flow.

Nutrients can be transported by grazing animals across landscapes and between patches (e.g., 76–78, 122, 124, 125, 171). Large animals are important because they typically graze (and remove nutrients) from patches containing high-quality forage and may return nutrients (by means of defecation) to areas in which they rest or sleep. However, research has not explicitly addressed the effects that different spatial arrangements of habitat have on nutrient transport by grazers.

The flux of gases between the atmosphere and the biota may be influenced by spatial heterogeneity. The source-sink relationship between soils, microbes, and plants potentially alter gas flux across the landscape (40). New technologies such as Long-path Fourier-Transform Infrared Spectroscopy (FTIR) offer new, powerful methods to study fluxes between ecosystems, potential patterning of biological processes, and scale-dependent processes (40).

Landscape position also influences redistribution processes. Landforms such as sediment deposits or landslide areas influence the temporal and spatial patterns of material fluxes carried across landscapes by surface water (137). Characteristics of water quality can vary with a lake's position in the landscape, as demonstrated in the Colorado alpine zone (11) and in Wisconsin forests (70). Lakes lower in the landscape had a higher specific conductance because their surface or groundwater supplies passed through more of the vegetation and soils, accumulating a greater concentration of dissolved material.

Ecosystem Processes at the Landscape Level

Landscape-level estimates of ecosystem processes (e.g. primary production, evapotranspiration, and decomposition) that are influenced by spatial heterogeneity are difficult to obtain. Frequently, sampling cannot be done at the appropriate spatial scale, and studies may need to rely on data collected for other purposes. For example, Turner (142) used agricultural and forestry statistics to estimate net primary production (NPP) of the Georgia landscape over a 50-year interval. According to her study, NPP of the Georgia landscape increased from 2.5 to 6.4 t/ha during the period from 1935 to 1982 (in comparison with a potential natural productivity of ~16–18 t/ha), but NPP varied among land uses and across physiographic regions.

Several recent studies have attempted to examine scale-dependent patterns of productivity, water balance, and biogeochemistry. Sala et al (116) demonstrated that the regional spatial pattern of aboveground net primary production (ANPP) in the grasslands region of the United States reflected the east-west gradient in annual precipitation. At the local scale, however, ANPP was explained by annual precipitation, soil water-holding capacity, and an interaction term. Sala et al concluded that, for a constant frame of reference, a model will need to include a large number of variables to account for the pattern of the same process as the scale of analysis becomes finer. This change in the ability of particular variables to explain variability as the spatial scale changes has also been demonstrated for other processes, such as decomposition (79, 80) and evapotranspiration (54). Regional trends in soil organic matters across 24 grassland locations in the Great Plains have also been predicted by using a few site-specific variables: temperature, moisture, soil texture, plant lignin content, and nitrogen inputs (100).

NPP has also been extensively studied at regional scales through the use of remote sensing technology (e.g. 139). Although a review of this literature is beyond the scope of this article, it is important to note that remote sensing technology offers considerable promise for the estimation of other ecological processes at broad scales. For example, evapotranspiration (ET) from forested landscapes can be estimated from remotely sensed data (e.g. 65, 66). Estimates of forest canopy ET that are based on data from the Thermal Infrared Multispectral Scanner (TIMS) compared well with estimates made through energy balance techniques (65).

Because the spatial heterogeneity of many ecosystem processes is not well known, the extrapolation of site-specific measurements to regional scales is difficult. Schimel et al (117) demonstrated that the spatial pattern of soil and forage properties influences cattle behavior and hence urine deposition in grasslands, making large-scale estimates of nitrogen loss challenging. King et al (58) tested two methods of extrapolating site-specific models of seasonal terrestrial carbon dynamics to the biome level. The first method, a simple extrapolation that assumed homogeneity in biotic, edaphic, and climatic patterns within a biome, was not adequate for biome-level predictions. The second method explicitly incorporated spatial heterogeneity in the abiotic variables that drive carbon dynamics, producing more reasonable results. Predictions were based on the mathematical expectation of simulated site-specific exchanges for each region times the area of the region. Four main ingredients were required to extrapolate the site-specific models across heterogeneous regions: (*a*) the local site-specific model, (*b*) designation of the larger region of interest, (*c*) the frequency distribution of model parameters or variables that vary across the region and define the heterogeneity of the region, and (*d*) a procedure for calculating the expected value of the model. Methods such as those developed by King et al (58) show promise for dealing with this difficult problem, so theory development and empirical testing should continue. The problem of extrapolation of site-specific measurements to obtain regional estimates of ecological processes remains a challenge.

CONCLUSION

Spatial pattern has been shown to influence many processes that are ecologically important. Therefore, the effects of pattern on process must be considered in future ecological studies, particularly at broad scales, and in resource management decisions.

Many land management activities (e.g. forestry practices, regional planning, and natural resource development) involve decisions that alter landscape patterns. Ecologists, land managers, and planners have traditionally ignored interactions between the different elements in a landscape—the elements are usually treated as different systems. Although this review has selectively emphasized the effects of spatial patterns on ecological processes, the landscape (like many ecological systems) represents an interface between social and environmental processes. Results from landscape ecological studies strongly suggest that a broad-scale perspective incorporating spatial relationships is a necessary part of land-use planning, for example, in decisions about the creation or protection of sustainable landscapes. A working method for landscape planning was presented by Steiner & Osterman (136) and applied to a case study of soil erosion.

The long-term maintenance of biological diversity may require a management strategy that places regional biogeography and landscape patterns above local concerns (93). With regional diversity and ecological integrity as the goal, the rarity criterion (for species management) may be most appropriately applied at regional/global scales (see also 120). Noss & Harris (94) present a conceptual scheme that evaluates not only habitat context within protected areas but also the landscape context in which each preserve exists. There

190   TURNER

remains a tremendous potential (and a necessity) for truly interdisciplinary cooperation among ecologists, geographers, landscape planners, and resource managers to develop an integrated approach to landscape management.

Landscape theory may have direct applications to the management of disturbance-prone landscapes. Franklin & Forman (32) presented a convincing argument for considering the ecological effects of spatial patterns of forest cutting patterns. The theoretical studies conducted by Turner et al (147) also have implications for landscape management. If a habitat type is rare (e.g. granite outcrops and remnant forests), management should focus on the frequency of disturbance initiation; disturbances with low frequencies may have little impact, even at high intensities of disturbance propagation, if there is insufficient landscape connectivity. In contrast, high frequencies of disturbance initiation can substantially change landscape structure. If a habitat type is common, management must consider both frequency and intensity. The effects of disturbance can be predicted at the extreme ends of the ranges of frequency and intensity, but effects may be counterintuitive for intermediate levels of frequency and intensity. For example, large tracts of forest can be easily fragmented and qualitatively changed by disturbances of low to moderate intensity and low to high frequency.

New insights into ecological dynamics have emerged from landscape studies and have led to hypotheses that can be tested in a diversity of systems and at many scales. Several studies have suggested that the landscape has critical thresholds at which ecological processes will change qualitatively. A threshold level of habitat connectivity may demarcate different sorts of processes or phenomena. The number or length of edges in a landscape changes rapidly near the critical threshold (34); this change may have important implications for species persistance. Habitat fragmentation may progress with little effect on a population until the critical pathways of connectivity are disrupted; then, a slight change near a critical threshold can have dramatic consequences for the persistence of the population. Similarly, the spread of disturbance across a landscape may be controlled by disturbance frequency when the habitat is below the critical threshold, but it may be controlled by disturbance intensity when the habitat is above the critical threshold. Hypotheses regarding the existence and effects of critical thresholds in spatial patterns should be tested through the use of a diversity of landscapes, processes, and scales.

Current research suggests that different landscape indexes may reflect processes operating at different scales. The relationships between indexes, processes, and scale needs more study to understand (a) the factors that create pattern and (b) the ecological effects of changing patterns on processes. The broad-scale indexes of landscape structure may provide an appropriate metric for monitoring regional ecological changes. Such an application is of particular importance because changes in broad-scale patterns (e.g. in reponse to global change) can be measured with remote-sensing technology, and an understanding of the pattern-process relationship will allow functional changes to be inferred.

A few variables may be adequate to predict landscape patterns. The relative importance of parameters controlling ecological processes appears to vary with spatial scale. Several studies suggest that, at the landscape level, only a few variables may be required to predict landscape patterns, the spread of disturbances, or ecosystem processes such as NPP or the distribution of soil organic matter. These observations could simplify the prediction of landscape dynamics if a significant amount of fine-scale variation can be incorporated into a few parameters. A better understanding of the parameters necessary to predict patterns at different scales is necessary.

It is important to identify the processes, phenomena, and scales at which spatial heterogeneity has a significant influence. For example, the effect of landscape heterogeneity on the redistribution of materials is not well known. The spatial patterning of habitats may be important to predict nutrient distribution in landscapes of small extent (e.g. the watershed of a lower-order stream) but less important as extent increases (e.g. an entire river drainage basin). The identification of instances in which spatial heterogeneity can be ignored is as important as the identification of the effects of spatial pattern. Neutral models of various types will continue to be helpful in the identification of significant effects of spatial patterns.

Future research should be oriented toward testing hypotheses in actual landscapes. Methods for characterizing landscape structure and predicting changes are now available, but the broad-scale nature of many landscape questions requires creative solutions to experimental design. Theoretical and empirical work should progress jointly, perhaps through an iterative sequence of model and field experiments. Microcosms or mesocosms in which spatial pattern can be controlled by the experimenter may also prove useful. Natural experiments, such as disturbances that occur over large areas or regional development, also provide opportunities for hypothesis testing. Of paramount importance is the development and testing of a general body of theory relating pattern and process at a variety of spatial and temporal scales.

ACKNOWLEDGMENTS

The comments and suggestions of V. H. Dale, R. T. T. Forman, R. H. Gardner, A. W. King, B. T. Milne, and R. V. O'Neill. improved this manuscript, and I sincerely thank them for their thoughtful reviews. Funding was provided by the Ecological Research Division, Office of Health and Environmental Research, US Department of Energy, under Contract no. DE-AC05-84OR21400 with Martin Marietta Energy Systems, Inc., and by an

192   TURNER

Alexander Hollaender Distinguished Postdoctoral Fellowship, administered by Oak Ridge Associated Universities, to M. G. Turner. Publication No. 3317 of the Environmental Sciences Divison, ORNL.

Literature Cited

1. Allen, T. F. H., Starr, T. B. 1982. *Hierarchy*. Chicago: Univ. Chicago Press
2. Baker, W. L. 1989. Landscape ecology and nature reserve design in the Boundary Waters Canoe Area, Minnesota. *Ecology* 70:23–35
3. Bormann, F. H., Likens, G. E. 1979. *Pattern and Process in a Forested Ecosystem*. New York: Springer-Verlag
4. Botkin, D. B., Janak, J. F., Wallis, J. R. 1972. Some ecological consequences of a computer model of forest growth. *J. Ecol.* 60:849–72
5. Bowen, G. W., Burgess, R. L. 1981. A quantitative analysis of forest island pattern in selected Ohio landscapes. *Rep. No. ORNL/TM-7759*, Oak Ridge Natl Lab., Oak Ridge, Tenn.
6. Brown, D. A., Gersmehl, P. J. 1985. Migration models for grasses in the American mid-continent. *Ann. Assoc. Am. Geogr.* 75:383–94
7. Buchwald, K. Engelhart, W., eds. 1968. *Handbuch fur landschaftpflege und naturschutz*. Bd. 1. Grundlagen. BLV Verlagsgesellschaft, Munich
8. Burgess, R. L., Sharpe, D. M., eds. 1981. *Forest Island Dynamics in Man-Dominated Landscapes*. New York: Springer-Verlag
9. Burke, I. C., Schimel, D. S., Yonker, C. M., Parton, W. J., Joyce, L. A. 1989. Regional modeling of grassland biogeochemistry using GIS. *Landscape Ecol.* In press
10. Burrough, P. A. 1986. *Principles of Geographic Information Systems for Land Resources Assessment*. Oxford: Clarendon
11. Caine, N. 1984. Elevational contrasts in comtemporary geomorphic activity in the Colorado Front Range. *Studia Geomorphol. Carpatho-Balcanica* 18:5–31
12. Casey, R. M., Jameson, D. A. 1988. Parallel and vector processing in landscape dynamics. *Appl. Math. Comput.* 27:3–22
13. Caswell, H. 1976. Community structure: a neutral model analysis. *Ecol. Monogr.* 46:327–54
14. Clements, F. E. 1916. Plant succession: an analysis of the development of vegetation. *Carnegie Inst. Wash. Publ.* 242
15. Clements, F. E. 1936. Nature and structure of the climax. *J. Ecol.* 24:252–84
16. Costanza, R., Sklar, F. H. 1985. Articulation, accuracy, and effectiveness of mathematical models: a review of freshwater wetland applications. *Ecol. Modelling* 27:45–68
17. Curtis, J. T. 1959. *The Vegetation of Wisconsin: An Ordination of Plant Communities*. Madison: Univ. Wisconsin Press
18. Dale, V. H. 1989. Wind dispersed seeds and plant recovery on the Mount St. Helens debris avalanche. *Can. J. Bot.* In press
19. Dale, V. H., Gardner, R. H. 1987. Assessing regional impacts of growth declines using a forest succession model. *J. Environ. Manage.* 24:83–93
20. Dale, V. H., Jager, H. I., Gardner, R. H., Rosen, A. E. 1988. Using sensitivity and uncertainty analyses to improve predictions of broad-scale forest development. *Ecol. Modelling* 42:165–78
21. DeCandolle, A. P. A. 1874. *Constitution dans le regne vegetal de groupes physiologiques applicables a la geographie ancienne et moderne*. Geneva: Archives des Science Physiques et Naturelles
22. Den Boer, P. J. 1981. On the survival of populations in a heterogeneous and variable environment. *Oecologica* 50:39–53
23. Dickman, C. R. 1987. Habitat fragmentation and vertebrate species richness in an urban environment. *J. Appl. Ecol.* 24:337–51
24. Fahrig, L., Merriam, G. 1985. Habitat patch connectivity and population survival. *Ecology* 66:1762–68
25. Fahrig, L., Paloheimo, J. 1988. Effect of spatial arrangement of habitat patches on local population size. *Ecology* 69:468–75
26. Forman, R. T. T. 1986. Emerging directions in landscape ecology and applications in natural resource managment. In *Proc. Conf. on Sci. Natl. Parks*, ed. R. Herrmann, T. Bostedt-Craig, pp. 59–88. Washington, DC: The George Wright Society
27. Forman, R. T. T., Galli, A. E., Leck, C. F. 1976. Forest size and avian diversity in New Jersey woodlots with some land use implications. *Oecologia* 26:1–8

28. Forman, R. T. T., Godron, M. 1981. Patches and structural components for a landscape ecology. *BioScience* 31:733–40

29. Forman, R. T. T., Godron, M. 1986. *Landscape Ecology*. New York: Wiley

30. Foster, D. R. 1988. Disturbance history, community organization and vegetation dynamics of the old-growth Pisgah Forest, southwestern New Hampshire, USA. *J. Ecol.* 76:105–34

31. Foster, D. R. 1988. Species and stand response to catastrophic wind in central New England, USA. *J. Ecol.* 76:135–51

32. Franklin, J. F., Forman, R. T. T. 1987. Creating landscape patterns by forest cutting: ecological consequences and principles. *Landscape Ecol.* 1:5–18

33. Freemark, K. E., Merriam, H. G. 1986. Importance of area and habitat heterogeneity to bird assemblages in temperate forest fragments. *Biol. Conserv.* 31:95–105

34. Gardner, R. H., Milne, B. T., Turner, M. G., O'Neill, R. V. 1987. Neutral models for the analysis of broad-scale landscape pattern. *Landscape Ecol.* 1:19–28

35. Givnish, T. J. 1981. Serotiny, geography, and fire in the Pine Barrens of New Jersey. *Evolution* 35:101–23

36. Gleason, H. A. 1917. The structure and development of the plant association. *Bull. Torrey Bot. Club* 43:463–81

37. Gleason, H. A. 1926. The individualist concept of the plant association. *Bull. Torrey Bot. Club.* 53:7–26

38. Gleason, H. A. 1939. The individualistic concept of the plant association. *Am. Midl. Natl.* 21:92–110

39. Gosz, J. R. 1986. Biogeochemistry research needs: observations from the ecosystem studies program of The National Science Foundation. *Biogeochemistry* 2:101–12

40. Gosz, J. R. Dahm, C. N., Risser, P. G. 1988. Long-path FTIR measurement of atmospheric trace gas concentrations. *Ecology* 69:1326–30

41. Gutierrez, R. J., Carey, A. B., eds. 1985. *Ecology and Management of the Spotted Owl in the Pacific Northwest. Gen. Tech. Rep. PNW-185.* USDA Forest Service, Pacific NW For. Range Exp. Sta., Portland, OR

42. Haggett, P., Cliff, A. D., Frey, A. 1977. *Locational Analysis in Geography*. New York: Wiley

43. Hall, F. G., Strebel, D. E., Goetz, S. J., Woods, K. D., Botkin, D. B. 1987. Landscape pattern and successional dynamics in the boreal forest. *Proc. Int.*

GeoScience Remote Sensing Symp., pp. 473–82. Ann Arbor, Mich.

44. Hansen, A. J., Di Castri, F., Naiman, R. J. 1988. Ecotones: what and why? In *A New Look at Ecotones*, ed. F. Di-Castri, A. J. Hansen, M. M. Holland. *Biol. Int. Spec. Issue* 17:9–46

45. Hansson, L. 1979. On the importance of landscape heterogeneity in northern regions for the breeding population densities of homeotherms: a general hypothesis. *Oikos* 33:182–89

46. Hardt, R. A., Forman, R. T. T. Boundary form effects on woody colonization of reclaimed surface mines. *Ecology*. In pres

47. Harris, L. D. 1984. *The Fragmented Forest*. Chicago: Univ. Chicago Press

48. Hayes, T. D., Riskind, D. H., Pace, W. L. III 1987. Patch-within-patch restoration of man-modified landscapes within Texas state parks. See Ref. 140, pp. 173–98

49. Henderson, M. T., Merriam, G., Wegner, J. 1985. Patchy environments and species survival: chipmunks in an agricultural mosaic. *Biol. Conserv.* 31:95–105

50. Hett, J. 1971. Land use changes in east Tennessee and a simulation model which describes these changes for 3 counties. *Rep. No. ORNL-IBP-71-8*, Oak Ridge Natl Lab., Oak Ridge, Tenn,

51. Hoover, S. R. 1986. *Comparative structure of landscapes across physiographic regions of Georgia*. MS Thesis. Univ. Georgia, Athens

52. Huston, M., DeAngelis, D., Post, W. 1988. New computer models unify ecological theory. *BioScience* 38:682–91

53. Isard, W. 1975. *Introduction to Regional Science*. Englewood Cliffs, NJ: Prentice-Hall

54. Jarvis, P. G., McNaughton, K. G. 1986. Stomatal control of transpiration: scaling up from leaf to region. *Adv. Ecol. Res.* 15:1–49

55. Johnson, W. C., Sharpe, D. M. 1976. An analysis of forest dynamics in the north Georgia piedmont. *For. Sci.* 22:307–22

56. Joyce, L. A., Hoekstra, T. W., Alig, R. J. 1987. Regional multiresource models in a national framework. *Environ. Manage.* 10:761–71

57. Kesner, B. T. 1984. *The geography of nitrogen in an agricultural watershed: a technique for the spatial accounting of nutrient dynamics*. MS Thesis. Univ Georgia, Athens

58. King, A. W., DeAngelis, D. L., Post, W. M. 1987. The seasonal exchange of carbon dioxide between the atmosphere

194   TURNER

and the terrestrial biosphere: extrapolation from site-specific models to regional models. *Rep. No. ORNL/TM-10570.* Oak Ridge Natl. Lab., Oak Ridge, Tenn.

59. Klopatek, J. M., Krummel, J. R., Mankin, J. B., O'Neill, R. V. 1983. A theoretical approach to regional environmental conflicts. *J. Environ. Manage.* 16:1–15

60. Knight, D. H. 1987. Parasites, lightning, and the vegetative mosaic in wilderness landscapes. See Ref. 140, pp. 59–83

61. Krummel, J. R., Gardner, R. H. Sugihara, G., O'Neill, R. V., Coleman, P. R. 1987. Landscape patterns in a disturbed environment. *Oikos* 48:321–24

62. Leopold, A. S. 1933. *Game Management.* New York, Scribners

63. Lin, C., Harbaugh. W. C. 1984. *Graphic Display of Two- and Three-Dimensional Markov Computer Models in Geology.* New York: Van Nostrand Reinhold

64. Lowrance, R. R., Todd, R. L., Asmussen, L. E. 1984. Nutrient cycling in an agricultural watershed: I. Phreatic movement. *J. Environ. Qual.* 13:22–27

65. Luvall, J. C., Holbo, H. R. 1989. Measurements of short-term thermal responses of coniferous forest canopies using thermal scanner data. *Remote Sens. Environ.* 27:1–10

66. Luvall, J. C., Holbo, H. R. 1990. Modeling forest canopy thermal response on a landscape scale using remotely sensed data. See Ref. 146. In press

67. MacArthur, R. H. 1972. *Geographical Ecology: Patterns in the Distribution of Species.* Philadelphia: Harper & Row

68. MacArthur, R. H., Wilson, E. O. 1967. *The Theory of Island Biogeography.* Princeton: Princeton Univ. Press

69. MacClintock. L., Whitcomb, R. F., Whitcomb, B. L. 1977. Island biogeography and "habitat islands" of eastern forest. II. Evidence for the value of corridors and minimization of isolation in preservation of biotic diversity. *Am. Birds* 31:6–12

70. Magnuson, J. J. Bowser, C. J., Kratz, T. K. 1984. Long-term ecological research (LTER) on north temperate lakes of the United States. *Verh. Int. Verein. Limnol.* 22:533–35

71. Mandelbrot, B. B. 1977. *Fractals. Form, Chance and Dimension.* San Francisco: Freeman

72. Mandelbrot, B. B. 1983. *The Fractal Geometry of Nature.* San Francisco: Freeman

73. McCoy, E. D., Bell, S. S., Walters, K. 1986. Identifying biotic boundaries along environmental gradients. *Ecology* 67:749–59

74. McIntosh, R. P. 1985. *The Background of Ecology.* Cambridge: Cambridge Univ. Press

75. McLellan, B. N., Shackleton, D. M. 1988. Grizzly bears and resource extraction industries: effects of roads on behavior, habitat use and demography. *J. Appl. Ecol.* 25:451–60

76. McNaughton, S. J. 1979. Grassland-herbivore dynamics. In *Serengeti: Dynamics of an Ecosystem,* ed. A. R. E. Sinclair, M. Norton-Griffiths, pp. 46–81. Chicago: Univ. Chicago Press

77. McNaughton, S. J. 1983. Serengeti grassland ecology: the role of composite environmental factors and contingency in community organization. *Ecol. Monogr.* 53:291–320

78. McNaughton, S. J. 1985. Ecology of a grazing ecosystem: the Serengeti. *Ecol. Monogr.* 55:259–94

79. Meentemeyer, V. 1978. Macroclimate and lignin control of litter decomposition rates. *Ecology* 59:465–72

80. Meentemeyer, V. 1984. The geography of organic decomposition rated. *Ann. Assoc. Am. Geogr.* 74:551–60

81. Meentemeyer, V., Box, E. O. 1987. Scale effects in landscape studies. See Ref. 140, pp. 15–34

82. Merriam, C. H. 1890. Results of a biological survey of the San Francisco Mountain region and desert of the Little Colorado, Arizona. *North Am. Fauna* 3:1–136

83. Merriam, C. H. 1898. Life zones and crop zones of the United States. *USDA Bull. 10.* Washington, DC

84. Milne, G. T. 1987. Hierarchical landscape structure and the forest planning model: discussants comments. In *FORPLAN: An evaluation of a forest planning tool,* pp. 128–32. USDA Forest Service

85. Milne, B. T. 1988. Measuring the fractal dimension of landscapes. *Appl. Math. Comput.* 27:67–79

86. Milne, B. T. 1989. Heterogeneity as a multi-scale characteristic of landscapes. In *Ecological Heterogeneity,* ed. J. Kolasa, S. T. A. Pickett. New York: Springer-Verlag. In press

87. Milne, B. T., Forman, R. T. T. 1986. Peninsulas in Maine: woody plant diversity, distance, and environmental pattern. *Ecology* 67:967–74

88. Milne, B. T., Johnson, K. M., Forman, R. T. T. 1989. Scale-dependent proximity of wildlife habitat in a spatially-

neutral Bayesian model. *Landscape Ecol.* In press

89. Mooney, H. A., Godron, M., eds. 1983. *Disturbance and Ecosystems.* New York: Springer-Verlag

90. Naveh, Z. 1982. Landscape ecology as an emerging branch of human ecosystem science. *Adv. Ecol. Res.* 12:189–237

91. Naveh, Z., Lieberman, A. S. 1984. *Landscape Ecology, Theory and Application.* New York: Springer-Verlag

92. Neilson, R. P., Wullstein, L. H. 1983. Biogeography of two southwest American oaks in relation to atmospheric dynamics. *J. Biogeogr.* 10:275–97

93. Noss, R. F. 1983. A regional landscape approach to maintain diversity. *BioScience* 33:700–06

94. Noss, R. F., Harris, L. D. 1986. Nodes, networks, and MUMs: preserving diversity at all scales. *Environ. Manage.* 10:299–309

95. O'Neill, R. V., DeAngelis, D. L., Waide, J. B., Allen, T. F. H. 1986. *A Hierarchical Concept of Ecosystems.* Princeton: Princeton Univ. Press

96. O'Neill, R. V., Krummel, J. R., Gardner, R. H., Sugihara, G., Jackson, B. et al. 1988. Indices of landscape pattern. *Landscape Ecol.* 1:153–62

97. O'Neill, R. V., Milne, B. T., Turner, M. G., Gardner, R. H. 1988. Resource utilization scale and landscape pattern. *Landscape Ecol.* 2:63–69

98. Orbach, R. 1986. Dynamics of fractal networks. *Science* 231:814–19

99. Paine, R. T., Levin, S. A. 1981. Intertidal landscapes: disturbance and the dynamics of pattern. *Ecol. Monogr.* 51:145–78

100. Parton, W. J., Schimel, D. S., Cole, C. V., Ojima, D. S. 1987. Analysis of factors controlling soil organic matter levels in Great Plains grasslands. *Soil Sci. Soc. Am. J.* 51:1173–79

101. Pastor, J., Post, W. M. 1988. Response of northern forests to CO_2-induced climate change. *Nature* 344:55–58

102. Peterjohn, W. T., Correll, D. L. 1984. Nutrient dynamics in an agricultural watershed: observations on the role of a riparian forest. *Ecology* 65:1466–75

103. Pickett, S. T. A., White, P. S., eds. 1985. *The Ecology of Natural Disturbance and Patch Dynamics.* New York: Academic Press

104. Pielou, E. C. 1975. *Ecological Diversity.* New York: Wiley-Interscience

105. Quinn, J. F., Hastings, A. 1987. Extinction in subdivided habitats. *Conserv. Biol.* 1:198–208

106. Risser, P. G. 1987. Landscape ecology:

state-of-the-art. See Ref. 140, pp. 3–14

107. Risser, P. G., Karr, J. R., Forman, R. T. T. 1984. *Landscape Ecology: Directions and Approaches. Special Publ. No. 2.* Ill. Nat. Hist. Surv., Champaign, Ill.

108. Robinson, G. R., Quinn, J. F. 1988. Extinction, turnover and species diversity in an experimentally fragmented California annual grassland. *Oecologia* 76:71–82

109. Romme, W. H. 1982. Fire and landscape diversity in subalpine forests of Yellowstone National Park. *Ecol. Monogr.* 52:199–221

110. Romme, W. H., Knight, D. H. 1982. Landscape diversity: the concept applied to Yellowstone Park. *BioScience* 32:664–70

111. Ruzicka, M. 1987. Topical problems of landscape ecological research and planning. *Ekologica-CSSR* 5:233–38

112. Ruzicka, M., Hrnciarova, T., Miklos, L., eds. 1988. *Proc. VIIIth Int. Symp. Probs. Landsc. Ecol. Res.,* Vol. 1. Inst. Exp. Biol. Ecol. CBES SAS, Bratislava, Czechoslovakia

113. Ruzicka, M., Kozova, M. 1988. Results achieved within the target-oriented project of the basic research, "Ecological optimization of the east Slovakian lowland utilization." See Ref. 112, pp. 349–58

114. Rykiel, E. J., Jr. 1985. Towards a definition of ecological disturbance. *Aust. J. Ecol.* 10:361–65

115. Rykiel, E. J., Jr., Coulson, R. N., Sharpe, P. J. H., Allen, T. F. H., Flamm, R. O. 1988. Disturbance propagation by bark beetles as an episodic landscape phenomenon. *Landscape Ecol.* 1:129–39

116. Sala, O. E., Parton, W. J., Joyce, L. A., Lauenroth, W. K. 1988. Primary production of the central grassland region of the United States. *Ecology* 69:40–45

117. Schimel, D. S., Parton, W. J., Adamsen, F. J., Woodmansee, R. G., Senft, R. L., et al. 1986. The role of cattle in the volatile loss of nitrogen from a shortgrass steppe. *Biogeochemistry* 2:39–52

118. Schlosser, I. J., Karr, J. R. 1981. Water quality in agricultural watersheds: impact of riparian vegetation during base flow. *Water Res. Bull.* 17:233–40

119. Schlosser, I. J., Karr, J. R. 1981. Riparian vegetation and channel morphology impact on spatial patterns of water quality in agricultural watersheds. *Environ. Manage.* 5:233–43

120. Schoener, T. W. 1987. The geographic distribution of rarity. *Oecologia* 74:161–73

121. Schreiber, K. F. 1977. Landscape planning and protection of the environment. The contribution of landscape ecology. *Appl. Sci. Dev.* 9:128–39

122. Schwartz, C. C., Ellis, J. E. 1981. Feeding ecology and niche separation in some ungulates on the shortgrass prairie. *J. Appl. Ecol.* 18:343–53

123. Sellers, R. E., Despain, D. G. 1976. Fire management in Yellowstone National Park. In *Proc. Tall Timbers Fire Ecology Conf., No. 14.*, pp. 99–113. Tall Timbers Res. Stat., Tallahassee, Fla.

124. Senft, R. L., Coughenour, M. B., Bailey, D. W., Rittenhouse, L. R., Sala, O. E., et al. 1987. Large herbivore foraging and ecological hierarchies. *BioScience* 37:789–99

125. Senft, R. L., Rittenhouse, L. R., Woodmansee, R. G. 1985. Factors influencing patterns of grazing behavior on shortgrass steppe. *J. Range Manage.* 38:81–87

126. Sharpe, D. M., Guntenspergen, G. R., Dunn, C. P., Leitner, L. A., Stearns, F. 1987. Vegetation dynamics in a southern Wisconsin agricultural landscape. See Ref. 140, pp. 139–58

127. Shugart, H. H. 1984. *A Theory of Forest Dynamics.* New York: Springer-Verlag

128. Shugart, H. H., Bonan, G. B., Rastetter, E. B. 1987. Niche theory and community organization. *Can. J. Bot.* 66:2634–39

129. Shugart, H. H., Michaels, P. J., Smith, T. M., Weinstein, D. A., Rastetter, E. B. 1988. Simulation models of forest succession. In *Scales and Global Change*, ed. T. Rosswall, R. G. Woodmansee, P. G. Risser, pp. 125–51. New York: Wiley

130. Shugart, H. H., Seagle, S. W. 1985. Modeling forest landscapes and the role of disturbance in ecosystems and communities. See Ref. 103, pp. 353–68

131. Shugart, H. H., Smith, T. M. 1986. Computer models of terrestrial ecosystems. In *Coupling of Ecological Studies with Remote Sensing: Potentials at Four Biosphere Reserves in the United States*, ed. M. I. Dyer, D. A. Crossley, Jr., pp. 71–81. *U. S. Dep. State Publ. No. 2504*, Washington, DC

132. Sklar, F. H., Costanza, R. 1986. A spatial simulation of ecosystem succession in a Louisiana coastal landscape. In *Proc. Summer Computer Simulation Conf.*, ed. R. Crosbie, P. Luker, pp. 467–72 Simulation Soc. Computer

133. Sklar, F. H., Costanza, R. 1990. The development of spatial simulation modeling for landscape ecology. See Ref. 146. In press

134. Stauffer, D. 1985. *Introduction to Percolation Theory.* London: Taylor & Francis

135. Steele, J. H., ed. 1978. *Spatial Pattern in Plankton Communities.* New York: Plenum

136. Steiner, F. R., Osterman, D. A. 1988. Landscape planning: a working method applied to a case study of soil conservation. *Landscape Ecol.* 1:213–26

137. Swanson, F. J., Kratz, T. K., Caine, N., Woodmansee, R. G. 1988. Landform effects on ecosystem patterns and processes. *BioScience* 38:92–98

138. Troll, C. 1939. Luftbildplan und okologische Bodenforschung, pp. 241–98. *Z. Ges. Erdkunde*, Berlin

139. Tucker, C. J., Sellers, P. J. 1986. Satellite remote sensing of primary production. *Int. J. Remote Sens.* 7:1395–1416

140. Turner, M. G., ed. 1987. *Landscape Heterogeneity and Disturbance.* New York: Springer-Verlag

141. Turner, M. G. 1987. Spatial simulation of landscape changes in Georgia: a comparison of 3 transition models. *Landscape Ecol.* 1:29–36

142. Turner, M. G. 1987. Land use changes and net primary production in the Georgia, USA, landscape: 1935–1982. *Environ. Manage.* 11:237–47

143. Turner, M. G. 1988. A spatial simulation model of land use changes in a piedmont county in Georgia. *Appl. Math. Comput.* 27:39–51

144. Turner, M. G., Bratton, S. P. 1987. Fire, grazing and the landscape heterogeneity of a Georgia barrier island. See Ref. 140, pp. 85–101

145. Turner, M. G., Dale, V. H., Gardner, R. H. 1989. Predicting across scales: theory development and testing. *Landscape Ecol.* In press

146. Turner, M. G., Gardner, R. H., eds. 1990. *Quantitative Methods in Landscape Ecology. The Analysis and Interpretation of Landscape Heterogeneity.* New York: Springer-Verlag. In press

147. Turner, M. G., Gardner, R. H., Dale, V. H., O'Neill, R. V. 1989. Predicting the spread of disturbance in heterogeneous landscapes. *Oikos.* 55:121–29

148. Turner, M. G., Ruscher, C. L. 1988. Changes in the spatial patterns of land use in Georgia. *Landscape Ecol.* 1:241–51

149. Turner, S. J., O'Neill, R. V., Conley, W. 1990. Pattern and scale: statistics for

landscape ecology. See Ref 146. In press

150. Urban, D. L, O'Neill, R. V., Shugart, H. H. 1987. Landscape ecology. *BioScience* 37:119–27

151. Van der Maarel, E. 1978. Ecological principles for physical planning. In *The Breakdown and Restoration of Ecosystems*, ed. M. W. Holgate, M. J. Woodman, pp. 413–50. *NATO Conf. Ser. 1 (Ecology)*, Vol. 3. New York: Plenum

152. Van Dorp, D., Opdam, P. F. M. 1987. Effects of patch size, isolation and regional abundance on forest bird communities. *Landscape Ecol.* 1:59–73

153. Vink, A. P. A. 1983. *Landscape Ecology and Land Use*. London: Longman

154. Von Humboldt, A. 1807. *Ideenzu einer geographie der pflangen nebat einem naturgemalde der tropenlander*. Tubingen

155. Walker, D. A., Webber, P. J., Binnian, E. F., Everett, K. R., Lederer, N. D., et al. 1987. Cumulative impacts of oil fields on northern Alaskan landscapes. *Science* 238:757–61

156. Warming, E. 1895. *Plantesamfund grundtrak of den okologiska plantegeographi*. Copenhagen: Philipsen

157. Watt, A. S. 1947. Pattern and process in the plant community. *J. Ecol.* 35:1–22

158. Wegner, J., Merriam, G. 1979. Movements by birds and small mammals between a wood and adjoining farm habitats. *J. Appl. Ecol.* 16:349–57

159. Weinstein, D. A., Shugart, H. H. 1983. Ecological modeling of landscape dynamics. See Ref. 89, pp. 29–45

160. White, F. C., Hairston, J. R., Musser, W. N., Perkins, H. F., Reed, J. F. 1981. Relationship between increased crop acreage and nonpoint-source pollu-tion: a Georgia case study. *J. Soil Water Conserv.* 36:172–77

161. White, P. S. 1979. Pattern, process, and natural disturbance in vegetation. *Bot. Rev.* 45:229–99

162. White, P. S., Pickett, S. T. A. 1985. Natural disturbance and patch dynamics: an introduction. See Ref. 103, pp. 3–13

163. Whitney, G. T., Somerlot, W. J. 1985. A case study of woodland continuity and change in the American midwest. *Biol. Conserv.* 31:265–87

164. Whittaker, R. H. 1956. Vegetation of the Great Smoky Mountains. *Ecol. Monogr.* 26:1–80

165. Whittaker, R. H. 1975. *Communities and Ecosystems*. New York: Macmillan

166. Wiens, J. A. 1976. Population responses to patch environments. *Annu. Rev. Ecol. Syst.* 7:81–120

167. Wiens, J. A. 1985. Vertebrate responses to environmental patchiness in arid and semiarid ecosystems. See Ref. 103, pp. 169–93

168. Wiens, J. A., Crawford, C. S., Gosz, J. R. 1985. Boundary dynamics: a conceptual framework for studying landscape ecosystems. *Oikos* 45:421–27

169. Woebse, H. H. 1975. *Landschaftsokologie und Landschaftsplanung*. Graz: R. B. Verlag

170. Wolff, J. O. 1981. The role of habitat patchiness in the population dynamics of snowshoe hares. *Ecol. Monogr.* 50:111–30

171. Woodmansee, R. G. 1979. Factors influencing input and output of nitrogen in grasslands. in *Perspectives in Grassland Ecology*, ed. N. R. French, pp. 117–134. New York: Springer-Verlag

172. Wrigley, N., Bennett, R. J., eds. 1981. *Quantitative Geography: A British View*. Boston: Routledge & Kegan Paul